J. Bernacchi

ANALOG-DIGITAL CONVERSION HANDBOOK

ANALOG DEVICES TECHNICAL HANDBOOKS

Published by Prentice-Hall

- Analog-Digital Conversion Handbook

Published by Analog Devices

- Nonlinear Circuits Handbook
- Transducer Interfacing Handbook
- Synchro & Resolver Conversion

ANALOG-DIGITAL CONVERSION HANDBOOK

by
The Engineering Staff of
Analog Devices, Inc.

Edited by Daniel H. Sheingold

PRENTICE-HALL, Englewood Cliffs, NJ 07632

Library of Congress Cataloging in Publication Data: 85-62726

ISBN: 0-13-032848-0

Printed in the United States of America

10 9 8 7 6 5 4 3 2 1

ISBN 0-13-032848-0 025

Prentice-Hall International (UK) Limited, *London*
Prentice-Hall of Australia Pty. Limited, *Sydney*
Editora Prentice-Hall do Brasil, Ltda., *Rio de Janeiro*
Prentice-Hall Canada Inc., *Toronto*
Prentice-Hall Hispanoamericana, S.A., *Mexico*
Prentice-Hall of India Private Limited, *New Delhi*
Prentice-Hall of Japan Inc., *Tokyo*
Prentice-Hall of Southeast Asia Pte. Ltd., *Singapore*
Whitehall Books Limited, *Wellington, New Zealand*

Contents in Brief

Table of Contents

Part V: SOURCES OF HELP

Preface

This book is about circuits and systems that interface between the analog variables of the real world and the digital world of processing, storage, communication, and display; almost invariably, such systems employ some form of analog-to-digital and digital-to-analog conversion.

Our principal objective is to provide engineers and scientists on both sides of the interface with the basic information they need to use conversion in a wide variety of settings. The range is from simple converters as circuit elements to converters with displays (digital panel instruments) to fully integrated intelligent data-acquisition systems employing converters "somewhere inside."

Since our readers will come from widely differing backgrounds, in terms of both the range of disciplines normal to the interface—from pure software to purely analog hardware—and, in the real world, from the gamut of disciplines in engineering, science, and physics, a key first step is to provide the common ground of understanding of the role of conversion in typical settings. This is accomplished in Part I, "Converters at Work (With and Without Microprocessors)."

In Part II, "A/D and D/A Converters," you can find basic information for an understanding of converter circuits, and how they communicate at both ends, inspect a sampling of the many available "mainline" converter products, and learn how they are designed, tested, specified, and applied for best results.

Part III treats of some converter forms designed for special areas of application. These include video converters, converters for synchros and resolvers, voltage-to-frequency converters, and intentionally nonlinear converters.

Part IV discusses some related circuits: analog references, sample-holds, switches and multiplexers, and a review of digital signal processing.

Depending on when it is consulted, Part V, a "Guide for the Troubled," will be an aid to either avoidance or cure (preferably the former) of problems often encountered by unsuspecting circuit and systems designers in the neighborhood of the interface.

For those desiring greater depth, we have provided a multidimensional bibliography, which is capable of fanning out to the many good in-depth sources of material on conversion. For those desiring to consult specific topics within this book after (or instead of) reading, we have provided a conscientiously detailed Index.

This book, a milestone rather than a culmination, is the third-generation outgrowth of a series of conversations with Ray Stata and Jim Pastoriza in the late 1960s and early 1970s. At that time (and increasingly since then), it was felt that the growing availability of data-processing facilities at low cost—especially minicomputers—would bring the analog-digital interface, in the form of modular a/d and d/a converters and accesories, out of the specialty houses and into the realm of the working design engineer.

Although there are books in print on digital, analog, and hybrid computing, on circuit design, and on digital communication theory and sampled-data systems, there were—and still are—few if any books that could serve as a guide to the engineer on the practical aspects of understanding, specifying, and applying the commercially available elements of conversion systems in these pursuits.

Lest any reader either expect or question our altruism in publishing this book, let us say that our viewpoint and credentials are those of a major producer of precision integrated circuits, modules, subsystems, and computer-based data-acquisition systems, for whatever level of system integration the user is comfortable with. Since we strive neither to hide nor to unduly emphasize our commercial motives, the reader may find that the resulting honesty will impart a down-to-earth sense of practicality and realism.

We have, however, attempted to restrain our temptations to crass commercialism to the extent of using model numbers and product specifications in the text for their flesh-and-blood illustrative effect only. Our catalogs, data sheets, hardware and software manuals, and other propaganda (and those of our competitors) are separately available in sufficient panoply, partisan quality, and timeliness— as well as depth of detail—to make any effort to outshine them in the present volume less than desirable, even if possible.

PREFACE TO THE 1986 EDITION

This volume is a successor to the *Analog-Digital Conversion Handbook*, first published in 1972, and its interim revision, the *Analog-Digital Conversion Notes*, published in 1977, which contained Parts I and II of the earlier book, updated to reflect the revolution in cost, size, and performance brought about by IC and hybrid technology—and incorporated two entirely new chapters to further reflect the changes in the structure of the technological marketplace brought about by the availability of both converters and computers as true components.

The current edition reflects the advances in technology that have led to an explosion in the use of converters and the tremendous variety of available converters that have appeared during the recent octennium.

Not only are more converters with increasingly improved performance available in monolithic form at rapidly decreasing prices, but, as a result of improvements in monolithic and hybrid circuit technologies, uses are burgeoning for such types as high-resolution video converters, which for years had been considered by many to be expensive laboratory curiosities.

Other major advances in conversion have occurred in level of data-acquisition system integration, to match corresponding advances in processing and memory. Complete data acquisition systems are becoming available in monolithic chip form, and giant steps have been made in remote data acquisition, permitting comprehensive intelligent interfaces between sensors and host computers.

It is probably not surprising that the basic principles have changed but little; on the other hand, it has been rewarding to observe that many application examples, described tentatively and prophetically in earlier editions, are now pretty much "old hat," and can be found, described in variety and depth, in a list of publications (many with then unknown names) that seems to grow at an uncontrollably explosive rate.

In this edition, we have retained the core of useful basic information and restored such invaluable features as the "Guide for the Troubled," the Bibliography, and the Index, which were omitted in the abbreviated 1977 "Notes" edition. Besides bringing all references to the design and application of available conventional conversion products up to date, we have added chapters on video conversion, synchro and resolver converters, analog-to-frequency converters, and intentionally nonlinear converters.

To hold this book to manageable size, however, some topics, which were treated in depth in earlier editions—including especially those that comprise the all-important field of analog signal conditioning—have been elevated to independent book-length treatment and have been published by us as the

Transducer Interfacing Handbook (1980), the *Nonlinear Circuits Handbook* (1974), and *Synchro & Resolver Conversion* (1980).

As with previous editions, it is our hope that this volume will successfully bridge the gap between the practicing engineer and the computer scientist, providing each with the complementary knowledge that will make possible a wider range of better designs for digital handling of real-world (analog) signals in this computer age. At the same time, we hope that the exposition of basic notions, the sampling of applications, and the descriptions of the wide variety of options for the interface will make the book equally attractive to teachers and students of modern electronic system practice.

We will always welcome the comments and suggestions of our readers for the benefit of the readers of future editions.

ACKNOWLEDGEMENTS

Contributions to this book have come from engineers in many departments of Analog Devices throughout the world, engaged in activities ranging from the design of monolithic integrated-circuit chips to the marketing of sophisticated high-level integrated systems for measurement, control, and test. It is one of the misfortunes of rapid growth that, while it was possible in the first edition to identify just about everyone who participated by name, it is impracticable to do so in the present volume because of the real danger of inadvertently omitting the names of many persons who have made significant contributions.

Nevertheless, the book would be incomplete without an expression of appreciation to a number of persons whose help was indispensible.

To begin with, much of the material in the first edition—especially in Part I—was so fundamental or universal in nature that it has survived with little change to benefit today's readers. Principal contributors to that volume included Walter Borlase, Cy Brown, Lew Counts, Bob Craven, Dick Ferrero, Stan Froud, Marty Gross, Barry Hilton, Mike Lindheimer, Wayne Marshall, Jim Maxwell, Frank McCormack, Berry Phillips, Al Sanchez, Rick Spofford, Dwight Wahr, Ivar Wold, and C. Peter Zicko.

New or heavily rewritten chapters of the present volume were spearheaded by Doug Grant (Chapter 8), A. Paul Brokaw (Chapter 9), Bruce Amazeen, Bruce Coleman, and Gerard T. Quilligan (Chapter 10), Stan Domanski (Chapters 11 and 12), Dale Zeskind (with Walter Kester and Don Brockman, Chapter 13—and Ed Friedman and Geoff Boyes, Chapter 14), Larry DeVito (Chapter 15), John Wynne (Chapter 16), John Sylvan and Scott Wayne (Chapter 17), David Duff (Chapter 18), Elwyn Davies (Chapter 19), Mike Stefani (Chapter 20), and Ted Dintersmith, (Chapter 21).

Substantial input, comments, and reviews came from all sides; just to name a few: Al Haun, Lisa Herbst, Steve Miller, John Mills, Jerry Neal, Rowan O'Riley, John Reidy, Al Ryan, Bill Schweber, Jim Surber, Don Travers, Russ VerNooy, Jerry Whitmore, Scott Wurcer. Feedback from numerous readers of earlier volumes was also useful. Encouragement, when needed, was provided at various times by Jim Fishbeck and Eric Janson.

The book was typeset by Joan Costa and Terri Dalton; the drawings were skillfully rendered by Ernie Lehtonen, Wendy Sheehan, and other members of the Analog Devices Publications Department, under the direction of Marie Etchells. Shelley Cohane designed the cover.

All of the above—as well as others too numerous to mention—contributed to the book's strengths and are responsible for any success it may achieve. Any weaknesses are the responsibility of the undersigned.

D. H. Sheingold

Norwood, Mass.

PART I

CONVERTERS AT WORK (WITH AND WITHOUT MICROPROCESSORS)

Chapter One

Introduction—
Data Systems and Components

The analog-to-digital and digital-to-analog converter are key elements of any system that uses digital techniques to process or communicate analog "real-world" electrical data. This book is basically about a/d and d/a converter circuits: understanding them, applying them, testing them, choosing them, and using them in systems.

When used in systems, they are often accompanied by a variety of other devices, both analog and digital, to measure input signals and perform intermediate processing with varying degrees of sophistication.

Depending on how eager the user is to wrestle with the details of electronic circuitry, converters for data-acquisition may be purchased within a wide variety of forms. They range from simple integrated-circuit conversion chips to systems that accept an electrical sensor's output, provide processing programmed by the user, and generate appropriate analog or digital output signals to control physical variables.

This chapter provides a brief thumbnail sketch of elements that tend to be used in systems with converters and are likely to be found in block diagrams in this book. General characteristics and aptitudes are summarized, and their roles in relation to converters are hinted at within the short discussions devoted to each entity.

Analog vs. Digital

As used here, "digital" refers both to electrical signals that represent numbers, control logic, and physical variables that can be measured by counting or identifying discrete states—and to related circuitry. Examples include event counts and binary* voltage.

"Analog" has to do with physical variables that are represented or measured by continuously variable aggregates, and to related circuitry. Temperature and electric current (as aggregates of molecular and electronic motion) and measurements of continuous quantities with linear scales (analogs) are examples of analog quantities.

In this book, by far the greater weight of discussion is given to the properties of *analog* circuits, in the performance of system functions. The reason for this is no mystery: the challenges to the analog circuit designer's ability are many, varied, and unrelenting.

The basic promise of analog circuits, in favorable environments, comprises functional simplicity (inherent parallel operation), high speed, and overall low cost, as well as the ability to mimic natural phenomena with electrical variables and parameters. The difficulties in dealing with analog circuits are a natural function of both the wide dynamic range that accompanies their extreme fine structure and the many degrees of freedom of interaction associated with them. In both concept and practice, designers must be concerned that analog circuits have to labor in the real world, where limits to resolution and accuracy are directly related to physical environment, electrical interference, signal magnitude, component tolerances, and the passage of time; and bandwidth adds one more dimension of complexity, being affected by all of the above.

Digital circuits, on the other hand, in dealing with binary quantities, have high (but by no means infinite!) noise immunity, no drift, high speed and low cost (individually); and the rules for using them are few and simple. With digital techniques, the principal challenges relate to the reduction of overall cost and complexity. They require ingenuity in pursuing optimal tradeoffs in the development of system architecture, avoidance of timing errors, the writing of foolproof software, and avoidance of problems arising from the nature of electronic parameters and circuitry—which are inherently analog!

Hence, designers of both analog and digital circuits, as well as software, must always anticipate where Murphy's Law will strike next and be prepared to debug when anticipation has failed.

With the exception of preamplification, a great many of the functions presented here in analog form may be performed digitally, after conversion. The choice of technology depends on tradeoffs of cost, speed, complexity, and requirements for adjustment and calibration. There has been a rapid increase, with no sign of a letup, in the the development of devices and subsystems that are intended to perform analog functions using digital components. Examples of these include analog function generation with read-only memories (ROM

*"Binary", in digital technology, has two meanings: *two-valued* (e.g., 1 or 0) and *base-2* (number system). The meaning is usually clear from the context.

lookup tables) and d/a converters, and the universe of arithmetic, logical, and control possibilities with microprocessors—including special-purpose μPs with analog capabilities and digital signal processors.

Figure 1.1 illustrates the relationships of the principal components of a data system in a "global" perspective; causality flows clockwise but can loop recursively. As a sampling of entities found in the loop, those elements or problems to be discussed in this chapter include:

Sensors
Operational Amplifiers
Instrumentation Amplifiers
Isolators
Analog Function Circuits
Analog Multiplexers
Digital Multiplexing
Sample-Hold Circuits
Analog-to-Digital Conversion
Digital-to-Analog Converters
Registers
Microprocessors
Counters
Filters
Comparators
Power Supplies
Digital Panel Instruments

Rarely will a system involve all of the above elements; but most systems will use many of them—and others, not mentioned here, as well. The shrinkage in size and cost of components—or increase of the level of system integration—has made possible the combination of many of these elements into integral parts of subsystems, in the form of chips, modules, or boards, having specified performance and even a modicum of intelligence.

THE MOST IMPORTANT ELEMENT

As Figure 1.1 indicates, there is one additional element that is always present in a data system but seldom shown on a block diagram: *homo sapiens*. These systems are inspired, designed, built, programmed, tested, perfected, and used by men and women to serve human purposes.

Much attention is given to the many ways in which humans can communicate with the system, read system data, and provide adjustments and instructions. These include: visual displays (indicator lights, cathode-ray tubes, light-emitting diodes, liquid crystals, printouts, X-Y plots), sound (computer speech, tones, beeps, clicks, etc.), keyboards and keypads (with display feedback), touch screens, mice, joysticks, switches, light pens, speech, and a growing variety of more-exotic means.

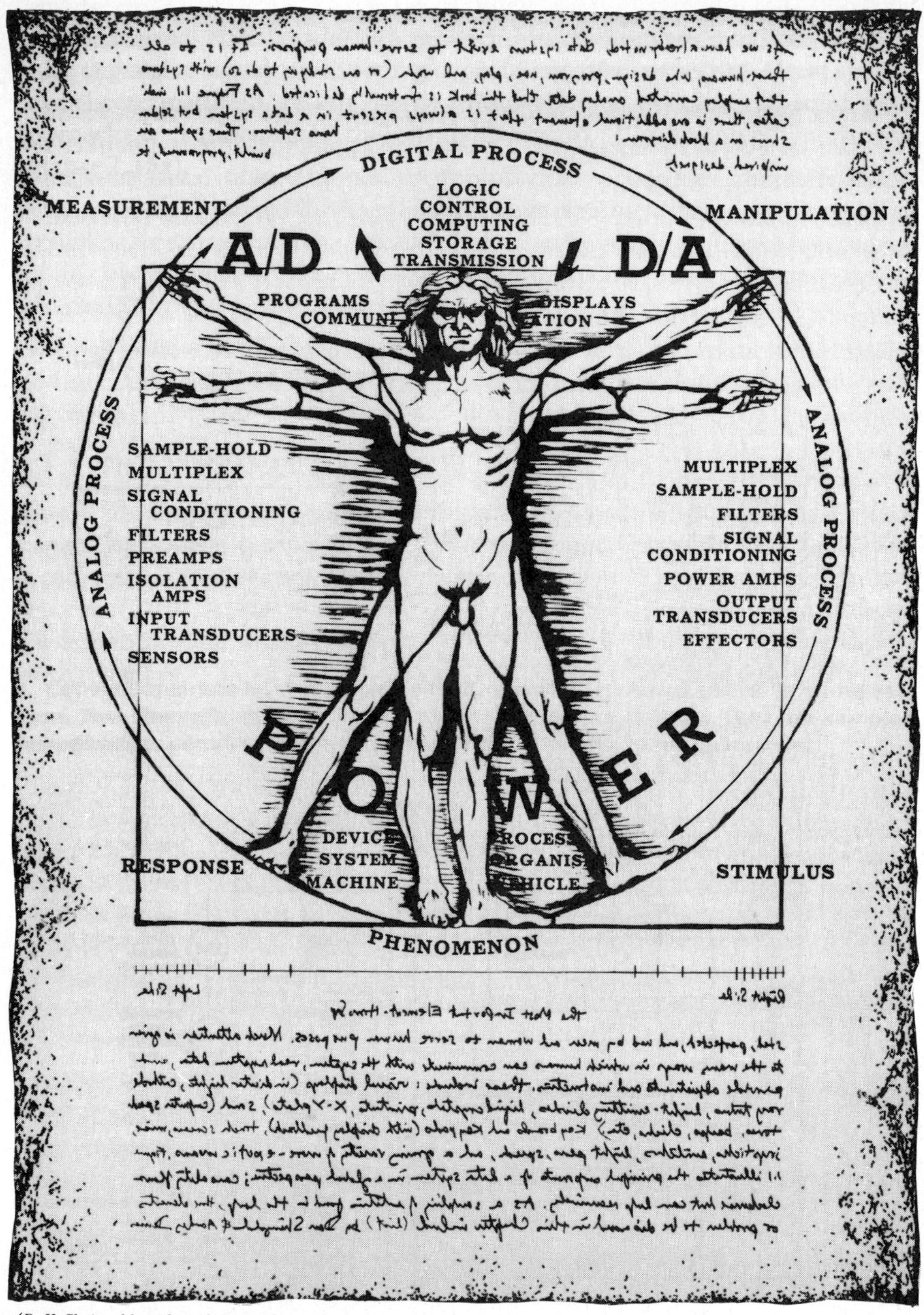

(D. H. Sheingold, with apologies to Leonardo da Vinci: Rule of Proportions, Academy of Fine Arts, Venice)

Figure 1.1 Relationship of functions and causality in a data system.

SENSORS

One might imagine that the electronic systems designer has very little say in the choice of sensor, that (s)he accepts whatever data signals exist without protest and gets on with the interface system design without further ado. However, the systems engineer who can have a say in the selection of the original transducer can go a long way towards easing the conversion-design task.

For example, in monitoring or controlling mechanical shaft rotation, the designer may be confronted with signals obtained by three radically different position-sensing approaches: optical shaft encoders, resolvers, and potentiometers, plus variations on all three. For a given task, different sensors with appropriate speed, accuracy, and reliability specifications will result in greatly differing interfaces.

Similarly, temperature measurements may be accomplished with thermocouples, RTDs (resistance temperature-detectors), thermistors, or semiconductor temperature sensors; while mechanical force may be measured directly by load cells and strain gages, or obtained indirectly by integrating the output from accelerometers, or even by counting interference fringes in an optical system.

Although it is not our rôle to recommend any particular type of transducer for a particular application, we thoroughly endorse the idea of getting the electronic systems-design engineer into the act before the signal sources are decided upon, instead of later, when it may be found that the designer is painted into a corner by the few options allowed.

OPERATIONAL AMPLIFIERS

Even if the transducer signals must simply be converted from current to voltage, or scaled up from millivolt levels to an a/d converter's 5- or 10-volt full-scale input range, signal conditioning is required.[1] Because of their low cost, one or more operational amplifiers in a suitable circuit with appropriate closed-loop gains and additive constant offsets may be the first (but not necessarily the best) choice. If the system involves many analog sources, a choice must be made between providing each transducer with its own signal conditioning and a central signal-conditioning facility that can handle a number of multiplexed inputs.

Besides gain scaling, operational amplifiers are used for a host of mathematical and signal processing functions: linear and nonlinear, static and dynamic. Although op amps are cheap, such dedicated devices as instrumentation and isolation amplifiers are always worthy of consideration in specialized applications; the design time they save often makes them the most cost-effective option.

[1]Considerable useful information about signal conditioning may be found in the Analog Devices *Transducer Interfacing Handbook* (1980). Consult the Bibliography for details.

INSTRUMENTATION AMPLIFIERS

If analog data must be transmitted over long distances (or often, even over quite short distances), differences in ground potential between signal site and data center will add spice to the interface system design problem. In order to separate common-mode interference from the signal to be recorded or processed, devices designed for the purpose (for example, instrumentation amplifiers) may be used.

Instrumentation amplifiers are functionally complete components characterized by good common-mode-rejection, high input impedance, low drift, and adjustable gain. IC types compete favorably with operational-amplifier circuit kludges in cost, as well as size and performance. Today's instrumentation amplifiers are generally monolithic ICs; gains are programmable in some by external resistors, others contain internal precision resistors and are programmed by jumpers or software.

ISOLATORS

In the event of high common-mode voltage levels or the need for extremely low common-mode leakage current, or both (as might be mandatory for many clinical applications in medical electronics), galvanic isolation is required to interpose a break in the common-mode path from the analog signal source to the data system. *Isolation amplifiers* may involve optical isolation or (more often) transformer-coupled carrier techniques; they usually have at least 1,000 volts of isolation, and they typically cost more than instrumentation amplifiers. Though most often used for isolating input data from system level, they may also be used for communicating system outputs to destinations at high common-mode voltage.

When isolation is necessary in the digital world, digital logic signals and voltage-to-frequency-converter output pulse trains are usually isolated by solid-state opto-couplers or fiber optics.

Amplifiers and analog signal conditioners are not the only analog devices that are isolated. For example, many line-powered digital panel instruments have isolation between analog inputs and digital output. A typical isolated digital-to-analog converter isolates digital equipment from its analog output (a 4-to-20mA current loop) and power supply. Thus, it can protect the analog output signal by preserving the information stored in its registers during system maintenance or crashes, as well as permitting manual updating.

ANALOG FUNCTION CIRCUITS

These "analog-to-analog" converters are analog computational circuits and special-purpose devices used to condition analog signals. Where their accuracy is adequate, they can simply, at low cost and with high speed, relieve a processor of an expensive and time-consuming software and computational burden.

The membership of this category is open-ended. Some of the more popular operations performed are multiplication; taking ratios; raising to powers; taking roots; performing special-purpose nonlinear functions such as linearizing transducers; performing rms measurements; computing trigonometric functions and vector sums; integrating and differentiating; transforming current to voltage or voltage to current, etc.

Some of these operations can be purchased in the form of such readily available devices as multiplier/dividers, log/antilog amplifiers, etc. Others represent but a sampling of the vast analog parallel number-crunching potential and program memory inherent in operational-amplifier circuitry, available to the competent designer at low parts cost, and limited only by human ingenuity.[2]

ANALOG MULTIPLEXERS

If data from many signal sources must be processed by the same computer or communications channel, via a single converter, a multiplexer is usually introduced to couple the input signals into the a/d converter in some preset or random sequence. An n-bit logic address input (2^n channels) determines which data source is to be coupled to the converter at any instant.

Multiplexers are also used in reverse, as distributors, or demultiplexers. For example, when the converter must distribute analog information to many different channels, the multiplexer, fed by a high-speed output d/a converter, can continually refresh the various output channels with updated information; generally, each channel must have analog storage to retain its information until the next update.

DIGITAL MULTIPLEXING

Digital systems often can do without a device that is specifically labeled "multiplexer" for parallel data. Such a device is cumbersome, requiring essentially a set of multipoint switches, one switch for each wire of the bus (16 multipoint switches for a 16-bit data bus) and a large number of wires converging on a single device. Instead, the digital multiplexing function is usually delegated to the devices being multiplexed, as they share a common input/output bus. They are connected to it via internal "three-state" switches.

When enabled, the switches connect a parallel set of individual 1's and 0's to the bus; otherwise, the data from the device is in effect disconnected (hence three states: 1, 0 or disconnected *). Addressed *read* commands from the pro-

[2]Considerable useful information about analog functional operations can be found in the *Nonlinear Circuits Handbook*. Consult the Bibliography for details.

*Three-state is a bit of a misnomer. Actually, there are two control states, *enabled* and *not-enabled*, and two data states, 1 and 0. The net result is three tangible states: 1, 0, and open.

cessor instruct the individual sources which one (and only one) among them must feed its burden of data onto the common bus, thence to its destination.

Input registers of all devices *receiving* data from the bus are connected to the bus. The device(s) chosen to receive the data appearing on the bus at a given time are strobed by a *write* signal, which latches the data into their registers.

SAMPLE/TRACK-HOLD CIRCUITS

In many interface systems, the analog signal varies quite rapidly. Since conversions take place at discrete (sampling) intervals, and an a/d converter cannot digitize the input signal instantaneously, substantial changes of the signal level during the actual conversion process could result in gross errors. The problem with the converter is that the conversion is completed at some appreciable (and not always constant) time following the repetitive, accurately timed conversion command, so that the final digital value would never truly represent the data level prevailing at the instant at which the conversion command was transmitted unless the analog signal was frozen at that instant.

Sample-hold devices make a fast acquisition of the varying signal, on a "sample" command and then—on a "hold" command—hold the signal constant at the output for the duration of the conversion process. Sample-hold circuits may also be used in multi-channel distribution installations, where they enable each channel to receive and hold its own signal level for activation of differing output processes. They are also used in "deglitching" to hold an output voltage steady while a large input transient is occurring during a d/a converter update, then quickly acquire the new data when the transient has subsided.

Typically, a sample-hold circuit used in data acquisition must acquire a signal rapidly (usually within microseconds), respond to the *hold* instruction within a fraction of a microsecond, with an uncertainty of a nanosecond or less, and hold the last value without significant "droop" for tens of microseconds. Most sample-holds also function as (and are often called) *track-holds*, i.e., once the analog signal is acquired, it is tracked until the hold command is received.

A/D CONVERTERS

These devices, which range from monolithic ICs to high-performance hybrid circuits, modules, and even boxes (such as digital panel meters), convert analog data—usually voltage—into an equivalent digital form. Key characteristics of a/d converters include absolute and relative accuracy, linearity, no-missing-codes, resolution, conversion speed, stability, and—of course—price. Other aspects open to choice include input ranges, digital output codes, interfacing techniques, presence of on-board multiplexing, signal conditioning, and memory.

Although the industry tends to converge upon the successive-approximations technique for a very large number of system applications, because of its inherently excellent compromise between speed and accuracy, other popular alternatives include integrating techniques (dual-ramp, quad-slope, and voltage-to-frequency), counting and tracking techniques (counter-comparator), and, for video-signal speeds, "flash" and digitally corrected subranging techniques.

Voltage-to-frequency converters can provide high-resolution conversion and such special features as long-term integration (from seconds to years), digital-to-frequency conversion, (with a d/a converter), frequency modulation, voltage isolation, and arbitrary frequency division or multiplication. Synchro-, resolver-, and Inductosyn® to digital converters are used where angular or linear position must be measured precisely and with high resolution, and converted into digital form.

A technique employed by some microprocessor users, to avoid the need for purchasing an a/d converter as such, is to use a d/a converter, a comparator, and the processor's logic to perform a tracking or successive-approximation conversion. While the financial cost is small, the software and program time burden is substantial.

D/A CONVERTERS

These devices reconstitute the original data after processing, storage, or even simple digital transmission from one location to another. The basic converter usually consists of an arrangement of weighted resistance values (or resistive or capacitive divider ratios), each controlled by a particular level or "significance" of digital input data, that is switched to develop varying output voltages, currents, or gains by selective summation in accordance with the digital input code.

The output of a d/a converter is proportional to the reference source used. Although most converters for data-handling applications are used with essentially fixed references, there is a special class of converter, capable of handling variable—and even bipolar ac—reference sources. These devices are termed *multiplying DACs*, because their output is the product of two variables—the number represented by the digital input code and the analog reference voltage; both may vary from full scale to zero, and even negatively.

Another way of looking at a multiplying DAC is to think of it as a digitally adjustable gain control. Some even have logarithmic conversion relationships, to permit digital inputs in decibels to control gain in steps having equal ratios.

For position outputs, the d/a converter may take the form of a digital-to-synchro or digital-to-resolver converter. And if the digital signal is a train of pulses at a given rate, it can be converted to analog by a frequency-to-voltage

®Trademark of Farrand Controls, Inc.

conversion circuit, often employing a voltage-to-frequency converter in a phase-locked loop.

REGISTERS

The digital register is a key component of digital systems. In this book's context, they are used to to hold information in readiness for passing it along from converters to computers, from computer buses to d/a converters, etc.

For example, a multi-channel interface system using an a/d converter for every input channel would store the parallel digitized values in an output register associated with each converter until called on by the computer to place the stored value on the common input bus. Conversely, in output multiplexing, a number of d/a converters provide different voltage levels for the independent output channels. Each DAC is fed by a storage register, which holds its digital input word (and the corresponding analog output variable) until the computer feeds in the new, updating digital value.

Like track-holds, registers may be transparent, allowing input data to appear continuously at the output until a *strobe* signal causes the data to freeze; or, like sample-holds, they may be opaque, holding the last data byte until the strobe causes the data currently on the input lines to replace the existing output data.

More than one rank of registers may be used, to make the output data independent of changes of data at the input—an especially useful feature for d/a converters that must acquire from 9 to 16 bits of information from an 8-bit bus in two bytes. In the first rank, for example, for a 12-bit digital word, an 8-bit byte is acquired, and then a 4-bit byte; then their outputs are strobed into a 12-bit register to update the DAC simultaneously.

Some converters contain memory registers. Such devices have both analog (conversion) capability and multiple internal digital registers to permit independent update and readout.

Shift registers are used where data is transmitted serially over a single data channel (e.g., pair of wires) instead of as parallel bits over many wires. Data may be strobed in in parallel and out in serial, or arrive in serial and be strobed out in parallel.

DIGITAL DATA PROCESSORS

Converters are used with processors at all levels of system integration. While the distinctions are not clear-cut, and barriers are tumbling continually as technologies march ahead, it might be worthwhile to indicate a few general categories.

At the lowest level are the rudimentary converters found on digital IC chips designed for direct processing and/or communication of low-resolution analog signals.

Microprocessors will be found in games, small computers, instruments, display terminals, and data acquisition boxes and boards, usually designed for specific purposes (perhaps slaved to a nearby or distant host computer) and characterized by "intelligence."

Microcomputer and minicomputer systems may be designed for host-computer status in systems involving higher levels of capability—large amounts of memory, complex programs—including multitasking—number-crunching capacity, and interfacing flexibility. The converters they interface with may be in instruments, often using such standard media as RS-232 or IEEE-488 interfaces, or on analog input/output interface cards designed to plug into the system's bus configuration.

Mathematical processing of large amounts of data by computers can be speeded up, with a greatly reduced CPU burden, by the use of coprocessors or auxiliary specialized number-crunching modules, for example *array processors*, which perform such tasks as digital filtering, fast Fourier transformations (FFT), and convolution. At the heart of these operations are hard-wired digital multipliers and multiplier-accumulators, which can multiply (for example) two 32-bit floating-point numbers at 5-MHz and faster rates.

At the highest level are mainframe computers, which handle very large amounts of data, perform computations at very high speeds, and perform a great many independent tasks simultaneously. Their dealings with analog signals are likely to be at second hand by the downloading of instructions to lesser entities.

UP-DOWN COUNTERS

These devices, analogous to ramp generators, are quite useful for performing a variety of tricks with a/d and d/a converters. They are used in forming electronic servo loops for automatic error correcting, offset adjusting, long-term sample-holds, time-function generation with ROMs and DACs, etc.

In electronic servo applications, the up-down counter accumulates pulses representing the variable being controlled, adjusted, or measured, in much the same way that a servo-motor shaft accumulates rotational angle. The counter is often used in conjunction with a d/a converter to develop an analog value proportional to the accumulated count. The process is also used in tracking-type a/d converters and resolver/synchro-to-digital converters. Counters are also used for updating multiplexer channels sequentially in response to pulses.

FILTERS

Low-pass filters are used on the input side of an a/d converter to remove unwanted high-frequency components of the input signal. Noise and line-frequency interference can also be attenuated by filtering, but at the expense of reduced response to fast input-signal amplitude variations. Filters (electrical or mechanical) are also used on the analog *output* from d/a converters, in order to smooth out the lumps created by discrete digital values. Filtering can be performed by digital techniques, using appropriate hardware and software, as well as by a wide range of active, passive, and sampling-type analog filters.

Pre-filtering is an important function, especially in high-speed information-processing systems, because the sampling frequency must be at least twice the highest frequency component of the signal input (Nyquist frequency). If higher frequencies are present, either within the signal or in the accompanying noise on the input channel, they can cause aliasing—intermodulation of unwanted high-frequency components of the signal (and input noise) with harmonics of the sampling frequency, to produce spurious signals at surprisingly low frequencies.

COMPARATORS

Analog comparators are important elements of data-acquisition systems. A high-gain analog comparator makes an elementary choice between the magnitudes of two inputs and decides which is the greater. This is the equivalent of a one-bit a/d conversion. Two comparators may form a decision "window". The a/d conversion process usually calls for a number of decisions; they may be made sequentially by a single comparator or simultaneously by a whole string of comparators, as in "flash" converters. Comparators may be free-running or latchable.

POWER SUPPLIES

Accuracy of interface systems is steadily rising, to the point where 12-bit resolution is quite routine, and 16-bit resolution is frequently needed for repeatability, resolution, linearity, and accuracy. Consequently, the design of the dc power system is no longer a trivial matter (it never really was!) since errors that remain third-order effects at 8-10 bits become menacing first-order effects at the 16-bit level. In many instances, careful separation of analog and digital grounds is required, demanding, in turn, considerable isolation between the various outputs that modern power supplies provide.

As with transducer selection, the priority of power-supply integration should be raised from the status of an afterthought. Too often, the power supply design (or choice) is left until last, where it is presumed to be able to take up all the slack or tolerances that other design stages create. Instead, power supplies (ac/dc or dc/dc) deserve at least as much initial attention at a competent engineering level as the selection of converters, amplifiers, sample-holds, multiplexers, and other devices.

A related question is whether to use power supplies and/or regulators in large, medium, or small chunks, for major portions of the system, for individual chassis, or perhaps even for mounting on individual boards. The issues involve space, cost, circuit independence vs. excessive lead length and wire size, connector technology, avoidance of ground loops, allowable local dissipation levels, etc., and must also include the question of interruptibility.

If continued operation despite loss of primary power is essential, either from considerations of overall system reliability or because of potential loss of data in volatile memory, the system design should include a determination of what level of operation or memory retention is necessary in case of an interruption—and some arrangement for continuity of supply, for detection when standby power is in use, and for contingency decisions or alarm.

DIGITAL PANEL INSTRUMENTS

These devices form a kind of self-contained digital processing system all by themselves, often comprising not only conversion and display, but also multiple-channel scanned inputs and signal conditioning (an example is the Analog Devices AD2036 6-Channel Scanning Thermocouple Thermometer). They can be either "dumb" or intelligent and can be used in conjunction with other digital equipment, since most DPIs have digital outputs available.

At its core, a DPI may be regarded as an (often self-powered) independent analog-to-digital converter—often including signal conditioning—complete with case, overrange capability, input protection, visual readout, and digital output, usually in parallel or multiplexed BCD (binary-coded decimal) format. Thus, the DPI can be used as a stand-alone converter that interfaces with both humans and machines.

DEDICATION

As we have already noted, data systems exist to serve human purposes. It is to all these humans, who design, program, use, play, and work (or are studying to do so) with systems that employ converted analog data, that this book is fraternally dedicated.

Chapter Two

Data Acquisition

Analog data is acquired in digital form for any or all of the following destinations:

Storage	Processing
Transmission	Display

Digital data may be *stored* in either raw or processed form; it may be retained for short, medium, or long periods. It may be *transmitted* over long distances (for example, to or from outer space), or short distances (from one part to another of a microprocessor-based instrument). The data may be printed on a printer or plotted on an X-Y plotter for a permanent hard copy, or it may be *displayed* on a digital panel meter, as part of a cathode-ray-tube presentation, as speech or other sounds, or in any other form that stimulates human senses.

Processing can run the gamut from simple comparisons to complicated mathematical manipulations. One might use it for such purposes as collecting information, converting data to a useful form, using the data for controlling a physical process, performing repeated calculations to dig out signals buried in noise, generating information for displays, simplifying the jobs of warehouse employees, controlling the color of paint, the thickness of a wrapper, maneuvers in a game, the speed of a subway train.

But it all starts with getting the data in digital form, as rapidly, as frequently, as accurately, as completely, and as cheaply as necessary.

The basic instrumentality for accomplishing this is the analog-to-digital (a/d) converter (ADC). It can be an IC chip, a shaft digitizer, a DPM with digital outputs, or a sophisticated high-resolution high-speed device; physically, it may take the form of a box, a card, a potted module, or an integrated circuit. It may be functionally integrated with other elements.

To accommodate the input voltage to the specified conversion relationship, some form of scaling and offsetting (signal conditioning) may be necessary, performed with an amplifier/attenuator. To convert analog information from more than one source, either additional converters or a multiplexer may be necessary. To increase the rate at which information may be accurately converted, a sample-hold may be desirable. To compress an extra-wide analog dynamic range, a logarithmic amplifier or conversion relationship may be found useful.

The properties which the data-acquisition system must have depend on both the properties of the analog data itself and what is to be done with it. This chapter deals with aspects of signal flow, from sensors through conversion. Chapters 4 and 6 have to do with integration into systems and instruments, Chapter 7 deals with the fundamentals of ADCs, Chapter 8 deals with some of the forms IC ADCs can take, and the Chapters in Part III deal with ADCs designed for special purposes, such as high-speed ("video") applications. A great deal more information on the nature and handling of analog signals from transducers in preparation for digital data acquisition can be found in the *Transducer Interfacing Handbook*[1].

In this chapter, we shall introduce some of the functional architectures that have proven useful and popular and discuss some of the considerations involved in the choice of configuration, components, and other elements of the system. Additional information will be found throughout the book.

2.1 THEN AND NOW

A quarter-century ago, a/d converters capable of 0.05% performance and 50,000-sample-per-second conversion rates cost about $8000, consumed about 500 watts, and occupied about one-quarter of a cubic meter.

Today, the completely self-contained monolithic Analog Devices AD573 requires less than 20 microseconds for a 10-bit 0.05% conversion, is available in quantity at less than 0.2% of the price, and is packaged in a 20-pin plastic DIP. And it is designed for easy interfacing with the modern microprocessor.

The space formerly occupied by the converter alone will now hold (for a similar order of dollar investment) a MACSYM 150: a complete multitasking minicomputer-based data-acquisition system including computer, keyboard, display, disk drive, converter, and a set of input-output cards.

In the past 25 years, as the above examples show, the processing power and complexity of data-acquisition and computer hardware have increased radically, thanks primarily to the semiconductor revolution. Sophisticated software and interactive terminals, which make computer techniques accessible

[1] Sheingold, ed., *Transducer Interfacing Handbook* (Norwood MA 02062: Analog Devices, Inc., 1980)

even to small children, permit interconnections, switching, and "knob twisting" to be performed under remote or programmable control. All of this, available at prices that were then undreamed of, has brought matters to the point that digital, rather than analog, "massaging" of information is a matter of routine, rather than exotic necessity.

What have not changed, however, are the fundamental system problems confronting every digital data-system designer. Of course, it helps to have small, quiet, low-cost, cool, low-current-drain components. But the designer is still up against the laws of Mother Nature, who often prefers to keep her secrets safely obscured by noise, EMI, ground loops, power-line pickup, and transients induced in signal lines from machinery. Separating the signals from these obscuring effects, then, becomes a challenge to ingenuity and imagination, coupled with a great deal of experience and persistence. Design is not merely a matter of purchasing fast, high-resolution a/d converters—but having them available at realistic cost is an incentive for giving them useful jobs.

2.2 ENVIRONMENT AND COMPLEXITY

Though there are many ways of starting to think about data-acquisition systems, a highly relevant approach has to do with environment. Some systems are intended to operate with modest accuracy in hostile environments (factories, vehicles, military surroundings, and remote installations); others are best suited to making high-precision measurements in such (probably fictitious) favorable surroundings as electrically quiet laboratories at room temperature; and yet others may be mixed, acquiring analog data generated in hot, noisy environments, processing it in the security of a quiet control room, and transmitting the resulting digital data through a world rife with interference.

Environment may give rise to such considerations as:

- Analog vs. digital signal transmission
- Signal accuracy vs. waveform recovery
- Isolation vs. direct wiring
- Subminiature vs. macroscopic size
- Simple vs. complex system architectures
- Integrated vs. distributed approaches
- Local vs. remote processing
- "Hi-rel" vs. "commercial" parts
- Choice of power supplies and physical hardware

Hostile environments manifest themselves in combinations of physical, chemical, and electrical challenges. *Physically* hostile environments may exhibit extremes of temperature, pressure, acceleration, humidity, and radiation (both ionizing and non-ionizing). *Chemically* hostile environments may involve such corrosive surroundings as salt spray, biological fluids, noxious atmospheres, dirt, and chemically active fluids and gases. *Electrically* hostile environments may include destructively high voltages and magnetic fields, as well as dc, ac, and transient interference over the whole spectrum.

In addition to these actively hostile forms of environment are "passively" hostile environments that must be disturbed as little as possible—physically, chemically, or electrically—by the introduction of data-acquisition equipment. Some environments are both actively and passively hostile.

Typical actively hostile environments might include the wide temperature range of aircraft engines, shock and vibration in railroad freight cars, moisture and corrosiveness in oceangoing equipment, high radiation levels in the vicinity of nuclear reactors, and combinations of many of these factors in geothermal wells. In passively hostile environments, the data-acquisition equipment (or portions of it) may be required to fit into small physical space, not raise the local ambient temperature substantially, not generate excessive electromagnetic interference, not be made of materials that would interact chemically with sensitive surroundings, and not use more than a small part of the power furnished to or generated within its surroundings.

On the other hand, for laboratory-instrument applications, the system designer's problems may be related more to the performing of sensitive measurements (usually under favorable conditions, with respect to electrical interference) than to the gross problems of protecting either equipment or the integrity of analog data. *However, with the increasing use of microprocessors and the promiscuous use of both analog and digital circuits in precision instruments, freedom from electrical interference cannot ever be taken for granted.*

Systems existing in hostile environments may require electronic devices capable of wide-temperature-range operation, excellent shielding, considerable design effort aimed at eliminating common-mode errors and preserving resolution, early conversion and digital data-transmission (perhaps via optical fibers), redundant paths for critical measurements, and—in some instances—considerable processing of the digital data to extract the maximum of information.

Measurements in the laboratory, with narrower temperature ranges and fewer sources of ambient electrical interference, may be easier to make and communicate, but higher accuracies (or resolutions) may require more sensitive devices, plus a still-considerable degree of effort to preserve appropriate signal-to-noise ratios.

2.3 KEY FACTORS

Environmental factors aside, the choice of configuration and circuit building blocks in data acquisition depends on a number of critical considerations, among them:

Resolution and accuracy
Number of analog channels to be monitored
Sampling rate per channel

Throughput rate
Signal-conditioning requirements
Intended disposition of converted data
The cost function

Besides the choice of appropriate component performance levels, careful analysis of the above factors is required to obtain the lowest-cost circuit configuration to obtain the desired overall performance.

Commercially available data-acquisition systems range from basic a/d converters to multichannel converters on monolithic chips to completely integrated systems on cards and in boxes, and they even include converters that are functionally inseparable from the digital processor. In later chapters, we will consider the various optional levels of integration available in a single package or piece of equipment. However, in this chapter, we will treat the functions peripheral to the a/d converter *as though they were embodied by separate components,* in order to make clear the architectural choices that a designer might have, with their characteristics, advantages, and disadvantages.

Typical configurations include:

Single-Channel possibilities
Direct conversion
Sample-hold and conversion
Preamplification
Signal-conditioning

Multi-channel possibilities
Multiplexing the outputs of single-channel converters
Multiplexing converter inputs
Multiplexing the outputs of sample-holds
Multiplexing the inputs of sample-holds
Multiplexing low-level data
More than one tier of multiplexers

Some of the more-interesting signal-conditioning options include:

Ratiometric conversion

Wide-dynamic-range options
High-resolution conversion
Range biasing
Automatic gain switching
Logarithmic compression
Logarithmic conversion
Digital correction of analog errors

Noise-reduction options
Filtering
Integrating-type converters
Digital processing

Finally, in evaluating tradeoffs, there are at least three types of "budgets" that should be considered: error budget, system time budget, and the relationship of the "make-or-buy" question to the cost budget at a given level of integration.

2.4 SINGLE-CHANNEL CONVERSION SUBSYSTEMS

Direct Conversion

Figure 2.1 represents the simplest digitizing system, a lone a/d converter performing free-running repetitive conversions at a rate determined by the time for a complete conversion. It has power inputs and an analog signal input. Its outputs are a digital code word—which may include overrange indication—in parallel, byte-serial, or serial form; polarity information (if the analog input is bipolar); and a "status" output that indicates when the output digits have become valid.

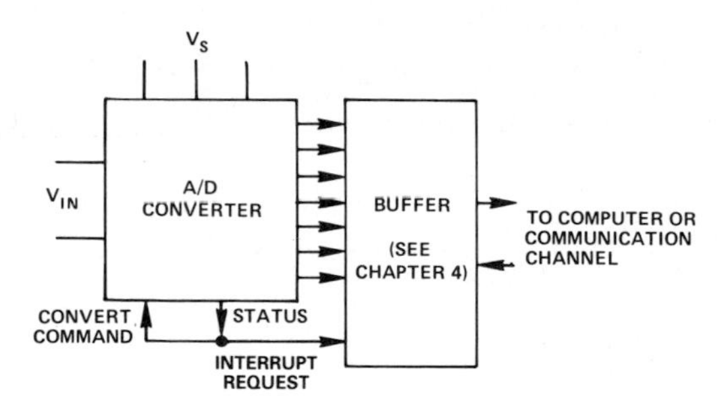

Figure 2.1. Single-channel free-running a/d converter.

Perhaps the best-known converter of this kind is the basic digital panel meter circuit, which consists of a basic a/d converter and a numerical display. For many applications, the sole purpose of digitizing is to obtain the display of the digits, i.e., to house the DPM circuit in a box and use it as a meter rather than as a system component.The DPM, however, is not necessarily the best way to digitize a single channel. Its two major shortcomings are: it is slow, and its BCD digital coding must be changed to binary if its output is to be processed by binary equipment. When free-running with a system, its output is strobed in following an Interrupt when the data becomes valid, rather than by a system interface command.

Converters designed for system applications (including many DPMs) can usually receive external commands to convert or hold. For dc and low-frequency signals, the converter is usually a dual-slope type (see Chapter 7), which has the advantage that it is inherently a low-pass filter, capable of averaging out high-frequency noise and nulling frequencies harmonically related to its integrating period. (For this reason, the integrating period is usually made equal to the period of the line frequency, since the major portion of system interference usually occurs at that frequency and its harmonics.)

The "actual" value of input that is converted by an integrating-type converter is represented by the average over the signal-integration interval. Since that interval is a fraction (about one-third, but not necessarily constant) of the total time required for the conversion cycle, what one can say about the value of the signal and when it occurred is that the digital output represents the most probable value during a significant portion of the conversion period.

For dc voltages or signals known to have constant values during the conversion interval, even if the changes occur rapidly (as long as they have settled prior to the start of a conversion), any new value will be converted to specified resolution and accuracy within the conversion interval.

For repetitive conversions of continuously changing inputs, however, the maximum rate at which the input signal can vary and still permit the converter to resolve 1 least-significant bit (LSB) of binary output, irrespective of the waveform, is

$$dV/dt = 2^{-n} V_{FS}/T \tag{2.1}$$

where V is the input, n is the number of bits of binary resolution, V_{FS} is the full-scale span, and T is the time between conversions. The maximum rate of change is thus 1 LSB per conversion period.

If $V = (V_{FS}/2) \sin 2\pi f t$, then

$$dV/dt = (V_{FS}/2)\, 2\pi f \cos 2\pi f t \tag{2.2}$$

and dV/dt_{max} is equal to the magnitude, $(V_{FS}/2)\, 2\pi f$. Thus,

$$2^{-n}/T = \pi f \tag{2.3}$$

and the maximum sine-wave frequency that can be converted with 1-LSB resolution is

$$f = 2^{-n}/(T\pi) \tag{2.4}$$

For example, if the conversion rate for a 12-bit binary integrating-type converter is 25 per second (T = 0.04s), f = 0.002Hz, corresponding to 0.12V/s of a 20-V span. For faster dV/dt's, changes cannot be resolved to within lLSB during a conversion period.

So far, the context has been that of the dual-slope integrating a/d converter, which spends about 1/3 of its sampling period performing an integration, and the remainder of the time counting out the average-value-over-the-integrating-period as a digital number, and resetting to initial conditions for the next sample. Though slow, the integrating a/d converter can be readily manufactured in integrated-circuit form and is quite useful for measurements of temperature, battery discharge, and other slowly varying voltages, especially in the presence of noise.

However, by far the most popular type of converter for system work is the successive-approximation device (Chapter 7), since it is manufacturable as an integrated circuit and is capable of high resolution (e.g., 16 bits), high speed (e.g., 1μs for 12-bit conversion), and quite reasonable cost.

The successive-approximation converter, used by itself, has the weakness that, at higher rates of change, it generates substantial linearity errors because it cannot tolerate change during the weighing process. The converted value will be at some value between the extreme values occurring during conversion, and the time uncertainty approaches the magnitude of the conversion interval. Figure 2.2 illustrates this point. Finally, even if the signal is varying slowly enough, noise rates-of-change (perhaps introduced by the signal itself) that are excessively large will cause erroneous readings that cannot be averaged, by either analog or digital means.

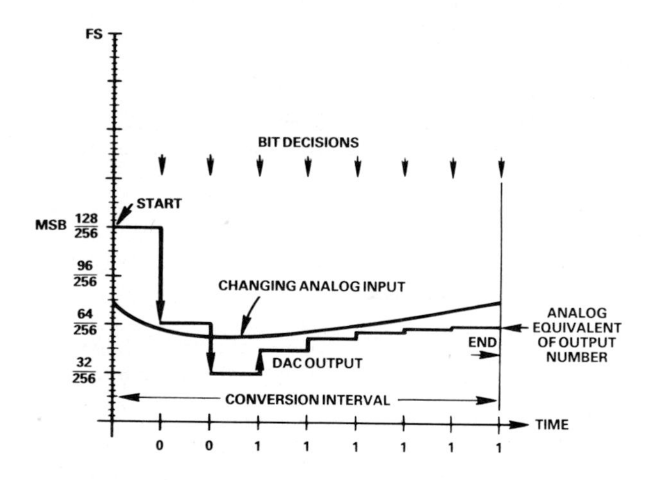

Figure 2.2. Error of successive-approximation 8-bit converter with changing input during conversion. The input is at the same level at the beginning and end of conversion, but the output value is about 16 LSBs low.

Since the final converted value occurs at an unknown time during the conversion interval, the time uncertainty corresponds to the conversion interval. In the example of Figure 2.2, a value equivalent to the output number occurs twice, at about 1/8 and 5/8 of the conversion interval. For a sinusoidally varying input, the relationships expressed in equations 2.1 through 2.4 apply. In the above example, if T, for a successive-approximation converter, is 1.5μs for 12 bits, the maximum allowable frequency for maintaining bit-at-a-time resolution of a sine wave becomes 52Hz, and the maximum rate of change for 1-LSB resolution of a 20-V span becomes about 1,600V/s—an improvement over the integrating converter, but far from sensational.

Sample-Hold and Conversion

A converter can be made to operate at considerably greater accuracies at high speeds with precise timing of samples, irrespective of the time required to complete a conversion—overcoming the weaknesses mentioned above—by introducing a *sample-hold* (or *track-hold*) between the input signal and the converter's input (Figure 2.3).* Between conversions, the device may acquire and track the input signal. Just before a conversion is to take place, it is switched to *hold* and remains in that state throughout the conversion.

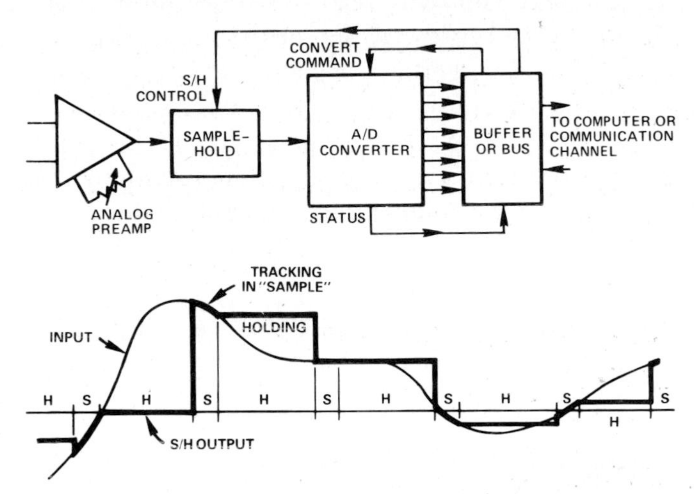

Figure 2.3. Sample-hold in single-channel data acquisition—block diagram and waveforms.

It can be seen that, if the S/H responds instantaneously and accurately, the converter can accurately convert signals having rates of change of any magnitude, at sampling rates up to the ADC's maximum conversion rate. In practical sample-holds, however (Chapter 18), there will be such time-related errors as acquisition time, tracking delay, and aperture time. Typical values of these parameters for track-holds designed to be used with fast a/d converters are 5μs acquisition time to 0.01%, 50ns tracking delay, and 25ns aperture time, with 0.5ns uncertainty (jitter). If the acquisition time is adequate, and aperture time and tracking delay: compensate one another, can be nulled by phase adjustment if necessary, or are unimportant in repetitive sampling as long as they are consistent, the principal (irreducible) source of time error in sampled-data systems is the *aperture uncertainty*.

When a sample/track-hold is used with an a/d converter, the signal is frozen as of the instant *hold* is achieved; thus, T in equations (2.1) through (2.4)—the uncertainty as to when the signal was sampled—represents the aperture uncertainty (instead of the much longer conversion time). If a track-hold with

*Strictly speaking, a *track-hold* (T/H) tracks the input until the sampling ("hold") command is received, then holds; a *sample-hold* (S/H) remains in hold until commanded to get a sample ("sample"), then quickly returns to hold. In most cases, the same device can be used in either fashion.

0.5ns aperture uncertainty is used with the 12-bit, 1.5μs converter mentioned earlier, the maximum-frequency sinusoidal signal that can be converted with 1-LSB resolution is $(2^{-12})/(0.5 \cdot 10^{-9} \cdot \pi) = 155$kHz (from 2.4), corresponding to a maximum rate-of-change of about 9.8V/μs. The sample-hold should have enough bandwidth to deal accurately with the signal amplitude (for stationary ensembles)—but also with phase for unique events or where phase between two channels is critical.

Figure 2.4 shows that—in contrast with Figure 2.2—at the end of each conversion interval, the converter, with a constant input applied by the S/H, will deliver an accurate digital representation of the input value as it was at the start of conversion. If the converter is accurate, any errors that are functions of time will be due to errors of the sample-hold, including the acquisition errors mentioned above, plus droop during the conversion interval, and any linearity, offset, and transient errors. Band-limited random noise present on regularly sampled signals, before sampling and conversion, may be susceptible to digital averaging by the processor.

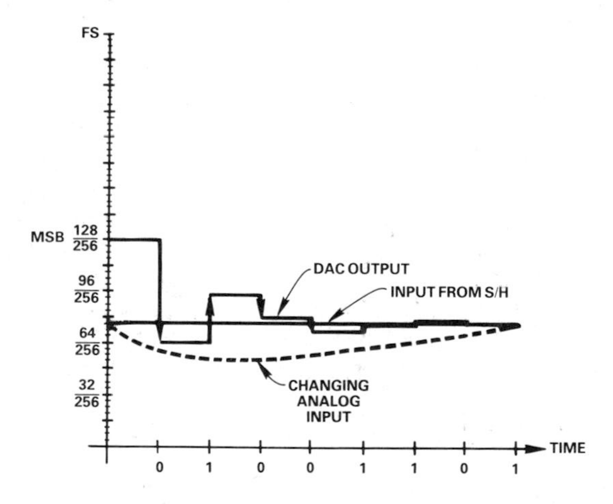

Figure 2.4. Same converter and input signal as in Figure 2.2, but with converter preceded by sample-hold, the changing input signal is ignored. Converter produces the correct output as of the instant of *hold*.

Sample, Signal, Noise, and Aliasing In order to avoid errors due to an insufficient number of samples, the Sampling Theorem tells us that regularly spaced sampling must occur at least at the *Nyquist rate*, twice the frequency of the highest-frequency signal or noise component; that is, either a sufficiently high sampling rate must be employed or else all components of signals and noise at frequencies equal to or greater than the *Nyquist frequency*, i.e., one-half the sampling rate, must be filtered out before sampling. Since practical filters require a compromise between attenuation in the pass band and transmission in the stop band, the sampling rate is often three or more times the filter cutoff frequency.

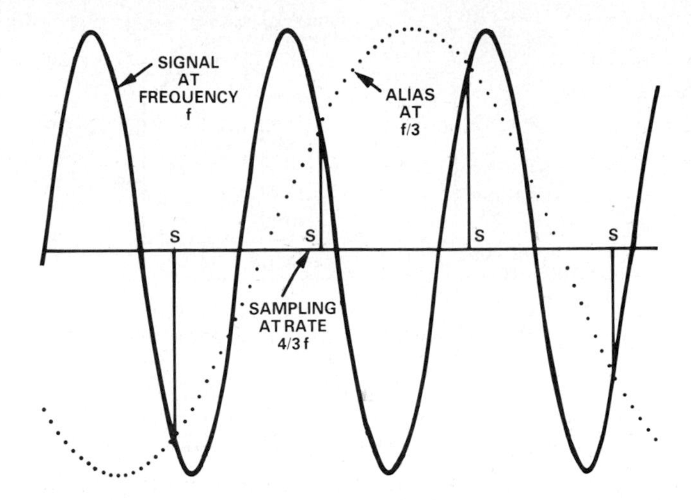

Figure 2.5. Example of aliasing. A sinusoid is sampled (S) at 4/3 of its frequency (f). The resulting set of samples form an alias at $(4/3 - 3/3)f = (1/3)f$, i.e., a train of pulses with amplitude and timing indistinguishable from those associated with a sinusoid at f/3.

If analog signals at higher frequencies are present, the sampling process will produce sum and difference frequencies with the sampling frequency and its harmonics; the difference frequencies, in particular, will produce spurious low-frequency signals, or *aliases*—in the signal passband—that cannot be distinguished from the signal. Figure 2.5 is a simple example illustrating aliasing.

Since sample-holds usually operate at unity gain*, with errors referred to full scale (which should be the same as the converter's full-scale range), scaling or preamplification should usually occur before the signal is applied to the sample-hold.

Preamplification (Figure 2.6)
In most cases, converters designed as components are "single ended" with respect to power common† and have normalized input ranges of the order of 5 or 10 volts, single-ended or bipolar. It makes sense to scale signal inputs up or down to the standard converter input level, to make fullest possible use of the converter's available resolution.

Figure 2.6a shows a typical preamplifier configuration. The preamplifier should have low dynamic output impedance, because the inputs of some types of a/d converters may have large current pulses, which will load the preamp's output and can cause errors. Sometimes these functions are combined with the converter. For example, in Figure 2.6b, we see a block diagram of

*Some types, however, do permit adjustment of gain and input configuration by a choice of external circuitry.
†Some converter designs and digital panel meters, have differential, and even isolated, floating inputs, and provide signal gain.

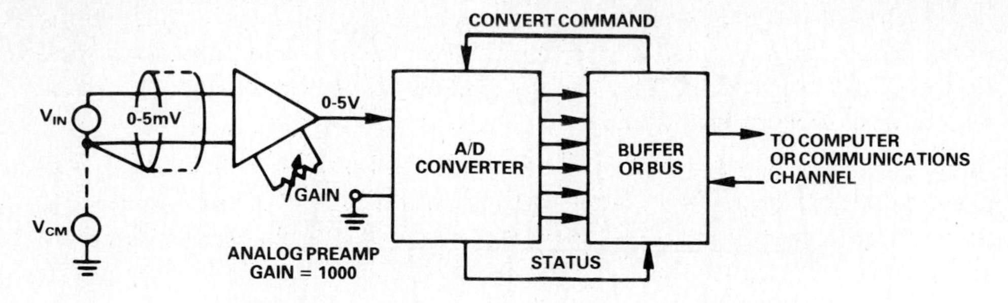

a. A/D converter with preamplifier/signal conditioner.

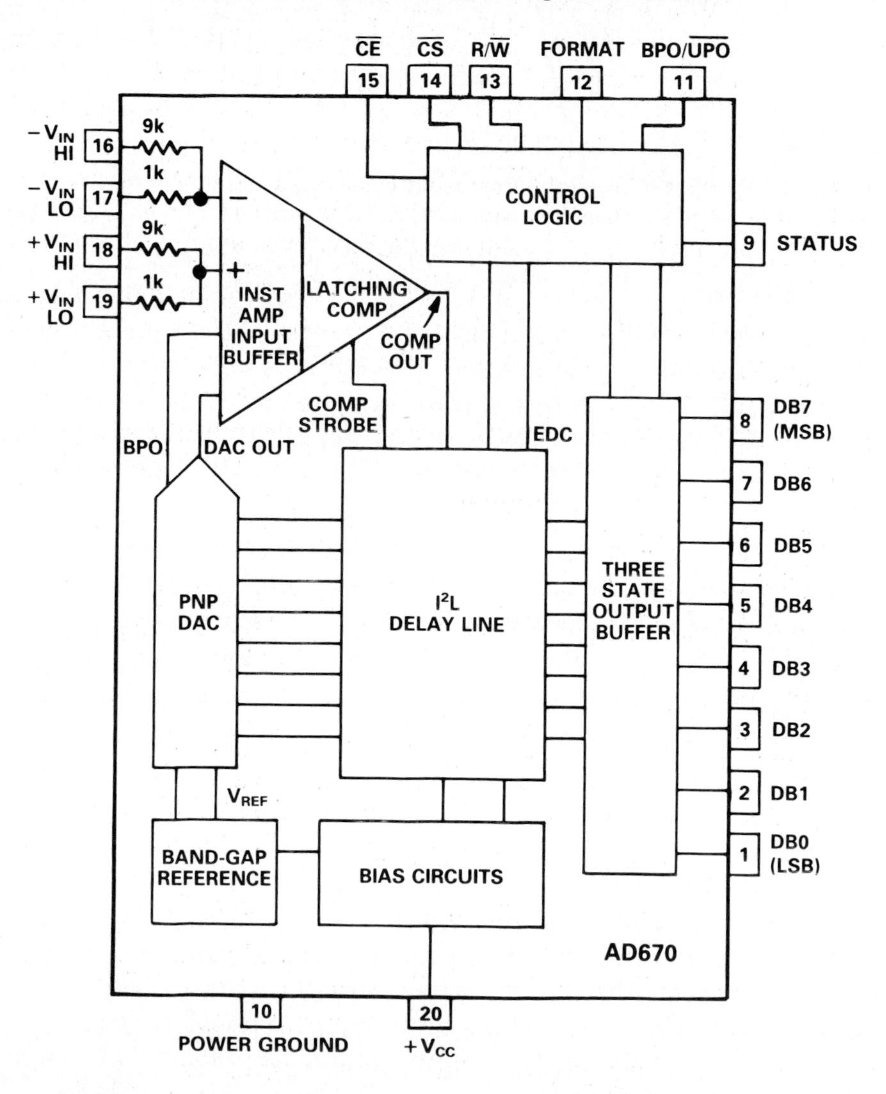

b. Block diagram of an 8-bit ADC with on-chip signal conditioning.

Figure 2.6. A/D Converter and preamplifier.

monolithic 8-bit a/d converter with a built-in gain-of-10 differential preamplifier and jumper-programmable choice of gains and input configuration (see Section 8.6.1). The on-chip preamplifier also buffers the input source from the conversion process.

Single-channel data-acquisition systems are available that include a programmable-gain amplifier (PGA)—with gain controlled by switching of resistors, either manually or under software control, plus a sample-hold function and an a/d converter in a single modular or integrated-circuit package.

If the signals are of reasonable magnitude (already preamplified), and already exist within a system referenced to a good-quality common "ground," the scaling may be simply accomplished with operational amplifiers in a single-ended or differential configuration. As is more generally the case, if the signals are from outside the system (or subsystem, or board, or neighborhood), or are quite small, or have an appreciable common-mode component, a differential instrumentation amplifier may be profitably used, with characteristics that depend on the gain required, the signal level, the needed CMR, bandwidth, impedance levels, and cost tradeoffs.

If the input signals must be galvanically isolated from the system, an isolation amplifier must be used to break all conductive signal paths. Such amplifiers generally employ optical or magnetic coupling. Isolation is mandated for protection of patients and subjects in clinical medical-instrument applications; it is also useful where common-mode spikes are encountered, as well as for industrial applications requiring *intrinsic safety* and for applications in which the signal source is at a high off-ground potential.

Signal Conditioning

This blanket term includes a wide variety of analog-to-analog possibilities. Many of the functions could also be performed digitally, under program control, depending on availability of processing capability, tradeoffs of cost, speed, accuracy, hardware vs. software, level of system integration, and the designer's personal orientation.

Signal-conditioning devices are available in a variety of packages and capabilities to meet the needs of the system designer. For example, instrumentation, isolation, and thermocouple amplifiers are available in IC packages, signal conditioners are available as board- or track-mountable modules, and whole families of modular signal conditioners for different purposes are available for mix-match mounting in expandable manifolds that include power supplies.

Here are some instances of signal conditioning. Scaling of input gains to match the input signal to the converter's full-scale span, using op amps or an instrumentation amplifier, is a simple, obvious example. One might also include dc offsets to bias odd ranges (for example 2-to-10 volts, derived from a 4-to-20mA instrumentation current loop via a 500-ohm load resistor) to levels more compatible with standard converters. Preamplification has already been mentioned.

Linearizing of data from thermocouples and bridges can be performed by analog techniques, using either piecewise-linear approximations (generated by biased-diode circuitry) or smooth series-approximations, using low-cost IC analog multipliers. It can also be done digitally, after conversion, by performing the necessary calculations with a microprocessor or by storing the inverse or complementary function in a read-only-memory (ROM) lookup table.

Analog differentiation can be used to measure continuously the rate at which the input varies; integration could be used to obtain total dosage from a rate of flow. Either could be used to produce a 90-degree phase shift; an op amp, connected as a simple all-pass filter, can be used to provide an arbitrary phase shift. Sums and differences could be used to reduce the number of data inputs (analog data reduction).

Analog multipliers canbe used to compute power by squaring voltage or current signals, or multiplying them together. RMS-to-dc converters compute rms directly. Analog dividers of various types could be used to compute ratios or the logarithms of ratios, or square roots. Devices that compute $Y(Z/X)^m$ can take ratios over wide dynamic ranges and perform analog function fitting.

Comparators can be used to make decisions based on analog levels (e.g., to convert only when an input exceeds a threshold or is within a "window"). Op amps and diodes may be used to perform simple "ideal-diode" functions.

And—what seems like getting "something for nothing"—logarithmic circuits can be used for range compression to permit the conversion of signals having wide dynamic ranges with converters having considerably less resolution than would be otherwise required.

Active filters are essential elements to minimize the effects of noise, carrier frequencies, and unwanted high-frequency components of the input signal. The increasing interest in filtering is reflected in the growing number of books and availability of hardware and software devoted to both analog and digital filter design. Analog filters can be either fixed or digitally programmable, using d/a converters.

One could go on and on but the basic point should have been made: that in system design, all data-processing need not be digital. Analog circuits can perform remote or local processing or data reduction effectively, reliably, and economically, and should be considered as alternative ways of reducing software complexity, noise, board space, and—quite often—cost. Figures 2.7 through 2.10 show a few examples.[2]

[2] For many more examples, see Sheingold, ed., *Nonlinear Circuits Handbook* (Norwood MA 02062: Analog Devices, Inc., 1974). Many additional examples can also be found in (1).

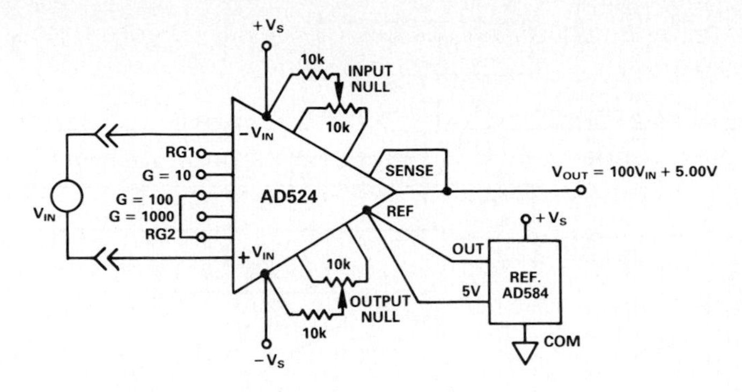

Figure 2.7. Instrumentation amplifier provides offset and scaling. Here, the gain is 100V/V, and precision offset is +5.00 V.

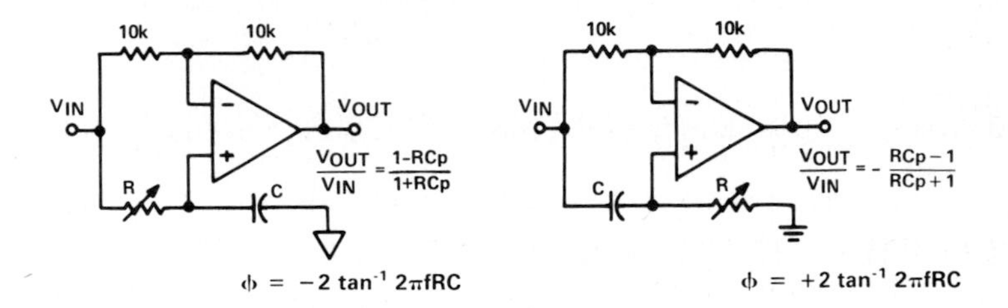

Figure 2.8. Operational amplifier generates an arbitrary phase shift. Gain = +1, all pass.

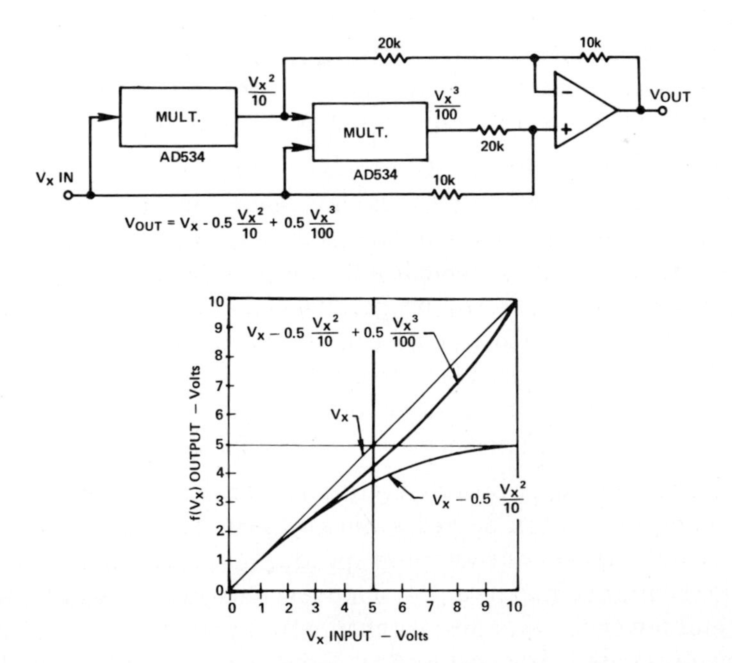

Figure 2.9. Using multipliers for nonlinear functional relationships, such as linearization.

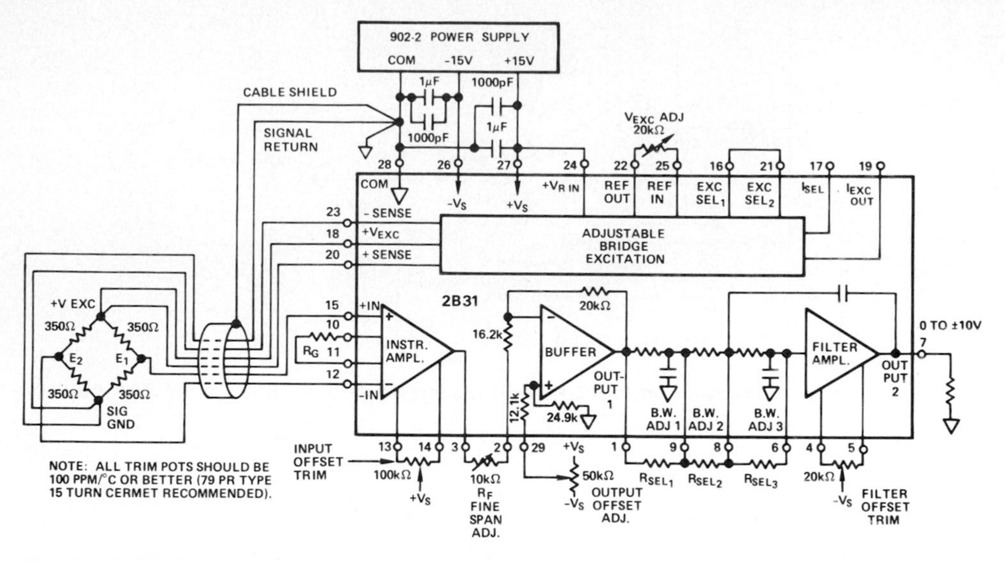

Figure 2.10. Using a signal-conditioning module for bridge excitation, preamplification, and filtering.

2.5 MULTI-CHANNEL CONVERSION

In multi-channel systems, elements of the acquisition chain may be shared by two or more input sources. This sharing may occur in various ways, depending on the desired properties of the multiplexed system. Large systems may combine several different kinds of multiplexing, as well as cascaded tiers of the same kind. Chapter 4 has more information about multiplexed data acquisition in practice at higher levels of integration, and Chapter 19 has further information about multiplexers and switches.

Multiplexing the Outputs of Single-Channel Converters

Although the conventional way to digitize data from many analog channels is to introduce the time-sharing process at the analog portion of the system by multiplexing the input of a single a/d converter among the various analog sources, in sequence, an alternative parallel conversion process is becoming increasingly practicable. Cost of a/d converters has dropped radically in recent years, and it is now possible to assemble a multi-channel conversion system, with the seeming extravagance of one converter for every analog source, yet considerably improved performance at reasonable cost (Figure 2.11).

This parallel conversion approach has its advantages. First, a desired overall digital throughput rate can be had with slower converters; alternatively, the converter-per-channel may run at top speed, providing a much greater flow of data into the digital interface. For a modest data rate, however, with more channels (and fewer conversions per channel), it may be possible for sample-holds to be eliminated, at a cost saving. Fewer conversions might also mean that a slower converter could be used, generally resulting in even further cost savings, especially since some channels may not require high resolution.

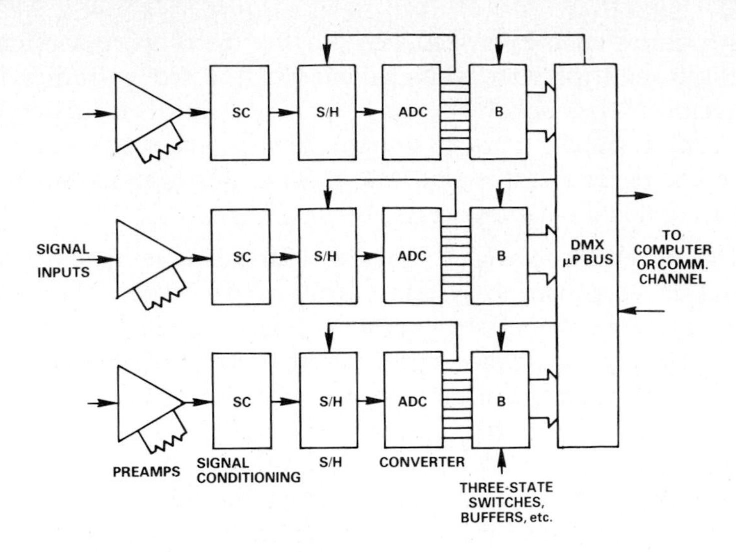

a. Basic multi-channel conversion scheme, using digital multiplexing.

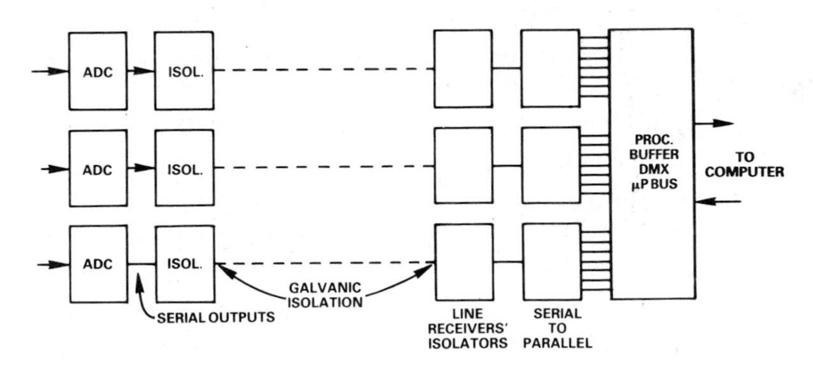

b. Multi-channel conversion using remote a/d converters.

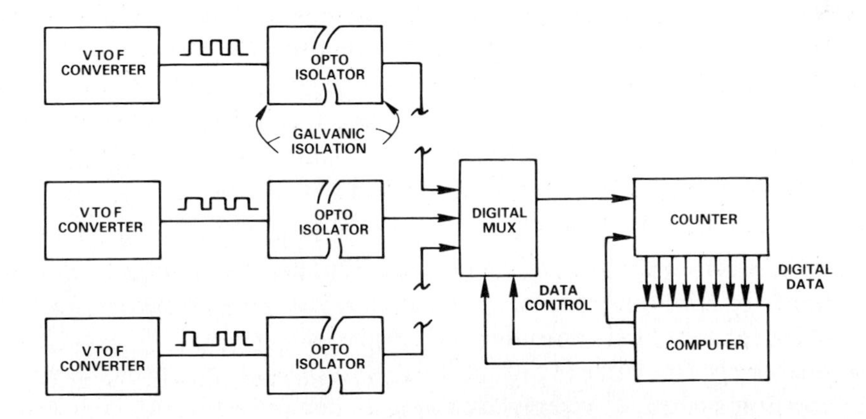

c. Multi-channel conversion using v/f converters and digital multiplexing.

Figure 2.11. Multi-channel conversion schemes.

The bus structures used by systems employing microprocessors encourage the use of digital multiplexing, with all devices connected to the bus via 3-state switches, enabled selectively by "chip select" logic signals from decoders and read/write control signals (*write* to initiate conversions, *read* to obtain the results). The converter's status line can provide *interrupt* signals indicating "conversion complete—data is ready".

The parallel-conversion approach provides a further advantage when applied to industrial data-acquisition systems, where strain gages, thermocouples, thermistors, etc., are strung out over a large geographical area. In essence, by digitizing the analog signals right at their source and transmitting serial digital data, rather than the original low-level analog signals (Figure 2.11b), a considerable immunity to line-frequency (50-60-400Hz) pickup and ground-loop interference is achieved. Among other factors, the digital signals can be coupled optically or even transmitted via fiber-optic links for complete electrical isolation and total indifference to electrical interference.

A multichannel array of voltage-to-frequency (v/f) converters is an interesting way of transmitting data generated by slowly varying signals, with dynamic ranges of up to 10^6—and requiring accuracies to within 0.01%. The outputs are TTL pulse trains, which may be easily isolated optically. The output of each VFC, in turn, is counted and read. As Figure 2.11c shows, the computer controls the multiplexer and acts as a time base for the counter.

Not least, among the subtle benefits of digitizing sensor signals at their source, is the ability to perform logical operations on the digitized data before it is fed into the computer. In this way, for example, mainframe involvement with data is streamlined and redundancies are minimized. More specifically, remote processing makes it possible, for example, to access data from slowly varying thermocouple sensors less frequently, while reading in data from rapidly changing critical sources at enhanced speed. In fact, the versatility of a digital subsystem may be exploited to make its own decision as to when a particular data channel should be brought to the attention of the computer by means of Interrupts. If certain signal sources remain constant or within a narrow range for long periods, then change rapidly later in the process, it is possible to ignore these data until the changes occur. (A local microcomputer can store the stationary values and make the decisions.)

In sum, a great deal of flexibility and versatility is gained by changing the interface process from analog multiplexing to digital multiplexing. Logic decision circuits or local microprocessors can exercise judgement on when and what data to feed the host computer and, in general, can give the overall interface a much larger measure of autonomy than is possible with an entirely analog conversion system. (Systems involving analog multiplexing have a Catch 22: The computer cannot make decisions about the data submitted by an analog multiplexing system until it has received the data upon which to base its judgements... this means that the data have been converted and interfaced before the computer can decide that a particular piece of information is redundant. And there is no guarantee that it will be redundant on the next pass.)

Finally it should be noted that if, for example, the data is being transmitted from a lunar vehicle to Earth, the channel is quite crowded, and the sort of *redundancy-reduction data compression* described above is absolutely essential to make sure that the items of data that get through are those having the highest priorities, by virtue of containing intelligence rather than redundant information.[3]

For each channel of the digitally multiplexed system, there could be the chain described earlier: preamplifier, signal conditioning, sample-hold, converter. It is also possible that, for one or more of the channels, there are a number of multiplexed subchannels, especially if they are carrying similar information.

Examples of Multiplexed Converter Inputs

In some scanning-type panel meters (such as the AD2037), the input channels undergo multiplexing, followed by signal conditioning and a conversion, in which the analog input circuitry is isolated from the digital logic circuits.

An unusual single-chip integrated-circuit data-acquisition system, the 8-bit AD7581 (Figure 2.12), multiplexes and converts 8 channels of analog data in sequence continuously, and stores the most recent value of each channel in a separate memory register, where it may be addressed and accessed via a microprocessor data bus at any time.

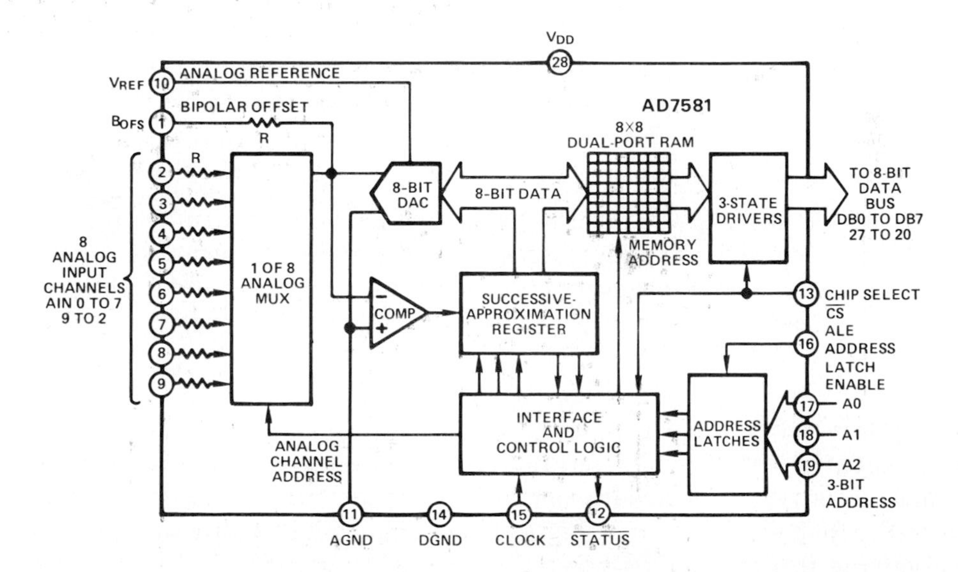

Figure 2.12. Block diagram of 8-channel 8-bit memory ADC.

[3] "New approaches to Data-Acquisition System Design," by T. O. Anderson, *Analog Dialogue* 5-1, 1971

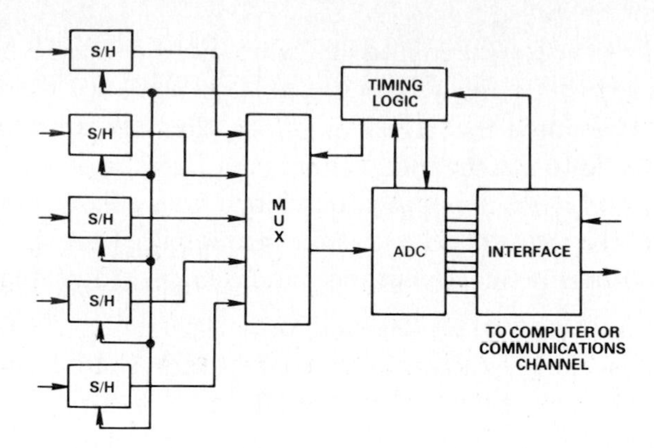

Figure 2.13. Multiplexed system with simultaneous sampling, sequenced conversion.

Multiplexing the Outputs of Sample-Holds

Working back from the interface (with a minimum number of shared elements, except for the bus) towards the more conventional situation, in which the number of shared elements is maximized, we consider the intermediate case of a shared a/d converter, with a multiplexer at its input, switching among the outputs of a number of sample-holds (Figure 2.13). This configuration is found where sample-holds are updated rapidly, perhaps even simultaneously, or at critical instants for individual inputs, then read out in some sequence. It is generally a high-speed system, in which all items of data delineating the state of the system must do so for the same given instant. Multiplexing may be done sequentially or, when required, by random addressing. The sample-holds must have sufficient freedom from droop to avoid accumulating excessive error while awaiting readout, which period may be considerably longer than in the case of the converter-per-channel. Increased throughput rate could be obtained by using additional converters, with fewer multiplex switch points and faster update rate.

Applications that might require this approach include wind-tunnel measurements, seismographic experimentation, or in testing complex radar or fire-control systems. Often, the event is a one-shot phenomenon, and the information is required in the neighborhood of a critical point during the one-shot event ... such as, for example, when a supersonic air blast hits the scale model.

Multiplexing the Inputs of Sample-Holds

The next step towards increased sharing is to share the sample-hold, as well as the a/d converter. Such subsystems may also include a programmable-gain amplifier for gain ranging (see Figure 2.17). Figure 2.14 shows a basic system embodying this idea, and a block diagram of an early form of device, with wire-programmed range. For most-efficient use of time, the multiplexer is seeking the next channel to be converted, while the sample-hold, in *hold*, is having its output converted. When conversion is complete, the *status* signal from the converter causes the S/H to return to sample (track) and acquire the next channel. Then, after the acquisition is completed—either immediately, or upon command—the sample-hold is switched to *hold*, a conversion begins, and the multiplex switch moves on.

This system is slower overall than the previous example, and the multiplexer could equally well be switching sequentially or in a random-access mode. For some older systems, a manual mode, for checkout, may also have been used (self-checking and keyboard random access are more typical nowadays). The random-access mode permits channels with more information, i.e., changes per unit time, to be accessed more frequently.

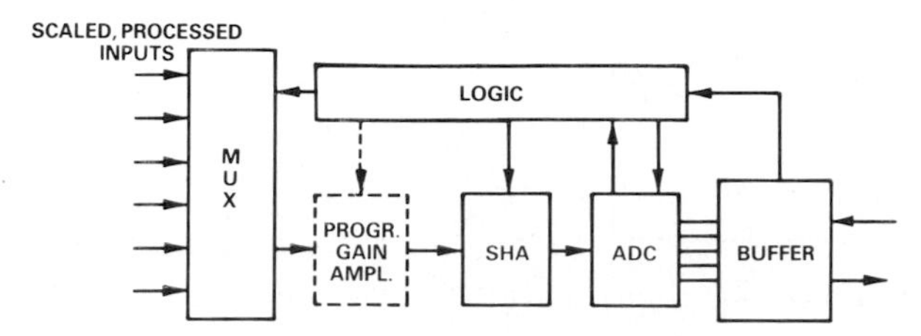

a. Data-acquisition system—basic architecture. The ADC, SHA, and even the PGA (if used) may be combined in one module or IC package to form a single-channel data-acquisition system (sampling ADC). Alternatively, the MUX, SHA, and amplifier are often packaged together to allow a choice of a/d converter.

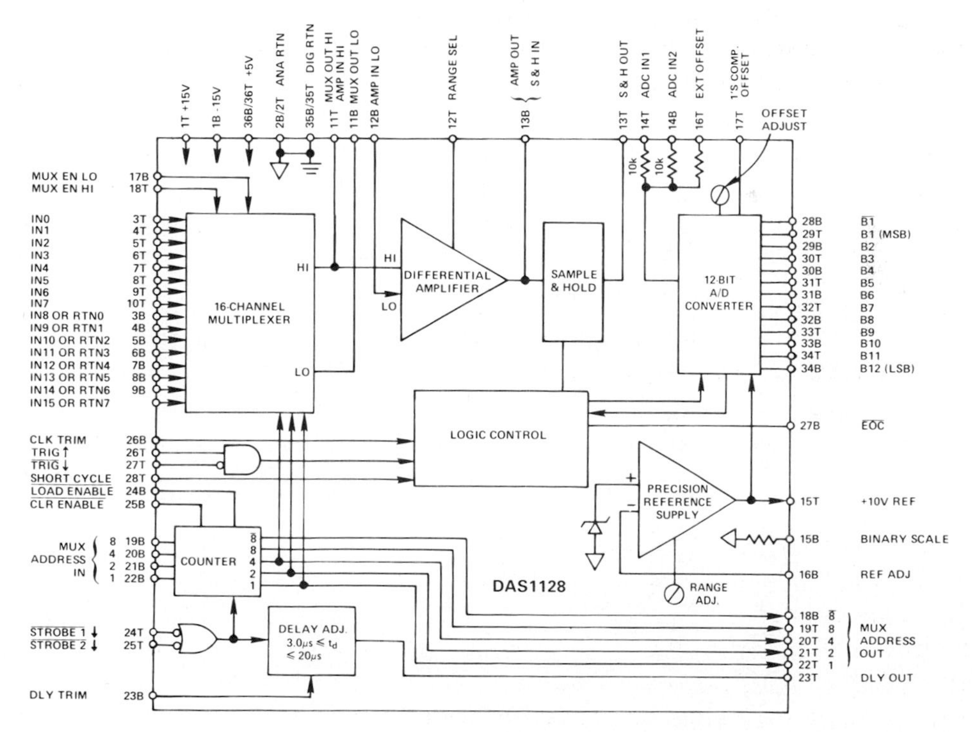

b. Block diagram of early data-acquisition module.

Figure 2.14. Conventional data-acquisition subsystem.

Multiplexing and Signal-Conditioning Low-Level Data

The idea here is that, in addition to sharing of the converter and the sample-hold, expensive signal-conditioning capacity must be conserved. Great strides have been made in recent years in developing effective all-solid-state approaches, supplanting the straightforward "brute-force" approach (Figure 2.15a).

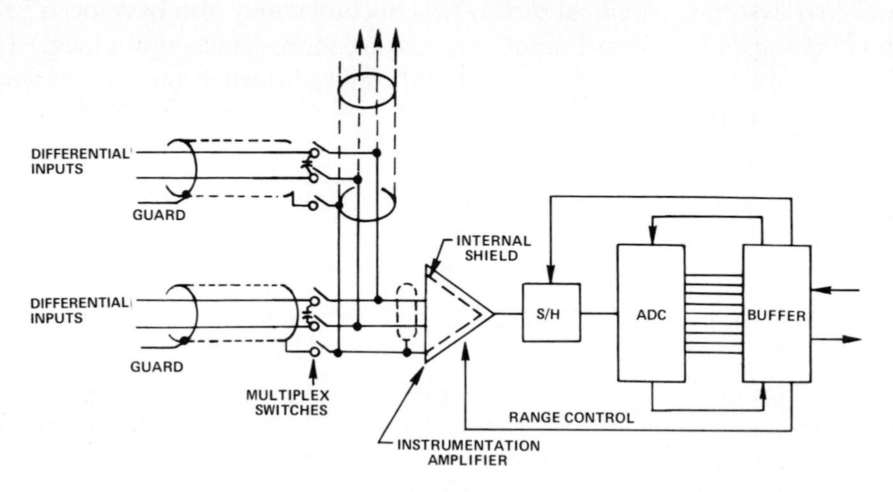

a. Conventional low-level multiplex circuit, showing switched guard. In some systems, two channels are reserved for zero and reference voltage. They are sampled periodically to permit software correction of amplifier offsets and gain errors.

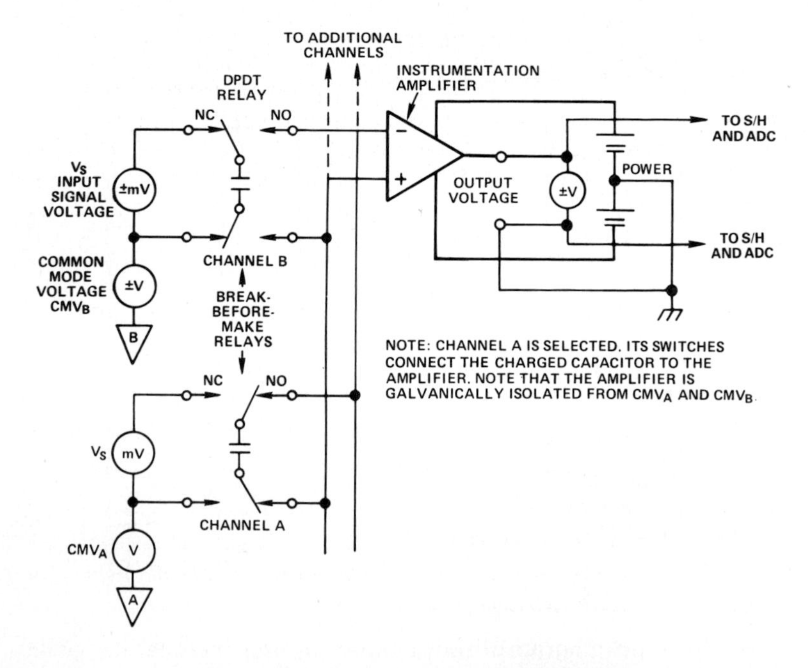

b. Example of flying-capacitor multiplexer.

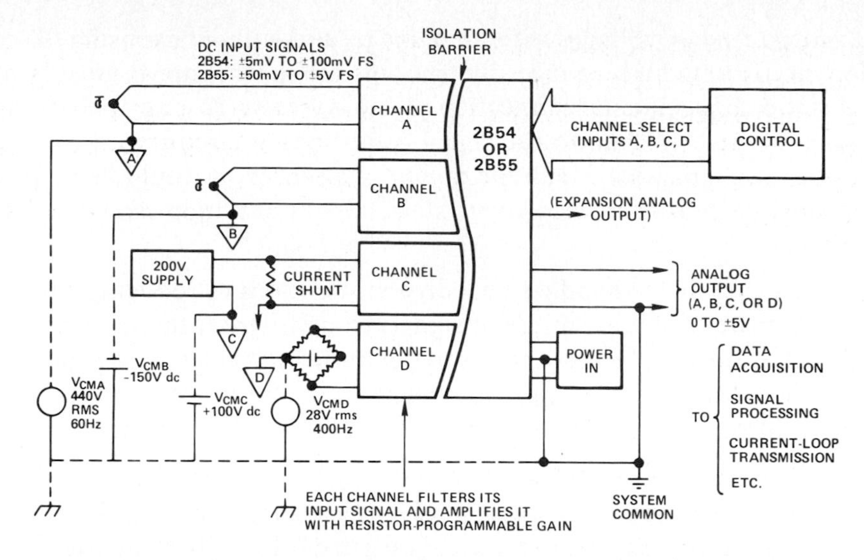

c. Block diagram of 4-channel multiplexed low-level signal conditioner.

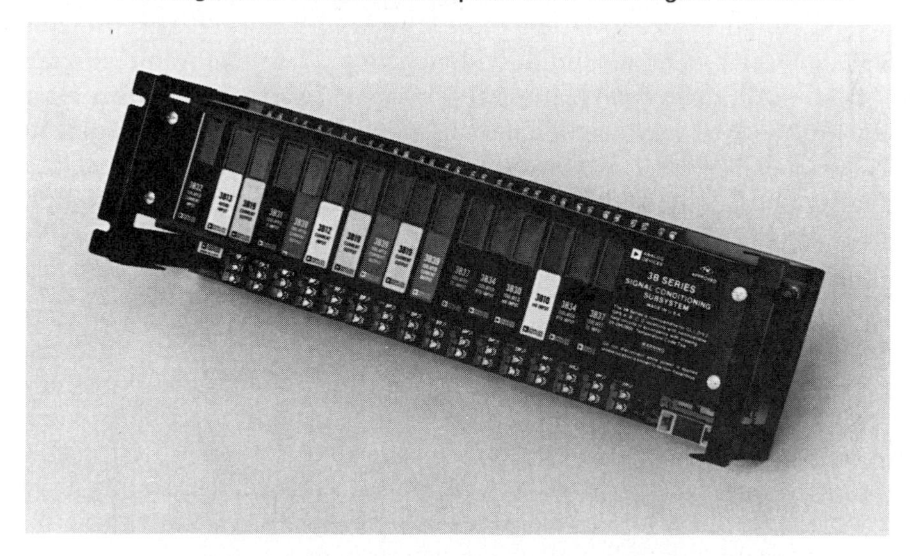

d. A rack-mounting 16-channel signal-conditioning subsystem.

Figure 2.15. Low-level data acquisition.

This has classically been a difficult function to perform. First, the circuitry must be capable of resolving millivolt-level voltages in the presence of large common-mode voltages, with low drift and nonlinearity. The common-mode voltage may be present on the signal's "ground" or it could be induced in the input leads inside a conduit in the vicinity of high-power mains. The signal itself may be afflicted with normal-mode noise.

Safety is a paramount consideration in many of the applications where this kind of circuitry is employed. The input circuitry must be able to withstand high common-mode voltages without damage, and without exposing the conversion circuitry to high voltage. By the same token, the inputs must be able to withstand the accidental connection of line voltage across any pair of input terminals without mishap to the signal-conditioning circuitry. If the input should be *open-circuited*, due to the failure of a sensor, it would be helpful if an appropriate indication were given, since false information can cause safety problems.

It should be possible to handle signal diversity without introducing complexity, whether the signals are millivolt signals from different thermocouples requiring different scale factors or mixtures of millivolt signals from strain gages and volt-level signals from potentiometers and current-transmitter loops. These conditions are—and have been—difficult to meet at reasonable cost.

Perhaps the most successful approaches are those exemplified by "flying capacitors" and circuits combining solid-state isolation and switching. A flying-capacitor multiplexer is illustrated in Figure 2.15b. Here, capacitors, are switched—generally by relays—between the signal sources and the system input bus; this provides isolation and multiplexing, as well as sample-hold, but it is not easily adaptable to individual gain adjustment. While not necessarily costly in terms of parts, the actual circuit construction requires great care, resulting in a custom installation at high overall cost, with a substantial software overhead for the calibration of individual channels. It is also potentially noisy, and common-mode range is limited by the voltage ratings of the capacitors and switch contacts; speed and life are limited.

An all-solid-state approach, as typically embodied by the Analog Devices Model 2B54 (Figure 2.15c), combines—in a single compact module—transformer-coupled isolation (±1000V peak max) for each input, differential input protection (130V rms @ 60Hz), signal conditioning, and multiplexing for four channels of millivolt-level input (e.g., from thermocouples). An expansion output provides for the multiplexing of additional groups of four channels *without the addition of external analog switches*. Channels can be scanned at a rate of up to 400 per second, minimum. Accessories are available for connecting sensor inputs directly to screw terminals.

For large numbers of channels having diverse input requirements, input and signal protection, direct connection to field wiring, and standard output formats (both voltage and 4-20-mA current-loop), subsystems are available consisting of powered manifolds of various sizes with assortments of plug-in modules (one per channel) that perform specific signal-conditioning functions. The Analog Devices 3B series (Figure 2.15d) is a typical example of such a system.

2.6 SIGNAL-CONDITIONING TOPICS

Discussed here are a few topics that keep coming up in connection with data-acquisition systems.

Ratiometric Conversion

Some a/d converters have a *ratiometric*, or external reference, connection, allowing the output digital number to represent the ratio of the input to an arbitrary (within specified limits) reference input. In effect, the device becomes an analog divider with digital readout.

Devices of this sort are useful in making precision measurements that ignore variation of a reference used in the measurement. For example, Figure 2.16 shows how a resistance ratio, which might represent a pressure, can be measured—independently of variations of the applied voltage—by applying the same voltage to the reference input of the converter.

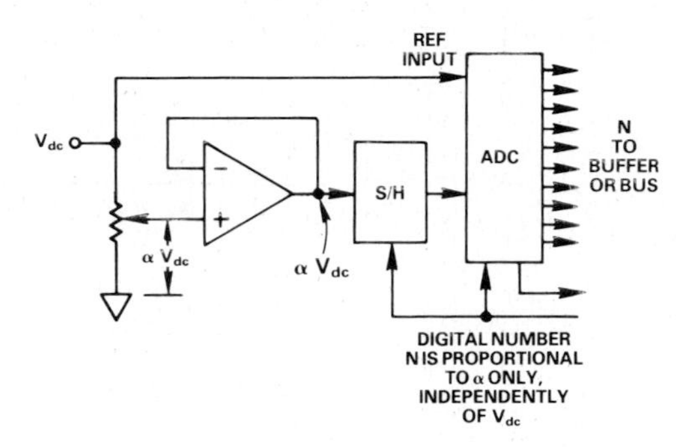

Figure 2.16. Measuring a resistance ratio, independently of the applied voltage—a ratiometric measurement. If common-mode rejection is necessary, the amplifier could be an instrumentation type (or a signal conditioner) instead of a simple single-ended unity-gain follower.

In a multiplexed system, where measurements may be taken from a number of similar devices with a common supply, for example, strain-gage bridges, the common bridge supply may be used as the converter's reference to eliminate normal-mode gain error caused by supply-voltage (or converter reference) variation.

Wide Dynamic Ranges (see also Chapter 17)

The need for wide-dynamic-range signal conditioning in a single channel may occur in two basic ways: Either it is necessary to resolve a voltage anywhere in the range to a high degree of accuracy, relative to full scale (for example in the measurement of position in a followup system); or it is sufficient to measure a quantity having a wide range of variation to modest accuracy, relative to actual value (for example, to within 1%, over a 10,000:1 range).

For signals in the first category, a high-resolution-and-linearity converter is the simplest answer.

To maintain wide dynamic range for small signals, it is feasible to use a moderate-resolution converter (say 12 bits), preceded by a software-programmable-gain amplifier, i.e., an amplifier with high-accuracy switched gain controlled from the digital interface. Software-programmable-gain amplifiers are available in several forms: They can be found in data-acquisition subsystems (Figure 2.17); purchased as complete modules or hybrids; and assembled from

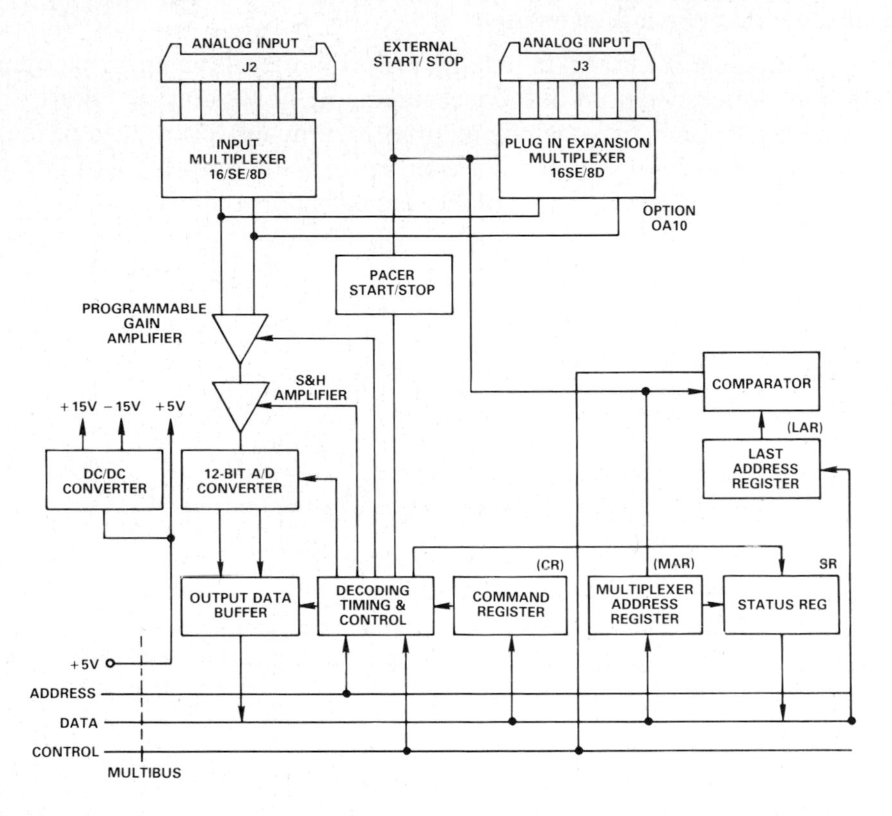

Figure 2.17. 1-2-4-8 programmable-gain amplifier extends converter dynamic range. Data-acquisition portion of the RTI-732 MULTIBUS-compatible analog I/O subsystem.

high-performance instrumentation amplifiers with external resistors and CMOS switches (Figure 2.18).

In one form of operation, a trial conversion is performed at the lowest gain; if the MSB is 0, the gain is doubled and another conversion is performed; if the MSB is still 0, the gain is doubled again, etc., until either the MSB is turned on or the top of the gain range is reached. Each doubling represents an additional bit of resolution for small signals. The scheme shown in Figure 2.17 is employed in a MULTIBUS-compatible analog I/O subsystem (see Chapter 4), for up to 15 bits of dynamic range and 12-bit resolution and accuracy.

Yet another possibility, when seeking accurate measurements of small variations about a fixed value of voltage, is to take the difference between the input and an accurately set voltage equal to the nominal fixed value. If the voltage

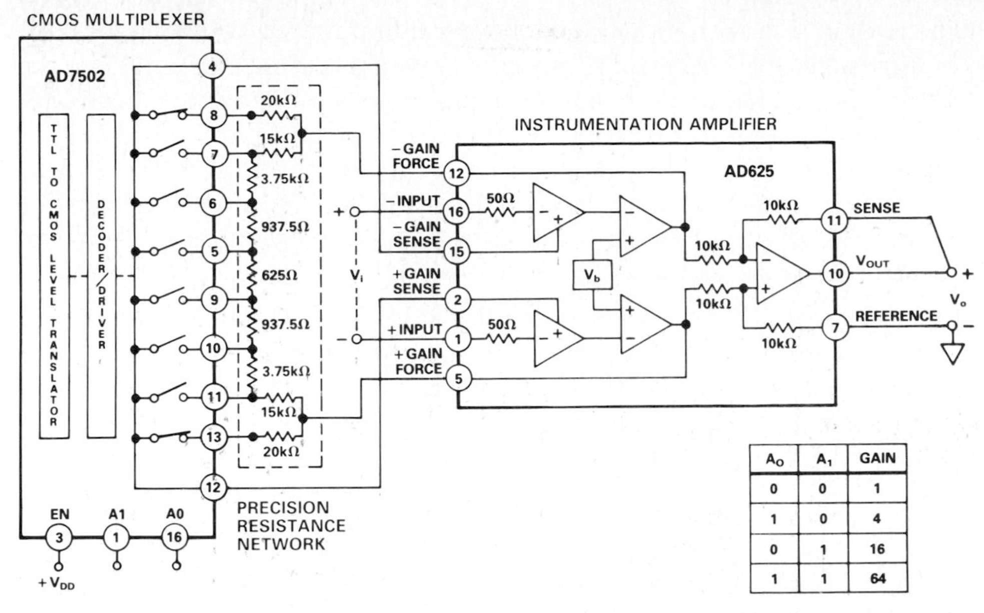

A_0	A_1	GAIN
0	0	1
1	0	4
0	1	16
1	1	64

Figure 2.18. Software-programmable-gain amplifier.

is applied via a high-resolution DAC (Figure 2.19), the interface can keep track, digitally, of both the initial value and the difference voltage, using an ADC of relatively modest performance. (The tradeoff here is the cost of a high-resolution DAC, plus logic and a modest 8-, 10-, or 12-bit DAC, vs. a 16-bit ADC.) This configuration also forms the basis of an excellent ADC test scheme (see Chapter 10).

Another approach to handling wide dynamic ranges with converters having limited resolution is to compress the data through the use of logarithmic techniques, in the form of either logarithmic converters or logarithmic processing of the analog signal (Figure 2.20). Logarithmic converters will be discussed in Chapter 16.

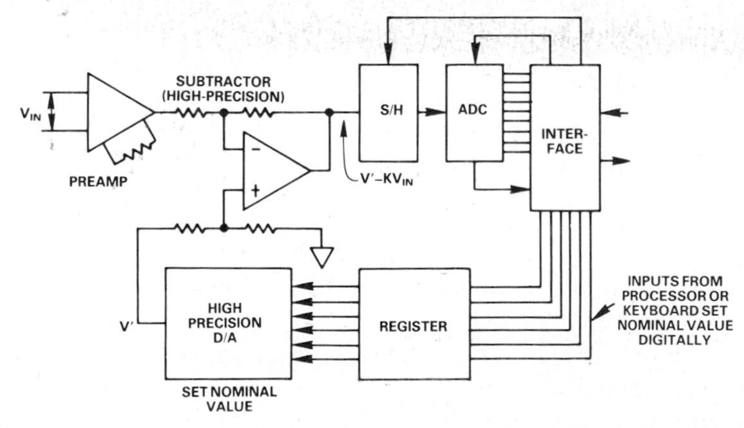

Figure 2.19. Use of a high-resolution DAC to measure small deviations about a precisely determined value.

The error of a logarithmic amplifier, after calibration, is a *log conformity* error (nonlinearity on a semilog plot), specified in terms of a maximum value at the output, or a maximum ratio to actual input over a specified range. For example, 1% log conformity error means that the error at the output, for 2V/decade scaling,* is 8.6mV, corresponding to an input uncertainty of $\pm 1\%$. Typical input voltage range (e.g., for Analog Devices Model 755) is 1mV to 10V. The corresponding output voltage range is ± 4V (i.e., ± 2 decades at 2 volts per decade, with respect to a 0.1-V reference level).

Since an error of 1%, referred to the 1-mV minimum input signal, is $1/10^6$ of full-scale input, and since the corresponding output error of 8.6mV is $0.0086/8 = 1.075 \times 10^{-3}$ of the output swing, the dynamic range of the signal has been compressed by a factor of 1,000, as a result of the logarithmic transformation. This means that a 12-bit converter (with suitable scaling) can be used to digitize the log-amplifier output, with a quite comfortable error margin.

Though it might appear that the representation of data having an inherent 20 bits of resolution ($10^6 \cong 2^{20}$) by a signal having 12 bits of resolution is getting "something for nothing," in violation of *some* Natural Law, the scheme really works. There are, however, some points to consider:

1. Compression is achieved by exponentially distorting the relative value of the least-significant bit. Thus, for a 10,000:1 signal range, represented by ± 4-V output, an LSB (of 12 bits, offset binary, suitably scaled) is worth about 23mV at 10V input (i.e., $0.1[\exp_{10}(2 - 8/8192)] - 10$V and 2.3μV for 1mV input. Therefore, while the approach is quite useful for compressing data requiring essentially constant *fractional* error (e.g., 1%) anywhere in a wide range, it is not at all suited to applications requiring high resolution (e.g., 0.01%FSR) *at any point* in the range.

2. Since the digital number is a logarithmic representation of the analog input signal, it must be dealt with as such in the digital process. If the number is to be used in computation, it should be antilogged, using either a lookup table or processor computing capacity—unless, of course, the computation is facilitated by the availability of a logarithmic relationship. If the data is to be stored or transmitted, and eventually returned to analog form unaltered, it does not require any further digital transformation, just an analog antilog operation following the output d/a conversion (unless logarithmic analog data is desirable).

3. Since a logarithmic function is inherently unipolar (the logarithm is real only for positive values of the argument—positive signals require a 755N, negative signals a 755P), it is far from ideal for signals that are inherently zero-centered. While it may be useful to bias some types of input signals into a single polarity, functions that demand symmetrical treatment may be badly

*A *decade* is a 10:1 range of input voltage or current.

distorted by the wide variation, in both resolution and speed, between zero and full-scale input. Such functions would profit by a type of compression that is symmetrical about zero. An example of an easily obtained form is a $\sinh^{-1}$ function (Figure 2.20), which involves two complementary antilog transconductors (752P and 752N) in the feedback path of an op amp. The resulting function is logarithmic for larger values of input, but it passes through zero, essentially linearly (and slowly).

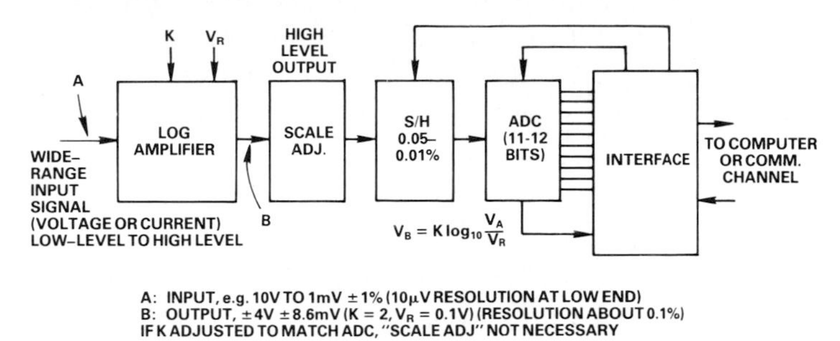

a. Log amplifier for range compression in a data-acquisition system.

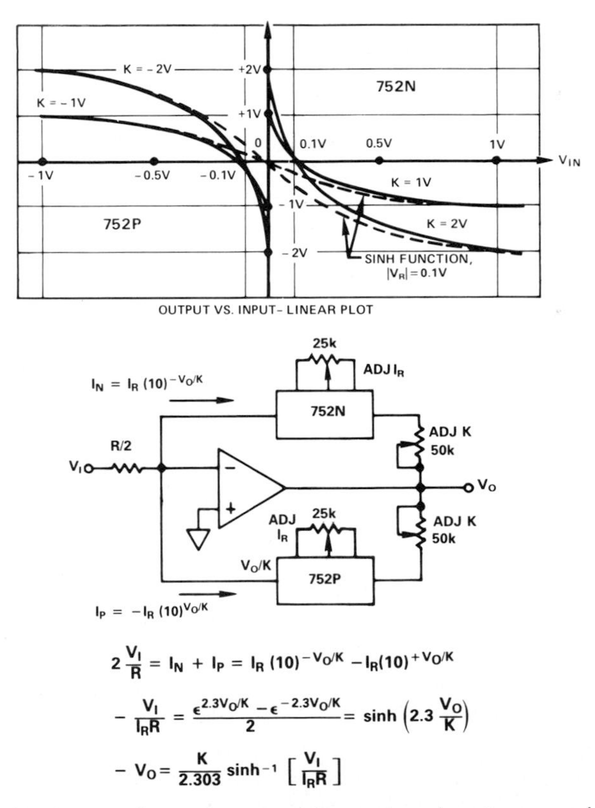

$$2\frac{V_I}{R} = I_N + I_P = I_R\,(10)^{-V_O/K} - I_R(10)^{+V_O/K}$$

$$-\frac{V_I}{I_R R} = \frac{\epsilon^{2.3V_O/K} - \epsilon^{-2.3V_O/K}}{2} = \sinh\left(2.3\,\frac{V_O}{K}\right)$$

$$-V_O = \frac{K}{2.303}\sinh^{-1}\left[\frac{V_I}{I_R R}\right]$$

b. Bipolar signal compression using complementary log transconductors to synthesize the $\sinh^{-1}$ function.

Figure 2.20. Logarithmic amplifiers in data acquisition.

Noise Reduction

Like diseases, noise is never eliminated, just prevented, cured, or endured, depending on its seriousness and the costs/difficulty of treating it.

Analog noise in data-acquisition systems takes three basic forms: *transmitted noise*, inherent in the received signal, *device noise*, generated within the devices used in data acquisition (preamps, converters, etc.), and *induced noise*, "picked up" from the outside world, power supplies, logic, or other analog channels, by magnetic, electrostatic, or galvanic coupling.

Noise is either *random* or *coherent* (i.e., related to some noise-inducing phenomenon within or outside of the system). Random noise is usually generated within components, such as resistors, semiconductor junctions, or transformer cores, while coherent noise is either locally generated by processes, such as modulation/demodulation (e.g., chopper-stabilization), or coupled-in. Coherent noise often takes the form of "spikes", although it may be of any shape, including—collectively from many sources—pseudorandom.[4]

In systems involving the conversion of analog signals, the finite resolution of the conversion process introduces "quantization noise," which may be thought of as either a truncation (or roundoff) error, or as a noise, depending on the context. See also Chapter 17.

Noise is characterized in terms of either *root-mean-square (rms)* or *peak-to-peak* measurements, within a stated bandwidth.* Random noise from a given source, within a given bandwidth, will give consistent *rms* measurements. For a typical gaussian amplitude distribution, and a sufficient number of measurements, one may expect a consistent relationship between the probabilities of obtaining peaks of a given size in relation to the rms, as shown in the Tables in Figure 2.21.

RMS values of noise from uncorrelated sources (e.g., from different devices, or from different portions of the frequency spectrum of the same device) add as the square-root of the sum-of-the-squares, and if the largest is more than 3 times as large as any other, the others may usually be safely ignored. However, if noise is dominated by picked-up spikes, root-sum-of-squares is of small comfort.

As we have indicated at the beginning of the chapter, there are ordinarily two basic forms of system-design problem: those involving essentially ordinary signal levels in unfavorable environments, and those involving extremely high-resolution measurements in favorable environments. (However, our readers should be aware that Murphy's Law would imply that their system design problems tend mostly to involve high resolution measurements in unfavorable environments.)

*For further references to noise, see the Bibliography.

[4]"Understanding Interference-Type Noise," *Analog Dialogue* 16-3 (1982), pp. 16-19.

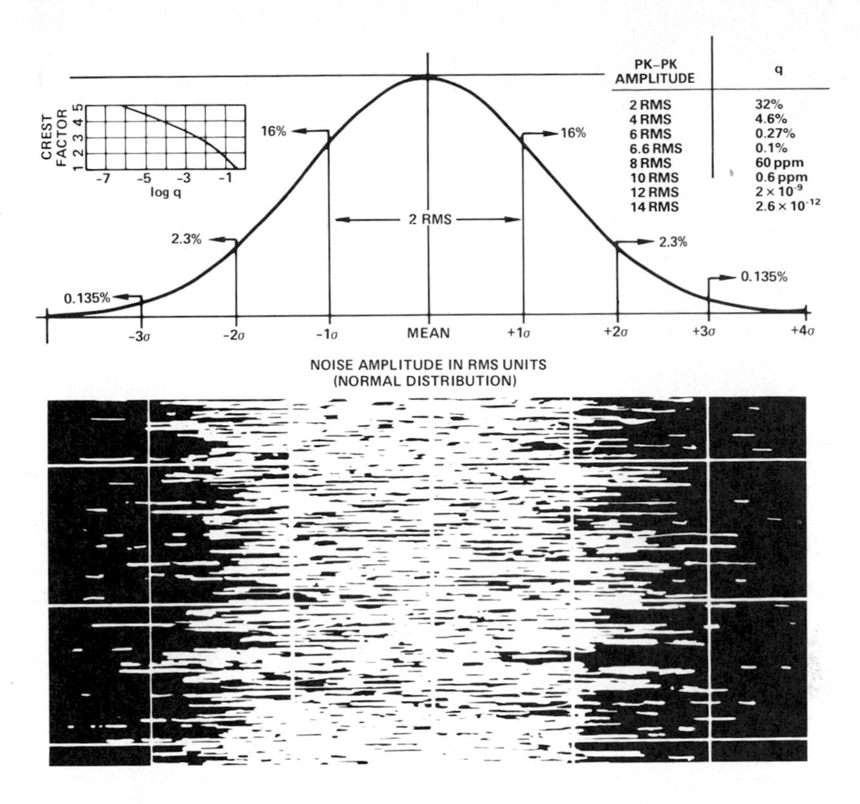

Figure 2.21. RMS vs. peak-to-peak amplitudes for gaussian noise.

For unfavorable environments, where the major sources of noise are *induced noise*, the designer must rely on early preamplification and conversion, isolation, shielding and guarding, signal compression and filtering, and—where possible—an information rate (via digital filtering, fast sampling or parallel paths) that has enough redundancy to allow the digital processor to retrieve data via digital filtering, correlation, and summation (see Chapters 6 and 21).

In favorable environments, where the measurement process and the processing hardware introduce the major portion of the uncertainty, the emphasis must be placed on measurement techniques, filtering, choice of data-acquisition hardware for best resolution, and—again—the use of high-speed digital processing for signal retrieval plus "intelligent" measurement techniques including automatic drift compensation and scale-factor adjustment.

Where noise is likely to have large spikes as a major component, the integrating-type converter usually provides additional filtering. For random noise, if there are sufficient samples taken of a given signal channel, the statistical properties of the noise are imparted to the digital output, which may be filtered by digital techniques.

Chapter Three

Data Distribution

After analog data have been converted to digital form and have been duly stored, transmitted, or processed, the results of this handling, as well as some newly created digital numbers, may be required once again to intervene in the "real world" of phenomena. In analog or digital form, they may be used to drive meters or motors, display information, stimulate devices under test, generate heat, light, or sound, modulate waveforms, sound the alarm, adjust an audio gain, or—in short—to manipulate energy according to a digital code.

The digital output words are made available fleetingly on an output bus—or for longer periods, at an output register—for distribution to their destinations. While an increasing number of real-world devices, such as numerical displays, stepping motors, printers, and the like, are operated more or less directly by digital numbers (perhaps with "decoding," but without the overt interposition of electronic analog variables), there is a widespread—and growing—use of electronic d/a converters in the redeployment of digital data in analog form. This chapter treats of systems that use d/a converters.

As with a/d conversion, but reversing the order, the basic objective is to get the data into the appropriate analog form, as rapidly, as frequently, as accurately, as completely, and as cheaply as necessary.

The basic instrumentality for accomplishing this is the digital-to-analog (d/a) converter (DAC). In response to a digital code, it may be used either to provide a voltage or current output (fixed-reference DAC), or to adjust the gain of an analog circuit (multiplying DAC). It can be a simple device on an IC chip, or a sophisticated high-resolution high-speed device with many "bells and whistles;" physically, it may take the form of a box, a card, a potted module, an integrated circuit—or even a portion of an integrated circuit. It may be functionally integrated with other system elements to form a subsystem.

To accommodate the analog output to the specified conversion relationship, some form of scaling and offsetting (signal conditioning) and energy translation (e.g., current-to-voltage) may be necessary, performed with amplifiers. To furnish analog information to more than one destination, either additional converters or a multiplexer and sample-holds may be necessary. To encompass an extra-wide analog dynamic range, an exponential amplifier or conversion relationship may be found useful.

The nature of the data-distribution system depends on the properties of both the digital and analog data, and what is to be done with it. This chapter deals with aspects of signal flow, from the digital data source through conversion. Chapters 4 and 6 have to do with integration into systems and instruments, Chapter 7 deals with the fundamentals of DACs, Chapter 8 deals with some of the forms IC DACs can take, Chapter 13 deals with DACs designed specifically for high-speed ("video") applications, Chapter 16 considers (intentionally) nonlinear DACs, and Chapter 17 discusses high-resolution converters.

Commercially available data-distribution systems range from basic d/a converters to multi-channel converters on monolithic chips to completely integrated systems on cards and in boxes, and even include converters that are functionally inseparable from the digital processor. In later chapters, we will consider the various optional levels of integration available in a single package or piece of equipment. However, in this chapter, we will generally treat most of the functions peripheral to the d/a converter *as though they were embodied by separate components*, in order to make clear the architectural choices that a designer might have, with their characteristics, advantages, and disadvantages.

3.1 FACTORS AFFECTING DISTRIBUTION-SYSTEM DESIGN

The configuration, choice of components and their specifications, the system timing, and location of multiplexing, depend, as with data acquisition, on

1. Number of channels
2. Update window per channel
3. Update rate
4. Bus width and word length
5. Output resolution
6. Output linearity and accuracy
7. Settling time per channel
8. The nature of the loads
9. The cost function

There are a number of additional areas for decision by the system designer:

Digital signal source: Parallel bus? port? register? serial data—bit-serial, byte-serial, ASCII?

Signal storage between or during updates: External or internal registers? Single-rank, dual-rank, n-rank? Analog storage (sample-holds, inertia)?

Multiplexing: μP bus? Digital switching? Multi-channel DACs? Analog multiplexing (sample-holds or multiplex switches)?

Update: Simultaneous? Sequential? Random?

Conversion: Near digital source or remotely? Many DACs—or few DACs with multiplexing?

Analog Output: Voltage, current, or gain? Discrete values or smoothed? Permissible level of switching transients, use of deglitching? Direct-wired or galvanically isolated circuitry?

Cost tradeoffs: Minimizing use of expensive components. Sample-holds vs. multiplexers vs. DACs. Inertial filtering. Using low-precision incremental "slave" DACs for accurate settings.

3.2 DIGITAL SIGNAL SOURCE

The basic d/a converter accepts parallel digital data at its input and continuously provides a representative analog output, usually based on a binary relationship. As soon as the digital code changes, the analog output seeks to follow it and—after a transient interval, of variable turbulence—comes to rest at the new value, within a period ranging from nanoseconds to microseconds.

Since there is a continuous input-output relationship, if a basic DAC must maintain a constant output after updating, it must be fed a continuous digital input. However, unless the DAC is wired so that its inputs are totally dedicated to a unique flow of information (for example, from a counter, as in Figure 3.1, or from a dedicated computer output port), it must get its inputs from a source which is rapidly changing and at the same time servicing other I/O (input/output)—or even internal—elements in the system.

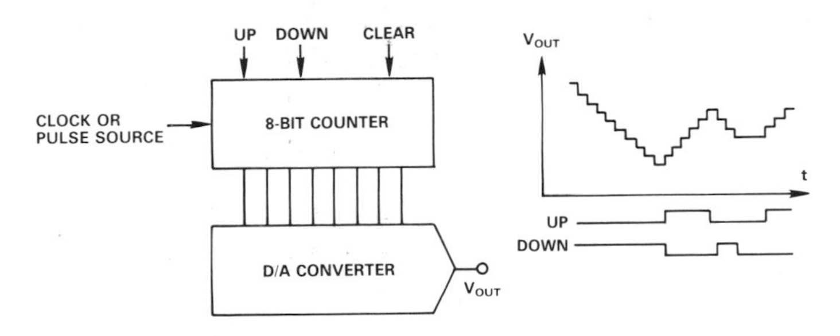

Figure 3.1. Converter output is a continuous analog representation of the counter's digital output (simplified block diagram).

3.3 REGISTERS

The d/a converter's input, therefore, almost always comes from a register, which is latched upon command ("clock" or "strobe") to preserve a fixed

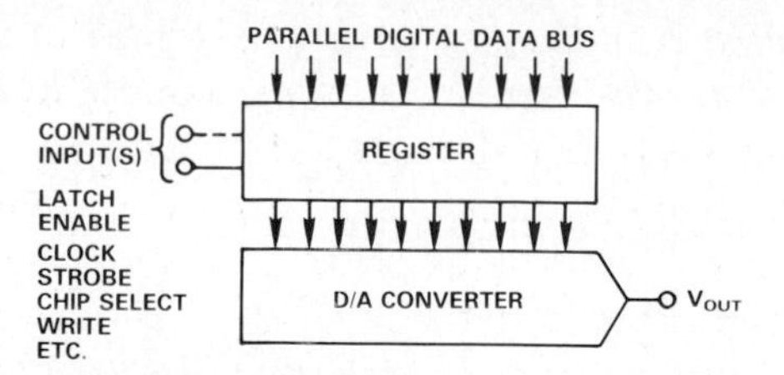

Figure 3.2. Basic d/a converter is buffered by a register. When it is latched, the output voltage is unaffected by the data on the bus.

digital code (Figure 3.2). For a single-stage latch, the output will follow the input while the control signal is in one state (transparent) and become latched when it changes state; for a two-stage latch, the output ignores the new value of input that is being acquired until it is latched in when the control signal returns to the inert state. The register may be external to the DAC chip (or package) but is often contained within it.

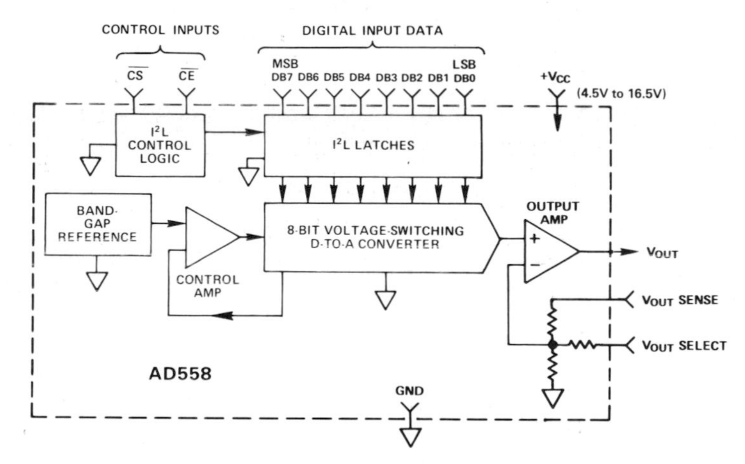

a. Voltage-output monolithic 8-bit DAC.

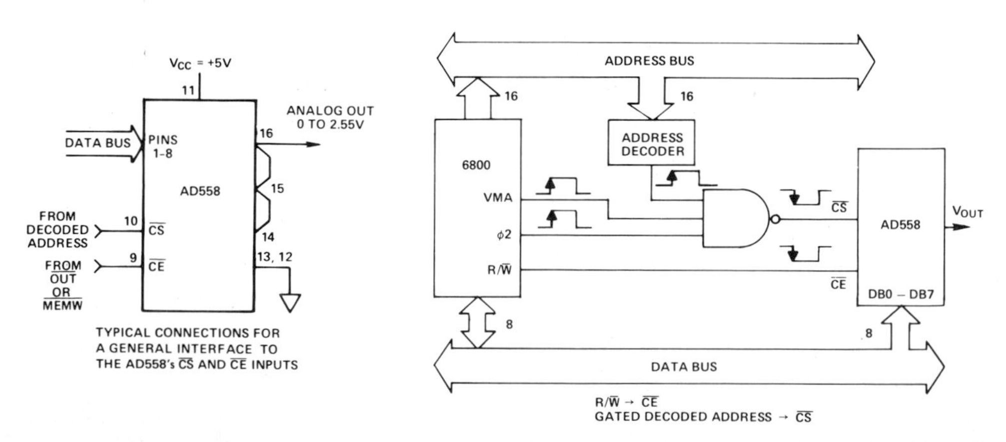

b. Control signal connections.

c. Interface to a 6800 microprocessor.

Figure 3.3. Interfacing an 8-bit DAC to a microcomputer bus.

The register is controlled by one or more latching signals. In microprocessor systems, the conjunction of a decoded addressing "chip select" and an enabling "write" command—within the window of time that valid data is on the data bus—will cause that data to be latched. Figure 3.3 shows a typical latching scheme used for a completely self-contained voltage-output monolithic 8-bit DAC having a self-contained register. The block diagram of the DAC is shown in (a), typical connections in (b), and the interface to a 6800 microprocessor in (c).[1]

The *chip select* input goes low when the clock ($\phi 2$) goes high, VMA (Valid Memory Address) goes high, and the address decoder selects the device in question by going high; the *chip enable* input goes low when $R/\overline{W}$ goes low to indicate that data is to be written to the device. When both are low, the DAC's register is "transparent" and open to updating by the data on the bus; as soon as either returns high, the data is latched. After one or two microseconds, the analog output has settled to its new value.

3.4 MULTIPLE RANKS

More than one rank of registers may be employed, in a manner determined by the way the data is arriving from the signal source. For example, if the data is placed on the bus at the convenience of the processor, but it is not yet time to update the DAC, the data may be latched by the first rank. It will be loaded into the DAC register (second rank), when the time arrives to update the DAC, either synchronously, as determined by the processor, or by some other (asynchronous) entity.

If all the data does not arrive simultaneously in parallel, two or more stages of latching may be required to prevent false data from appearing at the analog output. Figures 3.4 through 3.6 show examples of situations calling for more than one stage of latching.

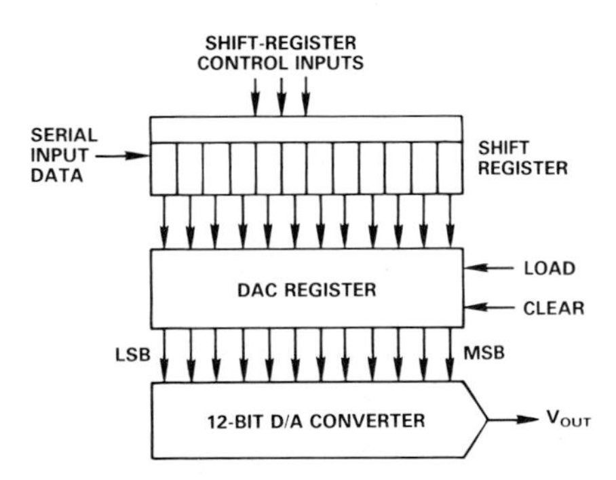

Figure 3.4. 12-bit DAC with serial input. Data is clocked into the shift register in serial (old data latched in DAC register), loaded into the DAC register in parallel at will.

[1]See "Putting the AD558 DACPORT™ on the Bus," by D. Grant. *Analog Dialogue* 14-2, 1980, pages 16-17.

In Figure 3.4, 12-bit data arrives—MSB first—at the serial input (SRI) of a DAC designed to accept bit-serial data. At a time when each bit can be expected to be stable, it is strobed into the shift register. The process continues, bit by bit, until all bits have been loaded, without affecting the DAC output. After all bits are loaded, the DAC register can be loaded at any time, to update the DAC output. The scheme is analogous to a 12-car train pulling into a station, car by car, then stopping and opening all the doors at once to detrain the passengers.

12-bit data generated for distribution on an 8-bit bus must be delivered as two separate bytes. The bytes can be presented in two ways, either "right-justified" or "left-justified." In the former scheme, with the data written in the fashion of whole numbers, the 12-bit word, 0100 1000 0001, would consist of two bytes, 0000 0100 and 1000 0001; in the latter scheme, written as a binary fraction, it would be 0100 1000 and 0001 0000. In either case, the two bytes must be simultaneously available at the input of the DAC register in order to obtain a correct analog output. Figure 3.5 shows the block diagram of a 12-bit DAC (the AD667), which accepts data in three four-bit nybbles, and the addressing scheme that would provide a left-justified 8-bit bus interface.

The 8 most-significant bits of the first rank, DB11 through DB4, are wired to the data bus, and the last four bits (the top four of the second byte) are wired to DB11 through DB8. When $\overline{\text{WR}}$ goes low, the first 14 bits of the address, A15 through A2, are decoded to enable the device, via the Chip Select input; when the last two digits of the address are 01, the 0 at A1 enables the last four bits, which are subsequently latched; then, when the last two address digits are incremented to 10, the 0 at A0 enables the 8 most-significant bits and at the same time loads the entire word into the DAC latch, updating the DAC's output.

It is easy to see that, with the facilities it has available, the same DAC could also be used to update the entire 12-bit word at once from a 12- or 16-bit bus, to update the word in three nybbles from a 4-bit bus, or to provide a right-justified 8-bit bus interface.

The forms of interfacing described above are often called *memory-managed interfacing*, since the d/a converter is communicated with in the same manner as a memory location (see also Chapter 4); it looks like a *write-only* memory. There also exist DACs with *readback* capability; they make the word that is currently providing the DAC output available to the bus in response to a *read* command—a helpful capability that makes it unnecessary for the processor to continually remember the last value, an especially useful feature at startup or after interruptions.

3.5 MORE THAN ONE CHANNEL

Since many I/O entities may be connected in parallel on the data bus, with each responding to a unique address code, digital multiplexing is inherent

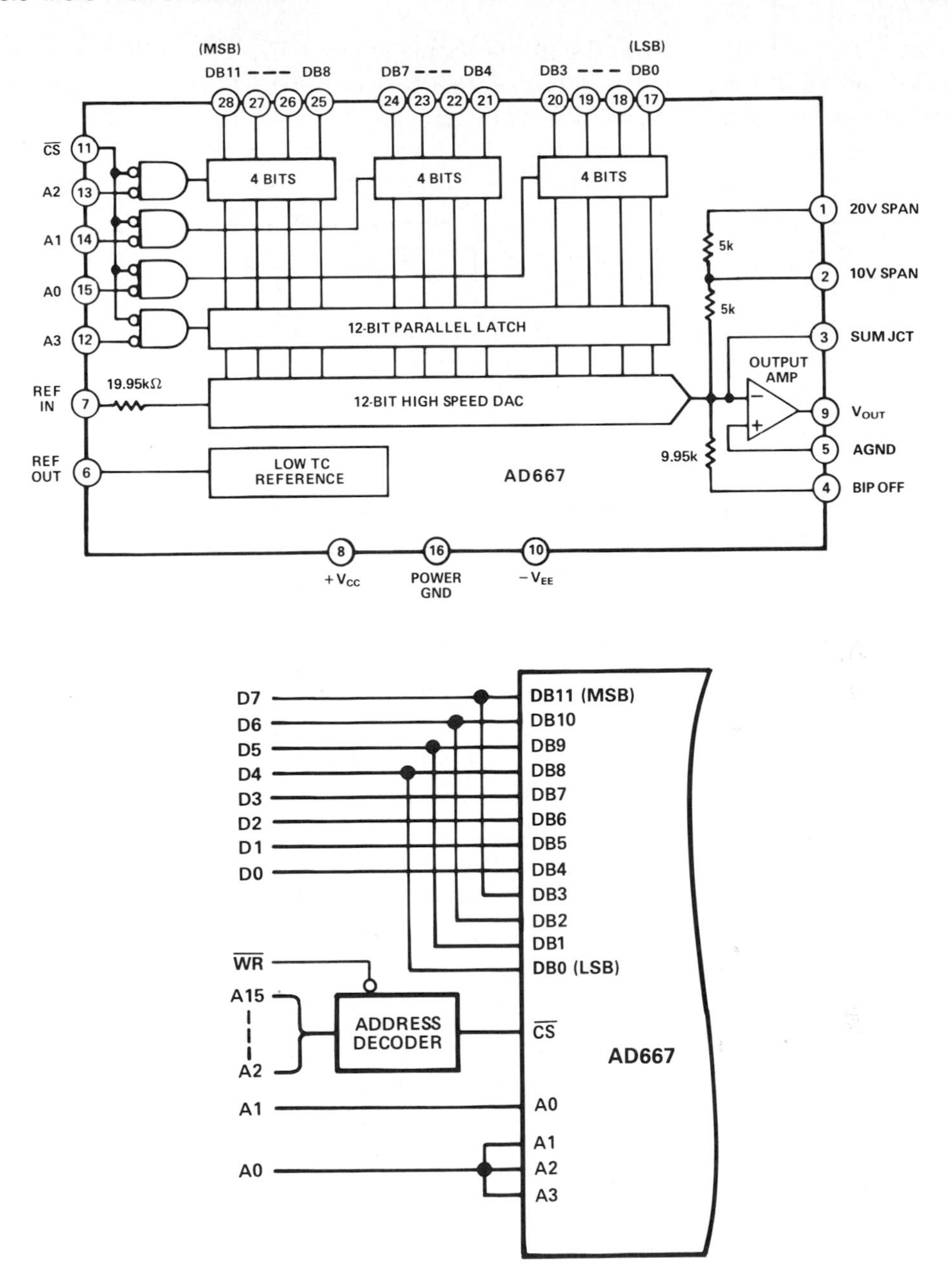

Figure 3.5. Double-buffered 12-bit DAC capable of interfacing with 4-, 8-, 12-, and 16-bit data buses. Left-justified connection to 8-bit bus.

(Figure 3.6a). If a number of DACs are connected to the bus, only those that are addressed will receive an update via their associated register(s) when the WRITE command is received. The converters may be addressed randomly (and repeatedly) or in a sequence. It is good practice to use two-stage latches or double buffering so that the basic DAC is never exposed to bus noise.

The multiple DACs may all be in the same package. For example, the quad DAC shown in Figure 3.6b contains four 12-bit double-buffered voltage-output DACs multiplexed on a 12- or 16-bit data bus.

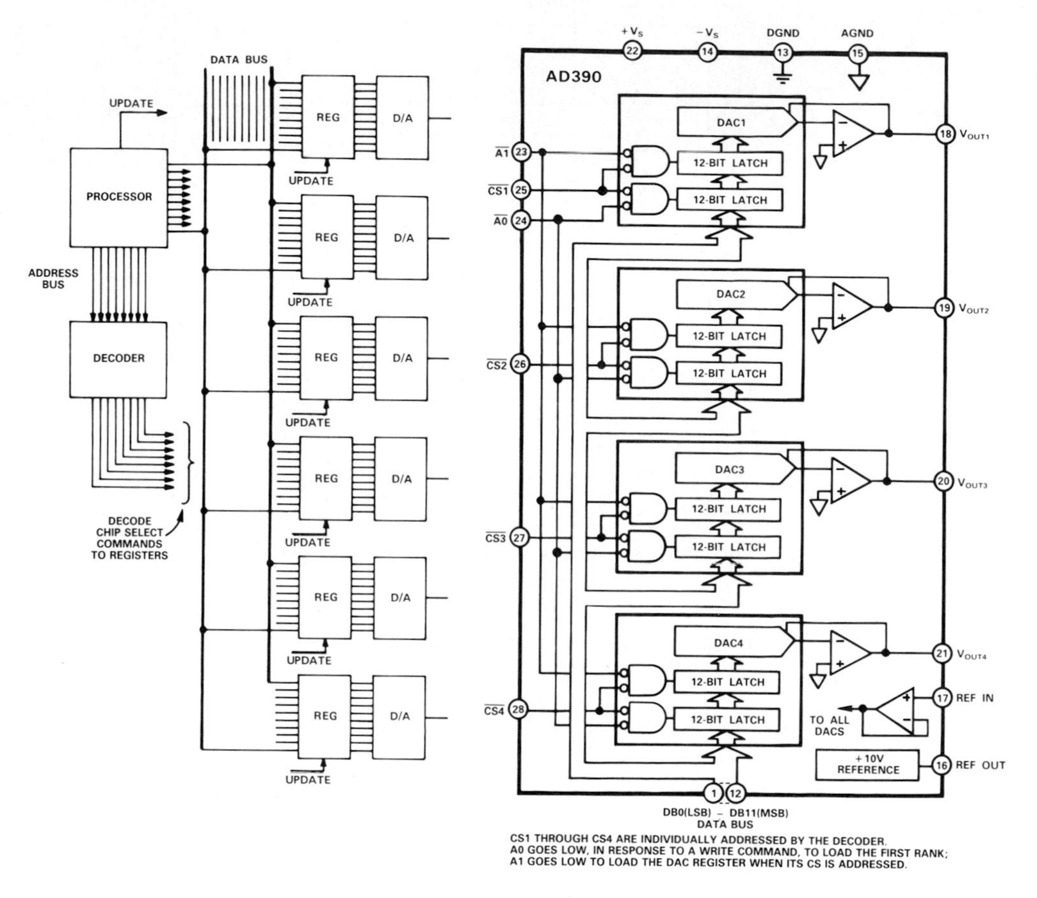

a. Converters share a common bus.

b. Quad DAC: 4 12-bit DACs in a single hybrid package sharing common bus.

Figure 3.6. Converter-per-channel distribution.

Digital multiplexing may also be performed with multipoint switching: For example, if four n-bit DACs with parallel inputs are to be updated, each bit is switched to one of four output lines (Figure 3.7). Though more complex, this approach has the advantage of reducing the load on the bus, since only the DAC currently being addressed is connected.

If there are many analog channels, and sample-holds are less costly than DACs for a given speed and resolution, the designer has the option of using a single d/a converter—multiplexing its output among many sample-holds, either with a multiplexing switch or by feeding the analog input to all of the sample-holds in parallel and decoding the address line to the sample-hold control inputs (Figure 3.8). A two-channel application of this technique might be found in digital stereo decoding, where a single 16-bit DAC updates the two audio channels alternately via sample-holds (see Chapter 17).

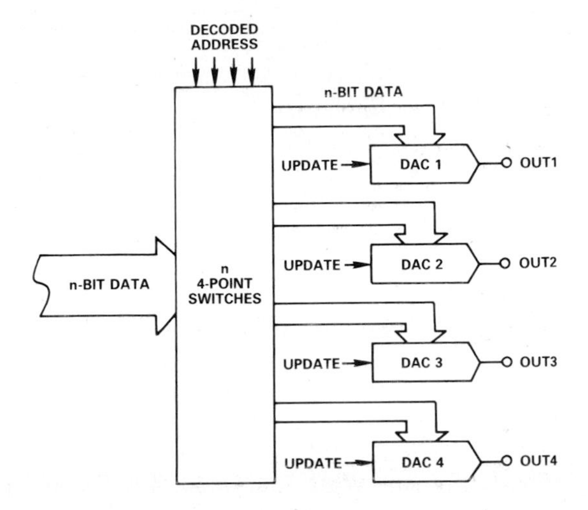

Figure 3.7. Multipoint switching of four d/a converters.

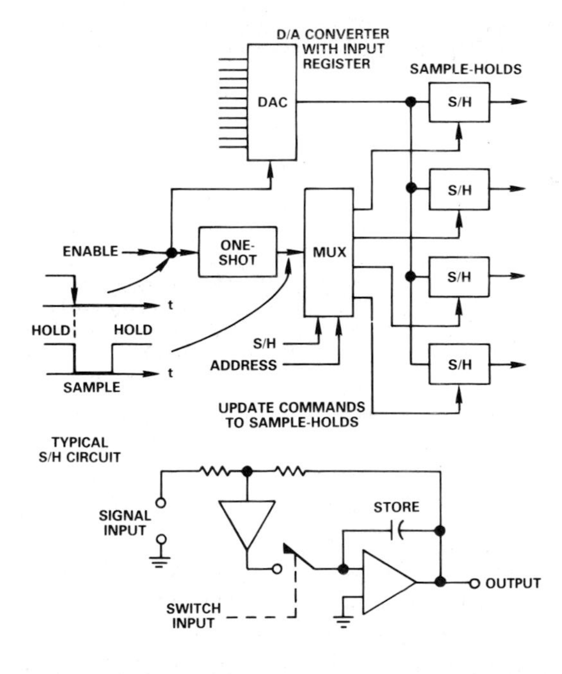

Figure 3.8. Sample-holds with common analog inputs and multiplexed control inputs. Desired sample/hold is addressed. Then, on next update, DAC is latched, and addressed S/H is switched to SAMPLE. One-shot switches it back to HOLD when sufficient time for settling has elapsed.

Comparing Sample-Holds and DACs

Once updated, a d/a converter-with-input register will store an analog value indefinitely, or at least for so long as the power is connected. By contrast, a sample-hold circuit, since it holds the analog data on a capacitor, is susceptible to a definite "droop" (positive or negative) in the analog output as the charge on the capacitor changes due to leakage across the switch, from the amplifier's summing point, or from the supplies (or perhaps even due to the capacitor's own leakage resistance or dielectric absorption).

Thus, even though the data may not change at all, if the same value must be maintained for a long period of time, it is necessary to update the sample-hold periodically to correct for output droop. On the other hand, so long as the data remains unchanged, a distribution system based on d/a converters (with registers) has no need to be periodically refreshed. In fact, the DAC's ability to store without error lays the foundation for saving time by "updating by exception," whereby the data channels are updated only if the data changes.

A further consideration in the use of d/a converters vs. sample-holds lies in the matter of allowing for acquisition and settling time. The data sheet for a typical general-purpose data-distribution sample-hold circuit at reasonable cost may call for acquisition periods ranging from 1μs to 26μs or more. Thus, the multiplexer must dwell at each channel for the duration of this acquisition period, and the update sequence must be arranged to avoid tying up the processor bus for long periods.

Offsetting some of the speed and flexibility of the DAC-per-channel method is the cost of interconnecting the DACs to the data source. Parallel data at the 10-bit (0.1%) level requires at least 12 conductors (10 data lines, common return, and command line). If the d/a converters are at any distance from the bus, or its buffer, installation cost for the cable may become the largest single economic factor, far outweighing the cost of the DACs. Cable and installation costs can be greatly reduced by introducing serial (bit-at-a-time), instead of parallel, transmission, but at a considerably reduced update rate.

In addition to allowing asynchronous timing of the loading and analog update for each DAC, systems employing double-buffered d/a converters permit simultaneous updating of a number of DACs. The first rank of registers is loaded as each item of data is made available, then the second ranks of all devices (the DAC registers) are updated simultaneously. (This is similar to the scheme used for a DAC that must accept data from the bus in more than one byte but must update the DAC register simultaneously.)

Analog Data Distribution

One of the approaches to sample-hold-circuit updating is shown in Figure 3.8. Analog data is sent over a common wire to all the sample-hold circuits. However, each s/h, normally in *hold*, remains oblivious to the input data until a command signal connects it momentarily to the analog data bus (*sample*). On

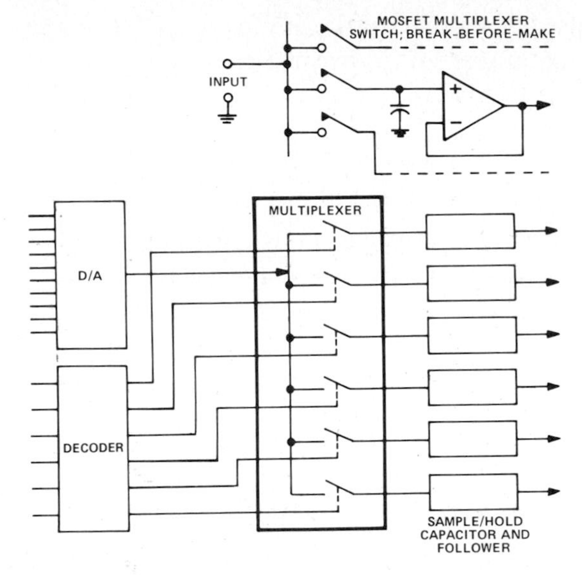

Figure 3.9. Sample-holds with multiplexed *analog* inputs. Multiplexer switch can do double duty in simplest configuration.

receipt of its update pulse, the sample-hold circuit acquires whatever analog information appears on the data line and holds this value until subsequently commanded to acquire a new signal level.

An alternative arrangement, potentially low in cost (and fraught with challenges for the designer), uses an analog multiplexer for distribution of analog data among individual channels, as shown in Figure 3.9. Here, the sample-holds respond to whatever signals are presented at their input terminals, and then hold this signal level when the analog input is disconnected. The multiplexer's switches serve double duty, both in multiplexing and as an interruptible path for charging the hold capacitor, though more complex switching arrangements may be used. It is more subject to leakage and crosstalk than the circuit of Figure 3.8, and requires some care in the timing of switch operations; but the idea is conceptually simple and low in cost.

3.6 ACQUISITION vs. DISTRIBUTION

As a rule, data *acquisition* poses more challenging problems than data *distribution*, but some of the problems assume different shapes. Since data distribution can take place at macroscopic power levels (volts and milliamperes), noise is not a great problem (except for induced noise in hostile environments). To the contrary, DAC outputs may be boosted, as in programmable power supplies; in such cases, it is useful for the DAC's output amplifier system to have remote sensing (force-sense, or Kelvin connections) to avoid errors due to voltage drops in the wiring. This is also good practice for high-resolution (16-18-bit) DACs, even at more-modest power levels.

Sample-Holds used in data acquisition must have short aperture time (or at least small aperture uncertainty) because they must either deal with the "instantaneous" value of a signal, or sample it rapidly at equal time intervals. Their *hold* time need be no longer than is necessary for the ADC to digitize the signal. In short, the usual emphasis in sample-hold circuits for data acquisition lies on rapid acquisition, followed by rapid conversion.

By contrast, sample-hold configurations used for data distribution usually permit relaxed update timing, but the analog values may have to be preserved for long periods without significant *droop*. Thus, sample-holds for distribution must have long *hold* times, and short acquisition-and-settling times. Where high resolution (12 bits or better) and large ratios of *hold* to *settling time* are necessary, multiple-DAC distribution—with register storage—becomes preferable; the decreasing cost of IC DACs makes the choice an easy one.

3.7 FILTERING AND DEGLITCHING

In data acquisition, analog filtering is used to remove (or at least reduce) analog transmitted, inherent, or induced input noise and to eliminate components of signal and noise at frequencies greater than one-half the sampling frequency to avoid aliasing. In distribution, filtering is used to reduce "noise" caused by quantization and sampling (finite increments of digital resolution and discrete output values due to sampling cause discontinuous analog outputs, which introduce unwanted frequency components) and to deal with coupled-in switching transients.

Small discontinuities are often tolerable, especially in dc-value testing; they occur at the application of test conditions, and readings are not taken until the system has settled. On the other hand, if the converter is producing an analog ramp in discrete steps, the discontinuities may have to be smoothed, and certainly any feedthrough transients and/or "glitches" must be minimized.

For reconstructing coarse sampled data, sophisticated analog interpolation techniques are used to overcome the limitations of simple filtering. An example is integration of the difference between two adjacent values so that the "points" are connected by straight lines or exponentials, and discontinuities become more-easily filtered changes in slope rather than steps.

A "glitch" is an insidious spike caused by intermediate codes introduced by asymmetrical switching times at major-carry transitions, such as from 0111 1111 to 1000 0000. In this example, where the DAC output is to change by one least-significant bit—from 1LSB below half scale to half scale—if the less-significant bits all switch off slightly before the MSB switches on, the DAC output will momentarily seek to go all the way towards zero (Figure 3.10a), then return to half-scale, creating a very large spike.

Linear filtering of glitches is impractical, because they have far-from-uniform magnitudes, and they do not occur at uniform intervals; hence, linear filtering can lead to badly distorted waveforms (Figure 3.10b). Glitches are minimized in high-speed "minimum-glitch" DACs by the use of latches and very fast switching, with the best-possible matching of rise and fall times, so that (if possible), the remaining error is a small, fast, filterable doublet pulse with near-zero average energy (c).

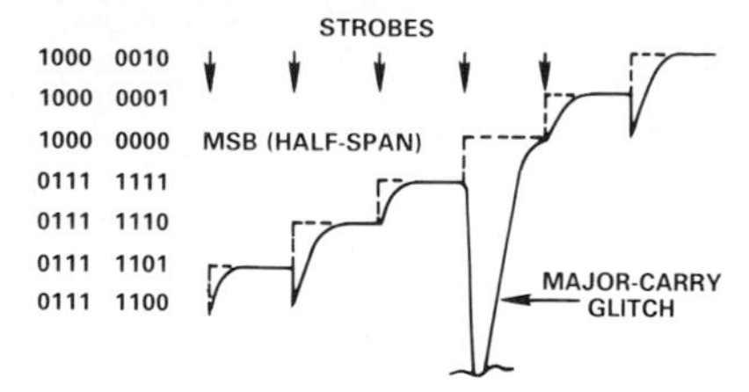

a. Transitions in the vicinity of the major carry, showing glitches.

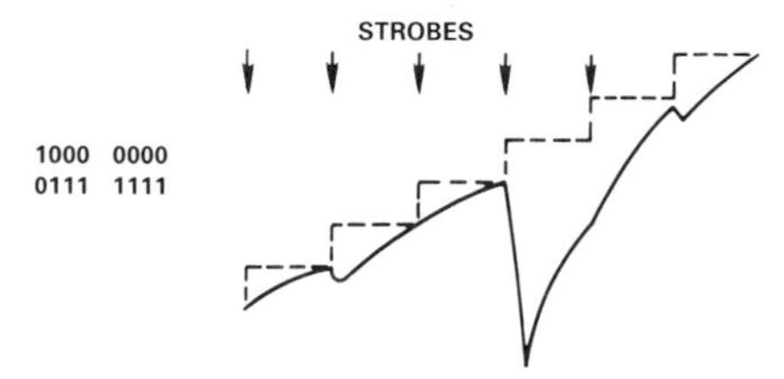

b. Filtering's distorting effect at the major carry.

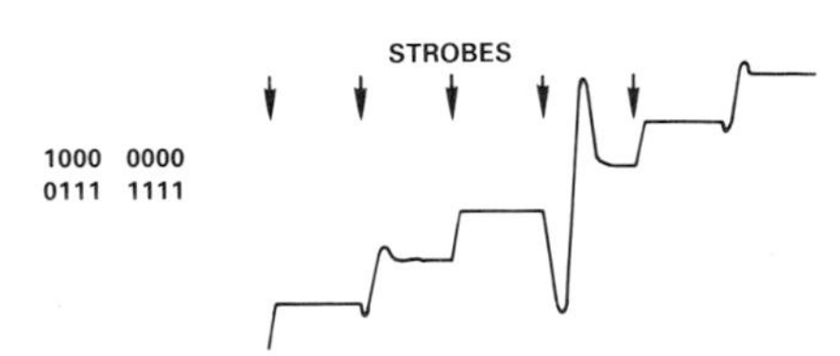

c. Typical response of DAC designed for minimal glitch.

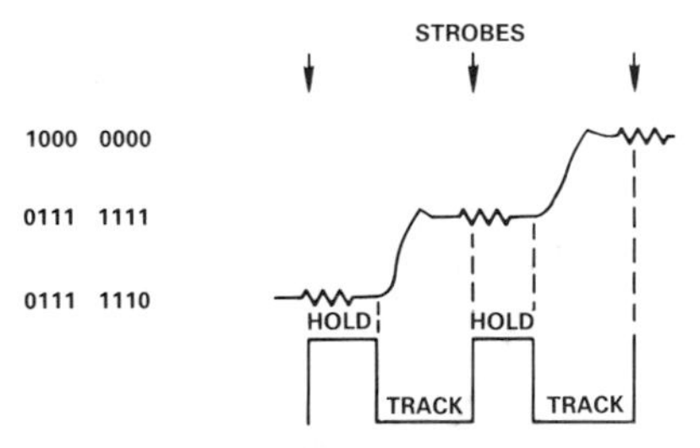

d. Effect of track-hold deglitching. Note slower update rate on same time scale as above.

Figure 3.10. Glitches in d/a converters.

In the more-usual case, track-hold *deglitcher* circuitry is used to cause the DAC's output circuit to ignore the glitch. The output circuit is switched into HOLD while the DAC is updated, then switched back to TRACK after a sufficient time (perhaps established by a one-shot) has elapsed for the glitch to settle out (Figure 3.10d). The deglitcher circuitry, though cleaning up the response, will result in a reduction of the update rate. Figure 3.11 is the block diagram of a complete 12-bit deglitched DAC capable of a 6-MHz update rate.

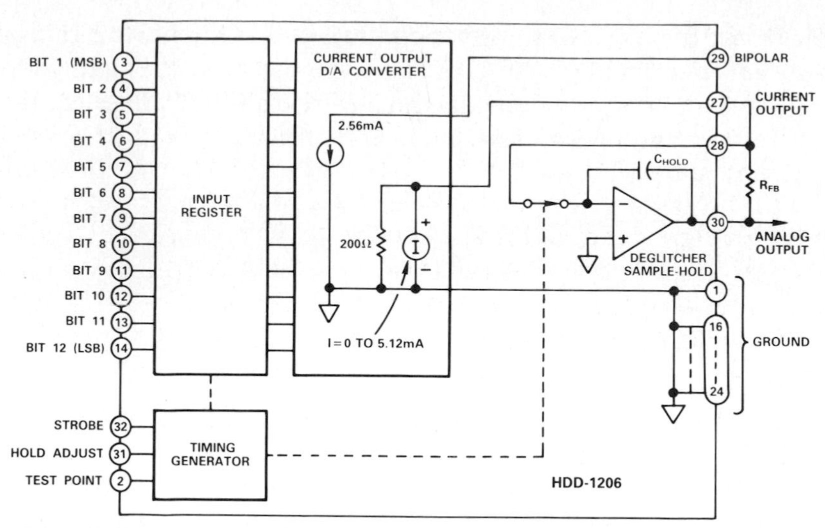

Figure 3.11. Block diagram of 12-bit deglitched DAC capable of 6-MHz update rate. Gain is set by external feedback resistor.

3.8 MINIMIZING CALIBRATION ERRORS BY SERVOING

In a test system, it may be necessary to set a number of parameters with high resolution and accuracy. It can be done using high-performance DACs, but where many channels of accurate analog output are required, it is possible to accomplish this at lower cost with a single high-performance DAC and sets of paired lower-performance converters. A representative scheme is shown in Figure 3.12.

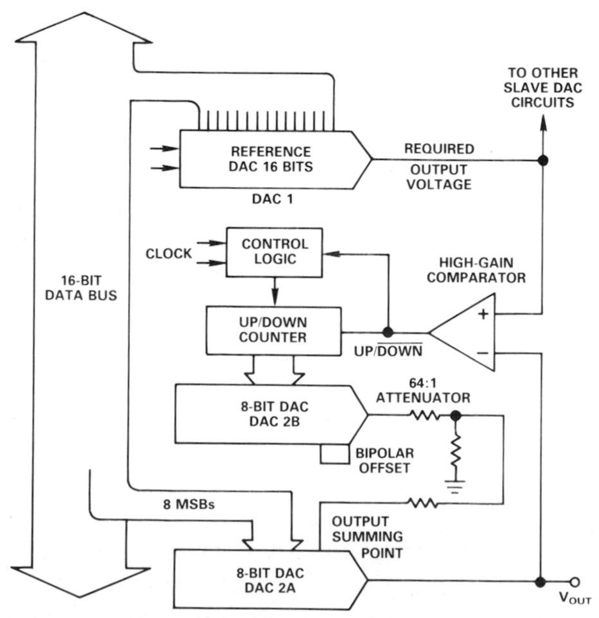

Figure 3.12. Master-slave precision d/a converter scheme, using 16-bit master and two 8-bit DACs per channel for 14-bit resolution, employing comparator and counter for correction.

The 16-bit DAC, D1, puts out an accurately set voltage equal to the desired output of D2, which consists of a two 8-bit DACs. D2A, with no attenuation, receives the eight most-significant bits. D2B's output is attenuated by a factor equivalent to 6 bits, and offset, so that it can add up to ±2 LSBs to the output of D2A. The outputs of the two DACs are summed and compared with the output of D1. D2B is driven by a an up-down counter whose direction, determined by the sign of the comparator output, tends to drive the error towards zero. When the comparator changes sign, the count stops, and D2's output has been set to an accuracy of better than 14 bits, regardless of the actual setting of the digital input to D2B. If the system has a number of such DACs, each is set, in turn.

This arrangement tends to be somewhat slow, especially for large step changes in output, it uses a fair amount of hardware, and it doesn't make best use of available software. An alternative method (Figure 3.13) uses a fast successive-approximation a/d converter to measure the output of a linear amplifying comparator (viz., instrumentation amplifier) which compares the output of the reference, DAC1, with the output of DAC2A (DAC2B output = 0). The

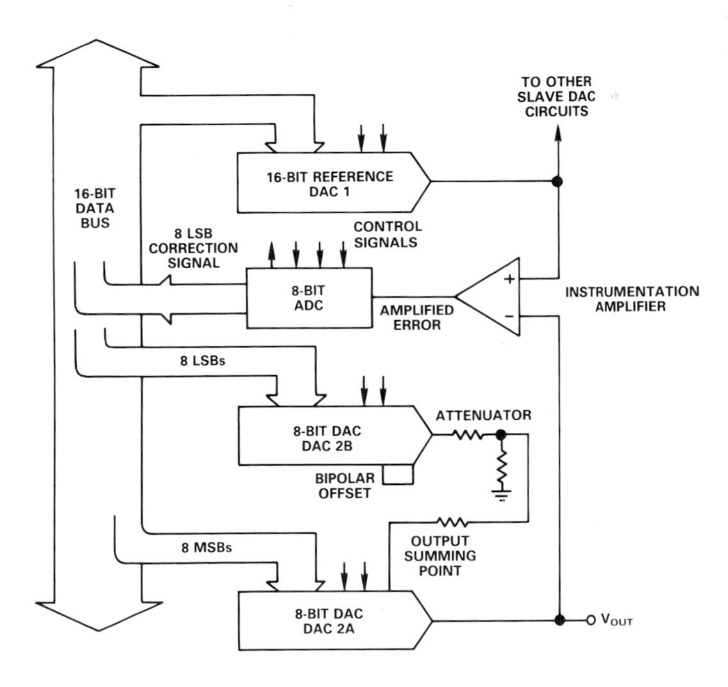

Figure 3.13. Master-slave scheme, using 8-bit ADC and software for correction.

digital output, which represents the difference of the analog values, is simply latched into the input of DAC2B. The total analog output of DAC2A and DAC2B is thus very nearly equal to that of DAC1, with a resolution of at least 14 bits; again, errors in DAC2A are taken into account in the setting of DAC2B.

For either of these schemes to be successful, the noise level and dc instability of DAC2A during the period between updates must be within a fraction of 1 LSB of the *overall* resolution.

3.9 ISOLATION

In any data-distribution system in which common-mode potentials pose a serious threat to the integrity of data, equipment, or organisms, it frequently becomes necessary to isolate the various analog loads from the digital data source. Otherwise, substantial differences in ground potential at the various locations could cause large ground currents, induced noise, or worse.

Isolation is accomplished by magnetic or optical coupling of either the digital or the analog signal. In one form of isolation, the analog output is isolated and then transmitted via a 4-to-20mA 2-wire current link, which is immune to voltage noise.

As a practical example of an isolated DAC, Figure 3.14 shows the architecture of a 10-bit systems DAC with 4-to-20 mA current output. The data, which might represent the setting of an actuator, is normally latched into the *preset* input of a latched 10-bit counter from an 8- or 16-bit data bus, in response to a WRITE command. The digital word is converted to an analog signal,

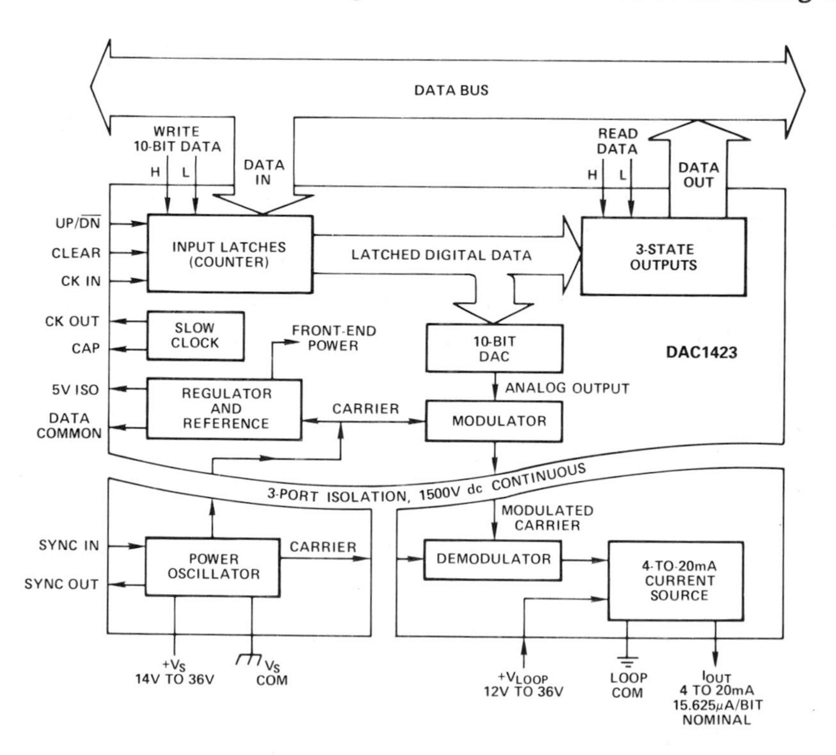

Figure 3.14. Block diagram of 10-bit System DAC having galvanic isolation, 4-to-20-mA output, data readback, power supply independent of bus system, and available external (e.g., manual) updating.

which is transmitted across an isolation barrier (via a modulated carrier), demodulated, and converted to a current signal with a 4-mA offset and a 16-mA span.

This DAC provides for preservation of the last setting and independent operation if—for any reason—useful data is not available from the bus. Under such conditions, pulses applied to the CLOCK input of the counter, while the UP/DOWN input is 1 or 0, can be used to increase or decrease the number stored in the register (and hence the analog output), one bit at a time. The pulses may be applied by manual switching, by an external backup processor, or gated in from an on-board slow clock generator.

As the Figure shows, there is also a three-state data-output register, which can place the information stored in the latches on the bus in response to a READ signal. This facility, which makes the DAC look like READ/WRITE memory, permits the computer to learn the state of the device (and the parameter it actuates) at any time—and especially when computer control is restored after an off-the-bus period (making possible "bumpless transfer" of control when the computer returns on-line). A CLEAR input can be used to set the latches to zero, e.g., during startup.

One of its most important attributes, the DAC has three divisions ("ports")—galvanically isolated from one another—with a breakdown rating of 1500V dc continuously or 1000V rms ac at line frequency for one minute. Its three sections might be termed: power, output, and front end. The *power* section contains a synchronizable high-frequency oscillator, the output of which is transformer-coupled to the other sections. The *front end* has a regulator and reference, a CMOS d/a converter, a modulator with transformer-coupled output, the slow clock for off-bus applications, and the digital logic circuitry. A small amount of isolated power is also available for external devices, such as logic gates for external drive in off-bus operation. The *output* section has the demodulator and current-output circuitry, with provisions for external offset and span adjustment. The current-loop power supply may be completely separate from the primary supply, or it may be the same supply.

Chapter Four

System Integration and Remote Data Acquisition

Now that we have seen a number of basic data-acquisition and data-distribution architectures in Chapters 2 and 3, it may be worthwhile to pause and consider some forms of system implementation. Areas of contemporary interest include:

1. Proprietary systems, subsystems, and components for interfacing and data communication.

2. Interfacing converters with nearby destinations, such as a microprocessor data bus, using parallel and byte-serial connections.

3. Using serial techniques to communicate sensor-based data with computers or distant destinations.

In Chapter 2, many of the configurations treat parallel data from the converter as an input to a nebulous "buffer" block. This buffer translates the converter's output into a machine data format, monitors the status of conversion, initiates conversions, addresses the multiplexer, controls sample-hold and gain-ranging, etc. These functions are achieved by purchasing proprietary interface products—ranging from ICs and modules to complete systems—and integrating them into the user's overall system.

In this chapter, we shall examine ways in which these forms of interfacing are achieved—first by outlining a number of proprietary systems and subsystems that form a hierarchy, then by a more-detailed consideration of interfacing techniques employing parallel and serial digital approaches, and the MACSYM ADIO (Analog-Digital Input-Output) bus.

As the preceding sentence may imply, we must be careful to limit the scope of this discussion, because any of these topics could itself justify a volume the size of this book for a thorough in-depth treatment of all possible cases. Our method will employ the following approaches: First, this book is oriented primarily towards data acquisition and conversion; we shall seek to maintain that focus in this chapter. For that reason, (and because they are widely documented elsewhere), we will avoid treatment of the popular IEEE-488 instrumentation bus and CAMAC concepts. Second, though we will tend to summarize in somewhat general terms, we will limit much of our discussion to ideas for which concrete embodiments can be found in the Analog Devices product lines. This, in turn, permits an abbreviated overview, with security in the knowledge that the reader who desires greater depth can find substantial amounts of detailed information about specific approaches (and products available to implement them) in our published literature.

4.1 SUBSYSTEMS FOR INTERFACING CONVERTERS TO THE ANALOG AND DIGITAL WORLDS

Briefly summarizing some salient ideas from earlier chapters, data acquisition is the process of transforming electrical voltages or currents, usually transducer outputs, into digital information to be received at some defined destination in a system, for storage, display, processing, or further transmission (Figure 4.1). The data-acquisition process typically involves these forms of activity:

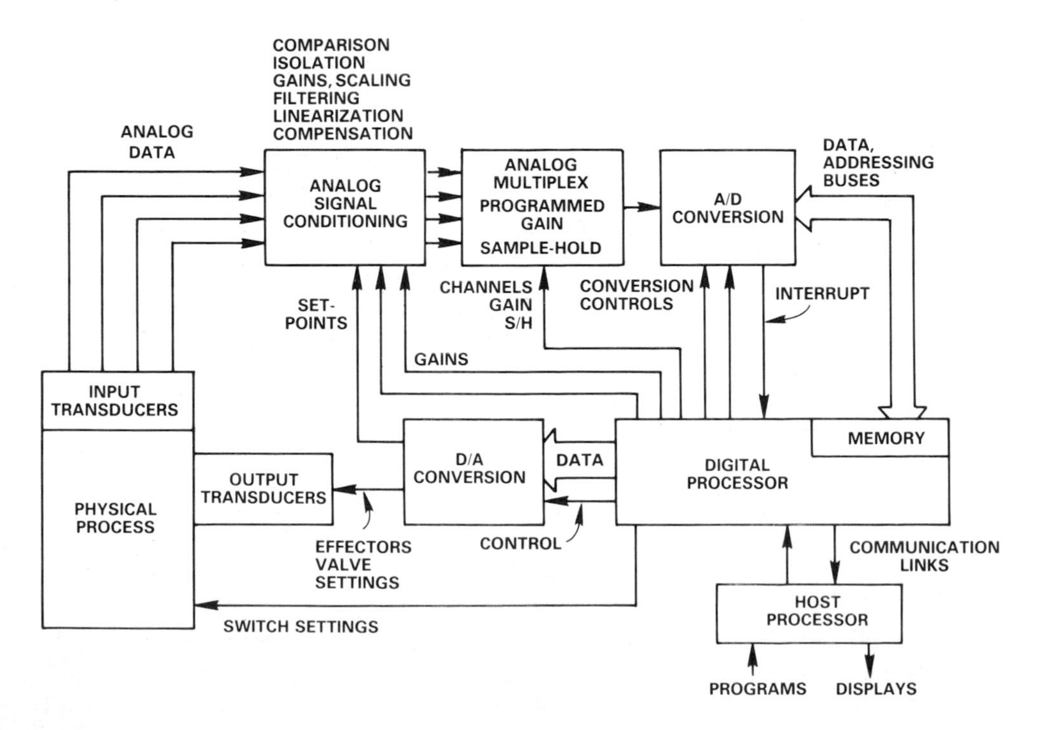

Figure 4.1. Data-acquisition function.

Analog signal-manipulation
Analog-digital conversion
Digital signal manipulation
Digital control manipulation, employing hardware and software

Analog signal manipulation includes such signal-conditioning operations as isolated preamplification, gain adjustment, linearization, algebraic functions (perhaps involving other inputs), sample/track-hold, and analog multiplexing.

An *analog-to-digital converter* produces a parallel or serial digital code that represents the ratio of an analog signal to a reference voltage or current. The digital code is usually—but not always—a binary or binary-coded-decimal number proportional to the ratio. Though widely and beneficially used, but not explicitly expanded upon here, a 1-bit a/d conversion can be the result of a go-no comparison of an analog signal with a set-point.

Digital SIGNAL manipulation might involve multiplexing, various arithmetic and logical operations—e.g., magnitude comparisons, algebraic operations, code or format conversions—storage, transmission to a slave or host processor, and deriving control signals for either digital handshaking or for operations on the "real-world" portion of the system.

Digital CONTROL manipulation includes control of all digital operations, programming of the analog functions (switching gains, channels, or circuit configurations), initiation of conversions, etc., and all of the associated software.

A *data-acquisition system* (for the present purposes) is an operationally self-contained subsystem, consisting of the conversion function (which involves at least one a/d converter and may involve d/a converters) and some portion of both the analog and digital manipulation circuitry. This definition (obviously quite flexible) permits a full functional range from a simple a/d converter (with a given analog span and digital controls) to an "intelligent" multichannel measurement-and-control subsystem—and a physical gamut from an integrated-circuit chip to a rack (or a room) full of equipment.

The desired properties of a given subsystem (in relation to the system) are determined by the application; their choice is affected by such factors as resolution, noise levels, system size and complexity, delegation of system tasks, frequency and level of interactions, the physical environment, software availability and compatibility, and (of course) cost—of hardware, of software, of wire, and of system development, prototyping, and manufacture.

When choosing the approach to take in designing a system, one system designer's component or subassembly may be another designer's turnkey system. In general, the design problem involves a classical "make-or-buy?" dilemma. The designer will purchase available components or subsystems that reflect the level of integration that is a best compromise between out-of-pocket cost and the many costs—both overt and hidden—of expending design and manufacturing effort in technological areas that are peripheral to one's primary mission.

The hierarchy of systems integration is especially deep at Analog Devices, ranging as it does from integrated-circuit converter chips to general-purpose computer-based data-acquisition subsystems, complete computer-based automatic test systems for integrated circuits and intelligent machine-vision systems. As examples of the levels of the hierarchy that are relevant for readers of this book, Table 4.1 (overleaf) lists some specific products that were available in 1985.*

We will review briefly the functional repertoire of these products in relation to the conversion interface, starting with the simplest and working our way up the chain.

4.1.1 THE AD574A INTEGRATED-CIRCUIT A/D CONVERTER

One of the simplest data-acquisition structures having a controllable interface capability is a bus-compatible a/d converter. A good example is the AD574A, a complete monolithic 12-bit a/d converter, furnished in a 28-pin dual in-line package (DIP). As the block diagram shows (Figure 4.2), it is a successive-approximation type with an internal 10-V reference. It accepts analog inputs with ranges of − 5V to + 5V, − 10V to + 10V, 0 to + 10V, and 0 to + 20V.

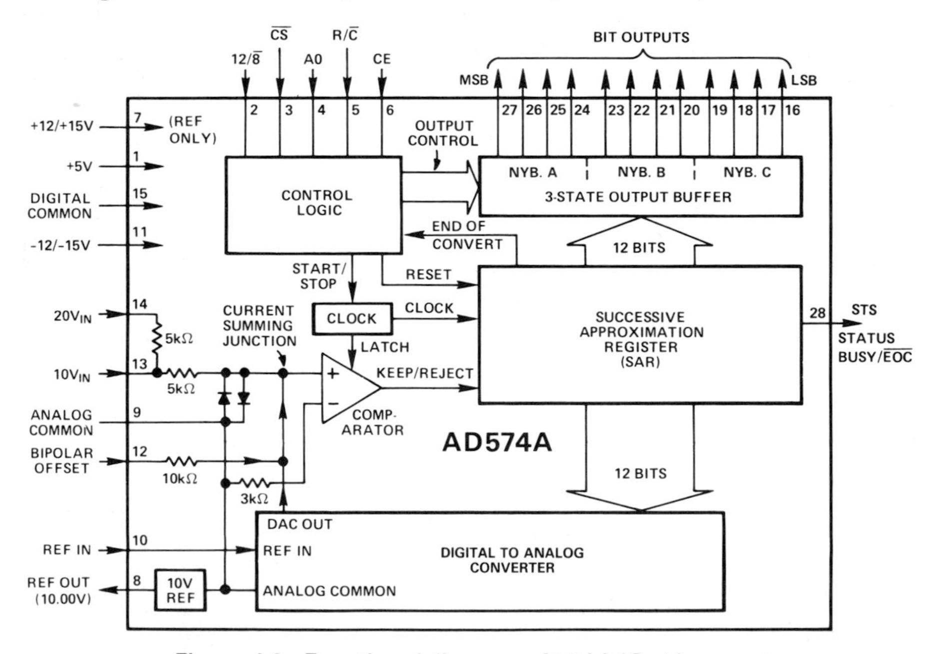

Figure 4.2. Functional diagram of 12-bit IC a/d converter.

*Since a complete discussion of their properties and applications is beyond the scope of this book, we will discuss several of these products in the context of their role in the conversion system. Complete information is available from the manufacturer, ranging from free data sheets and brochures to complete instruction books and software manuals at nominal cost.

Its digital output appears on three sets of three-state quad output latches and is left-justified. This means that the data represents the analog input as a fraction of full scale ranging from 0 to 4095/4096, implying a binary point to the left of the MSB (for example, .1100 0111 1001). The 12 bits of output data can be read either as one 12-bit word or as two 8-bit bytes—one with the 8 most-significant data bits (1100 0111), the other with 4 data bits and 4 trailing zeros (1001 0000). Conversions can be initiated either under program control, or in a stand-alone continuous-conversion mode. Its mode of interfacing will be discussed in Section 4.2.

4.1.2 THE AD364 DATA-ACQUISITION SYSTEM

The next level of simplification for the user (complexity for the device designer) adds a multi-channel analog front end. A graphic example of how this is done is the AD364, a complete microprocessor-compatible 12-bit data-acquisition system consisting of two complete functional blocks in hermetically sealed integrated-circuit packages (Figure 4.3). It essentially adds a multichannel analog front end to an AD574A a/d converter.

The analog input section, in a 32-pin dual in-line package, consists of two 8-channel multiplexers, a unity-gain differential buffer amplifier with high input impedance, a sample(track)-hold, channel-address latches, and control logic. The multiplexers can be connected to the subtractor in either an 8-channel differential or 16-channel single-ended configuration, under the control of a logic-operated mode switch. This means that the AD364 can perform in either mode without external hard-wired interconnections. Of perhaps

(continued on page 74)

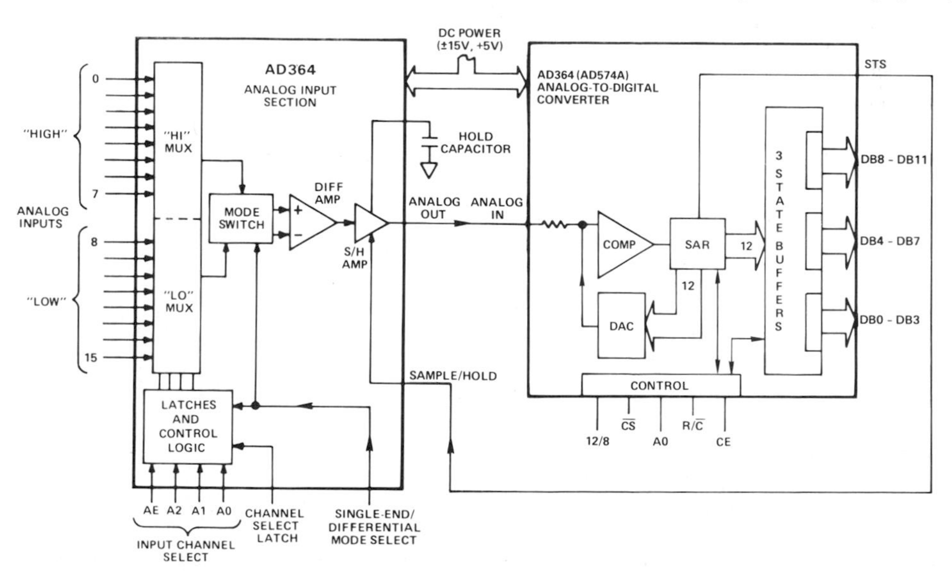

Figure 4.3. A/D converter and multichannel analog front end.

TABLE 4.1. REPRESENTATIVE EXAMPLES OF SYSTEMS, SUBSYSTEMS, AND SUBASSEMBLIES FOR DATA ACQUISITION MANUFACTURED BY ANALOG DEVICES IN 1985.

MACSYM 350 Fully Integrated Measurement And Control SYsteM

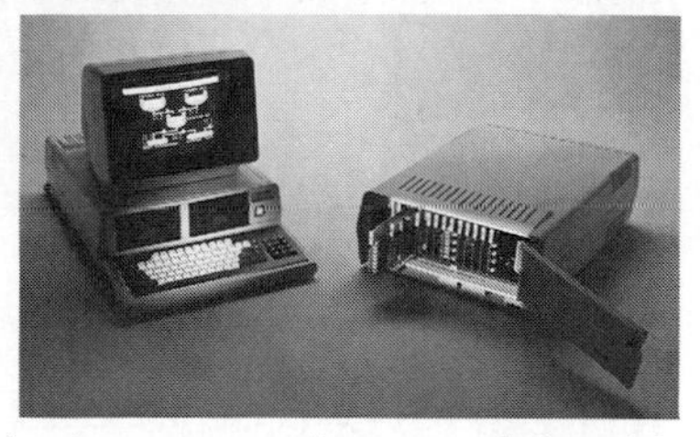

MACSYM 350 is a minicomputer-based measurement-and-control system used to automate the measurement, evaluation, and control of real-world phenomena, both analog and digital, while interfacing with human operators and other computers. It includes a keyboard, color display, and graphics, 5 ¼" floppy-disk storage, and a data-acquisition subsystem. It can hold up to 16 interchangeable analog-digital input-output (ADIO) cards—which may be field-wired to sensors. Its powerful multitasking real-time language, Measurement And Control BASIC (MACBASIC) is quickly grasped by users without extensive prior programming experience.

μMAC-5000 Programmable Measurement-and-Control Subsystem

μMAC-5000 is a single-board programmable measurement-and-control system. Combining direct connection to sensors—via screw terminals—with modular signal conditioning, conversion, digital inputs and outputs, a 16-bit microcomputer, an extended BASIC language for measurement and control, serial communication facilities, a power supply with uninterruptible features, and ruggedized construction, it provides a compact, easily expandable instrumentality for measurement and control, applicable in a broad range of stand-alone or distributed control systems. It is also physically and electrically compatible, serving the function of host, with μMAC-4000 subsystems, which have similar front ends, but limited computation and communication capability (Section 4.3.3).

AD2051 Microprocessor-Based Thermocouple Meter

The AD2051 connects directly to a switch-selected thermocouple (J, K, T, E, R, or S), corrects for the cold-junction temperature, amplifies and linearizes the thermocouple output, calibrates itself, provides a display in degrees Celsius (−165°C to +1760°C) or Fahrenheit (−265°F to +1999°F), and makes available a digital output, in the form of 7-bit character-serial ASCII. Also optionally available are a linearized analog output and facilities for a full-duplex 20mA isolated digital loop for communication with a computer or terminal.

RTI-1260 and RTI-1262 Microcomputer Analog I & O Subsystems

The RTI-1260 and RTI-1262 series of "Real-Time Interface" Analog I/O Subsystem Cards provide an analog input/output facility for microcomputer systems employing the popular STD bus. Interfacing as a block of memory locations, they fit easily into programs for microprocessors such as 8080A, 8085, 6800, 6809, and Z80. Capable of multiplexing 16 differential or 32 single-ended input channels, the RTI-1260 has a programmable-gain amplifier, sample-hold, and 12-bit a/d converter; the RTI-1262 has four channels of 12-bit d/a conversion. These cards interface with the STD Bus, but RTI-series Input, Output, and combined Input/Output boards, and others like them, are available for popular microcomputer bus structures, including the Multibus, TM990 Bus, LSI Bus, VME Bus—and as plug ins for PC slots.

AD7581 μP-Compatible 8-Bit 8-Channel Memory DAS

The single-chip CMOS AD7581 continuously scans 8 analog input channels, converts them to digital, and stores the data in bus-addressable RAM (read-write memory). The AD7581 interfaces directly with 8080, 8048, 8085, Z80, 6800, and other microprocessor systems. Data can be read at any time for any channel; on-chip logic provides interleaved direct memory access (DMA).

AD364 16-Channel 12-Bit Data-Acquisition System

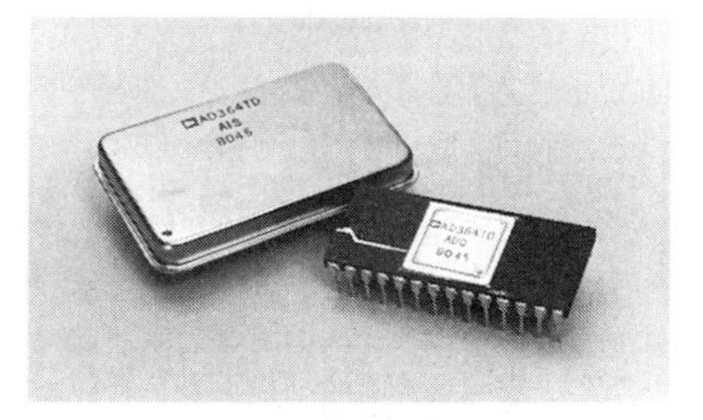

The AD364 is a complete data acquisition system in the form of two integrated-circuit packages. It includes a versatile multiplexer, differential amplifier, sample-hold, and 12-bit a/d converter. The multiplexer, and its associated mode switch, will handle from 8 differential to 16 single-ended channels, as well as intermediate combinations. The converter's 12-bit output will interface directly with 8- or 16-bit microprocessor buses, under software control.

AD574A μP-Compatible Analog-to-Digital Converter

The AD574A is an integrated-circuit 12-bit a/d converter. Accepting a variety of analog input voltage ranges, it interfaces to most popular microprocessor types having an 8-, 12-, or 16-bit data bus—without external buffers or peripheral interface controllers.

(continued from page 71)

greater significance, one AD364 can serve a mixture of both single-ended and differential sources under software control.

Multiplexer channel-address inputs are interfaced through a level-triggered (transparent) input register. With logic 1 at the Channel-Select Latch, the address signals feed through the register to directly select the appropriate input channel. This address information is latched into the register on the transition from logic 1 to logic 0 at the Channel-Select Latch input. The latching feature is useful when the user has no control over when input channel-address information may change—for example, when it is provided from an address, data, or control bus that may be required to serve many devices. Internal logic monitors the status of the differential/single-ended mode input and addresses the multiplexers according to an established scheme.

The sample-hold mode control input is normally connected to the status output from the a/d converter. When a conversion is initiated by applying a Convert Start control sequence (see AD574 Interfacing in Section 4.2), the Status goes High, putting the sample-hold into the Hold mode, freezing the information to be digitized for the period of conversion. When the conversion is complete, Status returns to logic 0 and the sample-hold tracks the input until the next conversion is initiated.

4.1.3 AD7581 8-BIT 8-CHANNEL MEMORY DAS

A quite sophisticated system capability on a chip is exhibited by the AD7581, a complete 8-channel, 8-bit data-acquisition system on a single monolithic chip. It consists of (Figure 4.4) an 8-bit ratiometric successive-approximation

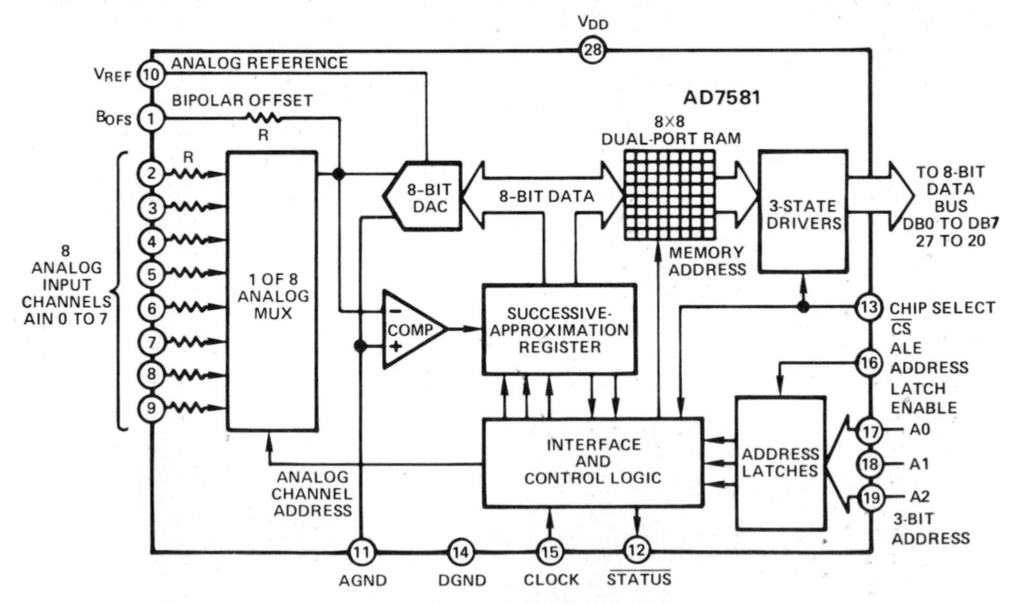

Figure 4.4. 8-channel, 8-bit data-acquisition system with memory.

a/d converter, an 8-channel analog multiplexer, an 8 × 8 dual-port random-access memory (RAM), three-state drivers (for interface), address latches, and microprocessor-compatible control logic. When used with appropriate references, it accepts either unipolar inputs (0 to +10V) or bipolar inputs (−5V to +5V, offset-binary output). It converts each channel in turn and stores the output; the digital value corresponding to any channel's input can be read from the AD7581's memory at any time. Its timing and operation with a microprocessor bus will be discussed in Section 4.2.

4.1.4 RTI-1260 AND RTI-1262 MICROCOMPUTER ANALOG INPUT AND OUTPUT SUBSYSTEMS

At the next level of complexity are "real-time interfaces" to standard microcomputer and personal-computer buses. When the intended function of a microcomputer system is measurement and control of real-world (i.e., analog) phenomena, analog I/O cards, which contain data-conversion components and bus interface logic, are necessary to interface the computer with the real world. They are designed to relieve the system designer of physical and electrical hardware problems relating to the analog-digital-processor interface (and many of the software considerations, too). RTI families, including input-only, output-only, and input/output boards, are available for popular buses, such as MULTIBUS, STD Bus and LSI-11 Bus.

The RTI-1260 Analog Input Subsystem and the RTI-1262 Analog Output Subsystem are representative of interface cards designed to work directly with

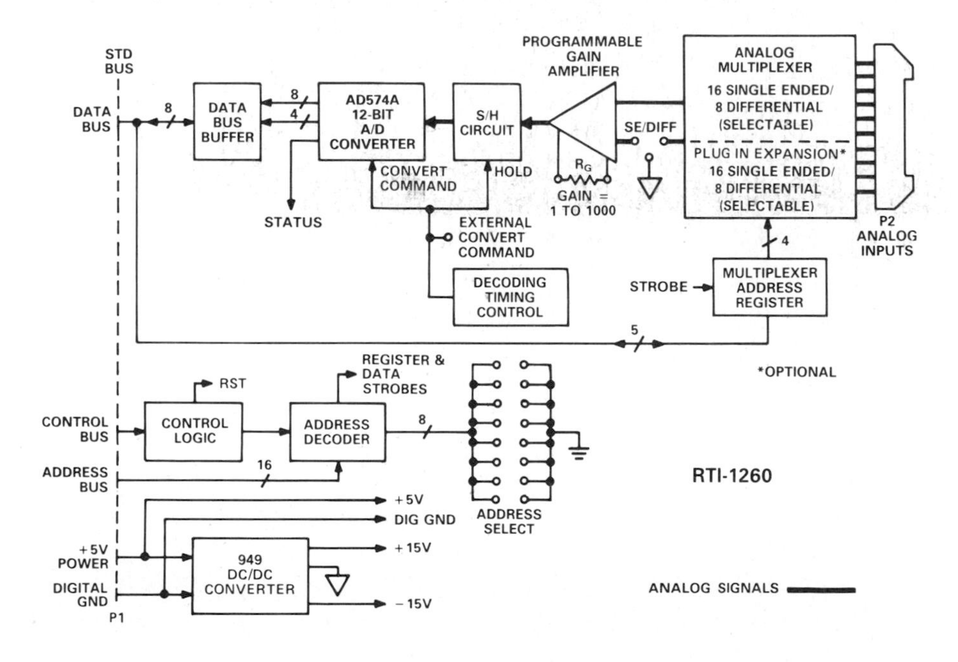

Figure 4.5. Functional diagram of data-acquisition card for STD bus.

microcomputer bus structures, interfacing in the same way as memory. They are designed for complete physical, electrical, and software compatibility with the STD bus, a popular bus that is compatible with many microprocessor types.

Figures 4.5 and 4.6 show the salient electrical features of the RTI-1260 Analog Input Card and the RTI-1262 Analog Output Card. The upper portions of the diagrams show conventional data-acquisition and distribution architectures. The lower portions detail the direct connections to the microcomputer buses: data, control, address, and power. A dc-dc converter converts the +5V bus supply voltage to the low-noise, isolated ±15V required for the analog circuitry.

Input cards for data acquisition. The basic architecture of an analog input card is shown in Figure 4.5. Analog inputs arrive at the terminals of a multiplexer, which selects one of 16 or 32 channels. The multiplexer can be configured for single-ended, differential, or quasi-differential modes. The single-ended mode is used when all signals are referred to a common ground—and are of sufficient magnitude in relation to noise to provide appropriate resolution; 16 or 32 channels are optionally available. For noisier environments, or where signals come from sources at differing common-mode levels, the differential mode can be used by the pairing of signal inputs to minimize the effects of common-mode noise; this halves the number of available inputs. If all signals have a common connection (not at system ground), it can be used as one side of the differential input; this quasi-differential connection takes advantage of the amplifier's differential inputs without sacrificing channel capacity.

The outputs of the multiplexers (which use dielectric isolation and can handle signals of up to ±35V without damage) feed a differential-input instrumentation amplifier, having gains programmable from 1 to 1000, to amplify the signal (±10mV to ±10V full scale) to the specified input range of the converter. The sample-hold tracks the signal and freezes it during a/d conversion. The converter produces an 8-, 10-, or 12-bit digital representation of the signal, and this result is made available to the microcomputer bus, via a set of three-state program-controlled registers.

The multiplexers accept software-determined commands from the microcomputer to select a specific analog channel and start an a/d conversion.

Output cards for control. Analog output cards (Figure 4.6) contain independent d/a conversion channels for driving chart recorders, servomechanisms, control valves, and output transducers. The analog output is set by writing a digital code to the appropriate address. If the resolution of the data exceeds 8 bits, the DAC requires two bytes of data. The DACs are double-buffered, so that both bytes may be separately loaded into the input register and then strobed simultaneously into the DAC, avoiding intermediate outputs and insuring cleaner transitions from one output value to the next.

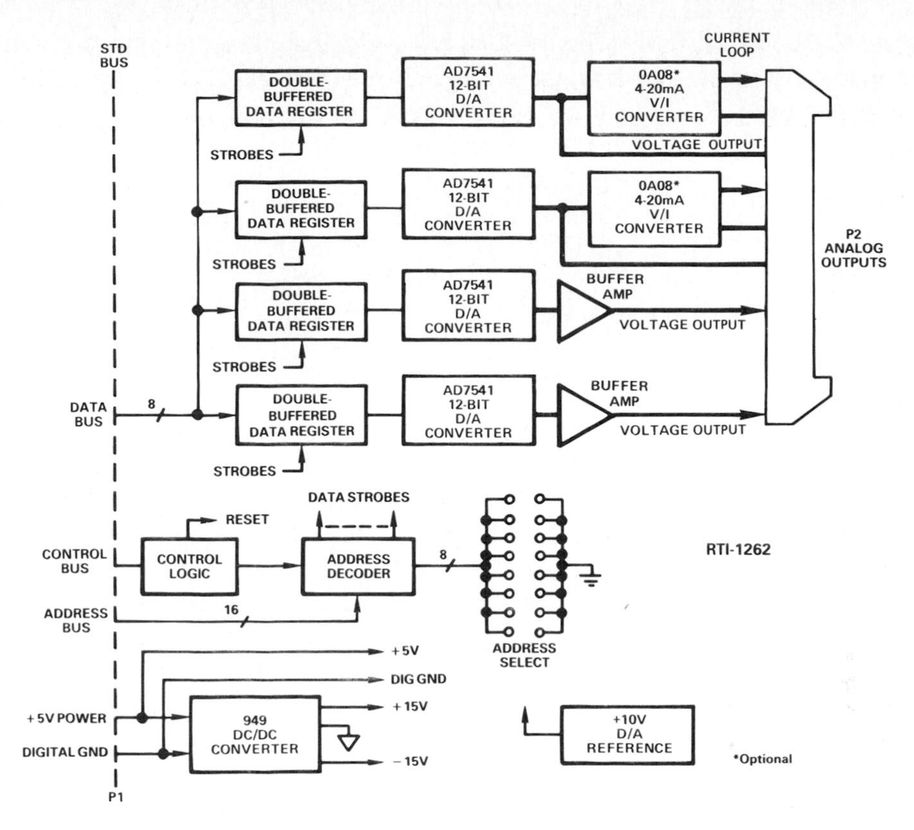

Figure 4.6. Analog output card for STD bus.

Usually, the cards offer single-ended or bipolar voltage output at standard levels. However, in some applications (e.g., automatic control), where the load may be at some distance from the microcomputer system, it is useful to transmit the analog signal as a current, rather than as a voltage, for immunity from voltage noise and IR drops in the output wiring. Current-output options typically provide an output current span from 4mA to 20mA. Digital interfacing of the RTI devices will be discussed in Section 4.2 of this chapter.

4.1.5 AD2051 μP-BASED THERMOCOUPLE METER

So far, we've discussed devices and subsystems that interface with high-speed multiwire microprocessor buses. Since long runs of parallel bus wire are high in noise, crosstalk, capacitance, and cost, it is usually imperative that the converter and its associated circuits be nearby—usually in the same card cage. For many applications, however, this may require that analog signals be carried over lengthy wire runs, with excellent prospects for signal degradation. The subsystems to be considered below can be operated closer to the signal source; they can communicate with computers by the use of serial transmission, and— because they have built-in processing capability and memory— they can stand alone in their transactions with analog signal sources.

The AD2051, for example (Figure 4.7), is a digital panel instrument that connects directly to a thermocouple. Switch-selected for the correct thermocouple type (J, K, T, E, R, or S), it corrects for the cold-junction temperature, amplifies and linearizes the thermocouple output, calibrates itself, provides a display in degrees Celsius (−165°C to +1760°C) or Fahrenheit (−265°F to +1999°F), and makes available a digital output, in the form of character-serial (7-bit parallel) ASCII. Also optionally available are a precision analog output and facilities for a full-duplex 20mA isolated digital loop for communication with a computer or terminal. These terms, and the ways in which the AD2051 interfaces, will be discussed in part 4.3.

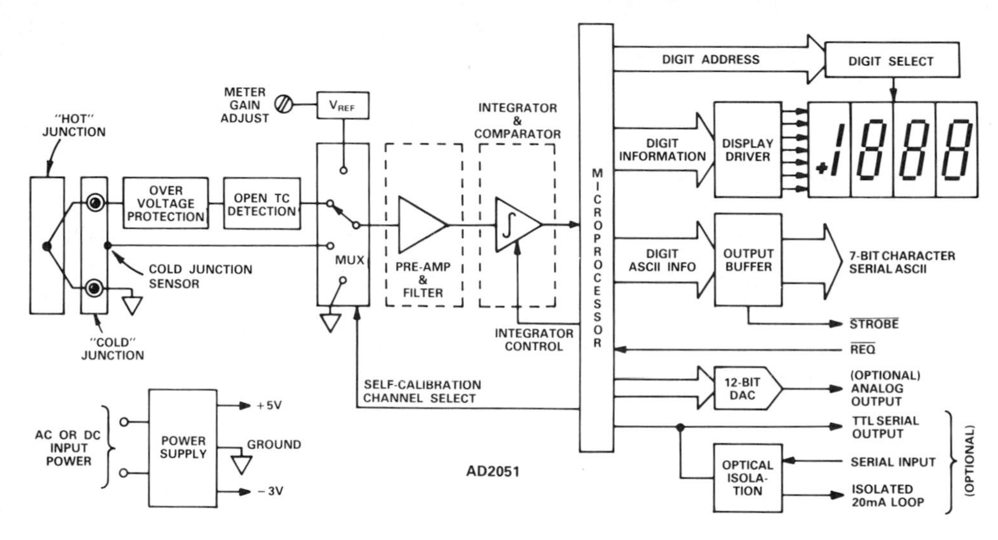

Figure 4.7. Versatile thermocouple thermometer.

4.1.6 μMAC-5000 PROGRAMMABLE SINGLE-BOARD MEASUREMENT-AND-CONTROL SYSTEM

The μMAC-5000 combines, on a single 9 ½″ × 13″ (241.3 mm × 330.2 mm) board, signal conditioning, multiplexing, and a/d conversion for 12 channels of analog sensor input (thermocouples, strain gages, etc.) It also has 8 digital outputs and 8 digital inputs, for communicating logic states, switch closures, etc. Additional channels are available via an expansion port and a family of analog and digital expansion boards that can be housed in the same card cage.

Fully programmable, employing μMACBASIC, a powerful, easy-to-use programming language, its 5-MHz, 16-bit 8088 on-board CPU processes the data collected from the input channels, sends out control signals, and comunicates with other equipment as directed by the program installed in its memory. Since the memory available over and above that required for the μMACBASIC operating system and internal operation is at least 16K bytes of both RAM and ROM, the user has the option to download programs into read-write memory from a host computer or to store them in non-volatile EPROM.

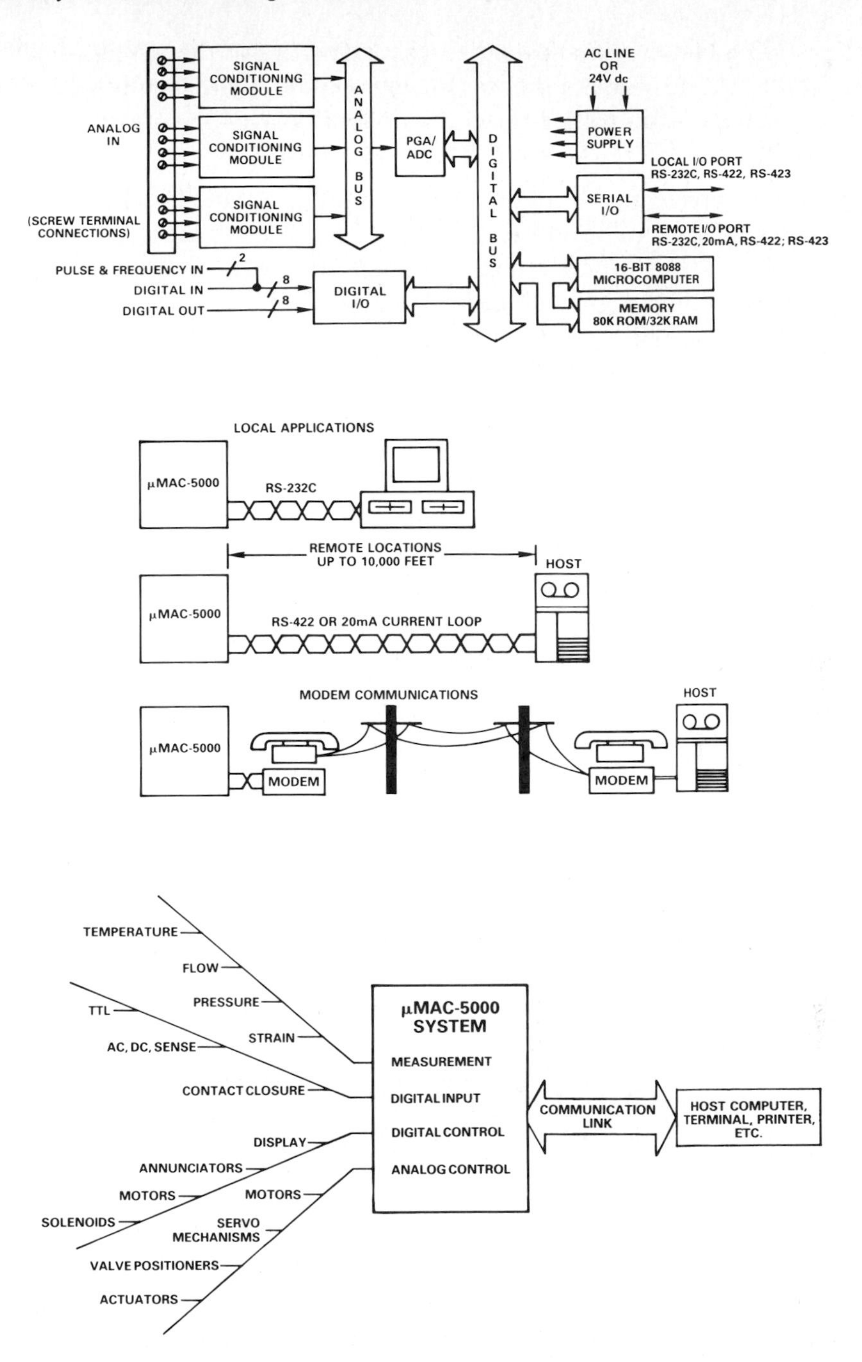

Figure 4.8. Functional diagram, typical communication modes, and system orientation for single-board measurement-and-control system.

The μMAC-5000 can be used as a stand-alone measurement-and-control system; it can provide remote intelligence for servicing a host computer; or it can act as a host for a μMAC-4000 slave measurement-and-control subsystem (see

Section 4.3.3). Figure 4.8 shows the basic structure of the μMAC-5000, its role in a system, and some of the ways it can communicate locally with terminals and printers—and either remotely or locally with host computers and other entities.

Analog signals arrive at the inputs of three plug-in quad signal-conditioner/multiplexers, which can be chosen from a variety of available types to provide such functions as cold-junction compensation for thermocouples, adjustable-gain preamplification, high-voltage isolation, and sensor fault indication. The function of each individual input is selectable in software. Each module's output is multiplexed onto the internal analog bus; the resulting signal is amplified by the programmable-gain amplifier (PGA) to fit the input range of the a/d converter, and the result of the conversion— appearing on the digital bus—is processed as required by the program. An integrating converter provides resolutions from 13-bits-plus-sign to 11 bits, depending on the desired number of conversions per second.

The μMAC-5000 software is easy to use. For example, on the command 'AIN(channel)', it automatically addresses the voltage on the specified analog input channel and converts it to digital; if identified as originating in a thermocouple, the data is linearized, compensated, and translated into engineering units. Programmers are relieved of writing these steps into their program. The use of an analog-input command AIN illustrates this:

```
10 TEMP1 = AIN(1)
20 PRINT "THE TEMPERATURE IS ";TEMP1;" DEGREES C."
```

When line 10 is executed, if the addressed, converted, and processed value of TEMP1 is 23.2, the displayed or printed response in line 20 would be:

```
THE TEMPERATURE IS 23.2 DEGREES C.
```

The first command measures the analog input on channel 1 (e.g., a thermocouple), computes the correct temperature in degrees Celsius, and assigns the value to the variable TEMP1. The second command formats and prints the data on a CRT or printer.

Besides such simplifications as these, μMACBASIC can further simplify the writing of programs by allowing the programmer to name Procedures or Functions. For example, in

```
10 IF AVG(3,10) > 700 THEN ALARM (2)
```

When line 10 is executed, a Procedure, AVG(3,10), is called (e.g., the averaging of 10 readings of analog channel 3), the result is compared with 700, and if the result is greater than 700, an alarm function (perhaps the outputting of a switch closure) is performed. The Procedure and Function may be written with any desired local line numbers and variable names without conflicting with the main program.

In Section 4.3.3, we will discuss the means by which the μMAC-5000 and other entities, such as personal computers, communicate with the response-only "intelligent" μMAC-4000 Measurement-and-Control Subsystem.

4.1.7 MACSYM 350 COMPUTER-BASED MEASUREMENT-AND-CONTROL SYSTEM

At the highest level of data-acquisition system integration, MACSYM 350 is a fully integrated Measurement And Control SYsteM developed specifically to acquire, reduce, store, present, and output real-time information in laboratory, process control, and discrete manufacturing applications. From architecture and packaging to software and documentation, the system is human engineered to minimize the time and experience required to configure, hook up, program, and operate in the user's environment.

The complete MACSYM 350 system, with integral signal conditioning, is packaged in two compact desktop units: the stand-alone MACSYM 150 workstation and the MACSYM 200 intelligent front end. The basic system, the MACSYM 150, includes a high-speed 16-bit 8086 processor and 8087 numeric coprocessor, keyboard, color display, and 5 ¼″ floppy-disk storage. Specialized I/O cards plug into a 6-slot internal backplane to provide analog, digital and communications input/output, and memory expansion.

It can be expanded to form MACSYM 350 systems, systems of much larger scope and range of functions, by the addition of one or more MACSYM 200 intelligent front ends, each of which provides for up to 16 additional slots for data-acquisition and control. A large library of analog/digital input/output (ADIO) cards—which can be interchangeably plugged into any of 16 card slots provided in the MACSYM 200's chassis—is available. Application programs can be written immediately in MACBASIC, a multitasking real-time BASIC, optimized specifically for measurement and control.

Figure 4.9 is a simplified overall block diagram of MACSYM 150 and 200, depicting the organization of the major components comprising such systems. The 16-bit central-processing unit has hardware floating-point and a minimum of 128K bytes of RAM. The system-control card includes a 24-hour real-time clock and console serial interface with RS-232C or 20-mA current loop (110 to 9600 baud).

A key element of MACSYM systems is the dual bus system. The processor communicates with memory and computer-type peripherals via the conventional computer bus structure. However, communication with the analog and digital I/O cards is established by way of the ADIO controller, an intelligent interface which provides a number of shared functions—including a/d conversion. The dual bus structure and shared functions minimize the complexity and cost of the individual cards and isolate the I/O bus from the noisy high-speed processor bus, permitting improved performance to be obtained with low-level analog signals. Since the ADIO controller deals with each card on

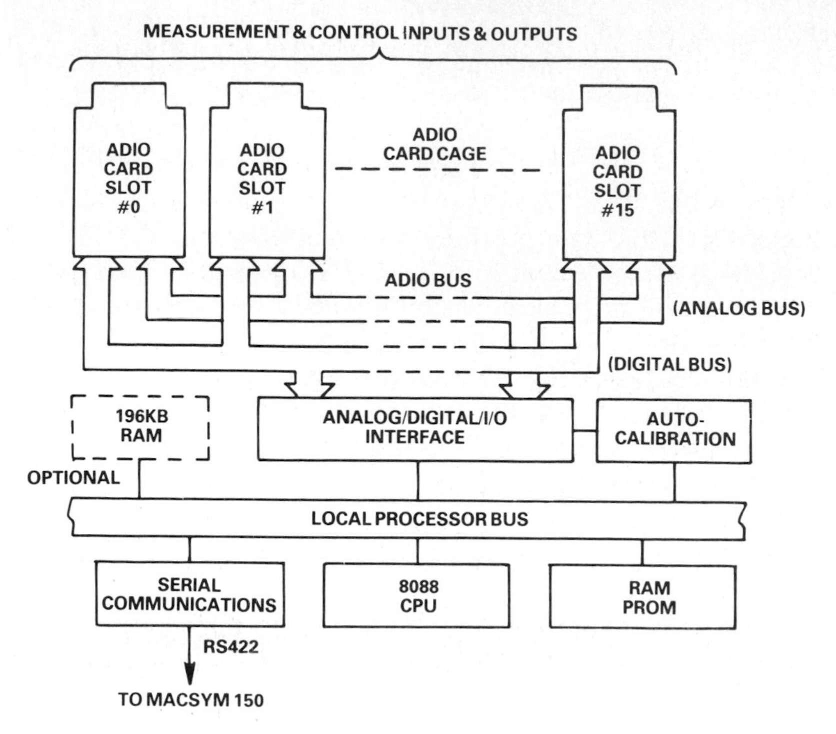

a. MACSYM 150.

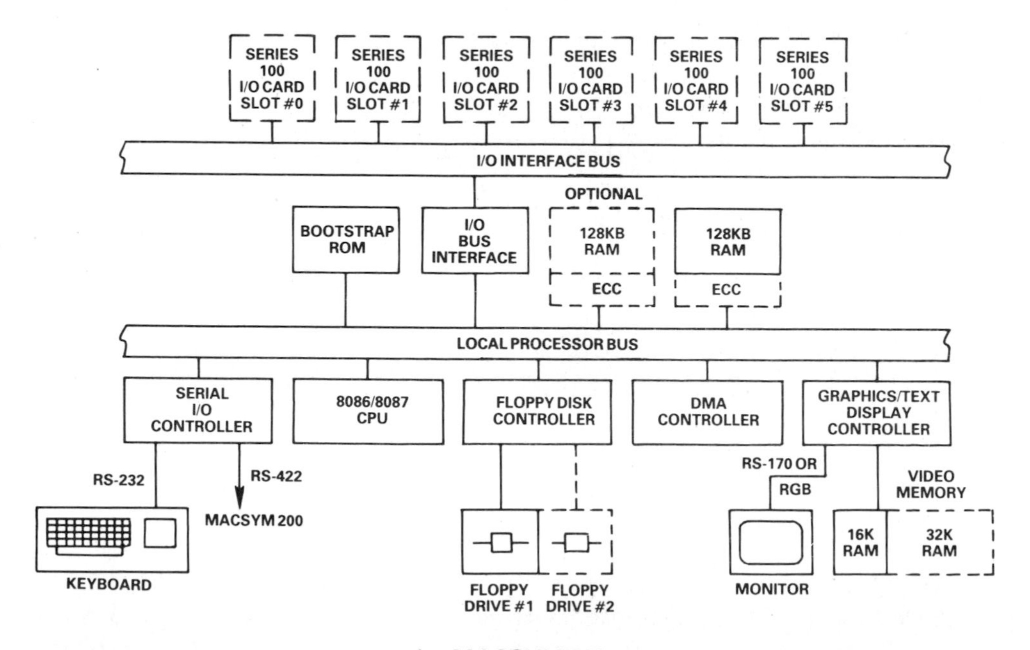

b. MACSYM 200.

Figure 4.9. Simplified block diagram of minicomputer-based measurement-and-control system.

the basis of its own identity, it is possible to utilize any assortment of cards, 16 per MACSYM 200 chassis.

Capabilities of the signal-conditioning card family include low-level analog in/analog out, direct sensor interfaces (thermocouple, strain gage, RTD, etc.), digital input/output, isolation, and a wide variety of special functions. Optional screw-terminal boards permit direct connections to sensor wires. The number of cards used in a a given application may be greatly increased by the use of extension chassis. A variety of peripherals, from disk drives and plotters to graphic terminals and air-conditioned NEMA cabinets are available.

MACSYM 350 can interface with the IEEE-488 (1978) general-purpose instrumentation bus (GPIB); it interfaces via the ACP100 I/O card and its associated software driver, which enables MACSYM 350 to send and receive data and operate as a bus controller or listener/talker, using BASIC, for supporting up to 15 external devices simultaneously, with 9 operational functions and 30 available programmable instrument addresses, at a 2000-byte-per-second maximum data rate.

A/D conversion in MACSYM 350 systems is achieved in software by the variable, AIN(card slot, channel); d/a conversion is effected by AOT(card slot, channel). Thus, the powerful MACBASIC statement,

$$\text{AOT}(8,3) = 5*\text{AIN}(2,5) + \text{AIN}(3,6) + \text{K}(5) \tag{4.1}$$

means that when the statement is executed, the voltage applied to channel 5 of the card in slot#2 is measured, multiplied by 5, added to the voltage measured at channel 6 of the card in slot #3, added to a variable identified as K(5), and that sum updates the analog output from channel 3 of the card in slot #8, all within milliseconds.

MACSYM 350 has been described here to illustrate the highest level of integration of conversion into a general-purpose data-acquisition system for measurement and control. Designed for end-user convenience, MACSYM's a/d and d/a conversion functions are internal, fully integrated, and buffered from the world by the ADIO bus and software. Because MACSYM functions essentially as a self-contained turnkey system, supported by a panoply of tutorial publications, instruction and software manuals, application briefs—and a field service organization—the details of its converter-interfacing methodology are beyond the scope of these pages.

4.2 INTERFACING CONVERTERS WITH MICROPROCESSORS, USING PARALLEL CONNECTIONS

Microprocessors, because of their low cost and ready availability, have become established as the prime arena for computer interfacing. It therefore makes sense to indicate how the general principles discussed in earlier chapters can be applied in fairly specific ways to microprocessors.

There is a plethora of detailed microprocessor architectures and software systems, differentiated by manufacturer, by generation, and by degree of integration. It would be futile and well beyond the scope of this chapter to explore even a few of them in detail. Instead, we shall seek to show the elements that most microprocessors have in common and to indicate how some of the conversion and data-acquisition devices mentioned earlier can be interfaced to μPs. Many of the underlying ideas can be found in several of the references (at the back of the book), as well as in manufacturers' literature.

A *microcomputer* is an operational computer system, with a specified amount of memory, based on a microprocessor (CPU) chip. A *microprocessor chip* (packaged integrated circuit) is something less than a microcomputer, and the difference between the two is simply a measure of a continually shrinking technological gap.

Figure 4.10 shows a functional diagram of the connections to an 8080 microprocessor. They include a 16-bit unidirectional latched address bus, which is used to address one 8-bit byte out of a possible 65,536 bytes (64K) of external memory; an 8-bit bidirectional data bus for transferring data to or from the processor; a set of power-supply terminals; a pair of clock terminals; and a set of incoming and outgoing control lines. The processor itself contains an accumulator, a set of registers, and the operational capability of carrying out up to 256 different instructions, coded in 8-bit words. The instruction groups

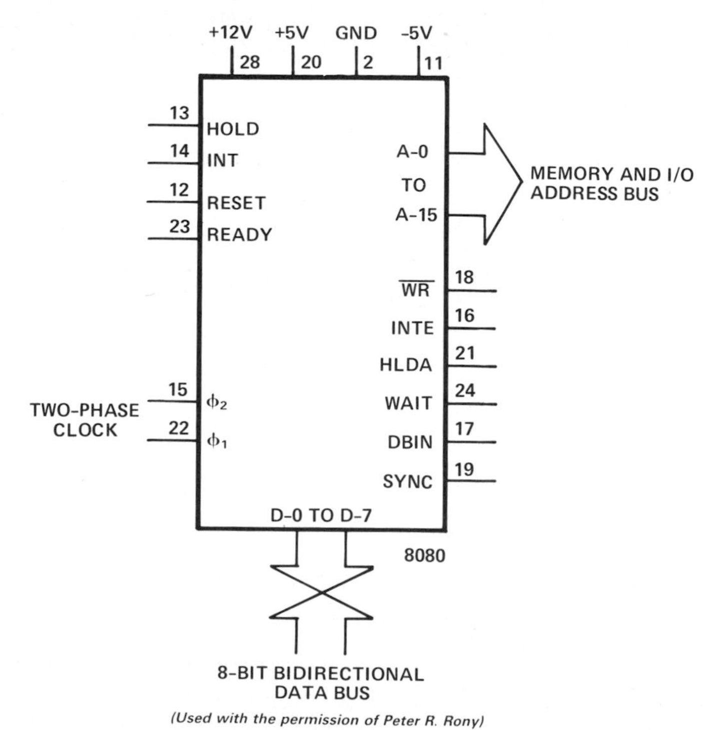

(Used with the permission of Peter R. Rony)

Figure 4.10. Typical microprocessor interface connnections.

include data-transfer; arithmetic operations; logic operations; branching operations; and stack, I/O, and machine-control operations.

Double-precision operations are inherent: a number of instructions string two 8-bit data bytes together as a 16-bit word. It will be seen that this is a useful feature in dealing with converters.

4.2.1 MICROPROCESSOR INTERFACING, I/O vs. MEMORY

To interface a converter or a data-acquisition system to a microprocessor, a number of requirements must be fulfilled:

It must be possible to address the converter subsystem, and if a MUX with random addressing is used, it must be possible to address specific analog channels.

The output of the converter must be transformed to a compatible format and to circuitry compatible with three-state busing.

Suitable software and control signals must be provided to initiate conversion, determine when conversion is complete, and transfer the data appropriately.

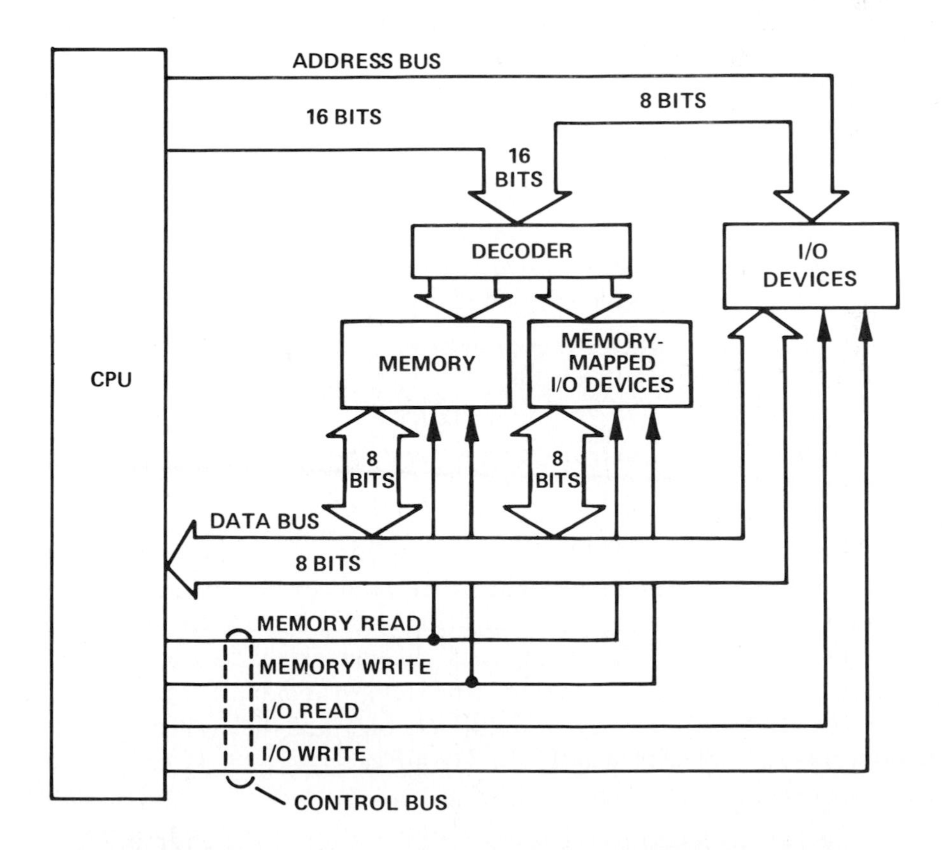

Figure 4.11. Accumulator (or isolated) I/O vs. memory-mapped I/O.

Analog-to-digital converters, like other I/O devices, may be interfaced tomicroprocessors by several methods. These methods include (but are not limited to) direct memory access (DMA), isolated, or accumulator I/O, and memory-managed (or memory-mapped) I/O. Direct memory access is the fastest, since conversions occur automatically, and data updates into memory are transparent to the processor. DMA logic is very processor-dependent and makes use of dedicated specialized hardware. Memory-managed and accumulator I/O are more-often used and somewhat easier to understand (Figure 4.11).

Memory-managed I/O assigns the I/O device to one or more locations in the memory space of the microprocessor. This technique has the advantage that the full range of memory-reference instructions may be used to operate on the data. The potential disadvantages include limiting the memory space available for program and data memory, somewhat more complex address decoding, and increased difficulty of isolating device-select pulses for system debugging. Many processors offer only memory-mapped I/O.

Accumulator I/O uses a set of control signals which are distinct and different from the memory control signals. These control signals, combined with use of a portion of the address bus, serve to define a totally separate I/O address space. This architecture is simpler from the hardware standpoint, since address-decoding requirements are less severe for the smaller space (for example, 256 inputs and 256 outputs may be used with the 8080), and distinct I/O Read and Write pulses are more easily located for system debugging purposes. However, processors using accumulator I/O generally can only send data to an output device from the accumulator. This can make the software more cumbersome, since processor-controlled transfers of I/O device data to a memory location cannot be accomplished in a single instruction.

Concrete examples of some of these concepts can be given in terms of some of the devices mentioned in this book.

4.2.2 INTERFACING AN IC 12-BIT A/D CONVERTER WITH MICROPROCESSORS

AD574A Controls. The AD574A contains on-chip logic for initiating conversions and reading out data, using signals commonly available in microprocessor systems. Referring to Figure 4.2 and Table 4.2, three of the logic inputs, CE, $\overline{CS}$, and R/$\overline{C}$, control the operation of the converter; the register-control inputs, A0 and 12/$\overline{8}$, control converted word length and data format. The state of R/$\overline{C}$ (read/convert), when CE (chip enable) and $\overline{CS}$ (chip select) are both asserted, establishes whether a Data-Read (R/$\overline{C}$ = 1) or a Convert (R/$\overline{C}$ = 0) operation is in progress.

The A0 line is usuallytied to the least-significant bit of the address bus. If a conversion is started with A0 low, a full 12-bit conversion cycle is initiated;

CE	$\overline{CS}$	$R/\overline{C}$	$12/\overline{8}$	A0	Operation
0	X	X	X	X	None
X	1	X	X	X	None
1	0	0	X	0	Initiate 12-Bit Conversion
1	0	0	X	1	Initiate 8-Bit Conversion
1	0	1	+5V	X	Enable 12-Bit Parallel Output
1	0	1	Dig. Com.	0	Enable 8 Most Significant Bits
1	0	1	Dig. Com.	1	Enable 4LSBs +4 Trailing Zeroes

Table 4.2. AD574A truth table.

if A0 is high during a Convert start, a shorter 8-bit conversion cycle results. During data Read operations, the state of A0 determines whether the 3-state buffers containing the 8MSBs of the conversion result (A0 = 0) or the 4 LSBs (A0 = 1) are enabled. The $12/\overline{8}$ pin determines whether the data is to be organized as two 8-bit words ($12/\overline{8}$ wired to Digital Common) or a single 12-bit word ($12/\overline{8}$ wired to VLOGIC).

The functions of the controls are summarized in Table 4-2. An output signal, STS, indicates the status of conversion. STS goes high at the beginning of a conversion and returns low when the conversion cycle is complete.

Microprocessor Interfacing. A typical a/d converter interface routine involves several operations. First, a Write to the ADC address initiates a conversion. The processor must then wait for completion of the conversion cycle, since most integrated-circuit ICs take longer than one machine instruction cycle to complete a conversion. Valid data can only be read after the conversion is complete.

The AD574A's STS (Status) signal indicates when a conversion is in progress. The processor can poll the signal by reading it through an external three-state buffer (or other input port). The STS signal can also be used to generate an Interrupt when the conversion is completed, if the system timing requirements are critical, and if the processor must perform other tasks during the time required for conversion (35μs for the AD574A) and must use the results of conversion as soon as they are ready. Another useful time-out method is to assume that the ADC will take 35 microseconds to convert, and insert a sufficient number of instructions to ensure that at least 35μs of processor time are consumed (or rely on an independent counter or clock, freeing the processor).

Once it is established that the converter has finished its cycle, the data can be read. If the ADC has 8 bits or less of resolution, a single data Read operation is sufficient. In the case of converters with more data bits than are available on the bus, a multi-byte data format is required, and multiple Read operations are needed. The AD574A includes internal logic to permit direct interfacing to 8- or 16-bit data buses, selected by connecting the $12/\overline{8}$ input low or high. In 16-bit bus applications (12/$\overline{8}$ high), the data lines (DB11 through DB0) may be connected to either the twelve most-significant or 12 least-significant bits of the data bus. The remaining four bits should be masked in software. The effect of the former is a left-justified fractional-binary format (1111111111110000, while the effect of the latter is a right-justified integer-binary format (0000111111111111). The interface to an 8-bit data bus (12/$\overline{8}$ low) is always done in a left-justified format with the AD574A. When the LSB of the address word is even (A0 Low), the eight most-significant bits (DB11 through DB4) are addressed; when the address is odd (A0 High), the output to the bus consists of 4 LSBs (DB3 through DB0), followed by four trailing zeros—bit-masking is unnecessary (Figure 4.12).

	D7							D0
XXX0 (EVEN ADDR):	DB11 (MSB)	DB10	DB9	DB8	DB7	DB6	DB5	DB4
XXX1 (ODD ADDR):	DB3	DB2	DB1	DB0 (LSB)	0	0	0	0

Figure 4.12. Left-justified data format for interfacing a 12-bit converter to an 8-bit bus.

Access time and data latency time for the AD574A's 3-state buffers are comparable to those for today's memory devices. Therefore, the AD574A can interface directly to many processor buses without the need for Wait states or external data buffers. We will show an example here of interfacing to 6800/6502 microprocessor systems. Interfaces to other systems, including the Apple II computer, may be found in the AD574A data sheet.

6800 Interfacing. The control signals and bus architecture of the 6800 and 6502 series of microprocessors are quite similar. In each, the state of the read-write (R/$\overline{W}$) signal at the rising edge of the $\phi 2$ (or equivalent) clock establishes whether a Memory Read or Memory Write is in progress. The memory address being exercised is signaled by decoding of the address bits to (usually) an active Low signal.

This control structure is directly compatible with the AD574A (Figure 4.13). The R/$\overline{W}$ (read/write) line can be used for R/$\overline{C}$ (read/convert); the active-low decoded base address (the AD574A occupies two memory locations) is applied to $\overline{CS}$ (chip select), and $\phi 2$ is used for CE (chip enable). The least-significant address line (even-odd) is tied to the AD574A's A0 input.

In this interfacing scheme, the processor initiates a conversion by performing a dummy Write to the converter, bringing read/convert low (the contents of the data bus during the Write are ignored). If A0 is low, a full 12-bit conversion will be started; if A0 is high, a short 8-bit conversion is started. After sufficient time has elapsed for the conversion to be complete, the processor can read the data (bringing read/convert high) in the two memory locations

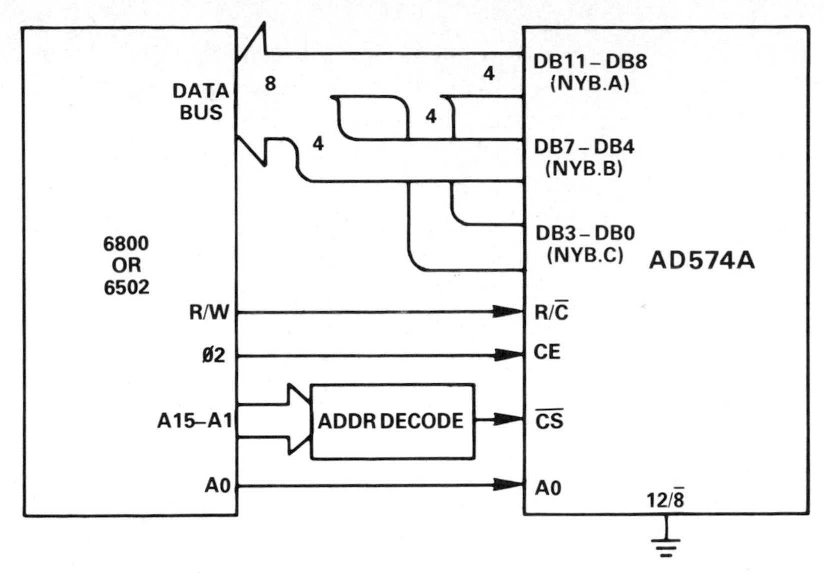

Figure 4.13. Interfacing a 12-bit ADC to 6800/6502 microprocessor.

occupied by the AD574A. The even location (A0 low) contains the 8 MSBs, and the odd location (A0 high), if used, contains the four LSBs and the four trailing zeros.

The AD574A may be used directly with 6800 series processors running at clock speeds up to 1.5MHz.

4.2.3 INTERFACING A DATA-ACQUISITION MODULE TO THE 8080 MICROPROCESSOR

Figure 4.14 shows how a DAS1128 data-acquisition subsystem (Figure 2.14b) might be interfaced with an 8080, using an 8255 peripheral interface chip. A typical sequence of events, slightly simplified, is this:

1. A setup byte, addressed to this channel (address decoded) is latched (written) into the 8255. It configures the 8255 as a set of two input ports (8 and 4 bits), which will receive the data from the converter, and one 4-bit output port, which will address the appropriate MUX channel. (The DAS1128 will have been configured for random addressing.)

2. A MUX-address byte, addressed to this channel (address decoded), appears on the data bus and is latched (written) into the MUX-address input of the DAS1128, causing the multiplexer to switch to the appropriate channel.

3. A conversion command, addressed to this channel (address decoded), is written into the DAS1128's Strobe input and initiates a conversion cycle, starting with sample-hold.

4. At some later time, when the conversion can be expected to have been completed, successive READ pulses, addressed to this channel (address decoded), cause data in the 8-bit and 4-bit input bytes of the 8255 to be transfer-

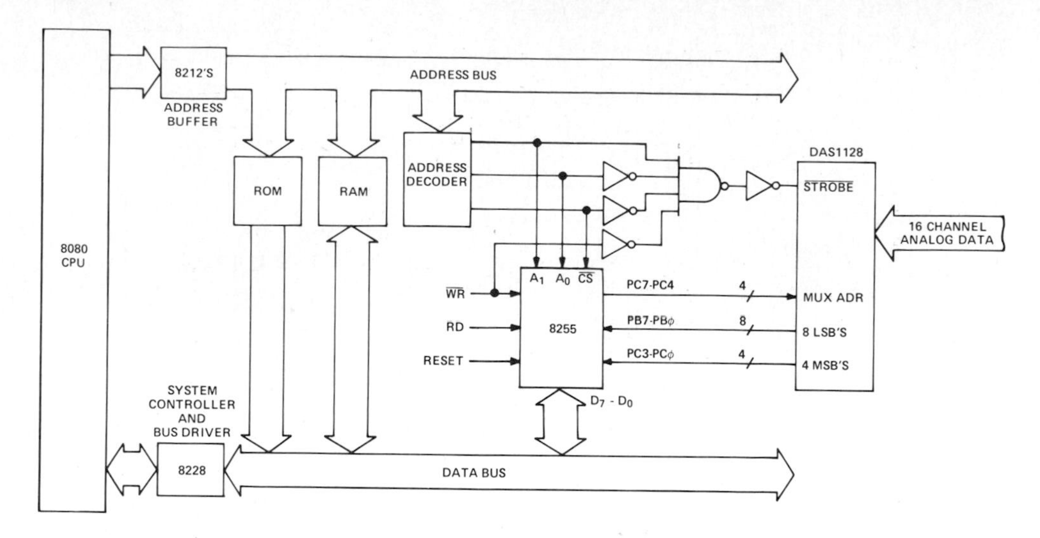

Figure 4.14. Interfacing a 12-bit DAS to an 8080 microprocessor.

red to the microprocessor. It is also possible (but not shown for the sake of simplicity) for the DAS1128's Status (Busy) line to have triggered an *interrupt* cycle when the converter's output became valid, in order to get fast handling without tying up the processor during the conversion. Interfacing could have been either I/O or memory-managed, trading off 8-bit (I/O) vs. 16-bit (memory) addressing for the sake of simpler software (2-byte instructions and faster machine handling).

It is easy to see that essentially the same technique could be used to interface any other simple 12-bit parallel-output (non-3-state) converter, such as the AD572, without the MUX, and with or without a sample/hold. If a separate sample-hold (e.g., the AD582) were used, an appropriate time delay must be used between initiating the Hold command and the start of conversion.

4.2.4 INTERFACING AN IC DAS TO THE BUS

Microprocessors. Since the AD364 data-acquisition system consists of an AD574 a/d converter plus an analog front end, the conversion control interface is the same as for an AD574. However, because there are from 8 to 16 channels of analog input, representing from 16 to 32 memory locations (for 12-bit conversion), which must be individually addressed, the decoding structure is different.

The lowest bit, A0, is still directly connected to the converter's A0 line. The next four bits, A1 through A4, are connected to Channel-Select inputs A0 through AE of the analog input section; they are internally decoded. Thus, only bits A15 through A5 on the μP's address bus need be decoded. The Channel-Select Latch is operated by the Write line; when the Write line goes low, the channel whose address is on A0 through AE of the analog input section is latched.

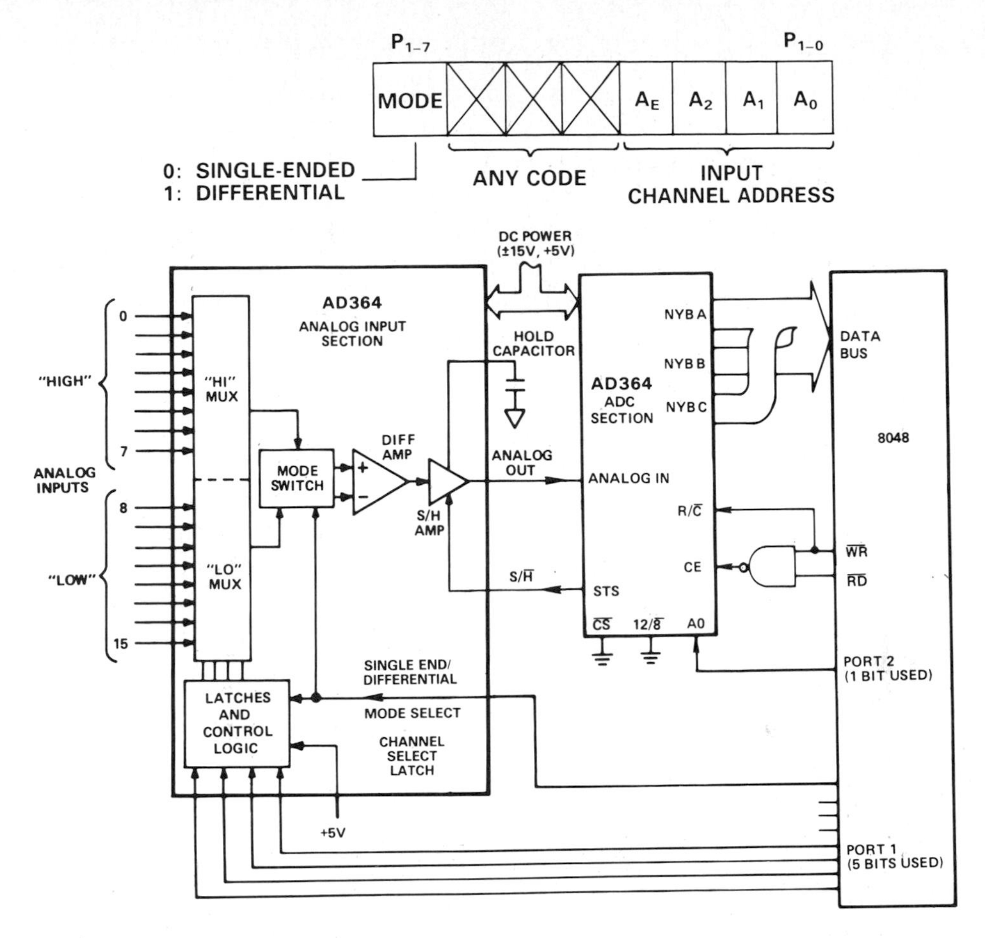

Figure 4.15. Interfacing an IC 12-bit DAS to an 8048 single-chip microcomputer.

Single-Chip Microcomputers. Single-chip microcomputers now available, such as the 8048, 6801, and 3870, include fully decoded I/O ports on the chip, as well as the central processing unit, RAM, and ROM. The fully decoded I/O ports sidestep the need for address decoding for I/O devices in many systems. For example, the 8048 contains 64 bytes of RAM, 1K bytes of ROM, and 2 programmable 8-bit I/O ports, which can be used either as inputs or outputs. A third 8-bit port, designated BUS, is a bidirectional port, which can be used for expanded I/O or memory.

As Figure 4.15 shows, the AD364 interfaces easily to an 8048 single-chip microcomputer, providing a complete data-acquisition system with minimal package count. In this system, 5 of the 8 bits of Port 1 drive the channel-select address inputs and single-ended/differential mode. Since the outputs of Port 1 are already latched, it isn't necessary to use the latch built into the AD364. The Latch input is tied to logic 1, which causes the latch to be transparent. The setup byte at Port 1 for the conversion takes on the format shown in the inset.

4.2.5 INTERFACING AN IC DAS WITH MEMORY TO THE BUS

The AD7581 (Figure 4.4) accepts eight analog inputs and sequentially converts each into an 8-bit binary word, using the successive-approximation technique. The result of each conversion is stored at an address in an 8-bit, 8-channel dual-port (input, output) RAM. Basic timing for the device is derived either from the microprocessor clock (in 6800-type systems) or from some suitable signal (ALE—Address Latch Enable, in 8085-type systems). Startup logic is included to establish the correct sequences within 800 clock periods when power is applied; required power is $-10V$ reference and $+5V$ excitation.

A complete conversion cycle for each channel requires 80 clock pulses; 640 pulses are needed for a complete scan through all eight channels (in decreasing order, from channel 7 to channel 0). When a channel's conversion is completed, the contents of the successive-approximation register are loaded into the proper channel location of the 8×8 RAM, and the Status line ($\overline{STAT}$) produces a negative-going pulse. When conversion of Channel 1 is complete, the status pulse lasts for 72 clock-periods (during the conversion of channel 0); for all other channels, the status pulse is only 8 clock periods in duration. Thus, an external pulse-width detector can be used to identify when channel 1 has been converted (and that channel 0 is being converted) and to derive conversion-related timing signals for microprocessor interrupts.

Each time the status line goes low, the multiplexer address is decremented; the next conversion starts eight clock periods later. Automatic interleaved DMA (Direct Memory Access) is provided by on-chip logic to ensure that memory updates take place at instants when the microprocessor is not addressing memory.

Memory locations are addressed via the three lines A0, A1, and A2. The input address latch is transparent when ALE is high and latched when ALE goes low. This address may be actively latched by ALE for systems which feature a multiplexed bus carrying both address and data information (Figure 4.16a); otherwise, for systems having separate address and data buses, the address inputs can be made transparent by tying ALE high (Figure 4.16b). $\overline{CS}$ (chip select) activates three-state buffers to place the addressed data on the DB0 to DB7 data-output pins.

4.2.6 INTERFACING ANALOG I/O BOARDS TO μC BUSES

The RTI-1260, described earlier, is one of a series of analog I/O boards, designed for complete plug-compatibility with the various microcomputer bus structures, utilizing memory-mapped architecture. In some cases, Port I/O is also available. Such cards offer the best results in the tradeoffs between hardware/software and cost/performance, with 12-bit data-acquisition throughputs of up to 30kHz.

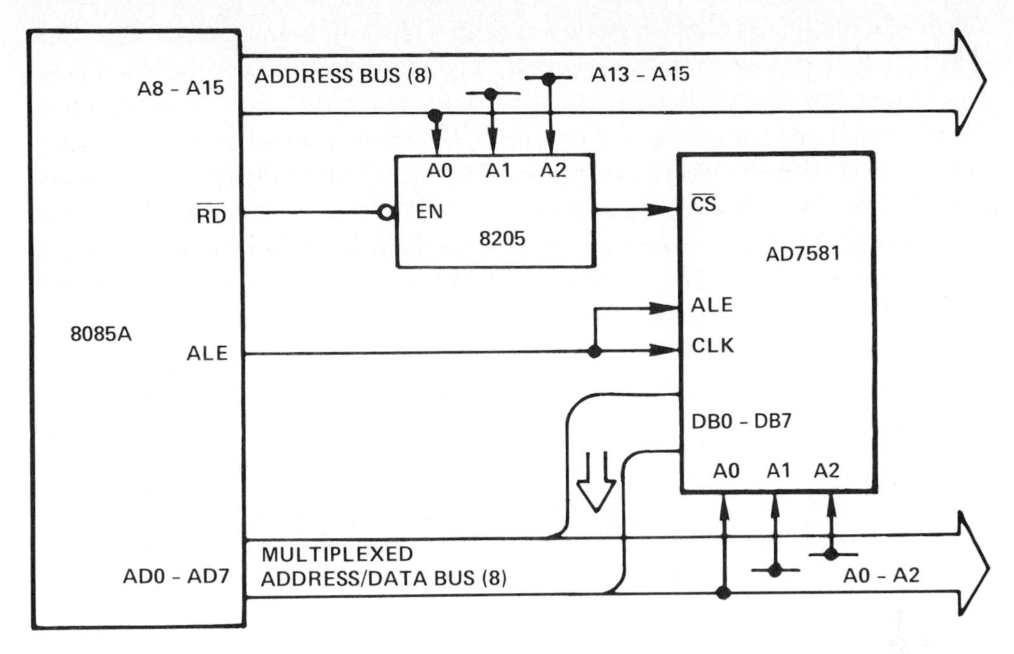

a. 8085 interface.

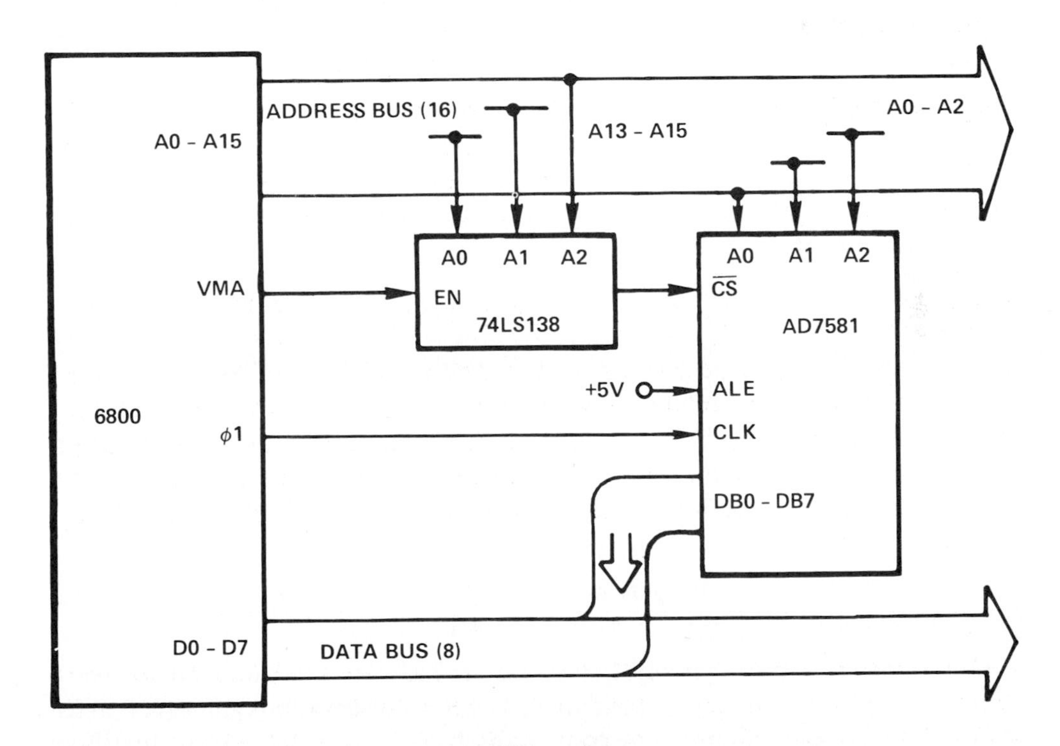

b. 6800 interface.

Figure 4.16. Interfacing a memory DAS to popular microprocessors.

Memory mapping treats an analog card as a block of memory locations. The cards, which decode the addresses placed on the address bus by the microcomputer, occupy a set of (e.g., 16) consecutive bytes within an unused 1K block of addresses in microcomputer memory. The choice of configuration is user-programmed by the use of on-board jumpers, which permit a choice among 256 locations in a 64K memory space.

When the RTI cards are addressed as memory, simple memory Read or Write instructions (STA, LDA) can be used. Memory mapping also allows programmers to access the RTI boards using any of the memory-reference instructions in the repertoire (SHLD; LHLD; MOV M, r; MOV r, M). Figure 4.17 shows an example of a data-acquisition subroutine for 12-bit a/d conversion.

```
THIS EXAMPLE USES 8085 ASSEMBLY LANGUAGE TO ADDRESS
CHANNEL 1, DO AN A/D CONVERSION AND STORE 12 BITS OF
A/D DATA IN REGISTER PAIR B AND C. BASE ADDRESS HAS
BEEN SET AT FFFB.

        LXI    H,FFFB
        MVI    M,Ø1      SELECT MUX ADDRESS
        LXI    H,FFFD
LOOP    MOV    A,M
        RLC
        JC     LOOP      TEST BUSY BIT
        MOV    B,M       READ ADC DATA HI
        DEC    H
        MOV    C,M       READ ADC DATA LO
```

Figure 4.17. Example of 12-bit a/d conversion program.

RTI-1260 Analog Input Card Memory Map. The RTI-1260 uses 3 bytes, which are assigned as shown in Figure 4.18.

MUX ADDR/CONV: Any one of 32 single-ended or 16 differential input channels can be selected at random by writing the channel code into bits D0 to D4 of this byte (00000 to 11111 single-ended, 00000 to 01111 differential). Writing to this byte also initiates an a/d conversion.

A/D DATA LO: The 8 lowest-order bits of the a/d converter's output word are available at this address.

A/D DATA HI: The Busy bit (D7) is used to indicate when an ADC conversion is complete, resulting in valid data. Logic 1 indicates busy, Logic 0 indicates conversion complete. The four highest-order bits of the right-justified ADC output are also available at this address (D3, D2, D1, D0). For twos complement bipolar coding of the ADC output, the inverted MSB is read at D6, D5, D4, and D3.

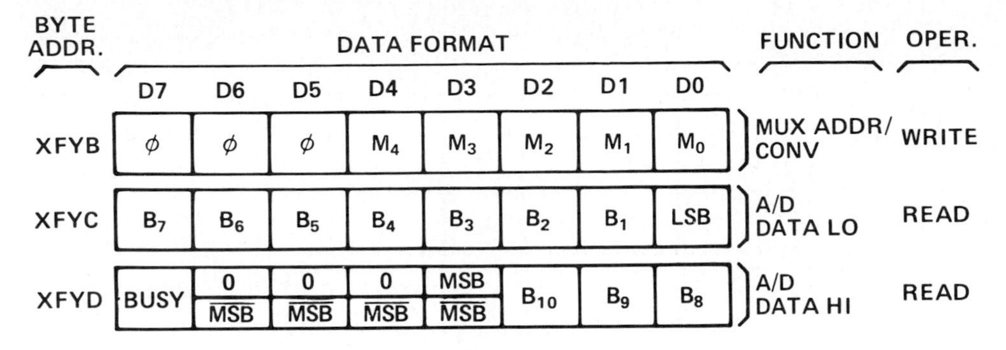

NOTES: 1. X AND Y ARE USER SELECTABLE.
2. BITS SHOWN AS ⊟ HAVE THE UPPER VALUE FOR UNIPOLAR CODES AND LOWER VALUE FOR 2's COMPLEMENT.
3. THE SYMBOL ϕ MEANS THE BIT IS IGNORED.
4. BUSY BIT EQUALS "1" DURING CONVERSIONS AND "0" WHEN DONE.

Figure 4.18. Memory map for microcomputer analog-input subsystem.

I/O Port Addressing. In the I/O port mode of operation, the RTI-1260 card occupies three consecutive ports in either an 8-bit or 16-bit port image. The port address is determined by on-board jumpers which can be configured to begin at any 16 port boundary. Since the RTI card is treated as a group of I/O ports, simple input and output instructions (INP, OUT) can be used. Port addressing eliminates the need to allocate memory when interfacing to the STD Bus.

RTI-1262 Analog Output Card Memory Map. The RTI-1262 uses 8 of the 16 contiguous bytes, assigned as shown in Figure 4.19. Since the byte addresses for the RTI-1260 and RTI-1262 are different, an input-output card pair require only a single block of memory.

D/A DATA: Two bytes are assigned to each of the four analog output channels. The digital data is written for each DAC in right-justified format, with the 8 least-significant bits in the lower address byte and the 4 most-significant bits in the upper byte. The DAC is double-buffered for one-step DAC updating; data for both bytes is loaded into the DAC only when the high byte is written to.

I/O Port Addressing. In the I/O port mode of operation, the RTI-1262 card occupies eight consecutive ports in either an 8-bit or 16-bit port image. The port address is determined by on-board jumpers which can be configured to begin at any 16 port boundary. Since the RTI card is treated as a group of I/O ports, simple input and output instructions (INP, OUT) can be used. Port addressing eliminates the need to allocate memory when interfacing to the STD Bus.

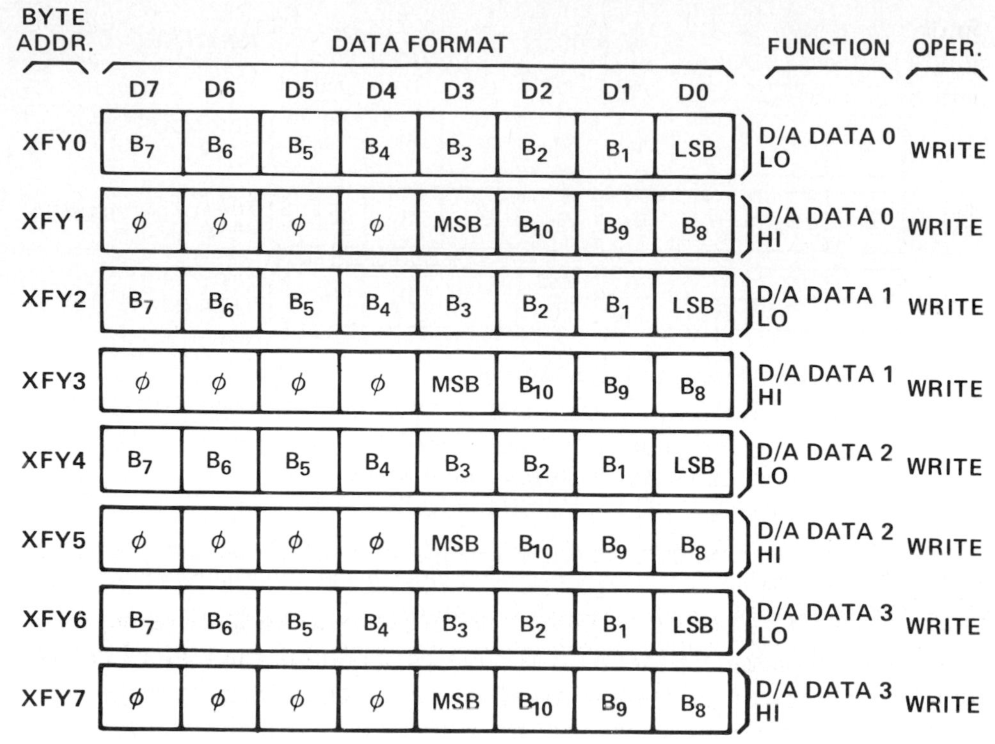

Figure 4.19. Memory map for microcomputer analog-output subsystem.

4.3 SERIAL INTERFACING

We've shown some of the ways analog information can be handled in its translation to digital and in the interfacing of the resulting digital information with a processor bus. This approach makes the most sense if the source of analog information is electrically and physically near the (host) processor. If, on the other hand, the data must be carried through an electrically noisy environment, if data and control signals must be transmitted over distances greater than a few meters, or if the data must be interfaced with terminals, communication links, or computer ports in a standard format, immediate conversion to some form of digital or pulse transmission employing a standard format and a minimal number of wires is strongly desirable.

Perhaps the simplest and most obvious approach is the use of a voltage-to-frequency converter (VFC), a device that produces a (usually asynchronous) output train of pulses or square waves at a frequency proportional to the input voltage or current (see Chapter 15). V/F converters offer high resolution at low cost, in common with other integrating methods. A v/f converter can continuously track the input signal without the need for clock pulses, convert-command signals, or any form of external logic. The direct count of its output pulses, over a measured time period, can produce a binary or BCD digital number, which represents the average value of the input during the counting period (Figure 4.20).

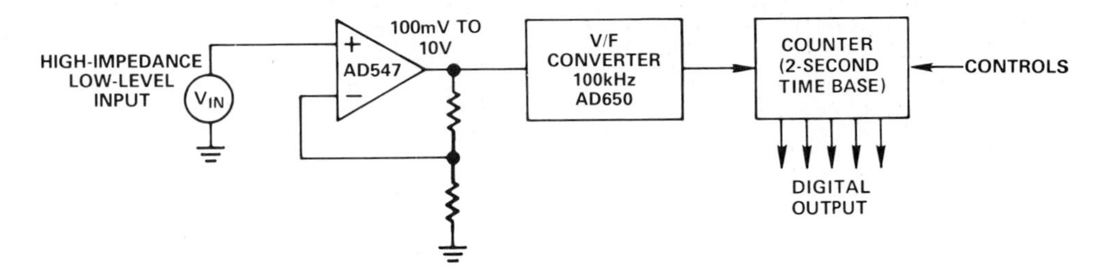

Figure 4.20. V/F converter, used as a nearly 18-bit binary (5 ½ BCD) a/d converter. Resolution is 1 pulse in 200,000, or 0.05% of smallest input signal (or 5ppm of full scale).

The VFC pulses require but a single wire-pair for transmission, unlike parallel converters, which—for n bits—require at least n + 1 wires, or synchronous serial converters, which require a form of clock signal. The v/f converter may share a local power source with a transducer and may be optically coupled for high common-mode isolation (Figure 4.21).

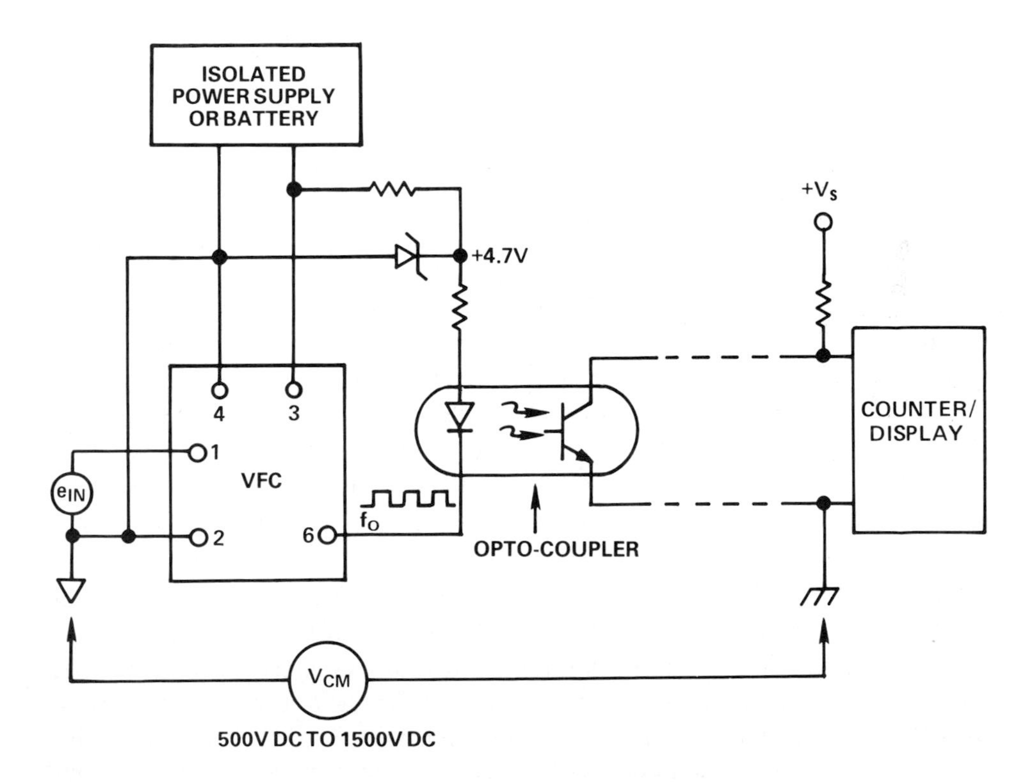

Figure 4.21. Optically isolated a/d conversion.

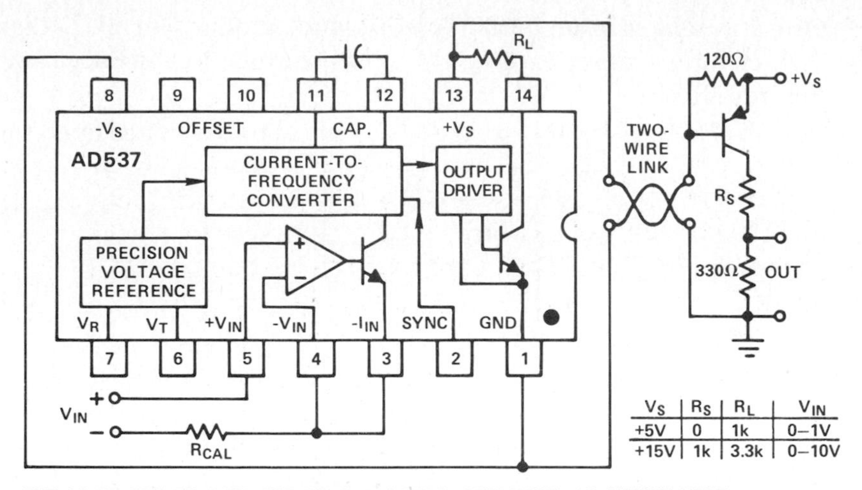

V_S	R_S	R_L	V_{IN}
+5V	0	1k	0–1V
+15V	1k	3.3k	0–10V

THE AD537 CAN BE USED FOR TRUE TWO-WIRE OPERATION, AS SHOWN HERE. THE FREQUENCY INFORMATION IS TRANSMITTED AS A CURRENT SIGNAL ON THE SUPPLY LINE TO THE DEVICE. THE SIGNAL IS CONVERTED TO A DTL/TTL OR CMOS-COMPATIBLE SIGNAL BY THE SINGLE-TRANSISTOR-TERMINATION CIRCUIT SHOWN. THE EXCELLENT SUPPLY REJECTION, HIGH OUTPUT-DRIVE CAPABILITY AND SQUARE-WAVE OUTPUT FROM THE AD537 ARE ALL ADVANTAGEOUS IN THE APPLICATION.

Figure 4.22. VFC two-wire operation.

Or a low-power-drain integrated-circuit v/f converter, such as the AD537, can use the two-wire link both to obtain its excitation voltage and to furnish an output-current pulse train, as Figure 4.22 shows. This avoids the need for local excitation. The current signal is converted to a DTL/TTL or CMOS-compatible signal by the single-transistor termination circuit shown. The excellent supply rejection, high output-drive capability, and square-wave output from the AD537 are all advantageous in this application.

The outputs from a number of AD537 VFCs may be multiplexed onto the same counter in random order by connecting their collectors together (sharing a single pullup resistor), and their emitters to the open collectors of a 1:N address decoder. Opening all gates but the one selected will cause its output pulse train to appear at the common collector terminal.

If a readily available pulse-to-fiber-optic cable driver, and the cable, are substituted for the photocoupler and transmission wires in the scheme of Figure 4.21, the VFC solves many interesting and difficult problems. These potentially include total isolation from up to millions of volts of common mode (HV transmission lines and atmospheric electricity), elimination of EMI (electromagnetic interference), privacy, light weight, small size, and—as time passes—lower cost.

VFC output pulses also lend themselves to acoustic transmission through water (or air), as well as to the modulation of RF or microwave carriers. A square-wave output, such as that of the AD537, is desirable. When using VFCs with narrow constant-width output pulses (e.g., the AD650 and ADVFC32, or the older Models 450 and 460), the output signal should be converted to a 50%-duty-cycle square wave with a divide-by-2 flip flop prior to driving the ultimate transducer or modulator. This maximizes noise immunity and minimizes pulse degradation.

Although VFCs are low in cost and simple to apply in uncontrolled operation, they have shortcomings that may be serious in data-acquisition systems employing serial transmission. Principal among these is the absence of "handshaking," that is, they are not readily controlled, and their format is not very suitable for the interchange of information. Furthermore, the time required for a complete conversion cannot easily be shared for transmitting other information over the line in either direction.

Much more desirable is a means of transmitting measurements and control signals in an economical format, at will, bidirectionally over (for example) a two-wire pair to permit interfacing with data teletypewriters or other human-operated data terminals, as well as minicomputers, microcomputers, etc. This can be accomplished by the use of a standard coding and serial asynchronous word format (ASCII), and serial communication via 20-mA loops and RS-232C systems.

4.3.1 ASCII

The key to such communication is ASCII, the American (National) Standard(s Institute) Code for Information Interchange. It is the most widely used code for transmitting alphanumerics and special characters, as well as control characters of undefined appearance. It may be found in teleprinting, computing, and instrumentation. There are 128 characters in the ASCII system, transmitted in serial or parallel as a 7-bit digital word. Table 4-3 lists all ASCII characters, with their decimal and hex equivalents.

Digital Panel-Instrument Communication. An example of the use of ASCII for transmitting the results of a measurement may be found in the communication capabilities of the AD2051 "smart" digital thermocouple thermometer (Figure 4.7). It has a character-serial ASCII output and optional TTL serial or isolated full-duplex 20mA serial output.

Character Serial: In the AD2051's character-serial transmission, each character is transmitted serially to a printer or terminal on a 7-bit parallel data bus—along with a Strobe line that carries a pulse that goes low when the data for each character becomes valid. A complete transmission consists of *nine* characters, as follows:

Polarity, + or −
Four consecutive characters representing temperature (or EEEE for over-range)
Character space
Temperature scale, C or F
Carriage return signifies end of data & left-justifies column
Line feed advances printer or terminal to next line.

Character	Value		Character	Value		Character	Value	
	Decimal	Hex		Decimal	Hex		Decimal	Hex
NULL	0	00	*	42	2A	T	84	54
SOH,CTRL A	1	01	+	43	2B	U	85	54
STX,CTRL B	2	02	,	44	2C	V	86	56
ETX,CTRL C	3	03	-	45	2D	W	87	57
EOT,CTRL D	4	04	.	46	2E	X	88	58
ENQ,CTRL E	5	05	/	47	2F	Y	89	59
ACK,CTRL F	6	06	0	48	30	Z	90	5A
BELL,CTRL G	7	07	1	49	31	[	91	5B
BS	8	08	2	50	32	\	92	5C
TAB	9	09	3	51	33	]	93	5D
LF	10	0A	4	52	34	∧	94	5E
VT	11	0B	5	53	35	←	95	5F
FF,CTRL L	12	0C	6	54	36	SPACE	96	60
CR	13	0D	7	55	37	a	97	61
SO,CTRL N	14	0E	8	56	38	b	98	62
SI,CTRL O	15	0F	9	57	39	c	99	63
DLE	16	10	:	58	3A	d	100	64
DC1	17	11	;	59	3B	e	101	55
DC2	18	12	<	60	3C	f	102	66
DC3	19	13	=	61	3D	g	103	67
DC4	20	14	>	62	3E	h	104	68
NAK,CTRL U	21	15	?	63	3F	i	105	69
SYN,CTRL V	22	16	@	64	40	j	106	6A
ETB,CTRL W	23	17	A	65	41	k	107	6B
CAN,CTRL X	24	18	B	66	42	l	108	6C
EM,CTRL Y	25	19	C	67	43	m	109	6D
SUB,CTRL Z	26	1A	D	68	44	n	110	6E
ESC	27	1B	E	69	45	o	111	6F
FS	28	1C	F	70	46	p	112	70
GS	29	1D	G	71	47	q	113	71
RS	30	1E	H	72	48	r	114	72
US	31	1F	I	73	49	s	115	73
sp	32	20	J	74	4A	t	116	74
!	33	21	K	75	4B	u	117	75
”	34	22	L	76	4C	v	118	76
#	35	23	M	77	4D	w	119	77
$	36	24	N	78	4E	x	120	78
%	37	25	O	79	4F	y	121	79
&	38	26	P	80	50	z	122	7A
'	39	27	Q	81	51	{	123	7B
(	40	28	R	82	52	|	124	7C
)	41	29	S	83	53	}	125	7D
						~	126	7E
						del	127	7F

Table 4.3. ASCII character chart.

In order for digital data to be transmitted on the bus, a Request input must bring the $\overline{REQ}$ line Low at a time when an output digital transmission is not occurring. If $\overline{REQ}$ remains Low, the results of each measurement will be transmitted continuously.

With the AD2051 connected to a low-cost printer, having a standard ASCII input format (Figure 4.23), an economical single-channel data logger can be assembled. A print-inhibit switch may be used to operate $\overline{REQ}$ so that only selected measurements are logged.

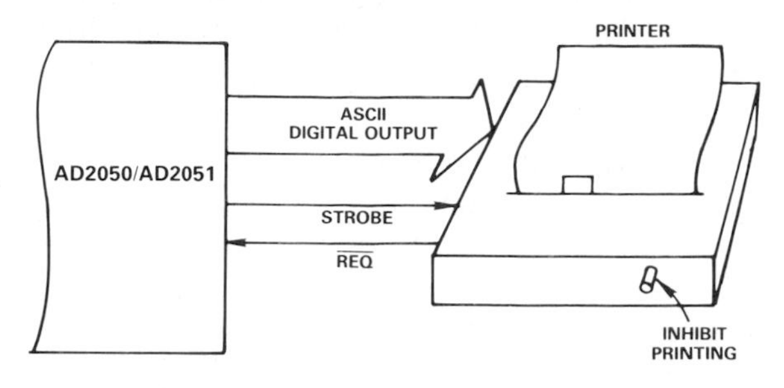

Figure 4.23. Interfacing digital thermometer to printer.

20-mA Current Loop, Bit-Serial, Full Duplex: A typical 20mA regulated half-duplex current loop is shown in Figure 4.24. A regulated 20-mA current source provides current for the loop. The current flows through receivers, transmitters, teletype machines, etc., so long as the switches are closed. When any of the devices on the line causes a switch to open, the current stops flowing, and all receivers on the line detect the level change. If a switch opens and

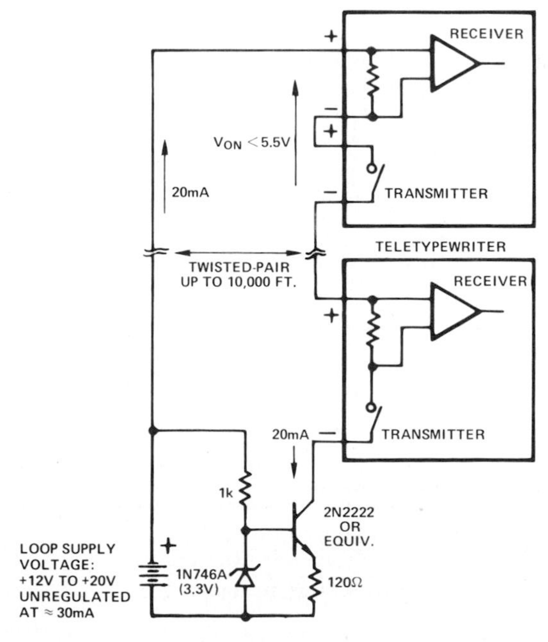

Figure 4.24. 20-mA current-loop data transmission.

closes a number of times, it will transmit a coded message to all enabled receivers; and the code, when decoded, will provide whatever information it represents and will cause whatever subsequent actions are appropriate.

A full-duplex loop consists of two current loops, one for transmitting, the other for receiving. This permits a device to receive a message while it is transmitting.

In the serial format, the bits that define an ASCII character are sent serially, as an asynchronous train of binary levels, LSB first. As noted above, the first bit of a transmission is initiated by turning off the current (0); this is the Start bit. Then, the seven ASCII data bits are transmitted (current on—1, current off—0). The data bits may be followed by an eighth (*Parity*) bit. Finally, two or more Stop bits indicate that the character transmitted has ended, and the output remains high until the next character is to be transmitted (Figure 4.25).

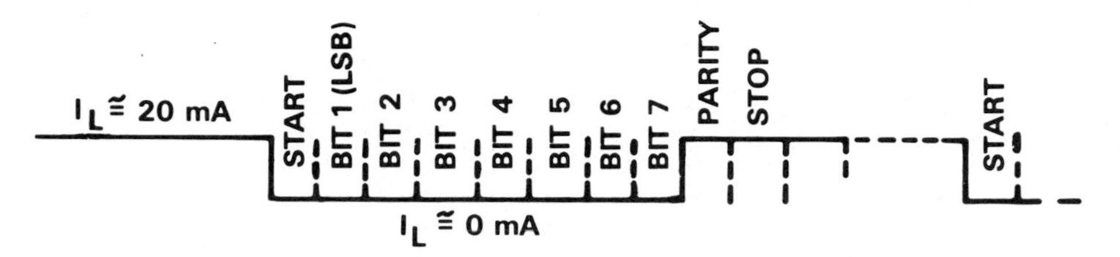

Figure 4.25. Typical ASCII code format for serial transmission.

Figure 4.26 shows how an AD2051 digital thermocouple meter communicates with a host computer or terminal using full duplex. A request for information is sent on the lower loop to the device's $\overline{\text{SERIAL INPUT}}$ terminals; the data transmission is sent back via the serial output terminals. A transmission can

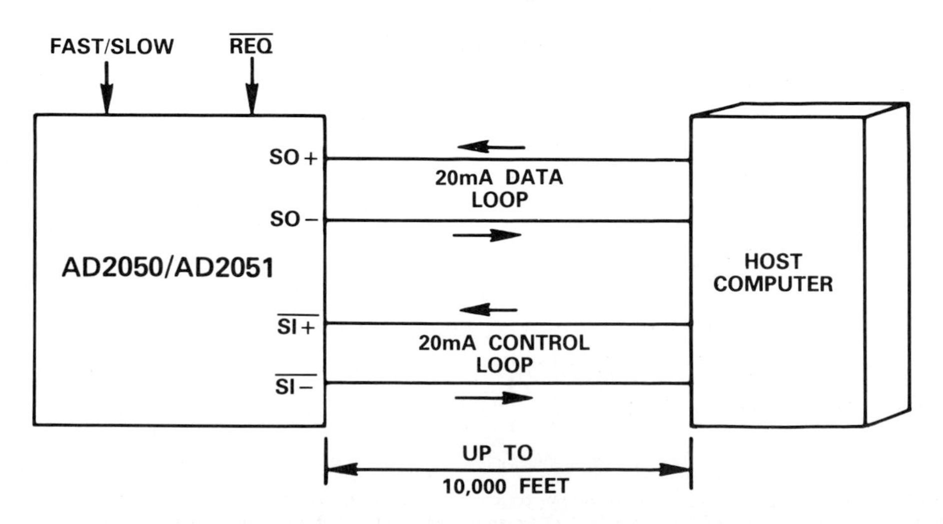

Figure 4.26. Four-wire full-duplex operation of digital thermometer.

also be initiated locally—at the meter—via the $\overline{REQ}$ line. There are two choices for the information rate, 300 baud (bits/second), or 1,200 baud. The tradeoff is information rate vs. maximum distance—10,000 feet at 300 baud, 4,000 feet at 1,200 baud.

4.3.2 RS-232C AND OTHER STANDARDS

In addition to current-loop transmission, over a minimum number of wires and fairly long distance, there is a form of voltage transmission of ASCII characters or other serial data, for distances up to 50 feet, involving a standard connector and the possibility of a variety of control and handshaking options, employing a number of additional wires.

For a peripheral device, such as a data-acquisition system, to send information back to a host computer, there are three things that must be consistent on both sides of the interface:

1. The method of encoding the information at the sending end and decoding it at the receiving end must be the same.
2. The signals used must correspond, with respect to voltage levels, timing, and sequence of signalling.
3. The connectors used must physically match.

The RS-232C Standard (C is the most recent revision), published by the Electronic Industries Association, is one of several standards* that specify some of the factors associated with these three areas; other factors are matters for agreement between companies supplying compatible equipment.

Further discussion of RS-232C is beyond the scope of this book. However, it is important for the reader to understand that, while the RS-232C standard is by far the most-common method of interconnection for computer-related electronic equipment, the term, "RS-232C-compatible" does not mean that all you have to do is plug the equipment together for everything to work. Problems can arise with the selection of parity, baud rate, and number of bits, and with the wiring of data and control lines with the up to 25 assigned wires. Syntax problems between more-complex RS-232C-compatible devices can also occur.

Compatibility with standards such as RS-232C is a desirable feature for all systems-level data-acquisition equipment that must interface with peripherals, such as terminals or printers, and host computers. MACSYM, the μMAC-5000, and the μMAC-4000, mentioned earlier, are RS-232C compatible, and the AD2051's isolated current-loop output can be translated to an RS-232C-compatible voltage format.

*RS-232C is applicable for short distances and low baud rates. For longer distances and higher baud rates, such standards as RS-422 and RS-423 are employed. For example, MACSYM 150 communicates with MACSYM 200 via an RS-422 link.

4.3.3 COMMUNICATIONS PROTOCOLS

We have seen that the AD2051 smart thermocouple thermometer communicates its output via a single 9-character word, in response to a simple digital stimulus. In the case of a multi-channel data-acquisition subsystem, which acquires analog and digital data, and can furnish analog and digital outputs to the outside world under the instructions of a terminal or a computer, the digital instructions it receives and the way it responds to them are a little more complex. In the case, of the μMAC-4000 single-board Measurement-And-Control subsystem, *communication protocols* are employed.

The μMAC-4000 measurement-and-control subsystem receives its instructions from a host computer, and communicates data upon request. Communication with the host is via a serial input/output port, which contains a full-duplex universal asynchronous receiver/transmitter (UART). The serial input-output may be jumpered for either 3-wire RS-232C communications or 4-wire full duplex.

A data-communication protocol specifies the format in which data is transferred. The μMAC-4000 employs two types of protocol ("C" and "T") to communicate with any host Computer or Terminal. The "C" protocol is designed to be used with computers and controllers, where communication efficiency, reliability, and adaptability to a wide variety of host systems are necessary. The "T" protocol, which uses simple, English-like commands, is designed for use with CRTs and teletypewriter terminals, for familiarization, debugging, system calibration, and manual control.

The μMAC-4000 will reply (only) in response to a command received over the serial link; this is known as command/reply (prompted) operation. Data is transmitted and received in standard ASCII format for each character, consisting of a *start* bit, 7 data bits, a parity bit, and one or two *stop* bits.

Command Sets. It is easy to operate the μMAC-4000, because the on-board microcomputer is programmed to respond to a simple command set. Through it, the host can delegate all measurement and control functions to the μMAC. The command set includes commands for transmitting analog and digital data, setting the digital output bits, activating channels, setting limits, and modifying the protocol.

"C" Commands Figure 4.27 shows typical examples of commands (a) and responses (b).

The command instructs the μMAC-4000 to scan and transmit the latest data from channels 0 through 3 (47 channels max) of cluster (μMAC and accessories) number 2 (7 clusters max).

The carriage return of the command initiates the response. The first word is an indication of system status (errors, exceeded limits, backup power supply, etc.) Then follow the data words requested, and finally a *checksum*, the sum of the numerical (hexadecimal) values of the preceding ASCII characters

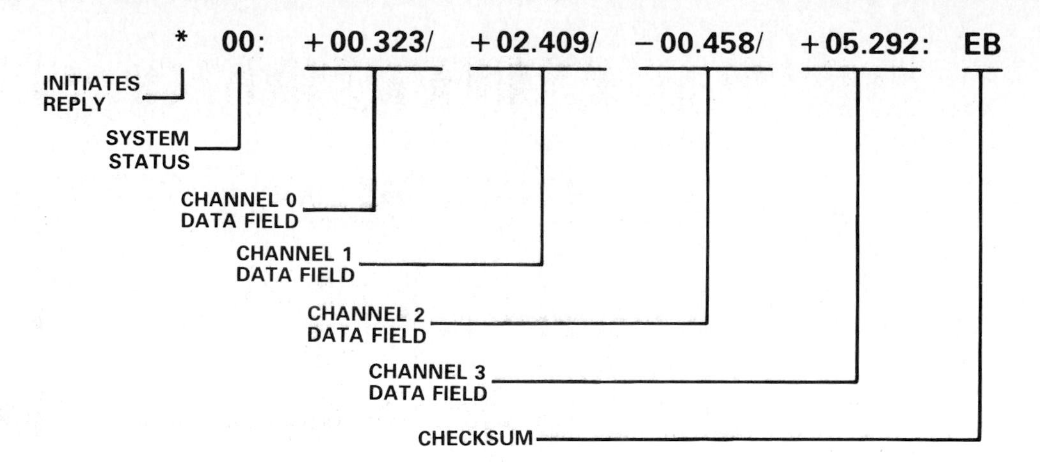

a. Typical command.

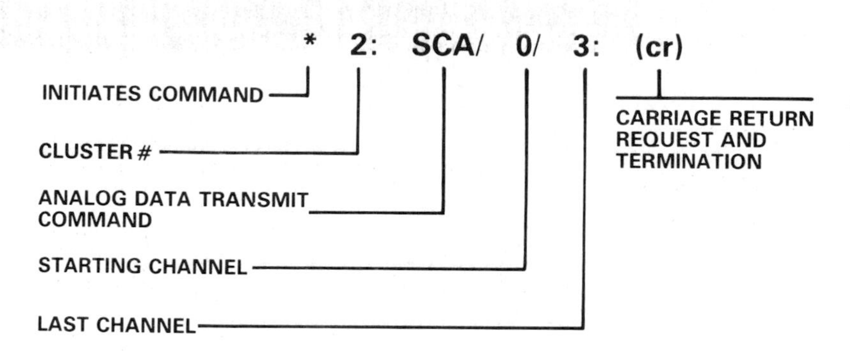

b. Typical response to the above command.

Figure 4.27. Command and response using "C" protocol.

modulo 100_H, (256). At the completion of the response, the μMAC-4000 generates a carriage return and line feed; in addition, a carriage return is also generated for each eight data fields.

"T" Commands A typical "T" command might be

CHANNEL 17 (cr)

This command requests the μMAC-4000 to transmit the latest data from Channel 17. A typical reply might be

CH 17 = +0024.7

If a thermocouple is connected to this channel, this response might indicate a temperature of +24.7°C.

The key feature of this protocol scheme is the small amount of software necessary to support an intelligent system. It permits easy implementation by hosts using high-level languages (FORTRAN, PASCAL, BASIC), and it is easily debugged. Software drivers, available for a variety of popular computers, make it unnecessary for the user to write special programs. For example, when the appropriate software drivers are used (Model AC1820) the list in Figure 4.28 contains (1) the entire APPLESOFT BASIC subprogram to read a temperature at one location and (2) the response to the RUN command.

```
LIST 80,130

80 TN$ = "0":TR$ = "3"
90  GOSUB 61080: REM  INITIALIZE VALUES
100 TN$ = "1": REM  SELECT CHANNEL 1
110  GOSUB 61260: REM  INVOKE CHANNEL 1 READ
120  PRINT "TEMP1= " + TQ$ + " DEGREES C":
     REM  DISPLAY RESULTS
130  END
]RUN

TEMP1= +0025.4 DEGREES C
```

Figure 4.28. Display of program LIST and results of RUN, using Apple II computer and software drivers with single-board measurement-and-control subsystem.

Since the μMAC-4000 retains the updated data in memory, the response appears immediately after the RUN is executed. A scan of a number of temperatures is almost as simple and as quickly executed.

4.4 CONCLUSION

We have summarized in this chapter a variety of means of implementing converter-interface functions in terms of standard readily available products, at the various levels of system involvement that are likely to concern our readers. More information on any aspect of this discussion is no farther away than the manufacturer's nearest applications engineer.

Chapter Five

Analog Functions with Digital Components

The world of analog system designers employs numerous circuit tricks to perform operations on voltages and currents—with op amps, multiplier/dividers, filters, phase shifters, function generators, etc. The availability of converters and fairly simple digital hardware and firmware greatly enlarges this bag of tricks (without even including the contributions of computers, software, etc., which are considered in great panoply elsewhere in the book).

The term "analog" is commonly understood in two contexts, one valid and one less-than-convincing: "analog" in the sense of dealing with measurable real-world quantities rather than abstract *digital* numbers; and "analog" in the sense of smooth, or at least *continuous* (derivatives existing nearly everywhere), vs. *discrete* (sampled or quantized), signals or relationships between inputs and outputs. We contend that, when used in the latter sense, the term *discrete* also connotes analog (measurable) quantities, but in a bridging state between analog and digital.

There have been a few excellent books on the applications of operational amplifiers, fewer on the applications of op amps and analog function modules, and virtually none on the use of digital and interface components (converters, counters, shift registers, etc.) in the service of analog relationships.

There are many excellent auguries favoring an intimate, long, and happy marriage between the two tribes. Analog devices are cheap, plentiful, and capable of a great deal of functional versatility; an increasing number of digital devices (including CPUs) are also cheap, plentiful, and capable of a great deal of functional versatility. The reasons there has been little apparent intercourse between them are twofold. Interface devices, such as A/D and D/A converters have heretofore been too expensive to be wasted as components (old-timers will remember the days of $227 op amps and $50 transistors). But more important—practitioners who volubly embrace the tricks of both trades have remained either extremely rare or well-hidden.

This chapter is in no sense intended as an encyclopedia (in either breadth or depth) of such connubial (i.e., "hybrid") circuits; that volume is yet to be written. Rather, the few representative items included here are intended to be suggestive of what is possible, and to stimulate the reader to bring creative faculties to bear on new ways of looking at problems that may have been conceived of as being strictly "analog" or "digital." For those already laboring in the vineyard, there will be no revelations, but perhaps there is something a little new or different to make a scan worthwhile. The circuits are presented in the form of independent modular panels that stand alone ("bite-size morsels," to aid digestion). The selected examples are:

SOURCES

- Digitally-Controlled Voltage Source
- Manual Digital Inputs
  - Thumbwheel BCD switch
  - Toggle-switch register
- Digitally-Controlled Current Sources
  - "Current-output" DAC
  - Current gain: floating load
  - Current gain: buffered load
  - Current to grounded load
  - 4-to-20-mA Current Generator

SCALE FACTORS AND MODULATIONS

- Digitally-Controlled Direct Gains
- Digitally-Controlled Inverse Gains
- Logarithmic Scale Factors
- High-Precision Analog Multiplication
- . . . or Division

FUNCTIONAL RELATIONSHIPS

- Analog Functions with Memory Devices
- Sinusoidal Input-Output Relationships

TRIGONOMETRIC APPLICATIONS

- Digital Phase Shifter
- Digital/Resolver Converter (Resolver Simulator)
- Coordinate Conversion

5.1 SOURCES

DIGITALLY CONTROLLED VOLTAGE SOURCE (or Precision Power Supply)

A well-calibrated internally referenced d/a converter is probably the simplest available source of arbitrary precision voltages. Turn on the power, set the digital input, and expect (and receive) the voltage you asked for. With a 10-bit converter, resolution is 0.1%; with a 12-bit converter, 0.024%; and with a 16-bit converter, 0.0015% (15 ppm).

Let it be driven by a computer, and you have a ready supply of voltage for fast or slow automatic testing, Set it manually (with a "toggle-switch or DIP-switch register," or with BCD thumbwheel switches), and it's a convenient "volt-box," or a handy reference source. Or set it permanently by hard-wiring its logic inputs. No resistors or pots necessary!

If its output op amp doesn't have adequate output current, follow it with an inside-the-loop current booster. Feedback to the built- in amplifier-feedback-resistor will make the output virtually independent of the booster's dc characteristics. It can be followed with an op amp having higher-voltage output and precisely set fixed gain, if high voltage is needed. Doing this outside the DAC's loop protects the converter's circuitry (including the low-voltage digital components) from accidental exposure to fault voltages.

Because the setting is done digitally, in either serial or parallel, the voltage can be set from a distant location, or in the presence of a fair amount of electrical noise, relying on the inherently high noise immunity of digital signals (at the cost of additional wire for parallel circuits). If noise pickup is not a major factor, it is interesting to note that in some cases the switches can be closed "passively," i.e., to the power-supply return for "0", left open for "1".* The serial-input AD7543 may permit remote voltage (or gain) settings, with minimal wiring, when appropriately pulsed.

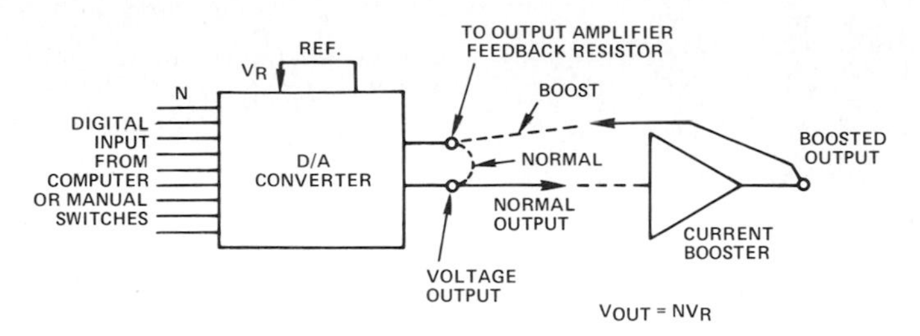

MANUAL DIGITAL INPUTS

All that is needed to obtain a given output voltage from a D/A converter is to close the appropriate switches. Human beings usually prefer base-10 numbers or BCD coding, despite the fact that it throws away inherent binary resolution at the rate of 2-bits-out-of-12 (12 BCD = 1/1000, 10 BIN = 1/1024).

Thumbwheel-Switch Encoder

A thumbwheel-switch encoder is the simplest way for the operator, especially one who is mathematically unsophisticated, since the base-10 number can be set directly, and all the appropriate switches are automatically closed. A D/A converter with BCD coding should be used. The switch points that are "0" (positive true) are connected to ground; those that are "1" are either left open* or connected to $+V_S$ (but be sure to use a break-before-make switch).

The first figure shows the principle for one decade of thumbwheel switchery ("1" open). If the converter has *complementary BCD* coding, the complementary switch connections should be used. A complete circuit arrangement employing a BCD d/a converter is also shown.

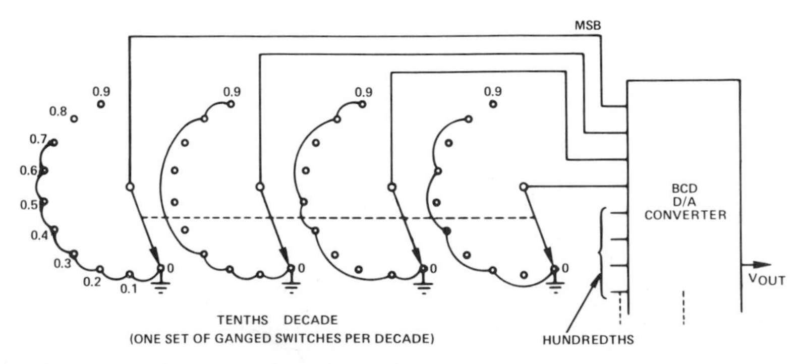

*"Pull-up" resistors, connected between each digital input and the positive supply, ensure that when the switch is open, the input to the DAC will indeed be "1".

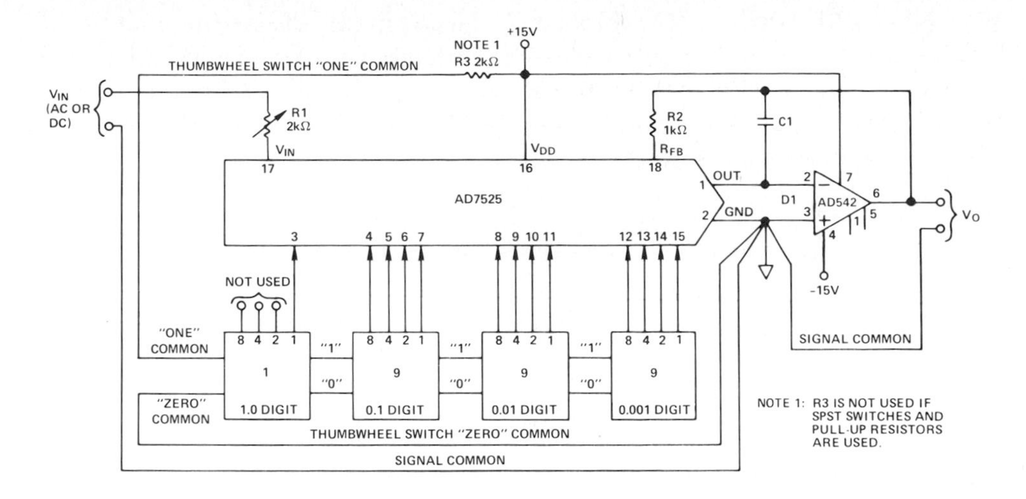

Toggle-Switch Register

The toggle-switch register is physically more elementary, and it may be used with either binary or BCD-coded DACs. It does require some calculations, though, especially for binary settings. As an aid to calculation, two tables are given, one for BCD (the same table is used for each digit), and one for binary equivalents of representative decimal fractions of full scale. Interpolation is performed by adding or subtracting an appropriate set of terms (binary rules) to form the desired sum (for example, $0.52 = 0.5 + 0.02 = 0.1000\,0101\,0001\,1110_2$.) Note that multiplication or division by 2 simply moves a number one place to the left or right: by 4, two places left or right, etc.

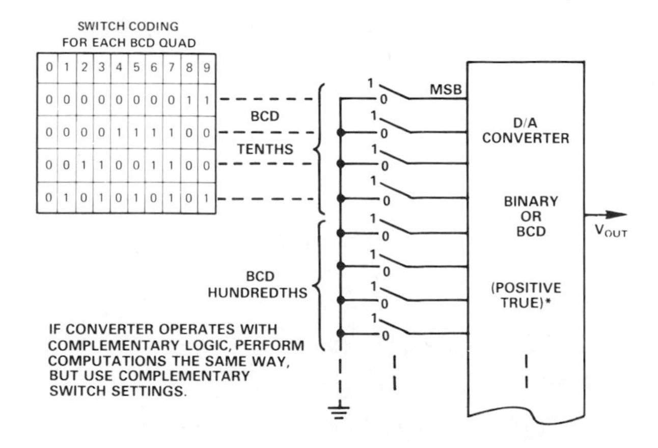

For unipolar binary coding, the digits to the right of the "decimal" point form the code, MSB leftmost. For bipolar 2s complement, divide the magnitude by two for the positive number, then complement all digits and add 1 LSB for the negative number. For offset binary, complement the 2s-complement MSB. (See Chapter 7 for a more-complete discussion of coding and conversion relationships in bipolar DACs.)

BINARY EQUIVALENTS OF DECIMAL FRACTIONS

	MSB ↓				
0.8	0.1100	1100	1100	1101	0
0.5	0.1000	0000	0000	0000	0
0.4	0.0110	0110	0110	0110	1
0.25	0.0100	0000	0000	0000	0
0.2	0.0011	0011	0011	0011	0
0.125	0.0010	0000	0000	0000	0
0.1	0.0001	1001	1001	1001	1
0.08	0.0001	0100	0111	1010	1
0.0625	0.0001	0000	0000	0000	0
0.04	0.0000	1010	0011	1101	0
0.02	0.0000	0101	0001	1110	1
0.01	0.0000	0010	1000	1111	0
0.008	0.0000	0010	0000	1100	1
0.004	0.0000	0001	0000	0110	0
0.002	0.0000	0000	1000	0011	0
0.001	0.0000	0000	0100	0001	1
0.0008	0.0000	0000	0011	0100	1
0.0004	0.0000	0000	0001	1010	0
0.0002	0.0000	0000	0000	1101	0
0.0001	0.0000	0000	0000	0110	1

Converting Base 10 Number to Binary Switch Setting — Two Examples (12-Bit Conversion):

1. +0.9FS (=0.5FS+0.4FS)

0.5	0.1000	0000	0000
+0.4	+0.0110	0110	0110
0.9	0.1110	0110	0110

Code: 1110 0110 0110, Straight Binary

2. −0.6FS, 2s Complement (Note: 0.6 = 0.5 + 0.1)

0.5	0.1000	0000	0000	
0.1	+0.0001	1001	1001	
0.6	0.1001	1001	1001	
Code:	1001	1001	1001,	Straight Binary
x 1/2	0100	1100	1100	Scale Expansion
Compl.	1011	0011	0011	Ones Complement
+1 LSB	1011	0011	0100,	Twos Complement

DIGITALLY CONTROLLED CURRENT SOURCES

Many analog current sources have been developed with the variations that provide such diverse advantages as low cost, simplicity, ability to transmit analog information over distance with reasonable noise immunity, ability to ground the load, etc. In conventional all-analog circuits, the original controlling input, if constant, is derived typically from a precision potentiometer, zener diode, or other reference—and if variable, from a sensor or signal conditioner. However, availability of versatile D/A converters now permits convenient digital control of current values, making, for example, programmable current supplies and digitally controlled current transmitters an inexpensive reality. As with voltage sources, the adjustments may be performed by either a computer or a human operator. They may be purchased as complete entities (see Figure 3.14 for a sophisticated example) or wired by the circuit designer. These are a representative few among the many ways of accomplishing current drive.

"Current-Output" DAC

An ordinary d/a converter specified as a current-output DAC would appear to be the simplest form of digital-to-current output source. However, most such devices are unsatisfactory as current sources because they generally have appreciable internal admittance "looking back," and this admittance (and the load) must be included in computations of the share of current reaching the load. In addition, there may be stringent limitations on voltage swing (*compliance* voltage). For this reason, current-output DACs are almost always used with the load in the feedback path of an op amp. The DAC drives the inverting input terminal, which is normally at zero potential, thus imposing negligible loading error. However, devices such as the AD561 (see Figure 9.14), with active-collector current outputs, may be treated as true current sources over the rated compliance-voltage range.

The output resistance of DACs made up of quad current sources is often introduced by the resistive dividers used for attenuation of less-significant-bit currents (see Figure 9.11 and accompanying text). For applications in which a restricted number of discrete values of current (say 16, at equal intervals) are required, one can construct a highly precise fast current-output converter with high internal resistance, using a quad current switch.

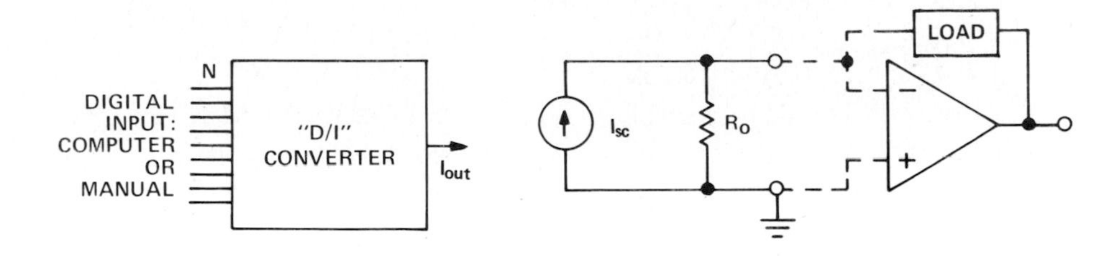

Current Gain – Floating Load

In this application, a load that has both terminals available is connected between the amplifier output terminal and the return lead of the feedback resistor. The attenuation introduced by R_M, if used, produces current gain. If the amplifier's output current is inadequate, a unity-gain current booster may be used, inside the loop (BF). For large currents, a separate booster supply should be used, with only the R_M pickoff point connected to the converter's analog ground.

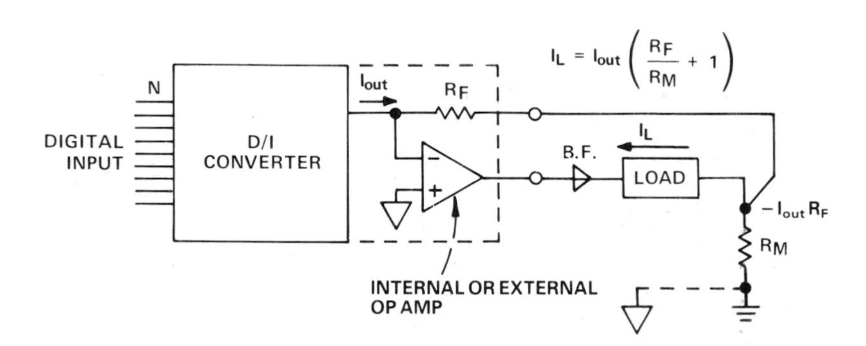

Current Gain – Buffered Load

For applications in which the amplifier's output range imposes serious restrictions on the kind of load that might be driven, a field-effect transistor (FET) with the load in its drain circuit allows a wide range of voltage swing (compliance voltage) for the load. Examples of loads that might be driven in this manner are CRT deflection coils, motor windings, chart-recorder pen drives, etc.

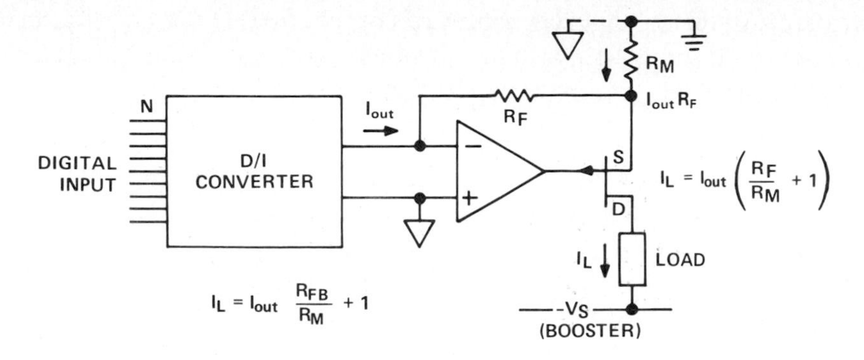

Current to Grounded Load

There are a number of ways of driving current to a grounded load, all of which employ both positive and negative feedback to measure and control the current. One example, using a voltage source and two operational amplifiers, is shown here. Amplifier A1 measures the difference voltage across R_M (direct from the top and inverted from the bottom via A2) and sets it equal to the DAC's V_{out}, thus forcing a current V_{out}/R_M through the load. In the general case, the resistor ratios can be adjusted for scaling, the drive could be from a current source, boosters could be used (at point "BF") etc. As with all operational-amplifier circuits having complicated (or even simple) dynamics, attention should be paid to dynamic stability: feedback capacitors may not be as helpful as capacitance shunting the load.

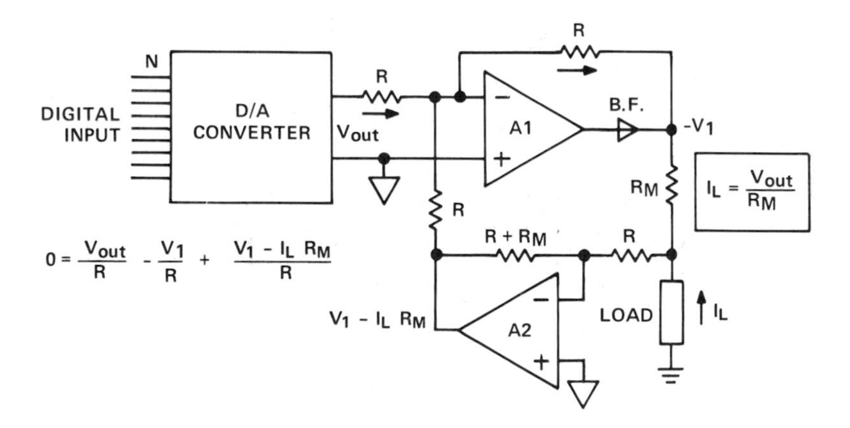

4- to 20-mA Current Generator

The current loop is a popular way to avoid the effects of voltage drops and minimize noise-voltage pickup due to interference when transmitting signals over moderate distances. The circuit on the next page illustrates one way to generate a digitally controlled 4-to-20mA current that is independent of the load resistance. The output current is equal to 4 mA + (D × 16 mA), where D is the fraction of full-scale output range represented by the digital input.

With a 10-bit DAC, such as the AD7533, the circuit provides an output range from 4 mA to 20 mA, with a resolution of 15 ⅝ μA. The maximum compliance voltage of the load is +25 V, equivalent to a load resistance of 1,250Ω maximum.

The DAC's output current flows through Q1 and R3. Operational amplifier A2 holds the lower end of R2 at the same voltage as the lower end of R3. Since the upper ends are connected together, they have equal voltage; therefore, the current through R2 is equal to the current through R3, amplified by R_3/R_2. The current through the Darlington follower and the load is very nearly equal to the current through R2.

Low-frequency voltage variation in the 30-volt supply does not significantly affect the current through the load, since A1 adjusts Q1's output conductance to keep the lower end of R3 at whatever voltage is necessary to maintain the current through Q1 equal to the output current of the DAC.

R1 is used to adjust the ratio of full-scale to zero-scale current (at OUT 1) in the ratio of 5:1. R3 adjusts circuit offset and span to provide proper values of 0 and full-scale current, set with a given value of R_L. Diodes D1–D4 limit the common-mode voltage to A2, and diode D5 protects Q1 during power sequencing.

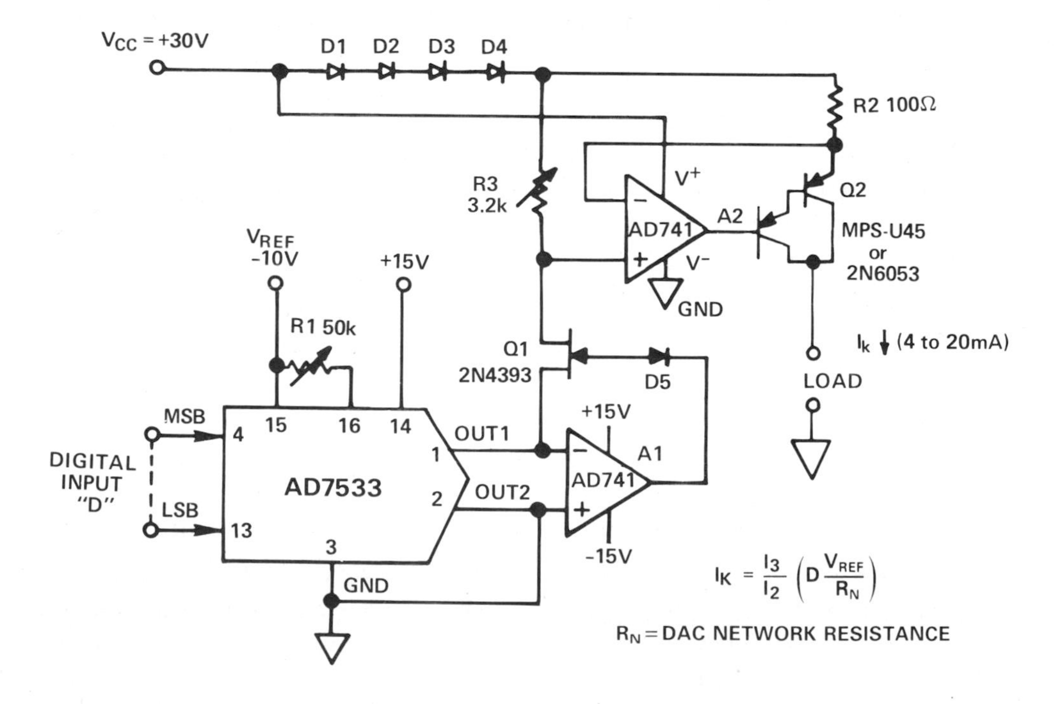

5.2 SCALE FACTORS

DIGITALLY CONTROLLED SCALE FACTORS

A D/A converter that accepts variable references (i.e., a *multiplying* DAC) can be thought of as a digitally-controlled potentiometer or a programmable-gain amplifier/attenuator. As such, it can be used for setting gains, either by a computer or a human operator. Computer-setting might be used, for example, in digital audio or adaptive control systems; manual setting might be employed where the device being controlled is remote (think of it as a potentiometer with a long shaft).

The multiplying D/A converter can also be thought of as a means of modulating a computer output by an analog signal. For example, if the computer is developing a square wave, the analog signal might be amplitude-modulating it (such an application might also be thought of as digitally weighted sampling).

The simplest device operates in one quadrant, with either a positive or a negative analog signal and straight binary or BCD coding. While CMOS DACs will do this easily, even many current-source DACs, such as the AD566A, which are usually unipolar, will accept a reasonably wide signal range (+1V to +10V) without excessive degradation of linearity.

For two-quadrant operations, there are two modes: bipolar analog and bipolar digital. Bipolar analog operation simply requires a bipolar analog input and straight binary or BCD digital coding. It also requires a converter that can accept analog signals of either polarity. CMOS DACs, such as the AD7545 in the current-switching mode, are well-adapted to this form of operation, providing wide bandwidth, good linearity, and low feedthrough. "Feedthrough" is the proportion of analog input signal that appears at the output when the digital input is calling for zero gain.

Bipolar digital operation can involve offset-binary (or twos complement) coding, with the output offset by one-half of the full-scale span; or sign-magnitude coding (unipolar DAC), where the sign-bit switches the output polarity between the direct output and an inverted version of it.

Four-quadrant operation involves a combination of circumstances: a DAC that can respond to both bipolar analog and bipolar digital inputs in the correct polarity, with appropriate speed and feedthrough performance. Again, CMOS DACs in the current mode, with an additional inverting op amp, perform this function with good linearity and speed, and low feedthrough.

Shown here are five basic ways (among many) that digital gain control can be used to perform useful functions.[1]

[1] See also *CMOS DAC Application Guide*, by Phil Burton, Analog Devices (1984).

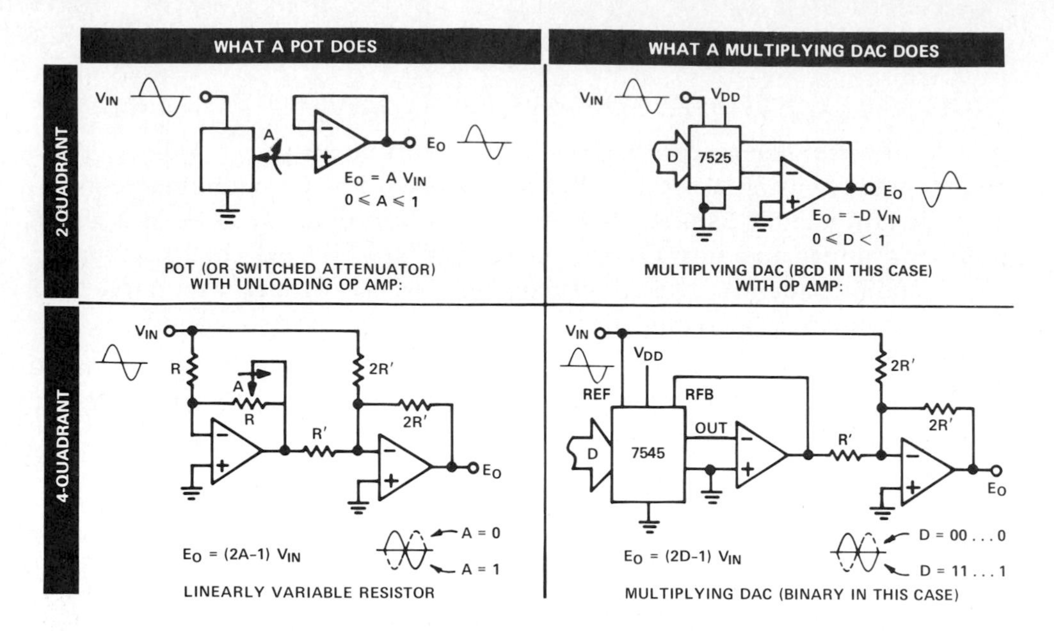

Direct Scale Factor

These "digital potentiometer" circuits provide simple linear digital scale adjustment, proportional to the unipolar or bipolar digital number. The comparison with analog potentiometer circuits is striking. As noted earlier, the digital number can be applied either by a computer signal or manually.

Inverse Scale Factor (Hyperbolic Gain Function)

With the DAC in the feedback loop of an operational amplifier, the gain is inversely proportional to the digital number (Gain = 1/D, where D is the fractional value of the digital number). The illustration shows a CMOS DAC, in the inverter connection. When the input is connected via the DAC's internal feedback resistor, the minimum gain is nearly unity, 1/(1 – LSB). Resistance may be added in series with the input for attenuation, so that normalized unity gain can occur at a mid-scale value (for example, if the total input resistance is $16R_f$), the gain will be 1/(16D).

This circuit will amplify signals; however, it will also amplify noise and gain errors, and bandwidth will be reduced as the gain increases. Also, there are constraints required by the nature of feedback: the digital input must be unipolar and must not go to all-zeros; however the analog input may be bipolar.

Logarithmic Scale Factors

For the many applications where logarithmic gain control is desirable (for example, dB steps of audio gain), a *logarithmic DAC* may be used. A logarithmic

D	NOMINAL $\frac{V_{OUT}}{V_{IN}}$
1111111111	$-\frac{1024}{1023}$
1000000000	-2
0000000001	- 1024
0000000000	OPEN LOOP

DAC (see also Chapter 16) is one for which the digital word is proportional to the logarithm of the gain, i.e., each one-bit digital change corresponds to a fixed ratio of gain change, hence a constant number of dB ($20 \log_{10} R$). For example, the AD7115 adjusts gain in 0.1-dB steps over a 20-dB range, while the AD7118 adjusts gain in 1.5-dB steps over a nominal 88.5-dB range. In the forward connection, attenuation is adjusted, and in the feedback connection, gain. The illustration shows the analog connections of a 2 ½-digit (0 to 199) BCD DAC that can be used for microprocessor controlled attenuation of up to 20dB in 0.1-dB steps.

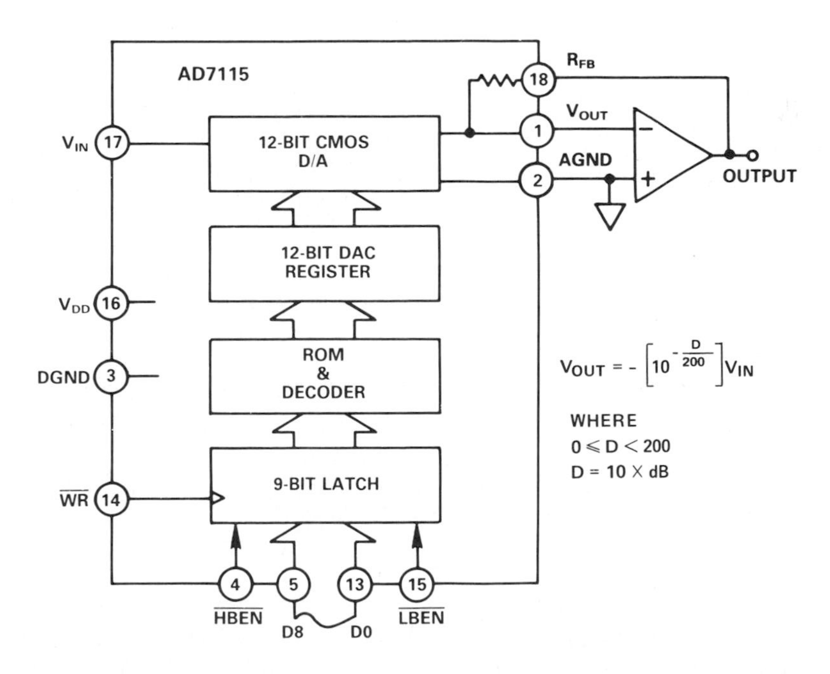

High-Precision Multiplication

Since a 12-bit multiplying DAC develops accuracies to within considerably better than 0.1%, it is possible to make an analog-to-analog multiplier having excellent accuracy by converting one of the inputs to digital form and using it to control the gain of a multiplying DAC. If the ADC is ratiometric, the output is a function of three variables. V_R should always be larger than V_1, or else overrange indication will be necessary. A sample-hold may be necessary if V_1 varies rapidly.

The availability of fast, high-resolution digital multipliers (e.g., 16 bits × 16 bits) may suggest the possibility of performing analog-to-analog multiplications at unheard-of speeds and accuracies. This is feasible in principle, but when the costs of sample-holds, conversion devices, and the multipliers themselves are added up, it makes more sense to consider such operations in a system context—as discussed under Digital Signal Processing—rather than as a restricted *ad hoc* circuit to perform a localized function with emphasis on simplicity and cost, which is the focus of this chapter.

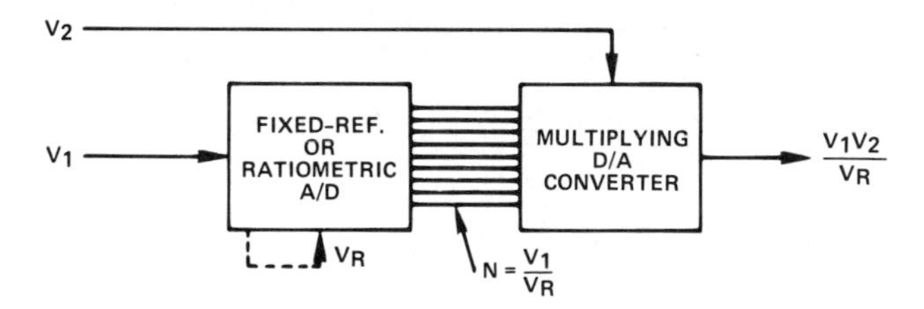

...or *Division*

There are at least two approaches. In the first, the same scheme is used as for multiplication, but the DAC is used in the feedback path of the output amplifier, thus dividing the multiplier input (V_2) by the digital number representing V_1. Alternatively, since an A/D converter digitizes the ratio of the "input" to the "reference", the digital word will be the quotient and a D/A converter will convert the ratio back to a voltage. Again, if the D/A is a multiplying type, the output is a function of three variables.

For both of these applications, the A/D may be connected for either clocked or free-running operation, and either the A/D or the D/A should have a register to store the previous value and buffer the D/A during the conversion process.

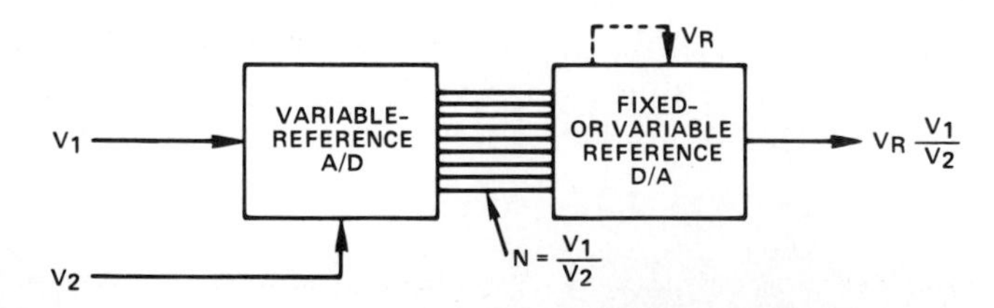

5.3 FUNCTIONAL RELATIONSHIPS

The term "functional relationship" (between two voltages/currents) implies a black-box causal operation, $y = f(x)$, f( ) being any single-valued linear or nonlinear realizable function. It is distinguished from a "function generator," which implies a *time function*; in a function generator, $y = f(t)$. By applying a linearly-increasing function of time to a device having a given functional relationship, one can create a function generator.

In analog circuitry, functions are traditionally embodied in three ways:[2]

1. Using a natural function (e.g., the inherently logarithmic diode characteristic for log and antilog circuitry, the transconductance relationships of transistors for transconductance multipliers, the ability of a capacitor to store charge for integration).

2. Using diode-resistor networks to form piecewise-linear approximations to a nonlinear function.

3. Using combinations of natural functions to approximate arbitrary relationships; for example, power series using multipliers to generate the x_2, x_3, x_4, etc., terms.

Now that converters and memories are available at low cost, a fourth approach becomes feasible:

4. Using memories (e.g., ROM's singly or in groups) to store a function digitally, and converting-in and -out with clocked or free-running A/D's and D/A's, as shown in the illustration. Typical applications include trigonometric transformations, thermocouple compensators, and arbitrary functions.

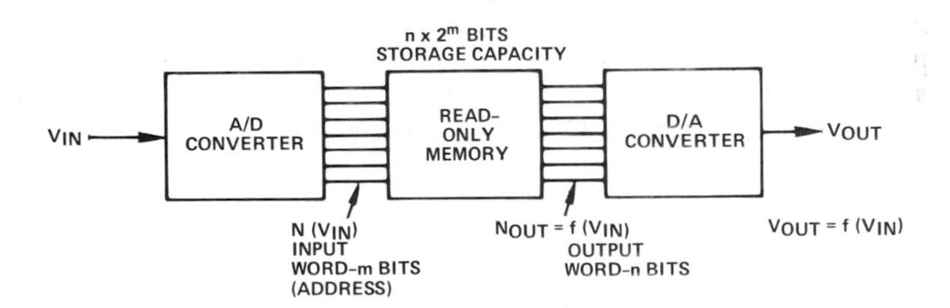

Arbitrarily Programmable Functional Relationships

Functions can be purchased in ROMs or programmed ("burned") into the various classes of programmable read-only memories (PROMs), and used in the relatively simple manner shown to provide continuous functional relationships in real time.

[2]Nonlinear Circuits Handbook, Analog Devices, Inc. (1974), has many details of these methods.

This approach may be favorably compared with a system approach that might go something like this: the same analog-to-analog functions, in wide variety, may be performed in software with computer programs written to acquire data (V_{IN}), using an ADC at an appropriate level of integration, retrieve the stored equivalent value from memory (or compute it on the spot), and output it to the DAC (V_{OUT}). If the ability to follow a rapidly varying input continuously in real time is desired, this can be unwieldy and costly in terms of time and software burden, even for a multitasking computer.

Sinusoidal Input-Output Relationships

An example of the approach is the use of a read-only memory that has the values of $sin\theta$ stored in it for $0° \leqq \theta \leqq 90°$. Two additional digits provide quadrant information, one to complement the input in the even-numbered quadrants, the other to provide the output sign-change for the 3rd and 4th quadrants. The input arrives from an angle-to-digital transducer, the corresponding sinusoidal number values are developed and applied to a D/A converter, and it in turn makes the sine function available as a voltage. If the D/A converter is a multiplying type, computations of the form $Rsin\theta$ are readily performed.

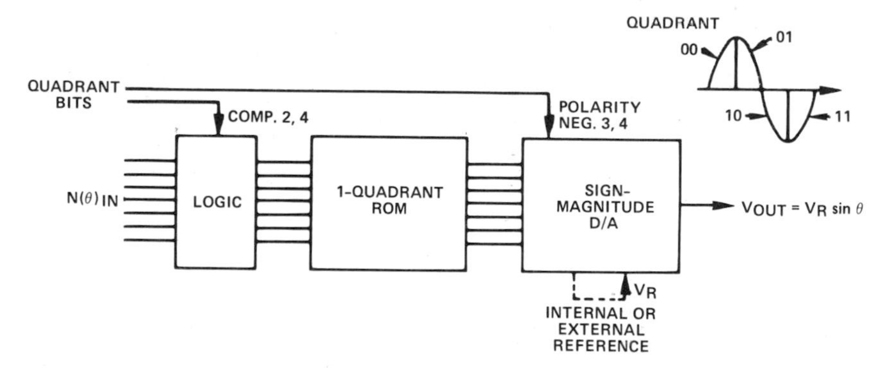

5.4 TRIGONOMETRIC APPLICATIONS

Digital Phase Shifter

The Figure shows two multiplying D/A converters used as digitally controlled attenuators multiplying the reference signals $V \sin\omega t$ and $V \cos\omega t$ by the vector components of θ. The difference of the outputs of the two converters is then the quantity $V \sin(\omega t - \theta)$, where the phase angle θ is set by the converter's digital inputs.

Digital/Resolver Converter (Resolver Simulator)[3]

Similar to the above configuration, but having the common reference input

[3]Information on synchro and resolver conversion can be found in condensed form in Chapter 14, and in considerable detail in the book, *Synchro & Resolver Conversion* (1980), available from Analog Devices, Inc.

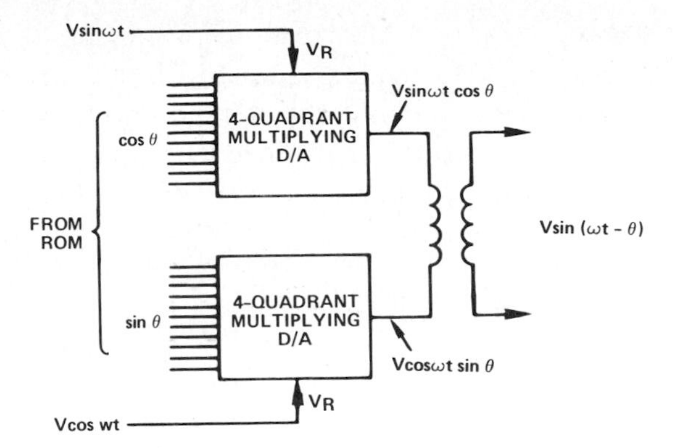

to both multipliers, V sinωt, this configuration obtains the two components, V sinωt sin θ and V sinωt cosθ, which express resolver data for angle θ. The resolver data can be converted into synchro format with a Scott-T transformer, or an equivalent network in which operational amplifiers provide the appropriate voltage ratios. This resolver simulator can be enclosed within a feedback loop to operate as a resolver-to-digital converter.

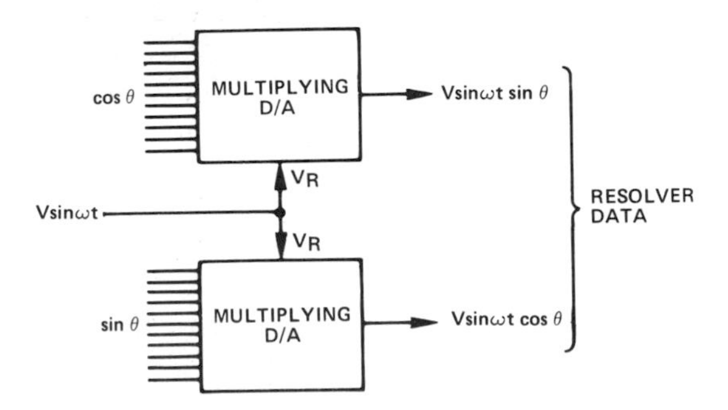

Using the *actual* resolver line voltages V sinωt sinθ and V sinωt *cos*θ as the converter reference inputs, and multiplying by digital equivalents to cosα and sinα, an output proportional to the angular error, $\theta - \alpha$ (for small angular errors) is developed. Operated in this mode, the configuration simulates a resolver control transformer.

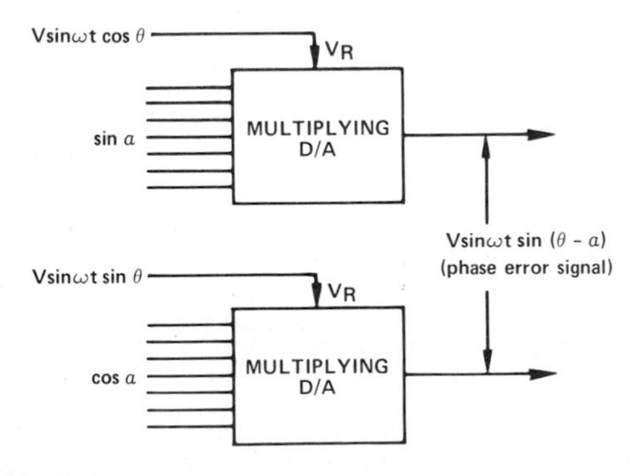

Coordinate Conversion

This interesting scheme takes a vector whose coordinates are X, Y, with angular equivalent $r\angle\theta$, and adds an angle of rotation, ϕ, to produce a vector, $r\angle\theta+\phi$, having a new set of coordinates, X', Y', and the same vector magnitude, with maximum error less than 1°.[4]

It employs two multiplying DACs, a multiplexer to deal with angular inputs in the proper quadrant, a set of precision resistors, and a handful of op amps. The cross-fed summations embody an algorithm that provides a sufficiently good approximation to keep the angle error within 1° and the total harmonic distortion on the order of 1%.

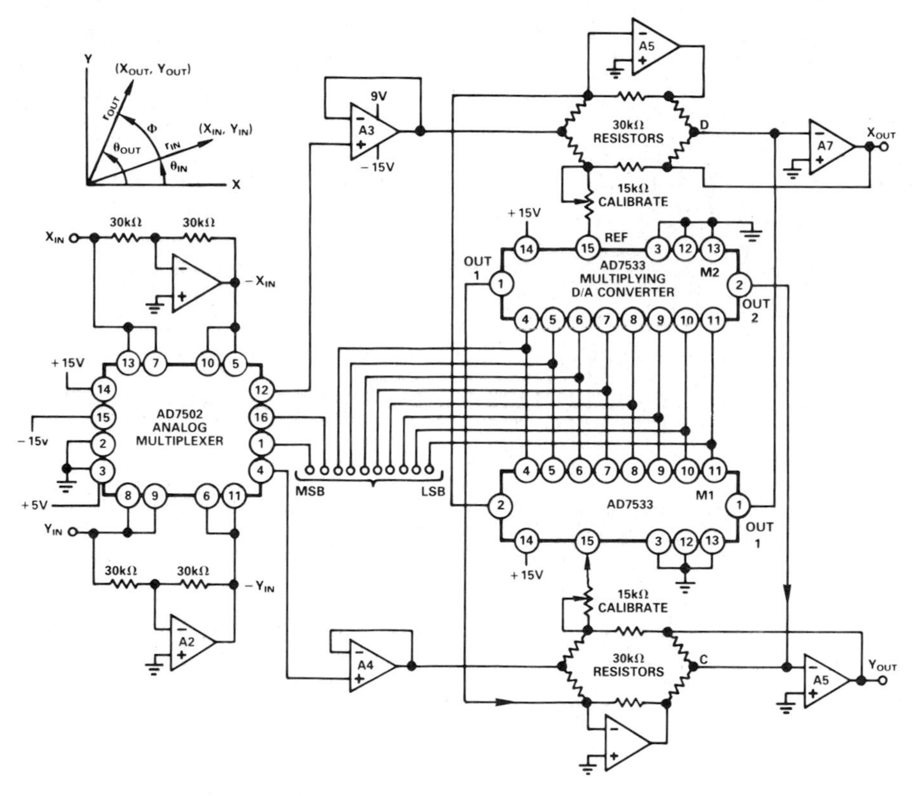

Courtesy of Arthur Mayer, U.S. Patent 3,974,367

5.5 WAVEFORM GENERATION

Linear time functions are generated digitally by clocks and counters, processed by ROM's or μPs for arbitrary wave shapes, and converted to analog functions of time by DACs.[5]

[4]See *Electronic Engineering Times*, July 9, 1979, Arthur Mayer, "Design a Multi-Purpose Network to Rotate Complex Numbers," and *Electronics*, Sept. 22, 1982, A, Mayer, "Low-cost coordinate converter rotates vectors easily," for extended discussions of the circuit, its applications, and PROM correction schemes for eliminating the residual errors.

[5]See also Chapter 6.2.

As long as the original function (with suitable dynamic range) can be created in digital form, then an analog output can be made to follow (within its speed limitations). The ease of manipulation and ability to lock timing operations to precise clocks give the digital approach considerable edge in versatility over many analog alternatives. Deglitching and filtering may be used as (and if) necessary to clean up the waveforms. Variable clock rates or arbitrary counting schedules may be used to obtain staircases having arbitrary, instead of uniform, duty cycles per step.

Sawtooth Generator

This sweep generator comprises a digital clock, a counter, and a DAC. The clock pulses increment the counter, and the sequential counter steps increment the DAC output. After the counter is full, it returns to its empty state and starts counting again. Both amplitude and period of the sweep generator are easily and precisely adjustable. The resolution is determined by the number of counts and choice of d/a converter, ranging from the 16-bit AD7546, with its 65,536 steps, down to 10- (or fewer) bit converters with 1,024 steps or less.

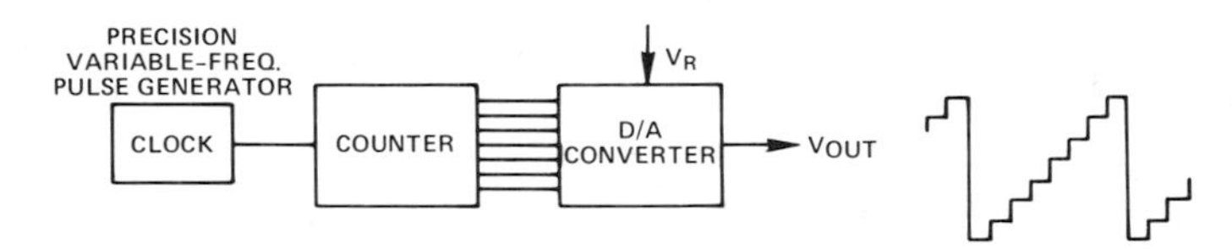

Triangular-Wave Generator

Instead of being allowed to overflow, the counter in this case is an up-down counter that is caused to change direction when it is full and again when it

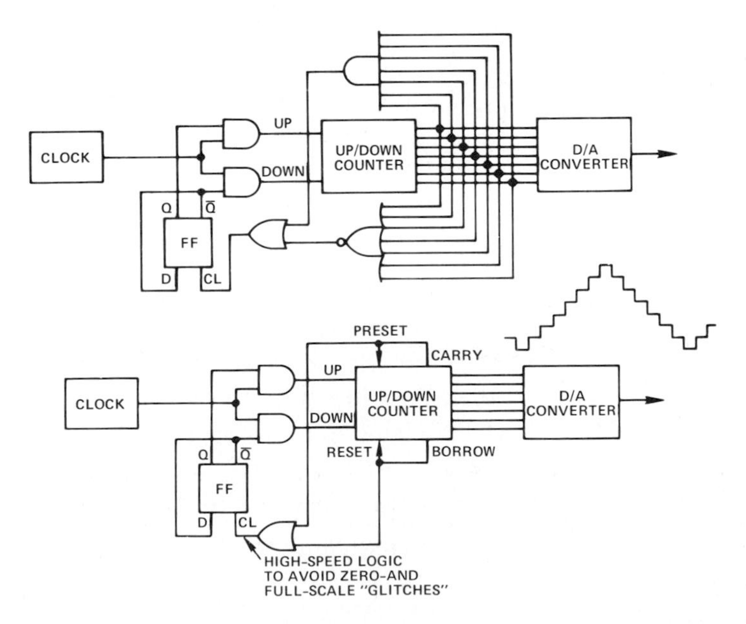

is empty. Two approaches to reversing direction are shown. In one, the reversal is generated during the full (and empty) states; in the other, it is generated by the carry(borrow) occurring at the leading edge of the next pulse. The result, at the DAC output, is essentially a triangular-wave of precise amplitude and frequency. With little additional logic, full-scale dwell (or dwell-and-reversal at any level) provides trapezoidal waveforms.

Sine-Wave Generator

If the digital count is fed to a sinusoidal ROM, and its output, accompanied by polarity information, is applied to a sign-magnitude-coded DAC, the output of the DAC will be an n-bit quantized sine wave. Its frequency is determined by the clock, and amplitude can be controlled at its destination or by the use of a multiplying DAC.

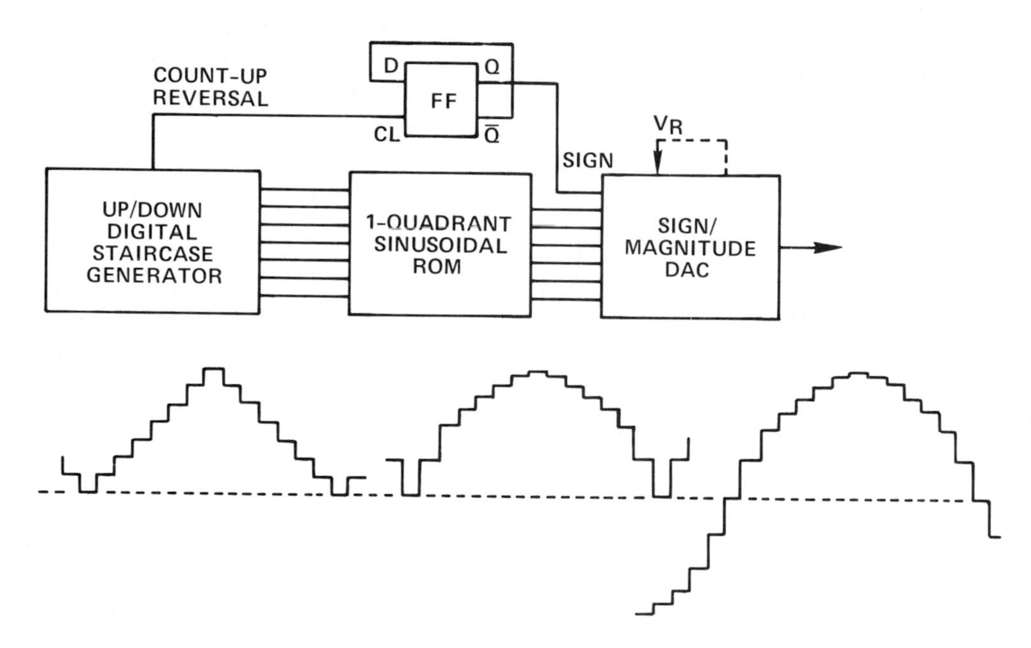

Digital-to-Frequency Conversion

A digital-to-analog converter and an analog voltage- or current-to-frequency converter (VFC) may be combined to generate a pulse train with frequency proportional to the magnitude of a digital word. If the DAC is of the multiplying type, the frequency will be proportional to both the digital input and an analog reference signal, for quadrants in which the VFC receives signals of proper polarity. The excellent resolution of available v/f converters will challenge the highest-resolution DACs.

The frequency of a signal may also be adjusted digitally to arbitrary (including non-integral) values by the use of a frequency-to-voltage converter, a multiplying DAC, and a VFC, as shown.

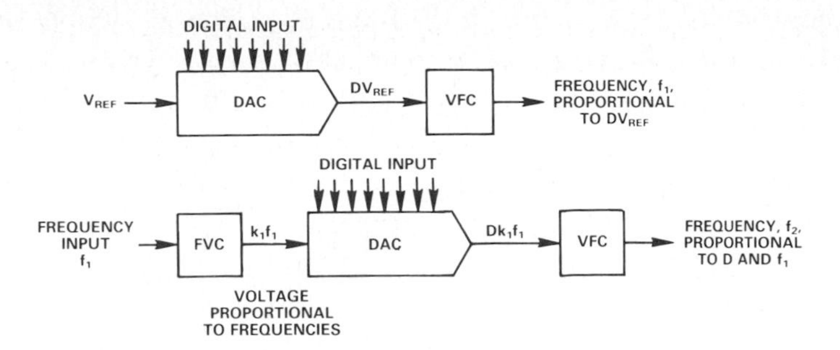

5.6 TIME RESPONSES

The ability of flip-flops to rapidly acquire and store digital information at rates (and for intervals) depending on precision crystal-controlled clocks, without degradation over time, and the continually decreasing cost of storage capacity, are strong motivations to seek ways of eliminating circuits employing capacitors as storage elements, with their leakage, dielectric hysteresis, parameter drift, and nonlinearity. "Distortionless" time delay, integration, and sample-hold are a few targets for such effort.

Precision Analog Delay Line

There are interesting applications for good analog delay lines: analog correlation, "distortionless" signal compression or expansion (e.g., "riding the gain" without missing a drumbeat), electronic echo-chamber effects, analog modeling of processes that incorporate pure time delay for predictive control, design of filters with arbitrary transfer functions, are a few.[6]

But there hadn't been a decent way of building a practical analog time-delay device that is variable over microseconds to minutes to months, without tying up a processor, until converters became available at low cost and FIFO (first-in, first-out) memories became available with reasonable word widths and stack lengths.

Active or passive filter-type delay lines were seldom "distortionless," analog "bucket brigades" had excessive leakage errors at low speeds, as well as a resolution-vs.-cost problem (this latter being solved by charge-coupled MOS high-speed bucket brigades), tape recording wasn't efficacious at high speeds, and the use of mainframe memory was too expensive (and bulky for portable instruments).

In the example shown, the delay is produced by FIFO registers (e.g., 9 bits $\times$ 128 stages). As the word resulting from each conversion is clocked into the FIFO, the earlier words are advanced; after 128 clock pulses, each input word emerges from the output port, with a delay of 128 clock pulses. For 9-bit conversion, signals that can be quantized into 512 discrete levels can be delayed

[6]See also Section 6.2.

with a resolution of 1/128 of the delay time (e.g., 2μs of 256μs, 1s of 128s, or 2.81° of a sinusoidal ac signal of period equal to the delay time, etc.). FIFOs can be stacked in parallel for increased bit-resolution, and in cascade for increased time-resolution.

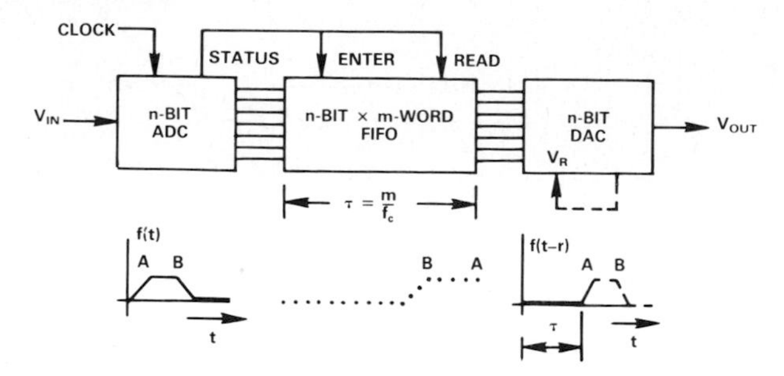

Tapped Delay Line

This device makes a number of points in the history of a waveform available simultaneously. It is simply the delay line with an increased number of discrete "chunks" of delay, and readout via DACs at each point. Multiplying

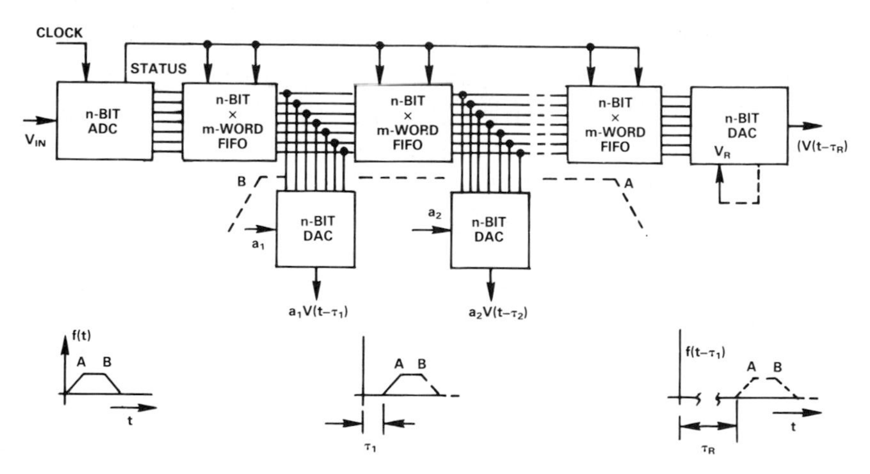

DACs allow such interesting functions as f(t) •f($t-\tau$) to be computed for a variety of values of τ. Hybrid IIR (infinite impulse response) and FIR (finite impulse response) filters with predictable indicial time responses may be constructed by this technique.

Serial Delay Line

For signals that do not require sampling at top speed, a considerable saving of the cost of FIFOs (or increase in the time-resolution of the delay) can be achieved by feeding the converted signal into the line serially, using shift registers for delay, and converting back to parallel information for the D/A conversion. Required shift register capacity depends on clock rate vs. delay, delay resolution, number of pickoff points, and length of data words.

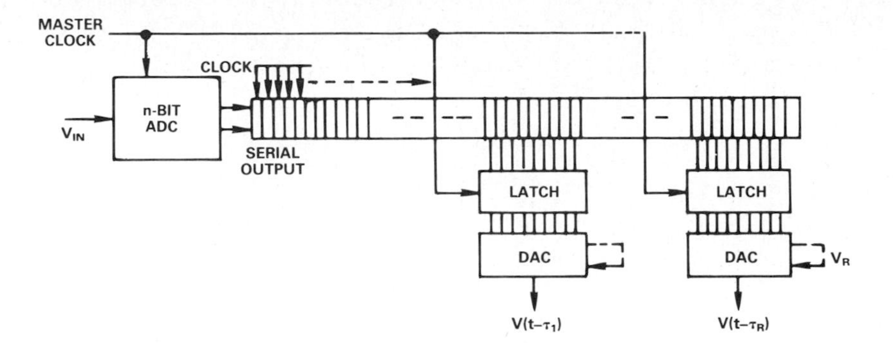

5.7 DIGITAL SERVO CIRCUITS

Most a/d converter designs involve feedback. Thus the very means of conversion implies that the combination of analog-digital interaction and the power of feedback can yield quite valuable results. A few examples are sample-hold, peak-detecting, and automatic zero-setting.

Tracking Sample-Hold (A/D Converter)

This circuit, also mentioned in the chapter on sample-holds, is especially useful as a track-and-infinite-hold device. It can acquire the analog signal within a minimum of 1 count and maximum of 2^{n-1} counts, and, upon command, *hold* it indefinitely without degradation, providing both digital and analog readout. Since it uses an up-down counter, it will track the analog signal at a constant rate (2^{-n} FS per count), and "hunt" between the two digital values that straddle the analog value, if it remains constant. Hunting may be avoided by the use of hysteresis, but a lag will be introduced.

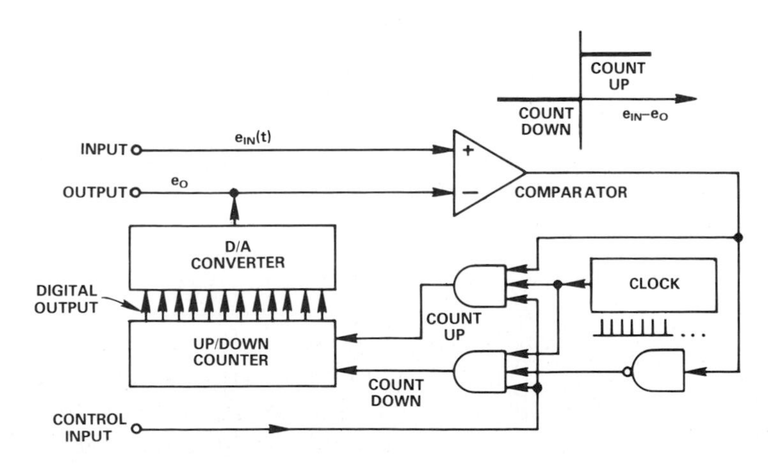

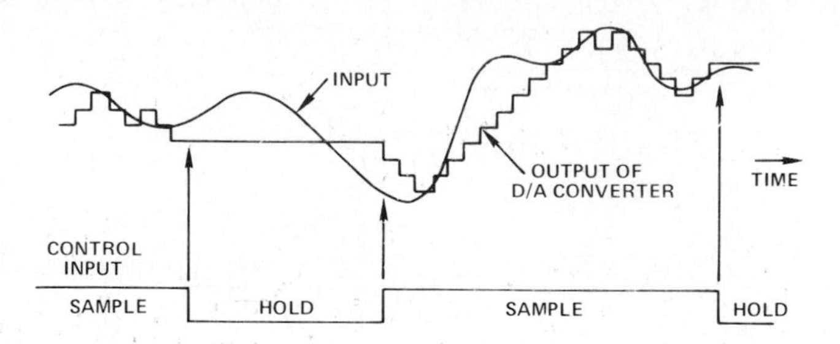

For analog signals without the sharp changes in slope, the tracking a/d converter is one of the lowest-cost ways to convert, since it eliminates the need for a sample-hold. However, its conversion time is variable, which introduces timing errors in sampled-data systems, since the most-recently acquired value may represent any value of signal during the interval between interrogations. Also, its response depends to a great extent on the amount and type of noise present. Tracking converters are discussed in Chapter 14.

Digital Pulse Stretcher

For extremely fast-acquisition-very-long hold, this circuit, consisting of a fast sample-hold and a fast successive-approximation A/D converter will provide the best results. Both analog and digital outputs are available. For single samples, if the internal D/A converter's output can be made available without slowing conversion, the output D/A shown in the Figure is unnecessary. The HTC-0300A is kept in *sample* at all times except during conversion. When switched to *hold*, it should have a "head start" of 200ns (acquisition time) for its transients to die down before the first conversion decision is made. Aperture time is about 6ns with 100-ps jitter.

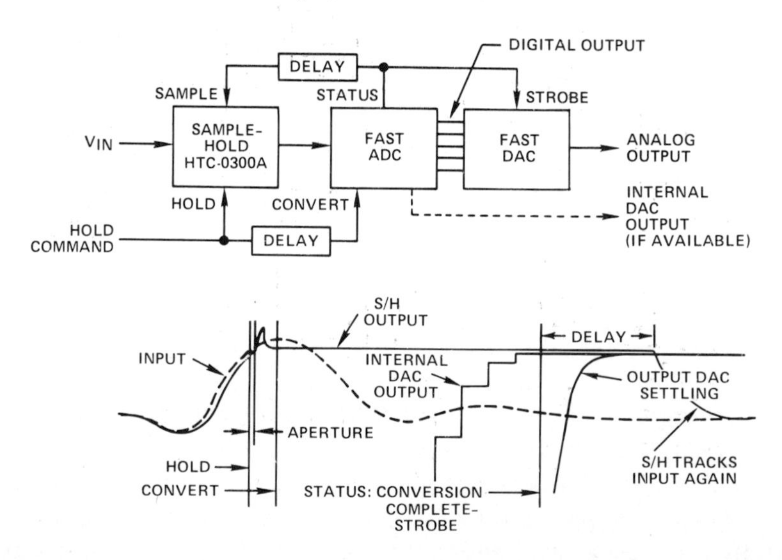

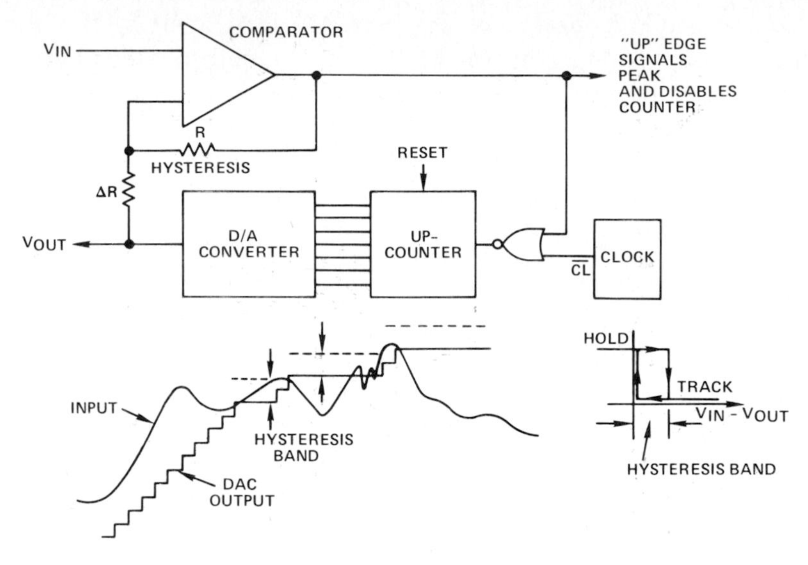

Digital Peak-Follower (with Hysteresis)

Similar to the tracking sample-hold, but using an up-counter (a valley-follower would use a down-counter), this circuit will ***hold*** the highest value of input that it has been able to track. However, to provide a small measure of immunity to noise, hysteresis makes the circuit insensitive to small changes; in order for the input to be followed, it must be higher than the stored value by a preset amount. A similar circuit can be used for valley following, and two such circuits with digital or analog subtraction will provide peak-to-peak measurement. As with all circuits employing comparators, care should be taken to avoid oscillations; ΔR is a very small fraction of R.

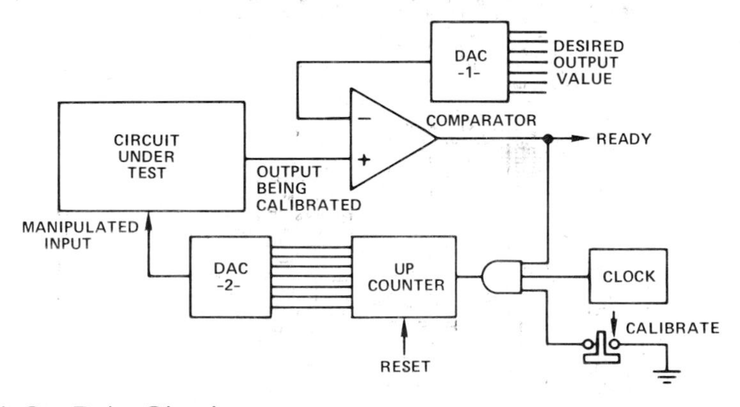

Automatic Set-Point Circuit

If a circuit under test is to be calibrated from time to time (e.g., each time some element, perhaps a device under test, is changed), the resetting and the level to which a test value must be reset, may be adjusted digitally. In the example, an output of the circuit must be set to a value equal to a calibrating value set by high-accuracy standard, DAC-1. The values are compared, and a clock increments a counter, which updates a DAC, setting the input that performs the calibrating adjustment. When the comparator changes sign, calibration is complete, and the sign change indicates a "Ready" condition. The calibration value is retained until a new calibration cycle is initiated by resetting the counter and gating the clock.

5.8 A FINAL NOTE: SOFTWARE vs. HARDWARE

The examples given here all involve hard-wired analog-digital circuitry. For applications in which microprocessors are available, it should be evident that functions involving memory, control logic, and digital data can be as well (and more flexibly if perhaps not quite as speedily) handled through the writing of appropriate microprocessor programs, at the cost of software development—and program running time.

Rather than viewing the techniques as competitive, the designer should consider that the three-way tradeoff between analog, hard-wired, and software approaches provides at least one more degree of freedom for developing cost-effective instruments, apparatus, and systems. The flexible designer will not arbitrarily exclude a given approach; on the other hand, the committed designer should know that other alternatives to one's own predilections do exist and may, on occasion, prove available to save the day.

Chapter Six

Applications of Converters in Instruments and Systems

Chapters 1-5 have introduced the basic hardware elements of systems and equipment that involve converters, shown the basic configurations of data-acquisition and data-distribution systems, and indicated a few examples of the uses of digital and analog elements in intimate combination.

In the real world, converters are necessary whenever data in analog form must be processed digitally—or devices that require analog inputs must get their input data from digital sources. Thus, any instrument, apparatus, equipment, or system involving real-world data is a potential home for one or more data converters.

This chapter will illustrate a few examples among the plethora of systems and equipment that have been conceived of or built involving converters. The examples are drawn from a variety of sources, but they share the ideas, hardware, and circuit structures that have already been touched upon.

The intent is to inform the reader of what has been done, to suggest what can be done, and to arouse thoughts of what *might* be done by adding the conceptual tools described in this volume to the fund of knowledge and experience already existing pertaining to the reader's own field of endeavor. To the digital expert, it should provide insights into the real-world connection; to the analog expert, it may suggest ways of doing analog jobs better with digital assistance; and to the person with much theory and little practice, it should give a more-concrete feeling for practical applications of digital techniques in the real world.

6.1 AUTOMATIC TESTING

> "Automatic testing of electronic devices has been a major factor not only in the overall improvement of product quality and reliability, but also in the dramatic lowering of product costs."[1]
>
> – Harold T. McAleer, General Radio Company

Although electronic devices are a major (and in some ways an obvious) market for electronic testing equipment, their makers and users are by no means the sole beneficiaries of automatic testing. Anyone whose blood has been tested recently, has flown safely in a modern jet aircraft, or has an automobile that has been inspected with modern equipment, has been exposed to the potential savings (and not just financial) inherent in automatic testing.

The cost savings, both immediate and long term, result from a number of characteristics of automatic testing:

Human resources are conserved. Fewer persons can conduct more (and more-thorough) tests of high complexity with minimal training.

Volume. Large numbers of tests can be performed in a short time, including individualized tests on complex devices and repetitive tests on large numbers of simple devices.

Reliability and consistency. A well-designed test program will perform identical tests leading to consistent results, with no aberrations due to misreading, fatigue, etc. If failure occurs in mid-test and repairs are made, the entire test cycle can be repeated, numerous times if necessary, with full confidence that the most recent test has "cut no corners."

Multiplexing of adjustments and readouts. An instrument designed for use in automatic testing bears little physical resemblance to conventional instruments, since it need have neither binding posts, knobs, readout, nor even "front panel;" it shares the system's readout devices; connections and adjustments are made by the system.

Automatic Calibration. Any necessary calibrations, zero adjustments, nonlinear-device compensations, or other predicted allowances, whether of the test subject, the sensors, or the test system itself, can be made under system command. Adaptive range-changing can be fully automated.

Measurement statistics. The system can retain in memory the results of all tests, the results of discrepant tests, and data on specific parameters; it can number-crunch the statistics and print the results upon request. Yield studies can lead to product improvements, elimination of sources of repeated rejections, and prediction or tracing of future failures.

[1] *IEEE Spectrum*, May 1971, "A Look at Automatic Testing"

In short, a well thought-out, well-designed, and well-implemented automated test facility can reliably perform large numbers of tests, around the clock, on a "100%" basis, consistently and without tiring, with accuracy and skill, and with feedback to the designer for the next generation of the product. Skilled test personnel can be freed for more-creative pursuits because they don't have to follow long, detailed procedures for routine manual testing of the ins-and-outs of complex systems, calibrate instrument dials, interpret go-no-go limits, and calculate odd ranges.

Then there are the important but less-measurable results, that pay off in human values as well as dollars-and-cents: the aircraft engine that didn't fail, the electrical chassis that didn't need field repair, the steel rolling mill that didn't run away, the hospital patient that survived, the vendor whose reputation remained consistently high.

6.1.1 USES FOR AUTOMATIC TESTING

The manufacturer of components, such as integrated circuits, benefits greatly, because testing is a far-from-negligible cost in the integrated circuits business. Besides delivering a higher level of acceptable quality to the customer, the manufacturer also develops more-accurate knowledge of yields and trends, and can develop specific selection categories for special orders. For the producer of high-performance specialty ICs, such as Analog Devices, it is an indispensable tool. Such devices as laser-trimmed low-offset op amps, high-accuracy monolithic multiplier/dividers, and—indeed—full-accuracy converters, with resolutions of 12 or more bits, would be so costly as to be infeasible if manual measurements and adjustments were involved. (Instead, the additional cost is a small fraction of the price of the untrimmed unit.)

The user of large numbers of identical components can also benefit: Machines can be used to weed out discrepant units in incoming inspection; measure, select, and grade units for different applications (freeing the user from paying the manufacturer extra to do the same job) and keep comparative statistics from lot-to-lot and vendor-to-vendor. It may be noted, as a matter of perspective, that an average saving of 10¢ on 100,000 units is $10,000.

The manufacturer of equipment and systems can test subassemblies in-process, or as received from subcontractors; (s)he can also test completed pieces of equipment thoroughly. In both cases, the test system can be programmed for GO/NO at points of discrepancy, and to either reject the device for later evaluation, or branch into a diagnostic mode, to isolate the portion of the circuit (or perhaps even the component or connection) that is faulty. Repaired units can be recycled and subjected to the same battery of tests as the new units, and serialized records of all such processing can be maintained.

Highly complex systems, such as jet aircraft and their various subsystems, can be tested thoroughly on the ground by a small number of persons in a short time, with a high probability of finding any faults, or the discrepancies that might indicate incipient faults. In addition, the on-board test and monitoring system can provide warning to the crew (using appropriate media) of anomalous subsystem behavior, provide automatic switching to backup systems, and, as the electronic portions become increasingly sophisticated, it can perform a degree of diagnostic testing, facilitating repair.

6.1.2 INGREDIENTS OF TEST SYSTEMS

For systems that test devices, the test begins with the *unit under test*. It must be handled, maneuvered into place, and connected to. Then a *stimulus* is applied, and a *response* must be measured. The response is compared with a set of possible responses, and a *decision* is made (accept, reject, grade-and-sort, perform an adjustment) and communicated (print, store, mark, analyze), and a new instruction is given (next test, next set of connections, next device, wait for manual instruction, etc.). An outline of such a system is shown in the block diagram of Figure 6.1.

The simplest devices have two leads (resistors, capacitors, diodes), but most devices or equipment subject to testing will have many more. For example, ICs often have more than 20 or 30 connection pads or pins, and printed-circuit boards subjected to all-node "bed-of-nails" testing may have more than 1,200 connection points. The program must call for connecting the appropriate

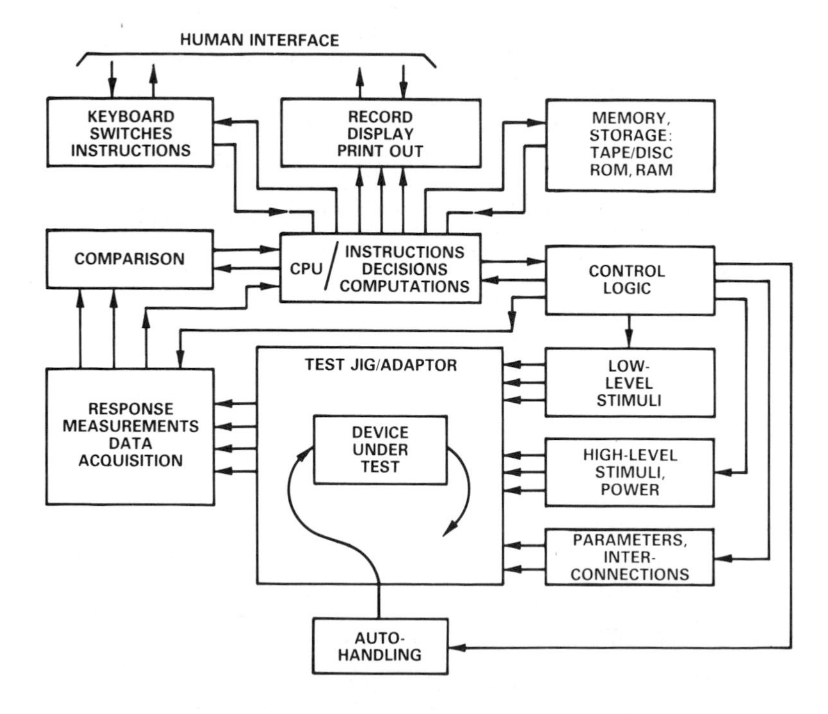

Figure 6.1. Test system ingredients in typical configuration.

stimulus generators and power sources to the appropriate terminals, and the appropriate measuring devices (bridges, amplifiers, etc.) to *their* appropriate terminals, and making all opens, shorts, "grounds," links, etc., as required for the test step. Some of these may be hard-wired in the adaptor; others must be called for by software or operator setting. It is absolutely essential that noise pickup and interference, as well as parasitic effects caused by lead resistance, capacitance, and inductance, be minimized.

A typical flexible, modular architecture employed in general-purpose electronic component and IC testers is shown in Figure 6.2. The device under test is plugged into an interchangeable socket board wired appropriately for the specific type. The socket connects to a socket assembly, wired for specified conditions; it in turn connects to a pluggable family board, which provides stimuli, gains, and special functions needed to exercise and test a particular group or class of devices (e.g., op amps, a/d converters, d/a converters, digital logic ICs, etc.)

Inside the system housing, the measurement section, which comprises the measurement card, source card, and digital I/O card, provides program-controlled measurement functions, voltage sources and references, and the digital I/O drivers and detectors required for performing device tests. It operates from a test-system bus, controlled by a 16-bit CPU. Besides performing the

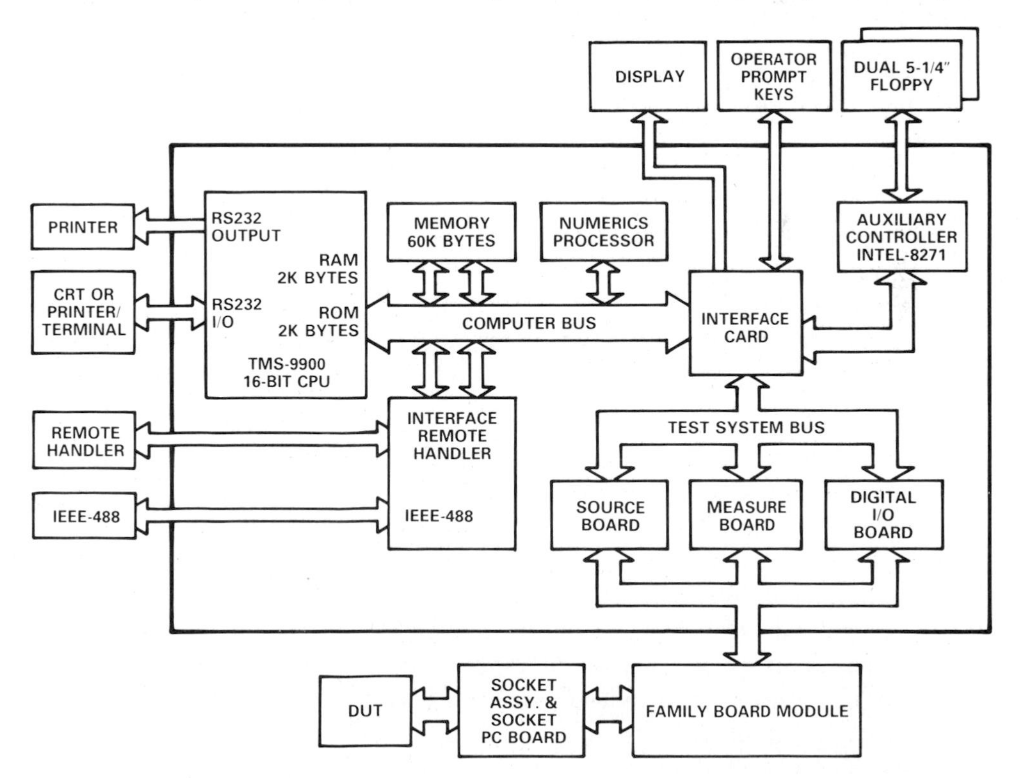

Figure 6.2. Block diagram of Analog Devices LTS-2015 device test system.

actual tests, the system interfaces with remote automatic handling devices, printers, and CRT terminals, and communicates locally via an alphanumeric keypad, operator prompt keys, a display, and one or two floppy-disk memories.

Converters in Test Systems

It may be fairly evident that much of the engineering and hardware cost of test systems goes into fixtures, switching devices, computers, peripherals, displays, wiring, and cabinetry. However, since the analog stimuli—and comparison-type measurements—are controlled digitally, and the analog responses must be returned to digital form for processing, it should be evident that converters and their accessories play a key role in ensuring test accuracy, speed, and reliability, yet represent but a small fraction of the cost of the system. For this reason, it may be false economy to use conversion devices that are anything but entirely adequate to do the job, or to seek to cut cost corners by risking marginal performance.

In the system shown in Figure 6.2, the reference system consists of a traceable temperature-controlled precision voltage reference and a 16-bit-accuracy 12-bit-resolution d/a converter. Figure 6.3 shows how DACs are used in the measurement section of that tester. Note the use of force-sense (i.e., Kelvin) connections for both the amplifier output and the ground return to enforce accurate test voltages at the device under test, irrespective of line drops.

Typical uses of D/A converters in testing include: programmable power supplies, pulse generators, sweep generators, waveform generators (with appropriate digital inputs). They may be used as offset and gain "potentiometers" in calibration loops, as bridge-balancing voltage sources, and as part of A/D converters, sample-holds, peak-followers, etc.

A/D converters, either with multiplexing or per-channel, return the measurements to digital form, often after processing by isolation or instrumentation amplifiers, by op amps as electrometers, and, in some cases, by multipliers, ratio devices, log devices, and all the other paraphernalia mentioned in Chapter 2.

An essential decision that must be made is the degree to which analog data reduction and/or digital signal-processing hardware will be used, as compared with the performance of similar functions by digital software. This consideration depends on such factors as the amount of number-crunching necessary and the time available for it, specified accuracies, cost and energy tradeoffs, as well as the background, experience, and inclinations of the designer. We suggest that analog-oriented designers not overlook the possibilities of software and digital signal-processing hardware for reliable routine computation, and that digital designers consider the decreasing cost of functions that can be performed with analog modules and linear integrated circuits at the front end, and the balance between too much and too little data.

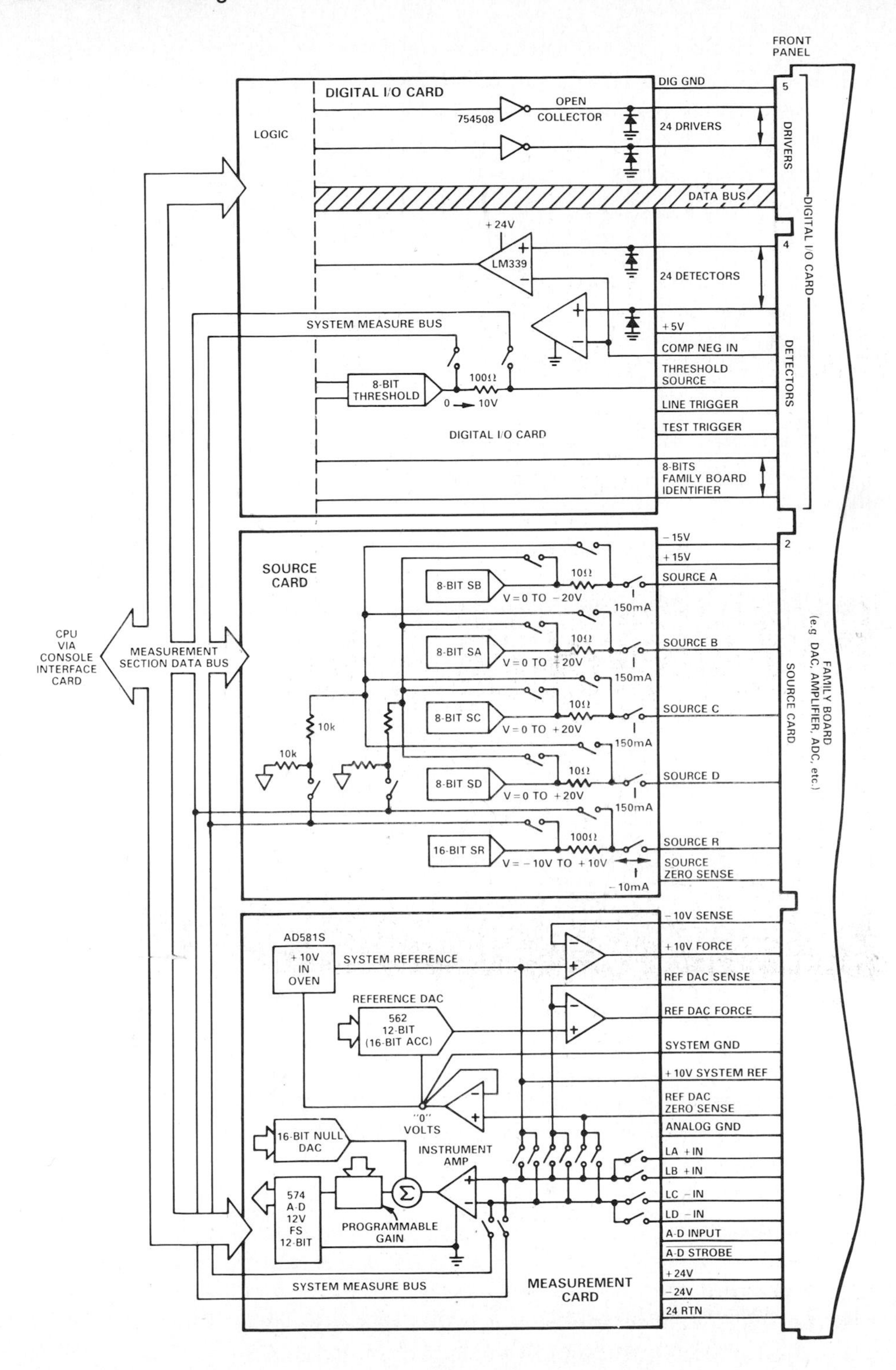

Figure 6.3. Measurement section of LTS-2000 tester, showing measurement, source, and digital I/O cards, with their connections to the family board.

Much that could be said about system optimization, in terms of getting the best-possible interference-free measurements of suitable accuracy, has already been mentioned in several places in this book (for example, Chapters 12 and 22); and test systems are probably the most representative class of design problems requiring active application of these principles. In general, it is best to keep the conversion function as close as possible to the analog measurements, especially in applications involving remote handlers.

In component tests, where the lead-runs to the unit-under-test (UUT) are controllable, as is the local environment, the main sources of interference arise from the proximity of input, output, power, and logic leads in the vicinity of the test adaptor. In large-system testing, long leads, including multiconductor cables and connectors; the presence of electrical noise (RFI, power line and switching-transient spikes); and possibly unfavorable environmental conditions (temperature, humidity, vibration), may combine to make the measurement problem extremely difficult.

It often turns out that, in the design of large systems, performing tests locally is an effective way of solving the interference problem, using a local μP-controlled self-contained measurement subsystem (or *instrument*) that communicates digitally with other such subsystems under the direction of an external system controller. The IEEE-488 bus structure was designed specifically for communication of data, instructions, and handshaking in tests involving sources (e.g., signal generators), measuring devices (e.g., precision voltmeters), display devices (e.g., printers), and controllers (e.g., keyboard/display terminals).

6.2 DIGITAL SIGNAL PROCESSING

In this section, we shall discuss briefly the class of digital applications that involve the generation, transmission, delay, recovery, processing, storage, characterization, and synthesis of analog waveforms. Conceivable applications include:

- Time expansion, compression, (relative) advance, and delay
- Transient storage and recording
- Synthesis and analysis of speech and music (and waveforms in general)
- Transfer-function synthesis and analysis
- Convolution
- Digital filtering
- Recovery of signals from noise by correlation techniques and fast Fourier transforms
- Scrambling and unscrambling of coded transmissions
- Generation of arbitrary signals and transfer functions

Digital methods, especially with microprocessors and digital signal-processing chips, can provide a powerful set of tools for dealing with analog functions and the transfer functions that are used to shape them in the time and frequency domains, as we have suggested in Chapter 5. These methods, and some components that make them feasible in a reasonable time frame are discussed briefly in Chapter 21 and at greater length elsewhere.

The key that unlocks the door is the A/D converter, which "freezes" a sample of the waveform and makes possible permanent storage without degradation. Thereafter, digital shift registers, multipliers and multiplier-accumulators, memories, comparators, microprocessors, and control logic can perform a virtually unlimited set of operations digitally at any time (*on-line* or *off-line*).

Errors are due to the discrete-time and quantized-amplitude nature of the sampled signal, and truncation or roundoff errors in computation (where necessary or permitted). If sampling occurs at an adequate rate, if the conversion has sufficient resolution, and if the computation carries enough significant bits, there is no loss of information, even though the signal be stored, multiplied, integrated, added, subtracted, correlated, or otherwise man(or machine-)ipulated. Via d/a conversion, the data can of course be returned to the analog domain and subjected to further processing there, but its attributes can be retained in digital memory for as long as desired.

Of the circuits and ideas that appear here, some are variations on the basic theme of the delay line, others are applications of basic digital signal processing techniques; they represent promising areas of application but are not necessarily new or original. Their purpose is to unleash the reader's curiosity and creativity, in the field broadly encompassed by the title of this section. We've tried to avoid, except where necessary, mathematical particularities (and the controversies they sometimes engender), since the purpose of this chapter is only to relate converters to the more interesting applications they are used in. The field is large enough that a book the size of this volume would be required to deal satisfactorily with it.

6.2.1 SHIFT-REGISTER DELAY LINE

The basic tool for performing many interesting functions is the shift-register delay line—with and without taps—mentioned briefly in Section 5.6, and shown here again (Figure 6.4) for further discussion. It should be noted that the delay line need not be an isolatable physical entity—it may be a sequence of memory locations in a microcomputer, incremented or decremented on appropriate clock pulses.

Suppose the analog signal is a one-shot occurrence, of which m samples have been taken, the sequencing clock has stopped, and the conversions have ceased. The signal is now stored in the delay line in digital form, and it will remain there until it is advanced or cleared or the power has been turned off. A number of interesting things may be done with the stored signal:

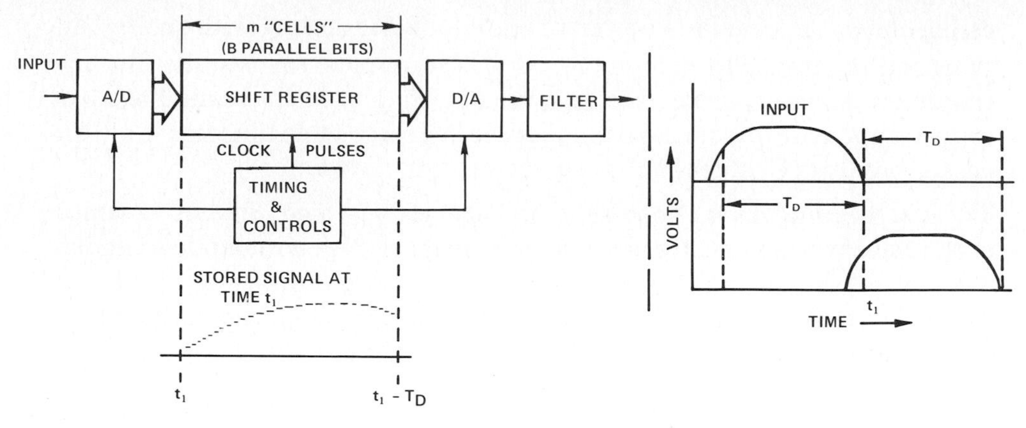

Figure 6.4. Digital delay line (word- or byte-wide shift register).

Read out into memory

The delay line can be considered as FIFO (first-in, first-out) buffer memory. The stored signal can be read out of the delay line, a word at a time, and stored elsewhere in memory, while the line awaits another transient (Figure 6.5).

Readout as an analog signal

The signal can be read out without further processing and converted to an analog signal with a DAC, each sample in turn—but at an arbitrary rate, determined by the choice of clock frequency. For example, the transient may have been quite rapid, but it is desired to plot it out on a chart recorder. Or, it may have been fed into the line slowly (perhaps even keyed in manually and asynchronously as an arbitrary waveform), to be used as a shaped stimulus for an analog process, and it is to be discharged at high speed. In applications such as this, digital techniques have a distinct advantage over purely analog switched-capacitor "bucket-brigade" delay lines, because there is no loss of accuracy with storage time or cumulative number of sections, hence no resulting constraint on either parameter.

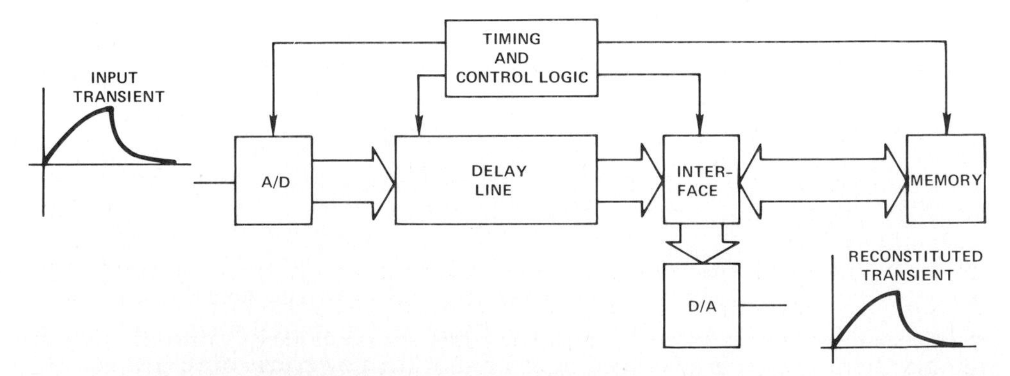

Figure 6.5. Digital delay line as FIFO buffer (transient recorder).

Recirculate

If the the data stored in a delay line is fed back repetitively from the output to the input end of the line (or the first of the memory locations), it becomes a recirculating delay line (Figure 6.6). The stored signal will then appear at the end of the line cyclically, allowing the signal to be displayed on an ordinary oscilloscope. By loading, or "charging", with an arbitrary or an analytic input signal (derived from either an analog or a digital variable), then providing rapid recirculation and D/A conversion, it is possible to create an extremely wide range of arbitrary repetitive analog waveforms, of controllable repetition rate and amplitude.

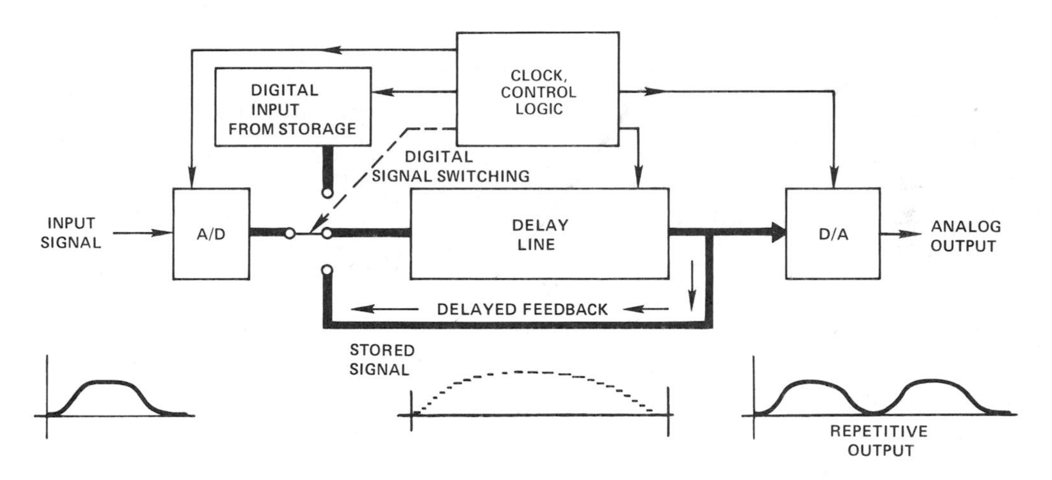

Figure 6.6. Recirculating delay line.

Perform waveform averaging by addition

If the same message is sent repeatedly but arrives at the converter accompanied by (and perhaps "buried in") noise, it can be recovered by summing all the versions of the message synchronously: the coherent portions will add directly with the number of items summed, while the rms noise will tend to be "averaged out" and will increase only as the square-root of the number of items. For example, with 100 repetitions, the signal will be increased in relation to noise by a factor of 10. This can be accomplished with a delay line or set of memory locations by summing the sample increment of the newest message in each position with the sum of the corresponding samples of previous messages, accumulated in the delay line. Thus, when the second message arrives, its first-position sample is summed with the already-stored sum of the previous first-position samples; its second-position sample is summed with the sum of the previous second-position samples; and so forth. Since the original messages are presumably identical, while the noise varies randomly, each iteration adds 1 unit of original signal to each position, while the noise components tend to be averaged out.

In practice, computing the simple sum

$$Y_{r,n} = \sum_{i=1}^{n-1} X_{r,i} + X_{r,n} \tag{6.1}$$

– where $Y_{r,n}$ = sum of n samples at the rth position, Σ = output of the delay line (n-1st sample at the rth position), and $X_{r,n}$ = input of the nth sample at the rth position—and averaging it by dividing by n afterwards leads to an open-ended stored-signal amplitude, which requires a large dynamic range (wide delay line). Thus some form of normalizing is desirable.

Figure 6.7 shows one scheme, based on the algorithm of (6.2):

$$Y_{r,n} = Y_{r,n-1} + \frac{X_{r,n} - Y_{r,n-1}}{n} \tag{6.2}$$

where $Y_{r,n-1}$ is the delay line's output (n – 1st sample at the rth position).

At the rth position, this is equivalent to adding the 1/n of the new input to the previous output, multiplied by (n – 1)/n. Thus, for n = 1, the output, $Y_{r,1}$, will be equal to $X_{r,1}$; for n = 2, the output will be equal to $(X_{r,1})/2 + (X_{r,2})/2$; for n = 3, the output will be equal to $(2/3)(X_{r,1} + X_{r,2}) + (1/3)X_{r,3}$, and—in general—the output is equal to the average over n inputs; The average value would be constant for n equal input values at position r. Since all variables are normalized, it is not necessary to carry large summation values or perform more than one division.

It is interesting to note that, as n becomes large, for identical inputs, Y_r, $Y_{r,n}$ becomes very nearly equal to $Y_{r,n-1}$, because each additional increment causes little change, being divided by n. For fast results, 1/n can be obtained from a lookup table[2] and multiplied by the difference in a digital multiplier-accumulator, which takes the product and sum.

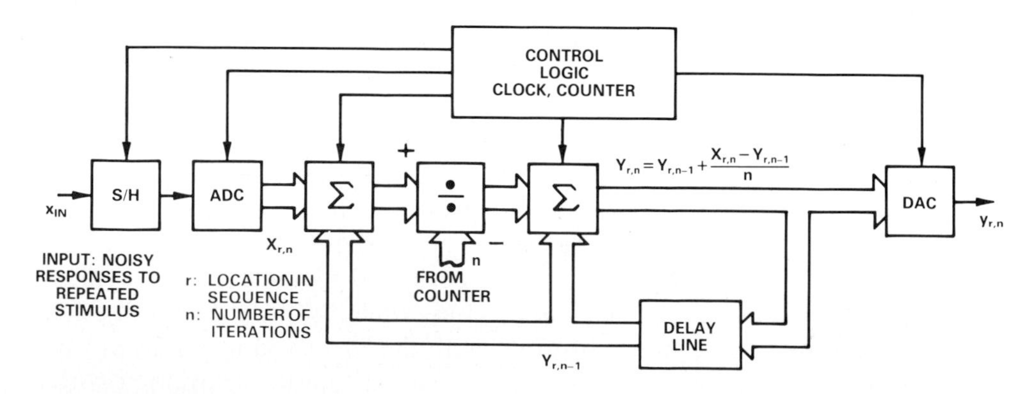

Figure 6.7. Waveform averaging scheme, in block-diagram form.

[2]The size of the lookup table and speed of computation can be minimized by the use of an approximation technique. See "Fast, Simple Approximation of Functions," by Matt Johnson, *Analog Dialogue* 18-1 (1984).

Time Compression by Sampling

This application employs a stroboscopic effect to apparently speed up a slowly varying signal.

In Figure 6.8, a shift register (or set of memory locations), is incremented at a high frequency, f_c, for example 513kHz. The converter is digitizing a slowly-varying signal at a rate f_s. Suppose that: the shift register has 512 steps, the line is full of previous values, and at a given instant, the 512th sample appears at the output and is fed back to the input. On the next step, starting the *m*th iteration, the a/d converter output, $X_{1,m}$, is fed onto the input bus to replace the existing value, its previous output, $X_{1,m-1}$. The line then advances for 512 steps. On the 512th step, the input is once again $X_{1,m}$; and $X_{2,m-1}$ appears at the end of the line, while $X_{2,m}$ is ready at the converter output. On the 513th step, the converter output is fed into the line to replace $X_{2,m-1}$. $X_{1,m}$ and $X_{2,m}$ are now indexed down the line, and on the 513th step, $X_{3,m}$ replaces $X_{3,m-1}$. By the time 512 conversions have occurred, in real time, each consecutive sampled signal (including new and previous values) has circulated 513 times, thus providing a 512-fold speeded-up version of the (for example, 1.96Hz) analog input waveform at the output of the D/A converter, at the equivalent of 1ms per sample.

If each cycle of the analog waveform is identical to the adjacent ones, and if the clock is synchronized to the analog signal, the output of the DAC, plotted on an oscilloscope screen, swept at 1kHz, will appear to stand still, plotting the low-frequency input, but *with no flicker*. Changes to the input signal, from iteration to iteration, will appear as progressively appearing changes to the stationary pattern. Since the compression ratio depends on the time required for each 511 samples, it is proportional to the clock frequency, which can be locked in at any convenient value.

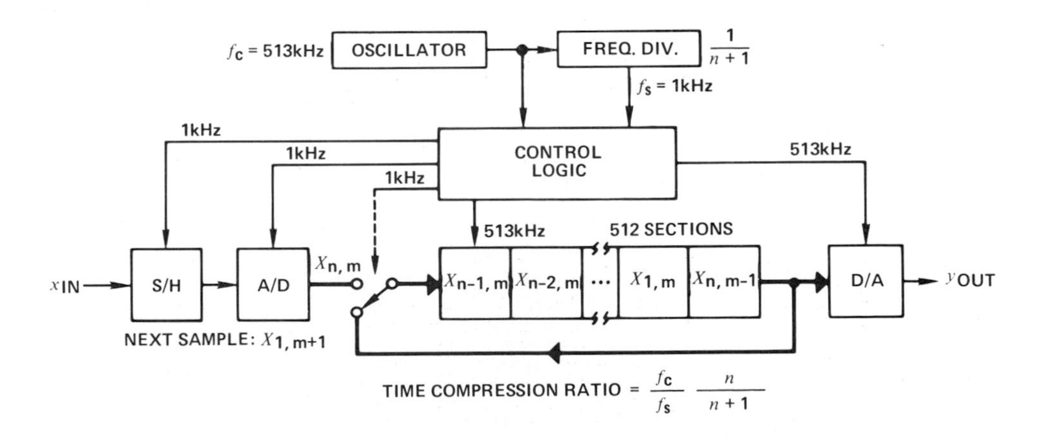

Figure 6.8. Time compression scheme, using a circulating delay line.

Real-Time Correlation

For an input function, f(t), the output of the delay line (or set of memory locations) over a complete circulation (in compressed time) is a set of values of $f(t - \tau_i)$. If the successive values are multiplied by the sampled value of another waveform, $g(t)$, which, with f(t), is updated after each circulation, and if each individual product is averaged with its synchronous counterparts from previous circulations, as described earlier, the output of the averager will represent a sample-by-sample cross-correlation of f and g at a real-time rate, delayed by the product of the sampling period and the number of samples circulated (Figure 6.9).

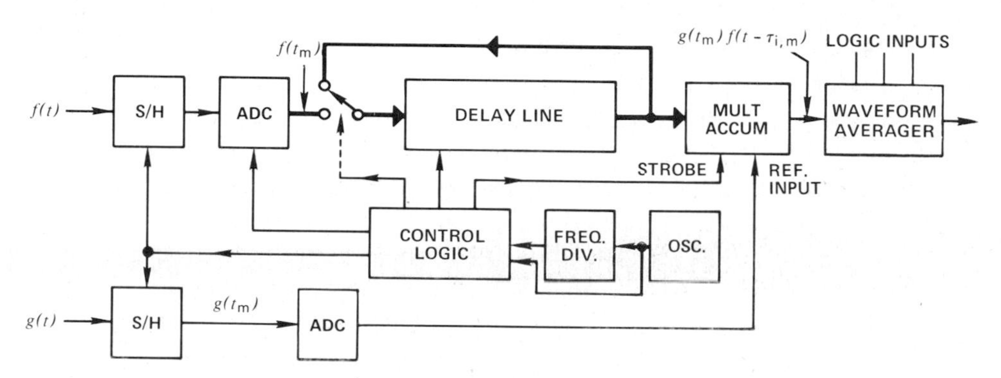

Figure 6.9. Real-time correlation scheme, employing a recirculating delay line and a waveform averager.

Incremental Delay Line as a Filter

If the delay line consists of a number of sections (or successively incremented memory elements), and the outputs at the taps are multiplied by arbitrary coefficients and summed (Figure 6.10), it is possible to synthesize arbitrary

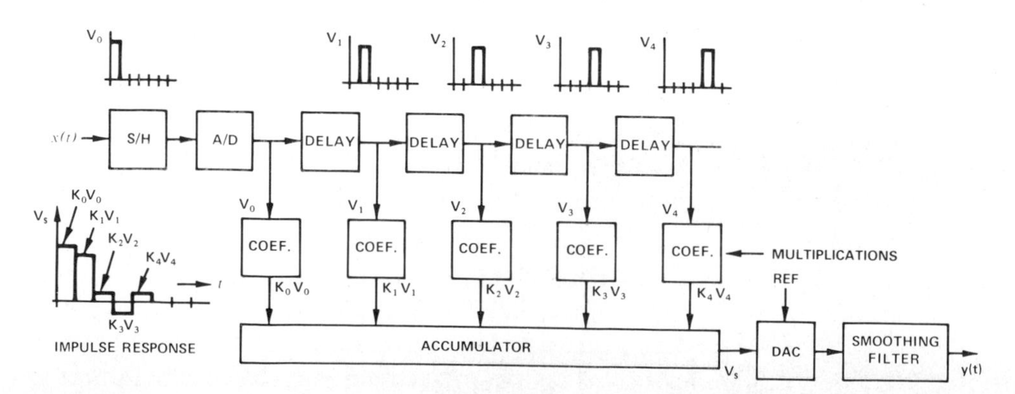

Figure 6.10. Delay line as a non-recursive filter with programmable real-time response (small number of sections shown for clarity). If fastest possible speed not necessary, fewer hardware multipliers are required; a single multiplier-accumulator could be adequate.

time-domain responses to steps, pulses, or other waveforms. Since the output bears a linear* relationship to the input, the resulting transfer function may provide amplitude and phase responses to other signal forms (over limited ranges of frequency) that can be expressed by transform integrals but were once otherwise formally considered "unrealizable."

In this case, a *non-recursive filter*, the output is a function of the input only and is inherently stable. In practice, if the signal is not too fast, a single multiplier-accumulator may be programmed to perform all the multiplications and the summation digitally, followed by a single output DAC, saving hardware. If the filter's response is defined in terms of frequency response, instead of time-domain response, the coefficients—although messy to compute from scratch—are handily computed by cut-and-dried techniques, such as the Remez Exchange Algorithm[3].

Recursive Filtering

When the output is a function of input only, the number of possible responses is limited, because—barring the introduction of recursion via an external feedback system—the output will settle within a finite time after the input has ceased to vary. However, by using recursion (i.e., feedback) to make the output a function of both output and input, a transfer function that is more general and more economically achieved (but more difficult to design and potentially unstable) becomes possible.

Recursion may be achieved by feeding forward (to the output) and back (to the input) from each tap point via individual coefficient multiplications (Figure 6.11a). A more elegant scheme uses step by step recursion, employing cells having identical form, in which each cell has two inputs—a direct and a recursion input, a delay, two multiply-and-adds, and two outputs, one going forward—the other going backward—forming a *lattice* structure (b).

6.2.2 CONCLUSION

Though it has been limited in scope, we hope that this section has provided the reader with an awareness of the power of digital techniques in signal processing, just through the use of the delay-line storage model. There are many more processing tricks available, if one is open to considering digital and analog, hardware and software, alone or in combination.

For use between the input a/d converter and the output DAC or display, the growing availability of digital components (see Chapter 21) of high complexity, increasing speed, and low cost (e.g., IC multiplier-accumulators and array processors), plus the availability of large amounts of memory and the possibility of overall control of the processing by CPUs and stored commands, has

*i.e., if the input is doubled, the corresponding output will be doubled.

[3]For an example, see Windsor and Toldalagi, "Digital FIR Filters without Tears", *Analog Dialogue* 17-2, 1983.

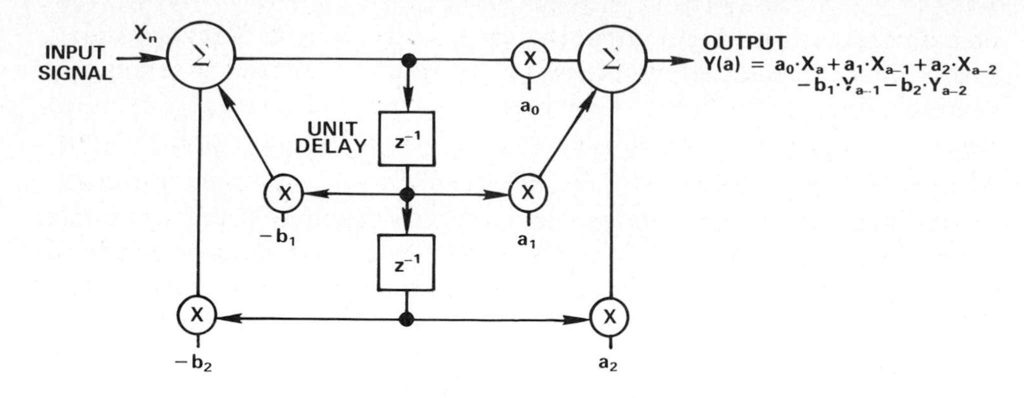

a. Feedback structure.

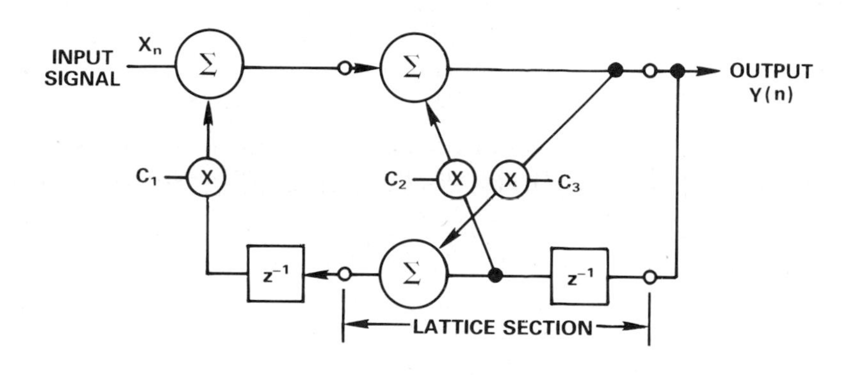

b. Lattice cell.

Figure 6.11. Two types of recursive filters.

made the outlook for analog waveform synthesis, analysis, and processing by digital techniques extremely bright, whatever the source. Speech, music, noise, gas chromatographs, electroencephalograms, image data from medical scans—whether NMR, ultrasound, or X-Ray—and mechanical vibrations are just a few.

6.3 CATHODE-RAY-TUBE DISPLAYS

In the industrial and scientific world, the close association of computer and cathode-ray tube provides an unparalleled method for speedy access to stored and real-time data, programs, and graphics. At the same time, it affords the opportunity for interactive dialogue with the computer, and through it, the system under test or control, for the purpose of actually instructing the computer or controlling real or simulated processes in real time. The recent growth of electron-beam recording (on film negative) poses a serious challenge to the centuries-old tradition of typesetting, while the ability to use computer power to adapt data to the needs of the human operator prior to presentation makes the computer-CRT display an all but indispensable combination.

While systems do exist for the sole purpose of display (i.e., monitors), the more general application of displays is in connection with data-acquisition systems and interactive systems involving computers. Some such systems deal with data that is purely alphanumeric by nature (e.g., accounting and word-processing systems). Others maintain contact with real-world physical variables through the use of a/d and d/a converters, either directly on-line in real time, or through recapitulation of data already captured. This category includes storage oscilloscopes and graphic-display scopes.

Whatever the display's purpose—or the source of the data—many cathode ray displays involve the use of D/A converters for generating sweeps, characters, and vectors, for positioning and intensification, relying on their inherent linearity, reproducibility, and controllability by entirely digital sources of command.

Since we are concerned here primarily with display systems that employ converters—with particular emphasis on the way they are used and the factors of importance in selecting and using them—the number of systems chosen will be limited and system descriptions will be brief.

The graphic-display oscilloscope can be found in increasingly widespread use—doubling every few years—in a growing variety of applications that were just wild dreams (if that) just a few years ago. Today's displays are characterized by high resolution, the promiscuous use of color, and a wide range of hardware and software options.

In general, a cathode-ray display system consists of a display-processing function and the CRT hardware, usually integrated into a single package. The processing function includes buffer storage to hold the information to be presented for update, the instructions for presenting it, the signals needed to activate the display elements, synchronization signals, and the digital-to-analog processing hardware; it may include a *refresh memory*. The CRT hardware comprises power supplies, CRT, circuitry for beam positioning intensification, nonlinearity correction, and focus.

Representative display techniques include:

- TV raster (picture and graphic displays)
- Stored-character display, e.g., Monoscope (alphanumerics)
- Dot-matrix (alphanumerics)
- Cursive: stroke and vector generators (alphanumerics and graphics)
- Rotating (PPI)

Reams of material have been written in recent years, debating the pros and cons of various CRT display systems. There is no "best" system for all purposes; they must be compared in terms of their advantages and disadvantages for specific applications.

Even the methods of deflecting the electron beam (or beams, for multicolor displays) have been subject to discussion. Systems with electromagnetic deflection use a magnetic yoke around the neck of the CRT to deflect the electron stream; systems with electrostatic deflection use voltages applied to sets of deflection plates built into the tube to steer the beam. Electromagnetic systems are slower and require more power than electrostatic systems, but they are cheaper and are widely used in TV-type raster-scan displays. Some systems use both—for example, electron-beam systems for integrated-circuit manufacture.

With computer processing, memory, and software widely available at low cost, computer-controlled displays have become popular in two basic forms: the *vector-refresh* (*alias* random-scan, calligraphic, stroke-writing, or directed-beam) display, which forms X-Y plots of digitally determined points or digitally programmed line-segments of random length and direction (in similar manner to oscilloscopes in the X-Y plot mode), and the TV-like *raster scan*, which places a large number of closely spaced dots of variable illumination—and color—along a raster of closely spaced horizontal lines every 1/30 or 1/60 of a second. Although usage of both techniques is growing, raster-scan is the more popular and growing at a much faster rate. Simplified traces characteristic of the two types of display are shown in Figure 6.12.

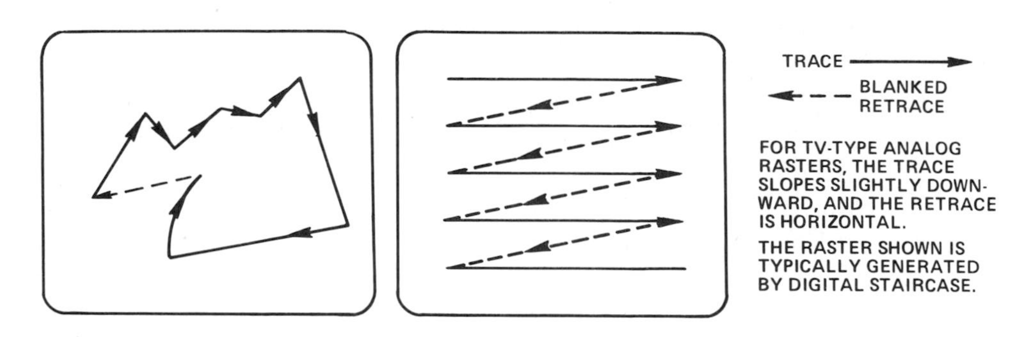

a. Randomly programmable vector-scan beam path.

b. Repetitive raster-scan beam path.

Figure 6.12. Display traces compared.

Vector-scan systems move the beam only to those portions of the screen forming the illuminated portion of the pattern, while the beam that illuminates the raster scans the entire screen, irrespective of the number and location of pixels to be brightened. In most applications, raster-scan displays use electromagnetic deflection (an inheritance from TV circuitry), while vector types use electrostatic deflection. The former has the advantages of simplicity and low cost; the latter requires less power, provides better resolution, and can display data changes with precision. Raster-scan systems are faster if large quantities of data are changing from frame to frame, while vector scan is often the choice where definition (i.e., precision and resolution) is important.

6.3.1 BASIC SYSTEM

Figure 6.13 shows the generalized system outline for an installation capable of accepting, processing, storing, and displaying information on a CRT screen. Conventional business data-processing systems do not normally in-

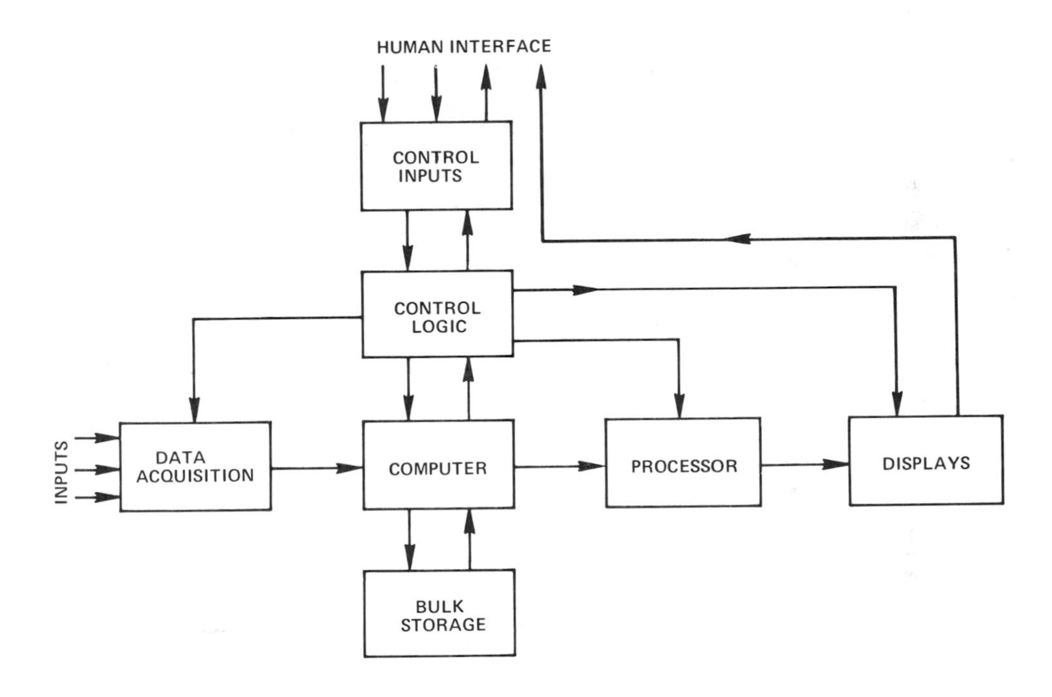

Figure 6.13. Display system outline.

volve sensors and A/D interfaces—but do involve other forms of peripheral data input. A computer game, employing a joystick for cursor position control, and an air traffic control system, based on radar data, are examples of ways CRT displays are used for interactive handling and presentation of information derived from the analog domain.

Further ingredients of the generalized display system are quite straightforward. The manual controls provide human interface, enabling the operator to call for a specific picture (or portion of a picture), to enter new information into the system, to command new modes of operation, and to initiate different data-processing and display functions. Bulk storage forms part of the data-processing capability; further auxiliary storage is often used for display refreshing at high speed to avoid annoying flicker. Control logic interfaces between computer data and the various peripheral devices, including displays, memories, communications links, the human operator, data-acquisition circuits, etc.

Driven to ever-higher scanning speeds to obtain increased resolution without flicker, the displays require extremely fast information transfer. For example, in a display with 1280×1024 fine structure, there are 1.31×10^6 picture elements (*pixels*). At 60 updates per second, each pixel must be displayed in less than 13 ns, if every data point in the field must be capable of being plotted, as is the case in raster graphics.

6.3.2 USES OF D/A CONVERTERS IN DISPLAYS

Raster Displays

Raster-scan graphic displays are very much like television pictures (and, in fact, often use TV hardware): as each vertical sweep scans linearly down the face of the tube, repeating its course every 1/30 or 1/60 of a second, the electron beam is repeatedly, rapidly, and linearly drawn across the face of the tube, forming a large number of equally spaced horizontal scans (lines), during which the signal information is modulating the intensity input of the cathode-ray tube. The major departure from TV is that, in high-resolution displays, there are many more horizontal lines, and more points per line. As in TV, the scans are synchronized, so that points brightened at the same time after the start of each horizontal sweep are directly above one another, and—if recurring on successive sweeps—will form a vertical line.

Because all points on the raster are scanned on each full cycle, the raster-scan cannot be as fast and sharp as the calligraphic display, especially for plotting simple figures; however the synchronization and standard nature of raster displays make it easier to obtain high-resolution shading, using the intensity gray scale, and to obtain a very large variety of color mixtures and hues, at low hardware cost.

One fast d/a converter channel provides intensity modulation for each electron gun: a single one for monochrome displays, three for color. Figure 6.14 is a block diagram of a typical computer-controlled raster-scan monochrome graphic display system.

A typical system comprises a MOS random-access memory buffer for storing display data in digital form, one or more memory controllers for managing the updating of the display and controlling the refresh cycle of the CRT, and a programmable microprocessor for generating display graphics and manipulating the image. The entire system operates as an intelligent peripheral

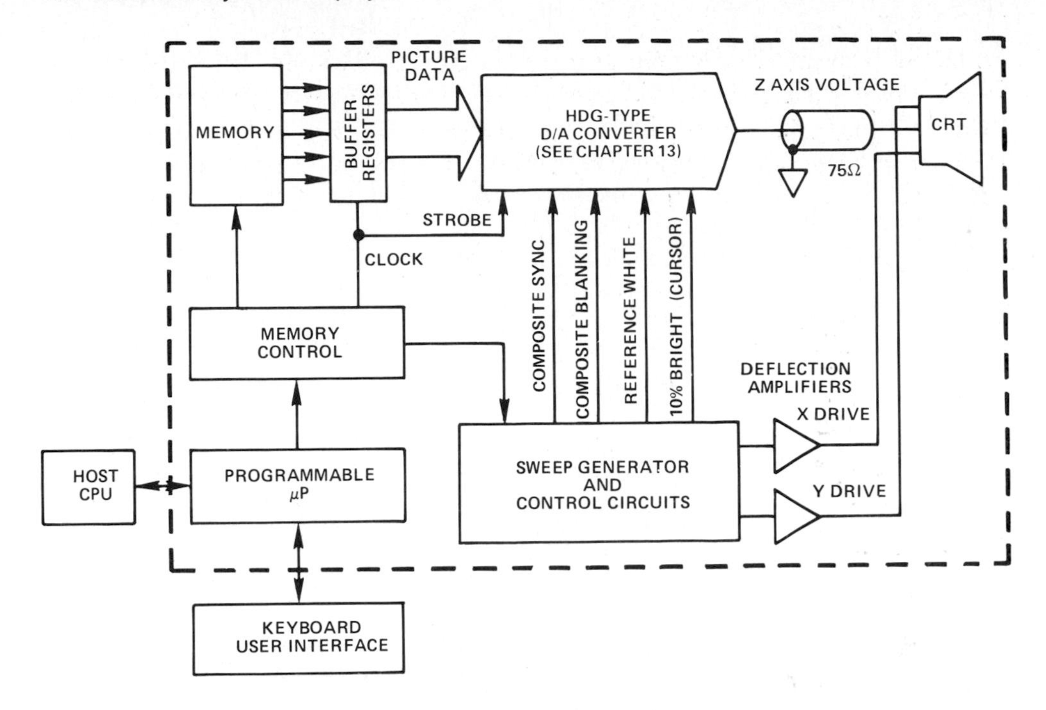

Figure 6.14. Raster-scan display system.

to a host computer; most of the processing associated with image and graphic display is down-loaded to the graphic subsystem.

If the picture resolution is specified as $1{,}024 \times 1{,}024$, there are 1,024 horizontal lines, each having 1,024 independent dots, and each dot has its own programmable intensity level; thus, there are 1,048,576 independent pixels. If the picture on the CRT is to be refreshed 60 times per second, then a new pixel must appear on the screen at least every 15.8 nanoseconds (exclusive of "overhead time" required for vertical and horizontal blanking during the sweep retrace portions of the cycle).

The d/a converter controls the Z-axis of the CRT, to modulate the brightness of the raster-scan beam. For the example mentioned above, the DAC must be capable of being updated at the pixel rate corresponding to 15.8 ns; it should be capable of settling to a new value in less than 10 ns. If a white dot is being plotted on a black background, the DAC's output must make a full-scale transition between adjacent pixels. The resolution of the DAC determines the number of finite intensity levels available. Typical DACs for this purpose have resolutions ranging from 4 to 8 bits, corresponding to from 16 to 256 levels of gray scale. For color displays, three memories and three DAC channels are required, one for each color gun of the CRT (red, green, blue). Triple DACs in monolithic form are becoming available for the purpose (e.g., the AD9702).

In addition to the programmed levels during the visible portion of the sweep, the electron beam must be blanked during the retrace. It is also useful to have an extra-bright level for the cursor in interactive applications. Besides a standard range of code-controlled intensity output levels, from reference black to reference white, special-purpose display DACs have control inputs that produce additional voltages to furnish standard levels of "blacker-than-black" for blanking and whiter-than-white for cursors.

Figure 6.15 shows the standard composite intensity waveform over 1 ½ cycles of the horizontal sweep. The controlled range of the DAC's full scale span (−643 mV) is from reference white (−71 mV) to reference black (−714 mV).

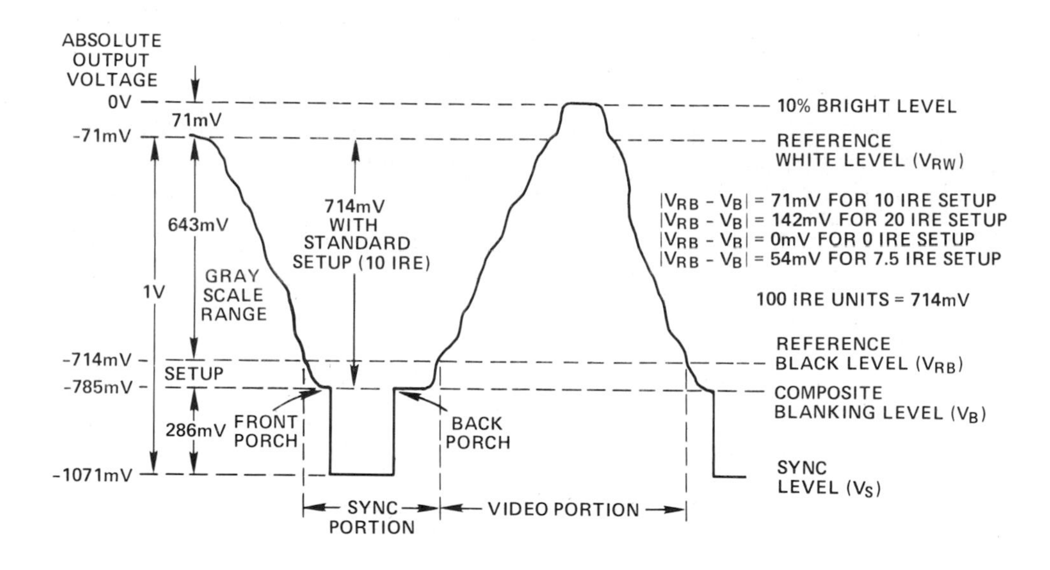

Figure 6.15. Composite DAC output waveform in raster-scan display system.

In the illustration, the intensity over the sweep interval is shown varying from full white to full black. At the beginning of the synchronized retrace portion of the cycle, the intensity signal drops to the blacker-than-black "front porch" (−785 mV), and then to the extreme black level (−1071 mV) during the horizontal retrace. As the next sweep starts, the intensity returns to the "back porch" (−785 mV) and, as the first element of the picture is triggered, to the controlled range of the DAC. During this scan, the cursor is displayed at the 10% "brighter than white" level (0 mV).

Sweeps may be produced either by analog ramps (integrating currents through capacitors) or by converting staircase outputs from digital counters to analog via DACs. D/A converters are especially well-suited to vertical sweeps, for a number of reasons:

- Timing, controlled by a clock and logic, is quite precise and uniform.
- Lines are horizontal (analog sweeps have slight tilt).
- Line-spacing uniformity depends on linearity, while maximum number of lines depends on DAC resolution. DACs having 10-bit-or-more resolution (1024+ lines) and 12-bit linearity (0.0125% linearity error) are readily available. In electron-beam recording, a 16-bit DAC can provide 4096 lines with less than 5% spacing error.
- DAC full-scale switching transients are blanked because they occur during the horizontal retrace interval.

In Section 5.5, a counter-driven D/A converter was suggested as a sawtooth sweep generator. When used for displays, such schemes can provide highly-repeatable, controllable, and linear sweeps of arbitrary resolution and accuracy.

For horizontal sweeps, the requirements on DACs are more severe, and analog sweeps win the cost tradeoff in many applications. For example, to resolve 500 points per line, at 500 lines per frame, at a 60-Hz frame rate, requires that each digital horizontal step settle well within 100ns, and that there be no "glitches," distorting the scan at major transitions and producing vertical stripes.

Vector Graphics

The general objective in vector graphic displays is to provide a flickerless high-resolution presentation of numerical, line-drawing, or pictorial information.

The usual problem is to start with the spot at a point having a given set of coordinates (which may be any point arrived at in the course of plotting the display, or it may be the starting point of a pattern), and plot a line to another point. A variety of methods have been used employing such analog techniques as modulating and summing ramps, performing integrations, etc. These methods, although using complex analog circuitry, are economical of digital data and resolution.

However, as fast deglitched d/a converters with high resolutions (12 bits and more) have become available, and as buffer memory becomes denser and cheaper, there is a trend towards defining all the data digitally and using point-by-point plotting.

Figure 6.16 is a block diagram of a typical vector-scan display system that plots point-by-point, using 12-bit DACs for the X and Y axes, and a fast DAC with coarser resolution to modulate the Z-axis (intensity). To obtain the perception of continuous lines when the display consists of discrete points, it is essential to use high-resolution d/a converters to drive the X- and Y-axis inputs. For example, a pair of 12-bit converters will provide a display of

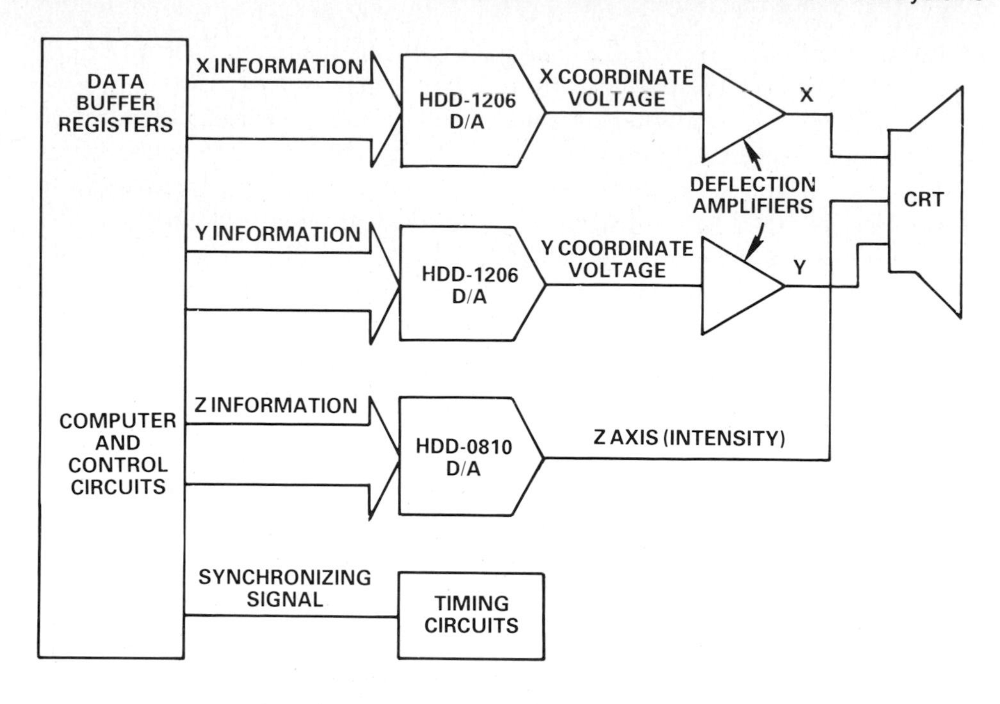

Figure 6.16. Typical vector display system.

4,096 × 4,096 bits, or about 16.8 million pixels. With a 21″ screen, resolution of about 0.1 mm would be available.

The high density of available pixels makes it mandatory that only high-resolution, high-quality "deglitched" DACs be used for plotting the digital data stored in memory. Any aberration in the DAC output, static or dynamic, can introduce an error in the output voltage and erroneously position the plotted point.

Figure 6.17 shows a simplified randomly programmable vector-scan beam path. The X-coordinate DAC moves the beam left-and-right; the Y-coordinate DAC moves it up-and-down. The beam is allowed to provide continuity between connected points, but it is blanked when it skips over a dark area. Errors in positioning in the X or Y direction produce lines with distorting lumps. D/A converter technology today is sufficiently advanced to ensure that problems with dc linearity and monotonicity are essentially nonexistent; so dc characteristics are not a significant source of errors in vector-scan displays.

However, besides high dc resolution, the d/a converters need a few other features. First and foremost is speed. The faster the DAC, the more points that can be illuminated within a given scan (usually 1/30 to 1/60 seconds, to avoid a flickering appearance). A converter—like the HDD-1206—with 60-nanosecond settling time for one-bit changes (while plotting a line) is compatible with a 6-MHz refresh rate; and only 2 μs are required for an accurate full-

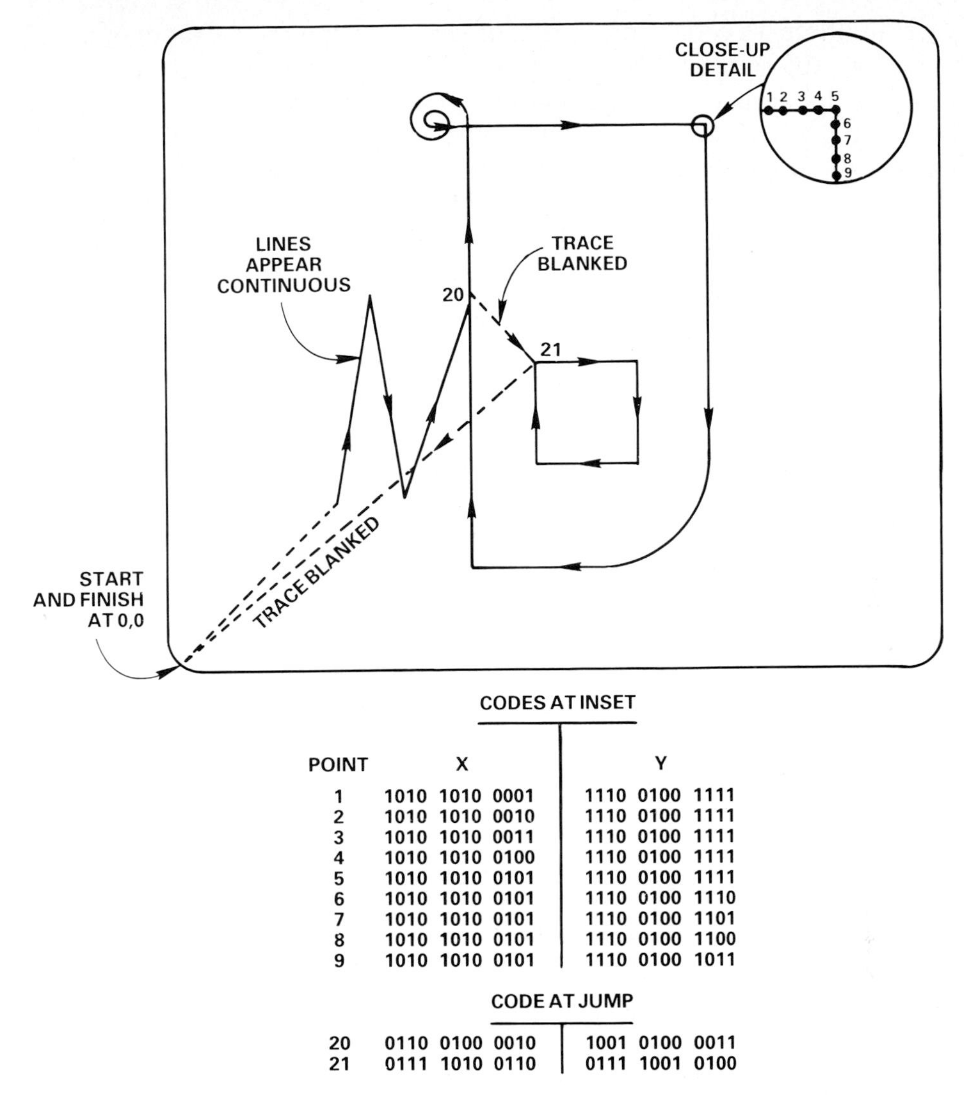

CODES AT INSET

POINT	X	Y
1	1010 1010 0001	1110 0100 1111
2	1010 1010 0010	1110 0100 1111
3	1010 1010 0011	1110 0100 1111
4	1010 1010 0100	1110 0100 1111
5	1010 1010 0101	1110 0100 1111
6	1010 1010 0101	1110 0100 1110
7	1010 1010 0101	1110 0100 1101
8	1010 1010 0101	1110 0100 1100
9	1010 1010 0101	1110 0100 1011

CODE AT JUMP

POINT	X	Y
20	0110 0100 0010	1001 0100 0011
21	0111 1010 0110	0111 1001 0100

Figure 6.17. Randomly programmable vector-scan beam path.

scale jump (less time for shorter jumps). Effective plotting strategies would call for many connected points and only short jumps, when feasible.

But high speed and high resolution are useless if the transition from point to point is not clean. A major problem in fast high-resolution DACs is the discontinuity, or "glitch" (Section 3.7), that can occur at major transitions, for example, in the one-bit transition from 0111 1111 1111 to 1000 0000 0000. If the less significant bits turn off faster than the MSB turns on, the output will swing rapidly towards zero, then rapidly swing back towards the original-level-plus-one-bit, causing the trace to swing wildly over the face of the CRT in the course of making that small change. Even at less-significant transitions, highly visible perturbations can occur.

Since the glitch is a code-sensitive nonlinear phenomenon, it cannot be simply filtered by linear techniques. It must first be minimized in duration and amplitude by designing the switching to be as symmetrical as possible; then the residual glitch can be eliminated by using a track-hold output amplifier circuit. When a new value of digital input is latched in, the output circuit switches to *hold* to retain the previous value until a time when the glitch can be expected to have settled out, then switches back to *track* to acquire the new value.

The "before" and "after" photographs in Figure 6.18 show the effects of glitches in a display and the great improvement that can be effected by deglitching. (Photos courtesy of The Foxboro Company.)

Figure 6.18. Effect of "glitches" on a display.

Dot-Matrix Displays

In vector graphics, elements that are repeated—for example, alphanumeric characters—need not have every point's complete address stored and traced out by the master X-Y DACs. Instead, a stored-character dot-matrix can be employed.

Each character is represented by a matrix of points, e.g., 4×7, with each point that is defined as part of the character intensified, by (for example) a character trace. The X and Y coordinates of each point of the character are located at addresses in two ROM's; the point is addressed by a word consisting of a format code for the character (e.g., ASCII) and a number from a counter indicating the order of the point in the writing sequence (i.e., i, in x_i,y_i).

In a typical system using this presentation (Figure 6.19), the outputs of two high-speed, low-resolution character DAC's (X and Y) are summed with the outputs of the main DACs. The main DACs establish the X and Y coordinates of the position of an index point on the character. The second set of DACs produce a sequential set of outputs that rapidly move the spot from one point to the next, dwell, and move on, until the character has been traced out.

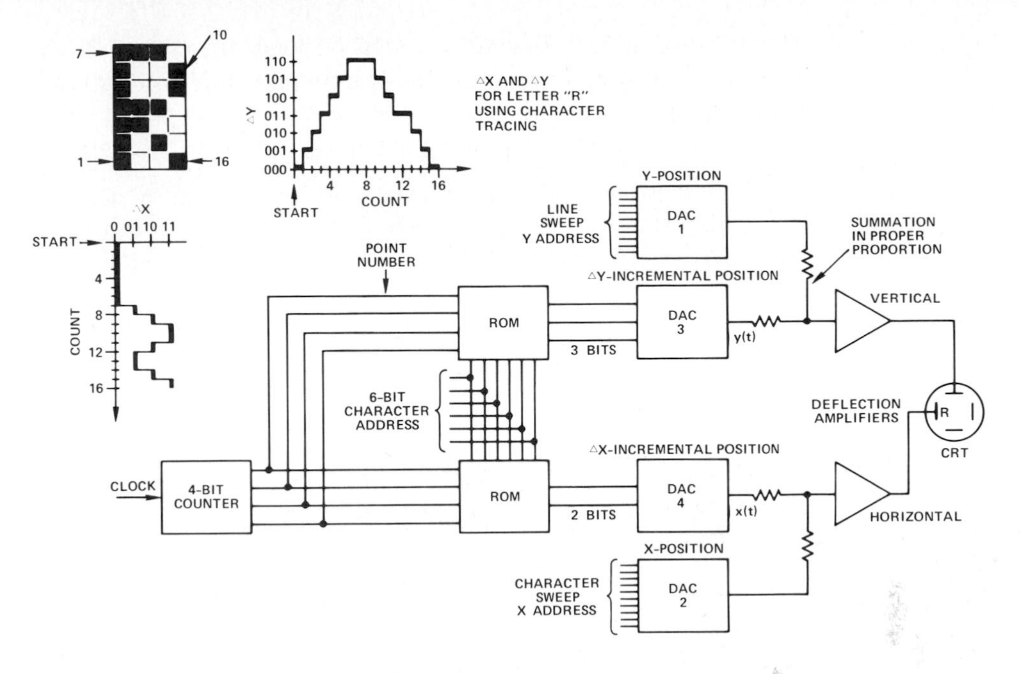

Figure 6.19. Dot-matrix display scheme.

An important advantage of this scheme is that the DACs that produce the characters need only have fast response (generally considerably faster than the main DACs), with very modest resolution and accuracy. In addition, the refresh memory needs to store only the character codes, rather than absolute point locations for each character.

Although 2 and 3 bits served to display the character adequately in the example, the D/A converter may have many more bits available for handling other forms of additive input. It is important to note, though, that the accuracy and resolution of the positioning DAC must be such that its errors are less than the relatively weighted value (taking differing scaling into account) of the least significant bits of DACs whose outputs are summed. Otherwise, overlapping or uneven spacing may result.

6.4 COMMERCE, INDUSTRY, AND ELSEWHERE

Because A/D and D/A converters were originally developed as computer interfacing devices, used primarily for getting data into and out of digital computers, the casual observer still tends to associate them with computer application alone. In reality, as Chapter 5 has demonstrated, A/D and D/A converters, as components, have followed the operational amplifier out of the computer laboratory and into the industrial world-at-large. But then, too, so has the computer, also as a component!

The reader who has arrived at this point (after presumably reading all of the material in Part One) has been exposed to a large variety of circuit configurations and application suggestions. It would not have been difficult to have noticed that some of the configurations looked more-or-less alike, though offered from somewhat different viewpoints.

In this section, closing the chapter and Part One, we will show just a few end-user applications, with the descriptive emphasis more on what they *accomplish*, rather than on how their circuits go together. The reader will not find anything especially different, from the circuit point of view, but it will serve as a microcosmic glimpse of end applications of conversion devices in the workaday world.

The applications include:

- Electronic instruments
- Medical imaging
- Industrial automation
- Oil-Well Monitoring

Electronic Instruments

Designers of test and measurement instruments, increasingly using automated measurement techniques that employ microprocessors, are finding a/d and d/a converters essential for direct measurements, setting gains, calibrating, and performing comparisons against standards. Electronic measurement techniques are used for accurately measuring virtually every conceivable parameter, including temperature, light, chemical composition, and pressure, including of course the familiar electrical parameters—as well as the performance characteristics of electronic equipment.

In performing measurements, an instrument's sources of error must, as a rule, be significantly less than other error sources in the measurement, or—for testing—in the device being tested. The concern for accuracy becomes particularly essential in the case of key components located within the critical measurement path.

A typical application where converter accuracy is a key to the desired measurement accuracy is found in sweep oscillators for wide-range testing of radio-frequency equipment. For example, Figure 6.20 is a simplified block diagram of Hewlett Packard's Model 8340A synthesized sweeper, a μP-controlled sweep generator that uses frequency synthesis techniques to generate signals ranging in frequency from 10 MHz to 26.5 GHz. Its output frequency can be swept over ranges up to the full frequency range of the instrument, making it adaptable to a wide variety of applications.

The value of the YIG-tuned output frequency range and offset, determined by the microprocessor, is established by a tuning voltage derived from an analog signal from a sweep-generator board and applied to the frequency generator. At the heart of that board (Figure 6.21) are three d/a converters—a 10-bit converter to control the gain of an integrator-generated sweep ramp

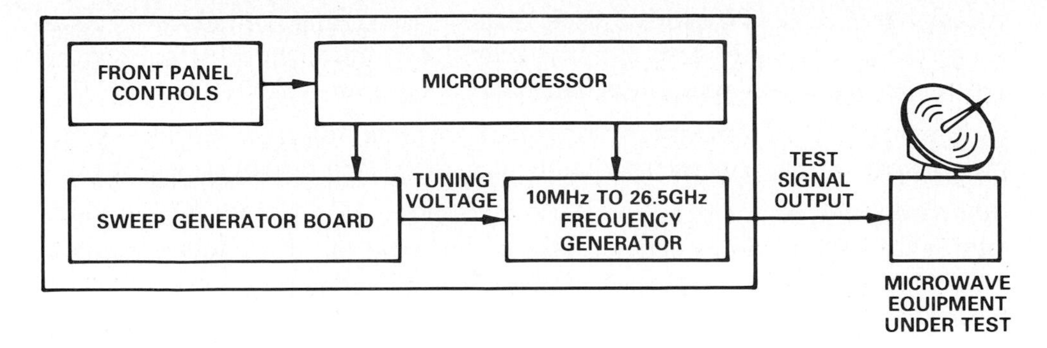

Figure 6.20. Synthesized sweeper using DACs to control the instrument's tuning voltage.

($\times 8$ and $\times 64$ gain switching provide an overall gain range of about 64,000); and a 12-bit (master) and an 8-bit (vernier) converter, used together—with 2-bit overlap—to provide an offset range of 262,000 values.

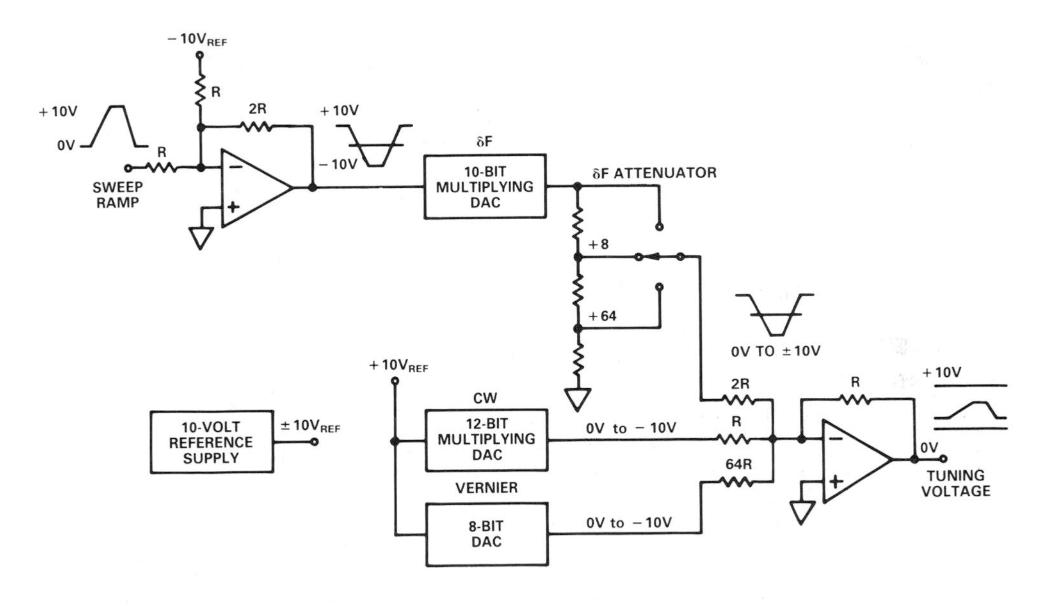

Figure 6.21. Sweep scaling and offset circuits condition the sweep ramp before it is sent to the frequency generator.

Medical Imaging

Since the time that German physicist Wilhelm Conrad Roentgen first performed his historic x-ray experiments in the late 1800s, medical innovators have sought continuously to increase the effectiveness of medical imaging methods for clinical diagnosis and therapy. Recent advances in imaging technology have provided diagnostic tools that are faster, safer, and more effective than ever before. In the area of x-ray technology, one of the most impressive recent developments is computerized axial tomography (CAT).

Computerized axial tomography is a technique for taking cross-sectional x-ray pictures of the body in planes perpendicular to the body's vertical axis.

Standard x-ray picture techniques, using photographic film, can diffuse x-ray energy throughout large portions of the body, including areas outside the area of immediate interest. The cumulative dosage over many pictures can be harmful to the patient. In addition, it is often difficult to interpret the pictures, because structures casting heavy shadows may be aligned with and block the images of other tissues that are of interest. This situation can be to some extent ameliorated if pictures could be taken from several different angles and then superimposed, eliminating redundant information, but the many pictures sometimes required could call for intolerable dosages of x-ray energy.

CAT scanning is a system approach that accomplishes the superposition of multiple images electronically, with considerably less irradiation of the patient—especially in portions of the body outside the area of interest. The major elements of the system are shown in Figure 6.22.

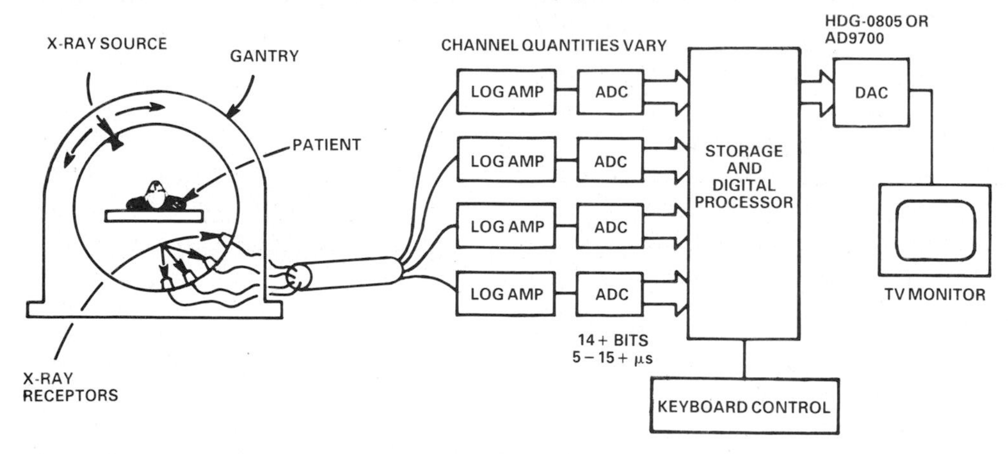

Figure 6.22. A/D conversion is the key to the effectiveness of CAT scanners. Once in digital form, the data can be processed to provide high-resolution pictures of cross-sections of internal body structures.

The x-ray source is mounted on a rotating cylinder within a large gantry, and the patient's body is centred within the cylinder. This positioning allows each successive exposure, measured with a set of sensitive x-ray receptors, to obtain data that can be accumulated and processed to produce a cross-sectional picture, or "slice," of the patient.

The scanner takes a picture in its initial position; the rotating cylinder, with source and receptors, is rapidly repositioned by 2° or 3°; a second picture is taken, and the process continues for a full 180°, requiring from 1 to 5 seconds for the complete cycle. After each "slice," the movable bed on which the patient is lying is incrementally advanced to a new position in the gantry. This allows multiple cross-section views to be stacked to form a three-dimensional picture of the part of the body being examined.

The analog data—collected by the receptors—for each picture are applied, via logarithmic amplifiers, to analog-to-digital converters. These converters often have 14+ bits of resolution and perform the digitizing function in 5 to 15 microseconds. The number of bits is dictated by the need for a wide dynamic range in order to observe small differences in image density that may be important in diagnosis.

After each 180° scan, the signal processor assembles the completed picture; it starts by noting the intercept points of the most dense tissue at each of the stopping points in the scan. Cross-correlation of these points allows the computer to assemble an extremely accurate picture of the interior of the body.

In addition to CAT scanning and various other approaches to medical imaging, there are many other occasions for data acquisition and conversion of real-world signals in health-care equipment, including patient-monitoring equipment, blood and body-fluid analysis systems, and other systems tht require measurements of such variables as body temperature, blood pressure, tissue structure, body chemical composition, and fluid flow.

Industrial Automation

Factory automation systems link automated, computer-controlled production- and materials-handling equipment with sophisticated communication networks. The continued viability of major segments of manufacturing industries worldwide will be dependent on the further development and implementation of factory automation products that assist in increasing productivity and quality, as well as lessening hazards in the workplace.

The current high level of interest in robotics technology is a direct result of the increased industrial manufacturing productivity it promises. Worldwide sales of industrial robots were expected to grow more than sixfold between 1983 and 1990.

Industrial robots are "intelligent" machines capable of independently performing various types of programmable factory tasks. They simplify many repetitive operations, such as sorting, inspection, and assembly, by putting those tasks under microprocessor control.

Typically, robots are required to perform two major types of tasks: those involving some kind of sensing, and those involving some form of locomotion or actuation. Whether the robot arm holds a drill, an arc welder, or a spray painter, the basic operation is the same: the robot must sense the object it needs to work on, it must locomote to that object, and it must quickly and efficaciously drill the desired hole, execute the desired weld, or coat the appropriate surface with a desired thickness of paint. Throughout this process, there is a feedback loop, which enables the robot to check its performance against the specified parameters. To effectively interface with its factory environment, the robot must operate with a very high degree of precision.

For example, the Mitsubishi Electric RW-1A is an arc-welding robot that has virtually the same manipulative ability with arc-welding tools as the human hand. In addition to the problem of positioning the arm, the robot must also maintain a desired profile of welding current. The welding current is sensed and converted to digital by a welding current sensor (Figure 6.23), using a

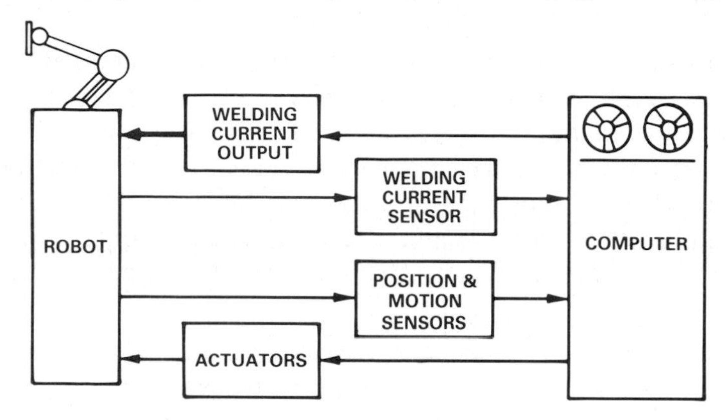

Figure 6.23. A 12-bit integrated-circuit ADC is used in the welding current sensor in monitoring and control of the robot's welding arm.

12-bit ADC, and the computer digests this feedback information and provides the adjustments necessary to maintain the specified values of current.

In addition to robotics, aspects of industrial automation providing opportunities for conversion devices include batch and continuous process control, and automatic inspection and test equipment.

Oil-Well Monitoring

High energy costs and growing concerns over the limits of the world's energy sources have led to the increasing use of automation as a means of using existing resources more efficiently and economically. This is especially true in the petroleum industry, where automation is being applied to all phases of production—recovery, refining, and distribution.

"Recovery" refers to the means by which oil and gas are brought to the surface once a producing well has been drilled. In most fields, the earlier wells produce by natural flow—that is, the initial pressure within the well is sufficient to cause the oil to flow to the surface without assistance. However, heavy crude oil does not flow readily at normal subterranean reservoir temperatures, and, as more oil is pumped from the ground, the pressure and natural flow of the well decrease. Natural recovery techniques are no longer effective for many of the mature oil fields in the United States; unenhanced processes typically recover only 15-30% of the oil in place.

The industry's push for increased production has led to enhanced oil recovery (EOR) projects, in which reservoir temperatures are raised through the use of steam injection. EOR is quite costly, so it is essential that field equipment be capable of effectively monitoring the pressure, temperature, and flow rates of the injected steam and the resulting fluids, to achieve the most efficient use of available material and economic resources. In the case of remote well sites, field equipment must also be capable of transmitting the data to a central location for analysis and control of the recovery operation.

As an example of this, a producing division of Texaco implemented remote monitoring on an EOR project involving 13 sites located more than 1 mile from the central computer. They wanted equipment at each well that could measure a broad range of critical parameters, and then convert the analog data into digital form for radio transmission back to the central computer. The means of performing the multiple-channel data acquisition—and remote control—was an integrated single-board measurement-and-control subsystem (see Table 4.1 and sections 4.1.6 and 4.3.3).

As Figure 6.24 shows, at each remote site, a μMAC-4000 accepts real-world signals from various sensors, such as thermocouples, flow meters, and pressure gages. These signals are filtered, amplified, and otherwise conditioned, then converted to a format which may be transmitted. Upon request from the host computer, the μMAC-4000 sends stored data via radio waves to the central location. Here, the data undergoes additional processing and analysis before the information is presented by the computer in printout form. If an engineer's review of the printout suggests that an adjustment in the EOR process is desirable, instructions can then be sent from the central computer back to the μMAC-4000, which will issue control commands to the appropriate valves and motors.

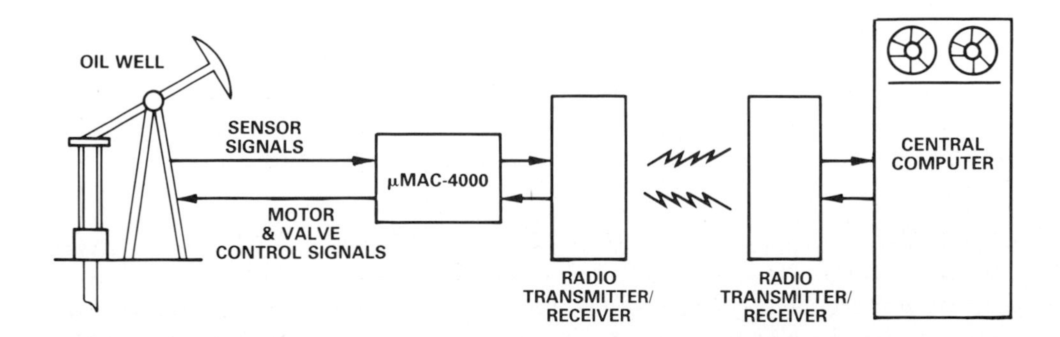

Figure 6.24. An integrated measurement-and-control subsystem provides effective monitoring of the enhanced oil-recovery process, ensuring efficient resource allocation.

The need for computer-aided measurement-and-control solutions, which involve conversion, in all levels of integration—from signal conditioning and conversion to complete stand-alone measurement-and-control systems, replete with field sensor connections and sophisticated software—exists wherever there is a need for an interface between the real-world process being measured and the information system. Applications for such subsystems include, not only energy conservation and development, but industrial process control, machine control, product test, laboratory research and development, and much more.

PART II

A/D AND D/A CONVERTERS

Chapter Seven

Understanding Converters

A/D converters translate from analog measurements, which are characteristic of most phenomena in the "real world," to digital language, used in information processing, computing, data transmission, and control systems. D/A converters are used in transforming transmitted or stored data, or the results of digital processing, back to "real-world" variables for control, information display, or further analog processing.

7.1 BINARY CODES AND CONVERSION RELATIONSHIPS

7.1.1 ANALOG QUANTITIES

Analog input variables, whatever their origin, are most frequently converted by transducers into voltages or currents. These electrical quantities may appear as fast or slow "dc" continuous direct measurements of a phenomenon in the time domain, as modulated ac waveforms (using a wide variety of modulation techniques), or in some combination, with a spatial configuration of related variables to represent shaft angles. Examples of the first are outputs of thermocouples, potentiometers on dc references, and analog computing circuitry; of the second, "chopped" optical measurements, ac strain gage or bridge outputs, and digital signals buried in noise; and of the third, synchros and resolvers.

The analog variables to be dealt with in this chapter are those involving "dc" voltages or currents representing the actual analog phenomena. They may be either wideband or narrow-band. They may be either scaled from the direct measurement, or subjected to some form of analog pre-processing, such as linearization, combination, demodulation, filtering, sample-hold, etc. As part of the process, the voltages and currents are "normalized" to ranges compatible with assigned converter input ranges. Ways and means of accomplishing appropriate pre-processing, including floating-point scaling, are discussed in the chapters on applications and system accessories. Analog output voltages or currents from D/A converters are direct and in normalized form, but they may be subsequently post-processed (e.g., scaled, filtered, boosted, etc.).

This chapter does not include the conversion of signals from resolvers and synchros—widely used in some control applications. Relevant material on this topic will be found in Chapter 14.

7.1.2 DIGITAL QUANTITIES

Information in digital form is represented by arbitrarily fixed voltage levels referred to "ground," either occurring at the outputs of logic gates, or applied to their inputs. The digital numbers used are all basically binary (in the sense of either-or); that is, each "bit," or unit of information has one of two possible states. These states are "off," "false," or "0," and "on," "true," or "1."

Words are groups of levels representing digital numbers; the levels may appear simultaneously in *parallel*, on a bus or groups of gate inputs or outputs, *serially* (or in a time sequence) on a single line,* or as a sequence of parallel bytes (i.e., "byte-serial") or nybbles (small bytes). For example, a 16-bit word may occupy the 16 bits of a 16-bit bus, or it may be divided into two sequential bytes for an 8-bit bus, or four 4-bit nybbles for a 4-bit bus.

Although there are several systems of logic, the most widely used choice of levels are those used in TTL (transistor-transistor logic), in which positive *true*, or 1, corresponds to a minimum output level of +2.4V (inputs respond unequivocally to "1" for levels greater than 2.0V); and *false*, or 0, corresponds to a maximum output level of +0.4V (inputs respond unequivocally to "0" for anything less than +0.8V). A unique parallel or serial grouping of digital levels, or a *number*, or *code*, is assigned to each analog level which is quantized (i.e., represents a unique portion of the analog range). A typical digital code would be this array:

1 0 1 1 1 0 0 1

It is composed of eight bits. The "1" at the extreme left is called the "most significant bit" (MSB, or Bit 1), and the one at the right is called the "least significant bit" (LSB, or bit *n*: 8 in this case). The meaning of the code, as either a number, a character, or a representation of an analog variable, is unknown until the *code* and the *conversion relationship* have been defined.

*In serial data transmission, if the levels return to ground between successive bits, they are denoted RZ (return-to-zero); if they change only when the leading or trailing edge of a clock pulse is present, and remain until the next such edge, they are denoted NRZ (non-return-to-zero).

7.1.3 BINARY CODE—INTEGERS AND FRACTIONS

The best-known code is *natural binary* (base 2). Binary codes are most familiar in representing integers; i.e., in a natural binary integer code having n bits, the LSB has a weight of 2^0 (i.e., 1), the next bit has a weight of 2^1 (i.e., 2), and so on up to the MSB, which has a weight of 2^{n-1} (i.e., $2^n/2$). The value of a binary number is obtained by adding up the weights of all non-zero bits. When the weighted bits are added up, they form a unique number having any value from 0 to $2^n - 1$. Each additional trailing zero bit, if present, essentially doubles the size of the number.

In converter technology, because full scale (i.e., the converter's reference) is independent of the number of bits of resolution, a more useful coding is *fractional* binary, which is always normalized to full scale. Integer binary can be interpreted as fractional binary if all integer values are divided by 2^n. For example, the MSB has a weight of ½ (i.e., $2^{(n-1)}/2^n = 2^{-1}$), the next bit has a weight of ¼ (i.e., 2^{-2}), and so forth down to the LSB, which has a weight of $1/2^n$ (i.e., 2^{-n}). When the weighted bits are added up, they form a number with any of 2^n values, from 0 to $(1-2^{-n})$ of full-scale. Additional bits simply provide more fine structure without affecting full-scale range. To illustrate these relationships, Table 7.1 lists the 16 permutations of 4-bits' worth of 1's and 0's, with their binary weights, and the equivalent numbers expressed as both decimal and binary integers and fractions.

When all bits are "1" in natural binary, the fractional number value is $1-2^{-n}$, or normalized full-scale less 1 LSB ($1 - 1/16 = 15/16$ in the example). Strictly

Decimal Fraction	Binary Fraction	Code: MSB (× 1/2)	Code: Bit 2 (× 1/4)	Code: Bit 3 (× 1/8)	Code: Bit 4 (× 1/16)	Binary Integer	Decimal Integer
0	0.0000	0	0	0	0	0 0 0 0	0
$1/16 = 2^{-4}$ (LSB)	0.0001	0	0	0	1	0 0 0 1	1
2/16 = 1/8	0.0010	0	0	1	0	0 0 1 0	2
3/16 = 1/8 + 1/16	0.0011	0	0	1	1	0 0 1 1	3
4/16 = 1/4	0.0100	0	1	0	0	0 1 0 0	4
5/16 = 1/4 + 1/16	0.0101	0	1	0	1	0 1 0 1	5
6/16 = 1/4 + 1/8	0.0110	0	1	1	0	0 1 1 0	6
7/16 = 1/4 + 1/8 + 1/16	0.0111	0	1	1	1	0 1 1 1	7
8/16 = 1/2 (MSB)	0.1000	1	0	0	0	1 0 0 0	8
9/16 = 1/2 + 1/16	0.1001	1	0	0	1	1 0 0 1	9
10/16 = 1/2 + 1/8	0.1010	1	0	1	0	1 0 1 0	10
11/16 = 1/2 + 1/8 + 1/16	0.1011	1	0	1	1	1 0 1 1	11
12/16 = 1/2 + 1/4	0.1100	1	1	0	0	1 1 0 0	12
13/16 = 1/2 + 1/4 + 1/16	0.1101	1	1	0	1	1 1 0 1	13
14/16 = 1/2 + 1/4 + 1/8	0.1110	1	1	1	0	1 1 1 0	14
15/16 = 1/2 + 1/4 + 1/8 + 1/16	0.1111	1	1	1	1	1 1 1 1	15

Table 7.1 Integer and fractional binary codes.

speaking, the number that is represented, written with an "integer point," is 0.1111 (= 1 – 0.0001). However, it is almost universal practice to write the code simply as the integer 1111 (i.e., "15") with the fractional nature of the corresponding number understood: "1111" ⟶ 1111/(1111 + 1), or 15/16.

For convenience, Table 2 lists bit weights in binary for numbers having up to 20 bits. The practical range for the vast majority of applications is about 16 bits; for numbers of bits than greater 20, continue to divide by 2.

BIT	2^{-n}	$1/2^n$ (Fraction)	"dB"	$1/2^n$ (Decimal)	%	ppm
FS	2^0	1	0	1.0	100	1,000,000
MSB	2^{-1}	1/2	–6	0.5	50.	500,000
2	2^{-2}	1/4	–12	0.25	25	250,000
3	2^{-3}	1/8	–18.1	0.125	12.5	125,000
4	2^{-4}	1/16	–24.1	0.0625	6.2	62,500
5	2^{-5}	1/32	–30.1	0.03125	3.1	31,250
6	2^{-6}	1/64	–36.1	0.015625	1.6	15,625
7	2^{-7}	1/128	–42.1	0.007812	0.8	7,812
8	2^{-8}	1/256	–48.2	0.003906	0.4	3,906
9	2^{-9}	1/512	–54.2	0.001953	0.2	1,953
10	2^{-10}	1/1,024	–60.2	0.0009766	0.1	977
11	2^{-11}	1/2,048	–66.2	0.00048828	0.05	488
12	2^{-12}	1/4,096	–72.2	0.00024414	0.024	244
13	2^{-13}	1/8,192	–78.3	0.00012207	0.012	122
14	2^{-14}	1/16,384	–84.3	0.000061035	0.006	61
15	2^{-15}	1/32,768	–90.3	0.0000305176	0.003	31
16	2^{-16}	1/65,536	–96.3	0.0000152588	0.0015	15
17	2^{-17}	1/131,072	–102.3	0.00000762939	0.0008	7.6
18	2^{-18}	1/262,144	–108.4	0.000003814697	0.0004	3.8
19	2^{-19}	1/524,288	–114.4	0.000001907349	0.0002	1.9
20	2^{-20}	1/1,048,576	–120.4	0.0000009536743	0.0001	0.95

Table 7.2 Binary bit weights or resolution.

The weight assigned to the LSB is the *resolution* inherent in numbers having *n* bits. The "dB" column represents the logarithm (base 10) of the ratio of the LSB value to unity (full scale), multiplied by 20, in the popular manner. Each successive power of 2 represents a change of 6.02dB [i.e., $20 \log_{10}(2)$] or "6dB/octave."

In natural binary, the normalized numerical value of 1 0 1 1 1 0 0 1, an 8-bit code, would be

INTEGER:

$$2^7 + 2^5 + 2^4 + 2^3 + 2^0$$

$$128 + 32 + 16 + 8 + 1 = 185 \qquad (7.1)$$

FRACTION:

$$\frac{2^7}{2^8} + \frac{2^5}{2^8} + \frac{2^4}{2^8} + \frac{2^3}{2^8} + \frac{2^0}{2^8}$$

$$\frac{128}{256} + \frac{32}{256} + \frac{16}{256} + \frac{8}{256} + \frac{1}{256} = \frac{185}{256}$$

$$2^{-1} + 2^{-3} + 2^{-4} + 2^{-5} + 2^{-8}$$

$$\frac{1}{2} + \frac{1}{8} + \frac{1}{16} + \frac{1}{32} + \frac{1}{256} = \frac{185}{256}$$

$$0.5 + 0.125 + 0.0625 + 0.0313 + 0.0039 = 0.7227 \qquad (7.2)$$

Bit numbering for microprocessor buses is based on whole numbers, not binary fractions. In such systems, the LSB is always Bit 0 (viz., 2^0), the MSB is always Bit $n-1$ (viz., 2^{n-1}). Setting up a correspondence between the bit numbers used in words in the two systems,

Integral
  Bit $n-1$: (2^{n-1}), Bit $n-2$: (2^{n-2}), . . . Bit $n-n$ (i.e., 0): ($2^0 = 1$)
Fractional
  Bit 1: (2^{-1}), Bit 2: (2^{-2}), . . . Bit n: (2^{-n})

Thus, the bit number is equal to the logarithm of its weight, base 2, in integral binary, and to the negative of the logarithm of its weight in fractional binary.

Since the binary point is to the right of the LSB in integral binary, additional bits are always added to the left; integral binary numbers are said to be *right-justified*. On the other hand, in fractional binary, additional bits are always added to the right; fractional binary numbers are *left-justified*. This is of importance when a number is represented by two bytes, for example, when a 12-bit number must be placed on or retrieved from an 8-bit bus. If the binary number is left-justified, the more-significant byte has the first 8 bits, starting with the MSB, and the less-significant byte has the last 4 bits and 4 trailing zeros; for right-justified numbers, the more-significant byte has 4 leading zeros and the four more-significant bits, while the less-significant byte has the 8 less-significant bits.

7.1.4 BASIC CONVERSION RELATIONSHIPS

Perhaps a graph is the most fruitful way of indicating in detail the relationship between analog and digital quantities in a conversion. Since there are two complementary conversion relationships to be discussed, two graphs must be plotted, one for A/D conversion, the other for D/A conversion.

Figure 7.1 shows the graph for an ideal 3-bit D/A converter. A 3-bit converter has 8 discrete coded levels, thus a total of 8 different inputs and 8 corresponding outputs, ranging from zero to ⅞ of "full scale." While full scale is not available digitally, it represents the reference quantity to which the analog

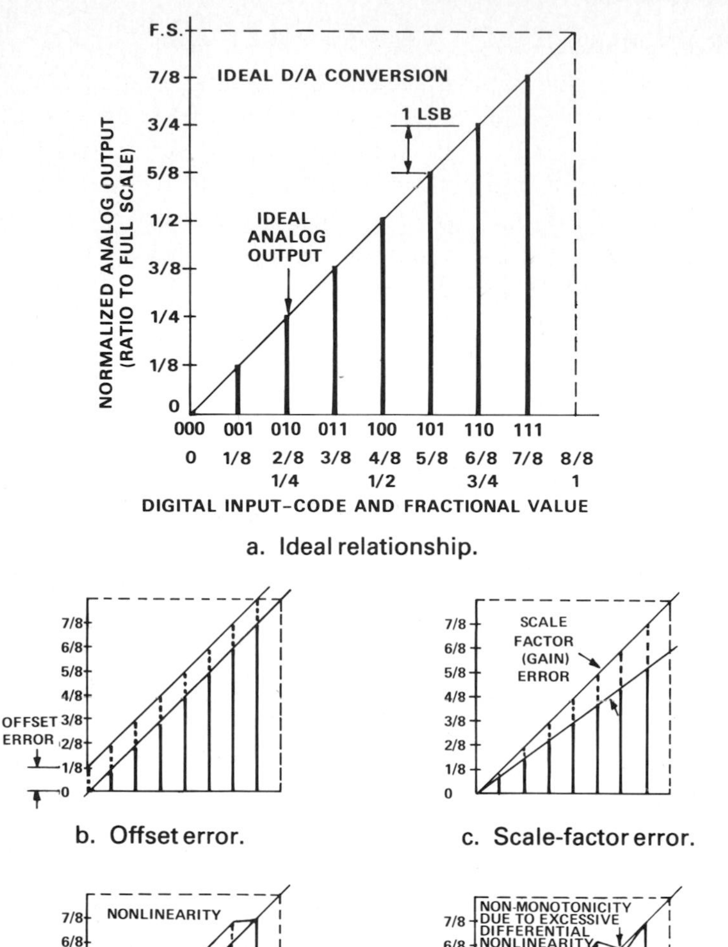

a. Ideal relationship.

b. Offset error.

c. Scale-factor error.

d. Linearity error.

e. Non-monotonic behavior (due to excessive *differential* nonlinearity).

Figure 7.1. Conversion relationships in a 3-bit d/a converter, showing ideal relationship and typical sources of error.

variable is normalized (the *full-scale range* (or *span*) is 0 to 1, not 0 to 7/8). Since only the eight coded levels can exist, Figure 7.1 is plotted as a column graph.

Practical D/A converters have errors. For example, the zero column may not be exactly zero, giving rise to *offset error*. The extrapolated range from zero to F.S. may not be exactly as specified; this scale-factor error is often called *gain error*. The differences between the heights of adjacent bars may not be equal or changing uniformly (*nonlinearity*), and—in fact—if the *differential* nonlinearity (difference between adjacent heights and 1 LSB) is sufficiently negative, the device may be *non-monotonic* (one or more values of analog output may actually be less than the values corresponding to codes having smaller weight). Even if differential linearity errors are within specification, there may be—for example—a gradually increasing nonlinearity error that becomes large between half scale and full scale; errors of this kind are called *integral nonlinearity* errors. The above errors (and others), the means of specifying and testing them, and some of the design techniques for keeping them small, are discussed in chapters 8, 9, 10, and 11.

To visualize the ideal performance of converters having larger numbers of bits, one may intensify this pattern by interpolating additional columns between the columns of this graph. For example, a fourth bit would require 8 additional columns with heights halfway between the levels indicated. The value of the LSB would be F.S./16, and the maximum value would be $7/8 + 1/16 = 15/16$ F.S. The next bit would interpolate 16 additional columns, the new LSB would be F.S./32, and the maximum value would be $31/32$, etc. The straight line connecting the tops of the columns is the locus of the *envelope* of the ideal conversion relationship.

Figure 7.2 shows the graph for an ideal 3-bit A/D converter. Since all values of the analog input are presumed to exist, they must be *quantized* by partitioning the continuum into 8 discrete ranges. All analog values within a given range are represented by the same digital code, which generally corresponds to the nominal mid-range value. These mid-range values correspond to the bar heights of the D/A converter.

There is, therefore, in the A/D conversion process, an inherent *quantization uncertainty* of $\pm 1/2$ LSB, in addition to the conversion errors analogous to those existing for the D/A converter. The only sure way to reduce this quantization uncertainty—which is like a roundoff or truncation error—is to increase the number of bits of resolution. (There are, of course, statistical interpolation tricks that may be performed in the digital processing or in analog filtering following subsequent D/A conversion, which will fill in probable analog values for large, rapidly varying or repetitive signals, but they will do nothing to indicate the variations within a quantum for an apparently constant digital number.)

Since it is easier* to determine the location of a *transition* than it is to determine a mid-range value, errors and settings of A/D converters are defined and measured in terms of the analog values at which transitions occur, in relation to the ideal transition values. Like D/A converters, A/D converters have offset error: the first transition may not occur at exactly $+1/2$ LSB; scale-factor (or gain) error: the difference between the values at which the first transition and the last transition occur is not equal to (F.S. – 2LSB); and *linearity* error: the differences between transition values are not all equal or uniformly changing.

*(using analog techniques)

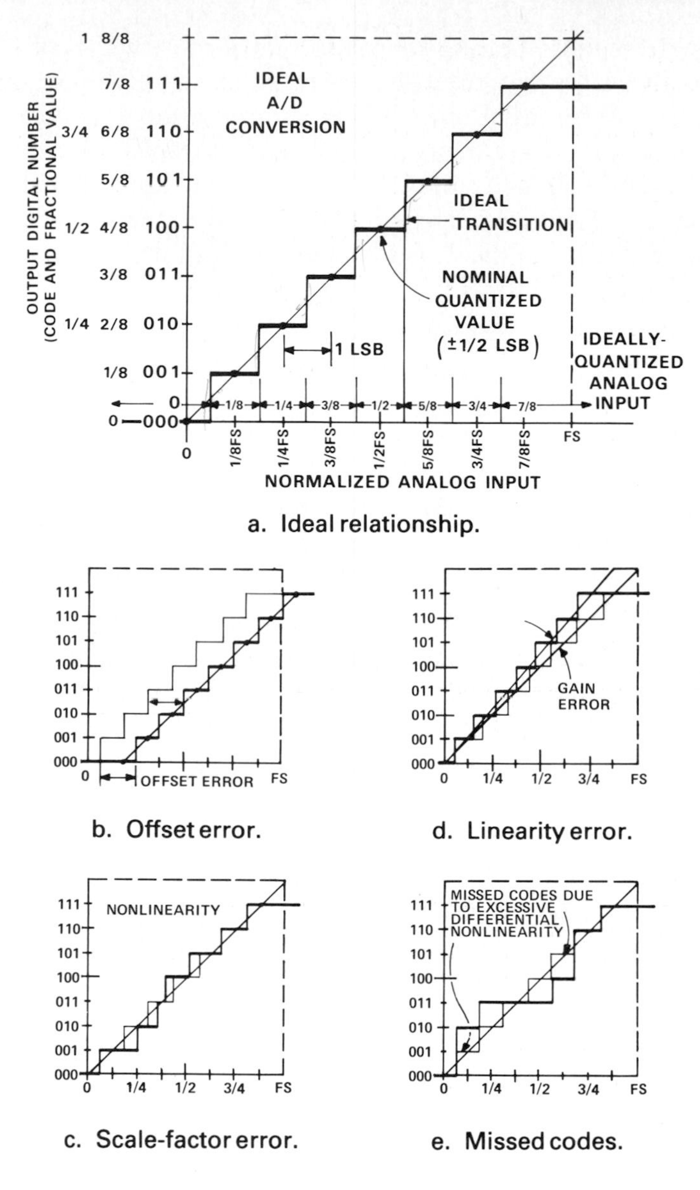

a. Ideal relationship.

b. Offset error.

d. Linearity error.

c. Scale-factor error.

e. Missed codes.

Figure 7.2. Conversion relationships in a 3-bit a/d converter, showing ideal relationship and typical sources of error.

If the *differential linearity* error is large enough, it is possible for one or more codes to be missed (the counterpart of non-monotonic D/A conversion).

An important factor in the conversion relationship is the choice of "Full Scale," the LSB magnitude, and the transition points. For a great many converters, full scale is in the vicinity of 10 volts: either exactly at 10V or at 10.24V. For 10V, the bit values are easily expressed as negative powers of 2, multiplied by 10; for 10.24V, the LSB can be expressed in "round" numbers, being a multiple or submultiple of 10mV.

Table 7.3 lists the LSB values, the "all 1's" value (i.e., F.S. – 1 LSB), and the A/D converter transition values at ½ LSB (for zero adjustment) and all

	10V Full Scale				10.24V Full Scale			
			A/D Transitions				A/D Transitions	
No. of Bits n	LSB	All 1's (Volts)	To LSB (1/2LSB)	To All 1's (Volts)	LSB	All 1's (Volts)	To LSB (+1/2 LSB)	To All 1's (Volts)
1	5V	5.0	2.5V	2.5	5.12V	5.12	2.56V	2.56
2	2.5V	7.5	1.25V	6.25	2.56V	7.68	1.28V	6.40
3	1.25V	8.75	625mV	8.13	1.28V	8.96	640mV	8.32
4	625mV	9.38	312mV	9.07	640mV	9.60	320mV	9.28
5	312mV	9.69	156mV	9.53	320mV	9.92	160mV	9.76
6	156mV	9.84	78.1mV	9.76	160mV	10.08	80mV	10.00
7	78.1mV	9.92	39.1mV	9.88	80mV	10.16	40mV	10.12
8	39.1mV	9.961	19.5mV	9.941	40mV	10.20	20mV	10.18
9	19.5mV	9.980	9.77mV	9.970	20mV	10.220	10mV	10.21
10	9.77mV	9.990	4.88mV	9.985	10mV	10.230	5mV	10.225
11	4.88mV	9.9951	2.44mV	9.9927	5mV	10.235	2.5mV	10.232
12	2.44mV	9.9976	1.22mV	9.9964	2.5mV	10.2375	1.25mV	10.2362
13	1.22mV	9.9988	610μV	9.9982	1.25mV	10.2388	625μV	10.2382
14	610μV	9.9994	305μV	9.9991	625μV	10.2394	312μV	10.2391
15	305μV	9.99970	153μV	9.99955	312μV	10.23969	156μV	10.23953
16	153μV	9.99985	76μV	9.99977	156μV	10.23984	78.1μV	10.23976
17	76μV	9.99992	38μV	9.99988	78.1μV	10.23992	39.1μV	10.23988
18	38μV	9.999962	19μV	9.999943	39.1μV	10.239961	19.5μV	10.239941
19	19μV	9.999981	9.5μV	9.999971	19.5μV	10.239980	9.77μV	10.239970
20	9.5μV	9.999990	4.8μV	9.999985	9.77μV	10.239990	4.88μV	10.239985

Table 7.3 LSB and (FS – LSB) values for 10V and 10.24V conversion.

1's (F.S. – 1 ½ LSB, for scale factor adjustment) for resolutions to 2^{-20}, for both 10V and 10.24V full scale. If full scale is 5V (also a popular value), simply divide the appropriate numbers by 2.

7.2 OTHER CODES

Although binary is the most commonly used code, there are a number of other popular codes used at system interfaces, depending on signal range and polarity, conversion technique, specially desired characteristics, and origin or destination of digital information.

7.2.1 BINARY-CODED DECIMAL (BCD)

This is a code in which each decimal digit is represented by a group of 4 binary-coded digits (or "quad"). In fractional BCD, the LSB of the most significant quad has a weight of 0.1, the LSB of the next has a weight of 0.01, the LSB of the next has a weight of 0.001, etc. Each quad has 10 permissible levels with weights 0 to 9. Group values in excess of 9 are not permitted. Table 7.4 gives examples of BCD coding for a variety of numbers between 0 and 0.99.

A/D converters with the BCD code are used primarily in digital voltmeters and panel meters, since each quad's output may be decoded to drive a numeric display using the familiar decimal numbers. If the display is of a BCD digitally

Decimal Fraction	BCD Code MSQ (×1/10) ×8 ×4 ×2 ×1	BCD Code 2nd Quad (×1/100) ×8 ×4 ×2 ×1
0.00 = 0.00 + 0.00	0 0 0 0	0 0 0 0
0.01 = 0.00 + 0.01	0 0 0 0	0 0 0 1
0.02 = 0.00 + 0.02	0 0 0 0	0 0 1 0
0.03 = 0.00 + 0.03	0 0 0 0	0 0 1 1
0.04 = 0.00 + 0.04	0 0 0 0	0 1 0 0
0.05 = 0.00 + 0.05	0 0 0 0	0 1 0 1
0.06 = 0.00 + 0.06	0 0 0 0	0 1 1 0
0.07 = 0.00 + 0.07	0 0 0 0	0 1 1 1
0.08 = 0.00 + 0.08	0 0 0 0	1 0 0 0
0.09 = 0.00 + 0.09	0 0 0 0	1 0 0 1
0.10 = 0.10 + 0.00	0 0 0 1	0 0 0 0
0.11 = 0.10 + 0.01	0 0 0 1	0 0 0 1
:		
0.20 = 0.20 + 0.00	0 0 1 0	0 0 0 0
:		
0.30 = 0.30 + 0.00	0 0 1 1	0 0 0 0
:		
0.90 = 0.90 + 0.00	1 0 0 1	0 0 0 0
0.91 = 0.90 + 0.01	1 0 0 1	0 0 0 1
:		
0.98 = 0.90 + 0.08	1 0 0 1	1 0 0 0
0.99 = 0.90 + 0.09	1 0 0 1	1 0 0 1

Table 7.4 Examples of 2-digit BCD weighting.

transmitted or processed number, or if the input is via a thumbwheel switch, a D/A converter that responds to BCD may be used to furnish a base-10 analog output from its digital input.

BCD is somewhat wasteful of bits, in the sense that each BCD quad has 10/16 the resolution of a comparable natural binary quad. Table 7.5 shows the relative resolution capability.

Number of Bits	Least Significant Bit Binary	Least Significant Bit BCD	Number of Binary Bits Needed For Same Resolution as BCD
4	0.062	0.1	4
8	0.0039	0.01	7
12	0.00024	0.001	10
16	0.000015	0.0001	14
20	0.000001	0.00001	17
24	0.0000002	0.000001	20

Table 7.5 Relative resolution of BCD and binary.

OVERRANGING

Many BCD A/D converters have an additional bit with weight equal to full scale, in a position "more significant" than the MSB.

This additional bit provides a maximum of 100% "overrange" capability. Additional "super-significant" bits would provide binary 300% (2 bits) and 700% (3 bits) overrange capability (or extend the range to nearly 800% of the

BCD "full scale"). Depending on how it is presented, the overrange bit is used in digital voltmeters and panel meters either to provide additional resolution or to indicate that nominal full scale has been exceeded and that the visual reading may be erroneous.

Overrange bits need not be restricted to BCD. They are useful as "flags" in any conversion process for which an overrange input would given an ambiguous reading, or where an overrange input indicates anomalous analog system behavior. The overrange bit must of course be of suitable accuracy, since it is, in effect, the MSB.

7.2.2 GRAY CODE

In Gray codes, each bit represents a binary-weighted segment of the range, and each code corresponds to a unique location in the range; but the bit weights do not readily combine to form a binary magnitude. However, Gray codes are easily translatable into natural binary (Table 7.6):

Decimal Fraction	Gray Code				Binary Code			
0	0	0	0	0	0	0	0	0
1/16	0	0	0	1	0	0	0	1
2/16	0	0	1	1	0	0	1	0
3/16	0	0	1	0	0	0	1	1
4/16	0	1	1	0	0	1	0	0
5/16	0	1	1	1	0	1	0	1
6/16	0	1	0	1	0	1	1	0
7/16	0	1	0	0	0	1	1	1
8/16	1	1	0	0	1	0	0	0
9/16	1	1	0	1	1	0	0	1
10/16	1	1	1	1	1	0	1	0
11/16	1	1	1	0	1	0	1	1
12/16	1	0	1	0	1	1	0	0
13/16	1	0	1	1	1	1	0	1
14/16	1	0	0	1	1	1	1	0
15/16	1	0	0	0	1	1	1	1

Table 7.6 Comparison of 4-Bit binary and Gray codes. Underlined bits indicate changes as number increases.

In Gray code, as the number value changes, the transitions from one code to the next involve only one bit at a time. The bits that change as the numbers increase are underlined in the table.

The conversion from binary to Gray code occurs as follows: If the binary MSB is zero, the Gray code MSB will be zero. Then, continuing to read from MSB to LSB, each change produces a "1," each non-change produces a "0." For example, binary 1011 becomes 1110 in Gray code (1 ⟶ 1, 1-to-0 ⟶ 1, 0-to-1 ⟶ 1, 1-to-1 ⟶ 0). Another example: the 12-bit binary number

101111000101 becomes 111000100111. Figure 7.3 shows one way in which binary to Gray code conversion may be mechanized.

The conversion from Gray code to binary is just the reverse of the conversion from binary to Gray code: the binary MSB will be the same as the Gray code

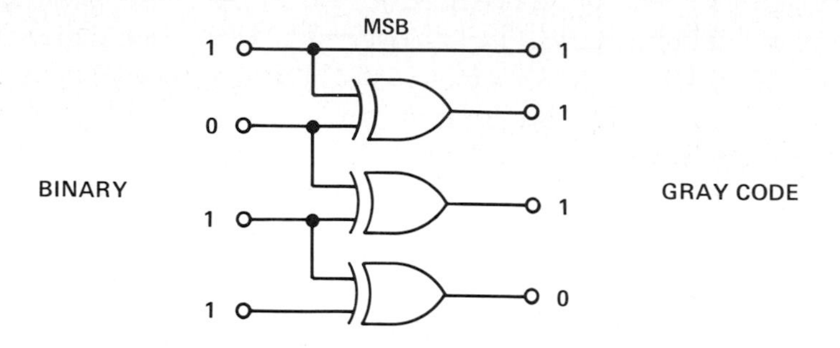

Figure 7.3. Binary-to-Gray code conversion using exclusive-or gates.

MSB. Then, continuing to read from MSB to LSB, if the next bit is 1, the next binary bit is the complement of the previous binary bit. For example, if the Gray code is 1110, the corresponding binary is 1011 (1 → 1, 1 → 1-to-0, 1 → 0-to-1, 0 → 1-to-1). Another example, the 8-bit Gray code 01110000 is 01011111 in binary. A mechanization of Gray code-to-binary conversion appears in Figure 7.4.

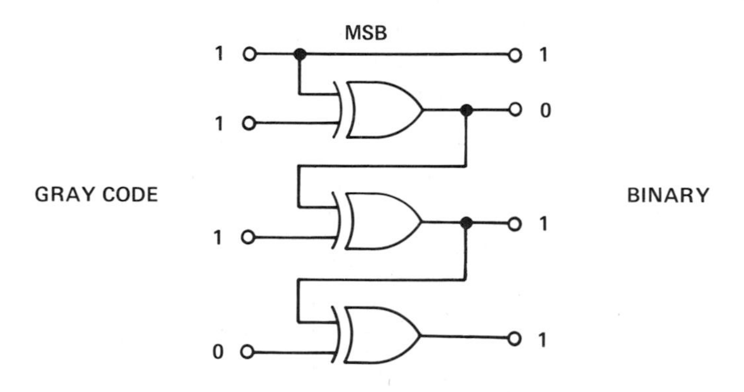

Figure 7.4. Gray Code-to-binary conversion.

Gray code is useful for shaft encoders (angle-to-digital converters) because the change of only 1 bit for each increment eliminates false intermediate codes that could occur in natural binary conversion. Here, for comparison, are Gray code and binary developed optical shaft encoders for 4-bit resolution.

Note that, with the Gray code converter, there is only one bit-change at each transition. If the edge of a shaded area is slightly out of line, the coding will be in error by a small fraction of an LSB. In the binary converter, *all four bits*

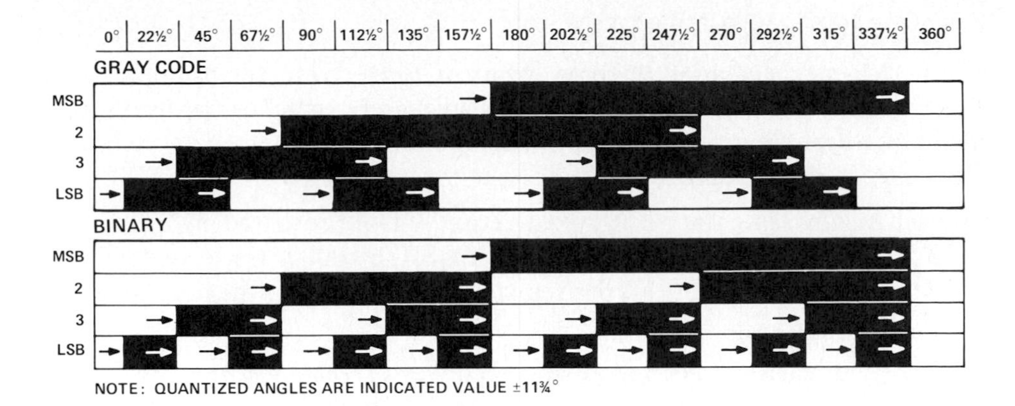

Figure 7.5. Gray vs. binary encoding.

change at once at the 180° and 360°transitions. If bit 2's shaded area were to end a little to the left of the 180° transition, the code, in a small region, would be 0011, indicating the 67 ½° range, or a fictitious progression from 157 ½° to 67 ½° to 180°. We leave the catastrophic implications of this to the reader.

The shaft encoder is a *simultaneous* converter: all bits appear at once and can be read in parallel at any time. An electrically equivalent form of simultaneous A/D converter, sometimes having a Gray code output, is the *flash* converter (see also Chapter 13). It employs a chain of biased comparators, the outputs of which provide a quantized indication of the analog input level: all comparators above it are 0, all comparators below it are 1. Multi-input gates then make the decisions necessary to obtain a parallel Gray code output. Such converters are quite fast, some being capable of producing 100 million or more meaningful conversions per second, but they require a number of comparisons that is a geometric function of the required resolution, (i.e., 2^n-1), as well as logic gates having large numbers of inputs.

A variation of this scheme, the cyclical converter, which also has Gray code output, uses fewer comparators, with more-accurate output states, but it requires more time to perform the conversion. It continuously tracks the analog input.

The use of Gray code in fast converters that provide continuous conversions has the same rationale as in the case of the shaft encoder. Any Gray code output value (for a 1-bit-accurate converter) that is latched into a register will always be within ± 1 LSB of the correct value, even if the latching occurs just as a bit is switching. With binary, however, where many bits can switch at a single transition, it is possible to latch in mid-flight, and, because of the "skew" between turn-on and turn-off speeds, lock in a false code. Sample-hold ahead of the conversion helps alleviate the situation in straight binary coding.

7.2.3 COMPLEMENTARY CODES

The actual mechanization of some forms of converters, (for example, early D/A converters using monolithic NPN quad current switches) required codes such as natural binary or BCD, but with all bits are represented by their complements. Such codes are called complementary codes.

In a 4-bit complementary-binary converter, 0 is represented by 1111, half-scale (MSB) by 0111, and full scale, less 1 LSB, by 0000. It can be easily obtained from the ["$\overline{Q}$"] outputs of a register, of which "Q" is the normal output sense.

Similarly, for each quad of a BCD-coded converter, *complementary* BCD is the code obtained by representing all bits by their complements. In complementary BCD, 0 is represented by 1111, and 9 is represented by 0110. As an example, Table 7.7 lists the equivalents for 1 through 11 in complementary binary and complementary BCD (with overrange bit).

Decimal Number						
INT	Fract. BIN	Fract. BCD	Natural Binary	Complementary Binary	BCD	Complementary BCD
0	0	0	0000	1111	00000	11111
1	1/16	1/10	0001	1110	00001	11110
2	2/16	2/10	0010	1101	00010	11101
3	3/16	3/10	0011	1100	00011	11100
4	4/16	4/10	0100	1011	00100	11011
5	5/16	5/10	0101	1010	00101	11010
6	6/16	6/10	0110	1001	00110	11001
7	7/16	7/10	0111	1000	00111	11000
8	8/16	8/10	1000	0111	01000	10111
9	9/16	9/10	1001	0110	01001	10110
10	10/16	10/10	1010	0101	10000	01111
11	11/16	11/10	1011	0100	10001	01110

Table 7.7 Complementary codes.

If a natural binary input were applied to a D/A converter coded to respond to complementary binary, the output would be in reverse order, i.e., zero output for all 1's, and F.S. – 1 LSB for all 0's.

The complementary codes discussed above involve complementing *all bits*, for convenience in implementing the conversion relationship using certain kinds of switches (i.e., those that respond to complementary logic). We could just as well have left the logic unchanged but redefined it as "negative true." However, for consistency in elucidation, we define all logic in terms of "positive true" TTL (or CMOS), as explained at the beginning of the chapter. It is important to understand that, for purposes of this discussion, these complementary codes have nothing to do with representation of the *analog polarity* (a matter that will be discussed next).

7.3 ANALOG POLARITY AND SCALING

So far, the conversion relationships mentioned have been unipolar (or nonpolar): the codes represent numbers, which in turn represent the normalized *magnitudes* of analog variables,* without regard to polarity. A unipolar A/D converter will respond to analog signals of only one polarity, and a unipolar D/A converter will produce analog signals of only one polarity.

For any application, a converter must be used whose reference and switches (and specifications) are compatible with the desired analog polarity. If, for reasons of economy or availability, a converter is available having a predetermined polarity different from that desired, the overall function's polarity may be modified by operating on the analog signal before A/D conversion—or after D/A conversion—to invert or double its polarity, and also to perform any necessary scale changes, if range must be adapted, too.

7.3.1 BIPOLAR CODES

For conversion of bipolar analog signals into a digital code that retains sign information, an extra bit, or sequence of bits, is necessary to indicate polarity. This extra "most-significant bit" doubles the analog range and halves the peak-to-peak resolution. In some cases, the sign bit is provided by re-interpreting the existing MSB, in which event the analog *range* may still be doubled, but the *resolution* is twice as coarse. For example, if a 10-bit converter's resolution is 1/1,024, for the range 0-10V, we may use a bipolar code having 11 bits,

Number	Decimal Fraction Positive Reference	Decimal Fraction Negative Reference	Sign + Magnitude	Twos Complement	Offset Binary	Ones Complement
+7	+7/8	−7/8	0 1 1 1	0 1 1 1	1 1 1 1	0 1 1 1
+6	+6/8	−6/8	0 1 1 0	0 1 1 0	1 1 1 0	0 1 1 0
+5	+5/8	−5/8	0 1 0 1	0 1 0 1	1 1 0 1	0 1 0 1
+4	+4/8	−4/8	0 1 0 0	0 1 0 0	1 1 0 0	0 1 0 0
+3	+3/8	−3/8	0 0 1 1	0 0 1 1	1 0 1 1	0 0 1 1
+2	+2/8	−2/8	0 0 1 0	0 0 1 0	1 0 1 0	0 0 1 0
+1	+1/8	−1/8	0 0 0 1	0 0 0 1	1 0 0 1	0 0 0 1
0	0+	0−	0 0 0 0	0 0 0 0	1 0 0 0	0 0 0 0
0	0−	0+	1 0 0 0	(0 0 0 0)	(1 0 0 0)	1 1 1 1
−1	−1/8	+1/8	1 0 0 1	1 1 1 1	0 1 1 1	1 1 1 0
−2	−2/8	+2/8	1 0 1 0	1 1 1 0	0 1 1 0	1 1 0 1
−3	−3/8	+3/8	1 0 1 1	1 1 0 1	0 1 0 1	1 1 0 0
−4	−4/8	+4/8	1 1 0 0	1 1 0 0	0 1 0 0	1 0 1 1
−5	−5/8	+5/8	1 1 0 1	1 0 1 1	0 0 1 1	1 0 1 0
−6	−6/8	+6/8	1 1 1 0	1 0 1 0	0 0 1 0	1 0 0 1
−7	−7/8	+7/8	1 1 1 1	1 0 0 1	0 0 0 1	1 0 0 0
−8	−8/8	+8/8		1 0 0 0	0 0 0 0	

Table 7.8 Commonly used bipolar codes.

*A/D converter input or D/A converter output.

with peak-to-peak resolution of 1/2,048 and range of ± 10V, or retain a code having 10 bits, but "stretch" the range to ± 10V, in which case the peak-to-peak resolution remains 1/1,024, which doubles the magnitude of the LSB.

The most-often-used binary codes in bipolar conversion are: twos complement, sign-magnitude (magnitude plus sign), offset binary, and ones complement. Table 7.8 shows each of these codes expressed for 4 bits (3 bits plus sign). Generally, if the bus that a converter is connected to is wider than a right-justified byte, the sign bit in twos complement is repeated as many times as required to fill the remaining spaces to the left (e.g., on an 8-bit bus, − 3/8 in twos complement should be processed as if it were 1111 1101).

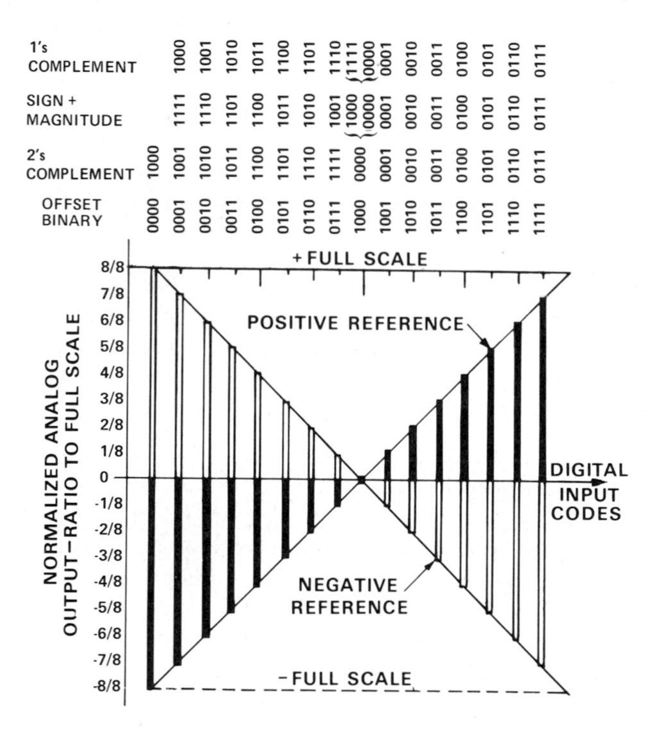

Figure 7.6. Ideal bipolar d/a conversion relationship for 4-bit (3-bit-plus-sign) offset-binary, twos-complement, sign-magnitude, and ones-complement codes.

Because the analog signal now has a choice of polarity, we must be careful about the relationship between the code and the polarity of the analog signal. "Positive reference" indicates that the analog signal* increases positively as the digital number increases. "Negative reference," on the other hand, indicates that the analog signal decreases towards negative full scale as the digital number increases. Conversion relationships for bipolar D/A and A/D converters are shown graphically in Figures 7.6 and 7.7.

*Gray code is an exception. Since it is not quantitatively weighted, it can represent any arbitrary range of magnitudes of any polarity.

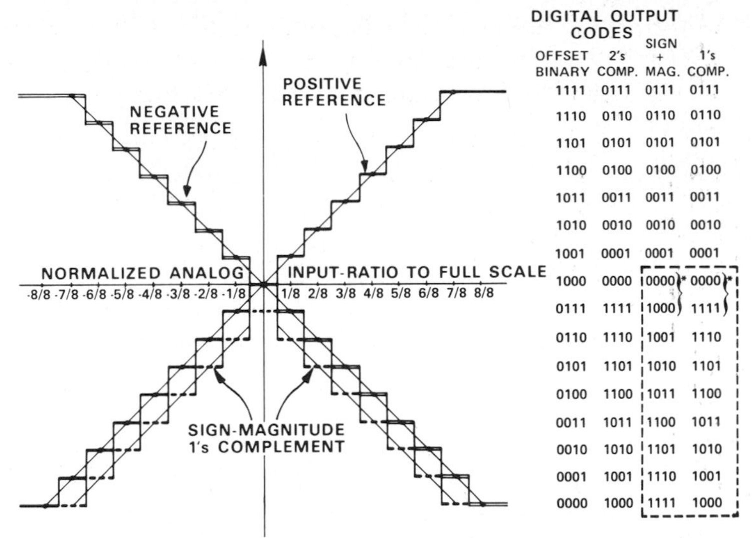

DIGITAL OUTPUT CODES

OFFSET BINARY	2's COMP.	SIGN + MAG.	1's COMP.
1111	0111	0111	0111
1110	0110	0110	0110
1101	0101	0101	0101
1100	0100	0100	0100
1011	0011	0011	0011
1010	0010	0010	0010
1001	0001	0001	0001
1000	0000	0000 }*	0000 }*
0111	1111	1000 }	1111 }
0110	1110	1001	1110
0101	1101	1010	1101
0100	1100	1011	1100
0011	1011	1100	1011
0010	1010	1101	1010
0001	1001	1110	1001
0000	1000	1111	1000

Figure 7.7. Ideal bipolar a/d conversion relationship for 4-bit (3-bit-plus-sign) offset-binary, twos-complement, sign-magnitude, and ones-complement codes.

Sign-Magnitude would appear to be the most straightforward way of expressing signed analog quantities digitally. Simply determine the code appropriate for the magnitude and add a polarity bit. It is used advantageously in D/A converters that operate in the vicinity of zero, where the application calls for smooth and linear transitions from small positive voltages to small negative voltages. As can be seen in the example in the table, it is the only binary code for which the three magnitude bits do not have a major transition (all 1's to all 0's, or equivalent) at zero. Sign-magnitude BCD is almost universally used for bipolar digital voltmeters (A/D converters).

It does have some shortcomings, though. In data-processing applications, the *other* codes are more-readily usable for computation with a minimum of translation. One of its problems is that it has two codes for zero. For this reason, sign-magnitude is harder to interface with digitally, because it requires processing by additional software and/or hardware.

Offset binary is the easiest code to embody with converter circuitry. An examination of the offset binary code for three bits plus sign will show that it is really a natural binary code for four bits, but the zero of the code is at negative full scale, the LSB is 1/16 of the bipolar range, and the MSB is turned on at analog zero. Therefore, to make an offset binary 3-bits-plus sign converter out of a 4-bit D/A converter having 0-to-10V full-scale range, we have only to double its scale factor (20V range), and offset its zero by one half of the full range (−10V), an operation which is neither difficult nor expensive. Similarly, for an A/D converter, one would attenuate the input by one-half, and add a bias of one-half the full range.

*When A/D converters with sign-magnitude or ones complement coding are required, the ambiguous zero must be handled appropriately.

Besides its ease of implementation, offset binary is compatible with computer inputs and outputs; it is easily changed to the more-computationally useful twos complement (just complement the MSB and any leading bits); and it has a single unambiguous code for zero. The all-zeroes negative full-scale code (0000), though not used in computing (because −F.S. +1 LSB is the most negative value defined in computing), is nevertheless useful as a converter checking and adjustment code.

The principal drawback of offset binary—unless the device is a sign-magnitude converter with translated logic—is that a major bit transition occurs at 0 (all bits change, from 0111 to 1000). This can lead to "glitch" problems dynamically (the difference in speed between bits turning on and off can lead to large spikes) and to linearity problems (the largest linearity errors are most likely to occur at major transitions, because the transition is essentially a difference between two large numbers). In offset binary, zero errors may be greater than with sign-magnitude, because the zero analog level is usually obtained by taking a difference between the MSB (½ full range) and a bipolar offset (½ full range)—again, two large numbers.

Twos complement, for conversion purposes, consists of a binary code for positive magnitudes (0 sign bit), and the twos complement of each positive number to represent its negative. The twos complement is formed arithmetically by complementing the number and adding 1 LSB. For example, the twos complement of ⅜ (0011) would be its complement plus 1 LSB, or 1100 + 0001 = 1101. If it were a right-justified number on an eight-bit bus (i.e., 0000 0011), its twos complement would be 1111 1101.

Twos complement is a useful code computationally because it can be thought of as a set of negative numbers. Therefore, addition can be used instead of subtraction. For example, to subtract ⅜ from 4/8, add 4/8 to −⅜, or 0100 to 1101 (i.e., 0000 0100 to 1111 1101). The result is 0001 (0000 0001), disregarding the extra carry, or ⅛.

If the twos complement code and the offset binary code are compared, it can be seen that the only difference between them is that the MSB of one is replaced by its complement in the other (Nature's way of helping converter manufacturers and users). Since both a digit and its complement are available from most flip-flops, an offset-binary-coded converter may be used for twos complement, just by using the complement of the MSB at the output of an A/D converter or at the output of a D/A converter's input register. And vice versa. Many converters are manufactured with both the MSB and its complement available.

Converters that produce (or respond to) twos complement directly have the same disadvantages as those coded for offset binary; the conversion process is generally identical. It is feasible (as noted earlier) to get improved analog performance by the use of a benign sign-magnitude conversion, with the external code translated to twos complement.

Ones complement is a common means of (or first step toward) representing negative numbers, because it is obtained arithmetically by simply complementing all of a number's digits. Thus, the ones complement of 3/8 (0011 or 0000 0011) is (1100 or 1111 1100). When a number is subtracted by adding its ones complement, the extra carry (that is disregarded in twos complement), if present, causes 1 LSB to be added to the total ("end-around carry.") Thus, subtracting 3/8 from 4/8, 0100 + 1100 = 0000 + 0001 = 0001 (or 1/8). Similarly, 0000 0100 + 1111 1100 = 0000 0000 + 0000 0001 = 0000 0001. A ones complement code can be formed by complementing each positive value to obtain its corresponding negative value, including—alas—zero, which is then represented by two codes, 0000 and 1111.

Besides its ambiguous zero, a disadvantage of this code in conversion is that it is not as readily implemented as twos complement. If it is not converted to twos complement before a D/A conversion with a twos-complement converter, by adding a 1 LSB increment digitally when the MSB is 1 (indicating a negative number), then the easiest way to implement the conversion is by performing a twos complement conversion, and—if the MSB = 1—adding the *analog* value of 1 LSB. The extra analog bit—in concept—can be added simply and elegantly by resistively dividing the digital MSB logic level down to the LSB's analog value and summing this attenuated signal, but it will not be free of errors and noise.

7.3.2 CODE CONVERSION

Code conversion may be desirable, either after A/D conversion or before D/A conversion, in order to make it possible to use a converter that produces the best results at the lowest cost (or one that simply happens to be available). For this purpose, the matrix of Table 7.9 succinctly outlines the relationships among the codes. For right-justified buses wider than the digital word, "Complement MSB" includes leading bits.

7.3.3 OTHER BIPOLAR CODES

The list of bipolar codes mentioned above may seem exhaustive, but it does not fully reflect the ingenuity and diversity of the computer and converter industries. There are a number of variations in more-or-less widespread usage that should be mentioned here because they will inevitably be encountered. Fortunately, they are based on codes we have already discussed and may be easily described.

Modified sign-magnitude: This is a version of sign-magnitude in which the polarity indication (i.e., the MSB) is complemented (1 for positive, 0 for negative).

Modified one's complement: Like modified sign-magnitude, a version in which the MSB is complemented (1 for positive, 0 for negative).

To Convert From To ↓ ☞	Sign Magnitude	2's Complement	Offset Binary	1's Complement
Sign Magnitude	No Change	If MSB = 1, complement other bits, add 00 . . 01	Complement MSB If new MSB = 1, complement other bits, add 00 . . 01	If MSB = 1, complement other bits
2's Complement	If MSB = 1, complement other bits, add 00 . . . 01	No Change	Complement MSB	If MSB = 1, add 00 . . . 01
Offset Binary	Complement MSB If new MSB = 0 complement other bits, add 00 . . . 01	Complement MSB	No Change	Complement MSB If new MSB = 0, add 00 . . . 01
1's Complement	If MSB = 1, complement other bits	If MSB = 1, add 11 . . . 11	Complement MSB If new MSB = 1, add 11 . . . 11	No Change

Table 7.9 Relations among bipolar codes.

Complementary everything: All of the above-mentioned codes may be completely complemented to form complementary sign-magnitude, complementary offset binary, complementary twos complement, and complementary ones complement. (These are, as explained earlier in this chapter, "negative true" versions.) Such codes, although they make life a little more complex, are the preferred coding for some now-popular "industry-standard" converter types based on early monolithic switching hardware. Users of monolithic and hybrid converters without registers should be prepared to adjust their thinking (and especially their test equipment) to include the possible application of complementary codes.

For the sake of completeness, Table 7.10 lists the codes mentioned above, for 3-bits-plus-sign.

7.3.4. ARBITRARY BIASING AND SCALING

The conversion relationships discussed so far have been either strictly one-sided (0 to full scale) or symmetrical (± full scale). The reason for this emphasis is that most commercially available converters are built that way—as general-purpose devices.

However, since the principal relationship between the analog variable and the digital number for linear converters is *proportionality*, the repertoire of codes corresponding to a given resolution may represent any portion of the analog voltage or current range.

Number	Modified Sign-Magnitude	Modified 1's Complement	COMPLEMENTARY CODES Comp. Sign-Magnitude	Comp. Offset Binary	Comp. 2's Complement	Comp. 1's Complement
+7	1111	1111	1000	0000	1000	1000
+6	1110	1110	1001	0001	1001	1001
+5	1101	1101	1010	0010	1010	1010
+4	1100	1100	1011	0011	1011	1011
+3	1011	1011	1100	0100	1100	1100
+2	1010	1010	1101	0101	1101	1101
+1	1001	1001	1110	0110	1110	1110
0+	1000	1000	1111	0111	1111	1111
0−	0000	0111	0111	0111	1111	0000
−1	0001	0110	0110	1000	0000	0001
−2	0010	0101	0101	1001	0001	0010
−3	0011	0100	0100	1010	0010	0011
+4	0100	0011	0011	1011	0011	0100
−5	0101	0010	0010	1100	0100	0101
−6	0110	0001	0001	1101	0101	0110
−7	0111	0000	0000	1110	0110	0111
−8				1111	0111	

Table 7.10 Modified and complementary bipolar codes.

For example, to encode the industrial transmitter current range from 4 to 20 mA in binary, using a 500-ohm resistor and the 2-to-10-volt portion of the range of a 0-10V A/D converter, simply apply the voltage without any transformation. However, a more range-efficient alternative would be to offset the input by −2 volts, amplify by 1.25, and apply the resulting 0-10V signal to the converter, thereby making use of the entire range of available codes and improving resolution by 25%. In a sense, the conversion relationship between the original input and the digital output is an *offset binary* code. The subsequent digital processing would take this transformation into account via the software.

Another sort of arbitrary scaling might result if the analog signal were proportional to a temperature range of (for example) 0° to 70°C, and one desired a direct readout of temperature on a "dumb" digital voltmeter. A typical approach might be to scale the voltage directly to the temperature numbers (e.g., 10°/V) and apply it to a DVM with a 10-volt scale, with the location of the decimal point re-interpreted. The DVM would then provide a readout from 0 V to 7.0 V scaled from 0 to 70 in *engineering units*.

7.3.5 DACs AS MULTIPLIERS AND ADCs AS DIVIDERS

As noted in Section 5.2, The D/A converter can be thought of as a digitally controlled potentiometer that produces an analog output (voltage or current) that is a normalized fraction of its "full scale" setting. The output voltage or current depends on the reference value chosen to determine "full scale" output. If the reference may vary in response to an analog signal, the output is proportional to the product of the digital number and the variable analog input. The polarity of the product depends on both the analog signal polarity and the digital coding and conversion relationship.

Four-quadrant multiplication is available, if the D/A converter accepts reference signals of both positive and negative polarities and the conversion relationship is bipolar. A typical conversion relationship for a 4-quadrant multiplying DAC having 3-bit-plus-sign twos-complement coding is shown in Figure 7.8, interpreting the multiplying DAC as a digitally controlled variable-gain amplifier.

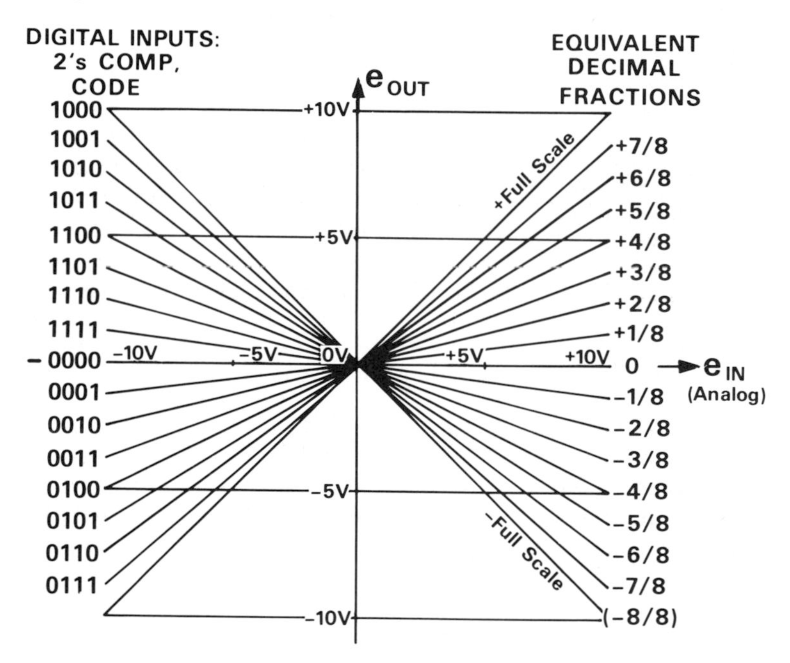

Figure 7.8. Digital-to-analog converter as four-quadrant multiplier of analog voltage and 3-bit-plus-sign, twos-complement digital number. Analog output vs. analog input as a function of digital input code.

In another interpretation, the envelope of the ideal bipolar D/A converter output in Figure 7.6 could be seen as proportional to the analog signal input, starting from full scale "Positive Reference," being attenuated as the analog signal is reduced, passing through zero, and increasing negatively to the "Negative Reference" envelope.

When the input plane is viewed from above (digital number on one axis, analog input on the other—output along the axis of viewing), multiplying D/A converters may be 4-quadrant, two-quadrant (single polarity of either analog or digital variable), or one quadrant. In a sense, they may even be fractional-quadrant, if the reference or the code cannot be varied to or through zero.

In analog-to-digital converters, the digital output number depends on the ratio of the quantized input to the "full-scale" reference. If the reference is allowed to change in response to a second analog input, the digital output will be proportional to the ratio of the analog signal to the reference signal. Thus, the "ratiometric" A/D converter can be thought of as an analog divider with digital output. Generally, in such devices, the reference input has more limited range and dynamic capabilities than the signal input.

7.4 ELECTRICAL INTERFACES WITH CONVERTERS

Converters may have associated with them six families of electrical inputs and outputs: analog signal(s), digital code, power, control, configuration, and reference. Table 7.11 indicates some of the properties of these interfaces, and the text that follows adds further detail.

	D/A Converters		A/D Converters	
ANALOG SIGNAL	Output:	Voltage or Current Polarity Magnitude	Input:	Usually Voltage Polarity Magnitude
DIGITAL CODE	Input: Output:	Buffered or Direct Coding Logic Levels Format: Serial Parallel Byte-Serial Readback	Output:	Coding Logic Levels Timing (Clock) Format: Serial Parallel Byte-Serial
CONTROL	Input	Strobe(s) Chip Select Chip Enable Write Reset	Inputs: Outputs:	Convert Command Chip Select Enable Clock Chip Status Overrange
CONFIGURATION	Serial/Parallel Short Cycle ( ) Bits Address – Multiple On-Chip Devices		Byte-Enable Short Cycle ( ) Bits Address – Multiplexed Inputs	
POWER	Analog: Digital:	Usually ± 15V or same as Digital + 5V (TTL) + 5V to + 15V (CMOS)	Analog: Digital:	Usually ± 15V or same as Digital + 5V (TTL) + 5V to + 15V (CMOS) – 5V to – 15V (CMOS V_{SS})
REFERENCE	Internal or External Fixed or Variable Polarity		Internal or External Fixed or Variable Polarity	

Table 7.11 Converter interfaces.

Figure 7.9 is a block diagram showing the relationships of typical connections to a parallel converter. There may be yet other connections, such as a clock synchronization input or output, complementary logic inputs or outputs, and

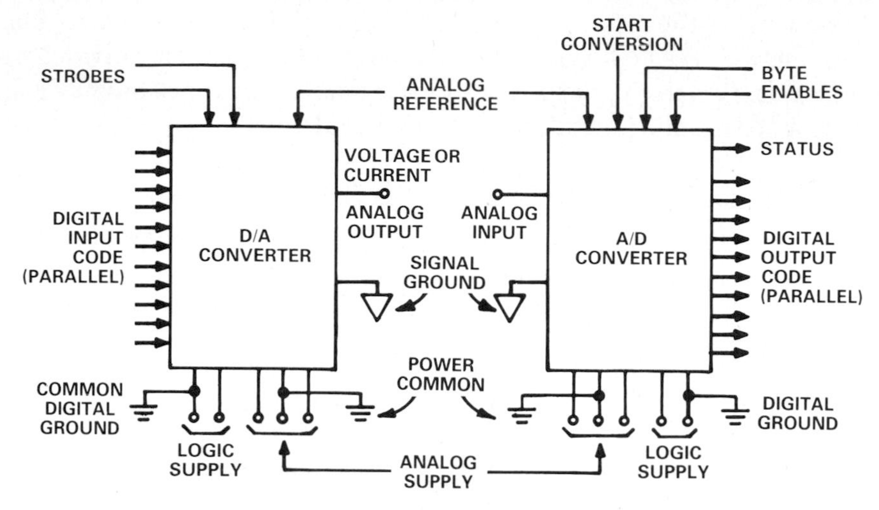

Figure 7.9. Typical classes of converter connection.

connections that are essentially internal but are brought out for the sake of optional flexibility, such as bipolar offset reference terminals. If the A/D converter is multiplexed, there will be additional analog inputs and one or more digital channel-select input lines. Figure 7.10(a) shows actual pin connections of a popular 8-bit IC DAC [AD558], and (b) shows the pin connections of a popular 12-bit IC A/D converter [AD574A].

Application block diagrams in this book (and much of the literature), for facility of communication, tend to depict only those interfaces that are of specific relevance to the point under discussion. For example, when the logic interface is discussed, the analog aspect of the circuit may be ignored—and vice versa; in many cases, the power and ground connections are not shown in detail. However, the reader should be continually aware that "out of sight" must never mean "out of mind."

7.4.1 GROUND RULE

The experienced circuit designer will recognize the feeling of wariness provoked by the presence of two supplies, and several classes of signals, all needing return "to ground." Grounding is indeed important to system performance; discussions of the essentials of grounding practice will be found (as appropriate) in several places in this book; however, for clarity in this present exposition, we will consider that all grounds are always at true zero potential with respect to all input and output signals. Accordingly, the discussions that follow will everywhere employ the inverted triangle, which represents ideal signal ground.

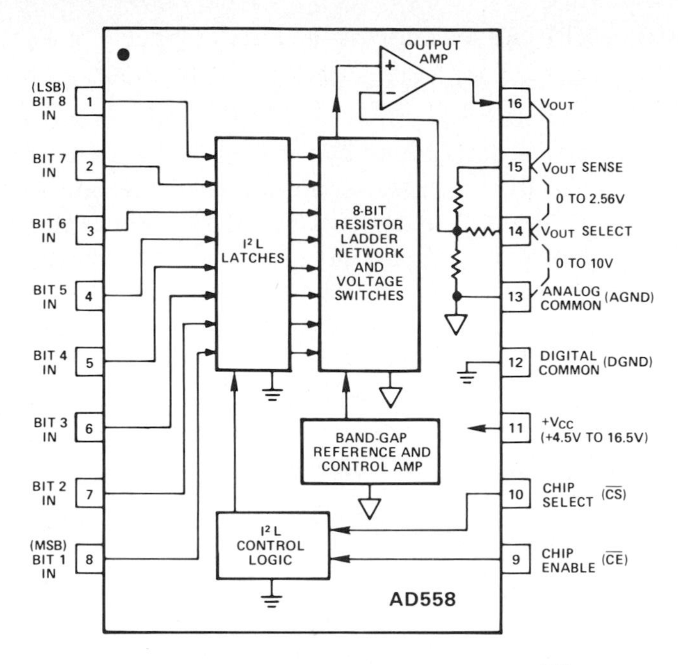

a. AD558 8-bit d/a converter (DACPORT™).

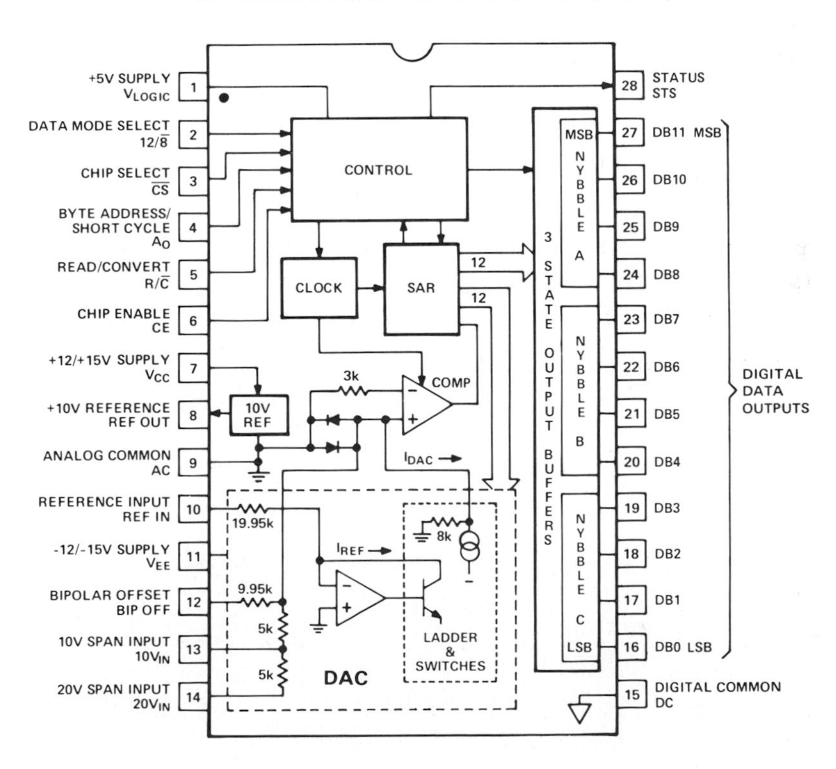

b. AD574A 12-bit a/d converter.

Figure 7.10. Pin-connection schemes of actual devices.

7.4.2 POWER SUPPLIES

The choice of power supplies for use with converters is governed by their effect on conversion accuracy, system noise, size and weight, reliability and cost. Supply capacity is determined by the choice of system philosophy: one main supply feeding all elements, vs. a number of satellite supplies or regulators sharing a common primary source (which might itself be a dc voltage derived from the ac mains). Generally, supplies that provide good operational-amplifier performance are sufficiently well regulated to provide rated converter performance, but adequate dynamic bypassing is essential because of the presence of fast digital edges. Converter performance as a function of dc variation of power supply voltage is a standard specification provided by the converter manufacturer. Conversion-system designers tend to avoid high-power switching-type supplies for converters.

7.4.3 DIGITAL LOGIC LEVELS

There are a variety of "standard" voltage levels and current-drive capacities corresponding to logic "0" and logic "1." This variety is a result of historical compromises between circuit/processing technology, the need for speed, reliable differentiation between the logic states, circuit complexity, and fanout capability. They are described by such sets of initials as TTL, DTL, HTL, ECL, CMOS, IIL.

In general, logic levels are associated with device technology; the most popular of these are CMOS, ECL, and TTL. However, there are varieties of specifications and large variations within each family of devices; for example, low-voltage CMOS voltage levels are more closely related to TTL than to high-voltage CMOS. In fact, most of the modern conversion system products are designed for some degree of compatibility with TTL, the most widely used logic system.

Logic levels are closely related to supply voltage; in fact, the ideal logic gate would switch between ground and supply voltage, but voltage drops in the devices and in the external circuits require specification of logic 1 as somewhat less than supply voltage and logic 0 as somewhat greater than ground. TTL and low-voltage CMOS are based on widely used single +5-volt supplies.

In classical TTL, as mentioned earlier, a gate must respond to "0" if the input to it is 0.8 volts or less, and it must respond to "1" if the input is 2.0 volts or greater, up to the maximum and minimum voltage ratings. In order to provide a measure of immunity to noise, including dc voltage drops, occurring in transmission, gate outputs (within their current ratings), must furnish a minimum of 2.4 volts to signify "1" and a maximum of 0.4 volts to signify "0."

Within the TTL system, there are further classifications by fanout (the number of gates that can be driven) and speed. For convenience, input or output currents are normalized in terms of the standard *TTL load*, which is a positive current of 40μA for "1" and −1.6mA (sink current) for "0"; however, an inspection of a random variety of "TTL-compatible" converter types (or even digital bit currents vs. digital control currents within the same converter) would often show substantial differences. To avoid possible difficulties at the interface, it is important for the user to read the manufacturer's data sheet and understand what "TTL-compatible" means for a given device.

When CMOS devices operate at higher voltages, they usually (but not always) have increased noise immunity. CMOS devices in general draw low current except when switching; they have low power dissipation and accordingly higher circuit-packing density and are especially useful in remote and portable system elements.

Products designed for one logic scheme can be used with other logic schemes by performing appropriate transformations (level shifting, gain-or-attenuation, sign-inversion). D/A converters designed for TTL logic will inherently accept DTL Inputs.

7.4.4 CONTROL LOGIC

The Status Output

In most applications, A/D Converters require a time interval, either fixed or variable, during which the system must wait for a conversion to be performed. During this time, the conversion data may be changing and may bear no relationship to the final result; if the input changes, it may cause erroneous results. Thus, the output of the converter must not be interrogated during the conversion, and the input track-hold must remain in the "hold" state until the input is ready to accept new data.

For this reason, the control output called *Status* (or "Busy," "Data Ready," "EOC"—*end of conversion*, etc.) changes state in response to the Convert command to define the conversion period; it does not return to its original state until a conversion is completed. It may be used as an interrupt, or to inhibit readout, or to update a *buffer* output register that holds the previous output word. It also serves to prevent another conversion from beginning, and to prevent the track/hold from changing state, until the converter's input is ready. In some high-speed converters (for example, digitally corrected subranging types—see Chapter 13), and in tracking types, a new conversion may start as soon as the previous conversion has cleared the input stage and an earlier conversion has been latched into the output to activate the Data Ready line.

Strobes

Most D/A converters have basic circuitry that responds immediately and continuously to whatever digital signals are applied. It is often desirable to buffer the basic circuitry from the source of digital information (for example, a busy bus) by a register, and update all bits simultaneously, upon command. The gating input is called the strobe (or "clock" or "enable.")

For use with microprocessors, the data word is often divided into bytes, typically having 8 bits (see Chapter 4), or 4-bit *nybbles*; bytes are *enabled* in sequence to transmit the information contained in the full word in *byte-serial* format. Input strobes to DACs might be called *high-byte strobe* (more significant bits), *low-byte strobe* (less-significant bits), and—to load the complete information into the DAC—*load-DAC strobe* (often asserted at the same time as the final byte becomes valid). Conversely, the parallel outputs of ADCs are placed on the microprocessor data bus in appropriate order by *high-byte enable* and *low-byte enable* strobes. In order for the *status* output to be treated as information appearing on the data bus, a *status-enable* strobe would be used with microprocessor-compatible ADCs.

7.4.5 ANALOG SIGNALS

Inputs to A/D converters are usually in the form of voltage. Outputs from D/A converters are often in the form of voltage, at low impedance, from an operational amplifier (an example is the AD667—Figure 8.8). However, many converters provide an active output *current* instead of a voltage (for example, AD567—Figure 8.6), and some simply provide an attenuation ratio; coupled with an op amp, they provide a digitally variable analog gain. As will become clear in the sections that follow, the basic conversion process may inherently develop a current output that is quite fast, linear, and free from offset. A built-in on-chip operational amplifier or an external op amp may be used to convert that current to voltage. As a result of the inevitable design tradeoffs, the amplifier will tend to limit converter performance, primarily by increasing settling time.

If the current is made available directly, the speed of response is under the control of the user, through the choice of an appropriate external output amplifier. The user can also choose the inverting or the noninverting mode. For example, in a 10-bit application calling for good accuracy and moderately high speed, at low cost, the full-scale settling time of the current output from the 10-bit AD561 to 0.05% (½ LSB) is 250 nanoseconds. The AD561 (Figure 7.19), followed by a general-purpose I.C. operational amplifier for voltage output, typically has settling time of 5 μs to the same resolution; but with a high-speed op amp, such as the AD509, for example, settling time can be reduced to 600ns.

Converters that have current outputs or "soft" voltage outputs (directly from resistive ladders) may be considered as either voltage generators with series resistance or current generators with parallel resistance (Figure 7.11).* They are used with operational amplifiers in either the inverting or the noninverting

*The impedance from summing point to ground is not always constant; a large class of converters—those using inverted ladders and connected directly to an op-amp summing point in the current-output mode—have code-dependent resistance and capacitance loading the summing point, affecting both linearity and dynamics.

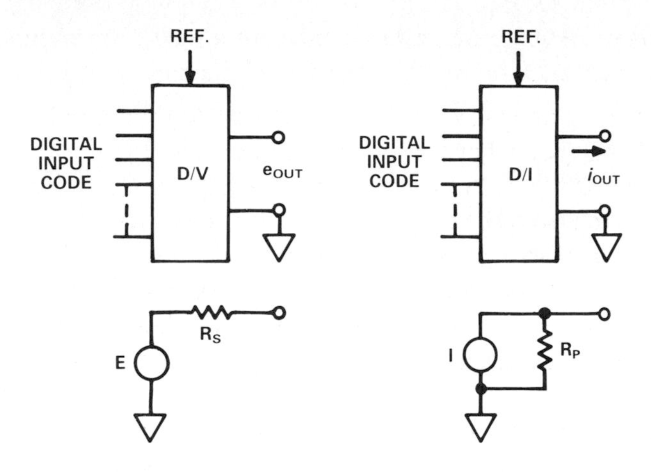

Figure 7.11. D/A converters as voltage or current generators.

connection (Figure 7.12). Some types have one or more internal feedback resistors (for appropriate output voltage scaling) that track the ladder resistors, to minimize temperature variations of gain in inverting configurations. Also present may be a terminating resistor, to develop passively a non-inverted output voltage, which may be amplified with a noninverting amplifier. The gain-determining feedback resistances (R_1, R_2) do not have to track the converter's internal resistors, only one another.

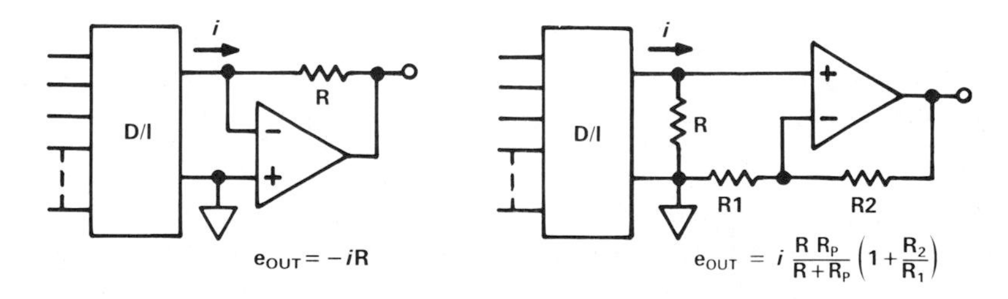

Figure 7.12. Current-to-voltage conversion, inverting and non-inverting, using operational amplifiers.

When current-output converters having active sources (see Figure 7.19 and Chapter 9) are used, the inverting connection is the preferred connection, for a number of reasons. With current-source output, the internal impedance of the D/I converter is usually high. Thus, the loop gain will tend to remain near unity, essentially independently of the value of feedback resistance, minimizing amplifier-contributed errors, such as voltage drift. Furthermore, the output swing of the D/I converter (at the amplifier's negative input terminal) will be negligible, minimizing loading of the current output—and any associated problems, such as voltage-dependent nonlinearity and variation of internal impedance with temperature. Finally, common-mode rejection is not important, since there is no common-mode swing.

However, if the DAC's basic circuit is that of a passively switched resistive attenuator (typical of CMOS DACs), a high-performance op amp must be used to minimize nonlinearity caused by voltage offsets and variable-impedance loading at the amplifier's summing point as codes are switched (see the next section and Chapter 12).

The conversion relationship of D/I converters is "positive reference" (Figure 7.6) if the current flowing *out of the converter* becomes more positive as the value represented by the digital code becomes more positive, irrespective of the actual polarity of the converter's reference element. If a noninverting amplifier configuration is used, the output voltage will have the same normalized conversion relationship as the output current. If an inverting connection is used, the voltage will have a conversion relationship of opposite output polarity, and will thus be "negative reference." Figure 7.13 illustrates this point, for both binary and complementary binary unipolar codes. On the other hand, if current flowing *towards the converter* increases as the value represented by the digital code increases, the relationship is "negative reference" for current, but "positive reference" for voltage in an inverting configuration.

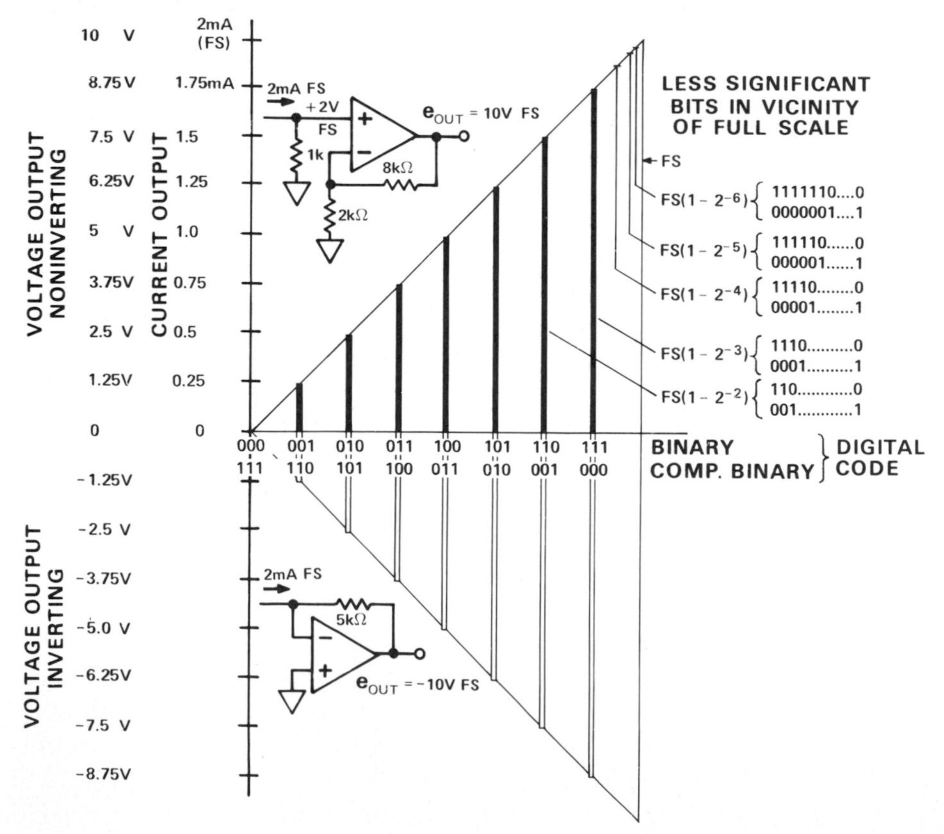

Figure 7.13. Ideal conversion relationships for the three most-significant bits in a positive-reference unipolar digital-to-current converter, with noninverting and inverting amplifier connections, and binary vs. complementary binary ("negative-true") codes.

7.5 D/A CONVERTER CIRCUITS

A basic D/A converter can be built with a voltage reference, a set of binary-weighted precision resistors, and a set of switches (Figure 7.14). An output buffer stage unloads passive elements, converts current to voltage or provides amplification, and furnishes a low-impedance voltage output.

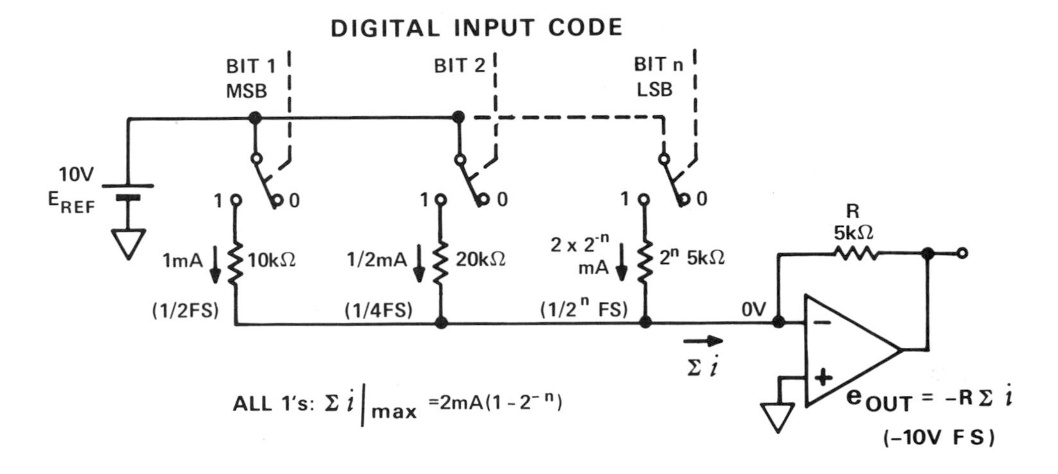

Figure 7.14. Simple digital-to-analog converter using binary-weighted resistors.

In this example, an operational amplifier holds one end of all the resistors in a set of n resistors at zero volts. The resistors are binary-weighted, i.e., each is weighted by 2^j; therefore, with equal voltage applied to all the resistors, the currents through them will be weighted by 2^{-j}. The switches are operated by the digital logic, open for "0," closed for "1." Each switch that is closed adds a binary-weighted increment of current E_{REF}/R_j via the summing bus connected to the amplifier's negative input. The negative output voltage is proportional to the total current, and thus to the value of one of the 2^n binary values represented by the input code. Thus, for an 8-bit converter and the code example of equation 7.2 (10111001), the output currents will be proportional to the terms in the equation, and the output voltage will therefore be proportional to the sum of the terms.

In general, for resolutions generally exceeding 4 bits, this scheme is not considered practical. For example, in 12-bit conversion, the required range of resistance values would be 2,048:1, or 20 megohms for the LSB to 10 kilohms for the MSB. If the resistors are to be manufactured in thin- or thick-film, or integrated-circuit form, such a range would be totally impractical. If discrete resistors are used, cost and size are increased, tracking advantages are lost, and inventory becomes a problem. Parasitic impedances (shunt and series) will also be difficult to deal with.

7.5.1 SWITCH DECODING

In the diagram of Figure 7.14, the selective summing of weighted currents provides the decoding. However, there is another approach, in which the binary code is digitally decoded, and N (from 0 to $2^n - 1$) equal increments of current, I, are summed. Figure 15a shows an example of this technique for

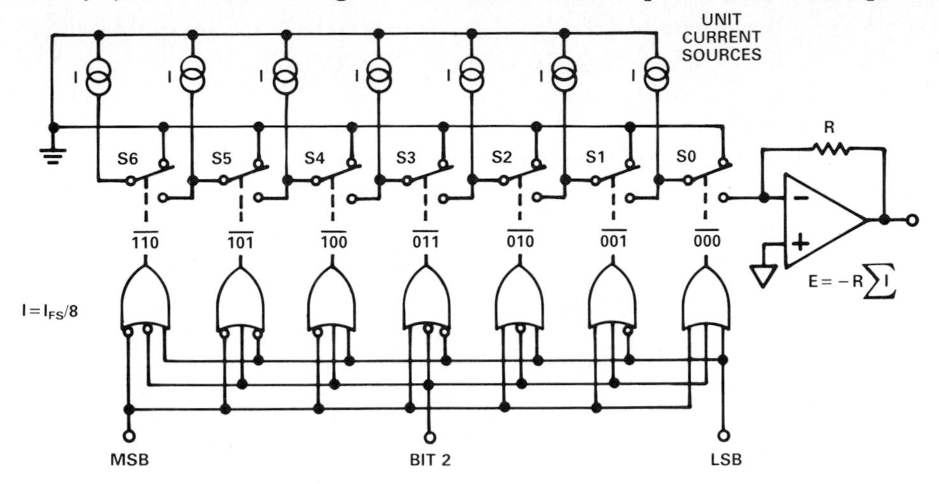

a. Current-switching.

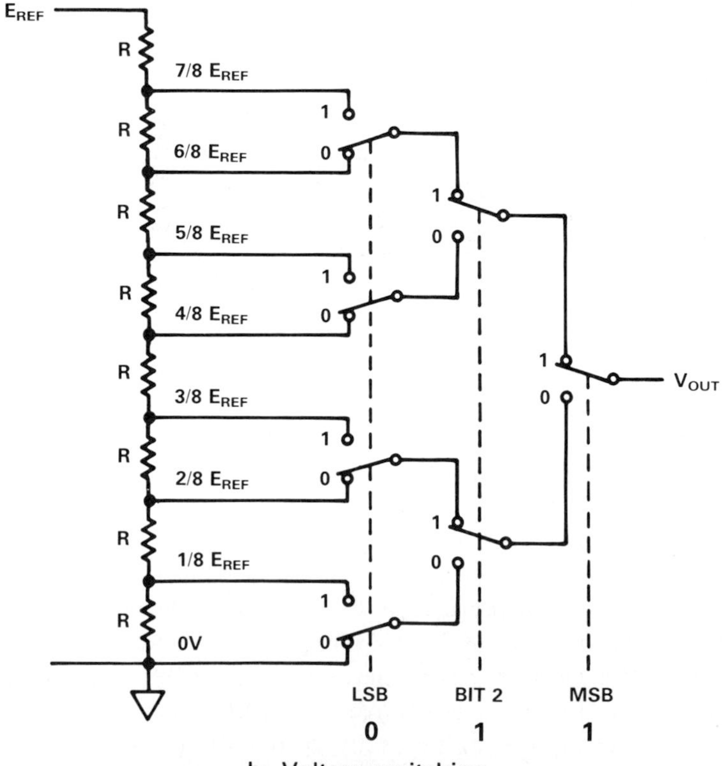

b. Voltage-switching.

Figure 7.15. Fully decoded 3-bit converter schemes.

3 bits, employing 7 equal current sources (or 7 equal resistors with a fixed reference voltage).

When all three bits are zero (000), switch S0 conducts its current to ground; no current flows through the op amp's feedback circuit. If one or more bits are 1, S0 will be closed, switching its current (as well as those from any other paths in a chain of consecutively closed switches) to the summing point. When the LSB only is 1 (i.e., 001), switch S1 will be open; for any other condition, it will be closed, allowing its source (etc.) to conduct toward the summing point—unless S0 is open. The position of any switch to the left of an open switch doesn't matter; its current flows to ground, not the summing point. Thus, corresponding to the binary chain of *j* codes, 000, 001, 010, etc., from 0 to 7, the sum of the currents flowing through the summing point will be *j*I. Figure 7.15b shows a corresponding scheme for a voltage-output DAC.

The principal advantage of fully decoded DAC architectures is that—assuming ideal switching—the output has to be monotonic, even if the resistor values or unit currents deviate substantially from their nominal values. Also, all resistors or current sources are identical. However, it is extremely difficult to carry out in practice, because $2^n - 1$ resistors or current sources and switches are required, plus decoding logic and a great many interconnections.

Although these configurations have tended to be impractical for complete high-resolution DACs, a limited number of fully decoded bits are used in designs that combine them with another form of DAC architecture to simplify the problem of obtaining high resolution and monotonic behavior. Thus, for example, in a 16-bit D/A converter, the four-most-significant bits might use full decoding to divide the range into 16 equal parts, with an easier-to-build 12-bit converter providing the required 4,096 levels of interpolation in each portion of the range. Examples can be seen in the descriptions of devices in Chapter 8; typical high-resolution devices employing partial decoding include the AD6012 12-bit DAC, and the AD7546 and AD569 16-bit DACs.

7.5.2 RESISTANCE LADDERS

A way to reduce both the number of resistors and range of resistance is to use a limited number of repeated values, in a configuration providing suitable attenuation. One convenient approach, shown in Figure 7.16, is to use a binary resistance quad, consisting of four binary-scaled values (e.g. 2R, 4R, 8R, 16R) for each group of 4 bits, with attenuation of 16:1 for each successive quad, down to the least-significant quad. Thus, the four most-significant bits are summed without attenuation, the next four bits are attenuated by 16:1 and summed, and the last four bits are either attenuated by 265:1 and summed with the most-significant quad, or (more likely) attenuated by 16:1 and summed with the next more-significant quad. A benefit of this scheme is that the proper relative quad weighting for BCD conversion can be achieved by using the same scheme with an attenuation between quads of 10:1.

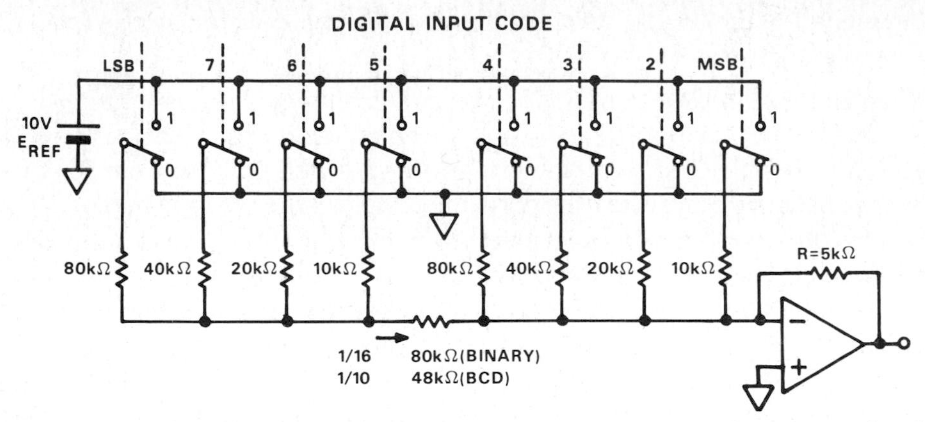

Figure 7.16. 8-bit digital-to-analog converter, using two equal-resistance quads with attenuation for the less-significant quad.

Carrying this reduction of resistance ratios farther, one arrives at the R-2R ladder, a convenient—and very popular—form. Figure 7.17 shows it with an inverting operational amplifier in a widely used configuration employing D/A converters that use CMOS switches.

If all bits but the MSB are "off" (i.e., grounded), the output voltage is $(-R/2R)\,E_{REF}$. For the second bit, the lumped resistance-to-ground of all the less-significant-bit circuitry (below the bit 2 leg) is 2R; the divider formed by series R and the two paralleled 2R elements has an attenuation of ½; therefore,

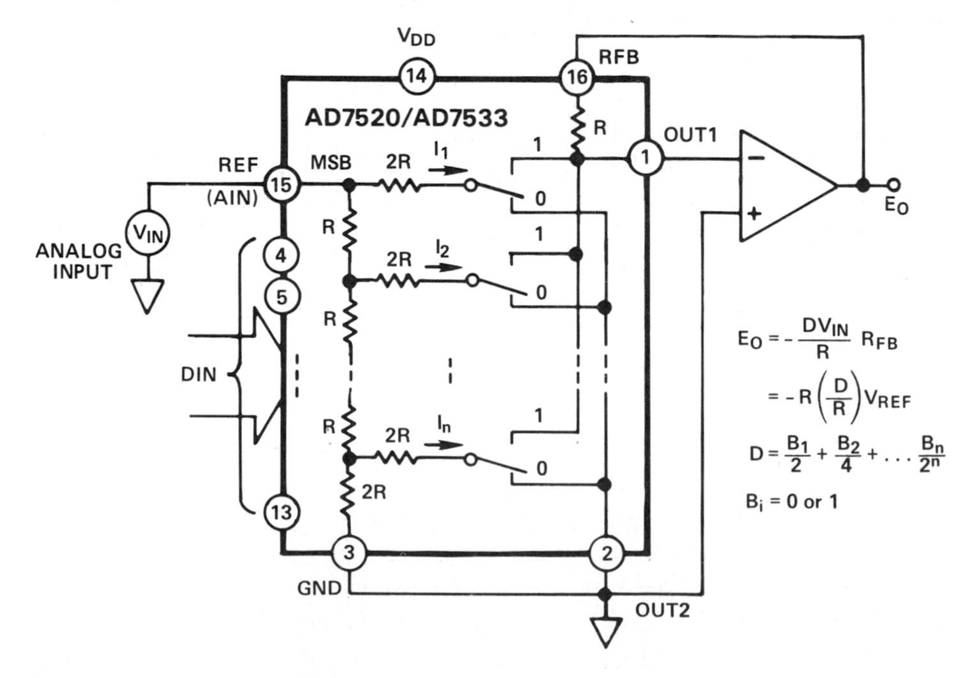

Figure 7.17. A typical CMOS DAC connected for conversion in current-steering mode. For clarity, not all resistance elements are shown.

if all bits but bit 2 are off, the current through the summing point will be one-half the MSB current, and the output will be $(-\frac{1}{2})(R/2R)E_{REF}$. If Bit 2 is off, the same current will flow to ground. Continuing down the ladder, each 2R resistor has one-half the voltage of the one above it, therefore it passes one-half the current. The output voltage is proportional (by superposition) to the sum of all the binary-weighted currents that have been switched on.

Because, in CMOS DACs, the switches are always very nearly at ground potential and will tolerate summing-point-type minuscule negative voltage swings, the reference can be either positive or negative, and the device can be used as digital gain control for ac signals or in four-quadrant multiplying DAC configurations.

The R-2R network can be employed to give unattenuated noninverting output *voltage*, simply by swapping the reference terminal and the output terminal. The reference terminal is driven at low impedance and the output terminal is connected to a high-impedance load, such as the input of a follower-connected operational amplifier (Figure 7.18). The effective resistance to ground of all resistors below a given node is 2R. Therefore, if the MSB alone is on, the output will be $(\frac{1}{2})\ V_{REF}$. When Bit 2 is on (all other bits grounded), its series 2R and the lower 2R form an effective generator of $E_{REF}/2$ in series with R. Since that R is in series with another R, and an effectively grounded 2R (via the MSB switch), the voltage at the output node is ½ the generator at node

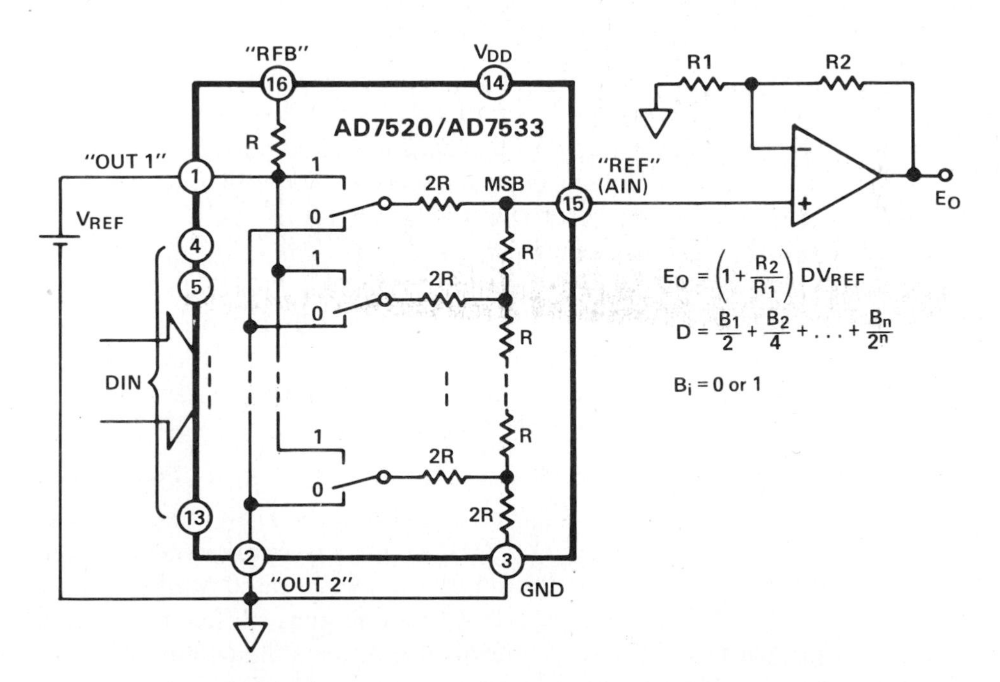

Figure 7.18. The same DAC as 7.17 connected for conversion in the voltage-switching mode. Generally, the magnitude of V_{REF} is limited in this configuration.

2, or (1/4) V_{REF}. The contributions of the subsequent bits form a binary progression, and superposition again provides an output proportional to the sum of the switched-on bits.

Since the entire network may be considered to be an equivalent generator having an output voltage D V_{REF} (where D is the fractional digital number), and an internal resistance, R, the output may be scaled down accurately by connecting precise resistance values to ground. Because of symmetry and self-duality R-2R networks may be used in other configurations. Some of these are discussed in Chapter 9.

7.5.3 SWITCHING

Needless to say, the switches used with the above networks are assumed here to be ideal. Switching may be performed in either the voltage mode or the current mode. A thorough description of the variety of voltage and current switches and switching schemes actually used in converters would be beyond the scope of this chapter. However, the use of monolithic bipolar transistor switches in converter design is addressed in Chapter 9. Information on popular converter designs will be found in Chapter 8, and a general discussion of CMOS switches will be found in Chapter 19. Considerable information about the characteristics of the switches used in CMOS DACs can be found in the *CMOS DAC Application Guide*, by Phil Burton, published by Analog Devices, Inc., 1985.

In the *current* mode, each leg of the ladder maintains constant current flow, which is steered either to an op-amp summing point or to ground. Two examples of current-mode converter block diagrams are shown in Figure 7.17 and 7.19. The first is that of a CMOS DAC, configured for current output. The bit currents are summed and converted to voltage via the feedback resistor, which is integrated on-chip, along with the ladder resistors, to maintain tracking of resistance with ambient temperature variations. This circuit is simple, and it can be used for 4-quadrant multiplication, but the presence of the switched resistors at the summing point of the op amp introduces problems with linearity in the presence of op-amp offsets, slowed dynamics due to switch capacitance, and noise coupling from the switch drive.

The second (Figure 7.19) is a DAC built on a bipolar chip. The design philosophy is described in some detail in Chapter 9. However, we can briefly describe here how it works: a buried Zener reference circuit develops −7.5 volts, which is scaled by the inverting amplifier to +2.5V. Applied to amplifier A2, with its 2.5-kΩ input resistor, it causes a 1-mA current to flow through the collector of reference transistor, Q1; the op amp adjusts the voltage across the 5-kΩ resistor to be whatever is necessary to maintain the feedback current at 1 mA. Since the voltage applied to that resistor is also applied at the input of the R-2R ladder, we can expect that the currents in successive legs will have

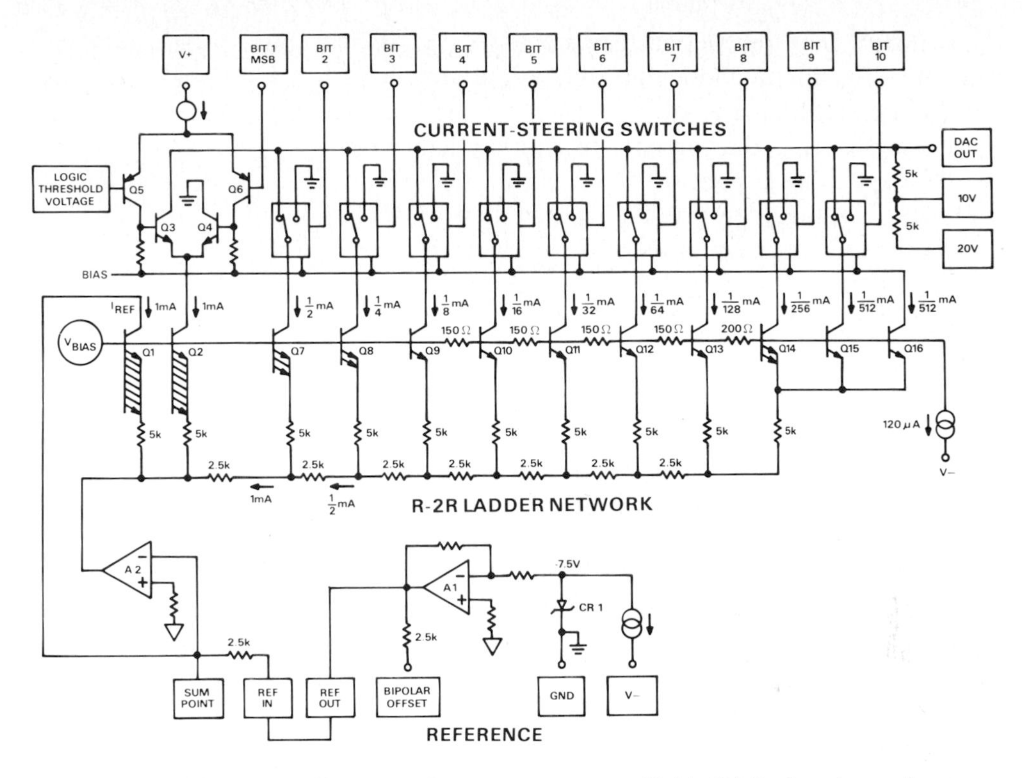

Figure 7.19. Schematic diagram of a current-output 10-bit DAC, showing reference, control amplifier, switching cell, and current-source arrangement.

a binary relationship; and, since Q1 is identical to Q2 (the MSB leg), the current that flows through its collector will also be 1 mA. The switches, which appear at the top of the diagram, switch the fixed currents either to ground or to the DAC OUT line, depending on the state of the control input for each bit.

While the reference in a DAC of this type is either fixed, or at best has a limited range of analog variation, the active reference circuit, compensated current sources, and fast-settling switches provide an excellent combination of speed and accuracy. A salient advantage of such current-mode converters is that the only significant voltage changes in the circuit appear at the output (in response to code changes); and the switch capacitances do not have to be charged through the ladder resistance.

Figure 7.18 shows a CMOS D/A converter being driven in the *voltage*-switching mode. In this mode, the ladder is used as a resistive attenuator; the switches alternate between a low-impedance reference voltage and ground. The magnitude of the current flowing through the switches is not important for precision, but the reference source and the switches must have sufficiently low impedance so that the current flowing through them does not cause significant voltage drops—or change in voltage drop as codes change.

For CMOS DACs in the voltage-switching mode, the constant resistance at the amplifier input (ladder output) eliminates linearity problems caused by modulation of the amplifier's offset voltage by code-dependent summing-point resistance in the current-steering mode of Figure 7.17. In addition, the switch capacitance is remote from the amplifier, and the charge is shunted to the input source or to ground, rather than to the summing point. Furthermore, the output capacitance of the network is considerably lower than in the current-switching mode. All of this results in cleaner and faster response of the circuit to code changes. As an additional feature, the system's output voltage is of the same polarity as the reference voltage; this makes it possible to operate the *DAC and its amplifier* from a single-polarity supply. Finally, only a single amplifier is required for bipolar digital operation, using offset binary or twos-complement coding.

The configuration has a few minor disadvantages. Performance is satisfactory for low values of reference voltage, but since the ON resistance of the FET switch is a function of the reference voltage, large values of reference voltage can produce significant nonlinearity. In addition, while current switching permits either polarity of input, only a single polarity of input is allowed in the voltage mode for CMOS DACs.

7.5.4 REFERENCES

Reference circuits for connection to converters are discussed in some detail in Chapter 20. Still popular as a reference device is the temperature-compensated breakdown ("Zener") diode, often used with operational amplifiers for operating-point stabilization, unloading, or transducing to current (Figure 7.20). It is being supplanted for many new designs by band-gap references, ICs that act like synthetic high-performance Zener diodes, and Zeners on constant-temperature substrates. On integrated-circuit chips, reference voltages are provided by stable, quiet, buried-Zener references—laser-trimmed to minimize error and temperature coefficient—and by band-gap circuitry.

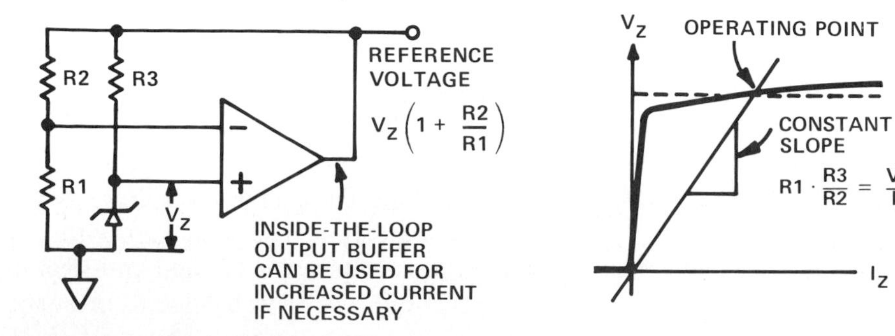

Figure 7.20. Stabilized diode reference. Amplifier adjusts feedback current to stabilize Zener-diode operating point independently of V_S or load variations.

In DACs employing active current sources, a reference must be provided that compensates for the characteristics of the current sources and switches. A powerful (patented) technique is described briefly in connection with Figure 7.19, and in some detail in Chapter 9.

7.5.5 BIPOLAR CONVERSION

For bipolar current-switching D/A conversion, using offset binary or twos complement codes, an offset current equal and opposite to the MSB current is added to the converter output. This may be accomplished with a resistor and a separate offset reference. More usually, it is derived from the converter's basic reference, in order to minimize drift of the output zero with temperature. It is usually jumpered externally, but there are converters in which the bipolar connection is programmed digitally.

The gain of the output inverting amplifier must be doubled, in order to double the output range, e.g., from 0-10V to ±10V. As indicated earlier (Figure 7.6), zero output corresponds to offset-binary 1 0 0 . . . 0 0, or twos-complement 0 0 0 . . . 0 0.

Figure 7.21 shows an example of a current-switching converter connected for bipolar output. Note that, because the amplifier is connected for sign inversion, the overall conversion relationship is "negative reference," i.e., +F.S. for all 0's (offset binary), −F.S. (1 − LSB) for all 1's.

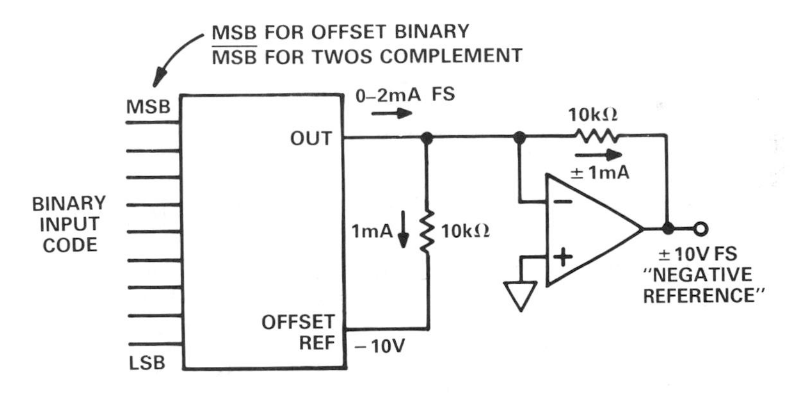

Figure 7.21. Bipolar connection of current-switching d/a converter for offset-binary or twos-complement codes.

For non-inverting applications, employing the output current of an active-current-source DAC to develop a bipolar output voltage across load resistance, the same values of offset voltage and resistance are used, but—for a given range of compliance voltage—the proper value of output voltage scale factor depends on the load presented by the parallel combination of the internal resistance, the offset resistance, the external load, and the gain of the buffering op amp. (Figure 7.22)

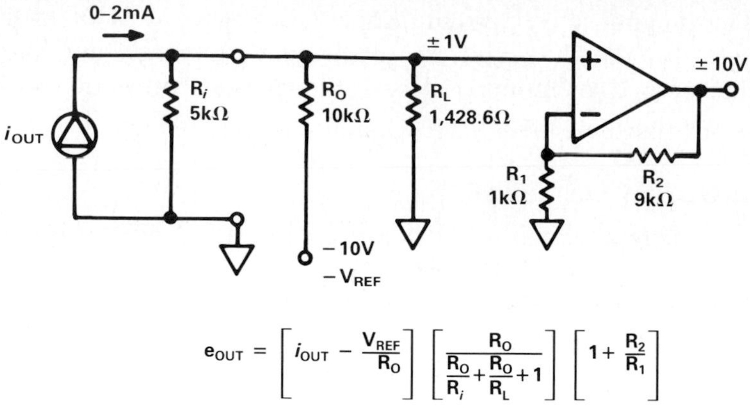

$$e_{OUT} = \left[i_{OUT} - \frac{V_{REF}}{R_O} \right] \left[\frac{R_O}{\frac{R_O}{R_i} + \frac{R_O}{R_L} + 1} \right] \left[1 + \frac{R_2}{R_1} \right]$$

Figure 7.22. Non-inverting bipolar output from current-switching d/a converter.

For bipolar D/A conversion using CMOS d/a converters in the voltage mode, the circuit of Figure 7.23 may be employed. It is similar to the circuit of Figure 7.18, but provides for subtracting the reference from the ladder output. If D is the digitally set DAC coefficient, the amplifier output is $2\,D\,V_{REF} - V_{REF}$. For $D = ½$, the output is zero; if $D = 0$, the output is $-V_{REF}$, and if D is all-1's, the output is $(1 - 2^{-(n-1)})\,V_{REF}$.

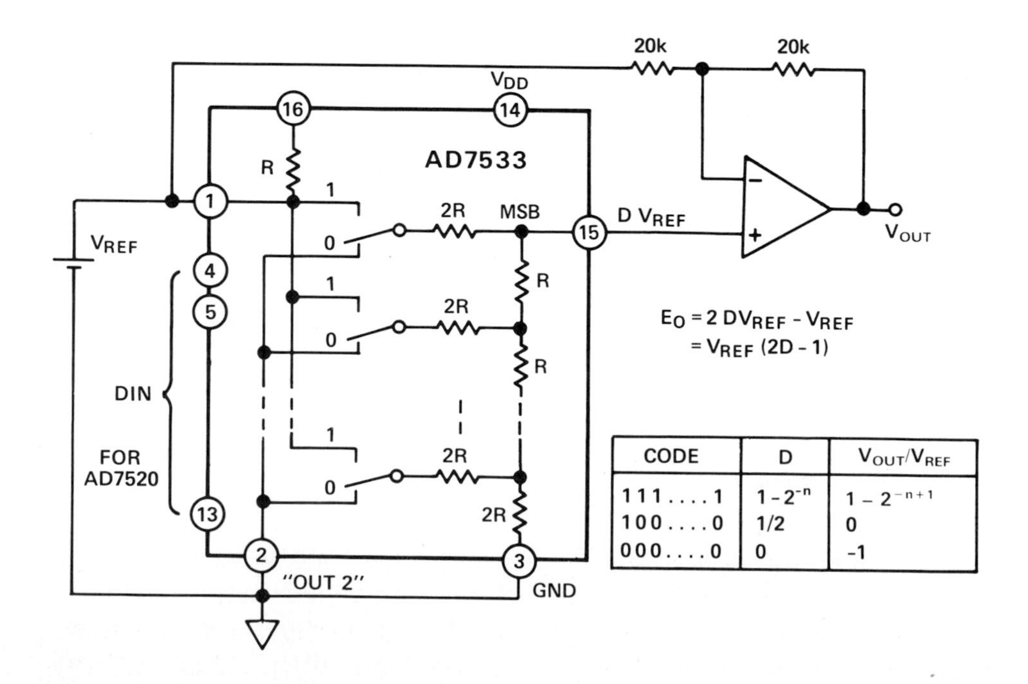

CODE	D	V_{OUT}/V_{REF}
111....1	$1-2^{-n}$	$1 - 2^{-n+1}$
100....0	1/2	0
000....0	0	-1

Figure 7.23. Connection of a CMOS DAC in the voltage-switching mode for bipolar operation. If V_{REF} is provided by a 2.5-volt source, such as the AD580, the nominal output swing is ±2.5V.

For full four-quadrant multiplying D/A conversion, employing CMOS DACs, the circuit of 7.24 provides a similar transfer function, except that V_{REF} may be a dc or ac voltage, positive or negative. The digital input determines the positive or negative value of gain applied to the signal.

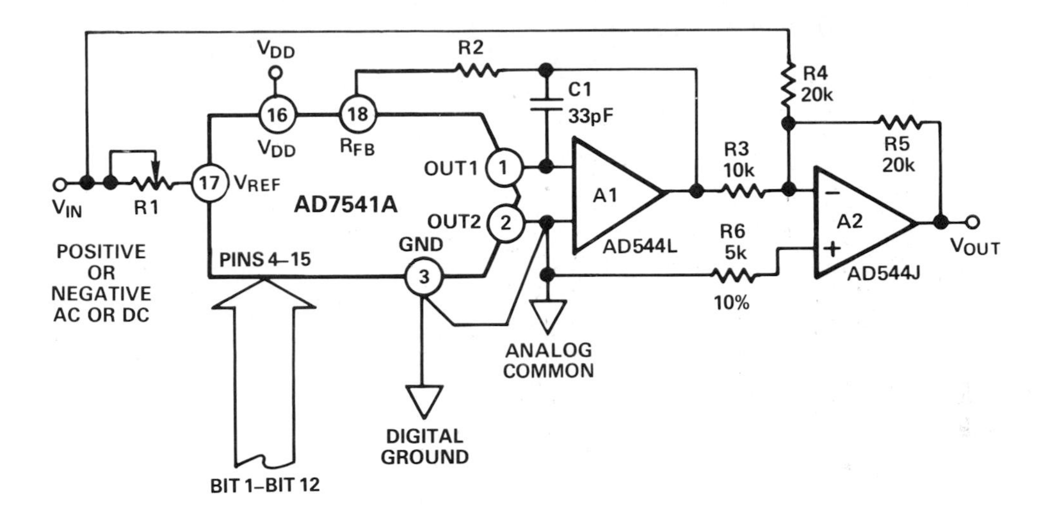

Binary Number in DAC MSB LSB	Analog Output, V_{OUT}
1111 1111 1111	$+V_{IN}\left(\frac{2047}{2048}\right)$
1000 0000 0001	$+V_{IN}\left(\frac{1}{2048}\right)$
1000 0000 0000	0V
0111 1111 1111	$-V_{IN}\left(\frac{1}{2048}\right)$
0000 0000 0000	$-V_{IN}\left(\frac{2048}{2048}\right)$

Figure 7.24. Connecting a CMOS DAC for 4-quadrant multiplication.

For sign-magnitude conversion in CMOS DACs, a scheme like that shown in Figure 7.25 may be used. The output of the current-switching DAC-and-amplifier is inverted for net positive gain, D (where D is the analog value of the digital gain setting), and added either to zero or to $-2\ D\ V_R$, for a net gain of either $+D$ or $-D$.

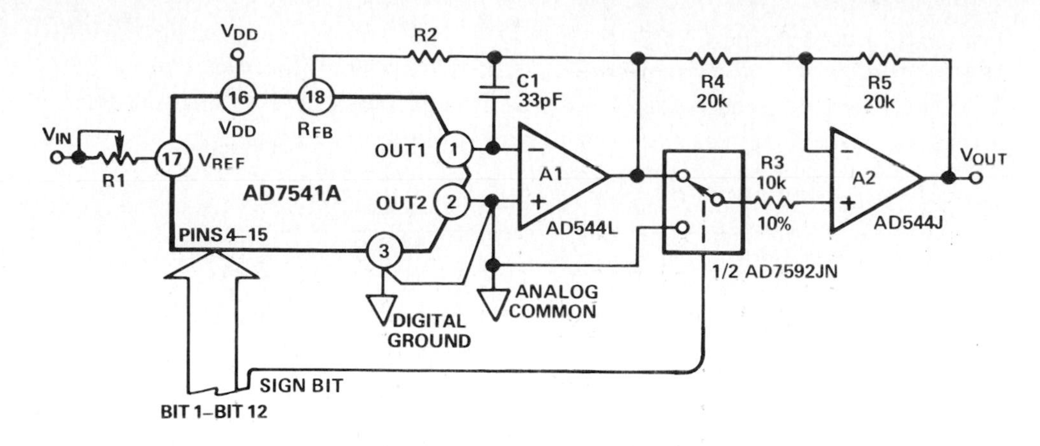

Sign Bit	Binary Number in DAC (MSB … LSB)	Analog Output, V_{OUT}
0	1111 1111 1111	$+V_{IN} \cdot \left(\frac{4095}{4096}\right)$
0	0000 0000 0000	0 Volts
1	0000 0000 0000	0 Volts
1	1111 1111 1111	$-V_{IN} \cdot \left(\frac{4095}{4096}\right)$

Note: Sign bit of "0" connects R3 to GND.

Figure 7.25. Connecting a 12-bit CMOS DAC for sign-magnitude bipolar operation.

The basic parallel-input D/A converter circuits considered so far have the common property that the analog output continually tends to reflect the state of the logic inputs. However, if the basic conversion circuitry is preceded by a register, either external or internal, the DAC proper will respond only to the inputs stored in its register. When the register is latched, the analog output is unchanging. This property is universally required in bus-type data distribution systems, where data is continually appearing, but it is desired that a DAC respond only at certain times, then hold the analog output constant until the next update. In this sense, a DAC with buffer storage may be viewed as a sample-hold with digital input, analog output, and (conceivably) infinite "Hold" time.

The register is enabled by a *strobe*, the net gating signal that results, in microprocessor systems, when the specific device has been both addressed and given a *write* signal, which causes it to update. The limiting rate at which the strobe may be allowed to update is determined by two factors: the settling time of the DAC, and the response time of the logic. In general, settling time of the analog portion of the D/A converter is at least an order of magnitude

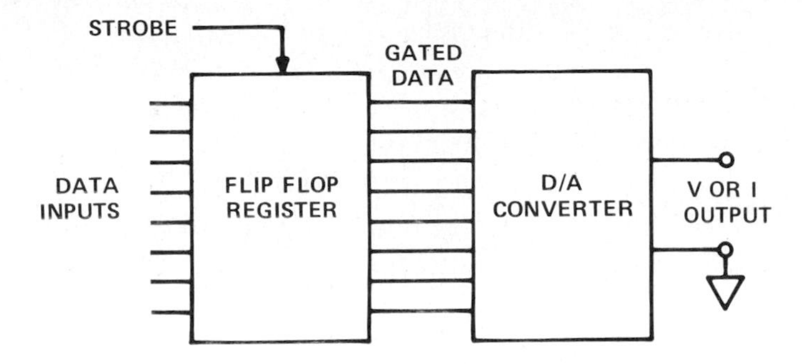

Figure 7.26. D/A Converter with buffer register.

slower than the response time of modern high-speed TTL logic circuits and is thus the limiting factor on update rate.

On-chip DAC register architectures vary. For example, a DAC may be single- or double-buffered—one or two registers—and the input register may accept data in more than one byte, or nybble, depending on the width of the bus to which it is to be interfaced. These structures are discussed in several places, notably in Chapter 3 and Chapter 8. If there are multiple DACs in the same package, they are generally connected to a single bus, which is an extension of the external bus; internal chip logic determines the manner of selection and updating.

The speed of the digital portion of a D/A converter is of especial importance when the "glitch" caused by unequal turn-on and turn-off times is an important factor in the application. The digital inputs to a DAC come from digital logic circuits, which may exhibit *skew*, or unequal turn-on and turn-off times. Skew is inherent in the on-off nature of saturated logic, such as TTL; symmetry is much more nearly achieved in ECL, which is always in the active range. The analog switches used in DACs also exhibit skew; however, even if the switch circuitry is specifically designed to minimize skew, the additional skew of the digital logic will constitute an irreducible minimum. In applications where it is an important factor, glitch energy (or *impulse*) can be reduced by using logic that is faster than would be necessary just to drive the analog switches.

7.6 A/D CONVERTER CIRCUITS

There are a vast number of conceivable circuit designs for A/D converters.[7] There are a much more limited number of designs available on the market in small, modular form at low cost, specifically designed for incorporation as components of equipment. The most popular of these are:

[7]See *Electronic Analog/Digital Conversions*, by H. Schmid (Van Nostand Reinhold, 1970), and more recently, *Data Conversion Integrated Circuits*, edited by Daniel J. Dooley (IEEE Press—Wiley, 1980)—for an encyclopedic panoply of A/D (and D/A) converter circuit designs.

Successive-approximation types
Integration (single-, dual-, & quad-slope and v-to-f) types
Counter and "servo" types
Parallel and modified-parallel types

Each approach has characteristics that make it most useful for a specific class of applications, based on speed, accuracy, cost, size, versatility.

7.6.1 SUCCESSIVE APPROXIMATIONS

Successive-approximation A/D converters are quite widely used, especially for interfacing with computers, because they are capable of high resolution (to 16 bits), and high speed (to 1 MHz throughput rates). Conversion time is fixed and independent of the magnitude of the input voltage. Each conversion is unique and independent of the results of previous conversion, because the internal logic is cleared at the start of a conversion.

Modern IC converters, such as the monolithic AD574A 12-bit ADC, include 3-state data outputs and byte controls to facilitate interfacing with microprocessors. A "three-state" output has, in addition to the normal "1" and "0" states, when enabled, a not-enabled condition, in which the output is simply disconnected via an open voltage switch. This permits many device outputs to be connected to the same bus—only the device that is enabled (one at a time) can drive the bus. Since many processor data buses are only 8 bits wide, 10- or 12-bit data must often be communicated in two steps, one "byte" at a time.

The conversion technique consists of comparing the unknown input against a precise voltage or current generated by a D/A converter (Figure 7.26). The input of the D/A converter is the digital number at the A/D converter's output. The conversion process is strikingly similar to a weighing process using a chemist's balance, with a set of *n* binary weights (e.g., ½ lb, ¼ lb, ¹⁄₁₆ lb (= 1 oz), ½ oz, ¼ oz, etc., for unknowns up to 1 lb.)

After the conversion command is applied, and the converter has been cleared, the D/A converter's MSB output (½ full scale) is compared with the input. If the input is greater than the MSB, it remains ON (i.e., "1" in the output register), and the next bit (¼ FS) is tried. If the input is less than the MSB, it is turned OFF (i.e., "0" in the output register), and the next bit is tried. If the second bit doesn't add enough weight to exceed the input, it is left ON ("1"), and the third bit is tried. If the second bit tips the scales too far, it is turned OFF ("0"), and the third bit is tried. The process continues in order of descending bit weight until the last bit has been tried. The process completed, the *status* line changes state to indicate that the contents of the output register now constitute a valid conversion. The contents of the output register form a binary digital code corresponding to the input signal's magnitude.

Figure 7.27a is a block diagram of a successive-approximations A/D conver-

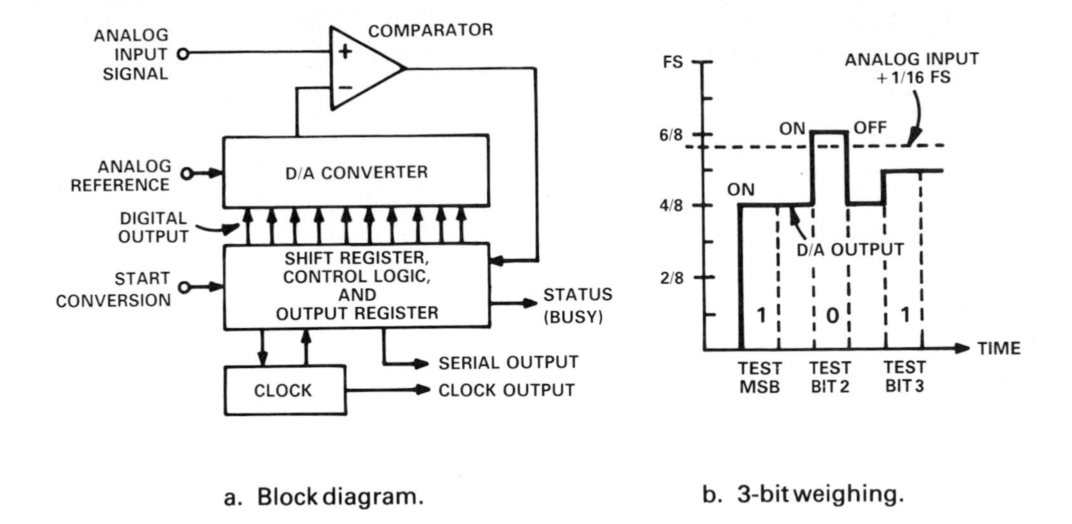

a. Block diagram.

b. 3-bit weighing.

Figure 7.27. Successive-approximation a/d converter.

ter, accompanied by a time history of a simple 3-bit conversion, in terms of the D/A converter output (the weight added to the balance pan). Note that, to place the D/A converter output in the center of each ideal output quantum, a ½-LSB "thumb" is placed on the scale (see Figure 7.2), in order to locate the transitions precisely at the ½ LSB points.

In the example of Figure 7.27b, the input does not change during conversion. If the input were to change during conversion, the output number could no longer accurately represent the analog input, for the same reason one would have difficulty using a chemist's balance with a changing unknown. However, even if the final weight were to match the final unknown, there would still be a question as to whether the final unknown was itself legitimate, especially if the weighing had to occur at a specific time. To avoid any problems of this sort, it is usual to employ a sample-hold device ahead of the converter to retain the input value that was present at a given time before the conversion starts, and maintain it constant throughout the conversion. The *status* output of the converter could be used to release the sample-hold from its *hold* mode at the end of conversion. A sample-hold may not be needed if the signal (by itself, or with filtering) varies slowly enough and is sufficiently noise-free that significant changes will not be expected to occur during the conversion interval.

Accuracy, linearity, and speed are primarily affected by the properties of the D/A converter (and its reference), and the comparator. In general, the settling time of the D/A converter and the response time of the comparator are considerably slower than the switching time of the digital elements. The differential nonlinearity of the D/A converter will be reflected in the differential nonlinearity of the resulting A/D converter. If the D/A converter is non-monotonic, one or more codes may be missing from the A/D converter's output range. Bipolar inputs are dealt with by using a D/A converter with bipolar output and offset binary coding, and appropriate input scaling.

7.6.2 INTEGRATION (RAMP AND V-TO-F TYPES)

This family of converters is also quite popular. Its members perform an *indirect* conversion, by first converting to a function of time, then converting from the time function to a digital number using a counter. Integrating types such as the *dual-ramp* and quad-slope types are especially suitable for use in digital voltmeters and those applications in which a relatively lengthy time may be taken for conversion to obtain the benefits of noise reduction through signal averaging.

Here's how the dual-ramp type works: The input signal is applied to an integrator; at the same time a counter is started, counting clock pulses. After a predetermined number of counts (a fixed interval of time, T), a reference voltage having opposite polarity is applied to the integrator. At that instant, the accumulated charge on the integrating capacitor is proportional to the average value of the input over the interval T. The integral of the reference is an opposite-going ramp having a slope V_{REF}/RC. At the same time, the counter is again counting from zero. When the integrator output reaches zero, the count is stopped, and the analog circuitry is reset. Since the charge gained is proportional to $\overline{V_{IN}}\, T$, and the equal amount of charge lost is proportional to $V_{REF}\, \Delta t$, then the number of counts relative to the full count is proportional to $\Delta t/T$, or V_{IN}/V_{REF}. If the output of the counter is a binary number, it will therefore be a binary representation of the input voltage. Converters of this type usually employ sign-magnitude coding for bipolar input ranges. However, if the input is attenuated and offset by half the reference voltage, the output will be an offset binary representation of a bipolar input, suitable as an input for computer systems (Figure 7.28a).

Dual-slope integration has many advantages. Conversion accuracy is independent of both the capacitance and the clock frequency, because they affect both the up-slope and the down-ramp in the same ratio. Differential linearity is excellent, because the analog function is free from discontinuities, the codes are generated by a clock and counter, and all codes can inherently exist. Resolution is limited only by analog resolution, rather than by differential nonlinearity; hence, the excellent fine structure may be represented by more bits than would be needed to maintain a given level of scale-factor accuracy. The integration provides rejection of high-frequency noise and averaging of changes that occur during the sampling period. The fixed averaging period also makes it possible to obtain "infinite" normal-mode rejection* at frequencies that are integral multiples of $1/T$ (see Figure 7.28b).

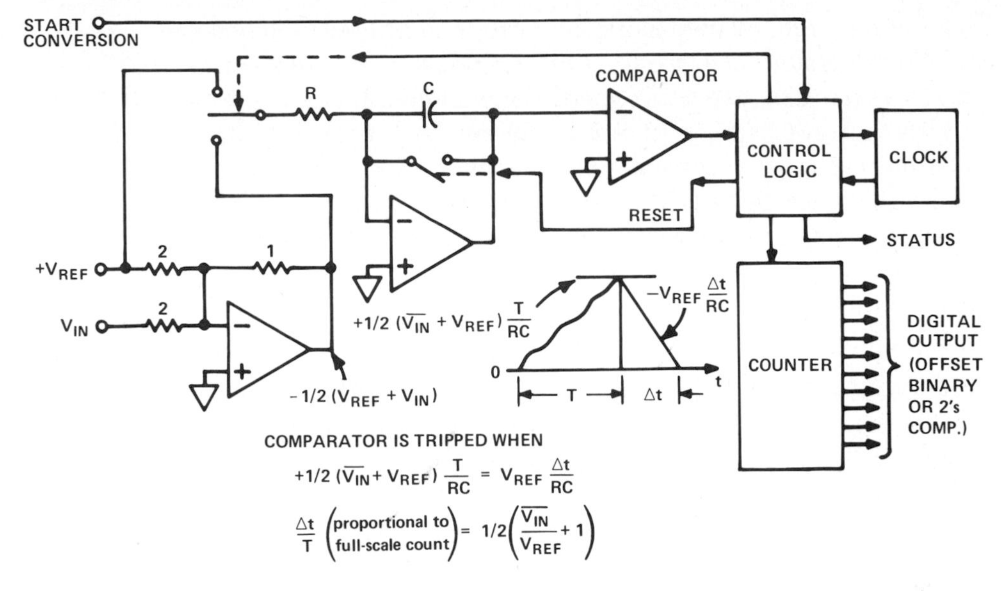

a. Conversion scheme for bipolar input.

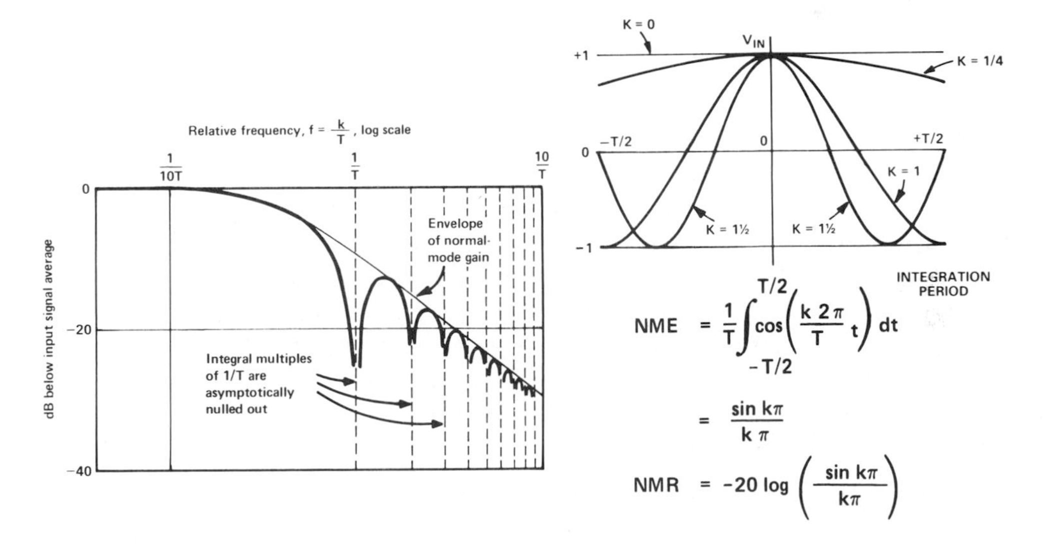

b. Worst-case normal-mode response of dual-slope ADC.

Figure 7.28. Dual-ramp converter (voltage-to-time).

***Normal-mode* noise consists of unwanted signals that appear on the input line, even if common-mode error is nil. If a low-frequency or dc quantity is to be converted in the presence of a high-frequency ripple, a successive-approximations A/D converter, even if preceded by a sample-hold, will convert the instantaneous values of signal-plus-noise, producing a noisy digital signal. On the other hand, an integrator will inherently attenuate high frequencies, producing smoothing, and, if combined with a fixed averaging period, will null out those frequencies that have whole numbers of cycles during the averaging period.

Throughput rate of dual-slope converters is limited to somewhat less than $1/(2T)$ conversions per second. The sample time, T, is determined by the fundamental frequency to be rejected. For example, if one wishes to reject 60Hz and its harmonics, the minimum integrating time is 16-2/3ms, and the maximum number of conversions is somewhat less than 30/s. Though too slow for fast data acquisition, dual-slope converters are generally quite adequate for such transducers as thermocouples; and they are the predominant circuit used in constructing digital voltmeters. Since DVM's use sign-magnitude BCD coding, bipolar operation requires polarity sensing and reference-polarity switching, rather than simple offsetting.

A shortcoming of conventional dual-slope converters is that errors at the input of the integrating amplifier or the comparator show up as errors in the digital word. Such errors are usually reduced by the introduction of a third portion of the cycle, during which a capacitor is charged with zero-drift errors, which are then introduced in the opposite sense during the integration, in order to (it is hoped) nullify them. An interesting scheme for nullifying most input errors is the patented *quad-slope* principle*; it stores the errors in the form of a digital count during a calibration cycle and subtracts them from the final count during the conversion cycle.

Other conversion approaches in this class include the single-ramp type and v/f converters. In the single-ramp converter, a reference voltage, of opposite polarity to the signal, is integrated (while a counter counts clock pulses) until the output of the integrator is equal to the signal input. At that time (Δt) the output of the integrator is $E_{REF}\Delta t/RC$. Therefore, Δt—hence, the number of counts and the corresponding digital number—is proportional to the ratio of the input to the reference. This process has the weakness that its accuracy depends on both the capacitor (extremely accurate and stable resistors are relatively easy to come by) and the clock frequency. In the v/f converter, a frequency is generated in proportion to the input signal; a counter measures the frequency and provides a digital output code, the value of which is proportional to the input signal. In both of the above schemes, offsetting may be used to obtain offset binary representation of bipolar analog inputs. V/F conversion is explored in some depth in Chapter 15.

7.6.3. COUNTER AND "SERVO" TYPES

Figure 7.29 is a block diagram of a counter-comparator A/D converter, which is analogous to the single-ramp type, but only conversion time (not accuracy) depends on the time scale. The analog input is compared with the output of a D/A converter, the digital input of which is driven by a counter. At the start of the conversion, the counter starts its count, which continues until the D/A

*U.S. Patent 3,872,466

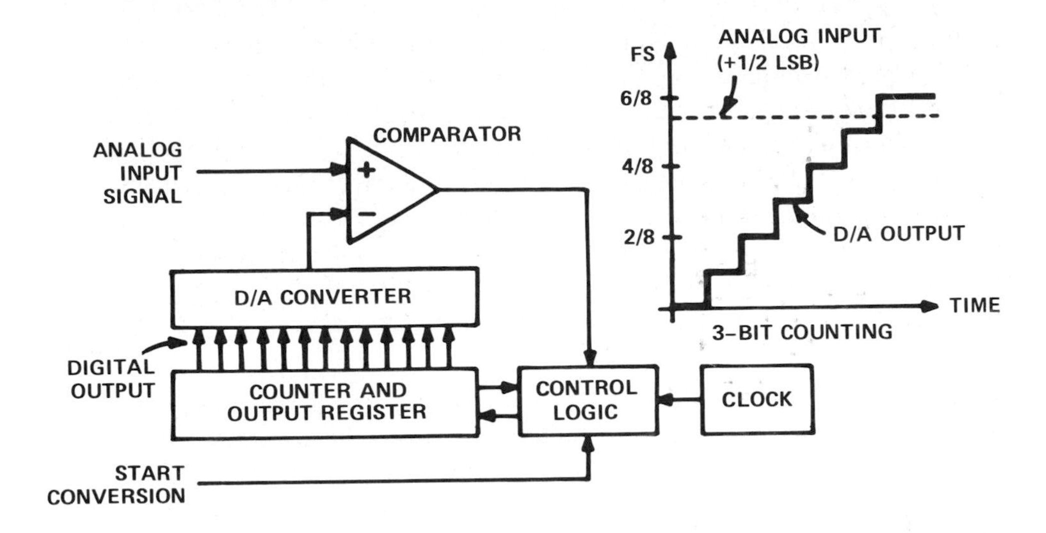

Figure 7.29. Counter-comparator a/d converter.

output crosses the input value. At that point, conversion ceases, and the converter is ready to perform the next conversion after the counter has been cleared and its output dumped into an output register. The number of counts appears in the output register. For bipolar inputs, a bipolar D/A converter is used, and the count is an offset binary representation of the input, starting from negative full-scale.

Though quite simple in concept, this converter has the disadvantage of limited speed for a given resolution, since the conversion time for a full-scale change is equal to the clock frequency divided into the maximum number of counts. For example, if the clock frequency is 10MHz, the maximum throughput rate for 10-bit resolution (1024 counts) is something less than 10kHz (100μs per conversion). A variation of this converter is the "servo" type, in which an "up-down" counter is used.

If the output of the D/A converter is less than the analog input, the counter counts up. If the D/A output is greater than the analog input, the counter counts down. If the analog input is constant, the counter output "hunts" back and forth between the two adjacent bit values. This "tracking"converter can follow small changes quite rapidly (it will follow 1 LSB changes at the clock rate), but it will require the full count to acquire full-scale step changes. The principle is widely used in resolver- and synchro-to-digital converters. Since it seeks to "home in" on the analog value, the analogy to a servomechanism is quite evident.

It seeks to convert continuously, which may be a disadvantage in tying it in with a fast data-acquisition system, since it can give a valid "conversion com-

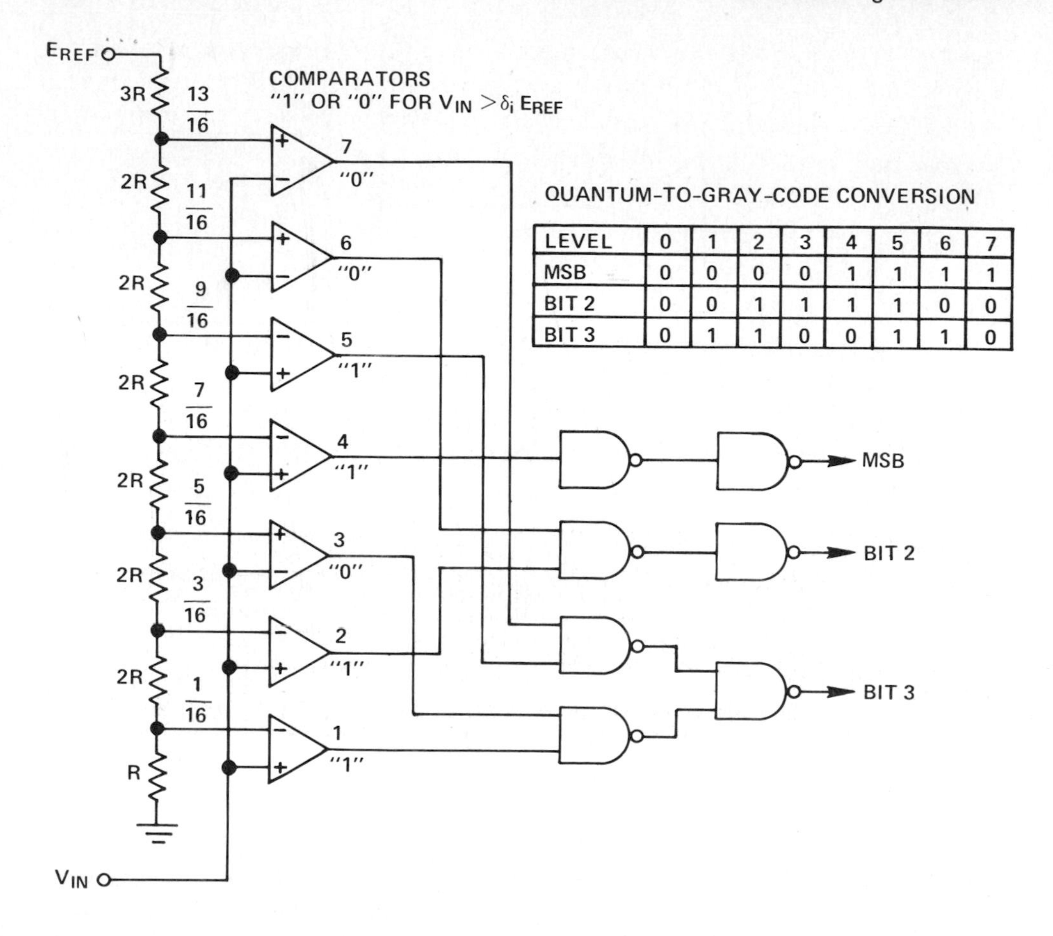

LEVEL	0	1	2	3	4	5	6	7
MSB	0	0	0	0	1	1	1	1
BIT 2	0	0	1	1	1	1	0	0
BIT 3	0	1	1	0	0	1	1	0

Figure 7.30. Parallel ("flash") 3-bit a/d converter with Gray-code output.

plete" report only during the clock period immediately following a change in state of the comparator (which in general occurs at irregular intervals). A buffer storage register may be used to store the previous count, while the counter is seeking the next value. By stopping the count (following a completed conversion) at an externally determined instant, the servo-type converter may be used as a sample-hold with arbitrarily long *hold* time (with no droop). If the "up" or the "down" count is disabled, the converter will act as a valley follower or a peak follower, counting in the appropriate direction only when the analog input exceeds the previous extreme value. Both the analog and the digital stored values are available.

7.6.4 PARALLEL TYPES

Figure 7.30 shows a parallel 3-bit "flash" converter with Gray code output. It has $2^n - 1$ comparators, biased 1 LSB apart, starting with + ½ LSB. For 0 input, all comparators are off. As the input increases, it causes an increasing number of comparators to switch state. The outputs of the comparators are applied to the gates, which provide a set of outputs that fulfill the appropriate conditions for Gray-code output. (Natural binary could be implemented in the same way, using an appropriate table).

The evident advantage of this approach is that conversion occurs in parallel, with speed limited only by the switching time of the comparators and gates. As the input changes, the output code changes. Thus, this is the fastest approach to conversion.

Unfortunately, the number of elements increases geometrically with resolution. As linear and digital integrated-circuit elements of increasing complexity become available, increased levels of resolution will tend to approach the threshold of practicality. But high resolution and the fastest speeds at low cost are still some time away.

By combining parallel conversion for moderate numbers of bits with iteration (to wit, 6 or 7 bits of flash in two successive conversions, employing *digitally corrected subranging*—see Chapter 13), it is possible to strike a compromise that gives considerably better resolution than a parallel approach, with less complexity, and improved speed over the successive-approximation approach. For example, the CAV-1210 has 12-bit resolution at a rate of 10 megasamples per second.

7.6.5 A NOTE ON SHARED LOGIC

In this book, we are concerned with the embodiment of the conversion function by means of ICs or other modules, in essentially complete form, with *completely defined specifications*. We should acknowledge, however, that when considering the tradeoffs between hardware and software, a software-oriented designer will be tempted to consider hardware savings inherent in using the control-logic capability of a microprocessor, along with basic precision analog functions (for example, a DAC with built-in reference and a precision comparator), to perform single or multiplexed conversions, employing the techniques mentioned here, but without using a piece of hardware identifiable as an "a/d converter" *per se*.

A decision to do this is in some respects equivalent to a decision to *design* a converter (analogous to the classical "make-or-buy" decision). It will tend to work well at low speeds and resolutions, in situations where the number of conversions required will not impose excessive time and software burdens on the processor. While there are applications for which the approach is eminently fruitful (e.g., dedicated instruments, to be manufactured in large quantity), the usual tradeoffs should be considered, lest the fascination of expending design- and manufacturing effort and software development in areas peripheral to one's primary mission, lead one down the "primrose path" of wasted resources.

CONCLUSION

In this chapter, we have attempted to provide the fundamentals for a basic understanding of converters. In the chapters that follow, we will discuss further some of the considerations faced and architectures employed by the converter designer, provide an understanding of and a guide to specifications of converters, and explore the elements of successful system design using converters.

Chapter Eight

Converter Microcircuits

This chapter endeavors to provide a thumbnail sketch of the wealth of integrated-circuit data converters available today, accompanied by historical insights into the technological influences affecting their development. Although no such summary from one source at one instant of time can be totally comprehensive or exhaustive, the reader should find the variety and diversity eminently satisfying.

8.1 INTEGRATED-CIRCUIT CONVERTERS

As the use of digital techniques in measurement, communication, and control grew by leaps and bounds, the size and price of processors and other LSI (and MSI) logic shrank in similar degree, with the inevitable result that further penetration of digital techniques into those fields became inevitable, in the regenerative fashion that is characteristic of the integrated-circuit era. Along with such other peripherals as keyboards, displays, and memories, converters have followed this spiralling trend—as a matter of necessity.

But it hasn't been easy. Linear IC's have always been more difficult to fabricate for reasonable degrees of resolution and accuracy than digital IC's—in part because the variables that are the input or output involve a continuum of voltage or current, rather than the easier-to-handle two-valued logic.

While the problems of implementing digital circuitry have involved questions of functions-per-chip, speed, and power dissipation, the analog (precision IC) problems have related more to simple existence and survival. Such matters as offset, bias current, drift, dynamic stability, common-mode errors, and open-loop gain—as well as slewing rate and settling time—have concerned both designers and users of op amps (the "representative" linear-IC product).

But converters are more difficult by at least an order of-magnitude. While IC op amps called for precision transistor circuitry and clever design, and IC analog multipliers added a need for precision resistors (and references)—but to-date have attained accuracies to within 0.1% (at best) and are hardly in the commodity class—*converters* call for all of these prodigies of linear design and processing, and *more*: on-chip switches, logic, and everyday resolutions of up to 16 bits. Furthermore, the downward trend in price for digital functions carries over to the converter world, pressing converter technology to produce higher performance at lower cost.

Although the technology still has a considerable distance to go, it is worthwhile to consider the progress made, just within the past fourteen years, as measured by entries within the Analog Devices catalog. In the 1972 *Product Guide*, the IC conversion product line consisted of just two families of monolithic quad switches (and compatible resistor networks) for constructing precision 8-10-12-bit A/D and D/A converters.

In comparison, the 1984 Analog Devices *Integrated Circuits Databook* contains over 700 pages of data on converter microcircuits, including A/D, D/A, V/F, and S/D converters as well as a wide variety of support components, such as switches, multiplexers, references, and sample-hold amplifiers. Most of these products are monolithic.

Since monolithic technology has evolved to the point where low-cost data converters are readily available, many applications have become economically attractive in much the same way that the low-cost monolithic computer (microprocessor/microcomputer) made digital computation applications feasible. As microprocessors expand into more and more applications, analog input and output capability become increasingly important. The advances in monolithic converter technology have thus opened up vast new application areas.

8.1.1 PROCESSES

Converters have been manufactured using nearly all of the same monolithic processes used for fabricating digital devices. However, most digital processes are designed for highest possible speed and density, and lowest power, which often leads to small geometries and hence low breakdown voltages, limiting both power-supply and signal voltage ranges. While this is acceptable in digital circuits, it limits the dynamic range and resolution of data-converter circuits.

Modern microcircuit converters are manufactured with five generic technologies: bipolar, bipolar/I^2L (integrated injection logic), CMOS, BiMOS, and hybrid.

The *bipolar* process used for converters is fundamentally the same process used to manufacture classical linear functional devices, such as operational amplifiers and voltage regulators. It has the advantages of being well-understood, due to its lengthy history, capability for high speed and low noise, and high breakdown voltages (leading to wide signal range). Its principal limitation is its generally poor logic capability, owing to the larger geometries used and relatively high power requirements.

Bipolar/I^2L processing allows the standard process to include more complex logic functions, at the expense of slightly reduced breakdown voltages. Many data converters use this process, since it retains most of the advantages of traditional bipolar technology. It is interesting to note that the use of I^2L is much more widespread in the data-converter world than in the digital world. This is because the speed-power product of MOS logic is more favorable than dedicated I^2L. Thus I^2L has evolved into a process with a specialized niche.

CMOS (Complementary Metal-Oxide Semiconductor) technology has evolved as the preferred MOS fabrication technology. It is superior to nearly all other processes in power requirements; and reduced-geometry devices are capable of speeds comparable to bipolar logic. In addition to these benefits, designers of data converters have found other advantages in CMOS. For example, one of the most common functions in data-converter designs is the analog switch. Switches can be produced in bipolar technology, but they are unidirectional, passing only currents of a single polarity. Bipolar transistors also exhibit a voltage drop (usually a junction—base-emitter—voltage) in series with the signal, and special circuit techniques must be employed in order to circumvent these shortcomings.

On the other hand, CMOS switches (also called transmission gates) have no V_{BE} drop to worry about. Since their switches act like resistors, they are also capable of passing current in either direction, which leads to the possibility of producing multiplying DACs that can operate with positive or negative, fixed or variable, reference sources. However, the limitation of the CMOS process that inhibits the use of CMOS for complete converters has been the lack of low-noise reference sources and gain (voltage-output) stages. Some operational amplifiers have been produced using CMOS technology, but they have fallen far short of the speed, stability, and low noise achievable with bipolar op amps. However, CMOS technology allows complex logic functions to be included on a converter chip, with very little additional power consumption.

BiMOS is a generic term applied to several manufacturing processes that combine bipolar and MOS transistors on one chip. This process is generally considered to be the most promising technology for future generations of data converters. It allows CMOS logic functions of low power and high density to be produced on the same chip as precision low-noise, high-gain, high-speed bipolar circuitry. Thus, it makes possible a complete converter with analog signal conditioning, conversion, memory, computation, and digital data communication—a "system-on-a-chip."

Hybrid technology has been an effective method for the merging of multiple IC technologies in a single design—in order to produce a more complete function in a small package. Many have believed that, as monolithic technology progresses, hybrid technology must outlive its usefulness. However, quite the opposite is true. As monolithic technology has advanced, it has provided the hybrid designer with an increasingly large inventory of devices to choose from. Thus, hybrid technology will always have a place in converter manufacture, because it will always make possible higher performance and more complete functions in a single small package.

8.2 BIPOLAR D/A CONVERTERS

The earliest integrated circuit d/a converters were manufactured using bipolar technology; they included only the basic core of a complete DAC—the array of switches and resistors to set the weight of each bit (Figure 8.1a). The 1408 and a later, higher-speed derivative—known as the DAC-08—fall into this category.

These converters are produced by several manufacturers and are available at very low cost. However, they require many additional external components in order to be usable in a system design (Figure 8.1). These external components include several resistors, a reference, a latch, an output operational amplifier, possibly a compensation capacitor, and usually one or more trimming potentiometers.

Converters like the 1408 were limited to 8-bit accuracy by the matching and tracking limitations of diffused resistors. The ability to match diffused resistors is limited by the definition of the resistor geometries in the photolithographic process, and post-fabrication adjustment is not possible. Furthermore, the temperature coefficient of diffused resistors is quite high, so that resistors which may match at room temperature could drift apart at other temperatures, causing degraded accuracy. When higher accuracy is desired, lower-tempco resistors are needed, and some means of post-fabrication adjustment is desirable.

Thin-film resistors exhibit low temperature coefficients and can be trimmed by use of a laser; they are well-suited for use in data converters. Thin-film resistors are manufactured using several different materials—silicon-chromium, nickel-chromium, and tantalum nitride are the most common. The resistor material is deposited on substates of ceramic, glass, or silicon, depending upon the manufacturer.

Hybrid converters have been manufactured with discrete switches and laser-trimmed thin-film resistor networks for quite some time. However, it was not possible to combine these components on a single monolithic chip (and thus produce a monolithic 12-bit DAC) until the late 1970s.

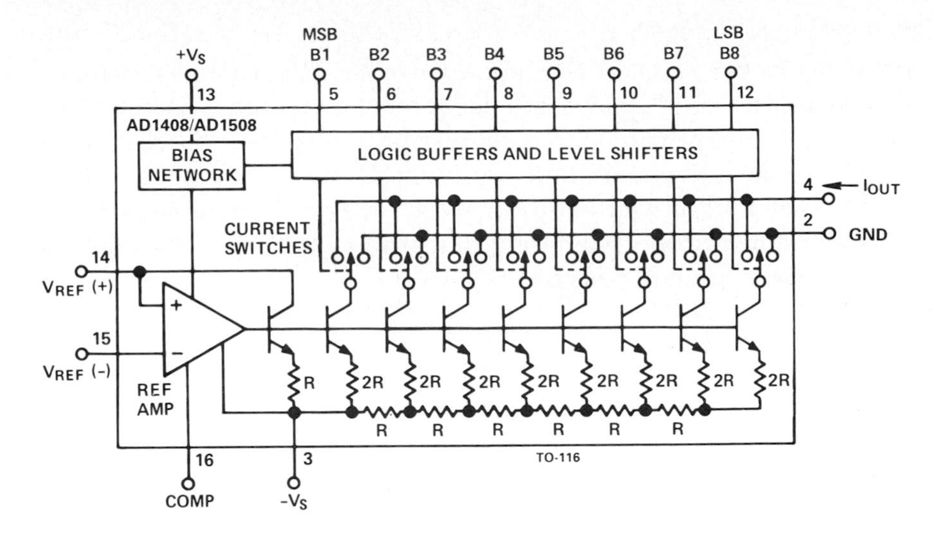

a. Functional block diagram.

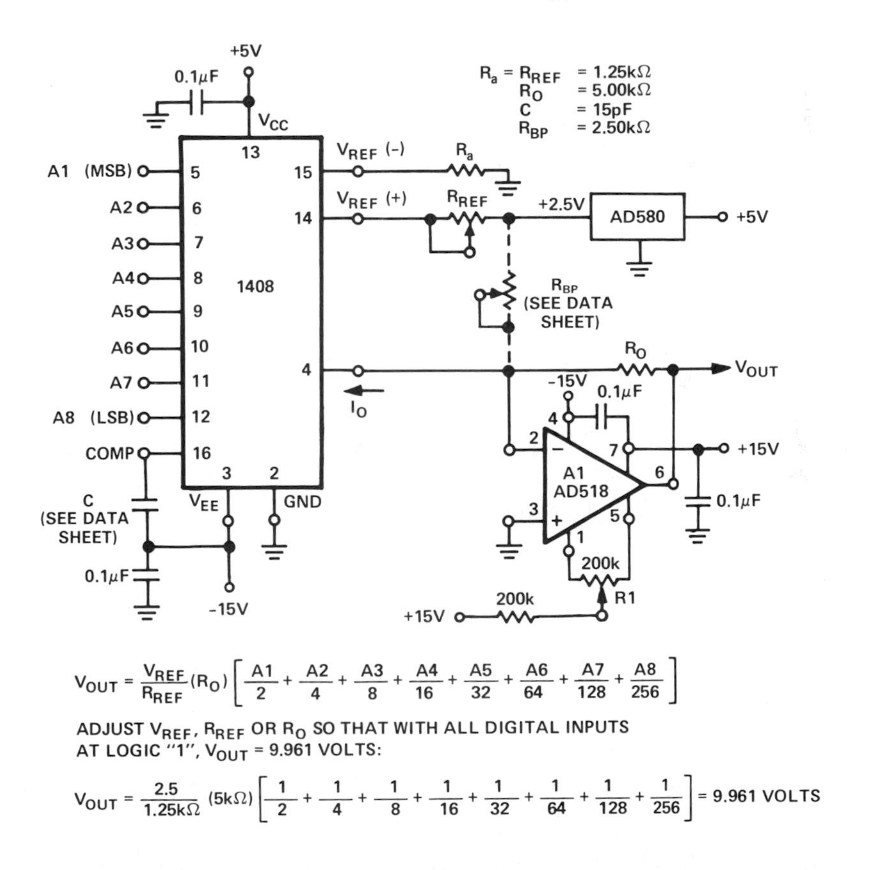

b. Implementing a 1408 application (voltage output, fixed reference).

Figure 8.1. An early bipolar-process d/a converter design.

12-Bit DAC

The Analog Devices AD562 was originally manufactured using a variation of hybrid manufacturing known as "compound monolithic integration," in which two IC chips were mounted in the same package without the traditional substrate for mounting and interconnection. Instead, the two chips were designed so that a set of wire bonds between the two chips (in addition to the usual ones to the package pins) were all that were necessary to assemble a 12-bit accurate DAC in an IC package.

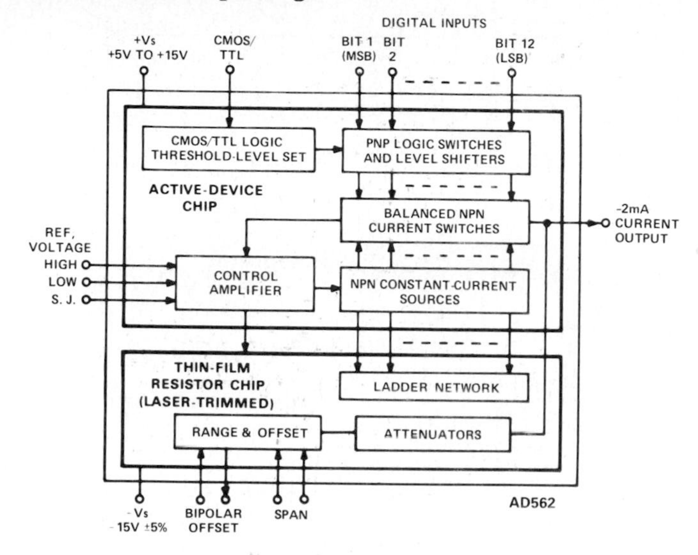

a. Functional block diagram.

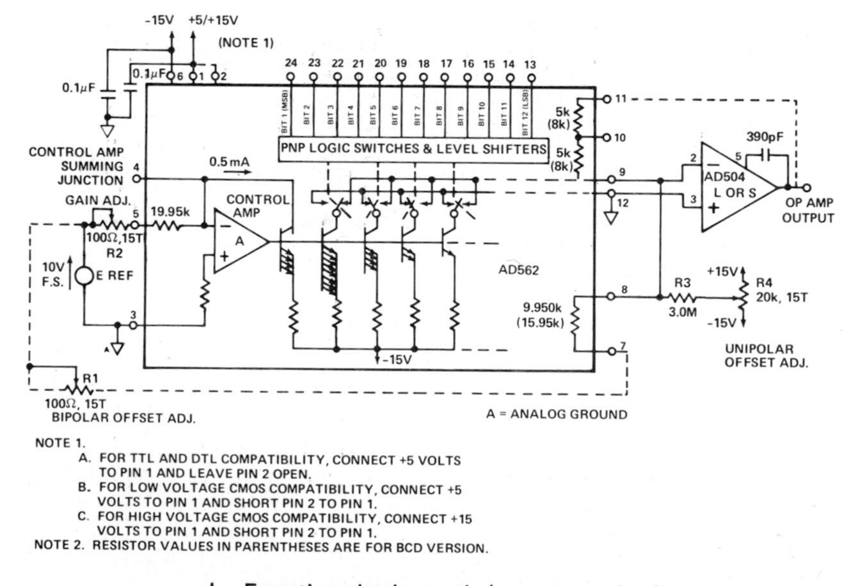

NOTE 1.
A. FOR TTL AND DTL COMPATIBILITY, CONNECT +5 VOLTS TO PIN 1 AND LEAVE PIN 2 OPEN.
B. FOR LOW VOLTAGE CMOS COMPATIBILITY, CONNECT +5 VOLTS TO PIN 1 AND SHORT PIN 2 TO PIN 1.
C. FOR HIGH VOLTAGE CMOS COMPATIBILITY, CONNECT +15 VOLTS TO PIN 1 AND SHORT PIN 2 TO PIN 1.

NOTE 2. RESISTOR VALUES IN PARENTHESES ARE FOR BCD VERSION.

b. Functional schematic (recent version).

Figure 8.2. First 12-bit 2-chip DAC.

In the original AD562 (Figure 8.2), one chip contained the resistor network (including bit weight-setting resistors and output gain-setting resistors), and the other contained the reference control amplifier and the current switches for the 12 bits. As the processing matured, the manufacture of larger chips became more practical. The two chips of the original AD562 were merged into a single-chip version, which has since become available from other sources (multi-sourced); it became the first 12-bit DAC qualified by the U.S. Department of Defense under MIL-M-38510.

Adding a Reference

While the AD562 was the first 12-bit IC DAC, and embodied the solution to some extremely difficult design problems, it was still really only a building block, since it lacked *latches*, a *reference*, and an *output amplifier*. Shortly after the two-chip AD562 was introduced, a version with a third chip was developed. The third chip was a 2.5-volt bandgap *reference*. This made the DAC function more complete. The resulting product, known as the AD563, also became quite popular and eventually made the transition to a monolithic device (Figure 8.3).

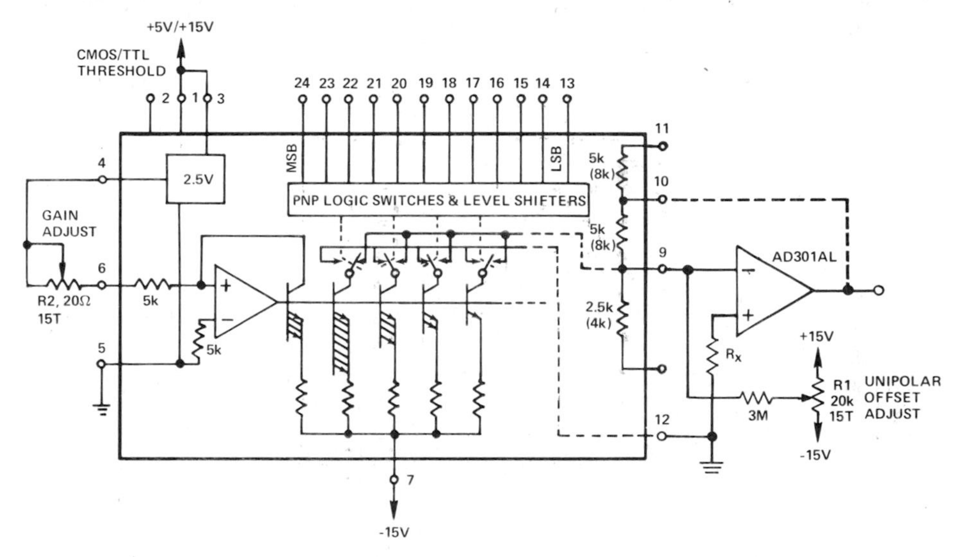

Figure 8.3. 12-bit bipolar-process DAC with additional 2.5-V reference.

Increasing the Speed

Another problem with the AD562 was that, while reasonably fast, it lacked sufficient speed for many applications; its settling time bordered on the slow side of the mystical 1-microsecond mark. Later advances in switch design and Zener-diode fabrication led to a higher-speed DAC, designated the AD565 (later followed by the AD565A). The bit switches used in this design are much smaller than those used in the AD562, allowing a substantial reduction in chip area and increasing the yield of good chips per wafer. The new switches also operate faster, with a factor-of-five improvement in settling time (Figure 8.4), from 1 microsecond to 200 nanoseconds. Furthermore, the switches have an internal threshold for 5-volt logic compatibility, reducing power consumption and eliminating the need for a separate logic power supply.

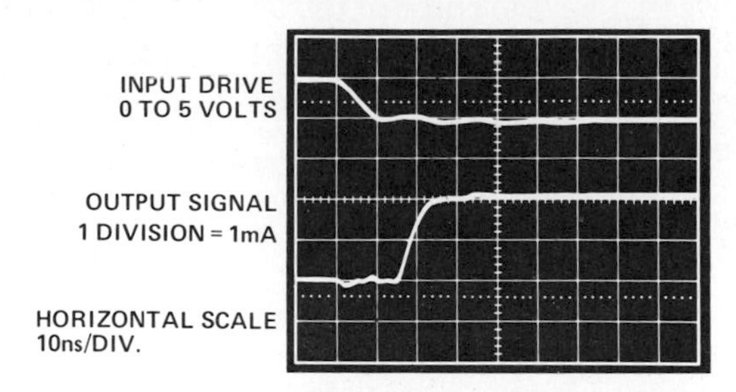

a. Transition time; note 10-ns/division time scale.

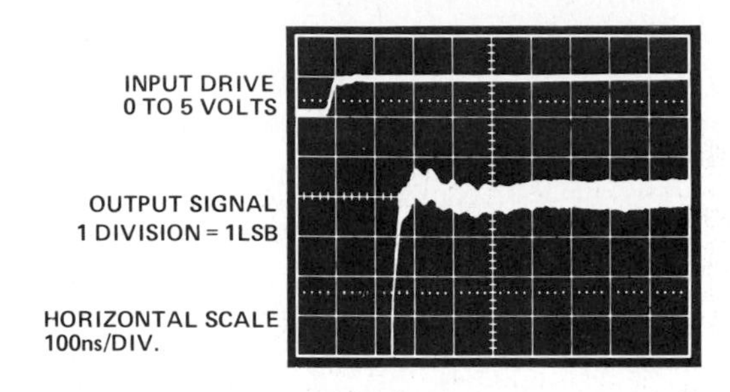

b. Settling time.

Figure 8.4. Full-scale transition and settling time of the AD565A.

Another improvement in the AD565 (Figure 8.5a) is the replacement of the bandgap reference by a buried-Zener type. The buried Zener differs from the conventional surface Zener in that the breakdown occurs below the surface of the chip. This frees the Zener from possible instabilities induced by migration of ions at the surface, as well as noise induced by imperfections at the silicon/passivation interface. The buried Zener also exhibits lower noise than the bandgap reference.

The AD565 retains the same pin configuration as the earlier AD563, allowing drop-in replacement in most applications with improved performance and lower price. An AD562-compatible version of the same chip is offered and is designated the AD566 (Figure 8.5b).

Adding Latches

While these products offer 12-bit resolution and linearity, additional external components are necessary to apply them in systems. Specifically, *latches* are needed for the digital bus interface, and an *output amplifier* is needed to convert the output current to a more universal—hence more useful—output voltage, without the use of external precision components.

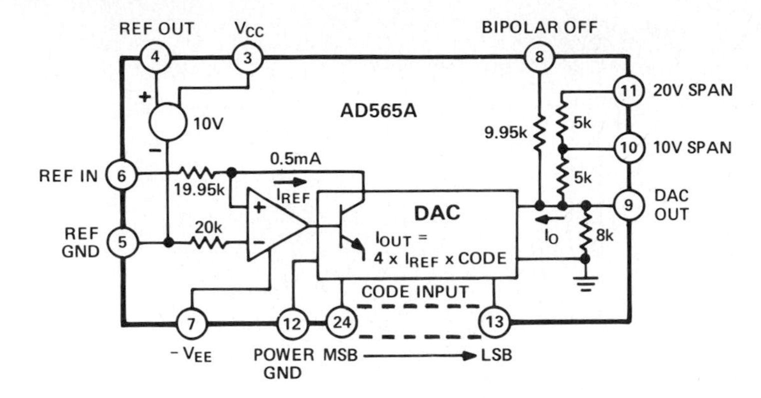

a. Complete version with reference (AD565A).

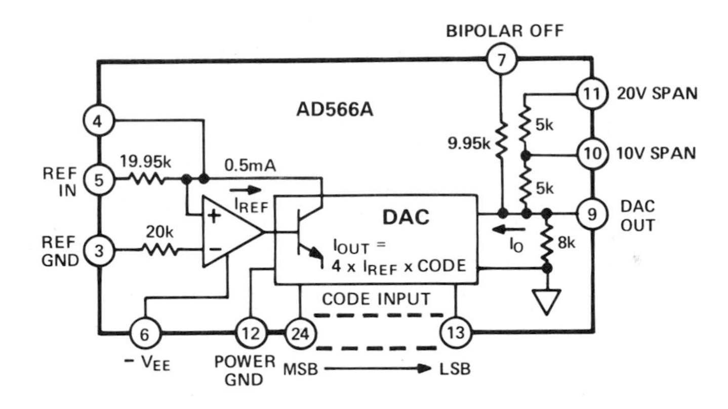

b. Speeded-up replacement for AD562 (AD566A).

Figure 8.5. Fast monolithic 12-bit bipolar-process DAC.

The evolution we are describing, towards the complete 12-bit bipolar DAC, progressed along two dimensions. The problem of adding the latches was solved in the AD567. This device is fundamentally similar to the AD565A, but includes latches implemented in an emitter-coupled logic form with TTL compatibility. These latches (Figure 8.6) are capable of accepting data in 4-bit nybbles, 8-bit bytes, or full 12-bit words from a bus.

The *double-buffered* arrangement of the latches allows the complete data word to be assembled in the first rank of registers, then transferred to the second rank, which drives the actual DAC inputs. This prevents invalid partial data (when using buses less than 12 bits wide) from reaching the DAC and generating spurious outputs during assembly of the complete data word. The latches are operated by independent address lines, gated with the Chip-Select and Write control inputs. The AD567 can operate with $\overline{CS}$ and $\overline{WR}$ pulses as short as 100 ns, and the current output settles to within ½ LSB in less than 500 ns.

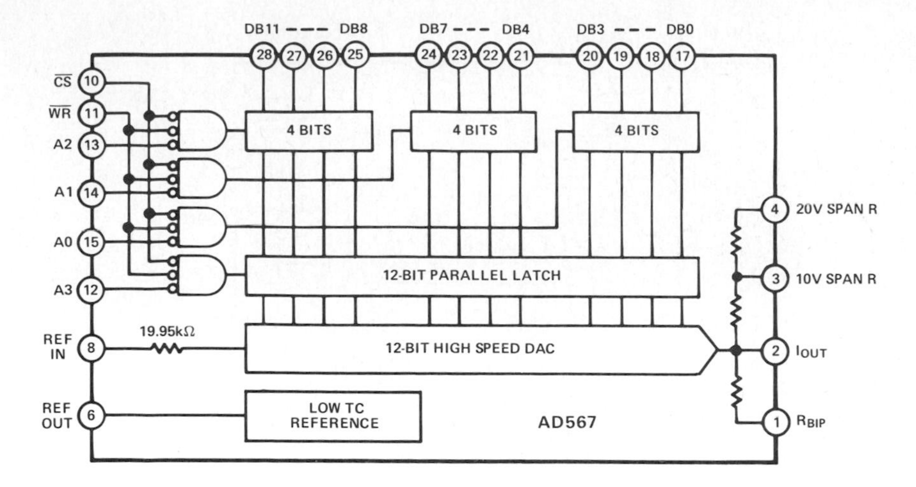

Figure 8.6. Functional block diagram of 12-bit current-output bus-interfaceable DAC.

Output Amplifier

At the same time that the AD567 was being developed, an output amplifier compatible with the process used to manufacture this family of DACs was developed, with the goal of integrating the amplifier on the same chip as the rest of the DAC. Certain design tradeoffs were possible in this amplifier. For example, since the inputs are always at ground potential, CMRR can be compromised. However, high open-loop gain is necessary in order to preserve 12-bit linearity, and dc parameters (offset voltage, offset drift, bias current, and bias-current stability) must be commensurate with 12-bit performance. Settling time should be as fast as possible with minimal overshoot and ringing.

Such an amplifier was first used in a monolithic voltage-output version of the DAC80 family of 12-bit DACs. The DAC80, a complementary-coded DAC, was originally produced in hybrid technology, with separate switch-, resistor-, reference-, and amplifier chips inside the hybrid package. It later was produced using only three chips—a 6.3-volt reference, a complementary-coded 562-type DAC chip, and an output amplifier. Now, the monolithic AD DAC80 is produced with a modified AD565-type chip, complementary logic, the 10-volt reference re-scaled to 6.3 volts, and the added output amplifier (Figure 8.7). The advent of the monolithic AD DAC80 also made possible low-cost plastic packaging of the device.

Putting It All Together

The DAC80, however, still cannot be considered a complete DAC, since it lacks latches for bus interface; and its complementary logic is inconvenient, though not difficult, to deal with. The AD667 (Figure 8.8), introduced in 1984, tops the evolutionary chain described here; it is a complete, general-purpose, fast, bus-compatible 12-bit DAC on a single chip. The monolithic

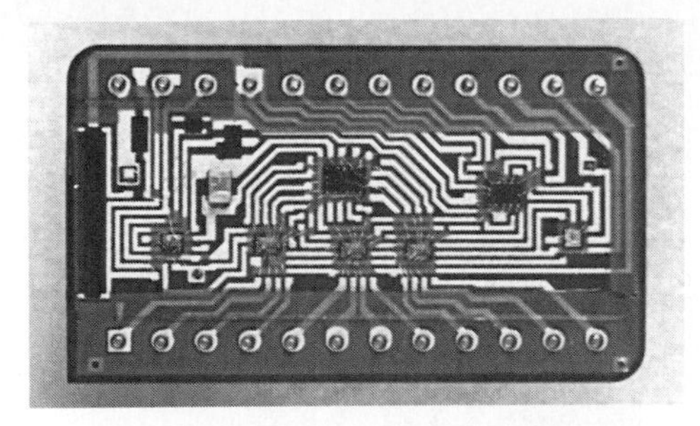

a. Early industry-standard version.

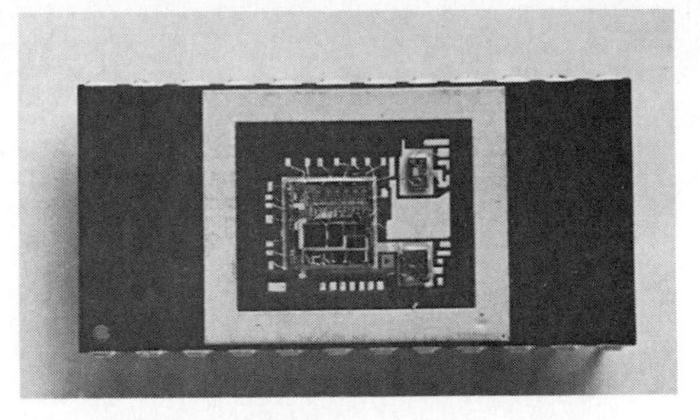

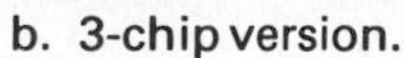

b. 3-chip version.

c. Completely monolithic version in ceramic and plastic packages.

Figure 8.7. Evolution of the DAC80.

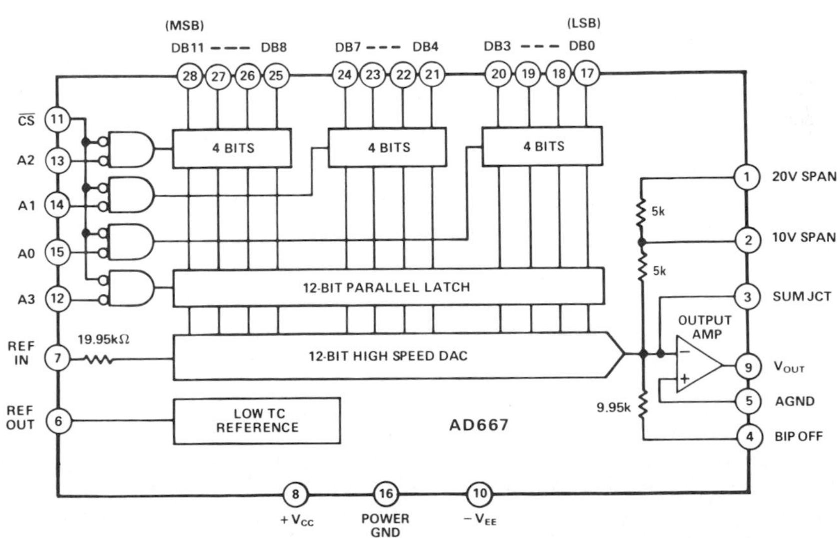

Figure 8.8. Functional block diagram of a complete, bus-compatible, voltage-output, monolithic 12-bit DAC.

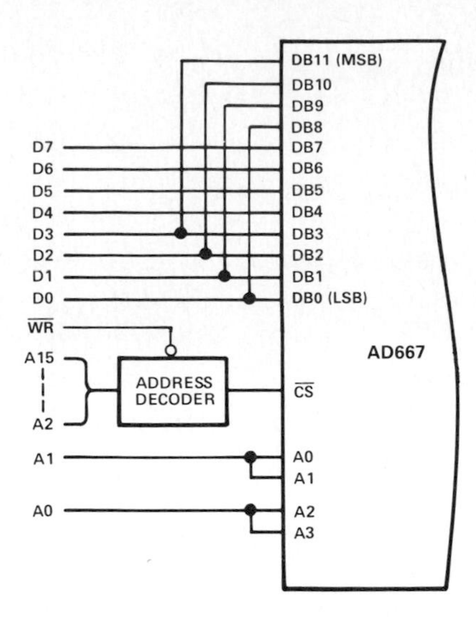

Figure 8.9. Connections of 12-bit DAC for right-justified interface to 8-bit bus.

chip bears a strong resemblance to the AD567, but includes the output amplifier that was used in the monolithic DAC80. Its applicability is enhanced by its availability in a variety of package forms: plastic and ceramic DIPs—and ceramic leadless chip carriers for surface mounting.

The AD667's positive-true digital data inputs are compatible with both TTL and 5-volt CMOS logic. Divided into 3 4-bit quads, they accept input data in a sequence of 4-bit nybbles from 4-bit buses; (8 + 4) or (4 + 8)-bit bytes (left- or right-justified—see Figure 8.9) from 8-bit buses; and 12 bits in parallel from 12-or 16-bit buses.

The inputs are double-buffered—this means that the converter may be updated when all 12 bits have been loaded, avoiding spurious analog output values. The control signals may be arranged for automatic updating when the full 12-bit word has been loaded. In addition, the double buffering makes it possible for a group of DACs to be updated simultaneously or in any desired sequence after having been loaded asynchronously by any of the above schemes. Since the latches are triggered by logic levels, rather than edges, they may be hard-wired in a transparent mode.

The on-chip 10-volt (±1%) reference, which the user may externally jumper to the device input, with typically 1 mA to spare, may also serve other devices or as a system reference. The output voltage may be pin-programmed for bipolar outputs of ±2.5V, ±5V, or ±10V, and unipolar outputs of +5V or +10V, at up to 5 milliamperes. A current booster may be connected inside the output op-amp's loop for high-current applications (e.g., line driving).

The use of precision high-speed bipolar current-steering switches and an on-chip high-speed output amplifier results in a 10 V/μs slew rate and output-voltage settling time of 3 μs maximum to within ± ½ LSB for a 10-volt change;

for 1-bit changes, typical settling time is 1 μs. The digital latch responds to strobe pulses as short as 100 nanoseconds, allowing the device to be used with fast microprocessors.

As an example of the specifications of a "universal" DAC, available in a number of performance options, all versions of the AD667 have guaranteed monotonic behavior over the specified temperature range. The AD667K (0°C to +70°C) and AD667B (−25°C to +85°C) guarantee maximum linearity error of ±¼ LSB at +25°C and maximum differential and integral linearity errors of ±½ LSB over temperature. Initial gain error is 0.2% of full scale (max) and offset is 2 LSB (max), while maximum temperature coefficients are ±15 ppm of full-scale range per °C for gain, ±3 ppm/°C for offset, and ±10 ppm/°C for bipolar offset.

8.2.1 COMPLETE 8-BIT DACS

Although the 12-bit DAC has arrived, it should not be forgotten that it once epitomized the goal of a supreme achievement in IC converter design, having the capability of resolving to 1 part in 4,096 and dealing in specifications expressible in tiny fractions of 1%, or even in parts per million. There are many applications that call for specifications that are far more modest in nature, easily satisfied by an 8-bit d/a converter capable of interfacing with an 8-bit bus, as long as it is complete, compact, low in cost, fast, easy to use, and of accuracy commensurate with its resolution.

The primitive 8-bit bipolar d/a converters mentioned earlier have historically been mere building blocks from which to build the DAC function. Devices like the 1408 and DAC-08 require many external digital and analog components, and adjustments, in order to perform what should be a relatively simple function. The need for a complete 8-bit DAC function was satisfied in 1980, with the introduction of the AD558 DACPORT™, an 8-bit DAC which can be considered truly complete (Figure 8.10).

It contains a precision voltage reference, an output amplifier, a latching register with *nor*'d CHIP SELECT and CHIP ENABLE inputs, for efficient microprocessor interfacing, and a precision DAC circuit.

Laser-trimming at the wafer stage eliminates any need for external adjustments; all versions are monotonic over temperature, and calibration accuracy is guaranteed over the full temperature range to within ±1LSB at full scale or zero ("K" and "T" versions).

Truly microprocessor-compatible, the device will run from the same single supply used by the host microprocessor, at any voltage from +4.5V to 16.5V, with a choice of two output ranges: 0 to 2.56V (10 mV/bit) and 0 to 10V (39.1 mV/bit, for $V_{CC} \geq 11.4$V). Settling time to full scale (Figure 8.11) is typically 0.8 μs to within ½ LSB (2.56-V range).

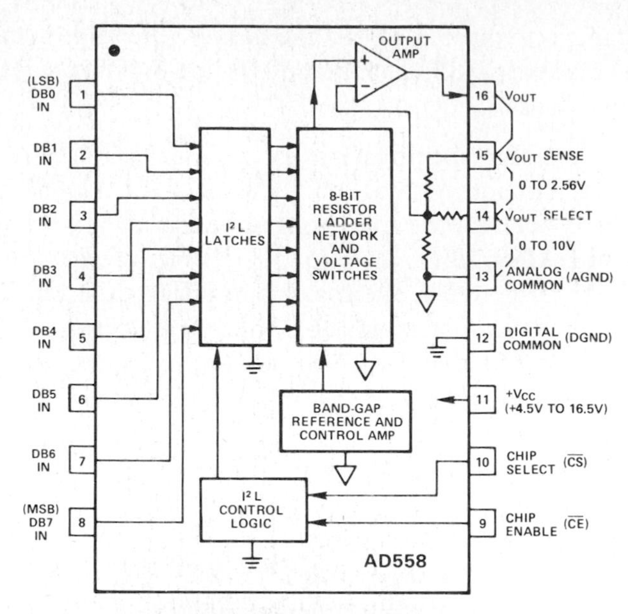

Figure 8.10. Block diagram of complete 8-bit bus-compatible d/a converter.

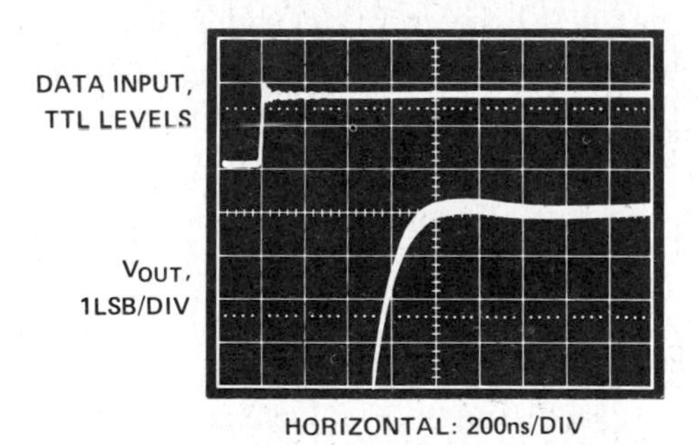

Figure 8.11. Detail of settling characteristic of 8-bit DAC for full-scale step, 2.56-V output range.

Keys to the Complete 8-Bit DAC

The block diagram (Figure 8.10) shows the elements of the AD558, which was—and remains—a triumph of circuit design, processing, and trimming technology, much of it patented.

First, its design is fully integrated, rather than a collection of stock circuits. This results in a small chip (hence better yield and lower cost), low dissipation (hence better performance over temperature, and a wider range of applications), and a compact pinout (only 16 pins, hence improved reliability and a smaller footprint).

Second, the use of Bipolar/I^2L, bandgap-reference, and thin-film-on-silicon technologies provide these important benefits: I^2L (integrated injection logic) permits efficient use of a single chip for digital and high-performance analog circuitry; bandgap reference provides tracking reference voltage with low tempcos, excellent long-term stability, and low-V_{CC} operation; and thin-film-on-silicon permits stable, linear, trimmable resistors to be fabricated for high-accuracy conversion.

Finally, the device is automatically laser-trimmed at the wafer stage. This technology results in converters that are fully calibrated—ending rejections in expensive packages and requiring no user trims, even when bought as chips for use in hybrid circuits—and monotonic over the entire operating temperature range.

Easy to Interface

The low-current logic inputs, set for TTL threshold voltage, can be operated by TTL or low-voltage CMOS over the entire operating V_{CC} range. The 100-μA maximum current minimizes bus loading.

The input latches simplify interfacing to 8- and 16-bit data buses. The latches are controlled by $\overline{CS}$ (Chip Select) and $\overline{CE}$ (Chip Enable) inputs (as mentioned earlier), internally *nor*'d so that the latches transmit input data to the DAC section only when $\overline{CS}$ and $\overline{CE}$ are both at logic zero. When either of the control inputs goes to logic 1, the input data is latched into the registers and held until both are again returned to zero. If the application does not involve control of inputs from a common data bus, both control inputs can be tied to 0 for transparency.

The AD558 acts like a "write only" location in memory. It can double up with a ROM slot, with no interaction; or, if doubled up with read-write memory, the memory will retain the word written into the DAC and can read it back without disturbing the DAC. Typical connections to a μP are shown in Figure 8.12.

8.3 CMOS DACS

As mentioned, CMOS technology has emerged as a superior process for producing low-power logic functions. In addition, CMOS transmission gates can be used as switches for the analog signals in DACs and ADCs.

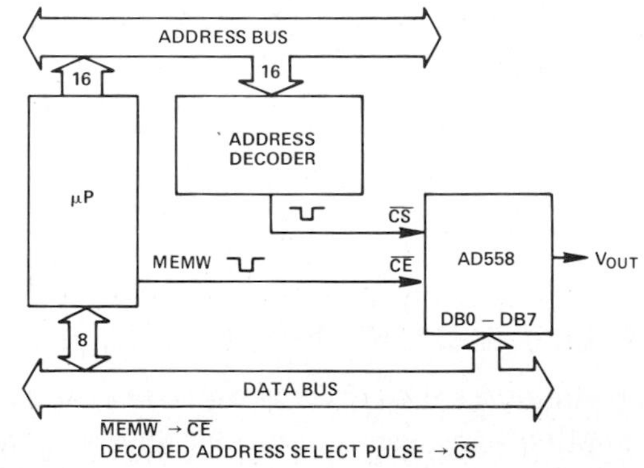

Figure 8.12. Typical microprocessor interface to 8-bit DAC.

The AD7520 10-bit multiplying DAC, introduced in 1974, was the first commercially produced CMOS DAC. It was the progenitor of an entire family of devices, from 8 to 16 bits in resolution, and including both D/A and A/D converters. The original design has since been revised to take advantage of smaller-geometry devices, and the AD7520 itself has been superseded by the AD7533.

8.3.1 CMOS D/A CONVERSION

Early commercially available monolithic d/a converters were principally processed by conventional bipolar linear processing techniques. Before 1974, when the AD7520 was introduced, 10-bit conversion had been difficult to obtain with good yields (and low cost) because of the finite β of switching devices, the V_{BE}-matching requirement, the matching and tracking requirements on the diffused-resistance ladders, and the tracking limitations caused by the thermal gradients produced by high internal power dissipation.

All of these problems were solved or avoided with CMOS devices. They have nearly-infinite current gain, eliminating β problems. There is no equivalent in CMOS circuitry to a bipolar transistor's V_{BE} drop; instead, a CMOS switch in the *on* condition is almost purely resistive, with the resistance value controllable by device geometry. The temperature-tracking problems of diffused resistors were solved easily: they weren't used.

The R-2R ladder is composed of 2-kΩ/square silicon-chromium resistors (a 10-kΩ resistor has a very manageable length/width of 5:1), deposited on the CMOS die. While the absolute temperature coefficient of these resistors is 150 ppm/°C, their tracking with temperature is better than 1 ppm/°C. The feedback resistor for the output amplifier is also provided on the chip to ensure that the DAC's gain-temperature coefficient is better than 10 ppm/°C—by sidestepping the absolute temperature coefficient of the network.

Finally, the low on-chip dissipation of only 20 mW (including the dissipation of the ladder network), in conjunction with the excellent tracking capabilities of the thin-film resistors, minimizes linearity-drift problems caused by internally generated thermal gradients. Low dissipation also helps to minimize the power and cooling requirements for circuitry that the AD7520/AD7533 is used in.

Figure 8.13 shows a functional diagram of the d/a converter, which employs an inverted R-2R ladder. Binary-weighted currents flow continuously in the shunt arms of the network; with 10V applied at the reference input, 0.5 mA flows in the first, 0.25 mA in the second, 0.125 mA in the third, and so on. The I_{OUT1} and I_{OUT2} output buses are maintained at ground potential, either by operational-amplifier feedback, or by a direct connection to common.

The switches steer the current to the appropriate output lines in response to the individually applied logic levels. For example, a "high" digital input to SW1 will cause the 0.5 mA of the most significant bit (MSB) to add to I_{OUT1}.

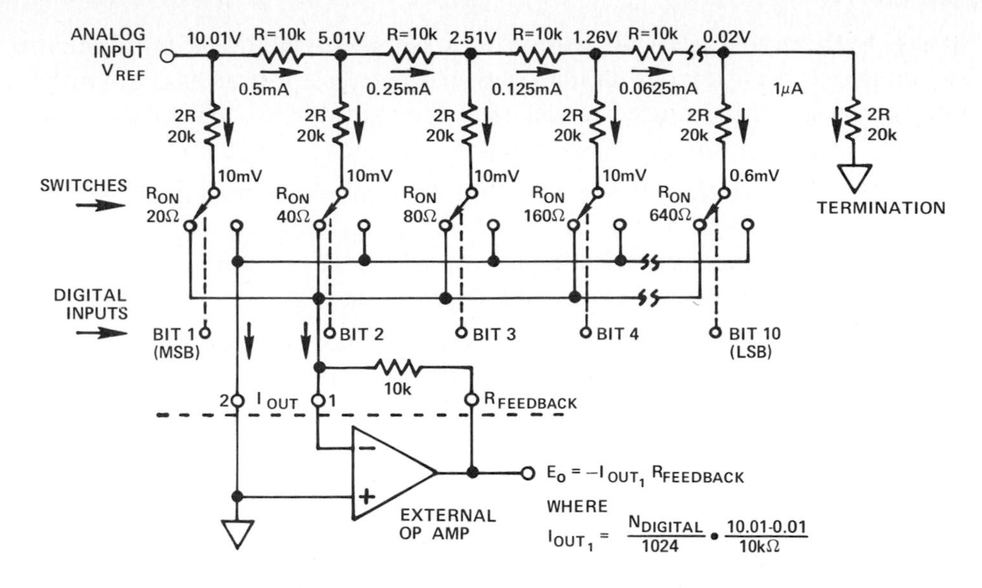

Figure 8.13. Functional diagram of CMOS d/a converter, with V_{REF} = 10.01 V. Bits 5 through 9 of 10-bit device are omitted for clarity.

When the digital input is "low," the current will flow through "I_{OUT2}." If I_{OUT1} flows through the summing point of an operational amplifier and I_{OUT2} flows to ground, then "high" logic will cause the nominal output voltage of the op amp to be $-(0.5\text{ mA})\times(10\text{k}\Omega) = -5\text{V}$, for a positive reference voltage of 10V, while "low" logic will make the contribution of Bit 1 zero. With all bits *on* (i.e., "high"), the nominal output will be -9.99V. With all bits *off*, the output will be zero.

Linearity errors, and—more important—their variation with temperature, are affected by variations of resistance in both the resistors and the switches. As we have seen, the resistance-network tracking is excellent. However, it is natural to expect that the switches, while tracking one another, will not track the resistance network. With identical switches having realistic resistance values (say 100 ohms), one would expect that, as temperature changed, the variation of resistance in the series legs would transform the network into an R-*n*R network, with *n* sufficiently different from 2 to destroy the binary character of the network and cause the converter to become non-monotonic.

The key to the linearity of the AD7520 is that the geometries of the switches are tapered so as to obtain *on* resistances that are related in binary fashion, for the first 6 bits. Thus, the nominal values of switch resistance range from 20 ohms for the first bit, 40 ohms for the second bit, through 640 ohms for the last 5 bits. The effect is, as can be seen in Figure 8.13, to provide equal voltages at the ends of the 6 most-significant arms of the ladder ($0.5\text{ mA}\times 20$ ohms $= 0.25\text{mA}\times 40$ ohms $= 10\text{mV}$). Since this drop is, in effect, in series with the reference, it causes an initial 0.1% scale-factor ("gain") error, which is well within the specifications but does not affect the linearity. Since the switches tend to track one another with temperature, linearity is essentially unaffected by temperature changes, and the gain drift is held to within the 10 ppm/°C specification.

Ten-bit linearity could, of course, have been obtained by scaling the *on* resistance of all the switches to a negligible value, say 10 ohms, but the switches would have required very large geometries, which would result in a 30 percent to 50 percent larger chip, at a substantial increase in cost.

Figure 8.14 illustrates one of the 10 current switches and its associated internal drive circuitry. The geometries of the input devices (1 and 2) are scaled to provide a switching threshold of 1.4V, which permits the digital inputs to be compatible with TTL and CMOS. The input stage drives two inverters (4 & 5, 6 & 7) which in turn drive the N-channel output switches.

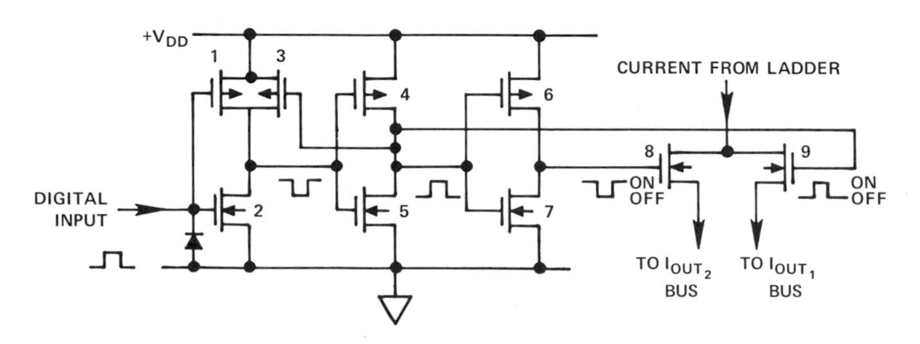

Figure 8.14. CMOS switch used in the AD7520 family. Digital input levels may be TTL or CMOS.

8.3.2 EQUIVALENT CIRCUIT

Figure 8.15 shows the equivalent circuit of the AD7533 at the two extremes of input, all inputs "high" (a), and all inputs "low" (b). V_{REF} (or I_{REF}, if a current reference is used) sees a nominal 10-kΩ resistance, regardless of the switch states. The current source, $I_{REF}/1{,}024$, represents a 1-LSB current loss through the 20-kΩ ladder-termination resistor, shown in Figure 8.13. R_{ON}, in this case, is the equivalent resistance of all ten switches connected to the I_{OUT1} bus (a) or the I_{OUT2} bus (b). Current-source I_{lkg} represents junction- and surface-leakage to the substrate. Capacitors C_{OUT1} and C_{OUT2} are the output capacitances-to- ground for the *on* and *off* switches. C_{SD} is the open-switch capacitance.

The 1,000:1 ratio between R_{ladder} and R_{ON} provides a number of benefits, all related to the small voltage drop across R_{ON}:

• V_{REF} can assume values exeeding the absolute-maximum CMOS rating, V_{DD}. For example, V_{REF} could be as large as ±25V, even if the DAC's V_{DD} rating were only +17V.

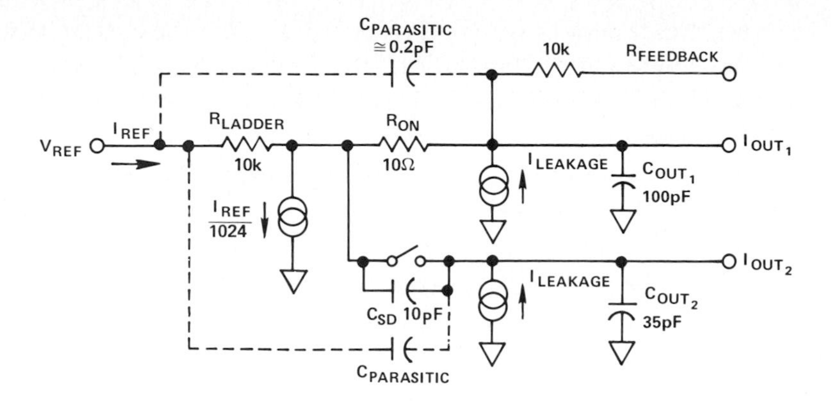

a. All digital inputs high.

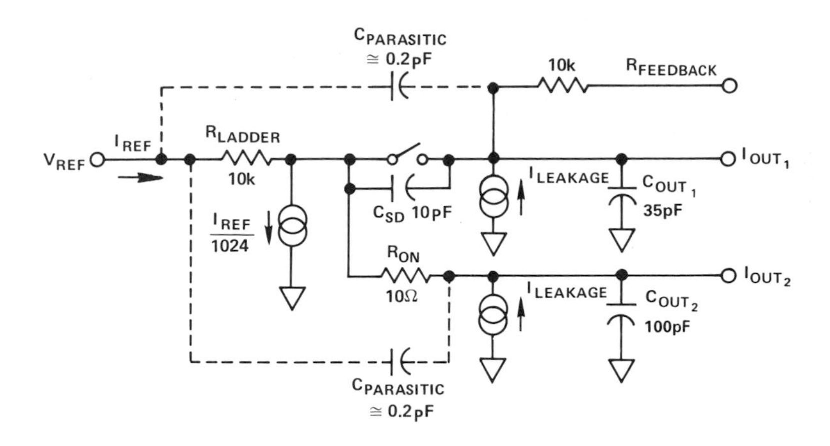

b. All digital inputs low.

Figure 8.15. Equivalent circuits of 10-bit CMOS d/a converter.

- The nonlinearity temperature-coefficient depends primarily on how well the ladder resistances track. Since R_{ON} is only a small fraction of R_{ladder}, any R_{ON} tracking errors will be felt only as 2nd- and 3rd-order effects.

- The same argument holds true for power-supply variations. Any change of switch *on* resistance, as the power supply changes, will be swamped by the 1,000:1 attenuation factor. Power-supply rejection is better than 1/3 LSB per volt.

- If V_{REF} is a fast ac signal, the feedthrough coupling via C_{SD}, the open-switch capacitance, will be negligible, again because of the 1,000:1 voltage step-down. The parasitic capacitances from V_{REF} to I_{OUT1} and I_{OUT2} comprise the major source of ac feedthrough. Careful board layout by the user can result in less than ½ LSB of ac feedthrough at 100 kHz.

Since the *on* resistance depends only on value of V_{DD}, not the current through the switch, and the resistance network is unaffected by V_{REF}, the full-scale output current (all bits "high") is nominally $V_{REF}/10.01\ k\Omega$, less the "constant" current losses shown in Figure 8.15. This means that I_{OUT} is almost perfectly proportional to V_{REF} over the whole range from $-10V$ to $+10V$. Equally important, the conversion linearity error (0.05%) is independent of the sign or magnitude of V_{REF}.

The extremely low analog-linearity error at constant digital input results in excellent fidelity to the input waveform, which suggests some interesting possibilities for the AD7520 family in the calibration and control of gain in signal generators, high-fidelity amplifiers, and response-testing systems.

The AD7520/7533 architecture is easily extended to 12-bit resolution by merely adding additional switch cells and resistors. However, in order to achieve consistent yields to 12-bit linearity, it is necessary to use error-correction techniques. The AD7541, the first CMOS DAC to offer 12-bit linearity and monotonicity, is schematically identical to the AD7520 and AD7533 with additional switches and resistors (Figure 8.16); but laser trimming at the wafer level adjusts the bit-weight ratios to the accuracy required for 12-bit performance.

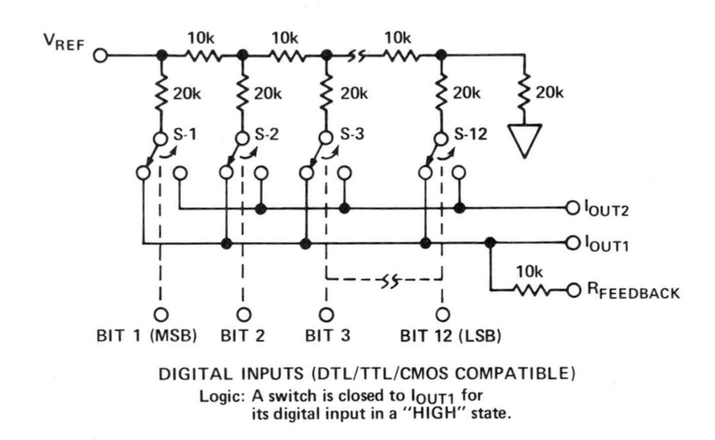

Figure 8.16. Functional block diagram of the AD7541 12-bit CMOS DAC.

8.3.3 DIGITAL BUFFERING

We have already pointed out that the great advantage of CMOS is the ability to embody complex logic functions without significantly increasing the chip's power dissipation. In a DAC, the most obvious logic function to add is the digital bus interface. As an example of what can be done, Figure 8.17 shows a family of bus-compatible 12-bit DACs based on the AD7541, each optimized for a particular bus architecture. The members of the family are:

DEVICE	INPUT WORD WIDTH No. of Bytes × Byte Width	BUFFERING Double/Single
AD7543	12 × 1 bit (bit-serial)	Double
AD7542	3 × 4 bits (nybble-serial)	Double
AD7548	1 × 4 bits and 1 × 8 bits (byte-serial)	Double
AD7545	1 × 12 bits (parallel)	Single
AD7549	Dual 3 × 4 bits (nybble-serial)	Double (dual DAC)

The AD7542 and 7543 are useful in applications where it is desirable to isolate the DAC from the rest of the system. With fewer input lines, the expense of wiring and optical isolators can be held to a minimum. Since these devices are housed in small 16-pin single-width DIPs, they require little board space, socket requirements are simplified, and system reliability is improved.

The AD7545 is designed to be used with 12-bit and wider buses. It accepts the 12-bit input word and applies it to the DAC inputs directly. In these applications, it is not necessary to use double-buffering on the chip.

The AD7548 is intended for use in systems where the data bus is 8 bits wide. The data can be presented to the DAC in either a right- or left-justified format in two bytes. This device, like the AD7542 and AD7543, uses double buffering to prevent the generation of spurious analog outputs from intermediate digital data.

A dual DAC, the AD7549, in which two DACs share the same bus, uses the same 4-4-4 loading format as the AD7542, to allow the device to be packaged in a 20-pin (0.3″) DIP.

8.3.4 ANALOG CONSIDERATIONS

The *analog* connections for any of the CMOS DACs discussed above are essentially the same. The output terminals are designed to be operated into ground potential (or the virtual-ground terminal of an output op amp). The current from each bit that is turned on is directed to the I_{OUT} terminal, which is connected to the inverting input of an op amp. The output current flows through the on-chip feedback resistor (which matches and tracks the ladder resistors) to provide a voltage output at the output terminal of the amplifier.

It is important to note that the output amplifier must have low offset voltage in order to preserve the linearity of the converter's transfer function. This is a rare example of two parameters which are usually independent actually interacting. Figure 8.18 illustrates the nature of the problem.

Inverted-ladder CMOS DACs—with characteristic ladder resistance, R—exhibit a code-dependent output resistance between ground and the amplifier summing point, varying from R to 3R (and actually to infinity in the trivial case of all bits turned off). This output resistance variation is not linear with

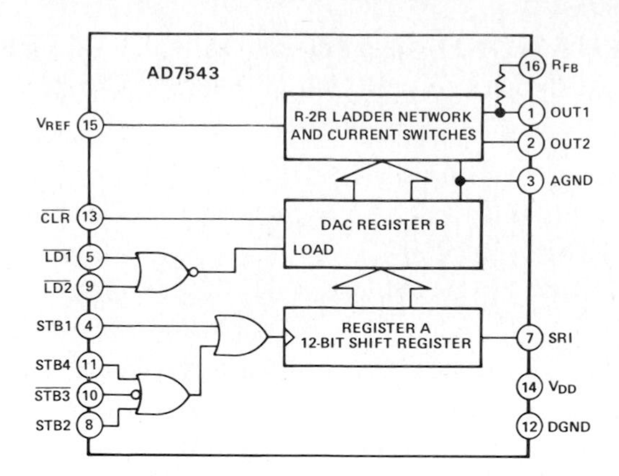

a. Serial-loaded, double-buffered DAC.

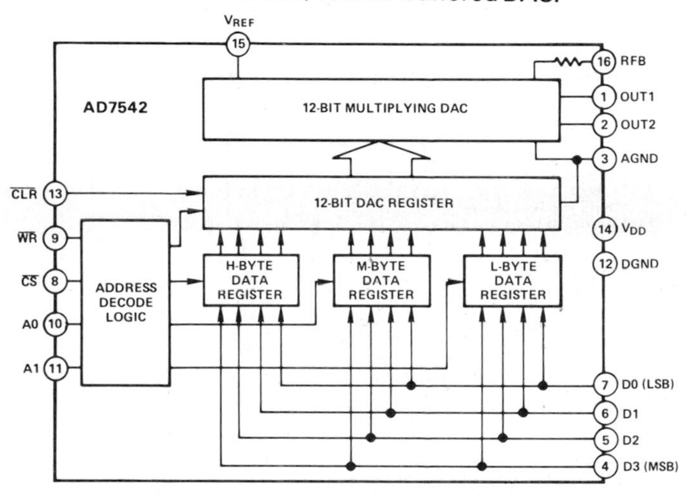

b. DAC with three 4-bit nybbles for 4-, 8-, or 16-bit buses.

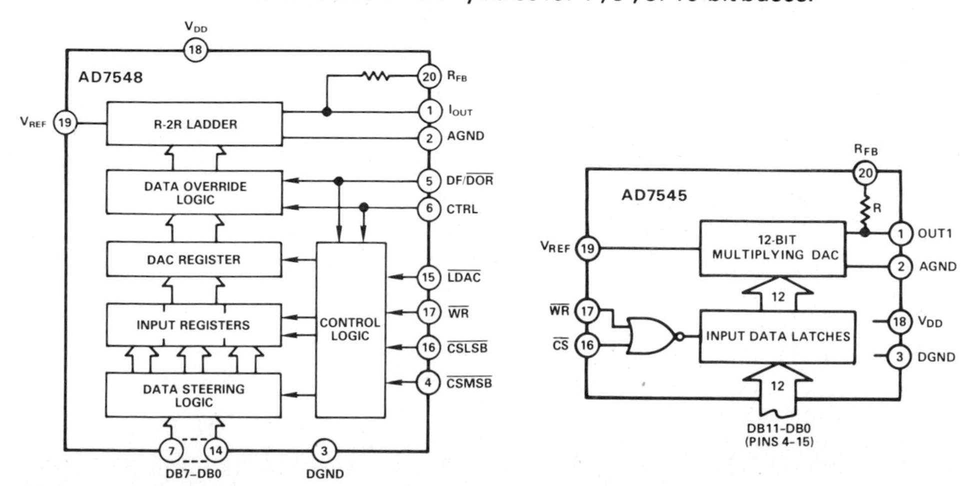

c. DAC with 2-byte loading, either left- (8,4) or right- (4,8) justified, for 8-bit buses.

d. DAC with single-byte 12-bit parallel loading for 16-bit buses.

Figure 8.17. Architectures of bus-compatible 12-bit CMOS multiplying DACs.

code, since a single-bit code—other than the MSB—approaches 3R output resistance, while codes with many bits turned on yield an output resistance which approaches R.

Since the amplifier is operating with a variable source resistance and a fixed feedback resistance (equal to R), the "noise gain," applied by the op amp to its own offset voltage, varies from 4/3 to 2. Since the worst-case gain changes occur at the same codes as the worst-case differential linearity, it is important to keep the error term (2/3 V_{OS}) much less than one LSB. Therefore, amplifiers with sub-millivolt offsets are required for 12-bit linearity.

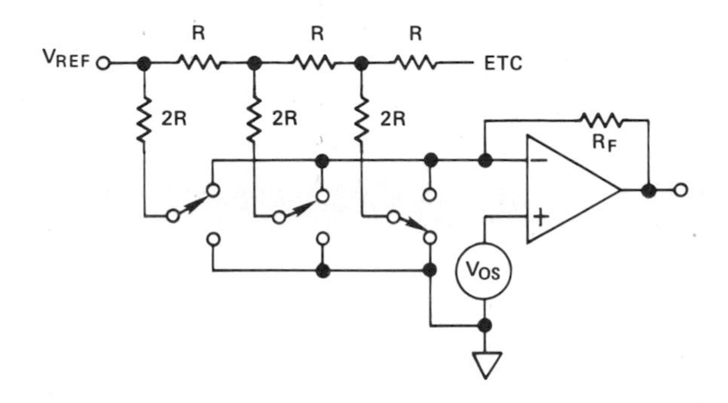

Figure 8.18. Modulation of op-amp circuit noise gain by CMOS DAC switches. In the case of V_{OS}, this can cause a code-dependent nonlinearity unless V_{OS} is negligible.

It is possible in some applications to circumvent the need for high-quality amplifiers by operating the DAC in a voltage-switching mode, rather than the current-switching mode (Figures 7.17 and 7.18). The AD7240 is designed to be used—and is specified for performance—in the voltage-switched mode (Figure 8.19).

In this mode, the terminal usually designated I_{OUT} becomes the reference input. The terminal usually designated REF IN becomes the voltage-output. This configuration removes the requirement for a low-offset output amplifier, since the DAC impedance seen by the amplifier input is constant (and equal to R). In applications where the load resistance driven by the DAC is high relative to R, no buffering is needed. However, if lower impedance loads must be driven, an amplifier configured as a voltage follower can be used. The voltage source connected to the reference input should have low dynamic impedance, since it must drive a switched load.

This mode of operation offers another benefit. It is possible to operate a CMOS DAC in the voltage-switching mode on a single power supply. However, the reference voltage range is small compared to that of current-switching DACs. The AD7240, for example, is rated for 12-bit linearity for a reference voltage of 1.2 Volts.

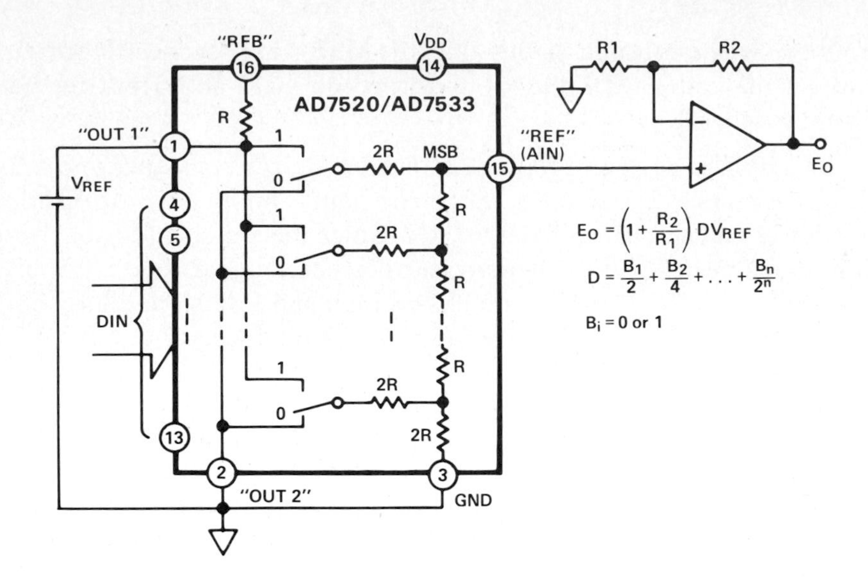

a. General principle.

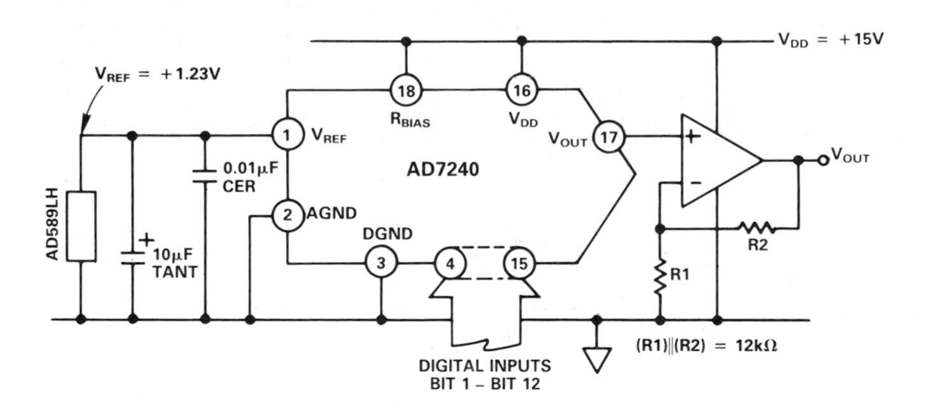

b. Application of the AD7240 voltage-switched DAC.

Figure 8.19. Using a CMOS DAC in the voltage-switched mode.

Higher reference voltages will cause degraded linearity due to the reduced drive available for the CMOS switches. The DAC is trimmed with the assumption that the gate-to-source voltage is large. Since the gate is driven to the positive-supply level, and the source is tied to the reference input, it can be seen that an increased reference will cause problems. Furthermore, only a positive reference can be used in this mode. A negative reference will forward-bias a parasitic transistor in the DAC and could result in damage to the device. This is not usually a problem in systems which use only a single positive power supply.

The CMOS DACs discussed above all require *external* op amps in the current-switching mode, and also—if buffering is required—in the voltage-switching mode. As noted earlier, the chief disadvantage of using CMOS technology to manufacture converter circuits is that the analog components (amplifiers, references, etc.) have traditionally been of poor quality. CMOS references, for example, generally exhibit higher noise and drift than bipolar devices. Amplifiers made from CMOS have suffered from poorer noise, drift, and output drive capability than bipolar units (the exception is the CMOS chopper-stabilized type, which has very low offset drift—but it still has more noise and lower drive capability than a bipolar type).

In many CMOS processes, a parasitic NPN transistor is formed, and great effort is expended in order to minimize its effect on the circuit. However, in the Analog Devices LC^{2}MOS (linear-compatible CMOS) process—specifically developed to permit both high-speed digital logic and precision analog circuits to be integrated on the same chip—the NPN transistor, used to great advantage, is the key to a DAC output amplifier with lower noise and higher drive capability than a pure CMOS device.

The single-supply 8-bit voltage-output DAC, shown in Figure 8.20, is an example of the performance achievable with the LC^{2}MOS process. The AD7224 is a monolithic 8-bit CMOS microprocessor-compatible d/a converter in an 18-pin dual in-line package. Designed for a variable or fixed external reference, it is a complete voltage-output device, with double-buffered data inputs and an on-chip output amplifier. Because of its low total unadjusted error—less than 1 LSB over temperature (L, C, and U grade)—it requires no adjustments.

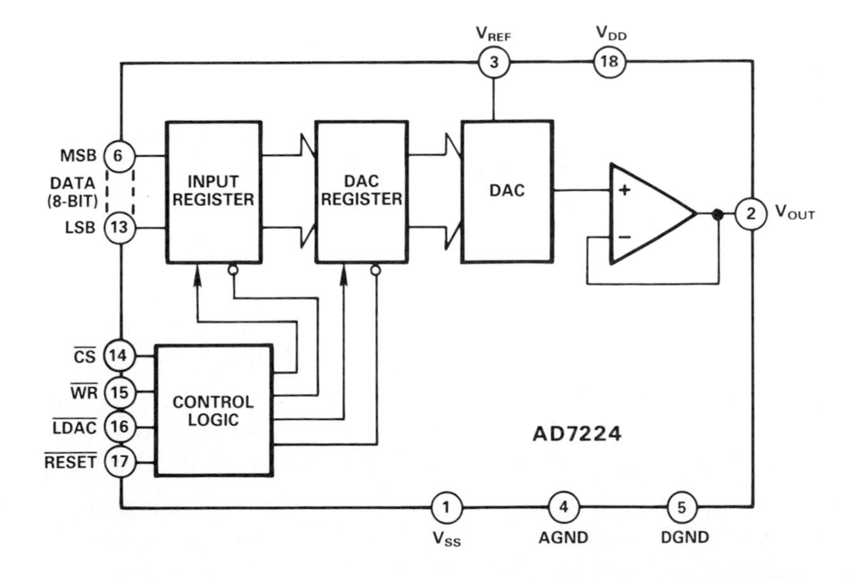

Figure 8.20. Functional block diagram of 8-bit, voltage-output CMOS DAC.

Its high-speed logic allows direct interfacing to most microprocessors. The double-buffered interface logic consists of an input register and a separately enabled DAC register. This arrangement allows the input register to be updated at the convenience of the microprocessor, while the DAC is updated whenever necessary. As a result, a number of DACs can be primed separately by the microprocessor, without changing their existing output levels, then enabled to deliver their new outputs simultaneously, or in a required sequence, a useful feature in (for example) test equipment.

A low-dissipation device (typically 35 mW with a single supply), it will operate with either a single positive supply and a +10-volt reference, or dual supplies and a +2-volt to +12.5-volt reference. The output amplifier can develop 10 volts across a load of 2 kilohms. The device is available in six grades, three temperature ranges, and 3 packages (including plastic).

8.3.5 MULTIPLE CONVERTERS

As low-cost converters have become increasingly more available, their use has grown and they have made many new applications possible. Often, it is cost-effective to add a converter in order to make an additional system feature available. On the other hand, it is desirable to reduce the physical size of systems as much as possible. New IC packaging technologies are evolving, but they are being primarily applied to digital circuits. Most linear and converter circuits have performance that is more sensitive to packaging, and they have remained in classical IC packages. Thus, the system designer is faced with the problem of adding increasing numbers of converters in relatively large packages.

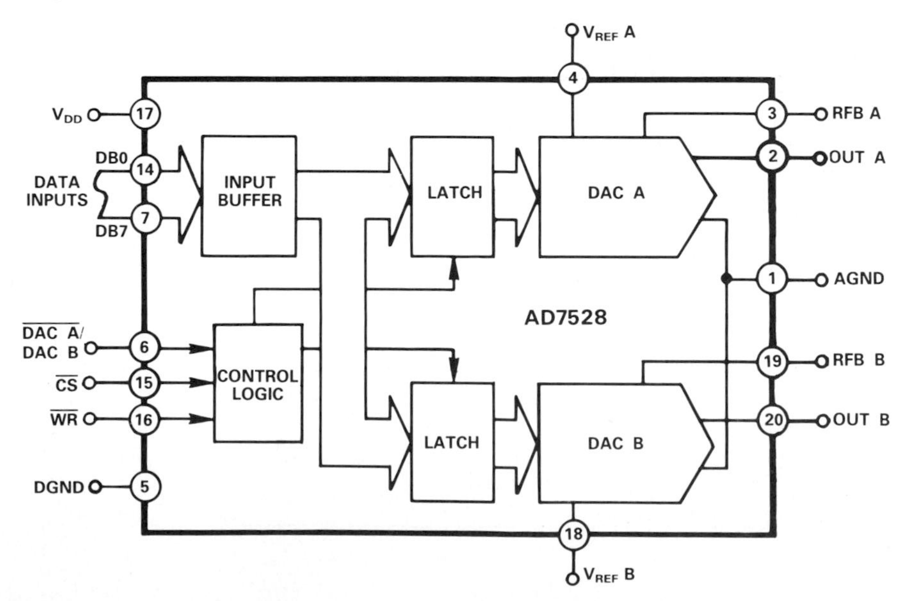

Figure 8.21. Functional block diagram of dual 8-bit monolithic DAC.

In the interest of improving circuit density in systems using many DACs, several multiple-DAC products have evolved. The task of packaging multiple devices on a monolithic chip or hybrid substrate has been made easier by the number of shared elements, from power supply to data and control buses. These have made it unnecessary to substantially increase the pin count of the chip or package each time a device is added.

The earliest, designated the AD7528 (Figure 8.21), is a dual 8-bit current-output device housed in a 20-pin single-width dual in-line package (DIP). The ladder resistances of the two DACs are tightly matched in this device, a useful feature in many applications. For example, it is possible to use the AD7528 in a digitally tuned state-variable filter.

Filter Application

The state-variable filter (or universal filter, as it is often called) is a convenient second-order filter block. It provides simultaneous low-pass, high-pass and bandpass outputs. All filter parameters can be readily adjusted. Figure 8.22 shows a typical filter circuit with expressions for center frequency, Q, and gain for the bandpass output.

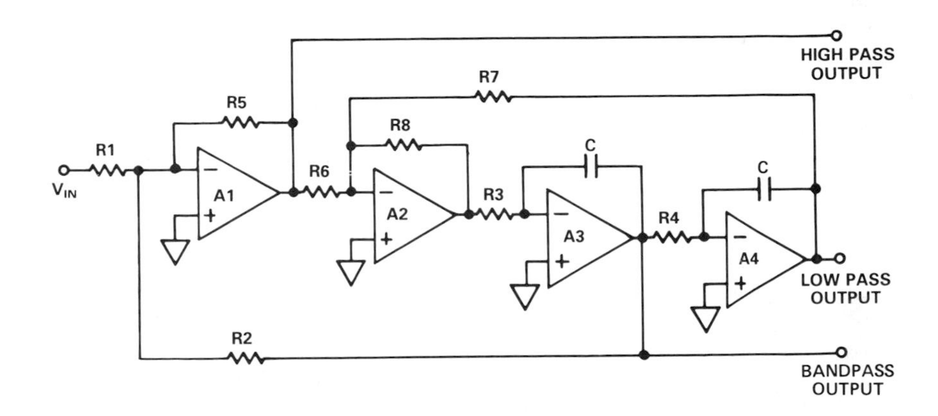

BANDPASS TRANSFER FUNCTION

$$\frac{V_{OUT}}{V_{IN}}(f) = \frac{A_O}{1 + jQ\left[\frac{f}{f_o} - \frac{f_o}{f}\right]}$$

Where f = input frequency of V_{IN}

A_O = gain at $f = f_o$

Q = circuit Q factor, i.e., $\frac{f_o}{\text{3dB Bandwidth}}$

f_O = resonant frequency.

$$f_o = \frac{1}{2\pi R_3 C} \cdot \sqrt{\frac{R_8}{R_7}} \quad (R_3 = R_4)$$

$$Q = \frac{R_6}{R_8} \cdot \frac{R_2}{R_5} \cdot \sqrt{\frac{R_8}{R_7}}$$

$$A_O = -\frac{R_2}{R_1}$$

Figure 8.22. State-variable filter.

DACs As Parameter-Control Elements. If R1, R2 and R3, R4 are functionally replaced with matched DAC pairs the filter parameters can be made programmable, as shown in Figure 8.23. DACs A1 and B1 control filter gain and Q, while DACs A2 and B2 control center frequency (f_o).

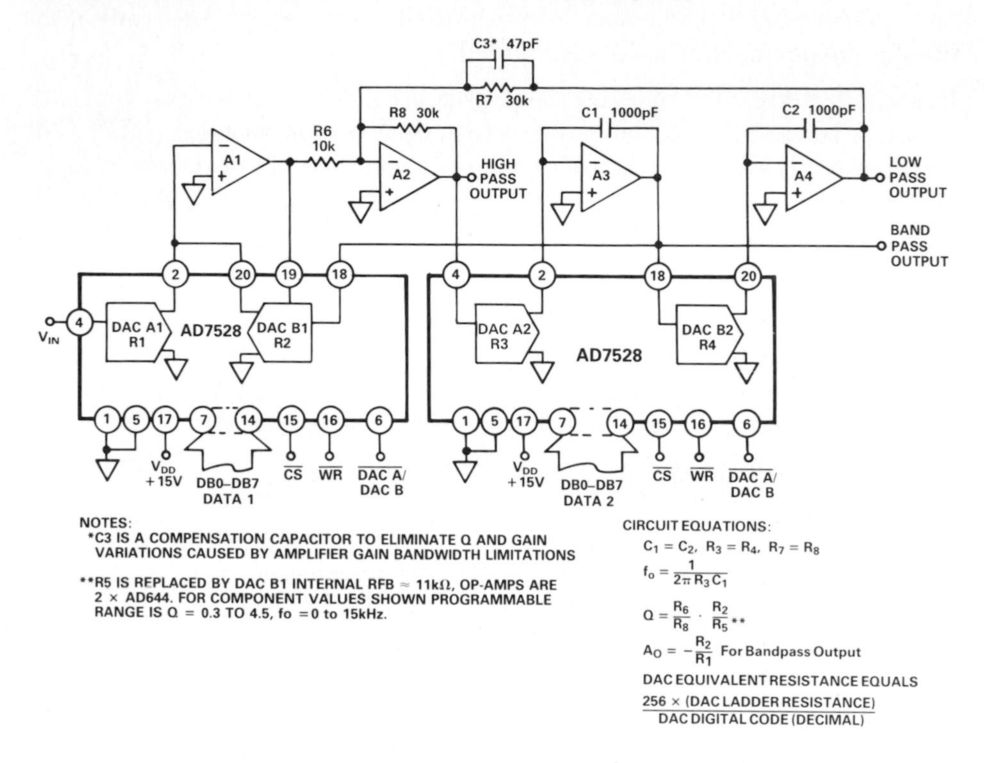

Figure 8.23. Digitally controlled state-variable filter.

For the component values shown, the programmable range of Q is from 0.3 to 4.5 and is independent of f_o (Figure 8.24a). Center frequency (f_o) is programmable from 0 to 15 kHz (Figure 8.24b) and is independent of Q.

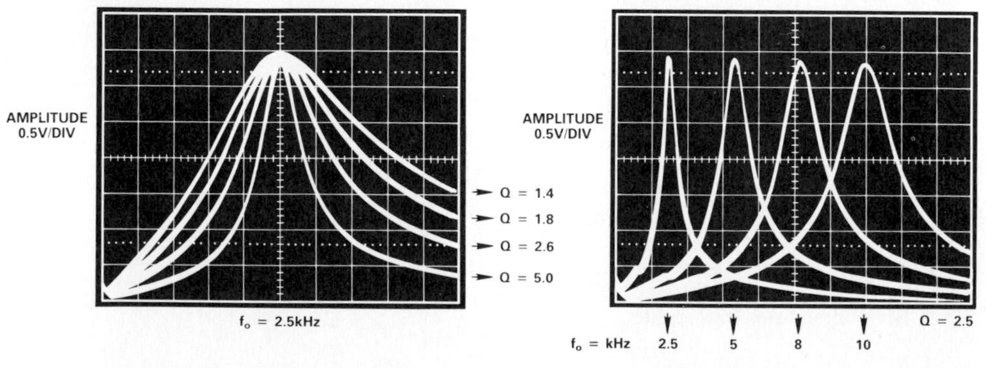

a. As a function of Q setting.

b. As a function of f_o setting.

Figure 8.24. Filter amplitude vs. frequency.

Programming the Parameters. Since maximum digital gain setting corresponds to minimum resistance, the input codes will be inverse with resistance value for a given parametric effect. The graph in Figure 8.25a shows how the circuit Q varies with DAC B1's input code (proportional to R_2, inverse with DAC B1 gain); and Figure 8.25b shows how the center frequency varies with DAC 2 (A and B) code for the component values given in Figure 8.23 (inverse with $\sqrt{(R_3 \times R_4)}$, direct with $\sqrt{(A2 \times B2)}$ gains.

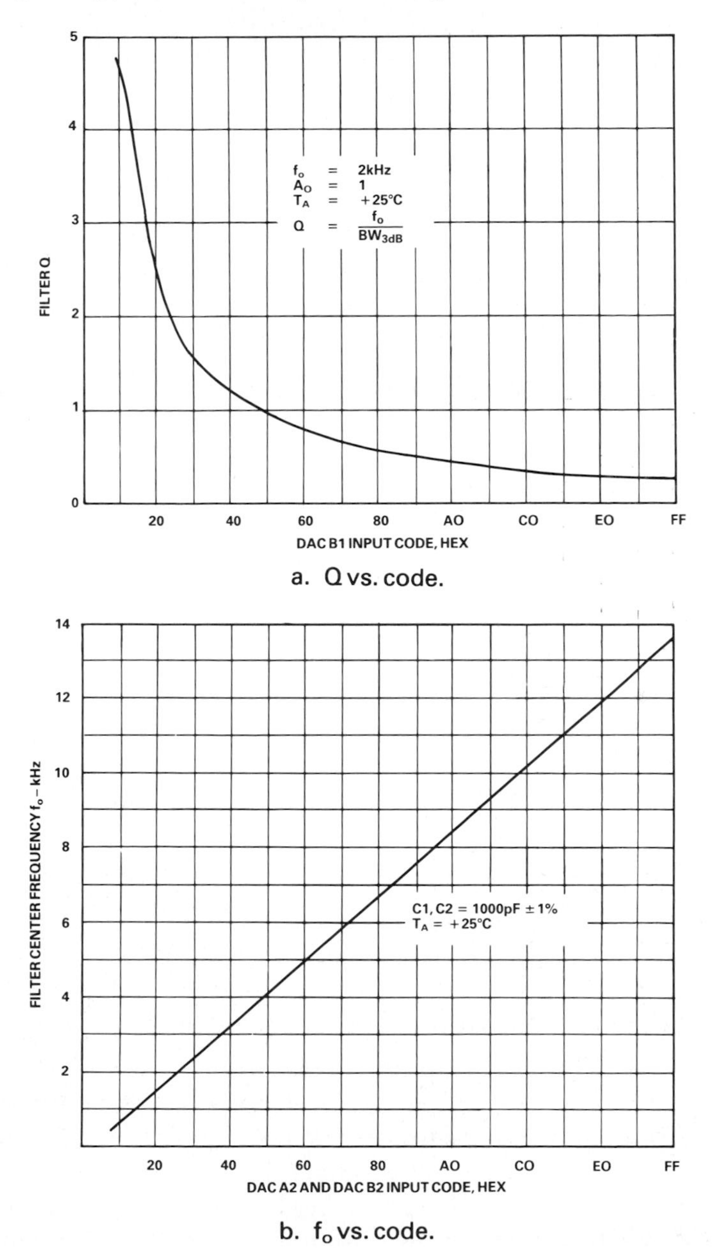

a. Q vs. code.

b. f_o vs. code.

Figure 8.25. Filter Q and f_o as a function of parameter values, expressed in hex code.

Gain variation alone, without affecting other parameters, is accomplished by changing DAC A1's input code. Unity gain occurs when the data in DAC A1 and DAC B1 latches is identical. Since the device's logic inputs are TTL or CMOS compatible, the DACs are readily interfaced to most microprocessors, thus providing an ideal microprocessor-control interface.

Quad Voltage-Output DAC

In applications where the DAC is being used strictly as a digitally controlled voltage source, rather than as a digitally programmable resistor replacement, a dedicated voltage-output unit is generally preferred. As an example of a multiple voltage-output 8-bit DAC, Figure 8.26 shows the architecture of the AD7226 quad 8-bit DAC, housed in a 20-pin single-width DIP.

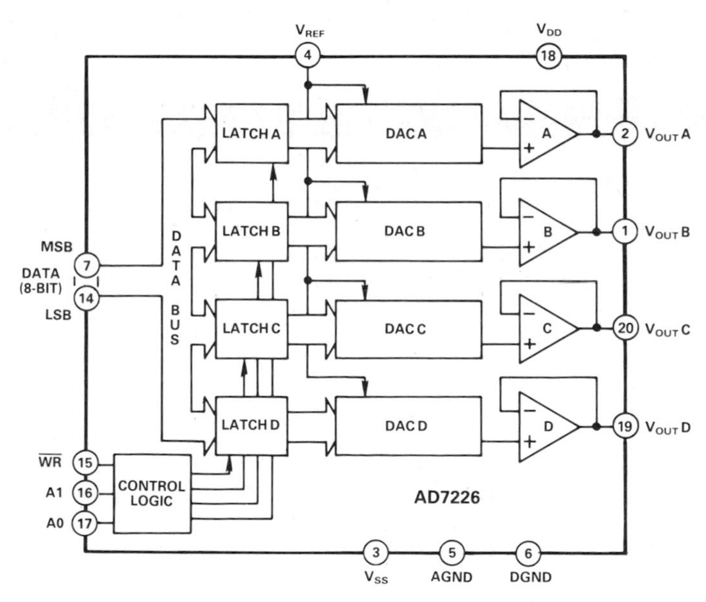

Figure 8.26. Functional block diagram of quad 8-bit CMOS DAC.

The four sets of latches are loaded via a common 8-bit data bus, under the control of two address bits (A0, A1) and an active-low WRITE pin ($\overline{WR}$). All logic inputs are level-triggered and compatible with both TTL and CMOS (5-volt). Because its logic-interface circuitry operates at high speed, the AD7226 is compatible with most 8-bit microprocessors; typically, all four channels can be updated at 2 MHz, and a single channel can be updated at 8 MHz.

Each converter consists of an 8-bit R-2R ladder and its associated switches, connected for operation in the voltage mode (Figure 8.27). The output of each DAC is buffered by a short-circuit-protected on-chip CMOS follower amplifier, capable of driving up to 5 mA of output current. Because the device operates in the voltage mode, with non-inverting buffers, a single supply (V_{DD}) from + 11.4 V to + 16.5 V may be used for unipolar output. The table shows the reference and output ranges for linear operation at each nominal supply-voltage level. Bipolar operation of the individual DACs is easily achieved with the addition of one external amplifier and 2 matched resistors.

V_{DD}	V_{REF} LIMIT		OUTPUT RANGE
	LOWER	UPPER	
15 V ±10%	+2 V	+10 V	0 TO +10 V
12 V ±5%	+2 V	+7.5 V	0 to +7.5 V

Multiple DACs of this type are useful wherever multiple voltages must be independently set by a digital source with resolutions of up to 1 part in 256. Examples include direct or incremental setting of test voltages, digital system-trim adjustments, setting of window widths in comparator applications, digital generation of multi-phase (e.g., 3-phase) sine waves, and setting of constants in analog computing circuits. Other applications might include adjustment of variable-capacitor voltage to tune multistage radio-frequency stages of VHF and UHF receivers optimally under microprocessor control.

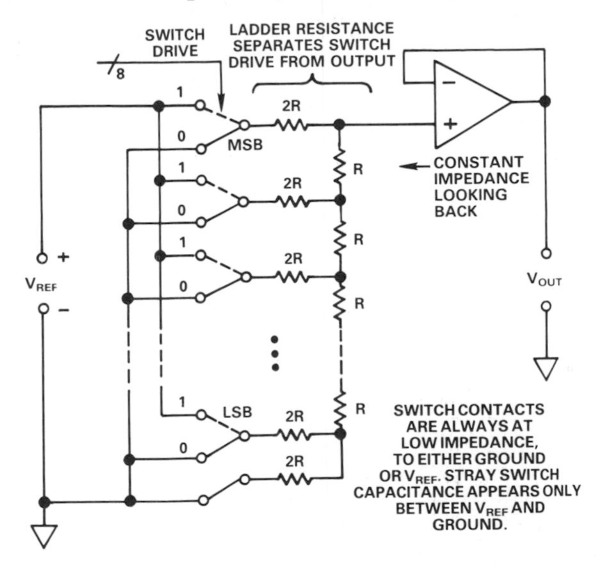

Figure 8.27. Functional diagram of one section of quad DAC.

8.3.6 PARTIALLY DECODED DACS FOR HIGHER RESOLUTIONS

All the CMOS DACs mentioned thus far use the R-2R ladder for setting the individual bit-weights, some of the DACs use the current-mode ladder; others use the voltage-mode ladder. Another topology which has become popular is the segmented architecture. In the R-2R ladder, it is necessary to have very tight matching between each bit and the sum of all the lesser bits in order to maintain monotonic operation. In the segmented design, however, these requirements are relaxed considerably, making monotonic high-resolution converters more practical.

Segmentation (see Figure 7.15) can be used in either the voltage or current mode. Its use in improving DAC resolution is somewhat easier to visualize when considered in the voltage mode.

In Figure 8.28, the 4 most significant bits of the input word are decoded to select one of 16 segments of the reference voltage, $V_{REFH} - V_{REFL}$, defined by a pair of adjacent taps on a string of 16 resistors. To V_{REFL} must now be added the fraction of the segment determined by the 12 lower bits. The chosen segment is buffered and applied to a second voltage division—this time by a 12-bit R-2R-type DAC, operated in the voltage mode; the position within the segment is selected by the lower 12 bits of the input digital word. The fraction of the segment thus chosen, added to V_{REFL}, gives the total output.

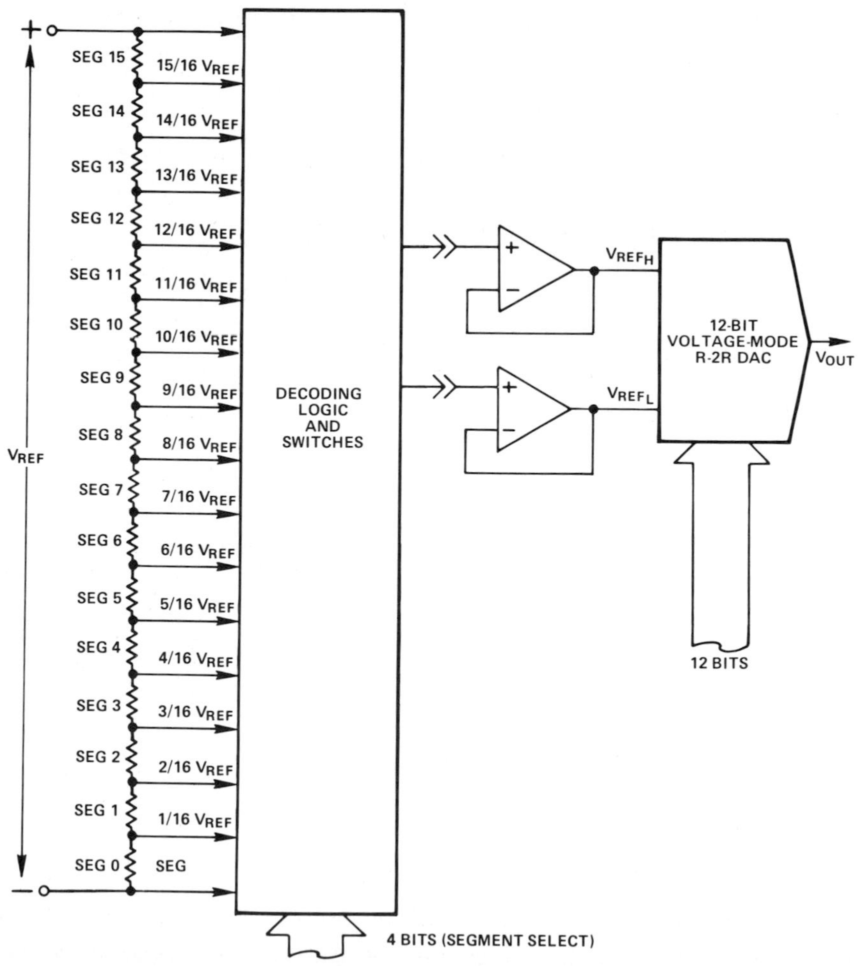

Figure 8.28. 16-bit segmented DAC architecture.

This voltage-segmentation technique was first used in the AD7546, a 16-bit monolithic voltage-output DAC with broadside input data latches for interfacing to 16-bit microprocessors (Figure 8.29). It employs a 12-bit R-2R DAC, operated in the voltage switching mode and supplied with a reference voltage from a 4-bit segment DAC under the control of the four most-significant bits. A monolithic CMOS device, the AD7546 offers outstanding differential-nonlinearity specifications, headed by ±0.0015% (16-bit monotonicity) for premium grades. An on-chip analog switch, synchronized with the latch loading signal, is provided for use with track/hold circuits for deglitching.

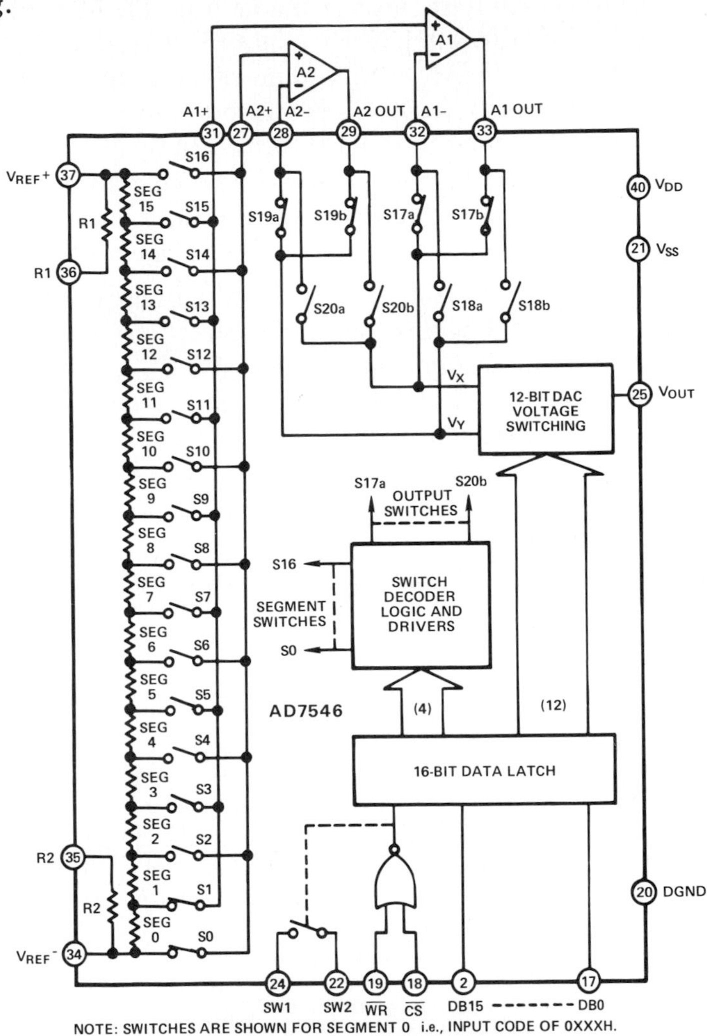

Figure 8.29. Functional block diagram of 16-bit segmented DAC.

The top four bits are decoded to select, via the segment switches, one of the 16 voltage segments available along the resistor chain. This voltage segment, $(V_{REF+} - V_{REF-})/16$, is buffered by external follower amplifiers and used as a voltage reference for a 12-bit R-2R-type d/a converter operating in the voltage switching mode. When the segment voltage is applied to V_X and V_Y, the output of the d/a converter may be expressed as follows:

$$V_{OUT} = V_Y + D(V_X - V_Y) \tag{8.1}$$

where D is the fractional analog value of the lower 12-bit digital code, V_X is the upper segment voltage and V_Y is the lower segment voltage. The 12-bit d/a converter's reference inputs, V_X and V_Y, are connected to the two buffered resistor-chain nodes, which define the segment of interest; the 12-bit DAC interpolates between the voltages at these two points.

In this way, the 65,536 output levels available from the 16-bit DAC are divided into 16 groups of 4,096 steps each. Since the largest output value from the 12-bit DAC doesn't exceed the minimum value of the next segment, the 16-bit DAC has to be monotonic—if the 12-bit DAC is itself monotonic. Thus, the monotonicity of the 16-bit DAC is limited by that of the resistance-ladder 12-bit converter; while these devices are reliably manufacturable with good yields, ladder-type converters for higher resolutions, approaching 16 bits are not yet fully feasible—hence the use of two-stage conversions.

The AD7546 has a 16-bit-wide internal latch to facilitate microprocessor interface. Signals $\overline{CS}$ and $\overline{WR}$ have the same interpretation (*chip select* and *write*) as in normal microprocessor systems. When both $\overline{CS}$ and $\overline{WR}$ are low the input latches are transparent, and the DAC output voltage follows the input data. With $\overline{CS}$ low, the input data is latched on the rising edge of $\overline{WR}$.

Also included on the chip is an SPST switch intended for use in a track-hold circuit to remove glitches from the DAC output and simplify low-pass filtering of the reconstructed output voltage. The switch is synchronized with the latch loading signals; it is open when both $\overline{CS}$ and $\overline{WR}$ inputs are low. The

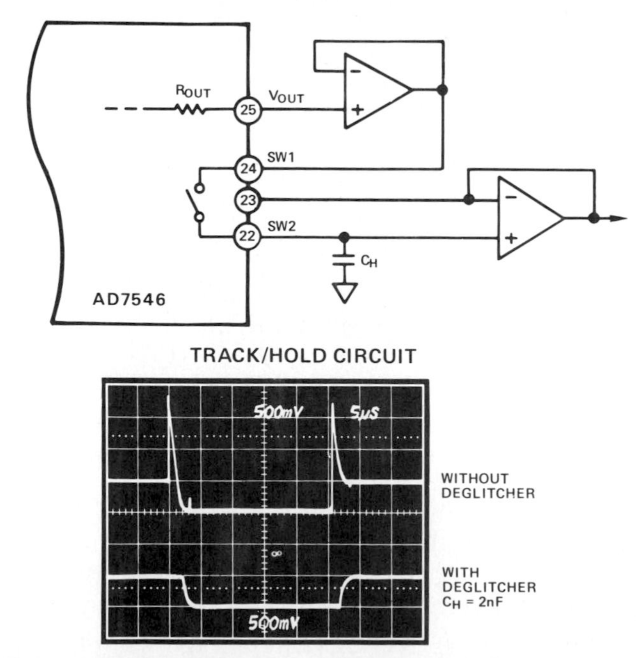

TYPICAL OUTPUT WAVEFORMS USING THE TRACK/HOLD CIRCUIT

Figure 8.30. Deglitching a 16-bit DAC.

internal logic of the AD7546 ensures that the switch opens before data to the latches can change.

To function as a track-hold, the switch is connected in series with the DAC output, as shown in the Figure 8.30, with pin 24 as the input and pin 22 as the output. Pin 23 is a pin with no internal connections; its purpose is to serve as a guard. It should be connected to the output to minimize any feedthrough resulting from stray capacitances at the two switch terminals. When the switch is open, the Hold capacitor stores the previous output voltage of the DAC. The WR pulse should be of sufficient duration to allow the DAC to settle to its new analog output and for all glitches to have settled out. Driving the $\overline{\text{WR}}$ input from a one-shot will ensure sufficient settling time.

It is interesting to note that the external amplifiers used with the AD7546 need not be particularly accurate to preserve monotonicity. Switches S17-20 are used to insure that the same amplifier that buffers the top of one segment is also used to buffer the bottom of the next segment. Thus, the transfer function of the DAC remains monotonic, even if the amplifiers have large offsets. However, integral linearity will be degraded by large amplifier offsets, since—depending on which segment is selected—the size of each segment will be equal to the nominal segment size, plus or minus the difference of the op-amp offsets.

In the AD7546, the amplifiers are left as external components to be added by the user, since amplifiers with sufficiently low noise for 16-bit applications, not possible with the CMOS process technology of the early 1980s, could not be integrated on the same chip as the rest of the DAC circuit.

8.4 BiMOS DAC TECHNOLOGY

In addition the use of standard bipolar and CMOS technologies in the manufacture of IC data converters, the new combinational technologies, BiMOS II and LC^2MOS, have been chosen to implement some of the new high-resolution converters at Analog Devices. The table outlines the relative strengths and weaknesses of the various technologies.

While development is under way on many general-purpose converters using LC^2MOS or BIMOS II, at this writing, existing designs are still implemented in bipolar and CMOS. Bipolar seems to be more capable of high accuracy and long-term stability, but CMOS is more versatile, in that it can operate more easily in a 4-quadrant multiplying mode. Bipolar devices are typically faster, but CMOS devices require much less power. The bipolar process has more versatile analog components available, for implementing reference circuits and amplifiers, but the CMOS process has high-quality digital components available for implementing on-chip logic, control, and storage functions.

The Analog Devices BiMOS process is an advanced, all-implanted, n- well process, which offers a combination of both high-speed-low-power CMOS logic and high-speed-low-noise analog circuitry. It is ideally suited to the manufacture of monolithic data converters, since complex digital functions and high-precision linear circuitry can both be included on a single chip. The first product manufactured with this process was the AD569 16-bit DAC. Like the CMOS AD7546, it uses the segmented architecture to achieve 16-bit monotonicity, but with less-stringent matching requirements in the resistor network (Figure 8.31).

COMPARISON OF IC DAC TECHNOLOGIES

	BIPOLAR	CMOS	BiMOS II	LC^{2}MOS
Switches				
Speed	Excellent	Good	Good	Good
Stability	Excellent	Fair	Good	Good
Accuracy	Excellent	Good	Good	Good
Power	High	Very Low	Very Low	Very Low
Resistors				
Diffused Silicon	Fair	Fair	Fair	Fair
Deposited Thin Film	Excellent	Excellent	Excellent	Excellent
References	Excellent	Poor	Good	Fair
Multiplying Capabilities	Poor	Excellent	Excellent	Excellent
Amplifiers				
Speed	Excellent	Poor	Excellent	Good
Accuracy	Good	Poor	Excellent	Good
Logic				
Speed	Good	Good	Good	Good
Power	Poor	Excellent	Excellent	Excellent
Size	Fair	Good	Good	Good

The AD569 has two cascaded 8-bit resistor strings, each with 256 taps. The 8 most significant input bits select the segment of the first string, and the lower 8 bits select the tap within that segment by use of the second resistor string. Since it is only necessary that each string be monotonic to 8 bits, resistor matching requirements are easily achieved without trimming of any kind. Thus the AD569's transfer function is monotonic to 16 bits without trimming of any kind. Integral linearity error is held to less than 0.02%.

Gain and offset errors are minimized in the AD569 by use of separate force-sense connections between the reference and the first resistance divider. Without these Kelvin connections, errors would arise from the parasitic resistances encountered in circuit-board tracks, package pins, and internal bond wires. While these errors may seem small, it is important to remember that, at 16 bits, 1 LSB on a 10-volt span is a mere *153 microvolts*.

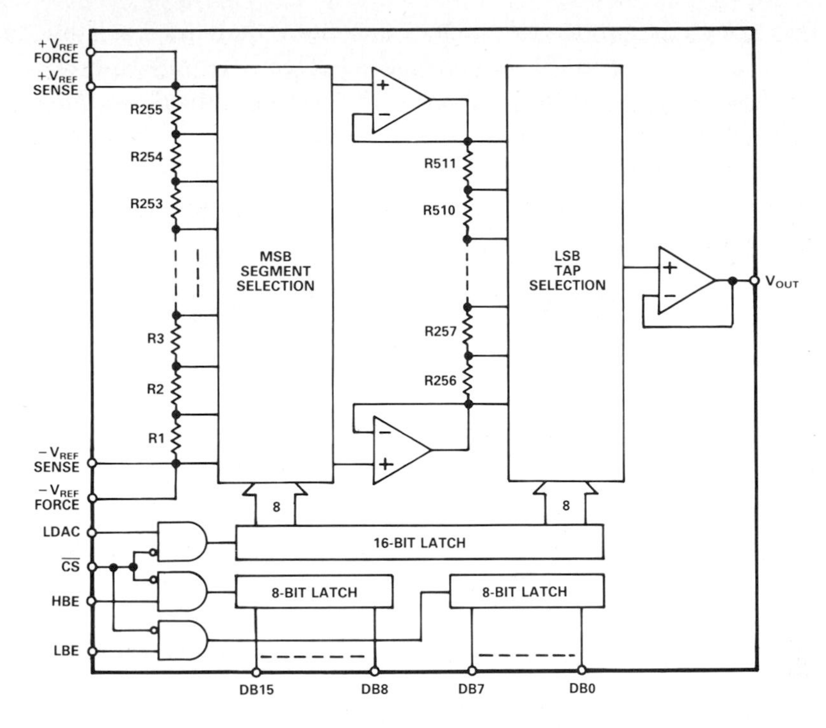

Figure 8.31. Double-buffered sixteen-bit monolithic d/a converter.

Unlike the AD7546, the AD569 includes high-precision buffer amplifiers on the same chip. These amplifiers reject common mode and achieve extremely high linearity in the follower mode by the use of a new circuit architecture which uses both bipolar and MOS transistors. Speed is not compromised in these amplifiers—output settling to 0.001% is typically 6 microseconds for a full-scale step. Furthermore, the AD569 can be used in a multiplying mode with reference signals of up to several hundred kilohertz.

The logic interface of the AD569 includes a set of double-buffered input registers. The control signals allow the first rank registers to be loaded from either an 8-bit or a 16-bit bus, followed by a transfer of the data from the first rank to the second.

The latches are controlled by four input signals: high-byte enable (HBE), low-byte enable (LBE), and load DAC (LDAC), all of which are internally gated with chip-select (CS). All control signals are compatible with all standard 5-volt logic families. The functioning of the control signals is shown here:

CS	LBE	HBE	LDAC	OPERATION
1	X	X	X	None—DAC Deselected
0	0	1	1	Load first-rank low-byte latch
0	1	0	1	Load first-rank high-byte latch
0	1	1	0	Load second-rank latch from first rank

The control signals can be tied together and more than one latch can be enabled at one time. For example, when loading directly from a 16-bit wide bus, HBE and LBE may be tied together and both latches enabled simultaneously.

8.5 HYBRID DACS

Digital-to-analog Converters have participated in a general progression from board (or module) form to hybrids, and then to monolithics. Hybrid and monolithic technologies share much in common—package size and manufacturing costs are similar and compare favorably with assembly of the DAC function in discrete form, as well as providing much higher reliability.

Hybrid construction allows several technologies to be combined in a single small package to provide a function which cannot be implemented in any existing monolithic technology. For this reason, it is unlikely that the advent of future IC technologies will bring about the demise of hybrids. Instead, hybrid suppliers will be able to combine these new technologies with other, older IC technologies in order to provide more-complex functions in a small physical area. Simply stated, the availability of more-advanced and more-powerful IC processes will make possible more-advanced and more-powerful hybrids.

For example, consider the evolution of the 12-bit DAC function. The first mass-produced modular 12-bit DACs consisted of a large number of discrete components to implement the various building blocks, such as switches, output and reference amplifiers, and resistance ladders. As monolithic technology progressed, operational amplifiers and arrays of matched switching transistors and level shifters became available in monolithic form. Thin-film resistor networks also became available. These were then used in modular converters to improve reliability and reduce manufacturing costs. Available in unpackaged chip form, these IC functional building blocks fostered economical manufacture of hybrid converters that were functionally as complete as the modules but with higher reliability and requiring much less physical space.

One of the most-popular 12-bit DACs is the DAC80, now available from several manufacturers in monolithic form (Figure 8.7). The original DAC80 was introduced in the mid-1970s as a hybrid device comprising 11 chips in a hybrid package. The 11 chips were: three 4-bit switch arrays, two operational amplifiers, two resistor networks, a Zener diode in chip form, two clamp diodes, and a chip capacitor. In 1978, when monolithic technology had progressed to the point where it was possible to combine the switch and resistor network functions on a single chip, a three-chip DAC80 was introduced. The three chips in this design included a reference chip, an output amplifier, and the switch/resistor/control-amplifier chip.

The newer design offered performance identical to that of the original DAC80, but with a tremendous improvement in reliability and at much lower cost.

Then, in 1983, the first single-chip DAC80 became available. It, of course, provided further cost reduction and reliability improvement, compared to the three- and eleven-chip hybrid versions. Finally, in 1984, this popular device was offered in a low-cost plastic DIP package. Thus, the DAC80 has evolved from a relatively high-cost hybrid to a high-volume commodity IC.

While this evolution is often used as an example to show the impending obsolescence of hybrid manufacturing, it is important to remember that the single-chip DAC80 can be used in chip form in hybrid designs. When combined with devices which are incompatible with the IC process used to produce the DAC chip (such as high-power drivers or MOS memory or logic functions), it is possible to produce more-complex functions in a package whose size is comparable to the monolithic, increasing the functional density available to the system designer.

A good example of the increase in functional density made possible by hybrid technology is found in the AD390 Quad DAC (Figure 8.32), which is based on the AD567 monolithic current-output DAC with its on-chip latches. The AD567 DAC chip requires only an op amp to convert the output current to a buffered output voltage. Normally, the AD567 is housed in a 28-pin DIP package. However, the AD390 contains *four of these DACs and their output amplifiers*, as well as an additional op amp to buffer the reference input, in a package which is the same size as a single AD567. Six address inputs allow any of the DAC registers (or any combination of registers) to be loaded from a 12-bit parallel bus.

The truth table indicates the functions available for various combinations of levels on the control bus.

$\overline{CS1}$	$\overline{CS2}$	$\overline{CS3}$	$\overline{CS4}$	$\overline{A1}$	$\overline{A0}$	Operation
1	1	1	1	X	X	No Operation
X	X	X	X	1	1	No Operation
0	1	1	1	1	0	Enable 1st rank of DAC1
1	0	1	1	1	0	Enable 1st rank of DAC2
1	1	0	1	1	0	Enable 1st rank of DAC3
1	1	1	0	1	0	Enable 1st rank of DAC4
0	1	1	1	0	1	Load DAC1 (second rank) from first rank
1	0	1	1	0	1	Load DAC2 (second rank) from first rank
1	1	0	1	0	1	Load DAC3 (second rank) from first rank
1	1	1	0	0	1	Load DAC4 (second rank) from first rank
0	0	0	0	0	0	All latches transparent

This device is especially useful in applications where many d/a converters are needed, but space is not available for a large number of IC packages. One such application is in automatic test equipment, where large numbers of DACs are used for waveform generation, level setting, and threshold setting, yet it is

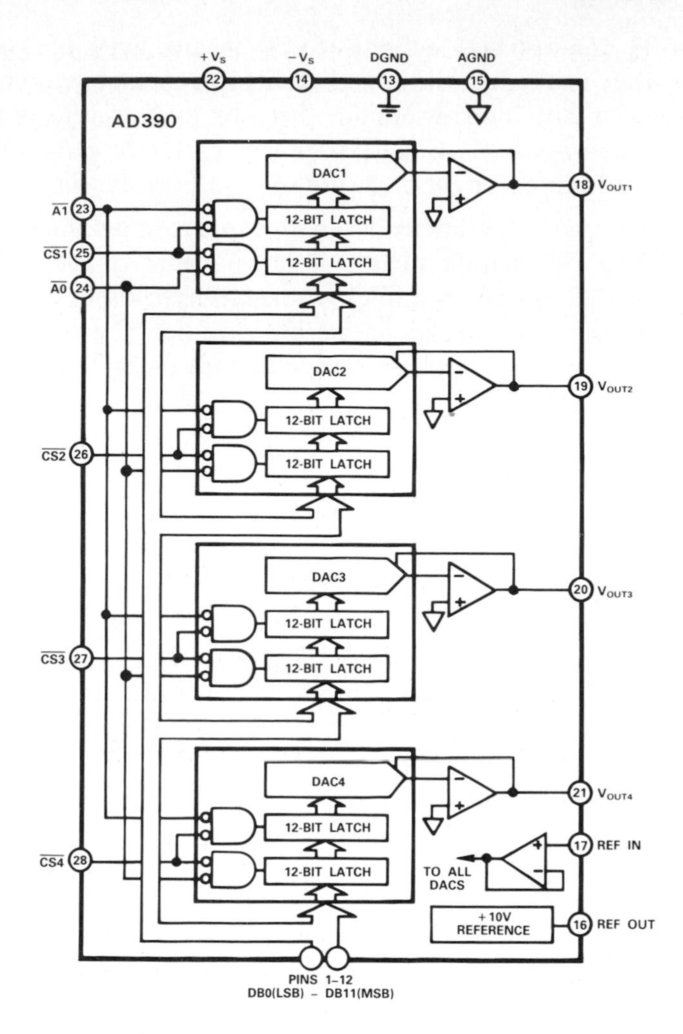

Figure 8.32. 4-channel 12-bit digital-to-analog converter.

desirable to fit the tester in as little floor space as possible. Another application is in automatic calibration of instruments and systems, where the cost of manual adjustment and/or maintenance would be prohibitive.

At the time of publication, no manufacturer of monolithic devices had yet demonstrated the ability to produce a component with functional density and performance comparable to those of the AD390 hybrid. Of course, should this occur, a hybrid manufacturer would be able to use four of these chips in a hybrid.

8.6 INTEGRATED-CIRCUIT ADCS

As has been pointed out in an earlier chapter, many circuit techniques have been developed, and several are in widespread use, for converting analog signals to digital form. Of these techniques, the successive-approximation method has become the preferred approach for performing this function in general-purpose applications, owing to its reasonable speed and resolution for a given amount of circuit complexity. Figure 8.33 is the block diagram of a successive-approximation converter:

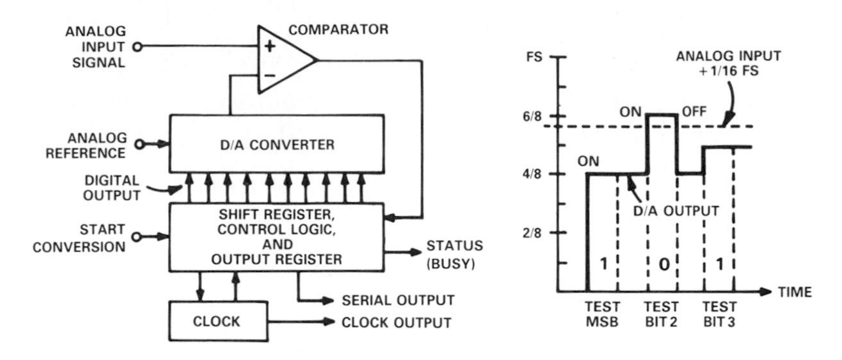

Figure 8.33. Successive-approximation a/d converter.

Both bipolar and CMOS technologies have successfully been employed to produce monolithic d/a converters. The addition of the logic, clock, and comparator blocks to complete the a/d conversion function has been accomplished in both technologies, but—as we shall see—with somewhat different approaches.

8.6.1 BIPOLAR PROCESSING WITH I^2L

The bipolar process is certainly capable of producing both the DAC and comparator functions. Bipolar technology is also used to fabricate many popular logic families, such as TTL and its derivatives. However, there are significant differences in the processes used for linear bipolar circuits and digital bipolar circuits. For example, bipolar digital circuits are generally designed for relatively low voltage operation (typically 5 volts), and the process is thus tailored for low breakdowns and high speeds. Linear circuits, on the other hand, need a wider supply spread in order to accommodate larger signal swings and achieve reasonable dynamic range. The higher breakdowns generally dictate larger geometries and lower speeds.

Since their objectives seem incompatible, it is therefore difficult to produce both high precision linear circuits and logic functions on the same chip using either the standard linear bipolar or the standard bipolar logic process by itself. A different process must be used.

An approach that has been very successful is the addition of the logic functions to a predominantly linear chip using the integrated-injection logic (I^2L) process. This process allows reasonably dense logic to be included on the same chip as high-breakdown precision linear circuitry. This process was used in the production of the AD571 monolithic 10-bit ADC in the late 1970s.

The I^2L process is particularly useful in manufacturing ADCs because only a single additional diffusion step is required beyond those used in the standard linear process. Furthermore, this diffusion does not interfere with the other steps in the process, so the analog circuitry is unaffected by the addition of the logic. The tradeoff between I^2L operating current and speed makes conversion times in the 2-3 microsecond-per-bit range possible with manageable amounts of supply current (and thus tolerable power consumption).

Unlike most logic families, the I^2L family uses multiple fan-out connections (multiple collectors) in a wired-or fashion rather than multiple fan-in connections (multiple emitters). This means that the IC designer must re-think the interconnection concepts, but the basic logic functions are still the same (Figure 8.34).

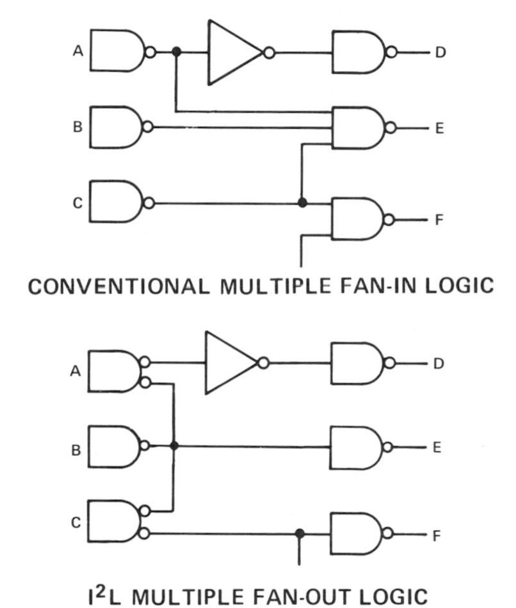

Figure 8.34. Comparing multiple fan-in logic with I^2L's multiple fan-out logic.

The simplicity of I^2L logic compared to conventional TTL logic is amply illustrated with a moderately complex function, such as a D-type flip-flop (Figure 8.35). The equivalent of six NAND gates is required to perform the function—in TTL, the four input gates require three components—two active and one passive—while the two output gates require nine components—five active and four passive—for a total of 30 components (a). The I^2L flip-flop requires only seven active components, since each multiple-collector transistor is a complete NAND gate (b). The reduction in area is even more impressive because of the simplicity of interconnection (c). I^2L technology is therefore ideal for implementing a complex logic function such as a successive-approximation register.

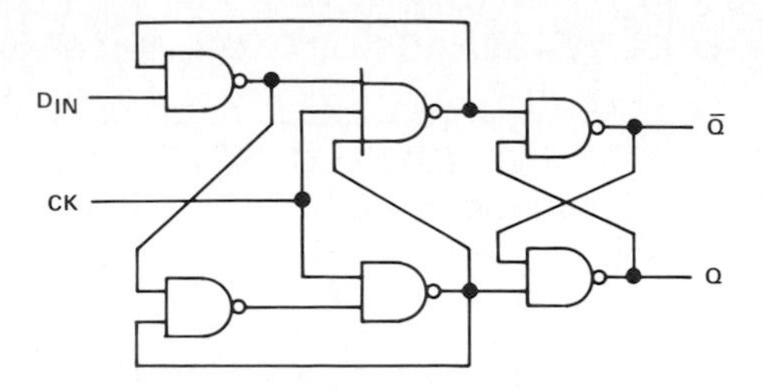

a. D-type FF using standard NAND gates.

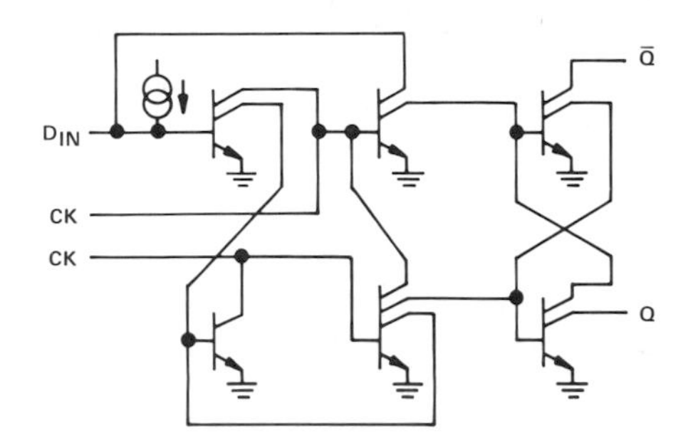

b. D-type FF using I^2L gates.

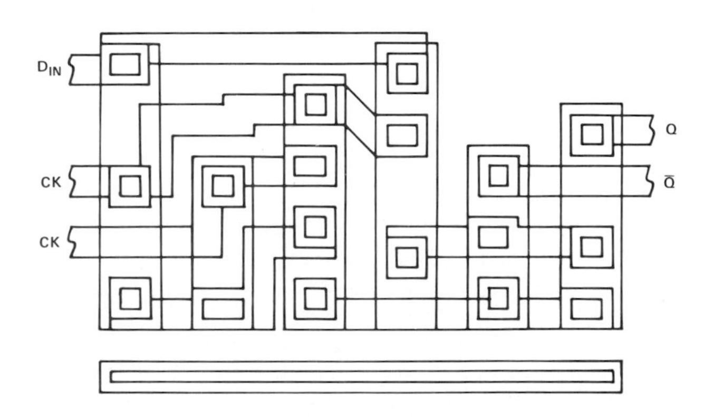

c. On-chip I^2L interconnections.

Figure 8.35. Comparison of TTL and I^2L logic in building a D-type flip-flop.

8- and 10-Bit ADCs

The 8-bit AD570 and 10-bit AD571 (Figure 8.36) employ the I^2L process to perform the SAR function. Each is a completely self-contained converter with internal clock, voltage reference, laser-trimmed DAC, and three-state output buffers. No external components are required to perform a full-accuracy conversion in less than 40 microseconds. The three-state output buffers are open whenever the blank/convert command is in Blank, or during a conversion; immediately after a conversion they come on and present data.

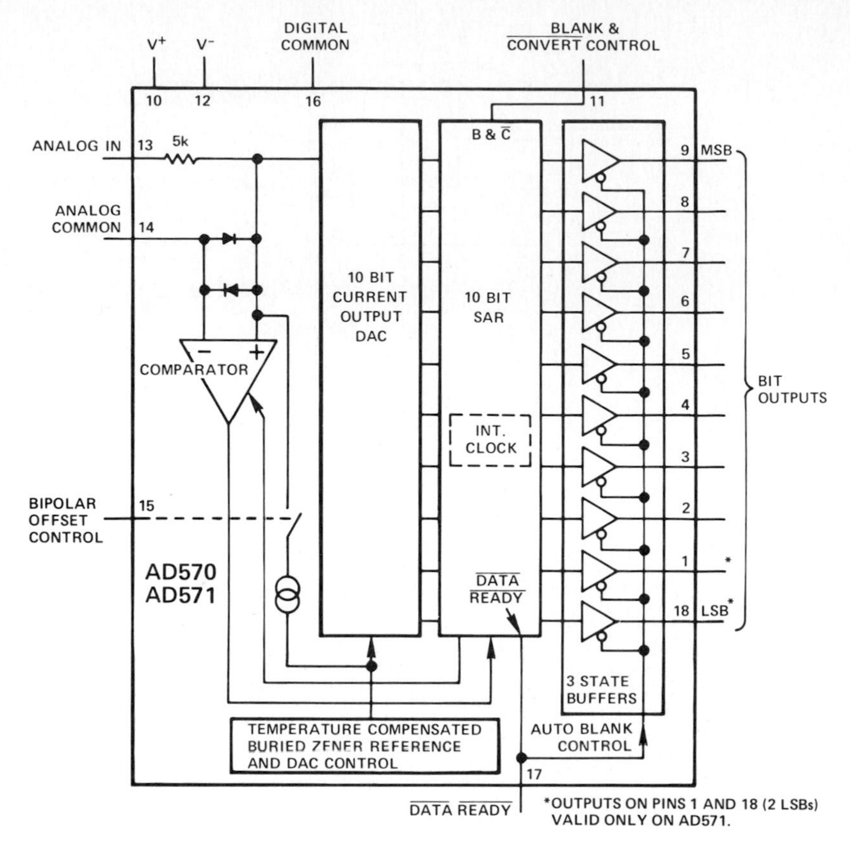

Figure 8.36. AD570/AD571 8/10-bit a/d converter with automatic 3-state outputs for single-line control.

The AD570 and AD571 are well-suited to applications where direct bus interface is not required. Where needed, the newer 10-bit AD573 and 8-bit AD673 (Figure 8.37), which use two additional package pins, provide bus-interface capability.

12-Bit ADCs

A logical extension of AD571-based technology is into the 12-bit area. But going from 10 bits to 12 bits requires more than adding two more switches, resistors and register states. In order to attain true 12-bit accuracy and stability, larger (and often many more) components must be used in critical locations. For example, trimming a resistor to ±0.006% (±¼ LSB at 12 bits) requires larger trim areas than would be necessary to trim to ±0.24% (±¼ LSB at 10 bits). Transistors need to have larger geometry for more predictable performance.

The AD574A is a complete 12-bit successive-approximation analog-to-digital converter with 3-state output-buffer circuitry for direct interface to an 8-, 12-, or 16-bit microprocessor bus. The AD574A design is implemented with a

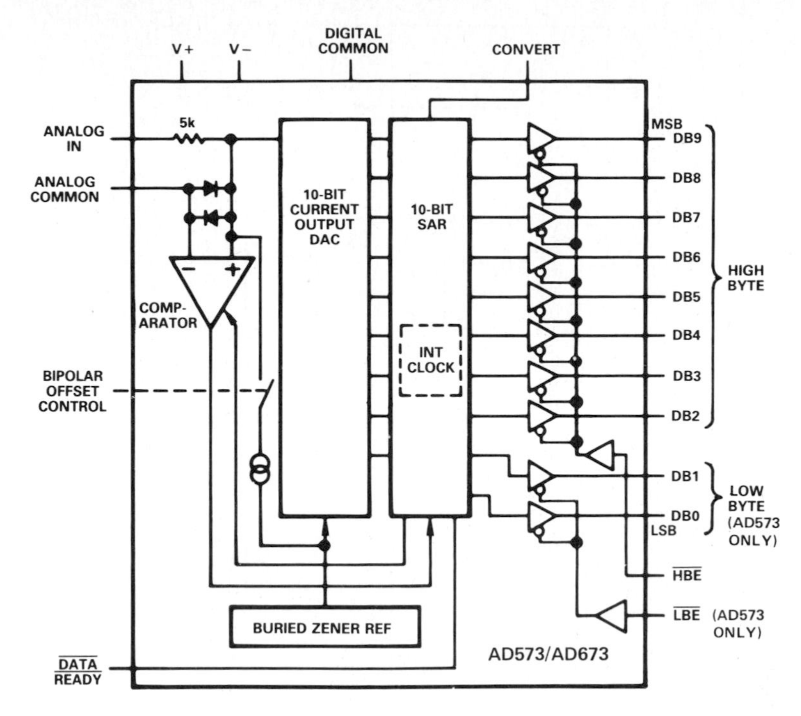

Figure 8.37. Block diagram of 10/8-bit a/d converter, capable of interfacing with 8- or 16-bit buses.

single LSI chip containing both analog and digital circuitry, resulting in maximum performance and flexibility at low cost.

Its introductory form, the AD574, which emerged as the industry-standard 12-bit ADC in the early 1980s, was manufactured using compound monolithic construction, based on two chips—one an AD565 12-bit current-output DAC, including the reference and scaling resistors, and the other containing the successive-approximation register (SAR) and microprocessor-interface logic functions, as well as the precision latching comparator. The block diagram is shown in Figure 8.38.

In 1985, the device became available in monolithic form for the first time; this made low-cost commercial plastic packaging possible. The transition from the two-chip version to the single-chip version was not made until manufacturing yields on the larger single chip reached economically viable levels. In performance, the monolithic version is a direct replacement for the two-chip device. A comparison of the physical appearance of the two devices appears in Figure 8.39.

At the time this is being written, CMOS technology cannot provide either a reference with comparably low drift or a high-speed comparator function. Likewise, bipolar processes without compatible I^2L are incapable of providing logic which is both fast enough and low enough in power to manufacture a 574-equivalent.

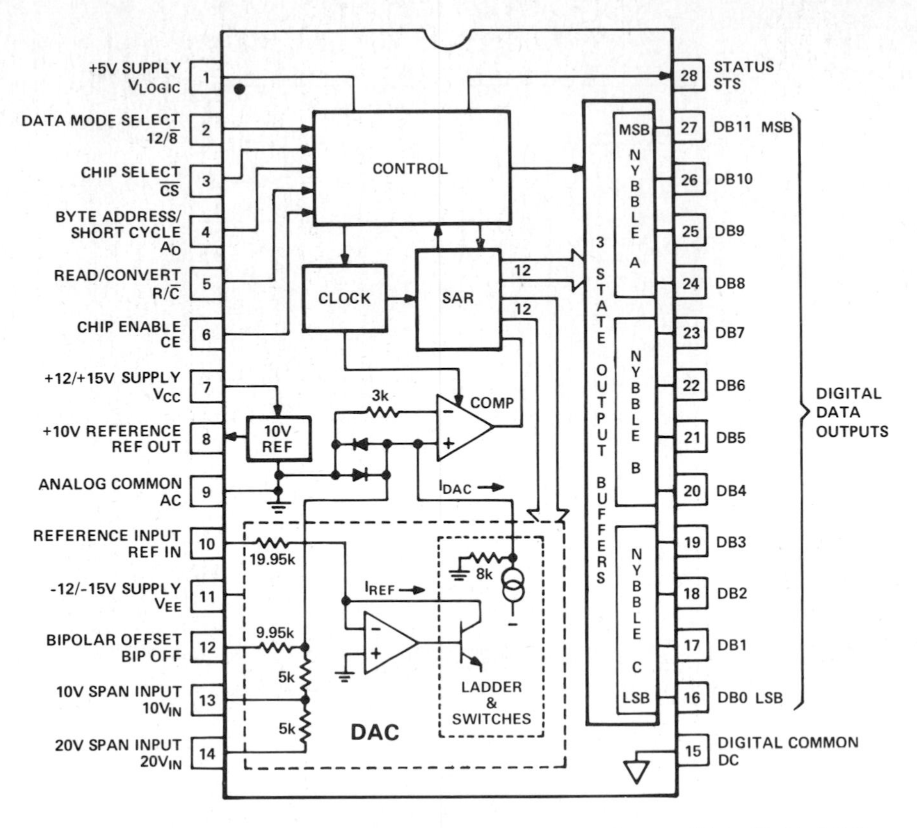

Figure 8.38. Block diagram and pin configuration of monolithic 12-bit ADC.

Figure 8.39. Monolithic and two-chip versions of 12-bit ADC compared.

8-Bit ADC with Simplified Logic and Instrumentation-Amplifier Input

Another bipolar-process approach to a general-purpose ADC is the AD670 8-bit, 10-microsecond unit. It uses a novel approach to perform the successive-approximation function. Rather than a string of D-type flip-flops connected as a shift register, the AD670 uses a delay line consisting of 80 I^2L gates to establish the timing sequences for the conversion. The delay-line SAR eliminates the need for a well-controlled clock oscillator on the chip. The timing sequence of the conversion is established by generation of a pulse at the beginning of a conversion and its propagation along the delay line, with taps on the delay line at points which allow appropriate timing for register reset, bit trials, DAC settling, and latching of data into the output registers. Figure 8.40 is a block diagram of the AD670.

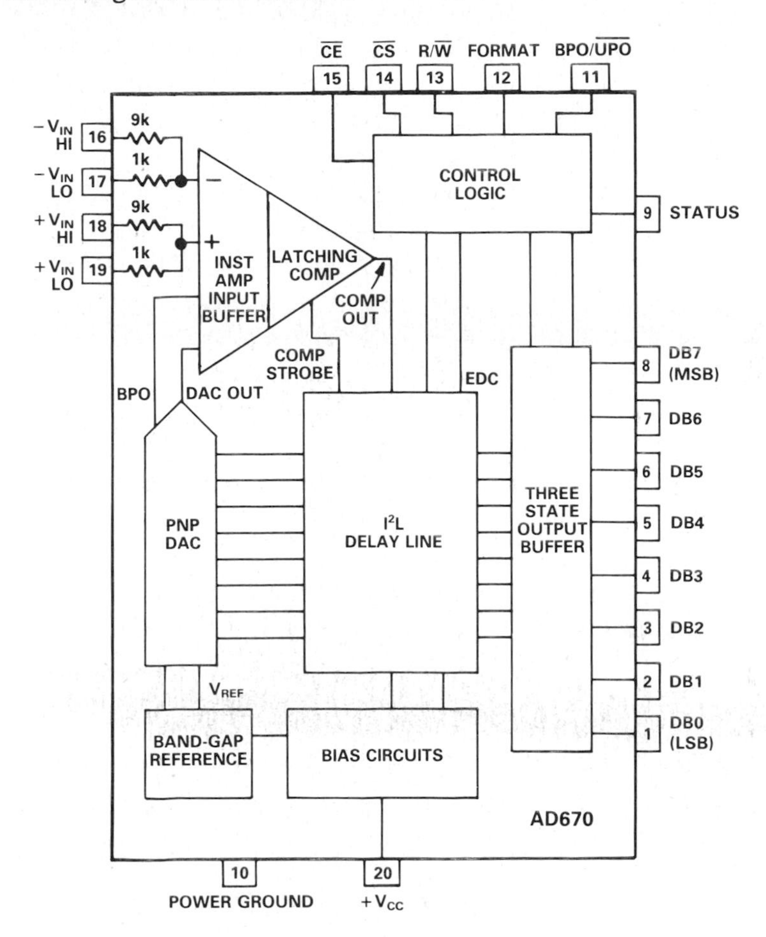

Figure 8.40. Functional block diagram of complete 8-bit ADC, including differential instrumentation-amplifier input.

The 80-gate delay line yields a total conversion time of 8-10 microseconds, fast enough to digitize signals in a wide range of bandwidths. The analog input section includes differential inputs capable of accepting input signals scaled at 1 mV/LSB, yet protected against overvoltages of up to ±30 Volts. A precision 10:1 attenuator allows an alternate input range of 2.56 Volts full-scale

(10 mV/LSB). Either unipolar or bipolar inputs are accepted, with output coding software-selectable, as indicated in the table:

BPO/$\overline{\text{UPO}}$	FORMAT	INPUT RANGE/OUTPUT FORMAT
0	0	Unipolar/Straight Binary
1	0	Bipolar/Offset Binary
0	1	Unipolar/Twos Complement
1	1	Bipolar/Twos Complement

The operation of the AD670 is controlled by the digital inputs $\overline{\text{CE}}$, $\overline{\text{CS}}$, R/$\overline{\text{W}}$, FORMAT, and UPO/$\overline{\text{BPO}}$. Conversions are initiated by writing to the AD670 ($\overline{\text{CE}}$, $\overline{\text{CS}}$, and $\overline{\text{W}}$ active). The states of the FORMAT and UPO/$\overline{\text{BPO}}$ inputs during the Write operation determine the input range (unipolar or bipolar) and the output digital data format (straight binary or twos complement). When the conversion begins, the STATUS line goes high to indicate that a conversion is in progress. The high-to-low transition of the STATUS line indicates that the conversion is complete and that data can be read.

The internal design of the AD670 is based on the AD558-type DAC (Figure 8.11). This architecture was chosen for its combination of speed and completeness as well as its ability to operate from a single +5-volt power supply. The circuit's reference is a low-drift bandgap type, chosen primarily for its low-voltage operation. A Zener-type reference would have precluded +5 V operation, since most Zeners produce output voltages higher than 5 Volts and thus would require higher input voltages.

The input stage of the AD670 is a bipolar-input differential buffer amplifier, with high common-mode rejection, even at high frequencies. This allows voltage to be measured in the presence of noisy grounds and dc common-mode offsets, where a single-ended input ADC might require external signal conditioning. Furthermore, many a/d converters in the same speed range as the AD670 require a buffer amplifier between the source and the converter to decouple the converter's changing input impedance during the conversion cycle (see Section 12.1.7). The AD670, however, provides this buffering on-chip, and presents to the source a resistive load of 10 kilohms (or several megohms on the 256-mV range).

The bipolar process is thus useful when the complete converter function is desired on a single chip. However, its versatility is limited by the availability of logic functions which may be required in more exotic (and logic-intensive) conversion algorithms. Although the linear functions (comparators and references) available in the bipolar process have been demonstrated to be superior to their CMOS counterparts, the digital functions are slower than CMOS, for a given power consumption.

Various versions of the bipolar process are also used to produce very high speed converters using the "flash" technique (Chapter 13). The flash a/d converter is implemented by comparing the analog input with voltage levels representing the boundaries of each of the codes. For example, an 8-bit flash converter uses a reference divider with 255 taps to establish comparison points for each of 255 high-speed comparators. In this type of circuit, speed is the most critical parameter for both analog and digital circuits, and power is a secondary concern. Therefore, bipolar processing is the most often used.

The higher-resolution converters using this circuit technique, along with specially developed high-speed fabrication processes, are available with up to 8 bits of resolution and many tens of MHz conversion rates. Present MOS processes do not lend themselves well to the flash conversion technique, since the usual power advantage of CMOS in the switching mode disappears when the circuitry is operating in its linear range most of the time.

8.6.2 CMOS A/D CONVERTERS

For lower speeds, however, CMOS converters using the successive-approximation method can be easily produced. The successive-approximation ADC requires logic to implement the SAR function, a DAC, a reference, and a comparator. The digital circuit capabilities of CMOS are well-understood, making the SAR and interface-control functions relatively easy to implement; also, the suitability of CMOS to DAC manufacture was described in some detail earlier in this chapter. The comparator and the reference are the circuit elements which provide the biggest design challenges in CMOS converters.

Most CMOS converters are designed for use with an external reference, since high-performance bipolar Zener and bandgap references are available at low cost and offer lower drift and noise than most CMOS references. Furthermore, there are many applications which call for either external system references or ratiometric conversion, for which an on-chip reference would be superfluous.

Early CMOS ADCs included neither reference nor comparator functions. Newer designs include the comparator function, and future designs, using improved linear-compatible CMOS processes, may include the reference.

8-Bit A/D Converters

A good example of a successive-approximation converter implemented in CMOS is the Analog Devices AD7574 (Figure 8.41a). It performs an 8-bit conversion in 15 microseconds and interfaces easily to microprocessors, while consuming only a few milliwatts of power.

The comparator used in the AD7574 is designed to resolve LSBs of a few tens of millivolts and is normally used with a 10-volt full-scale range, established by an external −10-volt reference, e.g., a low-cost bandgap type.

A newer 8-bit ADC, the AD7576 (Figure 8.41b), uses the linear-compatible

CMOS process (LC^2MOS) to implement a somewhat different comparator design, which allows the LSB size to be reduced—and both the input signal and the reference to be positive with respect to ground, permitting single-supply operation. With a very low-cost external 1.23-volt bandgap reference, the device accepts inputs with a 0-2.56-volt full-scale range.

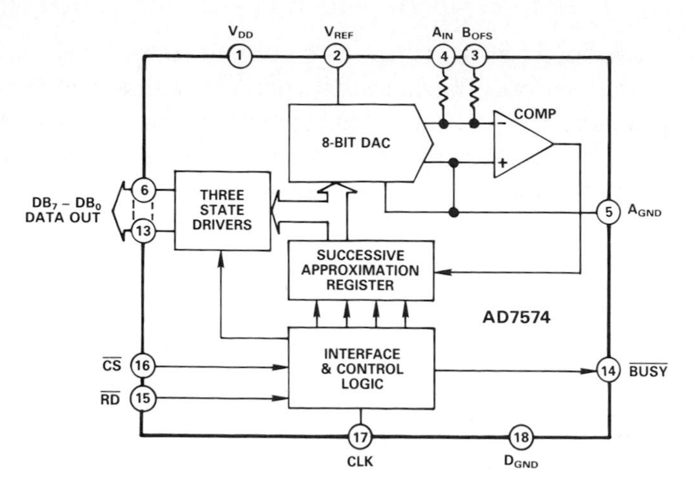

a. 10-volt-full-scale design.

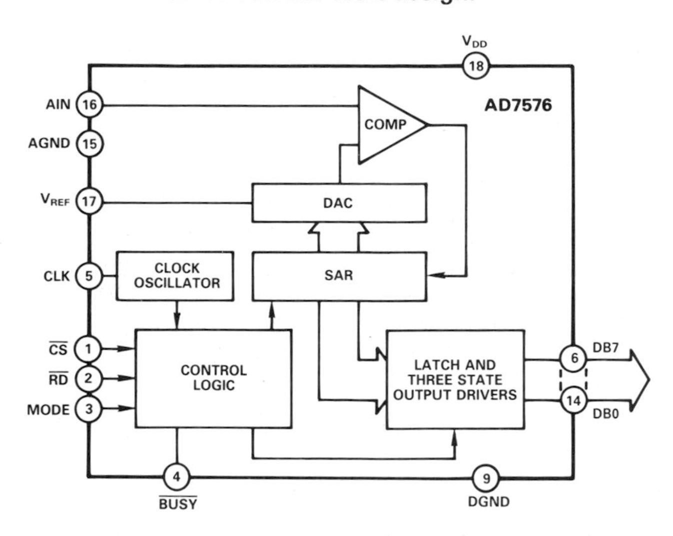

b. 10 mV/LSB design.

Figure 8.41. Functional block diagrams of uP-compatible CMOS 8-bit ADCs.

Like the earlier AD7574, the AD7576 uses an external resistor and capacitor to determine the frequency of the clock oscillator. Except for the reference—which may be a system reference or a voltage proportional to the signal's full-scale range (i.e., ratiometric)—no other external components are needed to perform conversions, which are typically completed in 5 microseconds.

Since the strength of CMOS lies in its ability to integrate logic functions, it is not surprising that the AD7576 features several microprocessor interface modes.

• **Timing and Control.** The AD7576 is capable of two basic operating modes which are outlined in the timing diagrams of Figures 8.42 and 8.43. These two operating modes are an asynchronous conversion mode and a synchronous conversion mode. The selection of the required operating mode is determined by the status of the MODE pin. When this pin is HIGH, the device performs conversions only when the required control signals ($\overline{CS}$ and $\overline{RD}$) are applied; with this pin LOW, conversions are performed continuously, and $\overline{CS}$ and $\overline{RD}$ are used only to access the output data.

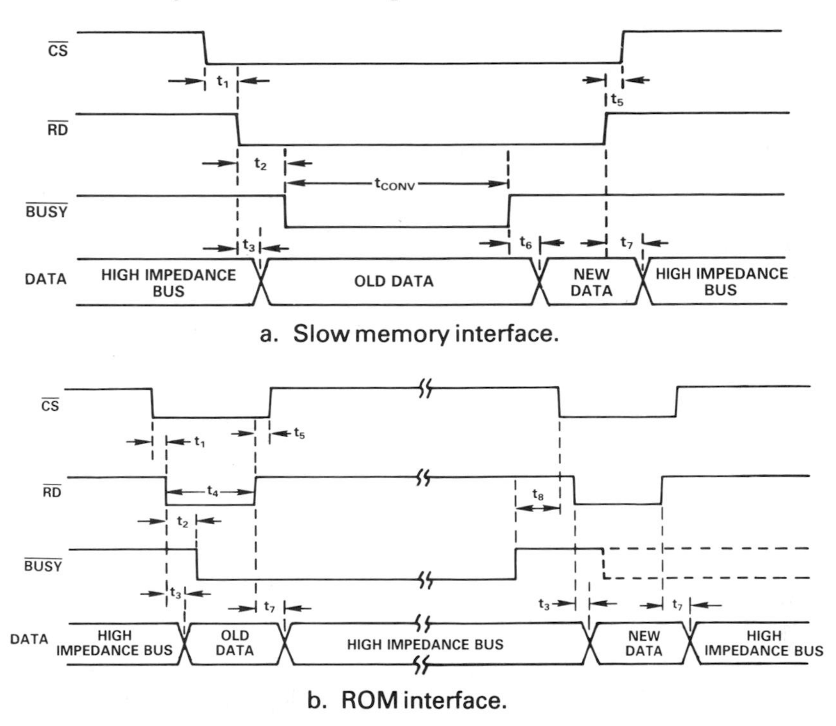

a. Slow memory interface.

b. ROM interface.

Figure 8.42. Synchronous conversion mode timing diagrams.

• **Synchronous Conversion Mode.** In the synchronous conversion mode (the MODE pin tied HIGH), the AD7576 will perform a conversion when requested to do so by the microprocessor. Once the conversion is performed, two interface options exist for reading the output data from the AD7576.

• **Slow Memory Interface.** The first of these interface options is intended for use with microprocessors which can be forced into a WAIT state for at least 5 μs. The microprocessor starts a conversion and is halted until the result of the conversion is read from the converter. Conversion is initiated by executing a memory READ to the AD7576 address. $\overline{BUSY}$ subsequently goes LOW, forcing the microprocessor READY input LOW, thus placing the processor in a WAIT state. When conversion is complete ($\overline{\text{BUSY}}$ goes HIGH), the processor completes the memory READ. The timing diagram for this interface is shown in Figure 8.42a.

The major advantage of this interface is that it allows the microprocessor to start conversion, WAIT, and then READ data with a single READ instruction. The fast conversion time of the AD7576 ensures that the microprocessor is not placed in a WAIT state for an excessive length of time.

Many processors test the condition of the READY input quite soon after the start of an instruction cycle. Therefore, in order for the READY input to be effective in forcing the processor into a WAIT state, $\overline{\text{BUSY}}$ of the AD7576 must go LOW very early in the cycle.

• **ROM Interface.** An alternative interface option in the synchronous conversion mode avoids placing the microprocessor into a WAIT state. In this interface, conversion is started with the first READ instruction, and a second READ instruction accesses the data and starts a second conversion. The timing diagram for this interface is shown in Figure 8.42b.

Conversion is initiated by executing a memory READ instruction to the AD7576 address. Data from the previous conversion is also obtained from the AD7576 during this instruction. This is old data; it may be disregarded if not required. $\overline{\text{BUSY}}$ goes LOW during conversion and returns HIGH when conversion is complete.

The $\overline{\text{BUSY}}$ line may be used to generate an interrupt to the microprocessor, indicating that conversion is complete. The processor then reads the newly converted data. Alternatively, the processor programming may be timed so that the delay between the Convert Start (first READ instruction) and the data READ (second READ instruction) is at least as great as the AD7576 conversion time. For the AD7576 to operate correctly in the ROM Interface mode, CS and RD should not go low before $\overline{\text{BUSY}}$ returns HIGH.

Normally, the second READ instruction starts another conversion as well as accessing the output data. However, if CS and $\overline{\text{RD}}$ are brought LOW within one external clock period after $\overline{\text{BUSY}}$ goes HIGH, a second conversion does not occur.

• **Asynchronous Conversion Mode.** When the MODE pin of the AD7576 is tied LOW, the device performs continuous conversions, and the control lines $\overline{\text{CS}}$ and $\overline{\text{RD}}$ are used only to read the data from the converter. The timing diagram for this operating mode is outlined in Figure 8.43.

Data is obtained from the AD7576 by executing a memory READ instruction to its address. The A/D process is completely transparent to the microprocessor and the AD7576 will behave like a ROM. Data may be read at any time, completely independent of the clock. This is especially useful in internal clock applications; the user does not have to worry about synchronizing the clock with the READ line of the microprocessor.

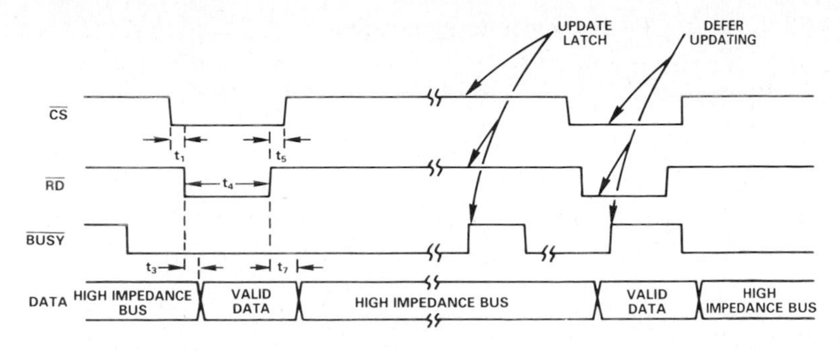

Figure 8.43. Asynchronous conversion-mode timing diagrams.

The data latches are normally updated by $\overline{BUSY}$ going HIGH. However, if $\overline{CS}$ and $\overline{RD}$ are LOW when $\overline{BUSY}$ goes HIGH, the contents of the data latches are frozen until $\overline{CS}$ or $\overline{RD}$ returns HIGH. This ensures that incorrect data cannot be read from the AD7576. The output latches are updated when $\overline{CS}$ or $\overline{RD}$ return HIGH and the converter is re-enabled. If $\overline{CS}$ or $\overline{RD}$ do not return HIGH the AD7576 will stop performing continuous conversions, and will not start again until either line goes HIGH.

The advantage of this mode is its simplicity. The disadvantage of this mode is that the data which is read is not clearly defined in time; however, it will not be older than one conversion period. If this uncertainty is a problem, it can be overcome by monitoring the $\overline{BUSY}$ line.

12-Bit CMOS A/D Converters

Extension of the performance of CMOS successive-approximation a/d converters to 12 bits and beyond is limited by comparator performance. A significant problem with linear CMOS comparators is that their offsets can be on the order of tens of millivolts; while comparator offset normally produces a simple offset in the converter, it is possible that anomalous behavior (i.e., missing codes) will result if offsets are as large as tens of LSBs in an open-loop circuit like a comparator.

Slower converter types, such as dual- and quad-slope units can tolerate slow response times in their comparators—and/or lower bandwidths (which reduces total noise). Successive-approximation converters, however, need relatively high comparator bandwidths if they are to convert in a reasonable length of time.

A good example of a medium-speed CMOS successive-approximation a/d converter is the AD7582 (Figure 8.44). A four-channel input device, it includes the SAR and microprocessor interface logic, a clock oscillator, and a high-precision autozeroing comparator. It performs 12-bit conversions in 100 microseconds with no missing codes.

The only passive components required are the autozero capacitor, C_{AZ}, and timing components, R_{CLK}, C_{CLK1} and C_{CLK2}, for the internal clock oscillator.

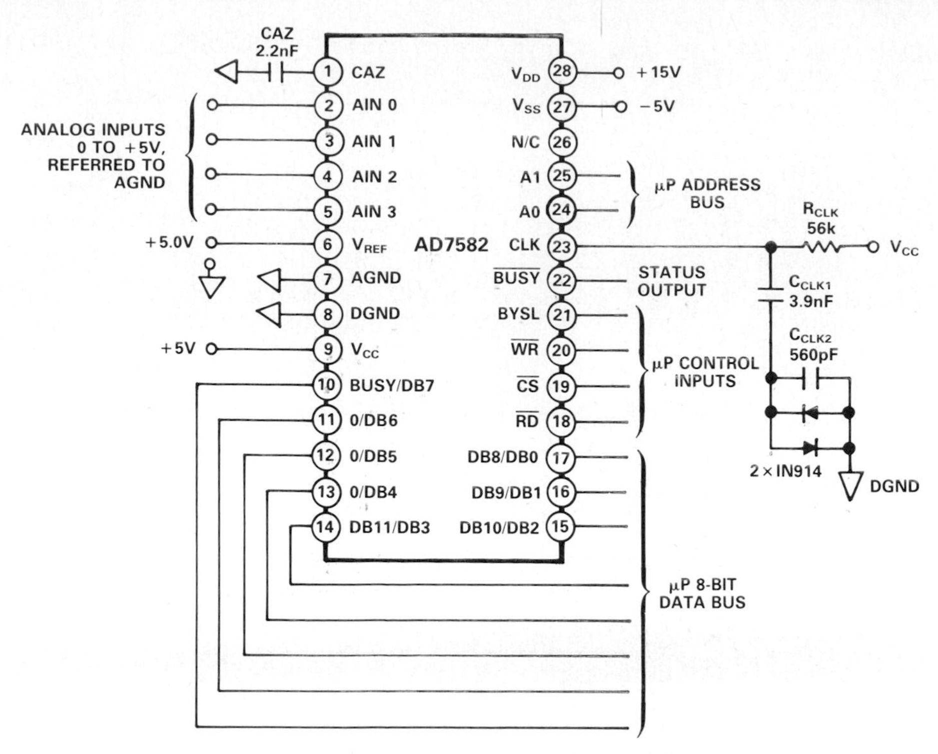

Figure 8.44. Connections to a 12-bit 4-channel CMOS a/d converter.

If the AD7582 is to be used with an external clock source, only C_{AZ} is required.

Between conversions ($\overline{BUSY}$ = HIGH), the converter is in the autozero cycle. When $\overline{WR}$ goes LOW (with $\overline{CS}$ LOW), to start a new conversion, the input multiplexer is switched to the selected channel, N, via address inputs, A0, A1. The autozero capacitor, C_{AZ}, now charges to $AIN_N - V_{OS}$, where V_{OS} is the input offset voltage of the autozero comparator.

A minimum time of 10 μs is required for this autozero cycle. In applications using the internal clock oscillator, it is not necessary for $\overline{WR}$ to remain LOW for this period of time, since a 10-μs delay is automatically provided by the AD7582 before conversion actually begins. This is achieved by switching a constant-current load across the clock capacitors, causing the voltage at the CLK input pin to slowly decay from V_{CC}. It occurs after $\overline{WR}$ returns HIGH; $\overline{WR}$ returning HIGH also latches the multiplexer address inputs, A0, A1 (see Figure 8.45).

The internal Schmitt-trigger circuit, monitoring the voltage on the CLK input, ends the autozero cycle when its LOW input trigger level is reached. At this point, the constant-current load across the clock capacitors is removed, allowing them to charge towards V_{CC} via R_{CLK}. When the voltage at the CLK input reaches the HIGH trigger level, the constant-current load is replaced across C_{CLK1} and C_{CLK2}. The MSB decision is made when the LOW trigger level is reached. This cycle repeats 12 times to provide 12 clock pulses for the conversion cycle.

The autozero capacitor should be a low-leakage, low-dielectric-absorption type. To minimize noise pickup, the outside foil of the capacitor should be connected to AGND. The offset voltage of the comparator is reduced to approximately 100 microvolts by the use of this autozero scheme. Input impedance of the four analog input channels is very high, and no buffering is required for source impedances up to 2 kilohms. Full-scale is normally 5.000 volts, established by an external reference, and total error is ±1 LSB, relative to this reference.

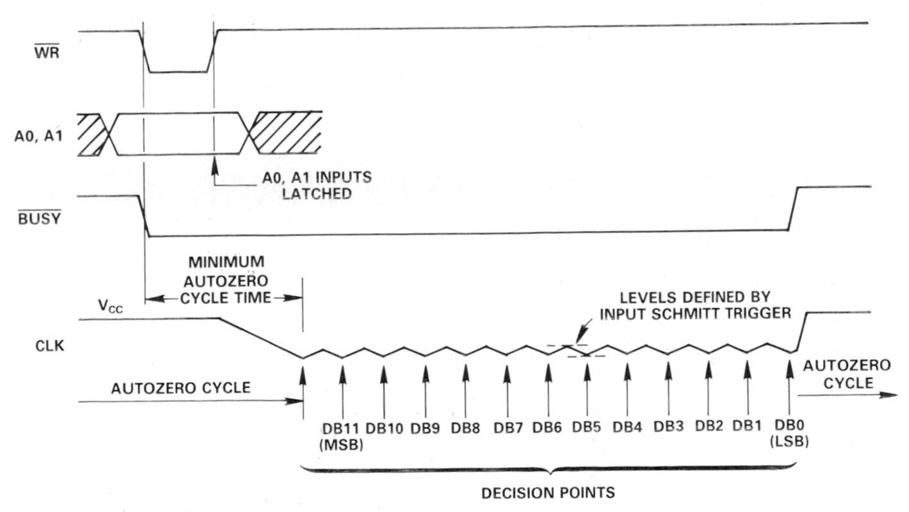

Figure 8.45. Operating waveforms – internal clock of AD7582.

8.6.3 HYBRID A/D CONVERTERS

Hybrid manufacturing technology is still quite often used for high-performance a/d conversion. As mentioned earlier, hybrid technology can serve as a means for combining several monolithic device technologies in a small package. Hybrid is particularly useful when the technologies to be combined are normally incompatible and cannot be integrated on a single chip.

An excellent example of the performance achievable with hybrid construction is the AD578 12-bit ADC (Figure 8.46).

The AD578 is a 12-bit successive-approximation a/d converter that uses monolithic devices from incompatible process technologies to obtain optimized performance. For example, the DAC (an AD565) is manufactured using an ion-implanted bipolar linear process—with laser-trimmed thin-film resistors, to provide high linearity; the SAR logic is implemented using high-speed CMOS for high speed at relatively low power; and the comparator is a combination of a standard bipolar device and several discrete devices, to provide fast response from low overdrive, while maintaining low noise and high accuracy.

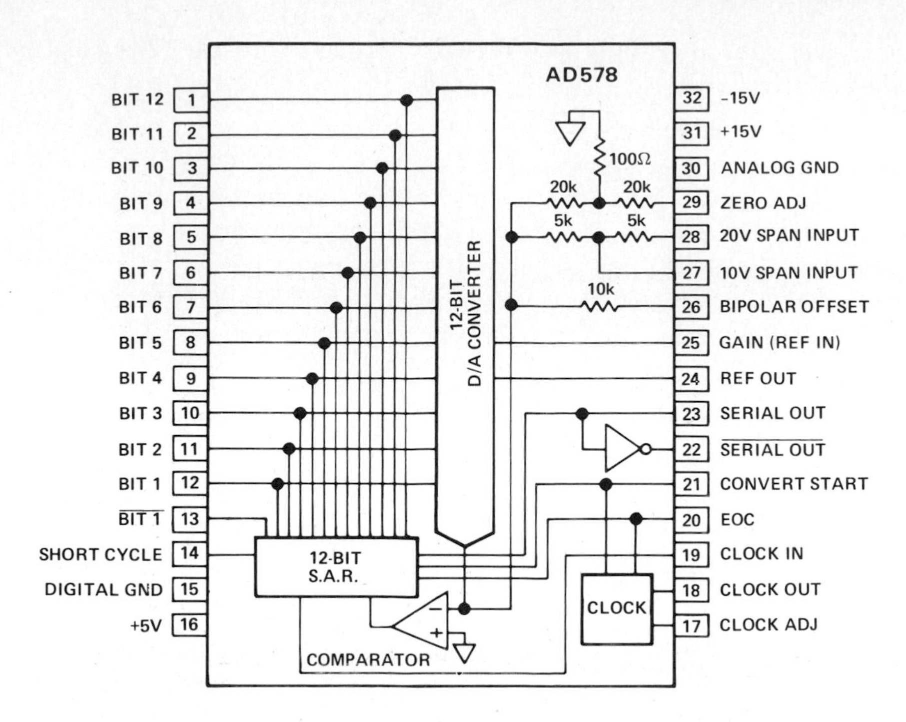

Figure 8.46. Block diagram of hybrid 12-bit ADC.

While it is possible for a skilled circuit designer to duplicate such a design on a board— as part of a larger circuit—using packaged DAC, logic, and comparator devices, or even using surface-mounted chips, commercial hybrid technology reduces such an assembly to the level of a single component, produced in large quantity, and thus achieving economies of scale. This reduces the circuit board area requirements and provides a tested, characterized function at reasonable cost.

Similar advantages arise in the manufacture of high-resolution, video-speed converters, since many components, which cannot be combined on a single chip, can be combined in a hybrid package—offering very high performance. Furthermore, it is generally easier to provide reproducible performance in a hybrid layout than in a printed-circuit board, since the parasitic resistances and capacitances tend to be lower (and more repeatable) in a hybrid package. Further details of the applications of hybrid technology to video converters will be found in Chapter 13.

Chapter Nine

I.C. Converter Design Insights

In Chapter 7, "Understanding Converters," there is a hasty survey of converter design principles. In Chapter 8, there is a discussion of characteristics, architecture, technology, and applications of a representative sampling of popular integrated-circuit converters. The intention of this chapter is to provide more-detailed information on factors the designer must consider in the design and construction of converters for high resolution and accuracy.

Why? In this day and age of monolithic 12-bit converters, it is unlikely (from the standpoint of both cost and engineering effort) that a user would design a converter, except to obtain not-readily available characteristics for a rather special application. Even then, the designer would tend to use available monolithic chips as critical design elements, to which would be added one's own resources of education and experience. Thus, it should not be our purpose to give our readers a blueprint for constructing a copy of a popular IC converter.

Rather, we are seeking here to create in the reader's mind—perhaps in somewhat simplified and general form—an understanding of the problems faced by designers of fixed-reference converters employing bipolar transistors, together with some of the answers that have since become classical (and often patented) approaches.[1]

9.1 REVIEW OF D/A CONVERTER TECHNIQUES

A current-output d/a converter in effect sums the digitally selected outputs of a set of binary-weighted current sources, as shown in the 6-bit example of

[1]Much of the material for this chapter has been adapted from the monograph, "Circuit Techniques for Monolithic DACs," (1979), by A. Paul Brokaw, originally published by Analog Devices and now out of print.

Figure 9.1. The DAC consists of a set of six binary-weighted current sources and six switches. The switches are digitally controlled, one switch for each bit. The switches channel the current flow either to the common output summing bus or to ground (in actuality, the grounded line could be used as a complementary output).

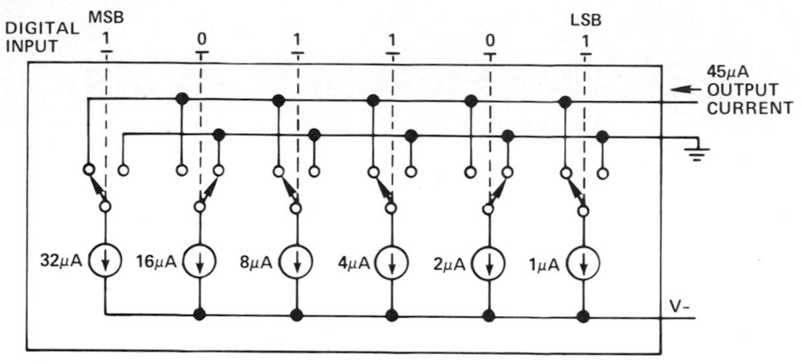

Figure 9.1. Functional schematic diagram of 6-bit current-summing DAC. Digital input 101101 results in 45μA current at output terminal.

The least-significant bit of the digital input (LSB) controls the current source with a weight of 1μA (2^0 μA). If the bit is 1, as shown, the current is switched to the output line; if the bit is 0, the switch diverts the current to ground, and zero current is added to the output line. Similarly, the second bit controls the 2μA (2^1 μA) weight by means of the second switch (shown at 0, or off), the third bit controls the 2^2μA current, and so forth.

The current sources are independent, so that the total output current is simply the sum of those currents switched to the output bus. The figure shows the switches responding to the binary code, $1\ 0\ 1\ 1\ 0\ 1_2 = 45_{10}$ (i.e., 32 + 0 + 8 + 4 + 0 + 1), transformed to 45μA. Each of the 64 unique code possibilities maps a binary number into one of 64 possible current values. The code, 0 0 0 0 0 0, would result in zero output current, and the code, $1\ 1\ 1\ 1\ 1\ 1_2$, would result in an output of 63μA (64μA full-scale minus 1μA).

Current, in Figure 9.1, is shown flowing from the bus toward the current sources. The choice is arbitrary, but there are are two good reasons for having done it that way. First, it is the preferred current direction for the actual structure to be developed, using NPN switching transistors; second, when used with an inverting op amp, to constrain the output bus voltage, this DAC will generate positive output voltage.

R-2R Ladder Networks

Although it is by no means the only method of producing them, the R-2R ladder is one of the most-often used starting points for generating a set of binary currents. Its popularity stems from the fact that although it can be used to produce currents spanning a range of more than 1000:1, it can be made with only two different resistance values (and even these can be constructed using a single resistance value singly or in pairs). This is an extremely important consideration in an integrated-circuit design, where only a limited range of dependable resistor materials are available in a given circuit. Because of the small range of basic resistivity, resistor values are determined largely by geometry. The range of values is limited on the one hand by photolithographic resolution and on the other by size, in relation to the size of both the chip and the conductors with which a resistor must make contact.

As one might expect, the R-2R ladder consists of a collection of resistors with values R and 2R. A single-element ladder, driven by excitation voltage, E, is shown in Figure 9.2. It illustrates that this circuit—which will be developed into a more complex ladder—has a resistance 2R, as viewed from the excitation, and that it divides evenly whatever currents flow into its input ($2I = E/(2R)$) between the leftmost 2R leg ($I = E/(4R)$) and the 2R effective resistance of the remainder of the circuit. Note that this is the same as the resistance which appears at the network input.

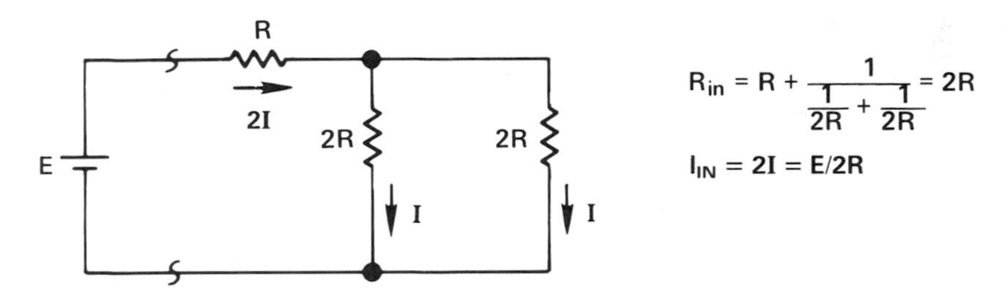

Figure 9.2. Primitive R-2R network. Parallel 2R resistors divide the input current equally; the series combination has a resistance 2R.

Since the resistance is the same, we could add another such network in place of the external 2R, without disturbing the current in the other resistors. Figure 9.3 shows such a circuit. The circuit also looks the same when viewed from the final 2R leg at the right toward the source; however, in order to keep I in the right-hand resistors constant, the input voltage, E, must be doubled.

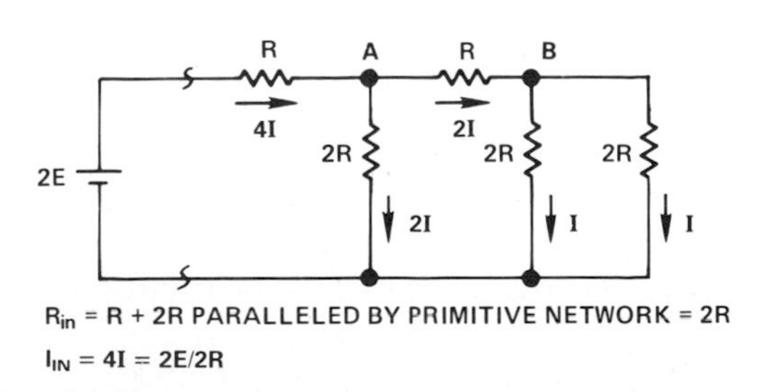

Figure 9.3. Two-stage R-2R network. An additional shunt 2R and series R cause input currents to divide evenly while preserving input resistance.

With the currents flowing out of node B both equal to I, the current through the right-hand R is $2I$. Since the currents through the 2R resistances at node A have divided equally, the total current through the left-hand R is equal to $4I$. Thus, the currents through the 2R legs at nodes A and B are related by a factor of 2. The extra I in the right-hand leg terminating the network is available for further division if additional stages are added. The network can be expanded in either direction. For example, if a shunt 2R is added at the left, a current of $4I$ wll flow in it; a series R will bring the network resistance back up to 2R, and doubling the input voltage to $4E$ will permit the same currents to flow throughout the rest of the network.

It should be noted that the voltages at the nodes ar also related by factors of 2, since each node is a division point between two equal resistances, equal to R. Thus, the voltage at node A, in Figure 9.3 is equal to $2E/2$, and the voltage at node B is equal to $2E/4$. These voltages can be used as the weighting for converters of types other than the ones to be described here.

The expansion process can be repeated as often as we wish by simply adding a parallel 2R resistor and series R resistor. After each addition, the input resistance of the network will remain unchanged. Each new 2R leg will carry twice the current of the leg to its right. When the network has been expanded to the desired number of 2R branches, it can be driven directy without the need for an additional series resistor, as shown in the generalized n-leg network in Figure 9.4, where the terminating resistor, at the extreme right, is considered to be an unused zero-order leg.

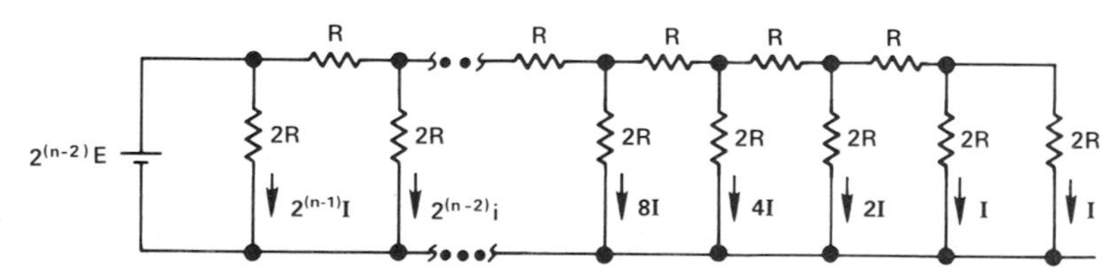

Figure 9.4. An n-stage R-2R ladder provides n binary-weighted currents.

The basic properties of the ideal R-2R network are that it has a constant impedance, regardless of the number of stages, and that each added stage operates at twice the current (and voltage) of the previous one. This ease of design and expandability, when coupled with its limited requirements on resistance range, further explain its popularity.

Getting the Currents Out

The network of Figure 9.4 does a fine job of producing binary currents; however, they're all locked up in a closed system. How do we bring them out for use in an application like that shown in Figure 9.1? In Figure 9.4, all the 2R legs are returned to the same voltage, using a common connection. For the network to make the currents available for switching and summation, the resistors must all be returned to the same voltage, *but not necessarily to the same point in the circuit.*

In the circuit of Figure 9.5, the 6-stage ladder is the same, but turned upside down (series legs at the bottom) and driven with negative voltage. The 2R legs have been returned to the emitters of a set of transistors having a common base connection. Neglecting (for now) small differences in base-emitter voltage among the transistors, we can assume that, since the V_{BE}'s are equal, the network is terminated with approximately the same voltage at the upper end of each leg. As a result, the network will behave as described earlier, producing binary-weighted currents which flow as the emitter currents of the transistors. The collector currents will be slightly less, being reduced by the finite common-base current gain (α). Nevertheless, if the transistors are all integrated on the same chip and have the same geometry, we can assume that their αs match, hence the emitter currents will be reduced in the same proportion at the collector for all transistors, and the collector currents will therefore retain their binary relationship.

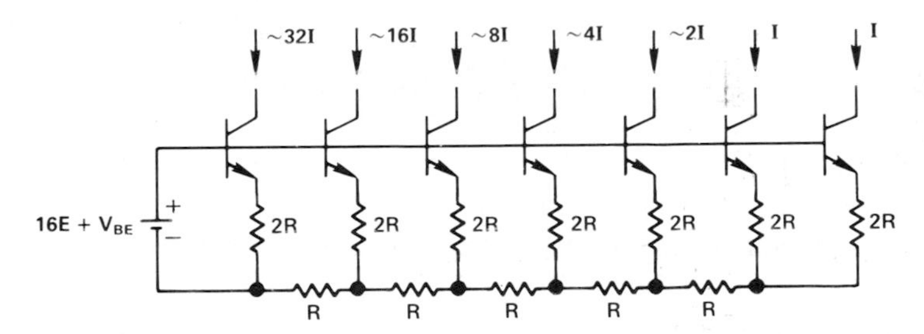

Figure 9.5. Common-base transistors terminate the ladder resistors with (approximately) equal voltages while making the binary weighted currents available as outputs.

Since the excitation voltage must be applied to the common base line instead of directly across the network, the desired excitation must be increased by the expected voltage drop between the common base line and the emitters, i.e., V_{BE}.

In practice, this circuit has several weaknesses. They will be discussed shortly. First, let us consider how the currents appearing at the transistor collectors are to be switched off and on and summed.

Switching

Having shown how the current sources of Figure 9.1 might be developed, we now consider how the switching can be accomplished. One of the most successful approaches is through the Craven cell*, consisting of a differential NPN current switch controlled by a differential-PNP level translator (Figure 9.6).

*U.S. Patent 3,961,326

Each of the current sources of Figure 9.5, except the rightmost terminating resistor, will have a switch similar to that shown in Figure 9.6. In the switching element (cell) shown, the lower NPN transistor is one of the current-source transistors; its collector current is one of the binary weighted currents. That current is directed to the common emitters of a pair of NPN transistors, operated so that one of them is normally conducting all the current from the weighting network (on) and the other one is off.

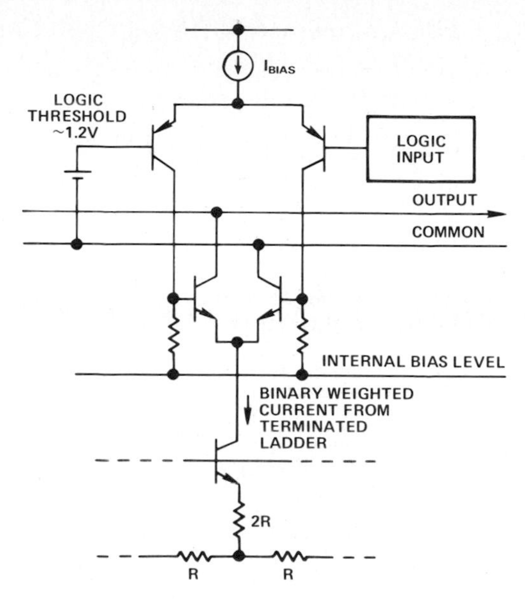

Figure 9.6. Craven cell switches bit-weight current under control of standard logic input.

When the transistor at the left is on, it conveys the weighted current to a line which is connected to all the switches and makes up the output signal, as in Figure 9.1. When the right-hand NPN is on, the ladder current is diverted to another line, which is common to all the switches, generally used as the "ground"—or signal return—line, but sometimes used as a complementary output (i.e., an analog output representing the complement of the digital input word).

The bases of the NPN switches are referred to an internal bias voltage level. Its value is somewhat critical; it must be more negative than the output-line compliance voltage, to avoid saturating the switches, but it must also be more positive than the base line of the current-source transistors, to prevent them from saturating. The NPN pair is switched by driving the base of one of them slightly (300mV to 600mV) positive with respect to the bias rail, while allowing the other base to be held at the bias potential by the associated resistor.

The drive voltage for the bases of the NPNs is generated by a positive bias current, which is controlled by the pair of PNP transistors. One of them is referred to the desired logic threshold (about 1.4V for TTL). The logic signal is applied to the other PNP. If it is substantially above the threshold, it is to be treated as a logic 1; if substantially less than the threshold, it is treated as a logic 0.

When the input is logic 1, the I_{bias} will flow through the left-hand PNP and develop a positive voltage at the base of the corresponding NPN, causing it to carry the current from the current-source transistor to the output bus.

When the input is logic 0, I_{bias} will flow through the right-hand PNP and cause it to turn on the right NPN, which will steal the weighted current from the output and direct it to the common "ground", or the complementary output.

In junction-isolated monolithic circuits, the PNP transistors will be lateral structures. Since they are included in the signal path, these reputedly slow devices might normally be expected to severely limit switching speed. In this application, however, they are greatly overdriven and are not required to settle accurately before switching of the NPNs can be completed. Craven cells of this general configuration can be made to switch in 50 to 60 nanoseconds.

Output Current Scaling

The last few figures have shown how binary-weighted currents can be obtained, how they can be made available from the network, and how they can be switched on and off the output bus. Until now, we've skirted the issue of how the excitation voltage can be set and maintained at just the proper level in a monolithic circuit. The problem is made challenging by the requirements to add the base-emitter (V_{BE}) voltage to the excitation voltage (Figure 9.5) and to compensate for the emitter current lost because $\alpha < 1$.

In addition it is worth noting that, when the Craven cell is used, the current passes through a second NPN transistor. Even if this transistor were to match the current-source transistor, current from the R-2R ladder is reduced to α^2 of its original value before reaching the output.

The circuit of Figure 9.7 addresses the problem of how the full-scale value of DAC output can be stabilized at a desired predetermined value. We continue to assume that the available output currents are in descending binary ratio, so that if the excitation voltage is set so as to adjust the MSB current to half the full-scale value, the other currents will automatically be at the proper level.

The key to accurate referencing is to use the ability of feedback circuits to compare a parameter (regulated variable) to a desired value and to adjust a related parameter (manipulated variable) electronically until the desired level is equalled.

In the case of the current-output DAC, it is inconvenient to make *direct* measurements of the individual output currents, due to the complications of switching and current summing on the output bus. However, since a series

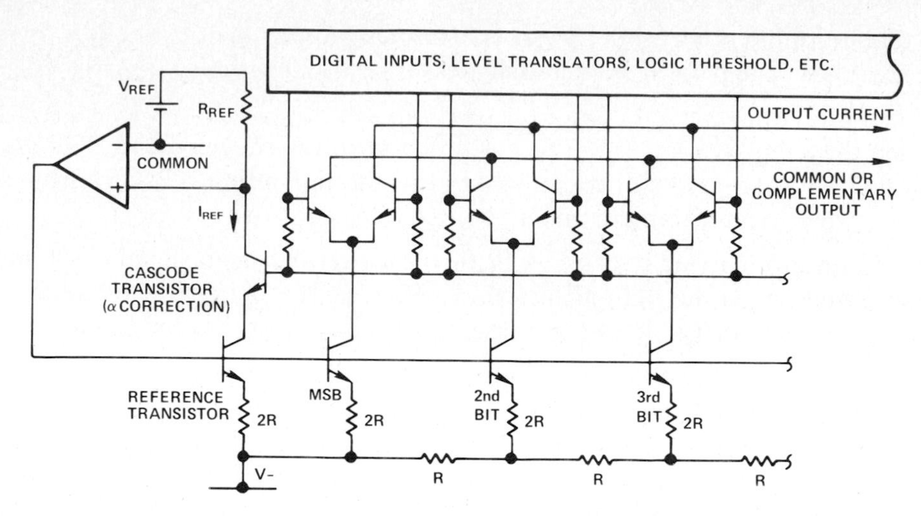

Figure 9.7. Switched-current DAC. Feedback loop precisely quantifies reference current, to which the output currents are ratiometrically matched.

of ratiometrically matched output currents is being produced anyway, very little incremental effort is required to provide an additional reference current, which will track the other currents and provide a representative feedback signal.* The circuit is arranged so that this current will accurately match the MSB current; and a feedback circuit will be used to adjust the excitation voltage until this reference current is equal to the desired value of the MSB.

In the circuit of Figure 9.7, the R-2R network has an additional 2R section at the left, connected to an additional common-base transistor to form a current source. This resistor-transistor combination is carefully matched to the resistor and transistor used to produce the MSB current. Since the excitation is applied to the two circuits in identical manner, the MSB current will be well-matched to the reference current, whatever the actual value may be.

The output of the reference transistor is passed through a second transistor, which is connected as a cascode and models the switching transistors; the two functions it performs are to insure that the collector of the reference transistor is at very nearly the same voltage as the collectors of the other current sources and to provide compensation for the current loss due to the α of the transistor that switches the MSB, so that the reference current, I_{REF}, will match that of the switched MSB.

The amplifier at the left is essentially an inverting op amp; its summing point, held by feedback at nearly 0 volts with respect to common, is at the (+) input because of the voltage inversion inherent at the collector of the reference transistor. The precision voltage, V_{REF}, and the associated resistor, R_{REF}, develop a current, $I_{REF} = V_{REF}/R_{REF}$, and the amplifier adjusts the base line to whatever voltage is necessary to duplicate I_{REF} through the collector of the

*U.S. Patents 3,803,590 and 3,978,473

cascode transistor, irrespective of α, V_{BE}, and their variations with temperature. As long as the MSB current through the switch closely tracks the current through the collector of the cascode transistor, the MSB current will also tend to be equal to V_{REF}/R_{REF}.

Variation in the initial resistance of the R and 2R resistors is of little consequence, so long as they remain in the proper ratio. This immunity extends to wafer-to-wafer variations due to semiconductor processing, so long as the entire circuit is fabricated uniformly.

Since the driven excitation voltage is the voltage between the base line and the bottom of the R-2R ladder, a viable variation of the circuit of Figure 9.7 is to fix the common base line of the transistors at a constant voltage and drive the bottom of the R-2R ladder from the amplifier output (reversing the inputs to the op amp). This arrangement requires that the amplifier supply the total R-2R ladder current, instead of just the transistor base currents, but it has certain advantages relating to the dynamic performance of the feedback loop, a "plus" with fixed reference and a vital necessity for a multiplying DAC (one that permits a variable V_{REF}).

Correcting V_{BE} Differences

By the use of the reference loop, we have corrected significant errors due to α and voltage drops in the excitation-voltage circuit. A remaining error source, which can be quite significant when good resolution or good multiplying performance is required, is inherent in the differences in base-emitter voltage between the current-source transistors of Figure 9.5.

The nature of the problem. For example, suppose that the current-source transistors are all well-matched to one another. Then their base-emitter voltages will differ, since they are operating at different current levels. That is, as we move from left to right in Figure 9.5, each transistor ideally operates at one-half the current of the one to its left. Since base voltage and collector current are logarithmically related, the transistors will have progressively smaller base-emitter voltages. The difference in the voltages means that the individual legs of the ladder do not terminate at the same voltage, and our analysis of the ladder breaks down. The output currents will not be in an exact binary ratio, and the DAC will have errors determined by the ratio of the offset voltage between the emitters to the voltage across the 2R legs.

It's quite possible to eliminate the offset without being required to examine it in detail. However, a detailed examination will lead to better understanding of the solutions, as well as—in fact—a better understanding of bipolar transistor circuitry—an appropriate tutorial goal for this chapter.

Consider two identical transistors, connected as shown in Figure 9.8, with base drive V_{BE1} and V_{BE2} and sufficient positive collector voltage ($+V$) to avoid saturation. We will want to keep tabs on the collector currents and the

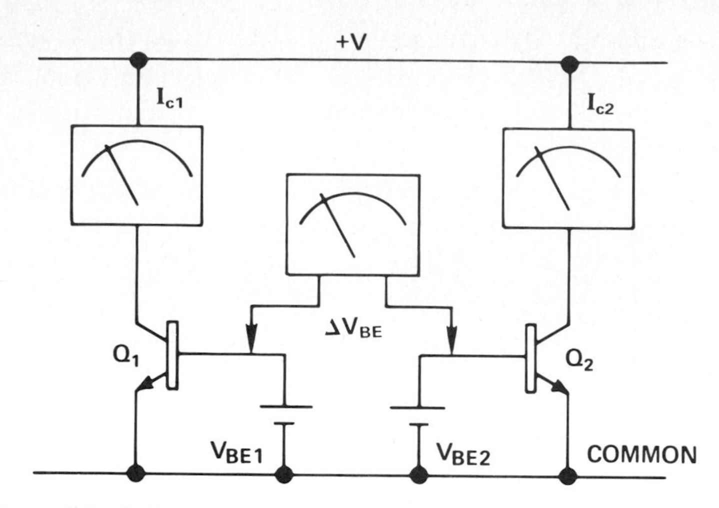

Figure 9.8. Collector currents and base voltages in transistor pairs.

difference in base voltage, symbolized by the meters. The transistors will be assumed to operate in the range of currents described by the relationship,

$$I_c = I_s(e^{qV_{BE}/kT} - 1) \tag{9.1}$$

where I_c is the collector current, V_{BE} is the base-emitter voltage, I_s is the saturation current for a transistor with a particular geometry and doping, T is absolute temperature, and q/k is equal to 11,605 kelvins/volt. Inasmuch as an integrated-circuit transistor, operating at—say—100μA, may have a V_{BE} of about 0.65V at room temperature, where q/kT = 39/V, the exponential factor in the equation will be of the order of 10^{11}, and the "– 1" term is negligible. This simplifies (9.1) to

$$I_c = I_s e^{qV_{BE}/kT} \tag{9.2}$$

Using this approximation, we can readily investigate the effect of operating matched transistors at different currents. If we establish the two collector currents at I_{c1} and I_{c2} by adjusting V_{BE1} and V_{BE2}, then the ratio of the two currents is

$$\frac{I_{c1}}{I_{c2}} = \frac{I_{s1}}{I_{s2}} \frac{e^{qV_{BE1}/kT}}{e^{qV_{BE2}/kT}} \tag{9.3}$$

Setting $I_{s1} = I_{s2}$ (since the transistors are matched), and taking the logarithms of both sides

$$\ln(I_{c1}/I_{c2}) = q(V_{BE1} - V_{BE2})/kT \tag{9.4}$$

Hence, the difference between the V_{BE}'s depends on the log of the current ratio,

$$\Delta V_{BE} = V_{BE1} - V_{BE2} = (kT/q)\ln(I_{c1}/I_{c2}) \tag{9.5}$$

To relate this expression to the problem of Figure 9.5, assume that the currents in the transistors are in a binary sequence. Then the ratio of collector currents in any two adjacent transistors will be 2, and the resulting difference in their base-emitter voltages can be calculated from (9.5) as

$$\Delta V_{BE} = (kT/q)\ln 2 \tag{9.6}$$

which, at room temperature, will be about 18mV. If the fraction obtained by dividing 18mV by the excitation voltage impressed across the R-2R ladder is comparable to the resolution of the DAC, a serious differential-linearity error will result. For example, suppose that the excitation voltage is 6.2V; then 18mV is about 0.003 of full scale, which would be a significant fraction of 1 LSB of 8 bits (for resolution of about 0.004).

In the design of integrated circuits, this error may be avoided or minimized in several ways. The oldest, and still one of the best—from the standpoint of performance—is to equalize the current density in the current-source transistors that terminate the R-2R ladder.

The multiple-emitter approach. In the equations associated with Figure 9.8 the transistors are assumed to be of equal area, so that the ratio, I_{s2}/I_{s1}, is equal to unity. If one of the transistors is larger than the other, this will not be so. For example, if Q2 has twice the emitter area of Q1, then it will have twice the saturation current ($I_{s2} = 2I_{s1}$), and equation 9.5 changes to

$$\Delta V_{BE} = (kT/q)\ln(2I_{c1}/I_{c2}) \tag{9.7}$$

If the currents are in a binary relationship, i.e, $I_{c2} = 2I_{c1}$, then ΔV_{BE} will be zero. This suggests that we can eliminate the difference between the emitter voltages of each pair of adjacent transistors by making the transistor with the greater current have a larger area.

It may be simpler to think in terms of passing current I through one transistor and sharing the $2I$ current in two parallel transistors. Assuming that the transistors all match, each of the parallel transistors will carry a current equal to I, and their base-emitter voltages will all be equal. Since the bases are common, the emitter of the transistor carrying I will be at the same potential as that of the pair sharing $2I$, and these legs of the ladder will be properly terminated. Continuing in this way, the next leg, carrying current $4I$, should be terminated in four parallel transistors, each operating at a current I, and so on.

In the transistors used for integrated circuits, extra emitters can be embedded in an enlarged base region. If these emitters are connected in parallel, the base, collector, and multiple emitters of the resulting device will behave like complete transistors in parallel. Figure 9.9 is a cut-away view of a typical integrated-circuit transistor made with four minimum-sized emitters. These emitters can be paralleled by the aluminum intraconnect to yield a transistor with a saturation current which is four times that of a single-emitter device. At a given V_{BE}, this device will yield four times the collector current of a single-emitter device.

Figure 9.10 illustrates the use of multiple emitters to obtain equal current density, and therefore equal base-emitter voltage, in the transistors terminat-

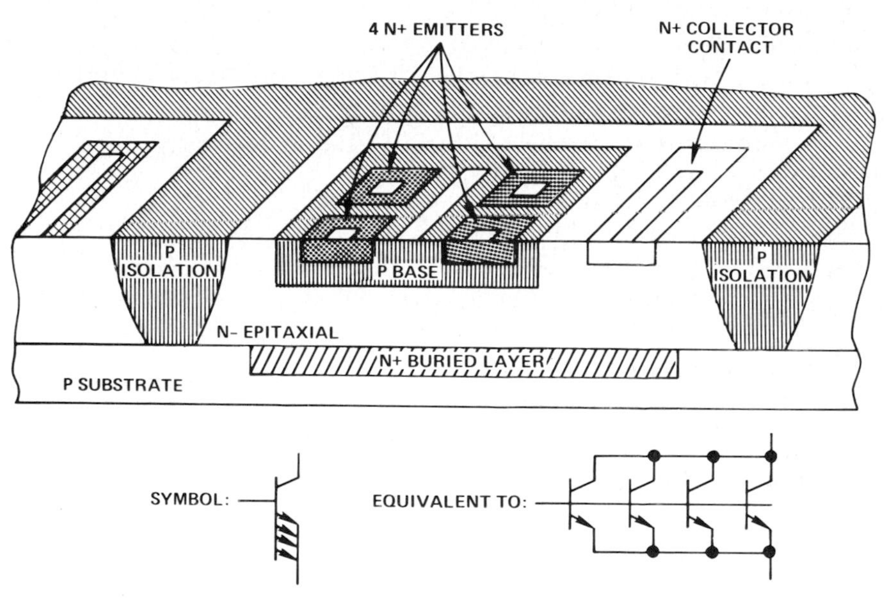

Figure 9.9. Junction-isolated NPN transistor with multiple emitter sites.

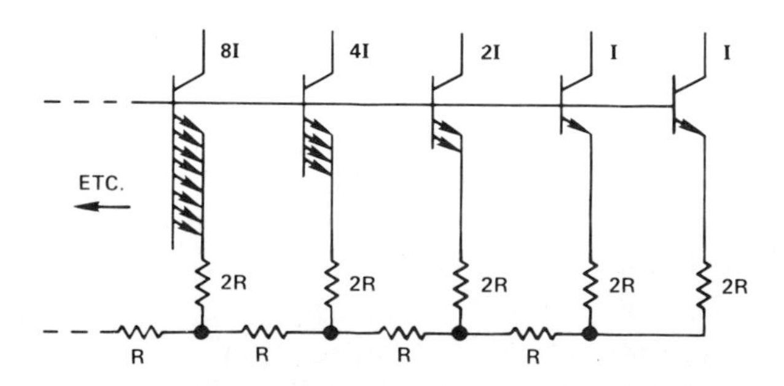

Figure 9.10. Equalizing current density equalizes emitter voltages.

ing a ladder. In the right-hand transistor, each emitter carries a unit of current. In the next transistor, two emitters share two units. In the following transistors, 4 emitters share 4 units of current, 8 emitters share 8 units, etc.

It's that "etc." that brings out the trouble with this solution. For example, in a 12-bit DAC, the weight of the MSB is 2,048 times that of the LSB. This means that the emitter area of the MSB transistor must be more than 2,000 times as large as that of the LSB transistor. Since the minimum emitter size is limited by photolithography, the large device will be very large indeed! This single transistor would be larger than many complete monolithic circuits. Moreover, although the other weighting network transistors would be *smaller*, they would not be *small*.

As a result of this problem, solutions involving current-density equalization by multiple emitters are not often used for DACs having more than 4 or 5 bits. A common practice is to divide a high-resolution DAC into quads (4-bit DACs)* and then combine their outputs with a second weighting network. For example, a 12-bit DAC function can be realized by making three DACs of 4 bits each and properly combining their output currents in a weighting network.

This solution is shown in simplified form in Figure 9.11. The DAC on the left supplies the four most-significant bits. The DAC in the middle supplies the next four bits. Its actual full-scale output is the same as that of the first four bits. To reduce the weight of these bits in the final output signal, the current is attenuated by a factor of 16. Similarly, the right-hand DAC controls the four least-significant bits. In order for them to have the proper weight in the overall current, the output of this four-bit DAC is attenuated 256 times.

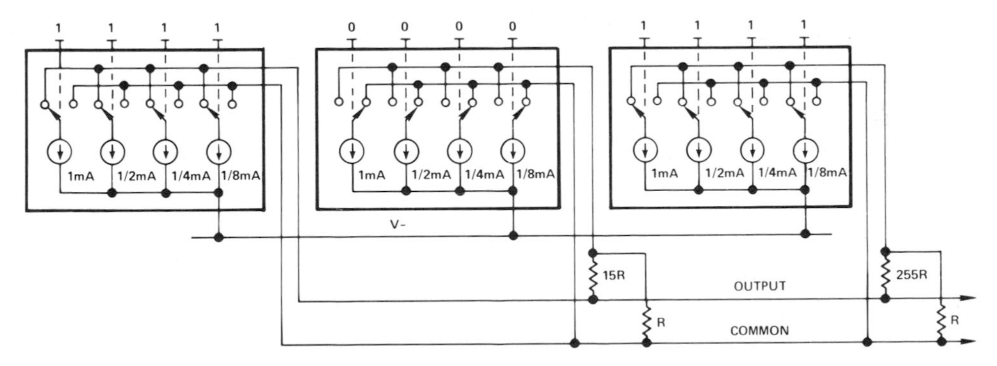

Figure 9.11. 12-bit DAC comprising 3 DACs of 4 bits each plus inter-quad dividers.

With the switches as shown, the current available at the output will be:

$$I_{OUT} = (1 + \tfrac{1}{2} + \tfrac{1}{4} + \tfrac{1}{8})\,\text{mA} + (\tfrac{1}{16})(0) + (\tfrac{1}{256})(1 + \tfrac{1}{2} + \tfrac{1}{4} + \tfrac{1}{8})\,\text{mA} = 1.8823\,\text{mA}.$$

*U.S. patent Re. 31850.

This is equivalent to

$$I_{OUT} = \frac{111100001111_2}{111111111111_2 + 1} \times 2mA$$

where 2 mA is the full-scale output of the DAC.

To avoid the 255:1 resistance ratio shown, the current in the right-hand DAC may be attenuated by a factor of 16, combined with the current from the middle DAC, attenuated again by 16, and combined with the output of the first DAC. The "interquad dividers," which mix the outputs of the three DACs in proper proportion, cause the DAC to have a relatively low output resistance. This may be advantageous in high-speed applications, but it requires the output to operate "shorted" (i.e., near ground potential) to obtain the full output current. This may be done by driving the virtual ground at the input of an inverting op amp. Such arrangements are common when the output from such a DAC is to be converted to voltage. The output common bus carries the surplus current from the lower-order DACs and may not be used as a complementary output.

Despite these limitations, the scheme shown in Figure 9.11 is capable of great accuracy and is probably the most widely used basis for high-resolution DACs made with bipolar monolithic and hybrid technologies.

There are several other methods used by designers of precision monolithic converters to equalize the emitter voltages of the transistors used to terminate the binary weighting network. Some of them involve other types of cascades of multiple-emitter transistors in arrangements which permit cycling of the 1, 2, 4, 8 area buildup, in ways comparable to Figure 9.11.

Offsetting the base voltages. A method which avoids these complications relies on the fact that the difference between base-emitter voltages of two matched transistors depends on the ratio of collector currents. This means that, in an array of matched transistors (of equal area) operating at currents in a binary sequence, the difference in base-emitter voltage is the same between any two adjacent transistors, as equation 9.6 demonstrates.

This suggests the possibility of driving the bases from a string of equal resistors carrying an appropriate value of current such that there are equal voltage increments between the bases to compensate for the V_{BE} voltage reduction with decreasing current. However, there is a problem: the ΔV_{BE} is proportional to temperature; therefore the correction voltage must also be proportional to temperature.

As a means of developing the correction, consider the circuit of Figure 9.12. It consists of a current mirror comprising a pair of PNP transistors, a pair of NPN transistors with a 2:1 ratio of emitter areas, a control resistance, R_M, and a measuring resistance, $R_M/2$.

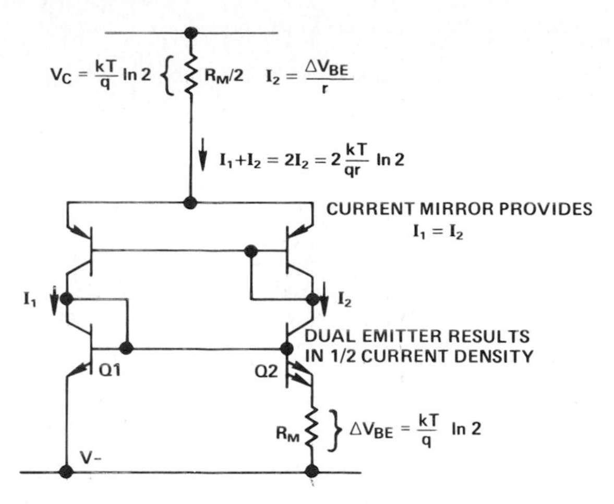

Figure 9.12. Repeater produces a voltage equal to ΔV_{BE} at a selected current-density ratio.

The current mirror forces the current in the two sides of the circuit to be equal. Since the two PNP transistors are joined at their bases and emitters, they should have equal collector currents. Neglecting base currents, the collector voltage of Q2 will drive both of the PNP bases to the point that the right-hand transistor absorbs Q2's collector current in its collector. Since the same base voltage is applied to the left-hand PNP, it too will operate at the same current.

Therefore, to a reasonable degree of approximation, $I_1 = I_2$, for any value of I_2. The NPN transistors are similarly connected: their common base connection is driven by the collector of Q1 until Q1 absorbs I_1 at its collector. The resulting base voltage of Q1 is applied to the base of Q2, causing it to conduct the current, I_2. This arrangement is regenerative at low currents, but the resistance, R_M, in series with Q2 permits the loop to stabilize as the currents increase and the loop gain drops to unity.

When the circuit reaches its stable condition, we can observe several things. First, to the extent that base current can be neglected, the PNP current mirror should force all four transistors to operate at the same collector current. The emitter current of Q2, which is twice the area of Q1, must divide between two emitters. Since the emitter current of Q2 is approximately equal to that of Q1, the current density is only half that of Q1 in each of the transistors of Q2. Since the transistors of Q1 and Q2 are all matched, the base-emitter voltage of Q2 must be smaller than that of Q1 by an amount

$$\Delta V_{BE} = (kT/q) \ln 2 \tag{9.8}$$

Since the bases are at equal voltage, the voltage difference must appear across the resistor, R_M. From this, we can calculate the sum of the currents, I_1 and I_2, in the equilibrium condition. I_2 must be equal to $\Delta V_{BE}/R_M$, and, since

$I_1 = I_2$, the total current flowing through $R_M/2$ must be equal to $2I_2$, and the correction voltage, V_c, developed across it is equal to $I_2 R_M$, or

$$V_c = (kT/q) \ln 2 \tag{9.9}$$

This is the same as the base-emitter voltage difference of matched transistors operating at a 2:1 current ratio, including the temperature proportionality. It has been achieved by generating a current, scaled by R_M, *proportional to absolute temperature* (PTAT). Circuits of this type are often called "PTAT generators." An important feature of this circuit is the available voltage compliance; the output of the four-transistor cell is a current which can be made to drive any load which provides more than about 1.5 volts of "headroom" and less than breakdown voltage. This means that the circuit could drive 2, 3, 4, or more resistors in series, and if each had the value $R_M/2$, each would independently develop the voltage, V_c, across itself.

Figure 9.13 shows how a string of series-connected resistors, each producing the correction voltage, V_c, can be used to equalize emitter voltages.* In this figure, a few sections of an R-2R ladder are shown with the 2R legs terminated in transistor emitters. The sequence of resistors and transistors is assumed to

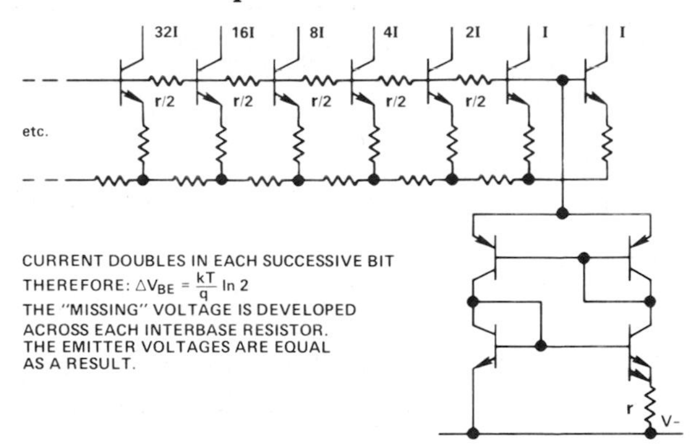

Figure 9.13. Equalizing emitter voltages with PTAT current and ratiometric resistors.

continue toward the left to the desired number of bits, providing very nearly complete correction for the problems mentioned above; and it is assumed to be driven by some suitable excitation circuit, such as the one in Figure 9.7.

Notice that the bases of the transistors are separated by resistances, r/2. The PTAT generator drives the series string comprising these resistors, and produces a voltage, $V_c = (kT/q)\ln 2$, across each of them, as we have seen. Since the base-emitter voltages of adjacent transistors should differ by this same amount, if they produce a binary sequence of currents, the interbase voltages should cause all the emitter voltages to be equal, so that the currents will indeed be in a precisely binary sequence.

*U.S. patent 3,940,760.

A feature of this scheme that makes it well suited to monolithic designs is that it relies upon ratiometric properties of the circuit elements, rather than upon absolute value. A voltage is produced by essentially the same mechanism which causes the problem. This voltage is then translated to the locations where it can correct the problem by a set of resistors. The absolute resistance values are not at all critical, so long as they have the proper ratio. This requirement is well met by monolithic techniques which may yield variations of as much as ±30% in initial resistance values, but can reliably produce resistors which match to within a fraction of 1%. Tight ratiometric tolerances can be met by using two $R_M/2$ resistors in series to form R_M in the PTAT generator. This and similar techniques are often used to avoid ratiometric errors due to contacting and edge effects in resistors with different geometries (e.g., in the R-2R network itself).

The PTAT generator shown in Figures 9.12 and 9.13 is incomplete, in that it lacks a starter (it also has a stable state in which I_1 and I_2 are zero). This problem is easily solved by providing a small leakage from other bias circuitry in the DAC. This bias can easily be large enough to cause the circuit to regenerate to the desired *on* state and small enough so that it introduces negligible error.

For high-resolution DACs, the elementary circuit of Figure 9.12 has some shortcomings; for example, the effects of base currents were ignored rather than compensated for. However, the principle is valid, and other, more-sophisticated PTAT generators are used in high-resolution converters. This generator was chosen because of the simplicity of illustrating the homomorphism between it and the problem it solves.

9.2 APPLYING THE TECHNIQUES TO THE FINAL PRODUCT

The techniques and circuits described here have been used in a number of monolithic converters. One of these converters, the 10-bit AD561, is illustrated in the schematic diagram of Figure 9.14. The R-2R ladder and its terminating resistors are prominent near the center of the circuit. The resistance ladder is terminated after bit 8, and the last two bits and the final LSB termination are produced by using transistors with matched 2:1 area ratios.

In the circuit of Figure 9.14, the base voltage is fixed, and the network excitation is provided by amplifier A2, which drives the resistors. The circuit combines the techniques of Figures 9.9 and 9.13, with multiple-emitter transistors in the three most-significant bits and interbase resistors correcting the remaining error voltages. The interbase resistors are driven by a properly scaled PTAT current with a nominal (room temperature) value of 120μA.

The output of Q1 develops a voltage across a 2,500-ohm resistor between the reference input and the summing point. Amplifier A2 adjusts the ladder voltage to set the current so that the summing point is at zero and the voltage across the 2,500-ohm resistor is equal to the 2.5-V reference. The reference

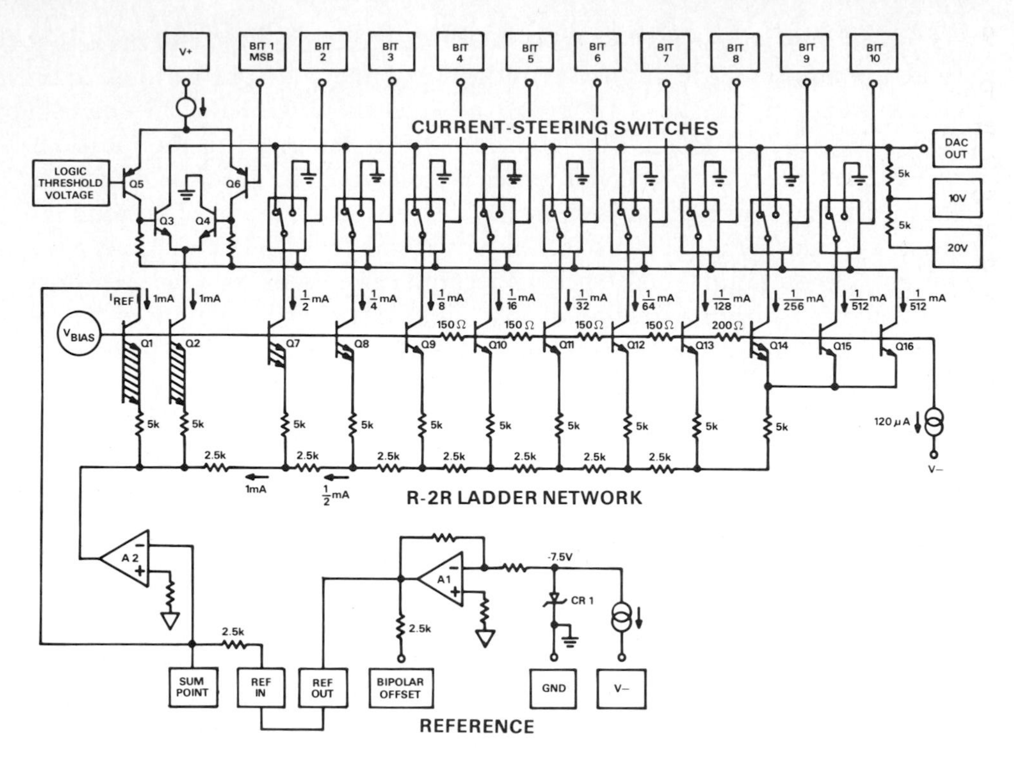

Figure 9.14. Circuit diagram of AD561 10-bit DAC, showing reference, control amplifier, switching cell, R-2R ladder, and bit arrangement.

voltage is generated with a temperature-compensated sub-surface, or "buried," Zener diode. The amplifier A1 conditions and scales this reference voltage to drive the reference input.

Ten Craven-cell switches are used to control the individual bit output currents. The current-output line also connects to two 5,000-ohm application resistors, ratio-matched to the reference-input resistor, for proper scaling when the DAC is used with an op amp for voltage output.

Since the full-scale output current is determined by the reference voltage and the input resistance, the voltage developed across the applications resistors by the DAC will be in an accurate ratio to the reference voltage. This arrangement ensures that the DAC output depends upon resistance ratios, which can be precisely controlled, rather than absolute resistor values, which are more difficult to control in monolithic circuits.

The full-scale output is, of course, directly related to the absolute value of the reference voltage. Although this voltage is more reproducible than the absolute resistance values, it is still subject to an undesirably wide range of variation in the course of many production runs. To correct this variation, the circuit shown in Figure 9.14 is laser-trimmed at the wafer stage.

9.3 LASER WAFER-TRIMMING

Power is applied to the circuit and it is operated via microprobes, and the reference voltage is measured. A fine, high-intensity spot, generated by a laser and steered under electronic control, is used to trim the feedback resistors associated with the reference-voltage buffer-amplifier, A1. These resistors—and many others in the circuit—are SiCr thin-film, deposited on the monolithic chip. The circuit is subjected to *active trimming*, i.e., it is trimmed while functioning, so that the reference-voltage output can be adjusted to its nominal value.

While the wafer is at the laser-trim processing station, other parameters can be adjusted. The temperature compensation for the Zener reference diode is adjusted to minimize the temperature coefficient of the reference-voltage output. In addition, the resistors of the R-2R ladder are adjusted to maximize the accuracy of the DAC. These resistors can be deposited with relatively high ratio accuracy; however the final yield of parts with overall accuracy to 10 bits and better can be substantially improved by laser-trimming.

The use of silicon-chromium thin-film resistors not only permits laser-wafer-trimming; it also ensures improved temperature stability and tracking compared to that obtainable with the commonly used diffused resistor. The temperature coefficient of diffused resistors is of the order of 1600 ppm/°C, which aggravates the ambient temperature sensitivity and—more importantly—results in irreducible errors due to minor temperature gradients on the monolithic chip. On the other hand, the tempco of Si-Cr thin-film resistors is typically less than 50 ppm/°C, with 1 ppm/°C tracking differences.

An interesting problem is encountered when trimming the R-2R ladder. Any of the individual bit currents can be adjusted by trimming its associated resistors. Subsequently, however, trimming *any other resistor in the network* will change the current in the previously trimmed bit. One might wonder if perhaps some interactive procedure could be derived whereby the network is repeatedly trimmed until all bit weights are within satisfactory limits. Fortunately, no such tedious method is required. Instead, there is a straightforward and simple—albeit proprietary—method which requires only one pass through the network and the trimming of no more than half the resistors.

This Chapter has touched on a few of the techniques used to implement bipolar current-summation DACs. Although the current-summation principle is probably the most widely used basis for monolithic DACs (and successive-approximation ADCs, too), it is by no means the only one. It was not our intention to produce here an exhaustive treatise on the various approaches, but rather to illustrate some of the considerations and techniques that have evolved, through the use of a representative example. Some other converter principles employed by Analog Devices are touched on elsewhere in this book. There is a burgeoning literature in technical journals, the trade press, and manufacturer publications, describing other aspects of converter design—at Analog Devices and elsewere. A number of references, which will provide fanout, can be found in the Bibliography.

Chapter Ten

Testing Converters

The purpose of this chapter is to illustrate common converter errors and deviations from ideal performance, discuss test principles, and outline schemes—for evaluating converter performance—that can be adapted to both manual and automatic testing.

The methods and test-fixture configurations needed to test DACs and ADCs are influenced by the prospective converter applications, nature and speed of tests to be performed, and skill of persons performing and interpreting the tests. The relative importance of the various converter performance specifications depends on the application; the converter user—unless performing a general evaluation—is more interested in testing parameters which significantly influence system performance than those which have little effect on performance.

These factors influence the choice of converter test-circuit configuration and degree of automation: the purpose of the test, e.g., engineering performance evaluation, incoming inspection, or functional checks only; the desired versatility of the test equipment; required measurement speed; data-reduction and display capability needed; and skill level of people intended to perform the tests.

Simple text fixtures designed to test relatively few converter parameters can be implemented easily and inexpensively. These generally must be operated by relatively skilled persons, and test data obtained usually must be number-crunched to extract meaningful performance information.

Although they are expensive when compared to a lashup of available laboratory equipment, general-purpose automatic testers, such as the Analog Devices

LTS-2000 series, perform tests quickly, can be operated by semi-skilled personnel, and are usually quite versatile—easily programmed, self-calibrating, and capable of providing printouts of test results and test statistics over many devices.

Converter performance parameters that are generally of importance are: calibration accuracy (both absolute and relative to full-scale), linearity (both integral and differential), offset, noise, conversion time, and, in the case of DACs, output-switching-transient *impulse* (amplitude-time product). Also of concern are stability of these parameters with variations in time and temperature.

In testing high-resolution converters, there are a potentially large number of data points to be examined to extract meaningful converter performance information. A 12-bit DAC or ADC, for example, has 2^{12}, or 4,096 possible input/output combinations. Fortunately, by knowing the types of converter errors, or deviations from ideal performance, that are commonly encountered, one can devise tests which permit useful performance data to be gained by investigating significantly fewer than the 2^n possible input/output combinations associated with an n-bit converter. Some of these short-cuts will also be discussed here.

10.1. D/A CONVERTER TRANSFER FUNCTION

A digital-to-analog converter converts binary numbers, represented by patterns of 1's and 0's, to discrete analog voltages or currents.

The input/output relationship of a DAC, or its *transfer function*, consists of a set of discrete points, corresponding to the number of digital codes, for which each output voltage is a fraction of a reference quantity. The fraction is determined by how the input binary number is coded. Depending on the way the DAC is configured, the transfer function can be unipolar (outputs having only positive or negative values, but not both) or bipolar (outputs can be either positive or negative). In some cases, the reference quantity is itself an input signal; converters used this way are called multiplying DACs (MDACs).

The transfer function of an ideal unipolar 3-bit DAC is plotted as a set of points in Figure 10.1a.

There are many input coding schemes: Figure 10.1a represents a DAC with the most common and best-known code, *natural binary*. In this case, there is a simple linear correspondence between the input codes and the output voltage levels, which—for an n-bit DAC—can be represented algebraically by the following.

$$V_o = V_{FS} \sum_{i=1}^{n} (b_i/2^i) \qquad (10.1)$$

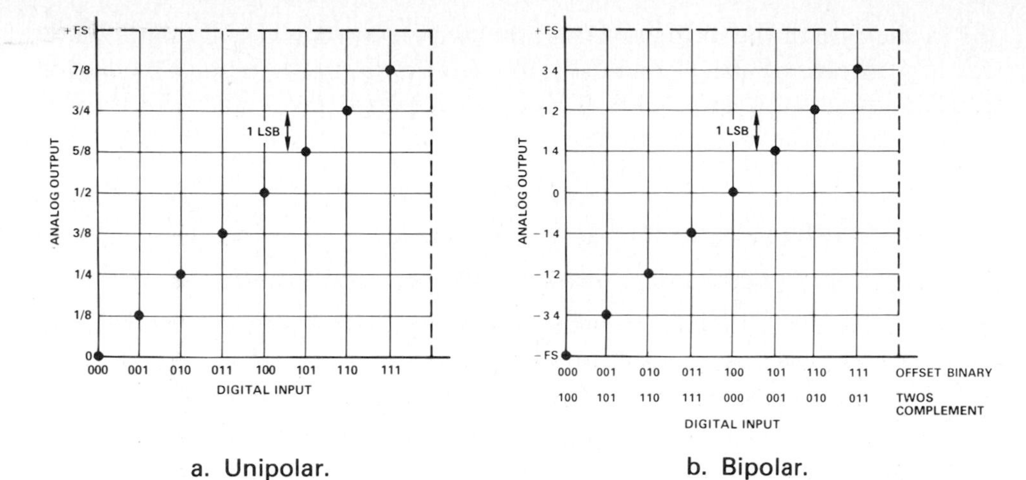

a. Unipolar. b. Bipolar.

Figure 10.1. Conversion relationships for an ideal 3-bit DAC.

The coefficients, $b_1 \ldots b_n$, represent the logic levels of the input bits, which can be either 1 or 0. V_{FS} is the reference quantity, which is usually a simple scalar multiple of the actual input reference voltage (but it does not have to be). *n* is the number of input bits: three, in this example.

Coefficient b_1 represents the most-significant bit (MSB), which has a weight of ½ V_{FS}, and b_n is the least-significant bit (LSB), which has a weight of 2^{-n} V_{FS} (in this case, ⅛ V_{FS}). Including zero, there are 2^n, or in this case 8, discrete voltage levels, corresponding to the 8 binary codes from 000 to 111.

Looking at equation (10.1), we see that the following conditions apply: (a) with all the input bits set to "0", the output voltage (V_o) is zero, and (b) with all the input bits set to "1", the all-1's output (not to be confused with "full-scale" output) voltage is:

$$V_{11} = V_{FS}\left(1 - \frac{1}{2^n}\right) \tag{10.2}$$

which is one LSB ($V_{FS}/2^n$) less than the full-scale range, or output reference quantity.

These input conditions define the end points of the transfer function; because the relationship is linear, all other points fall on a straight line drawn between them.

Figure 10.1b describes the transfer function of a DAC in the bipolar mode, where the alternate input codings shown are *offset binary* and—with the MSB complemented—the popular *twos complement*.

The bipolar transfer function can be described by the following equation:

$$V_o = -V_{FS} + 2\,V_{FS}\sum_{i=1}^{n}(b_i/2^i) \tag{10.3}$$

As in the case of the unipolar DAC, the two measurement end points occur for logic inputs of all "1"s and all "0"s. However, in the bipolar mode, for plus and minus the same value of V_{FS} (i.e., a *span* of 2 V_{FS}), an LSB is twice as big as it is in the unipolar mode.

10.2 DAC SPECIFICATIONS

The most important specifications of a DAC are *resolution* and *accuracy*.

Resolution refers to the number of unique output voltage levels that the DAC is capable of producing. For example, a DAC with a resolution of 12 bits will be capable of producing 2^{12}—or 4,096—different voltages at its output.

Inherent in the specification of resolution, especially for control applications, is the requirement for monotonicity. The output of a *monotonic* converter always changes in the same direction for an increasing digital code. The quantitative measure of monotonicity is the specification of *differential linearity* (step size).

The static absolute *accuracy* of a DAC can be described in terms of three fundamental kinds of errors: *offset errors*, *gain errors*, and *integral (non)linearity*.

Linearity errors are the most important of the three kinds, because in many applications the user can adjust out the offset and gain errors, or compensate for them without difficulty by building end-point auto-calibration into the system design, whereas linearity errors cannot be conveniently or inexpensively nulled out. But before we can understand the nature of linearity errors and how to test for them, the end-point errors must first be established.

10.2.1 END-POINT ERRORS

The transfer functions in Figure 10.1 are ideal—hence free from errors. The most commonly specified end-point errors associated with real-world, non-ideal DACs are offset error, gain error, and bipolar zero error:

Offset Error

Figure 10.2 illustrates the result of offset error only. The actual transfer function is offset from the ideal by two LSBs. Any such error—either positive or negative—that affects all codes by the same amount is an *offset* error.

In most DAC testing, the offset error is measured by applying the zero-scale code (e.g. all "0"s) and measuring the output deviation from 0 volts.

Although it is usually measured at zero, the offset error is not defined as being the error just at zero. It is just that usually no other errors are present at unipolar zero, so determining the offset error is easy. There are some DACs, though (e.g. AD558), where offset errors may be present, but not observable at the zero scale, because of other circuit limitations (such as zero coinciding with single-supply ground) so that a non-zero output at zero code cannot be read as the offset error. Factors like this make the testing of such devices a little more complicated; it will be discussed in more detail later.

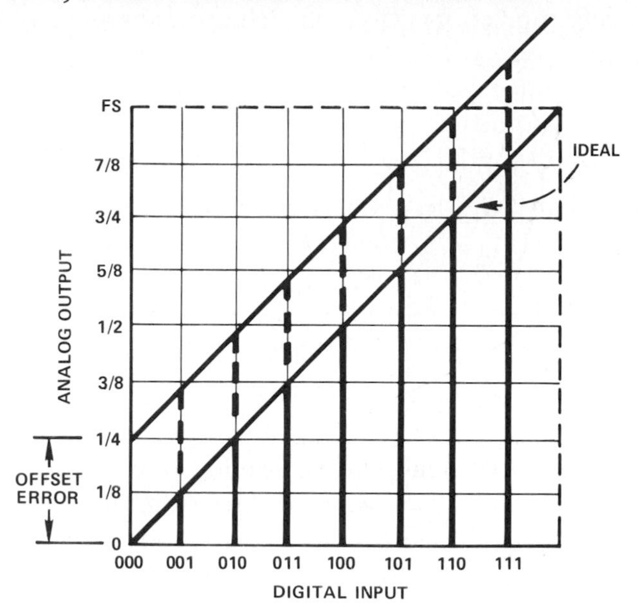

Figure 10.2. An offset error of +2 least-significant bits.

Gain Error

Figure 10.3 shows the effect of a gain error only. The ideal transfer function has a slope defined by drawing a straight line through the two end points. The slope represents the gain of the transfer function. In real DACs, this slope can differ from the ideal, resulting in a *gain* error—which is usually expressed as a percent because it affects each code by the same percentage.

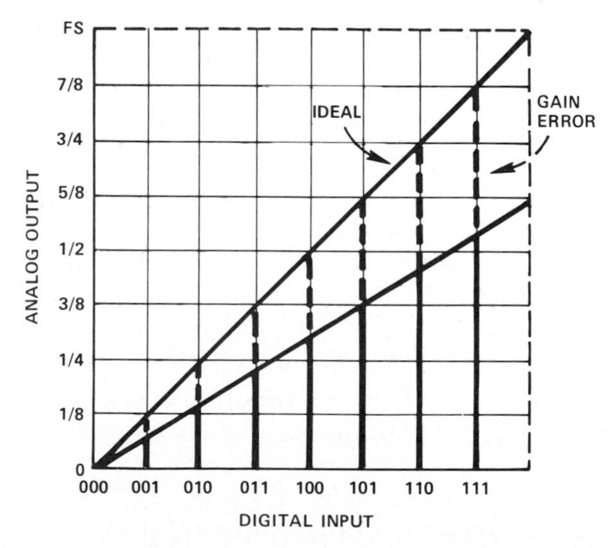

Figure 10.3. A gain error of −37.5%, or 3 least-significant bits at full scale.

Gain error is usually measured by first determining the offset error, then applying all "1"s to the DAC and measuring the error in the all-1's voltage. Since, by our definition of linearity, that voltage contains only offset and gain errors, the all-"1"s code, which is just 1 LSB less than the *full-scale* (reference) range, is the most convenient code to use to perform measurements to determine gain error.

Gain error is given by:

$$\frac{\text{Error}\,(\%)}{100\%} = \frac{V_{11} - V_{os}}{V_{FSR}\,(1 - 2^{-n})} - 1 \qquad (10.4)$$

Where V_{FSR} is the nominal full-scale range (the same as V_{FS}, for a unipolar DAC), V_{os} is the measured offset voltage, and V_{11} is the measured all-"1"s voltage.

Bipolar Errors

In the bipolar mode, the same two end-point errors should be measured. First, bipolar offset error (V_{os}), which is usually measured at negative full scale ($V_{os} = -V_{FSActual} - (-V_{FSNominal})$), and bipolar gain error, which is determined by measuring the positive full-scale error and subtracting the bipolar offset error (In Equation (10.4), $V_{FSR} = 2\,V_{FS}$, and the term "V_{os}" is the measured offset, $-V_{FSActual}$, which includes any bipolar offset error).

It is also common to specify and measure the *bipolar zero* error because of its importance in many applications. Refer to Figure 10.1b, which shows the ideal bipolar transfer function. The end points lie in the positive and negative regions, but what was mid-scale (½) in the unipolar mode is now zero scale in the bipolar one. This means that, with the MSB set to a "1", and all other bits set to "0", the output should ideally be zero. Any deviation from zero at this code is the *bipolar zero* error. To measure bipolar zero error, apply the code 10 . . . 00 to the DAC (in bipolar mode) and measure the output error from zero.

Bipolar zero error in DACs using offset-type coding is a derived, rather than a fundamental quantity, because it is actually the sum of the bipolar offset error, the bipolar gain error and the MSB linearity error. For this reason, it is important to specify whether this measurement is made before or after offset and gain have been trimmed or taken into account. Because of this error sensitivity, DACs that crucially require small errors at zero are usually unipolar types, with sign-magnitude coding (translated from twos complement, if necessary) and polarity-switched output amplifiers.

10.2.2 LINEARITY

Definitions

In a DAC, we are concerned with two measures of the linearity of its transfer function: integral linearity (or relative accuracy) and differential linearity.

Equation (10.1) describes a perfectly linear DAC by both of these measures, since the output is an exact binary-weighted fraction of the reference.

The transfer functions in Figure 10.1 also exhibit the corresponding straight-line relationship between input codes and output voltages. Note that the line has end points at the zero and full-scale voltages. This is consistent with our previous definitions of offset and gain errors and results in an "end-point" linearity specification (i.e., a specification of linearity with offset and gain errors corrected or taken into account). Occasionally, a "best-fit" linearity specification is used (i.e., by adjusting gain and offset arbitrarily to minimize linearity error), but that makes it difficult for a user to determine the actual error budget; fortunately, usage of this specification is becoming rare.

Relative-accuracy or *integral-linearity error*, or *integral nonlinearity* (INL), is the maximum deviation, at any point in the transfer function, of the output voltage level from its ideal value—which is on the straight line drawn through zero and full scale.

Differential-linearity error (DLE) is the maximum deviation of an actual analog output step, between adjacent input codes, from the ideal step value of +1 LSB (i.e, $+V_{FSR}/2^n$), calibrated based on the gain of the particular DAC. If the differential linearity error is more negative than −1 LSB (for DLE = −1 LSB, the step is equal to zero), the DAC's transfer function is non-monotonic.

10.2.3 SUPERPOSITION

Before proceeding with illustrations of DAC transfer functions showing linearity errors, it would be useful to consider the property of *superposition* and to be able to recognize its signature. Mathematically, superposition, a property of linear systems, implies that, if the influences of a number of phenomena at a particular point are measured individually, with all other influences at zero as each is asserted, the resulting total, with any number of these influences operating, will always be equal to the arithmetic sum of the individual measurements.

For example, let us assume that the DAC is ideal, except that each bit has a small linearity error associated with it. Equation 10.1 can be rewritten to include these errors, describing the performance of DACs that perform the decoding via a weighted resistor network with n binary-weighted taps:

$$V_o = V_{ref} \sum_{i=1}^{n} (b_i/2^i) + \sum_{i=1}^{n} b_i E_i \qquad (10.5)$$

where $E_1 \ldots E_n$ are the linearity errors associated with bits $b_1 \ldots b_n$.

Equation 10.4 does not apply to equal-resistor networks with 2^n fully decoded taps (segment architecture). In that case, we should write:

$$V_o = V_{REF} \sum_{i=1}^{2^n-1} (s_i/2^n) + \sum_{i=1}^{2^n-1} (s_i E_i), \tag{10.6}$$

where s_i is the decoded logic state, 1 or 0, of each segment and E_i is the linearity error associated with each segment.

Ideally, each bit- or segment error, (E_i) is independent; therefore, the linearity error at any code is simply the algebraic sum of the errors of each bit, or segment, in that code (i.e. superposition holds). In addition, by using end-point linearity, we have defined the linearity error in the vicinity of full scale to be zero. Thus, the sum of all the errors must be zero, since all bits, or segments, are summed to give the all-"1"s value.

The errors can be either positive or negative; therefore, if their sum is zero, the sum of the positive errors (positive summation) must be equal to the sum of the negative errors (negative summation). These two summations constitute the worst-case integral-linearity errors of the DAC.

Transfer functions can be drawn for DACs that have linearity errors for which superposition holds. Note that both segment-type and n-tap DACs have transfer functions that are symmetrical about the ideal straight line (Figure 10.4).

These two figures are significant because they illustrate two different kinds of behavior in the relationship between integral linearity and differential linearity.

In the transfer function of Figure 10.4b, the two codes with the worst-case positive and negative integral linearity errors are adjacent to one another, which results in the worst-case differential linearity error of twice the integral linearity error. This illustrates the consistency of specifying n-tap DACs as having ±½ LSB INL and ±1 LSB DNL. This two-to-one relationship is common for n-bit DAC's built with n binary-weighted circuit elements. Worst-case errors are often at half-scale, due to the sensitivity to element mismatches in that region, and the involvement of all bits in that transition.

Figure 10.4a, on the other hand, shows a much smoother transfer function, illustrating that the DNL error can be much less than twice the INL error. This relationship is typical of DACs built with "segment" architectures, which are much more tolerant of element mismatches for DNL performance.

For DAC's in which superposition holds, relative accuracy therefore bounds the worst case differential nonlinearity; often, as in the case of segmented architectures, the differential linearity is far better than the relative accuracy predicts for n-tap DACs.

Summarizing, if an n-bit DAC is linear to ½ LSB of n bits, we are certain that its differential-linearity error is no more than 1 LSB and that the response is monotonic. On the other hand, if the DNL error is less than 1 LSB, we cannot be assured that the DAC is linear to ½ LSB of n bits.

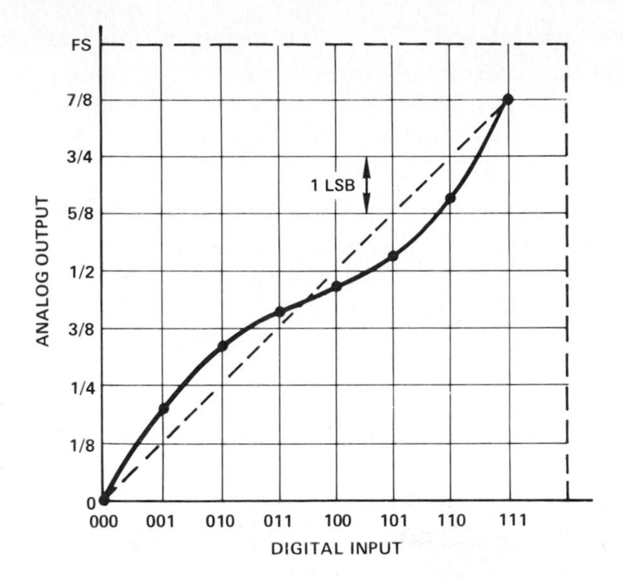

a. Linearity error of a 2^n decoded DAC, showing large integral linearity errors but low differential nonlinearity.

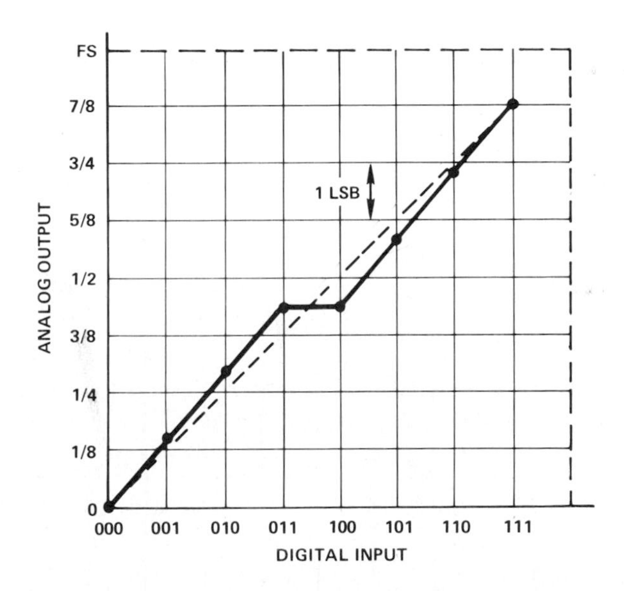

b. Linearity error of $n-$tap encoded-resistor-network DAC. Symmetrical integral linearity error with DNL $= -2 \times$ INL at major carry, showing DAC on the verge of non-monotonic behavior even though integral linearity is within ½ LSB.

Figure 10.4. Linearity errors without superposition error.

Having discussed the case of DACs for which superposition holds and for which there is thus no interaction between the bits—or segments—we must now say that completely independent or non-interacting errors is an ideal which cannot be entirely achieved, unless expensive steps are taken. In most well-designed DACs, reasonable compromises can be made to make the interactions negligible. Occasionally, DAC's appear which have significant interactions, producing non-symmetrical transfer functions like those illustrated in Figures 10.5a and 10.5b.

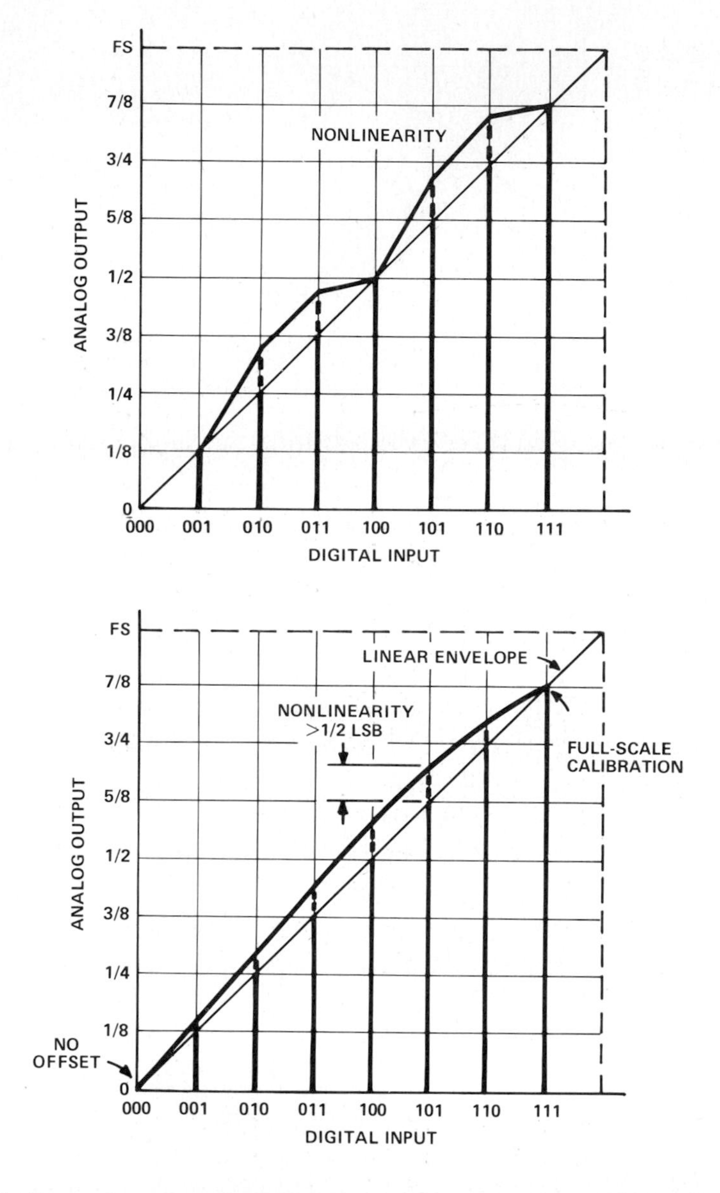

Figure 10.5. Examples of non-symmetrical integral linearity errors due to bit or segment interactions.

These non-symmetrical transfer functions can make testing for linearity errors more complicated, because there are no neat rules, like the one exemplified in equations 10.5 and 10.6, that make it possible to predict errors from a few sample test points.

10.3 DAC TESTING

We are now in a position to begin forming a test strategy for DACs. For characterization or small-quantity testing, there is no substitute for testing all codes and plotting the entire transfer function. Bench instrumentation is available in the form of meters, dc standards, and voltage dividers approaching 1 ppm, or 20 bits of linearity. However, as monolithic technology has evolved to the point that users routinely purchase large quantities of high-resolution and/or -accuracy (12 to 16 bit) DAC's, or evaluate DACs from a large number of possible sources, testing strategies have developed to guarantee performance with the least number of tests, using equipment that is as inexpensive as possible.

10.3.1 STATIC ERRORS

The minimum number of tests required must at least equal the number of unknowns being sought; so an efficient test strategy requires a prior knowledge of the DAC architecture and the degree to which superposition holds.

For example, taking the best possible case, consider an n-bit DAC built with n binary-weighted circuit structures and designed with sufficient skill such that bit interactions are negligible. We need to determine: offset error, gain error, and the integral nonlinearity (INL) error of each bit, i.e., a total of $n + 2$ unknowns. Therefore, we will need to perform at least $n + 2$ tests.

At the other extreme, an n-bit DAC built entirely with a "segment" architecture will have 2^n circuit elements with 2^n unknown errors; besides gain and offset error, it will require a maximum of 2^n tests to be sure of finding the worst-case error. In practice, we can take advantage of the fact that the INL changes slowly by design and will find that only a sampling of codes are necessary along the entire transfer function.

A DAC of any architecture with significant bit interactions is even worse than a "segment" DAC, because it cannot be counted on to have a smooth transfer function with slowly changing INL. An all-codes test, or at least an intelligent search routine based on an understanding of the particular bit interactions, will be required.

First, consider the case of a binary-weighted architecture with no bit interactions; what are the optimum $n + 2$ tests to perform? The measurement of offset error and gain error is straightforward and usually much easier to do than INL, because the offset is measured with respect to ground, and the tolerance on full scale is usually much looser than that on INL.

To determine INL, we could measure the linearity error of each of the n bits and calculate the positive and negative summation errors to determine worst case INL. This is an expensive—and often slow—procedure, because a meter with no more than ¼ as much INL as the DAC being tested is required.

Major-Carry Technique

A much better technique, in the case of no bit interactions, is to measure the n major carries. The three major carries for the 3-bit DAC example would be 001-000, 010-001 and 100-011.

If the DAC were perfectly linear, each carry difference would be exactly one LSB as determined by:

$$\text{LSB} = \frac{(V_{11} - \text{OFFSET})}{(2^n - 1)} \tag{10.7}$$

The errors measured for the major carries can be used to calculate the individual bit errors:

$$\begin{aligned} E_3 &= 001 - 000 - 1\ \text{LSB} \\ E_2 - E_3 &= 010 - 001 - 1\ \text{LSB} \\ E_1 - E_2 - E_3 &= 100 - 011 - 1\ \text{LSB} \end{aligned} \tag{10.8}$$

where E_3, E_2 and E_1 are the linearity errors of the LSB, Bit 2 and MSB, respectively. Notice that, since the offset is present equally in each code, it cancels from the calculation; and that any residual gain error, resulting in the wrong ideal LSB calculation, could be eliminated by calculating the sum of E_3, E_2 and E_1, for the total error, and apportioning any difference from zero back to the bit errors in binary fashion.

This is a quite powerful technique, because you can use a null-and-difference circuit with much less accuracy than the DAC being tested. For example, a 1% measuring error in the carry will only produce a 0.01–LSB error in the final INL determination.

This technique is, in fact, being used by manufacturers of sophisticated linear-device automatic test equipment to perform linearity measurements with up to 18 bits accuracy. They build "super DACs," which are very stable and non-interacting—but not necessarily very linear—and then software-calibrate them, using the major-carry technique.[1,2]

However, most test engineers who have the responsibility for developing DAC test packages for final test, outgoing QC, or incoming inspection, prefer not to assume that the DAC is free from bit interactions; they add linearity

[1]Robert B. Craven and E. Rachel Morris, "An 18-Bit Precision DC Measurement System," *Digest of Papers*, 1981 IEEE International Test Conference.

[2]Allan Ryan and John Chang, "Versatile System Console for Accurate Measuring," *Analog Dialogue* 14-3 (1980): 4-5

tests beyond the major carries. This decision is quite often necessary, but it is also expensive because it will require that the test-measurement scheme be significantly more accurate than the DAC being tested.

The most useful tests to add would be to measure the actual linearity errors at the two expected worst-case codes. Recall that if bit interactions are insignificant, these two summation errors will have different signs but equal magnitudes; therefore, the sum of these two errors can be used to estimate the degree to which superposition actually holds. In other words, if the sum of the positive and negative summation errors is significantly different from zero, additional tests will be needed to find the true worst-case INL error.

The most common method for determining the summation errors in non-segment DACs is to measure the INL error of each individual bit, then exercise (and measure) the code that contains only bits with positive errors, and then the code that contains only bits with negative errors. For example, if bits 1, 3, 5, 7 had positive errors, and bits 2, 4, 6, 8 had negative errors, the first code would be 10101010, and the second would be 01010101. This is more accurate than actually adding up the bit errors algebraically, because it avoids the accumulation of up to $n - 1$ measurement errors.

Measurement of INL errors is easily performed with an accurate, integrating-type voltmeter, but generally such meters perform their measurements too slowly for the accuracy required in testing DACs with resolutions of 10 bits and more. Another problem with purchasing such a meter for converter testing alone is the relatively high cost (for just the meter), compared to dedicated, customized measurement systems present in most of today's automatic test equipment. Finally, an instrument's ability to interface with intelligent controllers affects flexibility and—again—speed.

One of the most popular methods for testing DACs uses a precise reference DAC (REF DAC)—with much higher resolution and accuracy than the DAC being tested)—and a differencing circuit (using an op amp or instrumentation amplifier) to provide a measure of the error between the device under test and the REF DAC. Often, the error voltage is amplified (between 10 to 100 times) and applied to the vertical input of an oscilloscope, a chart recorder or the DC voltmeter in an automatic test system. The advantage of the differencing method is that it relaxes the accuracy requirement on the circuit that actually senses the measurement.

Bit-Scan Testing

Figure 10.6 shows the REF DAC technique implemented in a dynamic "bit scan" bench set-up for a 12-bit DAC, with the errors displayed on an oscilloscope. The control logic sequences through all-ones, then each bit individually, then zero, while the comparator and the shift register identify all bit with positive errors to determine which code has the worst-case positive error, and

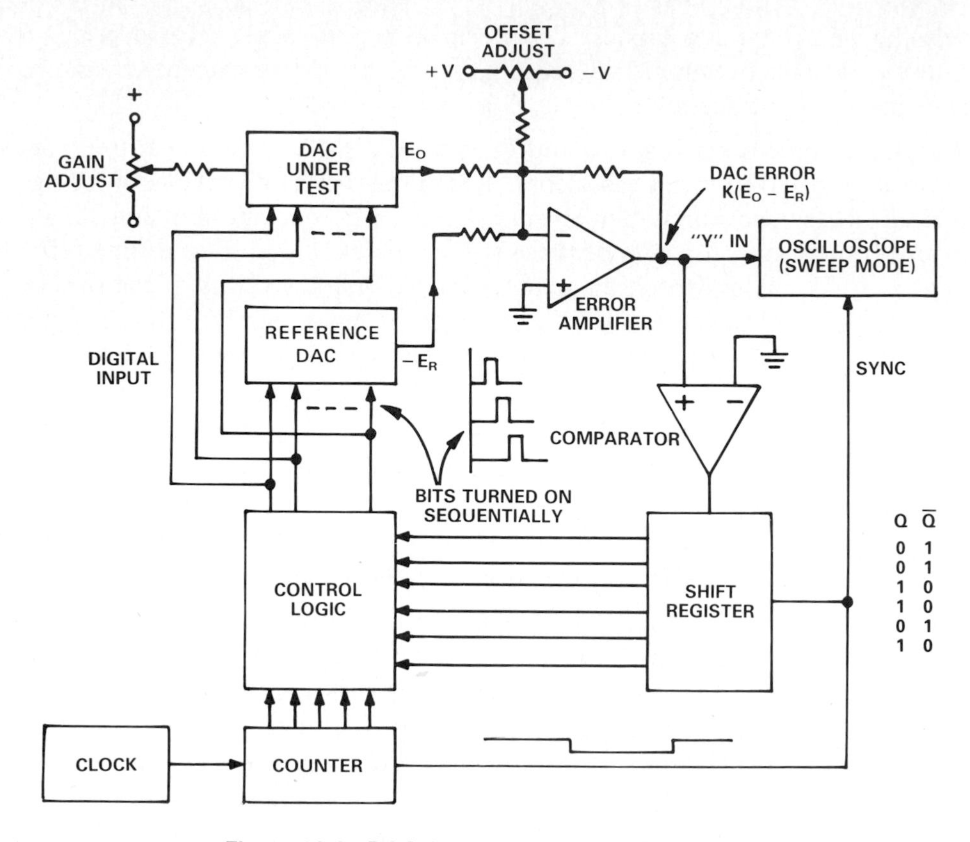

Figure 10.6. DAC dynamic test—bit-scan mode.

all bits with negative errors, to determine which code has the worst-case negative error; then these codes are tested.

Figure 10.7a shows a typical time-slot allocation for testing a 12-bit DAC in this mode. Typical error displays resulting from this test are shown in b, c and d. Figure 10.7b illustrates an error display for a DAC having the correct binary scaling weights for all bits, but full-scale gain calibration 1 LSB high. Since a 1-LSB full-scale gain error (with correct relative scaling) causes the MSB error to be ½-bit high, bit 2 to be ¼-bit high, bit 3 ⅛-bit high, etc., the error display is exponential in shape.

Figure 10.7c illustrates the error display for the case of a + ¼-LSB offset error, combined with a − 1 LSB full-scale gain error, with perfect relative weighting (i.e. linearity). This causes a reversal in fullscale error polarity from that shown in Figure 10.7b. In addition, the + ¼-bit Offset (zero) error shifts the complete display + ¼ bit from the zero baseline.

To find the linearity errors the device under test (DUT) must be calibrated in the following manner: Zero is adjusted to bring the bar representing the zero error (time slot T13 in Figure 10.7a) to the display baseline. The gain is then adjusted to bring the bar corresponding to fullscale (time slot TO) to

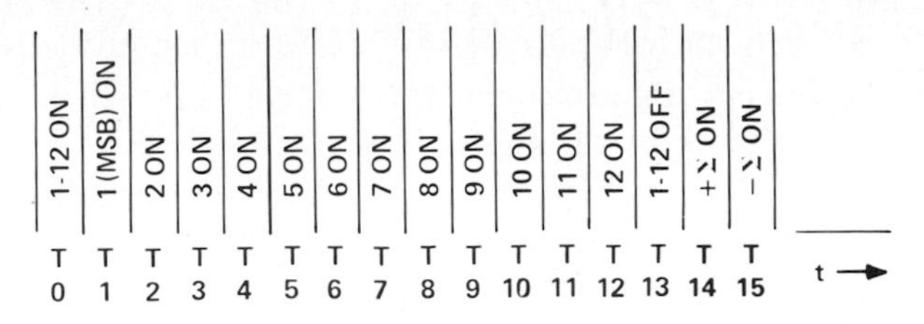

a. Commutator time-slot allocation.

b. DAC full-scale gain is 1-bit high, with zero offset.

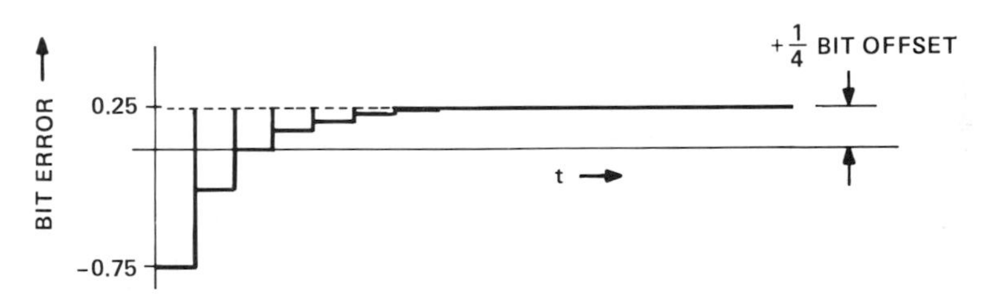

c. DAC full-scale gain is 1-bit low, with + ¼-bit offset.

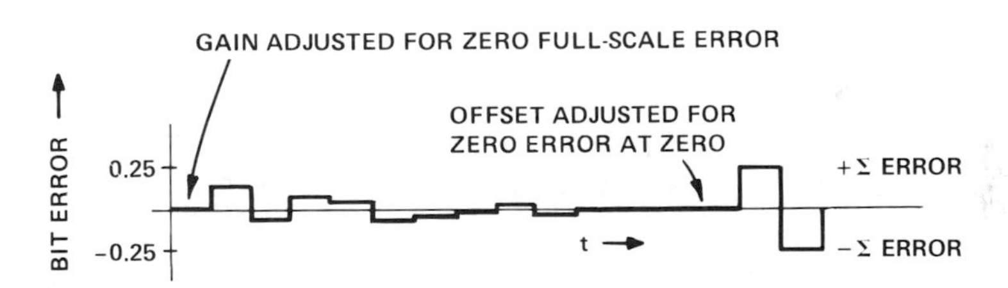

d. Typical bit-error distribution, with ¼-bit INL error.

Figure 10.7. Dynamic test waveforms of 12-bit DAC in bit-scan mode.

baseline. Zero and full scale of the DUT are now calibrated. (If the DUT does not have zero or full-scale adjustments, zero and full scale of the reference DAC can be adjusted instead, to normalize the display.)

A typical "bit-scan error" display after zero and full scale of the DUT (or reference DAC) have been calibrated, is shown in Figure 10.7d. If the DUT's bit interactions are negligible, the sum of all positive bit errors should equal the sum of all negative bit errors after zero and fullscale calibration; any difference indicates the presence of interactions.

A circuit similar to that in Figure 10.6 is found in many automatic testsystems, capable of measuring nonlinearities up to 14 bits. In a typical automatic test-equipment (ATE) application, calibration of the circuit's errors, excluding the DUT, would normally be provided for in software. This would include measurement of the error amplifier's offset- and gain errors, caused by mismatches of the amplifier's input-transistor base-emitter voltages & bias currents and mismatches in the feedback- and input resistors.

If an instrumentation amplifier is used instead of an op amp, there would be separate high-impedance inputs for the DUT output and the reference-DAC output and a saving of components. With an instrumentation amplifier, it's wise to measure its common-mode rejection, because it can be significant for DACs having resolutions of 12 bits and more (at 6 dB per bit, a 12-bit DAC would require at least 90 dB of CMRR). The reference DAC may also be calibrated by the system; this amounts to measuring its offset and gain errors (and sometimes the linearity errors, using the major carry technique), and storing them for future software correction.

10.3.2 DAC SETTLING-TIME MEASUREMENT

DAC settling time is a parameter of importance in high-speed applications. Settling time is defined as the time required for the output to approach a final value within the limits of a defined error band, for a step change in input. This fixed error band is generally expressed as a fraction of full scale, typically ± ½ LSB. If the device's step response overshoots or rings, so that the output swings through the defined error band before entering it for the final time*, the above definition requires that settling time is measured as the time taken for the output to enter the defined error band for the final time.

The above definition of settling time implies that, the greater the output step change, the longer the settling time (a non-overshooting 1-LSB output step change, for example, has settled to within ± ½ LSB when this change has reached only 50% of its final value.)

The accurate measurement of settling time for a high-resolution, high-speed DAC is fraught with practical difficulties. Measurement instrument bandwidth and thermal unbalance effects, coupled with the unavoidable presence of noise, can introduce significant measurement uncertainties when high-speed settling times to within error bands of the order of the order of 0.01% of final value are being measured. It is particularly easy to overlook a "long tail" in which the output continues to change for hundreds of micro seconds due to thermal gradients or dielectric absorption effects. It takes patience and significant analytical skills to develop, and verify the accuracy of, a high speed settling-time measurement setup.

*"Final time" does not mean just the final time within the time allotted for the measurement; if long tails (due to thermal or dielectric-absorption effects), orders of magnitude longer than the normal settling time, are suspected, the time allotted for the measurement should be long enough to detect their presence.

There are some important distinctions which must be mentioned when discussing speed and settling-time specifications. If the settling-time error band is described as ± ½ LSB, a DAC with 10-volt full scale range and 8-bit resolution has an allowable error band of ± 20 mV, a 10-bit DAC has a ± 5-millivolt error band, and a 12-bit DAC ± 1.22 mV. A given 12-bit DAC is not necessarily a slower design than a given 8-bit device; it is simply required to settle to within a more tightly specified error band.

Some converters settle faster in one direction than the other; if bipolar operation is considered, voltage output in the negative-going direction can settle much more slowly than in the positive direction. The current-output mode (the standard mode for most IC DACs), settles much faster than the voltage mode, since the currents are simply steered one way or the other, and there is no significant capacitance-charging. In the voltage mode, an output op amp is required, adding a delay, and stray capacitances must be charged by an amount corresponding to the voltage change. There are many ways to optimize settling time in the current-output-DAC/fast-op-amp connection; some of these are discussed in Chapter 12.

Settling time is often measured by looking at crossings on a trace that plots the difference between the output waveform and a voltage representing the final value of output—on an LSB-calibrated oscilloscope triggered by the initiation of the DAC input change. There are two major problems here. One is the narrow extent of the final settling voltage range; it calls for a very fast high-gain preamplifier. The other problem is the wide output swing in relation to the small range being depicted on the oscilloscope screen. It must invariably produce saturation until the trace comes into the final-settling window. Saturation of the oscilloscope preamplifier can cause large errors while the oscilloscope recovers; it can be avoided (or at least mitigated) by using a specially designed preamplifier with high gain, wide bandwidth, and controlled output swing that considerably reduces the amount of overdrive to the oscilloscope.

The circuit shown in Figure 10.8 is capable of measuring current-output 12-bit settling times as low as 200 ns. The unity-gain, high-bandwidth buffer formed by Q1-Q3 holds the DAC output voltage at ground while the current changes appear across R1. The pair of source followers, Q4 and Q5, isolate the scope's input capacitance from R1, and the Schottky diodes limit the signal swing into the plug-in to ± 400 mV, thereby reducing the problem of overload recovery.

When the magnitude of the LSB is in the neighborhood of 1 millivolt or less, detecting settling to beyond 12 bits requires limiting the input voltage excursion to the oscilloscope even further. Several "active clamping" techniques can be employed.[3] A fairly simple technique, not requiring synchronized

[3]Schoenwetter, H.R., "High-Accuracy Settling Time Measurements," *IEEE Transactions on Instrumentation and Measurement*, Vol. IM-32. No. 1, March, 1983.

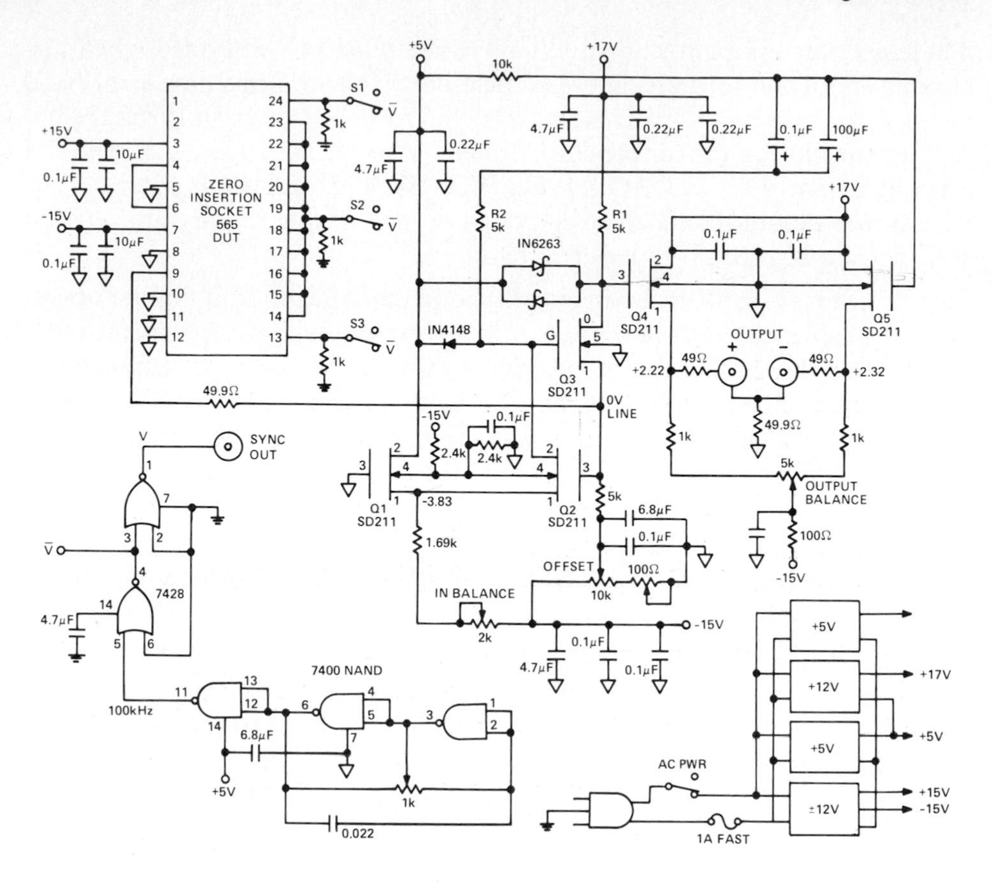

Figure 10.8. High-speed-settling-time test fixture.

switching, is illustrated in Figure 10.9. This circuit functions in a manner similar to the previous one, except that the reduced dynamic range of the output amplifier stage formed by Q7 and Q8 limits its output swing to 12 mV. Using this technique, settling time measurements to 14 bits in under 1 microsecond are possible.

Many applications which require fast settling time use the DAC in the current mode as a component of a fast, successive-approximation a/d converter. However, because in general the DAC's output is not tied to a virtual ground or a fixed voltage level in such applications, use of the current-output settling specification may prove overly optimistic. In such cases, settling-time test fixtures that replicate the actual conditions of the application should be used.

Figure 10.10 shows such an approach which, in addition, is more amenable to being automated than the previous circuits, because its outputs are a dc voltage and a pulse delay-time setting. Basically, the digital voltmeter can be made to trace out the settling waveform backwards, point by point, when the strobe timing of the latching comparator is varied with respect to the DAC input pulse.

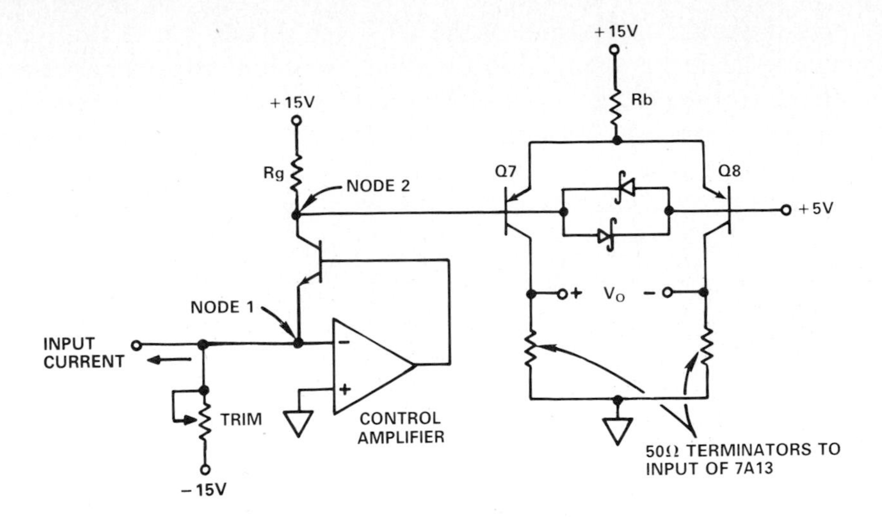

a. Simplified schematic illustrates the principle.

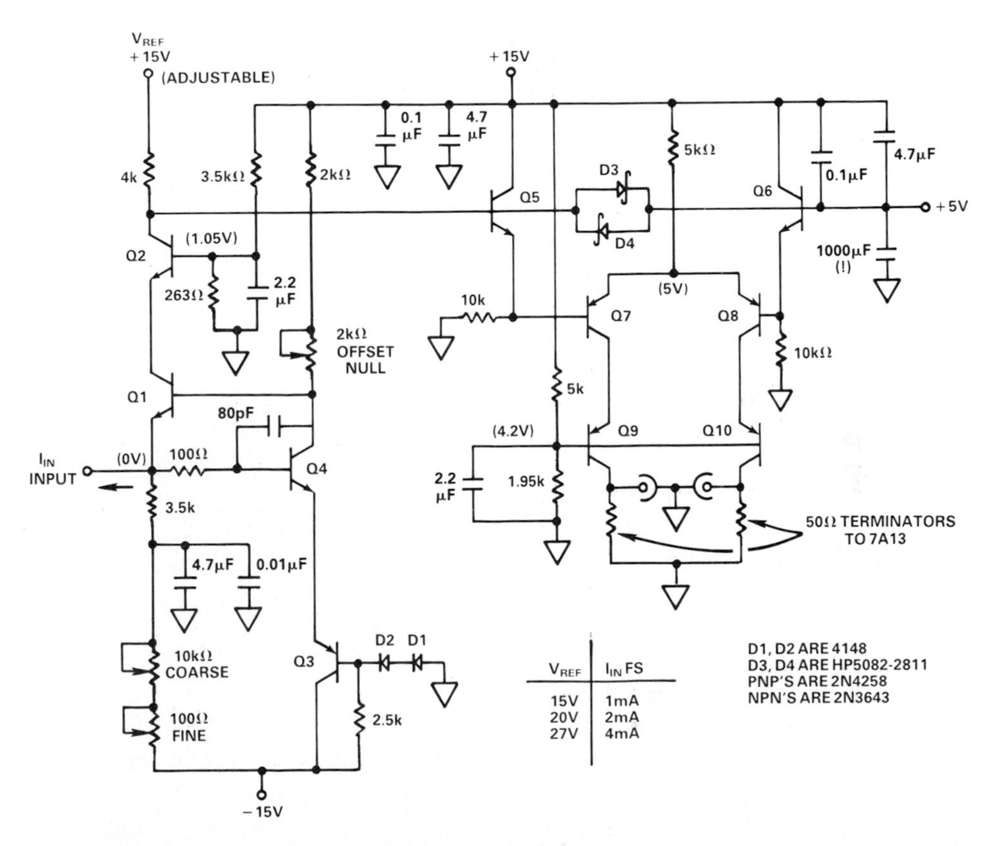

b. Complete schematic.

Figure 10.9. Preamplifier prevents overdrive recovery problems in the 7A13 oscilloscope plugin, because V_{OUT} max is only 12 mV. Transconductance from +IN to OUT is 1/4kΩ, or 2mV for a 0.5μA LSB.

Here's how it works: The output of the DUT (an AD569, in this case), is switched periodically between +5 and −5 volts, repeating its settling pattern continuously. In the configuration of Figure 10.10, we are concerned with positive step response, i.e., settling to +5V. The DUT's output is biased by −5 volts (V_{CANCEL}) so that the final value will be close to zero. A pair of Schottky diodes narrows the range being observed to a few tenths of a volt.

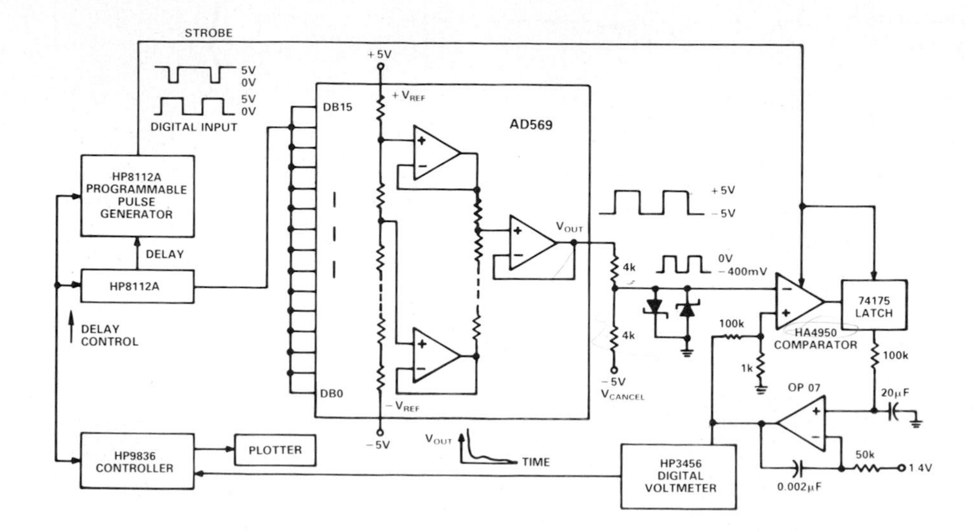

Figure 10.10. Tracking-loop scheme for measuring settling time.

The biased output is compared with the slowly varying output of an integrator, attenuated by 100. The comparator has very fast response, so the latched state will depend directly on the difference between the *instantaneous* value of the rapidly changing DAC output and the slowly varying integrator output. At the time the comparator is strobed, if the DAC output is less than 1/100 the integrator output, the latch output will be latched at TTL "0" (about 0.5 V); if greater, it will be latched at "1" (about 5 V).

The latched value is filtered and compared with a +1.4-volt threshold; if "0", the integrator will integrate downward, or if "1", the integrator will integrate upward, tending to follow the value of the DAC output at the time the comparator was strobed.

Thus, by repeatedly latching on the same value of DAC output at the same time in the cycle, while the integrator is tracking, the relatively slow loop will home in on that value, and make it available for reading as a dc voltage. In this way, any point on the response curve can be selected for reading as a dc voltage.

In practice, a measurement starts when the strobe signal is delayed long enough after the input signal to the converter to ensure that the device under test has settled. The loop offset value measured at this moment represents the final value. Next, the delay is shortened, and after the filtering loop has settled, the resulting offset change represents the converter's output change.

The strobe delay time is decreased in increments suitable for the resolution desired until the converter's instantaneous response is outside the desired error window. In the setup described, the generator can produce pulses as short as 1 ns. *The settling time is interpreted as the strobe delay period at which the offset change first becomes larger than the desired error band—typically ½ LSB.*

Schottky diodes limit the input to the high-resolution comparator to a few hundred millivolts. A clamped mode on that input is formed by two 4-kilohm load resistors, and a voltage (V_{CANCEL}) equal in magnitude but opposite in polarity to the expected final value of the converter's output for a transition of interest. Because the reference voltages to the converter must be very stable, precision power supplies are required.

10.4 A/D CONVERTER TESTING

10.4.1 ADC TRANSFER FUNCTION

In the ideal transfer function for a 3-bit A/D Converter (ADC), Figure 10.11, the analog input signal is on the horizontal axis and the digital output is on the vertical axis. Note that, unlike a DAC, where there exists a unique analog value for each digital code, the digital output of an ADC is valid over a range of input signals. The quantum of input for a given output code is called the "width" of the code. The ideal width is exactly 1 LSB (least-significant bit), but, in practice, each code-width is different from its neighbors; acceptable performance will typically occur with codes ½-LSB to 1 ½-LSB wide.

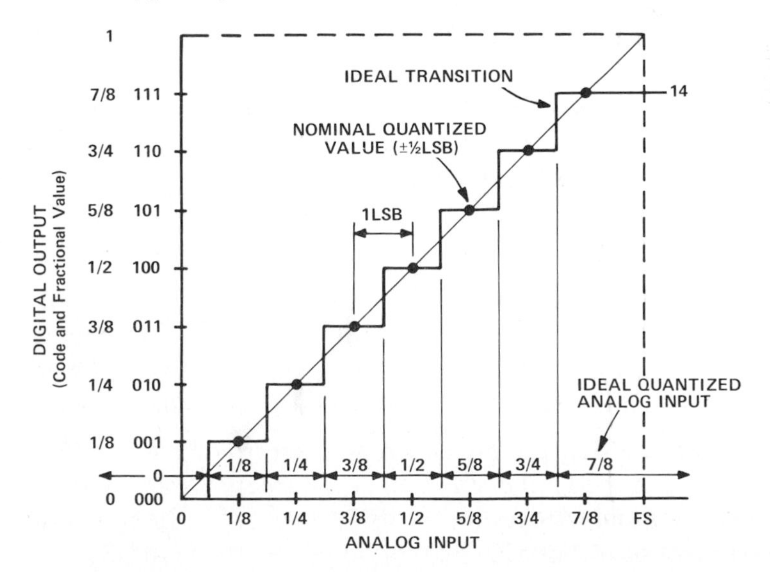

Figure 10.11. Transfer function of an ideal 3-bit ADC.

The transitions between codes occur at unique, measurable input voltages. In ADC architectures which employ a DAC, these transitions ideally occur when the input to the ADC equals one of the discrete DAC output values (comparator and noise can cause them to differ).

Determining the input voltages at which these transitions occur in an ADC is somewhat more difficult than measuring the output of a DAC at a given code. However, once instrumentation for identifying transitions has been implemented, the specifications and testing of ADC transfer functions are similar to those for DACs. That is, similar concepts of gain, offset and linearity apply with equal importance to ADCs.

The ideal analog values at which the transitions should occur in a n-bit ADC can be calculated as follows:

$$V_{in} = V_{FS} \sum_{i=1}^{n} (b_i/2^i) + V_{off} \tag{10.9}$$

where $b_1 \ldots b_n$ are the digits of the binary code, starting at the MSB, with values of 0 or 1. V_{FS} is the reference voltage or full-scale range and V_{off} is the offset. The analog input voltage, V_{in}, required to turn on each bit can be found by setting each bit, except the desired one, to zero in Equation 10.9.

Note that Equation 10.9 is identical to Equation 10.1, except for the inclusion of an offset (V_{off}) term. The reason for this is that in many ADC's, an offset is intentionally introduced to adjust the positions of the transitions to suit the particular ADC application.

For example, in Figure 10.11, an offset of $-\frac{1}{2}$ LSB has been introduced, moving the transitions to the left by that amount and aiming the ideal straight line through zero, full-scale range, and the center of each code. Thus, the position of the first transition (from Equation 10.9) is

$$2^{-n} V_{FS} - \tfrac{1}{2}\,\text{LSB, or } 1\,\text{LSB} - \tfrac{1}{2}\,\text{LSB} = +\tfrac{1}{2}\,\text{LSB}.$$

10.4.2. ADC GAIN AND OFFSET ERRORS

Unipolar Offset Error

The transfer function of an ADC is tested for errors in a predictable sequence. The first test to be performed is for offset error. An offset error is defined as a common deviation from the ideal transition voltages. It is usually tested by finding the error of the first (LSB) transition, because it is likely that the offset error is the only significant error present in that transition.

However, to ensure that the offset error is determined accurately, any gain or linearity error of the LSB itself must be subtracted first. To determine any such errors, it is usual to measure the difference in 1-LSB transition voltages between pairs of codes that will yield the actual width of the LSB, for example, code 0 . . . 0100 and code 0 . . . 0101, or some other such combination. The code-width error, or CWE, is equal to the measured width of an LSB step minus one LSB ($V_{ref}/2^n$). Once this error is determined, it is subtracted from the first transition voltage; the remaining deviation from the ideal value is the offset error (Figure 10.12). The offset error in the example of Figure 10.12 may be determined by the following equation:

$$V_{off(err)} = V_{trans(meas)} - CWE - V_{trans(ideal)} \tag{10.10}$$

where $V_{trans(meas)}$ is the actual first transition voltage and $V_{trans(ideal)}$ is the ideal first transition voltage. For Figure 10.12, where there are no apparent code-width errors, the offset error in LSBs, calculated from Equation (10.10), is:

$$V_{off(err)} = 1\,LSB - 0\,LSB - \tfrac{1}{2}\,LSB = +\tfrac{1}{2}\,LSB \tag{10.11}$$

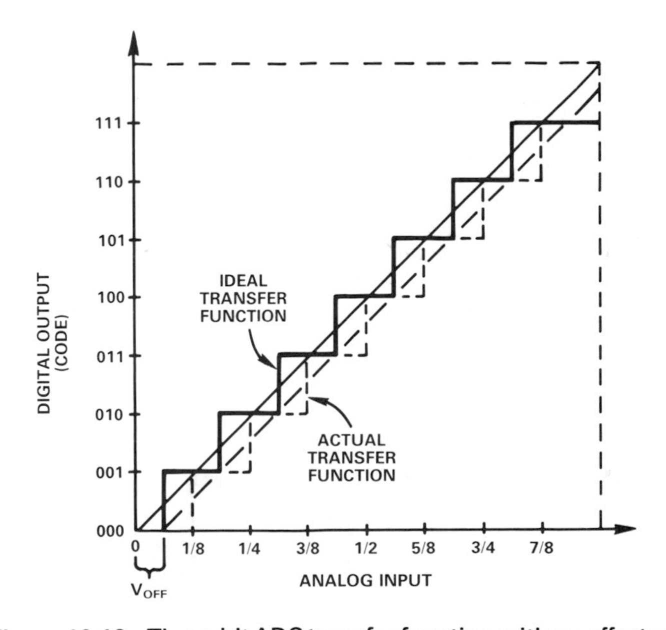

Figure 10.12. Three-bit ADC transfer function with an offset error.

Unipolar Gain Error

The next test performed is for the gain error. Gain error appears as a change in slope of the transfer function. Thus, gain error is the same as full-scale error, except that the offset error is subtracted. Gain error affects each code in an equal ratio (Figure 10.13).

To determine the gain error of an ADC, measure the final transition voltage and subtract the first transition voltage. Since this interval is ideally equal to (V_{FS} − 2 LSB), the deviation of the difference of the measured values from the value of (V_{FS} − 2 LSB) is the full-scale gain error. The gain error can be expressed as %FSR or in LSBs. The Gain Error of Figure 10.13 may be determined by the following equation:

$$\text{G.E.} = (V_{FS} - 2\,\text{LSB}) - (V_{11(meas)} - V_{zs(meas)}) \tag{10.12}$$

where $V_{11(meas)}$ is the last transition voltage, V_{zs} is the first transition voltage, and $(V_{FS} - 2\,\text{LSB})$ is the full-scale range of the ADC minus 2 ideal LSBs. For Figure 10.13, the gain error in LSBs is calculated from Equation (10.12) to be:

$$\text{G.E.} = 8\,\text{LSB} - 2\,\text{LSB} - (7\tfrac{1}{2}\,\text{LSB} - \tfrac{5}{8}\,\text{LSB}) = -\tfrac{7}{8}\,\text{LSB} \tag{10.13}$$

at V_{11}, or -1 LSB at full scale ($12\tfrac{1}{2}\%$ of any code).

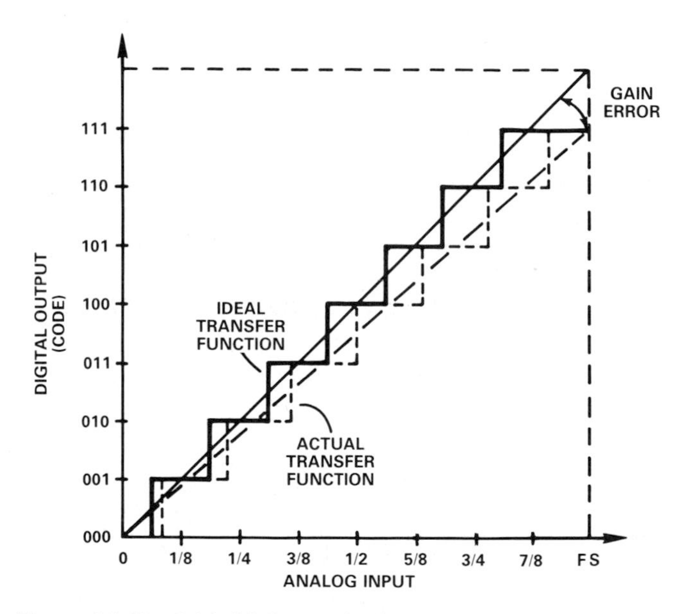

Figure 10.13. 3-bit ADC transfer function with a gain error.

Bipolar Offset Error

The transfer function for an ideal bipolar ADC resembles the unipolar transfer function, except that it is offset by $-V_{FS}$ (Figure 10.14). In addition to offset and gain, there is typically one additional parameter specified, bipolar zero.

Bipolar offset error is usually measured as the deviation of the first transition from the ideal, which is usually designed to occur at $(-V_{FS} + \tfrac{1}{2}\,\text{LSB})$; 1 LSB is equal to $2\,V_{FS}/2^n$. As with unipolar offset, the bipolar-offset error is common to all codes, and any nonlinearity or gain error of the LSB must be accounted for in its calculation. The offset error of Figure 10.15 may be determined using Equation (10.10). In the case shown in Figure 10.15,

$$V_{off(err)} = -V_{FS} + 1\,LSB - 0\,LSB - (-V_{FS} + \tfrac{1}{2}\,LSB) = \tfrac{1}{2}\,LSB \quad (10.14)$$

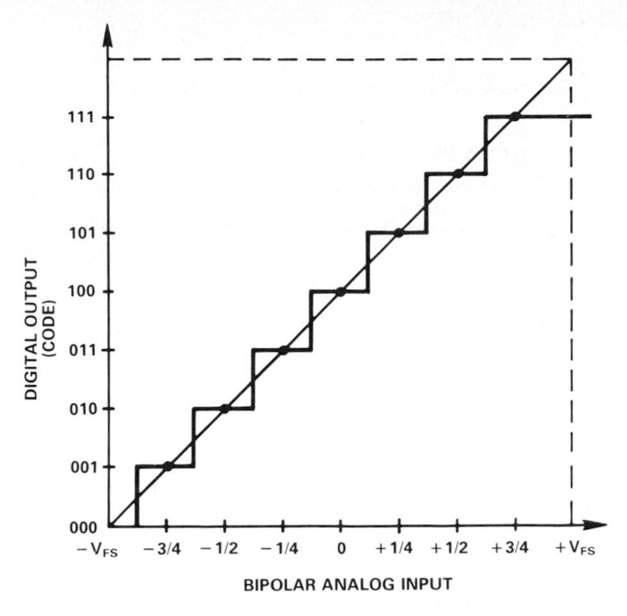

Figure 10.14. Transfer function of an ideal 3-bit bipolar ADC.

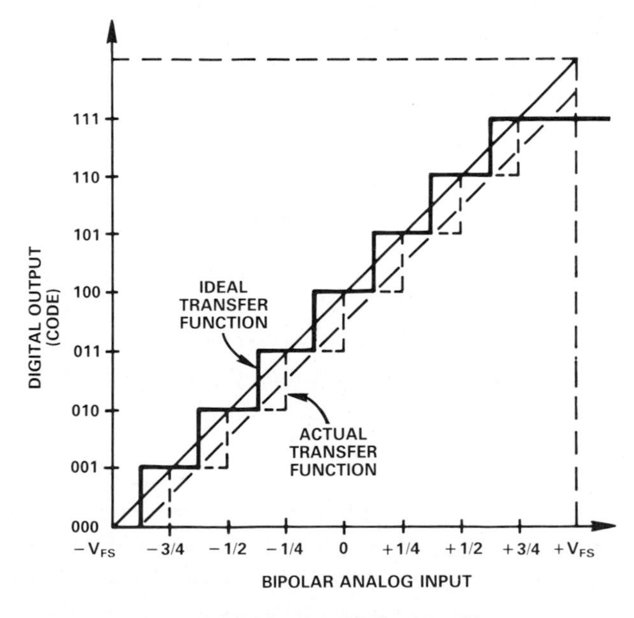

Figure 10.15. 3-bit bipolar ADC with offset error.

Bipolar Gain Error

Bipolar full-scale error, or gain error, is measured in the same way as unipolar full-scale (gain) error, except that the initial point corresponding to zero for the unipolar case is $-V_{FS}$ (Figure 10.16). The gain error of a bipolar-input

ADC is calculated using Equation (10.12). The gain error, in LSB's, of Figure 10.16 is calculated as follows:

$$\text{G.E.} = 8\,\text{LSB} - 2\,\text{LSB} - (7\tfrac{1}{2}\,\text{LSB} - \tfrac{5}{8}\,\text{LSB}) = -\tfrac{7}{8}\,\text{LSB} \qquad (10.15)$$

at V_{11}, or 12.5% at any code.

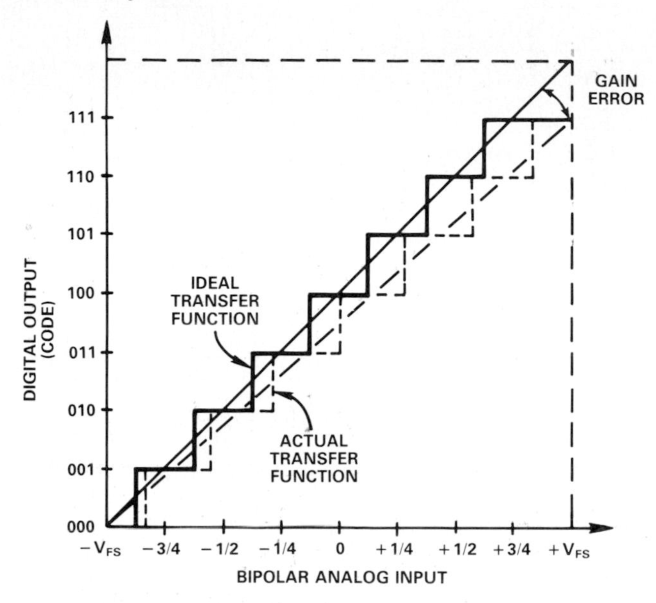

Figure 10.16. 3-bit bipolar ADC with gain error.

Bipolar Zero Error

The transition that is next most-often measured for bipolar input ADCs is for determining ***bipolar zero error***, which is defined as the deviation of the mid-scale transition voltage from the ideal in the vicinity of 0 volts (Figure 10.17). The bipolar zero error may be determined by the following equation:

$$\text{BZE} = V_{\text{zero(meas)}} - V_{\text{zero(ideal)}} \qquad (10.16)$$

where $V_{\text{zero(meas)}}$ is the measured mid-scale (MSB) transition voltage and $V_{\text{zero(ideal)}}$ is the ideal mid-scale transition voltage. The bipolar zero error of Figure 10.17 is calculated using Equation (10.16) to be:

$$\text{BZE} = -\tfrac{1}{4}\,\text{LSB} - (-\tfrac{1}{2}\,\text{LSB}) = +\tfrac{1}{4}\,\text{LSB} \qquad (10.17)$$

Bipolar zero error is not an independent variable; it is the sum of the bipolar offset error, one-half the gain error, and the MSB linearity error. It is usually measured, rather than calculated from the gain and bipolar offset measurements, to avoid errors due to tolerance buildup.

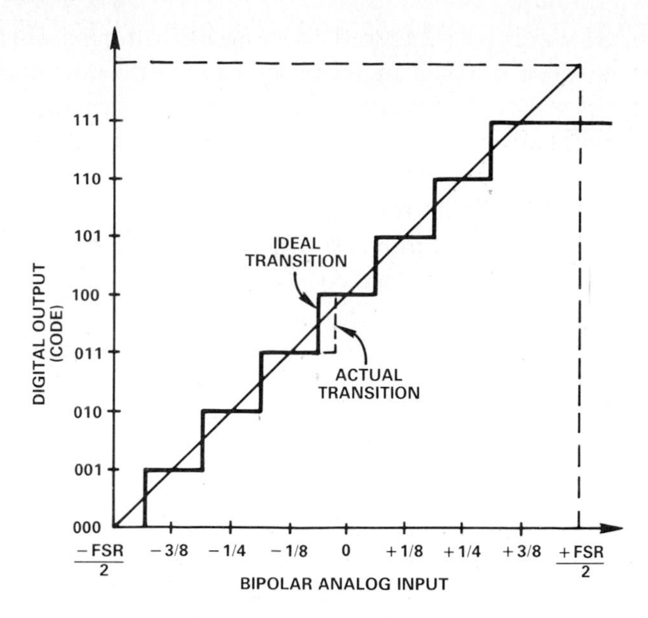

Figure 10.17. 3-bit bipolar ADC with error at analog zero.

10.4.3 ADC LINEARITY ERRORS

With most ADC's, the gain and offset specifications are not the most critical ones that determine an ADC's usefulness in specific applications. Typically, offset, gain and zero errors can be calibrated out, in either hardware or software. The most important specifications for the bulk of ADC applications, because they represent irreducible errors, are *differential nonlinearity* (DNL) and *integral nonlinearity* (INL).

Differential Linearity

Differential nonlinearity in an ADC is defined as the deviation in code width from the value of 1 LSB (i.e., $V_{FS}/2^n$). If DNL errors are large, the output code widths may represent excessively large and small ranges of input voltages; and if the worst-case DNL is more negative than −1 LSB, the code-width will vanish entirely and the ADC will have at least one missing code. This means that there will be no voltage in the entire full-scale range that can cause that code to appear.

Integral Nonlinearity

Integral nonlinearity is the deviation of the transfer function from the ideal straight line. Most ADC's are specified with *end-point* INL, i.e., INL specified in terms of deviations from a straight line between the end points of the transfer function (rather than a "best straight line") because it conservatively specifies the worst deviation that will occur for the transfer function; as the description below suggests, it is also easier to measure.

End-point linearity can be measured most easily in terms of the transition points; but it is often defined in terms of the ideal code midpoints. The straight line for *low-side-transition* (LST) linearity is drawn from the offset transition (i.e., from all-0's to the LSB on) to the last transition of the transfer function. The deviation of any transition from its corresponding point on that straight line is the INL of that transition. When *center-of-code* (CC) linearity is specified, the ideal straight line is shifted by ½ LSB, and the deviation of the center of a code is the INL of that code.

Best-straight-line integral nonlinearity is the deviation of any code from a straight line calculated to minimize the worst-case INL errors. The straight line is typically calculated using the least-squares method. This basically involves testing the codes of interest, reducing the data to determine the best straight line, and then retesting to the new straight line.

10.4.4 ADC Test Strategy

Minimum Number of Tests

Before testing an ADC for DNL and INL, the test engineer must determine which codes must be tested to guarantee performance over the entire transfer function. The problem is similar to that of DACs, in that different ADC architectures will require that different code patterns be tested. In addition, dynamic errors and noise are often significant and unpredictable and require additional tests in the same way bit interactions affect DAC testing.

For example, multi-comparator, or flash, ADCs are constructed using 2^n matched resistors and $2^n - 1$ comparators. Their architecture requires testing all codes; fortunately (for the test engineer), they are typically low in resolution, seldom having resolutions that exceed 8 bits.

Tracking ADCs are constructed using an internal DAC, a comparator, an up/down counter, and support circuitry. The codes to be tested for DNL and INL are primarily influenced by the DAC design, as dynamic errors are not usually present for moderate clock frequencies. One note of caution to consider when testing tracking ADCs is their susceptibility to noise, as compared to integrating ADCs.

Integrating ADCs are typically high-resolution, high-accuracy devices. They have very good DNL by design. The major sources of INL occur at the inputs of the integrating amplifier and comparator or are due to non-ideal behavior of the integrating capacitor, making it difficult to predict the location of the worst case INL codes. Fortunately, changes in INL will be very gradual, so a simple sampling of codes along the transfer function is usually sufficient to find the worst-case error.

Successive-approximation ADCs have both static and dynamic sources of DNL and INL. These errors come from static nonlinearities of the DAC, long DAC settling time, slow comparators, parasitic capacitance at the summing junction, and noise. They can require a large number of tests, often repetitively, if noise is significant.

End-Point Tests

For all ADC's, regardless of architecture, the transfer function's end points must be measured and normalized before linearity testing can begin. Normalization can be done in either hardware or software. Once the appropriate straight line is determined, all that remains is to measure the required transitions or codes and compare them to the ideal.

Major-Transition Testing

In some rare cases, an all-codes test routine may be required—perhaps even repeatedly, to take into account the effects of noise—to guarantee a tight INL specification in a stringent application. With modern ATE techniques, it is feasible to do this on a sampling basis as part of the production process, after a 100% screen using an abbreviated test routine. Most abbreviated routines are based on the major transitions, plus-and-minus three codes.

The major transitions consist of all the major and minor carries, plus the sums of the most-significant-bits (MSBs). The reason for testing to plus and minus three codes from the major transitions is to check the adjacent codes that may be affected by dynamic conditions of the device under test (DUT).

Figure 10.18 compares the transfer function of an ideal 4-bit ADC with the transfer function of a 4-bit ADC that has DNL and INL errors. The INL in

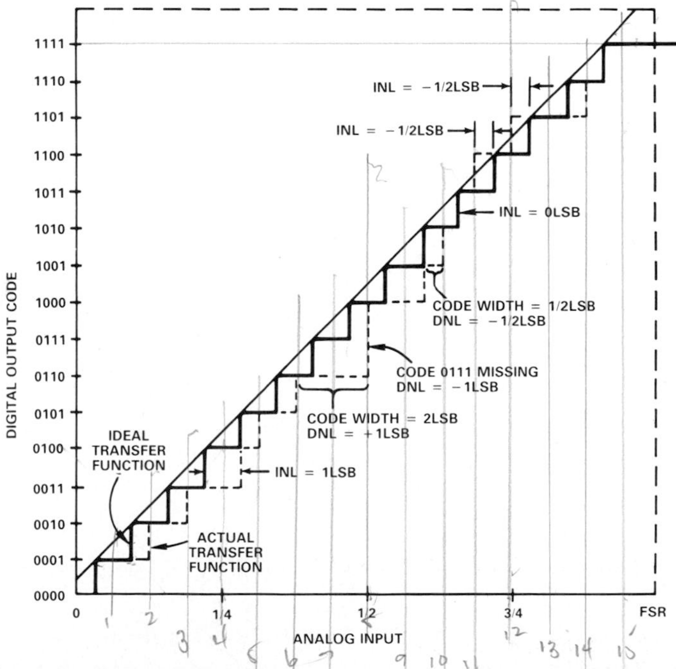

Figure 10.18. 4-bit ADC with linearity errors—both differential and integral (low-side transition).

this case is specified as low-side-transition (LST). The transition to code 0100 is shifted to the right by 1 LSB; this means the LST of code 0100 has integral nonlinearity of + 1 LSB.

The transition to code 1101 is shifted left by ½ LSB; this means the LST of code 1101 has INL of − ½ LSB. The code-width of code 0110 is 2 LSBs; it means that code 0110 has differential nonlinearity of + 1 LSB. The code width of the code 1001 is ½ LSB; thus, code 1001 has DNL of − ½ LSB.

Note that code 0111 does not exist for *any* input voltage. This means that code 0111 has − 1 LSB DNL and the ADC has at least one missing code.

Figure 10.19 shows the same transfer function, but drawn for a center-of-code (CC) integral-nonlinearity specification. The DNL of all codes remains the same, but notice the change in the values of INL. Code 1101 had − ½ LSB of low-side-transition INL, but it has 0 LSB of CC INL. Similarly, code 1011's 0 LSB of LST integral nonlinearity becomes − ⅛ LSB of center-of-code INL. The same phenomena can be observed for codes 1100, 0100, etc.

Some ADCs are specified with CC and some with LST integral nonlinearity; the choise depends on the ADC design and the intended end use. Users should consult the manufacturer if there is any doubt about which type should be employed for a given device.

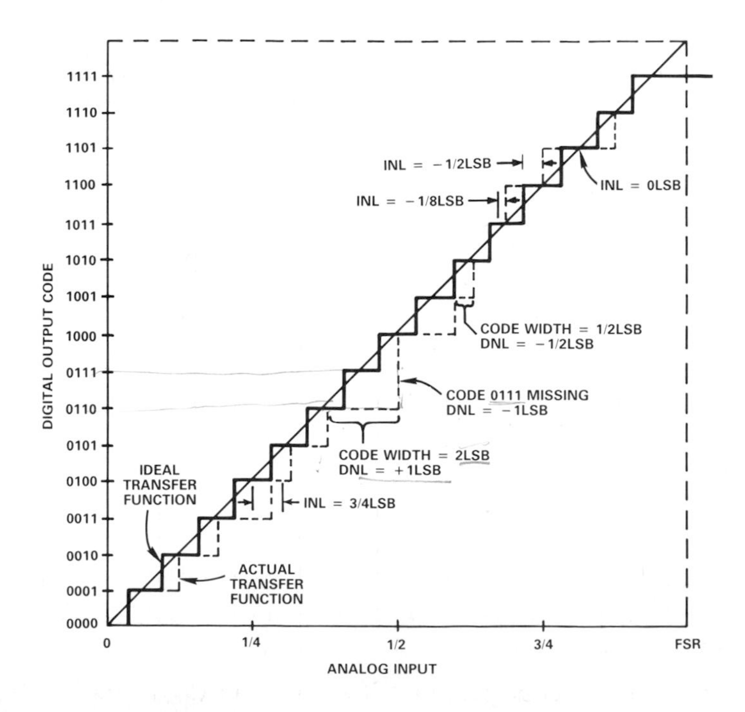

Figure 10.19. 4-bit ADC with linearity errors—both differential and integral (center-of-code).

10.4.5 BENCH-TEST HARDWARE

The most common method of determining the DNL and INL on the bench is by using a test circuit that performs a crossplot (Figure 10.20). If a small asynchronous sinusoidal or triangular ac signal—varying at a fast enough rate to provide a persistent image on an oscilloscope, yet slow enough to permit a large number of conversions—is summed with an analog dc voltage and applied to the input of the ADC under test, the ADC's output will be "dithered" about through several codes either side of the code representing the dc voltage. The output of an elementary 2-bit DAC, which decodes the two least-significant bits, is plotted vertically, and the ac analog input is plotted horizontally, producing the repetitive short staircase through codes (. . . 0)00, (. . . 0)01, (. . . 0)10, (. . . 0)11, (. . . 1)00, etc., as shown in the illustration.

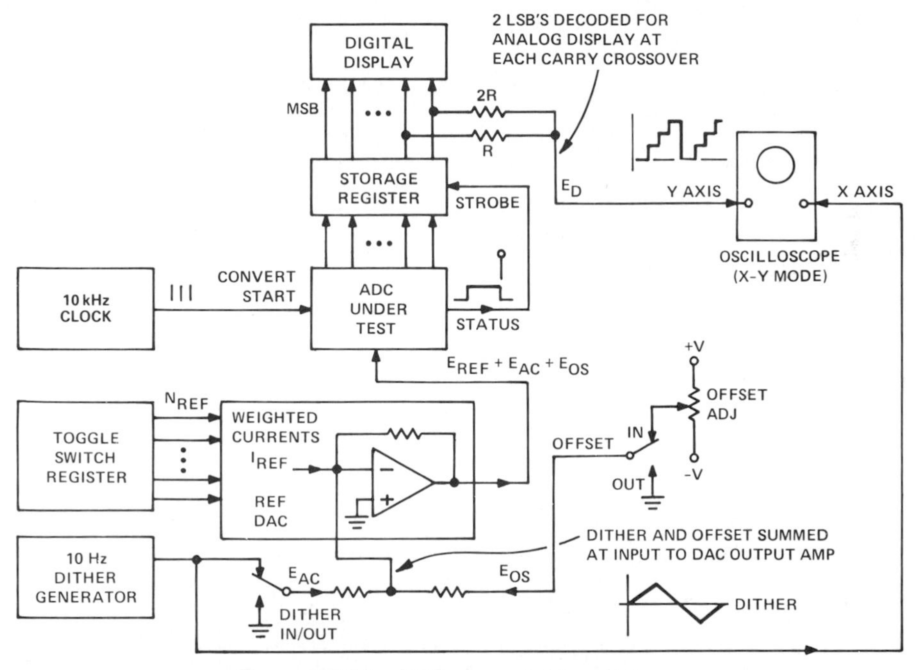

Figure 10.20. ADC crossplot test fixture.

Since the ac voltage serves, over time, to scan through all input voltages within the range of its amplitude, the oscilloscope display is an actual display of the code widths over a short portion of the input voltage range. If the dc voltage is equal to, say, the center of a code, this permits the analog voltages corresponding to the transitions and the center of each code, as well as the adjacent codes, to be readily displayed; this direct display permits determination of device INL and DNL to high accuracy and resolution; it also permits an "eyeball" estimation of noise.

Here's how the dynamic crossplot circuit works: The digital code to be tested, N_{ref}, is entered into the reference DAC via the toggle-switch register, thereby applying E_{ref}, the analog equivalent of N_{ref}, to the analog input of the DUT.

The ac dither, E_{ac}, and an adjustable dc offset, E_{os}, are summed with the reference DAC's output. The dither signal has a low frequency in comparison to the conversion rate of the DUT, allowing the digitized output of the DUT to track its analog input to within the ± ½ LSB quantization limits, i.e., without introducing dynamic errors. A digital register stores the results of each conversion. The 2-bit DAC is formed, using resistors with weights of 2R and R, to sum the LSB and the adjacent bit of the stored ADC output. The resulting 4-step analog output, corresponding to the ADC's two least-significant bit states, is applied to the Y axis of the oscilloscope. The ac dither signal is applied to the X axis of the oscilloscope. Figures 10.21b to 10.24f show a number of waveforms obtained using the dynamic crossplot test circuit.

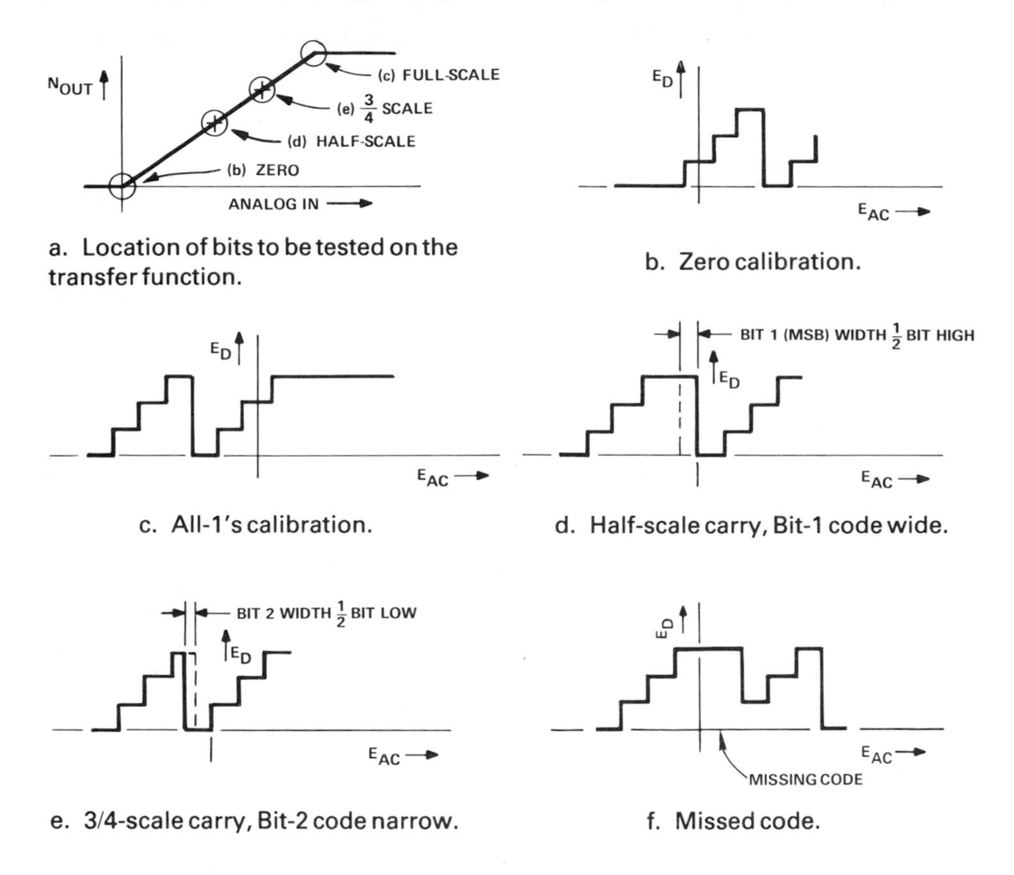

Figure 10.21. ADC crossplot testing.

The device under test is calibrated thus: The CRT beam is first positioned in the center of the screen with the X and Y axis drive signals grounded. Then, with all bits of the reference DAC off except the LSB, adjust the DUT's zero to center the first step of the decoded output staircase waveform, corresponding to the digital code 00 . . . 01 as shown in Figure 10.21b. Next, all bits except the LSB of the reference DAC are turned on, corresponding to the digital code 11 . . . 10, and the gain of the DUT is adjusted to center the next highest step of the decoded staircase waveform, as shown in Figure 10.25c. Zero and full scale are now calibrated.

Differential nonlinearity and integral nonlinearity at each code transition can be seen and measured visually by examining the width of the codes and the displacement of the transitions, to the left or right, as successive bits are turned on. Transition noise can also be assessed; it appears as a jitter in the location of each code transition.

Some additional points about the crossplot test circuit are worth noting:

- Since only the two least-significant-bits of the DUT's digital output are decoded, the crossplot waveform repeats every four steps. For this reason, the DUT should be originally calibrated to less than 2-bit error before the crossplot is used, so that one can be assured that the desired code transition—and not one four LSBs away—is being displayed.

- The dither waveform is shown in Figure 10.20 is triangular—but it could just as well be in a sine-wave, since a linear time relationship is not required in the X-Y display mode for a linear Y vs. X presentation.

- The external storage register shown in Figure 10.20 can be eliminated (at the expense of minor crossplot display degradation) if the conversion rate is reduced so that the time between conversions is large, compared to the conversion period.

- For the configuration of Figure 10.20, typical dither frequencies of 4 to 40 Hz, and conversion clock-frequencies from 10 kHz to 100 kHz have been found useful for crossplot analysis of-high speed successive-approximation ADC performance.

Static Errors

Figure 10.21d illustrates the waveform that would appear at the major carry code transition (011 . . . 1 to 100 . . . 0) of a successive-approximation ADC if the MSB of the internal DAC had + ¼ LSB INL and the summation error of all the lower bits was − ¼ LSB. Notice that the transition representing the code (100 . . . 0) is shifted ¼ LSB to the right, and the transition resulting from combinations of the lower bits is shifted ¼ LSB to the left. The result is that the major carry code is too wide by ½ LSB.

Similarly, in Figure 10.21e, the plot is representative of an ADC where bit two of the internal DAC has − ¼ LSB INL and all the other bits sum to + ¼ LSB INL. The result is a narrow code produced by the transitions on both sides shifting towards one another. Figure 10.21f shows a typical missed-codes signature.

Dynamic Errors

In addition to the errors due to static relationships within the converter, dynamic errors may arise—including missed codes—if the frequency of the clock that times the bit decisions is too high to permit correct decisions to be made. The effects of these errors will also appear in a crossplot. If the ADC's internal timing is controlled by an external clock, dynamic errors may be found by first checking the static errors, then increasing the clock frequency until it affects the crossplot. Experienced converter designers and test engineers can gain considerable information about the internal static and dynamic behavior of the ADC from the nature of the errors revealed by the crossplot.

10.4.6 ATE METHODS

Crossplot methods of testing ADCs work well for engineering analysis or low-volume production testing, but they are slow and tend to be specialized to particular device types. There are many differences among ADCs, and the job of testing devices in outgoing or incoming inspection could become a complicated—if not hopeless—task, even on a sampling basis, without automatic test equipment (ATE). A truly universal ADC test setup should be able to test any of the various types of converter with minimal hardware changes. Here are some of the many differences:

• *Input ranges*: 0 to 10V, −5 to +5V, −10 to +10V, and −2 to +2V are typical. A single ADC may be capable of a combination of input ranges.

• *Output codes* may be binary, offset binary, ones complement, twos complement, sign-magnitude, seven-segment decoded format, or a complement of any of these. What's more, an ADC can have more than one output format (for example, binary is converted into two's complement simply by inverting the MSB).

• *Output modes*: An ADC's output also may be read in different modes. A device may have a serial (non-return-to-zero) data output or a (12-bit) byte/nibble format for 8-bit-microprocessor-compatible output; and it may be necessary to check all digital input and output lines for meeting loading and timing specifications.

A variety of ATE techniques have evolved; the best method to use generally depends on the ultimate end-use of the ADC and the practical considerations of implementing the technique. For example, statistical techniques have the advantage of not requiring accurate analog measurements. Digital signal processing techniques, or the analysis of reconstructed waveforms, may be relevant to the end-use in radar, video, or audio applications.

The technique in most widespread use, evolved to handle converters for the large medium-speed data acquisition, instrumentation, and control markets, is a digital feedback approach, in which the input of the ADC is driven to selected output code transition voltages in a fashion similar to the crossplot scheme.

Digital Feedback Loop

A block diagram of a simple digital feedback loop that can be used on a tester with relatively slow digital I/O and high-accuracy analog measurement capability is shown in Figure 10.22. It consists of the converter under test, a decision maker, which compares the ADC's digital output with the programmed digital setpoint, and an integrator with switched polarity and controllable gain.

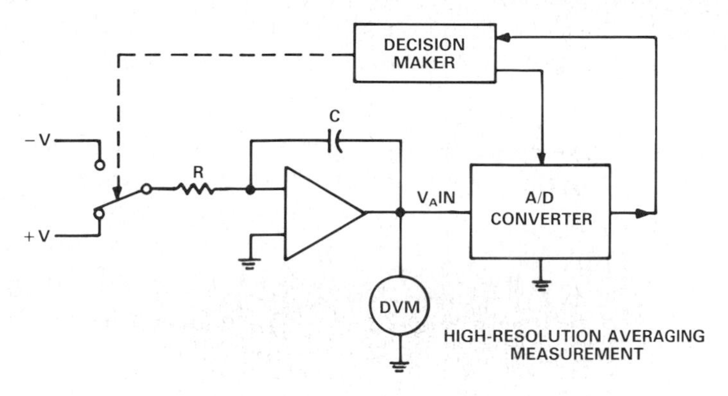

Figure 10.22. Feedback loop to measure transition voltage.

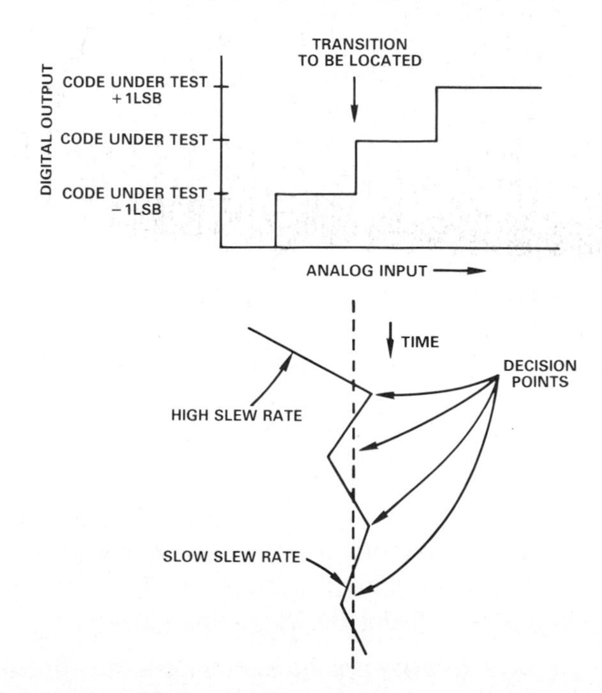

Figure 10.23. The integrator homes in on an average analog output voltage that corresponds to the desired code.

The analog input is made to lock onto a transition voltage by controlling the polarity of a ramping integrator output (Figure 10.23). In a typical sequence, suppose the digital code is below the transition being tested. The decision maker switches the integrator to the negative input, causing the integrator output to increase in the positive direction. On each successive conversion, a new decision is made; when the code is too high, the integrator is switched to ramp in the reverse direction; the integrator speed is also reduced, in order to increase the sensitivity of the measurement.

The decision-maker could be a simple digital comparator. It controls the ramp rates and polarities so the analog input voltage eventually dithers around the transition being sought. A high-resolution dc voltmeter reads the average value of the integrator output to determine the mean transition voltage.

Besides simplicity, this approach offers the advantage that the input voltage has infinite resolution because it is analog. Also, it integrates out any noise in the transition by naturally seeking the 50% point, if the positive and negative ramp rates are tightly matched. But there are drawbacks: besides requiring equal ramp rates, the analog measurement must have high resolution and accuracy (16 bits); and the low-level triangular wave must be averaged in the measurement of the DUT input level. Such measurements are not fast enough for many uses, even with modern measurement systems.

A further improvement to this technique combines the infinite resolution of analog dithering with the high accuracy of a 16-bit reference DAC, allowing the integrator to be used for a low-resolution difference measurement, instead of controlling system accuracy (Figure 10.24). The reference DAC sets the desired transition voltage, and the integrator homes in on the difference be-

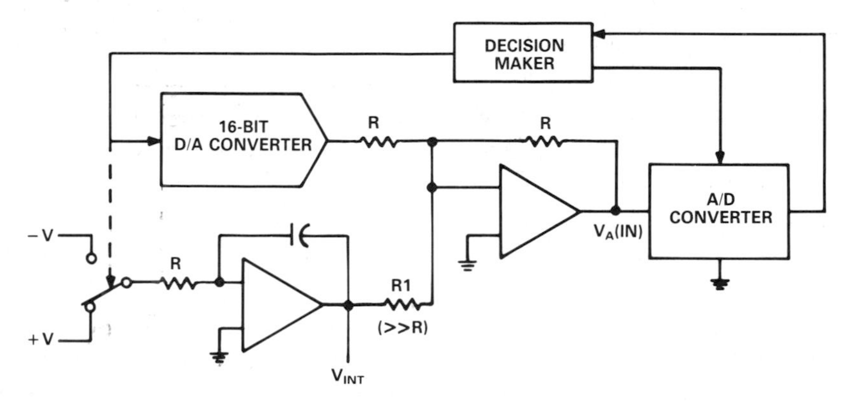

Figure 10.24. Error loop with a 16-bit reference DAC to provide an analog set point to correspond to the digital set point. The integrator reads out the ADC's transition error.

tween the desired and actual transition voltages. The difference is measured by a digital voltmeter (requiring less resolution than that in 10.22). The constraints on—and limitations of—the integrator still apply, however, and limit the tester's speed and flexibility.

These limitations disappear if the integrator is replaced by a 12-bit "dither" DAC, to form a digital tracking loop (Figure 10.25). This approach offers many advantages; the most important is that the dither DAC provides resolution comparable to using an analog integrator but with much greater speed and flexibility. In addition, the analog input presented to the DUT is completely defined digitally; no analog measurement is needed.

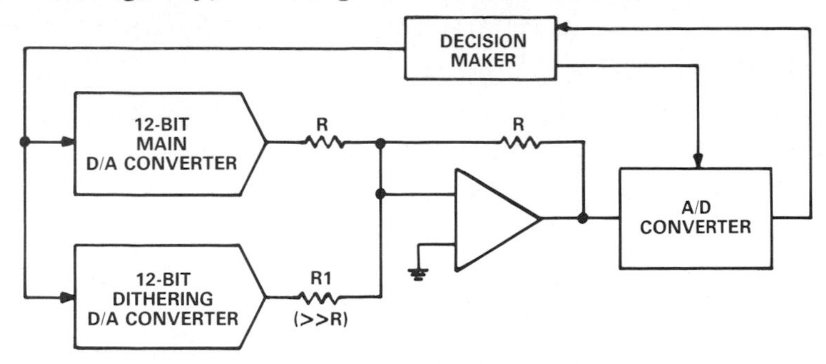

Figure 10.25. Set-point DAC and dither DAC combine in digital control loop to find ADC transition error.

The output of the 12-bit dither DAC sums with the main DAC's output to produce the ADC's input voltage. If $R_1 = 100\,R$ (as in this example), the LSB of the dither converter has an effective resolution of 0.01 LSB (of 12 bits), which is more than adequate for 12-bit testing. Actually, R1 may be further reduced to optimize the resolution for different applications. At maximum attenuation ($R_1 = 500$ R), the effective resolution of the dithering DAC becomes 10 microvolts. Since 10 μV corresponds to 0.06 LSB on a 16-bit 10-volt-span ADC, converters having greater than 12-bit resolution can be tested for all parameters, with INL limited only by the accuracy of the main DAC.

When this scheme is used in a computer-based automatic test system, the decision maker can be more "intelligent" than just a digital comparator, and the function could, in principle, be taken over by the test-system CPU. However, the CPU is generally programmed in some high level language, such as BASIC, which would slow down the execution of linearity routines or the search for all 4,095 code transitions of a 12-bit ADC. Instead, if the processing is distributed— with a slave microprocessor as the decision maker— many further advantages in speed and flexibility will accrue. For example, the slave processor, programmed in machine language, would determine all transition locations, while the CPU need only tell the slave how many approximations to use to create an average digital value for the transition of the "code under test" (CUT).

Figure 10.26 is the functional block diagram of a family test board dedicated to ADC testing for use in the Analog Devices LTS-2000-series of general-purpose benchtop component testers. Its operation is controlled by a slave processor, which receives instructions downloaded from the mainframe's central-processing unit.

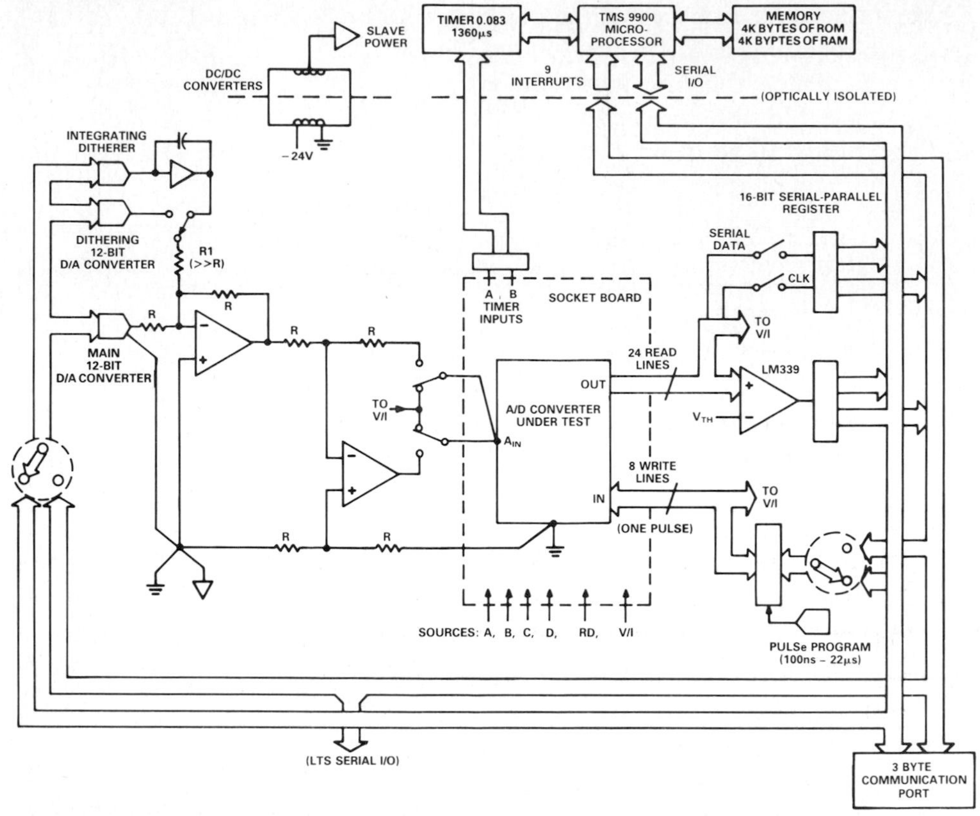

Figure 10.26. A/D Converter family test board for an LTS-2000-series benchtop automatic test system.

The slave processor, therefore, offers the advantage of raw processing speed. Because it is dedicated to testing, test time is independent of the system overhead and hardware settling time. A distributed-processing system also maximizes flexibility for testing the linearity of ADC's. For example, the slave processor may be told to report to the CPU values for all codes tested, or only for codes that fail, or only for the two codes having the greatest positive and negative linearity errors. Bit values in the variable responsible for code reporting can control whether linearity is to be tested at the transitions or at the center of the codes.

With the slave processor, a 12-bit, 5-µs ADC can be tested for linearity of all codes (reporting only worst-case codes) in less than 15 seconds to a sigma (rms deviation) repeatability of 0.04 LSB.

The slave processor also has the ability to map a transition. A digital value may be entered and the slave engaged to perform a great many conversions and report the percentage of conversions into the code under test for voltages in the vicinity of the transition. (See Figures 10.27 and 10.28) The input may be stepped incrementally by the programmer and the resulting values plotted on a CRT to present the transition graphically as a function of voltage and percentage conversion into the code being tested for voltages in the vicinity of the transition.

The circuit also contains an integrating dither that may be used to create a crossplot of the DUT (much like the crossplot circuit of Figure 10.20). The slave uses the integrator to sweep across a selected number of codes around a selected code. The two LSB's of the DUT are decoded to create four levels at the DAC dither. The outputs of the DAC dither and the integrating dither are connected to an oscilloscope's Y and X inputs, respectively, to produce the crossplot.

No-missing-codes is the most important requirement for most ADC applications. Until recently, the manufacturer's ability to guarantee the presence of all codes by actual 100% test has been hampered by the test time required to find all codes. A statistical technique sometimes used is called a ramp-input, fill-the-buckets method.[4] In this scheme, a free-running, large-signal, low-frequency waveform is fed into the DUT. The CPU keeps track of the number of times each binary code appears in the output. This technique, while faster than the crossplot or analog integrator methods, cannot operate at the speed of the distributed-processor ADC test circuit.

By virtue of the slave processor and the dither DAC, the existence of all codes of the DUT can be determined very quickly: on the order of 2 seconds for a 12-bit ADC. This routine works by a controlled search for the codes. It starts by making an intelligent guess as to the location of the code. If the code is not found, the processor either increments or decrements the dither DAC by counting or by a successive-approximation routine, depending on the output code just found. It will seek a code down to a resolution of 0.03 LSB before the code is considered missing. Once a code is found, the main DAC's output is increased by one LSB of the device under test, and the process is repeated for the new code.

10.5 PRACTICAL CONSIDERATIONS

Now that the basics of ADC testing have been discussed, it's time to consider some of the practical problems encountered when trying to test an ADC. These problems usually fall into one of four categories: nonlinearities in signal-conditioning circuits, grounding and decoupling errors, inattention to dynamics, and incorrect definition of parameters.

10.5.1 TRANSITION UNCERTAINTY

In the preceding text, except for an occasional mention of noise, a transition has been treated as if it were sharp, occurring at a unique voltage. In fact, transitions appear wide or blurry on a crossplot display, due to noise and/or finite internal comparator gain.

[4] D. Philip Burton, "Checking A/D Converter Linearity," *Analog Dialogue* 13-2 (1979): 10.

The effects of noise (occurring in either the signal or the converter, or picked up in the wiring) are to introduce an uncertainty in the precise determination of the analog input values at which the output code transitions take place, and to, in effect, increase or reduce the quantization band. The nature of the effects of quantization and noise uncertainty errors is shown in Figure 10.27.

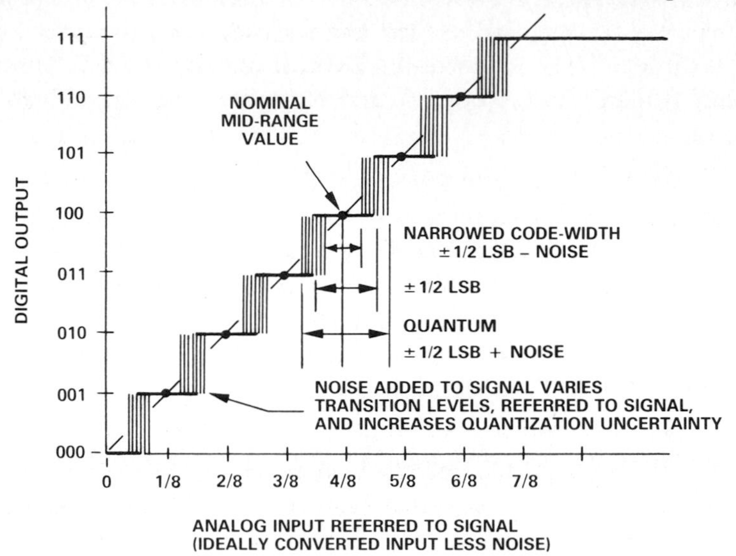

Figure 10.27. Quantization and noise uncertainty error.

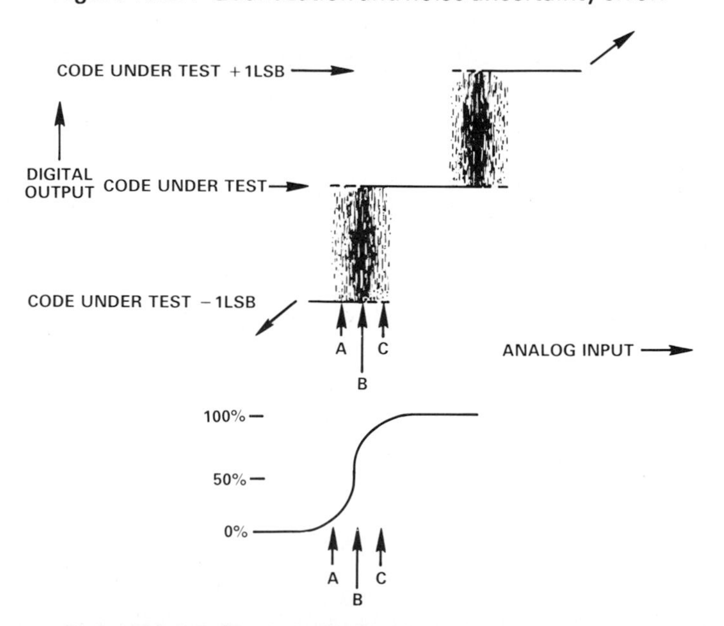

Figure 10.28. Statistical definition of code transition.

The model of a transition can realistically be drawn as a probability function (Figure 10.28). Starting (say) near code center, all (or nearly all) conversions are to the nominal code. Then, as the input voltage is slowly increased toward and through a transition, the ADC converts more and more frequently into the next code. The most-common definition for the location of the transition is that input voltage which results in 50% probability of conversion into each of the two adjacent codes.

Taking this into account, servo-loop and test hardware—and software—must be designed to insure repeatable readings consistent with the definition. For example, if the ramp size of Figure 10.23 is small enough—and given enough time—the analog integrator output will settle to the voltage at which the average current in the integrating capacitor is zero. This will be the 50% transition voltage if the input current sources are exactly equal.

In digital dither loops, more flexibility is available in choosing the search algorithm. It has been found more efficient to mathematically average the results of multiple successive approximations, rather than imitate the analog integrator by simulating a linear triangle wave.

10.5.2 HARDWARE CONSTRAINTS

It is inherently harder to design and build a successful test circuit for an ADC than for a DAC because of the dynamic nature of the ADC. In testing a DAC, the analog measurement can be delayed until the DAC and all the analog support circuitry have settled to their final states. An ADC, on the other hand, converts at its own speed without regard for the condition of the signal or the support circuitry. This makes the need for proper grounding and power-supply decoupling critical and places severe demands on the analog input buffer.

Grounding

It is good practice to establish one point as the analog reference point. All analog supply decoupling should connect to this single point via separate PC board foils or wires. When testing ADC's, every attempt should be made to arrange circuit topology in such a way that the analog ground reference point and the analog ground connection of the ADC are located as close to the ADC as possible (Figure 10.29).

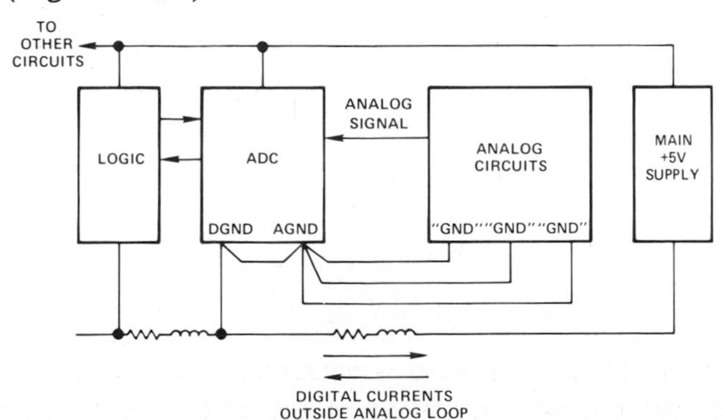

Figure 10.29. Proper grounding arrangement.

Input Buffer

The most common problem encountered is the inability of the buffer amplifier to reproduce the outputs of the reference DAC and the dither circuit at the input of the ADC. The amplifier must have the rare combination of excellent dc precision and fast dynamic settling.

The input impedance of many ADCs is low and—worse yet—changes abruptly during the conversion process. For example, in successive-approximation converters, the input current is compared to the internal DAC's output current. The comparison point (summing junction of the comparator) is diode-clamped, but it may swing plus and minus several hundred millivolts. This gives rise to a modulation of the input current (Figure 10.30).

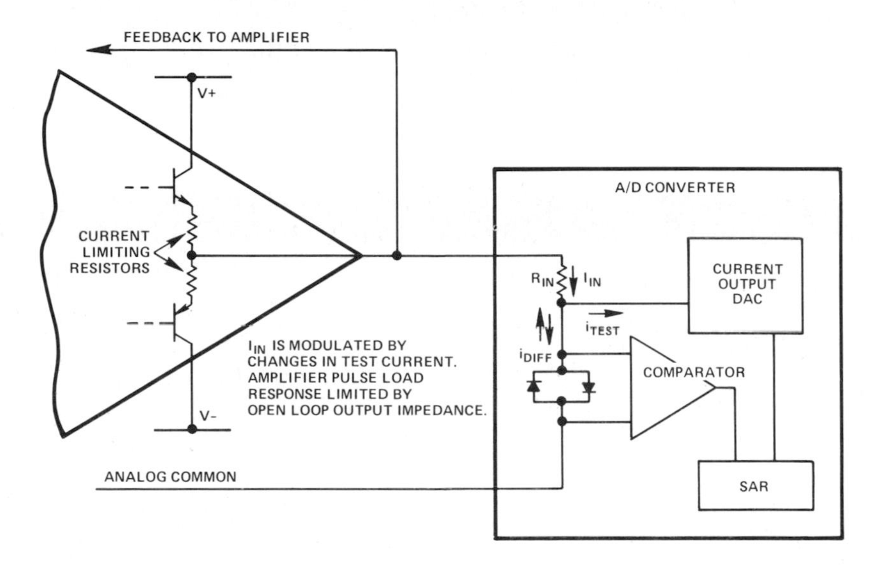

Figure 10.30. The problem of driving a successive-approximation ADC; transient load currents produce changes in output voltage.

For example, an amplifier supplying 2.5 volts to an ADC in the 10-V unipolar range will experience a 32% increase in load current when the ADC tries Bit 2. The amplifier will have to re-establish 2.5-V output—to well within the LSB—before the internal comparator is strobed, or an error in the output code will result. This requires an output impedance of under 10 ohms at all frequencies up to 2 MHz for a modest-speed (12-μs) 12-bit ADC with a 5kΩ input resistor.

It is well known (and expected) that the output impedance of a feedback amplifier is reduced by its loop gain. However, at high frequencies, where the loop gain is low and phase shift affects feedback polarity, the amplifier output impedance rises from its dc value to approach or exceed its open-loop value. In the case of most IC amplifiers, the open loop output impedance is typically 100 to 200 ohms because of the current-limiting resistors. Even a few hundred microamperes, reflected from the change in the load presented by the converter, can introduce errors in the instantaneous input voltage. If the bandwidth of the amplifier is sufficient to handle the ADC's conversion speed, the output will return to the nominal voltage before the converter makes its comparison, so that little or no error is introduced.

However, many precision amplifiers have relatively narrow bandwidth. This means that they recover very slowly from output transients. Naturally, precision amplifiers are more likely to be used in testing high-resolution ADCs, where small dc errors are less tolerable. As a result, fast, high-resolution test systems may suffer from amplifier output-transient errors. A wideband IC amplifier, such as the AD509, which does not include output current-limit resistors, is often sufficient, as is the more accurate, slightly slower AD OP-27. A hybrid or discrete unity-gain buffer amplifier may be added inside the amplifier loop (Figure 10.31) to improve the load transient response. High-linearity sample-hold amplifiers, such as the AD585, also are designed to have low output impedance and fast loop response.

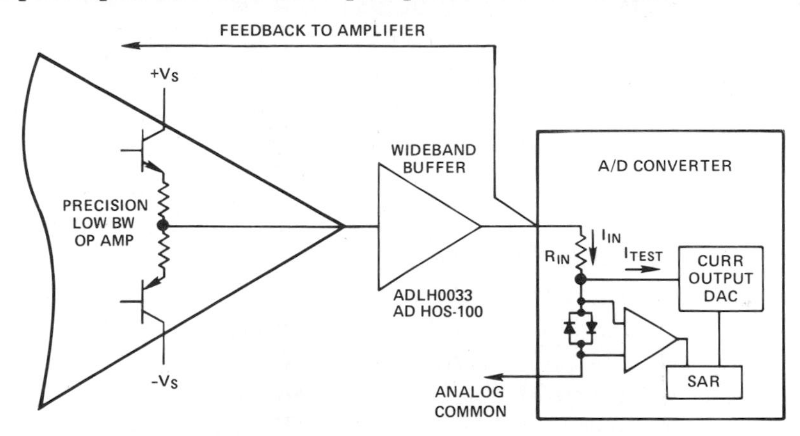

Figure 10.31. Wideband buffer inside-the-loop will drive the ADC.

Another dynamic problem often arises when testing ADCs that have multiple input ranges. Depending on the architecture of the inputs of the ADC, there are different ways of connecting unused input pins. Some ADCs require the unused pins to be grounded. This is typically to eliminate noise from being introduced into the summing junction through the unused input resistors. Other ADCs require that the unused input pins be left unconnected. This is typical because the input resistor is still part of the actual analog input. Any parasitic loading on this pin can result in unpredictable device performance. In *either* case, inattention to these details can result in inaccurate conversions or even missing codes, rendering testing basically useless.

10.5.3 TESTING UNDER DYNAMIC CONDITIONS

The bench version of the crossplot tester, and its ATE counterpart—the digital feedback circuit—test the ADC with essentially a dc signal at its input.

However, static testing is not sufficient for devices intended to be pushed to their limits at high speed, such as video ADC's; converters can demonstrate virtually ideal digitizing transfer functions under dc or low-frequency ac stimulation, but under dynamic stress, i.e., ac test signals approaching the Nyquist sampling-rate limit, $f_s/2$, they often exhibit otherwise hidden shortcomings, such as missing codes, non-monotonic conditions or linearity errors.[5]

Therefore, some form of dynamic testing is required, albeit it employs standards and procedures differing from those of industry-accepted static testing for lower speeds. Users tend to focus only on those ac specifications that highlight performance in a particular video-converter application. For example, *differential gain and phase* are paramount in digital-video uses, and *signal-to-noise ratio* dominates evaluation of devices aimed at data-communication tasks. More on this subject will be found in Chapter 13.

Analog Waveform Reconstruction

Traditionally, dynamic testing took the form of a waveform reconstruction technique, which connects a high-performance d/a converter to the video a/d converter under test and permits dynamic performance measurements with conventional analog test equipment. To obtain valid test results, however, the DAC's static and dynamic performance must exceed that of the a/d converter by two bits or more.

Fortunately, because many DACs achieve such higher performance, back-to-back a/d-to-d/a testing proves successful (Figure 10.32). A deglitcher connected to the DAC's output removes unwanted harmonics caused by DAC glitches or analog-output discontinuities. Essentially a track/hold amplifier, the deglitcher switches to its *hold* mode immediately before the DAC gets updated.

Probably the most powerful indicator of an A/D converter's dynamic performance is its signal-to-noise ratio (SNR). When stimulated by a spectrally

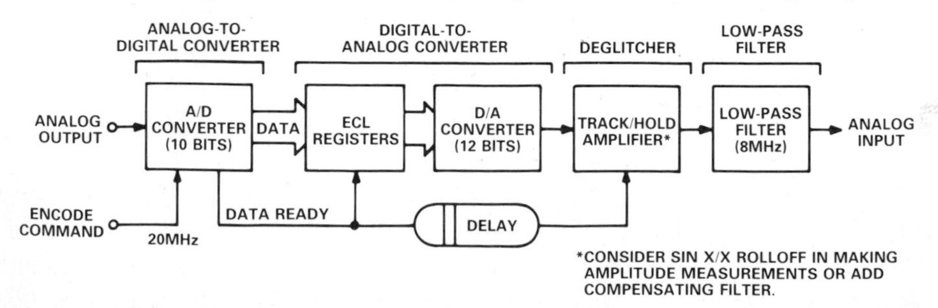

Figure 10.32. Testing an a/d converter by the use of a higher-performance back-to-back DAC.

[5]Walter Kester, "Test Video A/D Converters Under Dynamic Conditions," EDN, August 18, 1982.

pure sine wave, an ideal a/d converter generates $q/\sqrt{12}$ rms quantizing noise in an $f_s/2$ bandwidth, where q equals the weight of the least-significant bit (LSB). This rms noise level is independent of the input sine wave's level and frequency as long as the level lies within the a/d converter's operating range.

Theoretically, the logarithmic expression of the ratio of a full-scale sine wave's rms level to an ideal N-bit a/d converter's rms quantizing noise (measured over an $f_s/2$ bandwidth) equals

$$SNR = 6.02\,N + 1.8\,dB \tag{10.19}$$

Practically, though, an a/d converter's noise floor increases with the full-scale sine wave's increasing input frequency; the corresponding SNR thus decreases. Conversely, holding the input frequency constant but reducing the sine-wave amplitude decreases the a/d converter's noise floor.

A/D-converter SNR test results can be displayed and analyzed in several ways. One way is to plot SNR versus sine-wave frequency for a constant-amplitude full-scale sine-wave input. SNR can also be plotted versus input-signal amplitude for a fixed sine-wave frequency. The two plots can be combined (Figure 10.33) to form a set of curves that permit the number of effective bits of resolution for various amplitudes and frequencies to be determined and compared with the theoretical values.

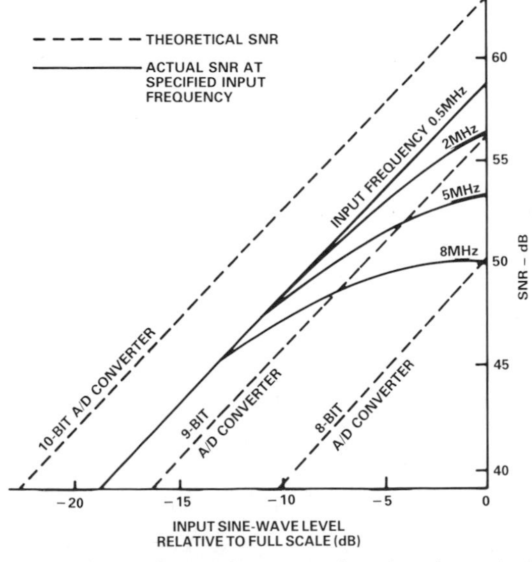

Figure 10.33. Signal-to-noise ratio vs. sine-wave level and resolution of an ADC. These plotted curves help analyze a video a/d converter's theoretical and actual SNR test results.

Note that the SNR can be measured for input signals greater than the Nyquist limit, $f_s/2$. However, it is important to be aware that the fundamental sine wave then appears as an alias component within the $f_s/2$ bandwidth. For example, a 12-MHz sine wave's in-band component when sampled at 20 MHz occurs at 8MHz.

Digital Waveform Analysis

In modern ATE, digital processing is cheaper to provide than a precision dynamic AC measurement capability. In these systems, the digital output stream of the ADC under test is sent directly to a high speed buffer memory for later analysis by a computer or array processor. The computer may simply create a histogram of the output codes, which can be analyzed for gain, offset, missing codes and linearity errors. DSP techniques, such as the fast Fourier transform, may be employed to do spectral analysis of the digital output table to determine signal-to-noise ratio, differential phase and differential gain, in addition to the above characteristics.

Chapter Eleven

Specifying Converters

The applications for digital data-handling equipment and the products of the conversion-and data-acquisition industry have spawned a multiplicity and diversity of companies, product lines, and products. We find it sobering,* though not a little gratifying, to discover that (as a major manufacturer, with a reasonably complete line of monolithic, hybrid, and modular products) we can deliver hundreds of distinct converter types—of which a large number are in the "recommended-for-new-designs" category, and that the line is growing substantially each year.

Thus, the very large number of converter products available in the marketplace, even from a single manufacturer, can overwhelm even the most informed engineer, when faced with the problem of selecting a device, or a group of devices, for a given application.

Interpretation of the specifications adds another dimension to the task, which is further complicated by the difficulty of finding standardized definitions of specifications that all manufacturers can agree upon.

To remedy this situation, and attempt to make the system designer's job of finding the "right" converter a little easier,† this chapter lists some of the elements of the decision—and steps a user can take to help "home in" on a near-optimum selection. In this chapter are also summarized interpretations of the

*It's even more disconcerting to realize that this paragraph appeared in the first edition of this book in 1972, when the Analog Devices product line was much smaller, with the words, ". . . we can deliver some 250 distinct D/A[!] converter types, . . . growing by 75 types per year."

†It's possible that some of the points raised here, if previously unanticipated by the reader, may actually make the initial selection more involved, with the benefit that problems will be fewer at a later (and more expensive) stage.

specifications, consistent (it is to be hoped), not only with the previous three chapters and with engineering practice at Analog Devices, but also with interpretations that may become accepted as standard within the industry.

For the convenience of the engineer who may seek orientation to some of the categories of devices available off-the-shelf, Tables 1 and 2 are abbreviated selection charts containing a sampling of popular general-purpose integrated-circuit converter-product families manufactured by Analog Devices, Inc., in 1984. They are based on the a/d and d/a converter sections of the 1984 Analog Devices *Databook*, which contains comprehensive selection guides, complete specifications, descriptions, and applications information for converters and a wide variety of converter-related products. The latest edition available should be consulted when specific designs are being considered. The reader is invited to request a copy from the manufacturer's home or local sales offices. The conscientious engineer will also seek comparable data from other manufacturers.

Finally, a brief example of a data-acquisition design process is given, based on the suggestions in this chapter.

11.1 TWO BASIC FACTORS

The two key factors in choosing the right device are:

Completely define the design objectives. Consider all known objectives and try to anticipate the unknowns that will pop up later. Include such factors as signal and noise levels, required accuracy, throughput rate, characteristics of the signal and control interfaces, environmental conditions and space factors, and anticipate budgetary limitations that may force performance compromises or a different system approach.

Understand what the specs mean. It is essential to have a firm understanding of what the manufacturer means by his set of specifications. It should not be assumed (in 1985) that any two manufacturers mean the same thing when they publish identical numbers defining a given parameter. In most cases, the manufacturer has honestly attempted to provide accurate information about his product. This information must be interpreted, however, in terms meaningful to the user's requirements, which requires a knowledge of how the terms are defined. Two examples that give an insight into how differences arise are included and discussed at length in the Specifications section: *linearity* and *temperature coefficient*.

11.2 DEFINING THE OBJECTIVES—APPLICATION CHECKLISTS

General Considerations

A. Accurate description of input and output
   1. Analog signal range; source or load impedance

2. Digital code needed: Binary, twos complement, BCD, etc.
3. Logic-level compatibility: TTL, CMOS, etc., logic polarity (unless otherwise noted, logic levels mentioned in Analog Devices publications are standard TTL, positive-true)

B. Data throughput rate
C. Control and data-interface details or constraints
D. What does the system error budget allow for each block?
E. What are the environmental conditions: temperature range, supply voltage, re-calibration interval, etc., over which the converter should operate to the desired accuracy?
F. Are there any special environmental conditions that must be coped with? EMC, high humidity, shock and vibration, and cramped space are a few.
G. What are the bounds of integration for the purchased portion of the system? Turnkey system, real-time interface, data-acquisition subsystem, subassemblies, components? What are the hardware/software, analog/digital tradeoffs?

In addition to the above general considerations, there are specific items to consider when choosing each block in a system.

Considerations for D/A Converters

A. What resolution is needed? How many bits (e.g., 8, 10, 12, etc.) of the incoming data word must be converted? To what degree of accuracy, linearity, etc.?
B. What logic levels and codes can be provided to the DAC? (The most popular logic system is TTL, and the most-frequently used codes are binary, twos complement, offset binary (twos complement with a complemented MSB), as outputs of systems, and BCD, usually derived from digital voltmeters or thumbwheel switches.) Is digital input serial or parallel?
C. What kind of output signal is needed: a current or a voltage? What is the desired full-scale range? (Most DACs are available with either current output—at very high speed—or voltage output, with the added delay of an internal operational amplifier. Voltage-output DAC's are the more convenient to use but—unless designed specifically for high speed—will serve only in applications not calling for submicrosecond settling times. Current-output DAC's are used in applications where high speed is more essential than stiff voltage output, such as circuits with comparators (e.g., A/D converters), or where fast amplification is to be provided externally (e.g., via CRT deflection amplifiers).
D. What kind of reference is needed, fixed (internal or external) or variable (multiplying DAC)? For multiplying DACs, how many quadrants are needed, and how arranged (1-quadrant, 4-quadrant, 2-quadrant digital, 2-quadrant analog)?

E. What is the nature of the digital interface? What are the speed, requirements? What is likely to be the shortest time between data changes? After a change in the digital input data, how long can the system wait for the output signal of the DAC to settle to the desired accuracy for a full-scale change? For a 1-bit change at the major carry? Are switching transients of any consequence? Can they be filtered? Must they be suppressed (i.e., deglitched) within the DAC? What is the analog signal feedthrough requirement for multiplying DAC's at low frequencies? At high frequency?
F. Over how wide a temperature range (at the device, including its internal temperature rise) must the converter operate? Over how much of this range must the converter perform essentially within its specifications with readjustment? What degradation of specifications is permitted (gain vs. linearity, etc.)?
G. How stable are the terminal voltages of the power supplies that will power the DAC? How stable should they be? Is the power-supply sensitivity specification adequate to hold errors from this source within reasonable limits? Are there constraints on converter dissipation?

Though no list can be complete, the above items will be the minimum consideration in any more-complete tabulation.

Considerations for A/D Converters

The process of selecting an a/d converter is similar to that involved in the selection of d/a converters. Some of the following considerations are analogous to those for DACs, and others are unique to ADCs.

A. What is the analog input range, and to what resolution must the signal be measured.
B. What is the requirement for linearity error, relative accuracy, stability of calibration, etc.?
C. To what extent must the various sources of error be minimized as ambient temperature changes? Are missed codes tolerable under any conditions?
D. How much time is allowed for each complete conversion?
E. Is the reference to be fixed, adjustable, or variable (ratiometric measurement)?
F. How stable is the system power supply? How much error due to power-supply variation is tolerable in the conversion system? Are there constraints on converter dissipation?
G. What is the character of the input signal? Is it noisy, sampled, filtered, rapidly varying, slowly varying? What kind of pre-processing is to be (or can be) done that will affect the choice (and cost) of the converter? What conversion circuit philosophies are acceptable for—or indicated by—the application? (e.g., successive-approximation, dual-slope inte-

gration, counter-&-comparator, etc. As a rule, integrating types are best for converting noisy input signals at relatively slow rates, while successive-approximation is best suited to converting sampled or filtered inputs at rates up to 1MHz. Counter-comparator types provide lowest cost but may be both slow and noise-susceptible; they are useful for peak followers and sample-holds that employ digital storage.)

H. What is the bus width and format of the digital interface? Parallel, byte-serial, serial? Buffered, three-state? What kind of logic? ECL, CMOS, TTL?

Considerations for Analog Multiplexers and Sample-Holds

When a sampled-data-system is to be assembled, in which one a/d converter is time-shared among many input channels by the use of a multiplexer and sample-hold, their contribution to system performance errors must be taken into account. These accessory devices are discussed elsewhere, but they are also discussed briefly in this chapter because of their relevance to the converter selection process.

Multiplexers

A. How many input channels are needed? Single-ended or differential? High-level or low-level? What dynamic range?
B. What kind of hierarchy is used, if a great many channels are involved? What is the addressing scheme?
C. How much time is needed for settling to desired accuracy when switching from one channel to another? Maximum switching rate?
D. How much ac crosstalk error between channels is allowable? At what frequencies?
E. What error is produced by the leakage current flowing through the source resistance?
F. What will be the multiplexer "transfer" error, produced by the voltage divider formed by the *on* resistance of the multiplexer and the input resistance of the sample-hold. Is the multiplexer active or passive (i.e., does it have an output amplifier, or is it simply a set of switches?)
G. Is the channel-switching rate to be fixed or flexible? Continuous or interruptible? Should it be capable of stopping on one channel for test or calibration purposes?
H. Is there danger of damage to active signal sources when the power is turned off? MOSFET multiplexers are inherently "safe" (at least in this sense), since the switches open when power is removed. JFET multiplexer switches can conduct when power is removed, making it possible to interconnect, and therefore damage, active signal sources.

Sample-Holds

A. What is the input signal range? Gain?

B. Considering the slewing rate of the signal and the multiplexer's channel-switching rate, what is the sample-hold's allowable acquisition time to within the desired error band?
C. What accuracy is needed (gain, linearity, and offset errors)?
D. What aperture delay and jitter are allowable, going into *hold* (The delay component of aperture time is considered to be correctible, since the switching operation can be advanced to compensate. The uncertainty (jitter) cannot be compensated, and a random jitter of 5ns applied to a signal slewing at, say, 1V/μs produces an uncertainty of 5mV. In sampled-data systems, operating at a constant sampling rate, with data that is not correlated to the sampling rate, delay is of no importance if fixed, but jitter modulates the sampling rate.
E. How much droop is allowable in *hold*?
F. What are the effects of time, temperature, and power supply variation?
G. What offset error is caused by the flow of the sample-hold's input bias current through the series resistance of the multiplex switch and the signal source?

11.3 DEFINING THE SPECIFICATIONS

Figures 11.1 and 11.2 depict the specifications of typical d/a and a/d converters. Though the specs probably mean "what you think they mean," it is important that their meaning and implications be spelled out. The following list, in alphabetical order, should prove helpful. (Additional definitions will be found in Chapter 10 and in chapters on specific kinds of devices.)

Accuracy, Absolute. Absolute accuracy error of a *d/a converter* is the difference between actual analog output and the output that is expected when a given digital code is applied to the converter. Error is usually commensurate with resolution, i.e., less than $2^{-(n+1)}$, or "½ LSB" of full scale. However, accuracy may be much better than resolution in some applications; for example, a 4-bit reference supply having only 16 discrete digitally chosen levels would have a resolution of 1/16, but it might have an accuracy to within 0.01% of each ideal value.

Absolute accuracy error of an *a/d converter* at a given output code is the difference between the actual and the theoretical analog input voltages required to produce that code. Since the code can be produced by any analog voltage in a finite band (see *Quantizing Uncertainty*, and also Figure 7.2), the "input required to produce that code" is defined as the midpoint of the band of inputs that will produce that code. For example, if 5 volts, ±1.2 mV, will theoretically produce a 12-bit half-scale code of 1000 0000 0000, then a converter for which any voltage from 4.997 V to 4.999 V will produce that code will have absolute error of (½)(4.997 + 4.999) − 5 volts = +2 millivolts.

Sources of error include gain (calibration) error, zero error, linearity errors, and noise. Absolute accuracy measurements should be made under a set of

SPECIFICATIONS ($T_A = +25°C$, ±12V, ±15V power supplies unless otherwise noted)

Model	AD667J Min	AD667J Typ	AD667J Max	AD667K Min	AD667K Typ	AD667K Max	AD667S Min	AD667S Typ	AD667S Max	Units
DIGITAL INPUTS										
Resolution			12			12			12	Bits
Logic Levels (TTL Compatible, T_{min}–T_{max})[1]										
V_{IH} (Logic "1")	**+2.0**		**+5.5**	**+2.0**		**+5.5**	**+2.0**		**+5.5**	V
V_{IL} (Logic "0")	**0**		**+0.8**	**0**		**+0.8**	**0**		**+0.7**	V
I_{IH} (V_{IH} = 5.5V)		200	**300**		200	**300**		200	**300**	μA
I_{IL} (V_{IL} = 0.8V)		50	**100**		50	**100**		50	**100**	μA
TRANSFER CHARACTERISTICS										
ACCURACY										
Linearity Error @ +25°C		±1/4	**±1/2**		±1/8	**±1/4**		±1/8	**±1/2**	LSB
$T_A = T_{min}$ to T_{max}		±1/2	**±3/4**		±1/4	**±1/2**		±1/2	**±3/4**	LSB
Differential Linearity Error @ +25°C		±1/2	**±3/4**		±1/4	**±1/2**		±1/4	**±3/4**	LSB
$T_A = T_{min}$ to T_{max}	**Monotonicity Guaranteed**			**Monotonicity Guaranteed**			**Monotonicity Guaranteed**			LSB
Gain Error[2]		±0.1	**±0.2**		±0.1	**±0.2**		±0.1	**±0.2**	%FSR[3]
Unipolar Offset Error[2]		±1	**±2**		±1	**±2**		±1	**±2**	LSB
Bipolar Zero[2]		±0.05	**±0.1**		±0.05	**±0.1**		±0.05	**±0.1**	% of FSR
DRIFT										
Differential Linearity		±2			±2			±2		ppm of FSR/°C
Gain (Full Scale) T_A = 25°C to T_{min} or T_{max}		±5	±30		±5	±15		±15	**±30**	ppm of FSR/°C
Unipolar Offset T_A = 25°C to T_{min} or T_{max}		±1	±3			±3			**±3**	ppm of FSR/°C
Bipolar Zero T_A = 25°C to T_{min} or T_{max}		±5	±10			±10			**±10**	ppm of FSR/°C
CONVERSION SPEED										
Settling Time to ±0.01% of FSR for										
FSR Change (2kΩ‖500pF load)										
with 10kΩ Feedback		3	4		3	4		3	4	μs
with 5kΩ Feedback		2	3		2	3		2	3	μs
For LSB Change		1			1			1		μs
Slew Rate	10			10			10			V/μs
ANALOG OUTPUT										
Ranges[4]		±2.5, ±5, ±10, +5, +10			±2.5, ±5, ±10, +5, +10			±2.5, ±5, ±10, +5, +10		V
Output Current	±5			±5			±5			mA
Output Impedance (dc)		0.05			0.05			0.05		Ω
Short Circuit Current			40			40			40	mA
REFERENCE OUTPUT	**9.90**	10.00	**10.10**	**9.90**	10.00	**10.10**	**9.90**	10.00	**10.10**	V
External Current	0.1	1.0		0.1	1.0		0.1	1.0		mA
POWER SUPPLY SENSITIVITY										
V_{CC} = +11.4 to +16.5V dc		15	**25**		15	**25**		15	**25**	ppm of FS/%
V_{EE} = −11.4 to −16.5V dc		3	**10**		3	**10**		3	**10**	ppm of FS/%
POWER SUPPLY REQUIREMENTS										
Rated Voltages		±12, ±15			±12, ±15			±12, ±15		V
Range[4]	**±11.4**		**±16.5**	**±11.4**		**±16.5**	**±11.4**		**±16.5**	V
Supply Drain										
+11.4 to +16.5V dc		5	**8.5**		5	**8.5**		5	**8.5**	mA
−11.4 to −16.5V dc		18	**25**		18	**25**		18	**25**	mA
TEMPERATURE RANGE										
Specification	0		+70	0		+70	−55		+125	°C
Operating	−55		+125	−55		+125	−55		+125	°C
Storage	−65		+125	−65		+125	−65		+150	°C

NOTES
[1]The digital input specifications are 100% tested at +25°C, and guaranteed but not tested over the full temperature range.
[2]Adjustable to zero.
[3]FSR means "Full Scale Range" and is 20V for ±10V range and 10V for the ±5V range.
[4]A minimum power supply of ±12.5V is required for a ±10V full scale output and ±11.4V is required for all other voltage ranges.

Specifications subject to change without notice.

Specifications shown in boldface are tested on all production units at final electrical test. Results from those tests are used to calculate outgoing quality levels. All min and max specifications are guaranteed, although only those shown in boldface are tested on all production units.

Figure 11.1. Typical microcircuit d/a converter specifications (AD667).

standard conditions with sources and meters traceable to an internationally accepted standard.

Accuracy, Logarithmic DACs: The difference (measured in dB) between the actual transfer function and the ideal transfer function, as measured after calibration of gain error at 0 dB.

Accuracy, Relative. Relative accuracy error, expressed in %, ppm, or fractions of 1 LSB, is the deviation of the analog value at any code (relative to

SPECIFICATIONS (typical @ +25°C with V_{CC} = +15V or +12V, V_{LOGIC} = +5V, V_{EE} = -15V or -12V, unless otherwise specified)

Model	AD574AJ Min	AD574AJ Typ	AD574AJ Max	AD574AK Min	AD574AK Typ	AD574AK Max	AD574AL Min	AD574AL Typ	AD574AL Max	Units
RESOLUTION			12			12			12	Bits
LINEARITY ERROR										
25°C (max)			±1			±1/2			±1/2	LSB
T_{min} to T_{max}			±1			±1/2			±1	LSB
DIFFERENTIAL LINEARITY ERROR (Minumum resolution for which no missing codes are guaranteed)										
25°C	11			12			12			Bits
T_{min} to T_{max}	11			12			12			Bits
UNIPOLAR OFFSET (max) (Adjustable to zero)			±2			±2			±2	LSB
BIPOLAR OFFSET (max) (Adjustable to zero)			**±10**			±4			±4	LSB
FULL SCALE CALIBRATION ERROR (with fixed 50Ω resistor from REF OUT TO REF IN)										
(Adjustable to zero) 25°C (max)			**0.25**			**0.25**			**0.25**	% of F.S.
T_{min} to T_{max} (Without Initial Adjustment)		0.47			0.37			0.30		% of F.S.
(With Initial Adjustment)		0.22			0.12			0.05		% of F.S.
TEMPERATURE RANGE	0		+70	0		+70	0		+70	°C
TEMPERATURE COEFFICIENTS (Using internal reference)										
T_{min} to T_{max}										
Unipolar Offset			±2			±1			±1	LSB
			10			5			5	ppm/°C
Bipolar Offset			±2			±1			±1	LSB
			10			5			5	ppm/°C
Full Scale Calibration			±9			±5			±2	LSB
			50			27			10	ppm/°C
POWER SUPPLY REJECTION										
Max change in Full Scale Calibration										
$+13.5 \le V_{CC} \le +16.5V$ or $+11.4V \le V_{CC} \le +12.6V$			±2			±1			±1	LSB
$+4.5 \le V_{LOGIC} \le +5.5V$			±1/2			±1/2			±1/2	LSB
$-16.5 \le V_{EE} \le -13.5V$ or $-12.6V \le V_{EE} \le -11.4V$			±2			±1			±1	LSB
ANALOG INPUT										
Input Ranges										
Bipolar		−5 to +5			−5 to +5			−5 to +5		Volts
		−10 to +10			−10 to +10			−10 to +10		Volts
Unipolar		0 to +10			0 to +10			0 to +10		Volts
		0 to +20			0 to +20			0 to +20		Volts
Input Impedance										
10 Volt Span	**3**	5	7	**3**	5	7	**3**	5	7	kΩ
20 Volt Span	**6**	10	**14**	**6**	10	**14**	**6**	10	**14**	kΩ
POWER SUPPLIES										
Operating Range										
V_{LOGIC}	+4.5		+5.5	+4.5		+5.5	+4.5		+5.5	Volts
V_{CC}	+11.4		+16.5	+11.4		+16.5	+11.4		+16.5	Volts
V_{EE}	−11.4		−16.5	−11.4		−16.5	−11.4		−16.5	Volts
Operating Current										
I_{LOGIC}		30	**40**		30	**40**		30	**40**	mA
I_{CC}		2	5		2	5		2	5	mA
V_{EE}		18	**30**		18	**30**		18	**30**	mA
POWER DISSIPATION		390	725		390	725		390	725	mW
INTERNAL REFERENCE VOLTAGE	**9.9**	10.0	**10.1**	**9.9**	10.0	**10.1**	**9.9**	10.0	**10.1**	Volts
Output current (available for external loads) (External load should not change during conversion)			1.5[1]			1.5[1]			1.5[1]	mA
PACKAGE OPTION (D28A)		AD574AJD			AD574AKD			AD574ALD		

NOTES
[1]The reference should be buffered for operation on ± 12V supplies.
Specifications subject to change without notice.
Specifications shown in boldface are tested on all production units at final electrical test. Results from those tests are used to calculate outgoing quality levels. All min and max specifications are guaranteed, although only those shown in boldface are tested on all production units.

Figure 11.2. Typical microcircuit a/d converter specifications (AD574A).

the full analog range of the device transfer characteristic) from its theoretical value (relative to the same range), after the full-scale range (FSR) has been calibrated (see *Full-Scale Range*; see also Chapter 10).

Since the discrete analog values that correspond to the digital values ideally lie on a straight line, the specified worst-case relative-accuracy error of a linear

ADC or DAC can be interpreted as a measure of end-point nonlinearity (see *Linearity*).

The "discrete points" of a D/A transfer characteristic are measured by the actual analog outputs. The "discrete points" of an A/D transfer characteristic are the midpoints of the quantization bands at each code (see *Accuracy, Absolute*).

Acquisition Time. The acquisition time of a sample/track-hold circuit for a step change is the time required by the output to reach its final value, within a specified error band, after the sample (or track) command has been given. Included are switch-delay time, the slewing interval, and settling time for a specified output-voltage change.

Aperture (Delay) Time, in a sample/track-hold, is the time required after the hold command, for the switch to open fully. The sample is, in effect, delayed by this interval, and the hold command would have to be advanced by this amount for precise timing.

Aperture Uncertainty (or Jitter) is the range of variation in the *aperture time*. If the *aperture time* is tuned out by advancing the *hold* command by a suitable amount—or if the signal is being sampled repetitively—this spec establishes the ultimate timing error, hence the maximum sampling frequency for a given resolution.

Automatic Zero. To achieve zero stability in many integrating-type converters, a time interval is provided during each conversion cycle to allow the circuitry to compensate for drift errors. The drift error in such converters is substantially zero.

Bias Current is the zero-signal dc current required from the signal source by the inputs of many semiconductor circuits. The voltage developed across the source resistance by bias current constitutes an (often negligible) offset error.

When an instrumentation amplifier performs measurements of a source that is disjoint from the amplifier's power-supply, there *must* be a return path for bias currents. If it does not already exist and is not provided, those currents will charge stray capacitances, causing the output to drift uncontrollably or to saturate. Therefore, when amplifying outputs of "floating" sources, such as transformers, insulated thermocouples, and ac-coupled circuits, there must be a high-impedance dc leakage path from each input to common, or to the driven-guard terminal (if present). If a dc return path is impracticable, an *isolator* must be used.

Bipolar Offset. See *Offset*.

Channel-to-Channel Isolation, in multiple d/a converters. The proportion of analog input signal from one DAC's reference input that appears at the output of the other DAC, expressed logarithmically in dB. See also *crosstalk*.

Character-Serial BCD. Multiplexed BCD data outputs, where the 4-bit BCD code for each digit is gated in sequence onto four common output lines.

Charge Transfer (or *Offset Step*), the principal component of *sample-to-hold offset* (or *pedestal*) is the small charge transferred to the storage capacitor via interelectrode capacitance of the switch and stray capacitance when switching to the *hold* mode. The offset step is directly proportional to this charge, viz.,

$$\text{Offset error (volts)} = \frac{\Delta Q}{C} = \frac{\text{Incremental charge (picocoulombs)}}{\text{Capacitance (picofarads)}}$$

It can be reduced somewhat by lightly coupling an appropriate-polarity version of the *hold* signal to the capacitor for first-order cancellation. The error can also be reduced by increasing the capacitance, but this increases *acquisition time*.

Code Width. This is a fundamental quantity for a/d converter specifications. It is defined as the range of analog input values for which a given digital output code will occur (the stair treads in Figure 7.2). The nominal value of a code width (for all but the first and last codes) is the voltage equivalent of 1 least-significant bit (LSB) of the full-scale range, or 2.44 mV out of 10 volts for a 12-bit ADC. Noise modulates (and narrows) the effective code width. Code width should generally not be less than ½ LSB or more than 1½ LSB. Because the full-scale range is fixed, the presence of excessively wide codes implies the existence of narrow—and perhaps even missing—codes.

Common-Mode Range. Common-mode rejection usually varies with the magnitude of the range through which the input signal can swing, determined by the sum of the common-mode and the differential voltage. *Common-mode range* is that range of *total* input voltage over which specified common-mode rejection is maintained. For example, if the common-mode signal is ± 5V and the differential signal is ± 5V, the common-mode range is ± 10V.

Common-Mode Rejection (CMR) is a measure of the change in output voltage when both inputs are changed by equal amounts of ac and/or dc voltage. Common-mode rejection is usually expressed either as a ratio (e.g., CMRR = 1,000,000:1) or in decibels: CMR = $20 \log_{10}$ CMRR; if CMRR = 10^6, CMR = 120 dB. A CMRR of 10^6 means that 1 volt of common mode is processed by the device as though it were a differential signal of 1-microvolt at the input.

CMR is usually specified for a full-range common-mode voltage change (CMV), at a given frequency, and a specified imbalance of source impedance (e.g., 1 kΩ source unbalance, at 60 Hz). In amplifiers, the common-mode rejection ratio is defined as the ratio of the signal gain, G, to the common-mode gain (the ratio of common-mode signal appearing at the output to the CMV at the input.

Common-Mode Voltage (CMV). A voltage that appears in common at both input terminals of a device, with respect to its output reference (usually "ground"). For inputs, V_1 and V_2, with respect to ground, CMV = $\frac{1}{2}(V_1 + V_2)$. An ideal differential-input device would ignore CMV. *Common-mode error (CME)* is any error at the output due to the common-mode input voltage. The errors due to supply-voltage variation, an internal common-mode effect, are specified separately.

In *isolation amplifiers*, the rating, *CMV, inputs to outputs*, is the voltage that may be safely applied to *both* inputs, with respect to the outputs or power common. This is a necessary consideration in applications with high CMV input or when high voltage-transients may occur at the input.

Compliance-Voltage Range. For a current source (e.g., a current-output DAC), the maximum range of (output) terminal voltage for which the device will maintain the specified current-output characteristics.

Conversion Time and Conversion Rate. The time required for a complete measurement by an analog-to-digital converter is called *conversion time*. For most converters (assuming no significant additional systemic delays), conversion time is essentially identical with the inverse of *conversion rate*. However, in some high-speed converters, because of pipelining, new conversions are initiated before the results of prior conversions have been determined; thus, for example, the Analog Devices MOD-1205 can provide 12-bit output data at a 5-MHz word rate (200 ns/conversion), even though the time for any one conversion, from start to finish, is two clock periods plus 275 ns, or 675 ns, at 5 MHz.

In digital panel instruments, *conversion rate* is the frequency at which readings may be processed by the instrument. Specifications are typically given for internally clocked rates and maximum permissible externally triggered rates. *Conversion time* is the maximum time required for the instrument to complete a reading cycle—it is specified for the full-scale reading

Crosstalk. Leakage of signals, usually via capacitance between circuits or channels of a multi-channel system or device, such as a multiplexer, multiple op amp, or multiple DAC. Crosstalk is usually determined by the impedance parameters of the physical circuit, and actual values are frequency-dependent. See also *channel-to-channel isolation*.

Multiple d/a converters have a *digital crosstalk* specification: the spike (sometimes called glitch) impulse appearing at the output of one converter due to a change in the digital input code of another of the converters. It is specified in nanovolt-seconds and measured at $V_{REF} = 0$ V.

Deglitcher (See Glitch). A device that removes or reduces the effects of time-skew pulses in d/a conversion. A deglitcher normally employs a sample-hold circuit, often specifically designed as part of the DAC. When the DAC is updated, the deglitcher holds the output of the DAC's output amplifier constant at the previous value until the switches reach equilibrium, then samples and holds the new value.

Digital-to-Analog Spike (also Glitch) Impulse. For CMOS multiplying DACs, this is a measure of the charge injected from the digital inputs to the analog outputs when the inputs change state. It is usually specified in terms of the area of the spike in nanovolt-seconds, and is measured with V_{REF} at analog ground and a fast operational amplifier as the output amplifier.

Droop Rate. When a sample-hold circuit using a capacitor for storage is in *hold*, it will not hold the information forever. Droop rate is the rate at which the output voltage changes (by increasing or decreasing), and hence gives up information. The change of output occurs as a result of leakage or bias currents flowing through the storage capacitor. The polarity of change depends on the sources of leakage within a given device. In integrated circuits with external capacitors, it is usually specified as a (*droop* or *drift*) current, in modules or ICs having internal capacitors, a rate of change. Note: dV/dt (volts/second) = I/C (picoamperes/picofarads).

Dual-Slope Converter. An integrating A/D converter in which the unknown signal is converted to a proportional time interval, which is then measured digitally.

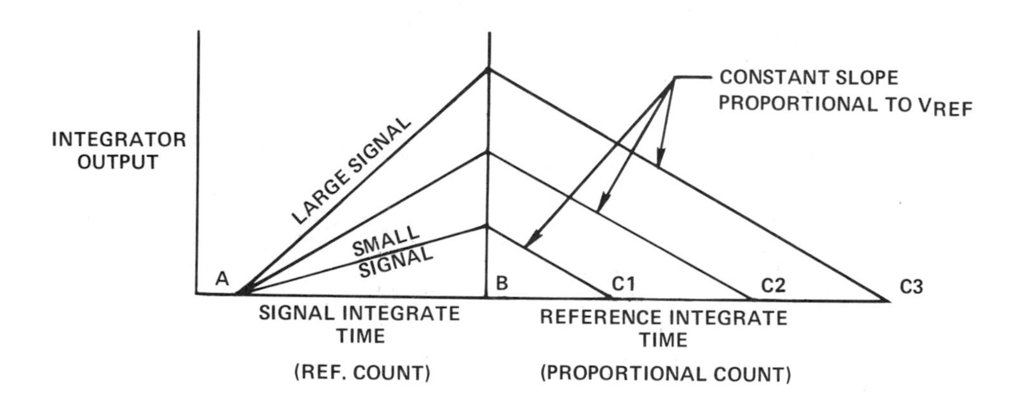

Figure 11.3. Voltage-time relationships in dual-slope conversion.

This is done by integrating the unknown for a predetermined length of time. Then a reference input is switched to the integrator, which integrates "down" from the level determined by the unknown until the starting level is reached. The time for the second integration process, as determined by the counter, is proportional to the average of the unknown signal level over the predetermined integrating period. The counter provides the digital readout.

Feedthrough. Undesirable signal-coupling around switches or other devices that are supposed to be turned off or provide isolation, e.g., *feedthrough error* in a sample-hold, multiplexer, or multiplying DAC. Feedthrough is variously specified in percent, parts per million, fractions of 1 LSB, or fractions of 1 volt, with a given set of inputs, at a specified frequency.

In a multiplying DAC, *feedthrough* error is caused by capacitive coupling from an ac V_{REF} to the output, with all switches off. In a sample/hold, *feedthrough* is the fraction of the input signal variation or ac input waveform that appears at the output in *hold*. It is caused by stray capacitive coupling from the input to the storage capacitor, principally across the open switch.

"Flash" Converter. A converter in which all the bit choices are made at the same time. It requires $2^n - 1$ voltage-divider taps and comparators—and a comparable amount of priority encoding logic. A scheme that gives extremely fast conversion, it requires large numbers of nearly identical components, hence it is well-suited to—and really only feasible in—integrated-circuit form. Flash converters are often used in pairs for two-stage conversion in *sub-ranging converters*, to provide high resolution at somewhat slower speed than pure flash conversion.

Four-Quadrant. In a multiplying DAC, "four quadrant" refers to the fact that both the reference signal and the number represented by the digital input may be of either positive or negative polarity. Such a DAC can be thought of as a gain control for ac signals ("reference" input) with a range of positive and negative digitally controlled gains. A four-quadrant multiplier is expected to obey the rules of multiplication for algebraic sign.

Frequency-to-Voltage Conversion (FVC). The input of a FVC device is an ac waveform—usually a train of pulses (in the context of conversion); the output is an analog voltage, proportional to the number of pulses occurring in a given time. FVC is usually performed by a voltage-to-frequency converter in a feedback loop. Important specifications, in addition to the accuracy specs typical of VFCs (see *Voltage-to-Frequency conversion*), include *output ripple* (for specified input frequencies), *threshold* (for recognition that another cycle has been initiated, and for versatility in interfacing several types of sensors directly), *hysteresis*, to provide a degree of insensitivity to noise superimposed on a slowly varying input waveform, and *dynamic response* (important in motor control).

Full-Scale Range (FSR). For binary ADCs and DACs, that magnitude of voltage, current, or—in a multiplying DAC—gain, of which the MSB is specified to be exactly one-half—or for which any bit or combination of bits is tested against its (their) prescribed ideal ratio(s). FSR is independent of resolution; the value of the LSB (voltage, current, or gain) is 2^{-n} FSR.

There are several other terms, with differing meanings, that are often used in the context of discussions or operations involving full-scale range. They are:

Full Scale—similar to full-scale range, but pertaining to a single polarity. Thus, full-scale for a unipolar device is twice the prescribed value of the MSB and has the same polarity. For a bipolar device, *positive or negative full scale* is that positive or negative value, of which the next bit after the polarity bit is tested to be one-half.

Span—the scalar voltage or current range corresponding to FSR.

All-1's—*All bits on*, the condition used, in conjunction with *all-zeros*, for gain adjustment of an ADC or DAC, in accordance with the manufacturer's instructions. Its magnitude, for a binary device, is $(1-2^{-n})$ FSR. *All-1's* is a *positive-true* definition of a specific magnitude relationship; for complementary coding the "all-1's" code will actually be all zeros. To avoid confusion, all-1's should never be called "full scale;" FSR and FS are independent of the number of bits, all-1's isn't.

All-0's—*All bits off*, the condition used in offset (and gain) adjustment of a DAC or ADC, according to the manufacturer's instructions. All-0's corresponds to zero output in a unipolar DAC and negative full-scale in an offset bipolar DAC with positive output reference. In a sign-magnitude device, all-0's refers to all bits after the sign bit. Analogous to "all-1's," "all-0's" is a *positive-true* definition of the *all-bits-off* condition; in a complementary-coded device, it is expressed by all ones. To avoid confusion, all-0's should not be called "zero" unless it accurately corresponds to true analog zero output from a DAC.

The best way of defining the critical points for an actual device is a brief table of critical codes and the ideal voltages, currents, or gains to which they correspond, with the conditions for measurement defined.

Gain. The "gain" of a converter is that analog scale factor setting that establishes the nominal conversion relationship, e.g., 10 volts full-scale. In a multiplying DAC or ratiometric ADC, it is indeed a gain. In a device with fixed internal reference, it is expressed as the full-scale magnitude of the output parameter (e.g., 10 volts or 2 milliamperes). In a fixed-reference converter, where the use of the internal reference is optional, the converter gain and the reference may be specified separately. Gain and zero adjustment are discussed under *zero*.

"Glitch" (see Figure 11.4). Transients associated with code changes generally stem from several sources. Some are spikes, known as digital-to-analog feedthrough, or charge transfer, coupled from the digital signal to the analog output, defined with zero reference. These spikes are generally fast, fairly uniform, code-independent, and hence filterable. However, there is a more-insidious form of transient, code-dependent, and difficult to filter, known as the "glitch."

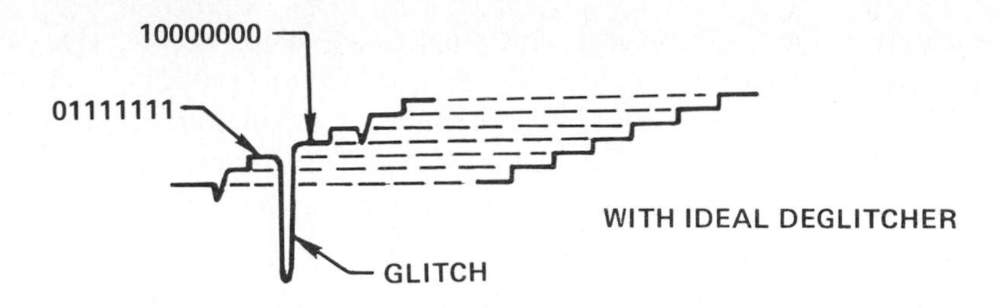

Figure 11.4. Glitch at a major carry.

If the output of a counter is applied to the input of a DAC to develop a "staircase" voltage, the number of bits involved in a code change between two adjacent codes establish "major" and "minor" transitions. The most major transition is at ½-scale, when the D/A switches all bits, i.e., from 011. . .111 to 1000. . .00. If, for digital inputs having no skew, the switches are faster to switch *off* than *on*, this means that, for a short time, the DAC will seek zero output, and then return to the required 1 LSB above the previous reading. This large transient spike is commonly known as a "glitch." The better-matched the input transitions and the switching times, the faster the switches, the smaller will be the energy contained in the glitch. Because the size of the glitch is not proportional to the signal change, linear filtering may be unsuccessful and may, in fact, make matters worse. (*See also Deglitcher*.)

The severity of a glitch is specified by *glitch impulse*, the product of its duration and its average magnitude, i.e., the net area under the curve. This product will be recognized as the physical quantity, *impulse* (electromotive *force* × Δ*time*); however, it has also been termed "glitch energy" and "glitch charge." Glitch impulse is usually expressed, for fast converters, in units of picovolt-seconds (equivalent to the more-readily visualized millivolt-nanoseconds).

The glitch can be minimized through the use of fast, non-saturating logic, such as ECL, matched latches, and non-saturating switches. Very fast DACs, such as those used in high-resolution raster-type displays, often include arrangements for trimming the glitch to take the form of a small, filterable doublet pulse, which has near-zero net area and a doubled fundamental frequency (see Section 13.2.1).

Glitch Charge, Glitch Energy, Glitch Impulse. See *Glitch*.

Leakage Current, Output. Current which appears at the output terminal of a d/a converter with all bits "off." For a converter with two complementary outputs (for example, some CMOS DACs), output leakage current is the current measured at OUT 1, with all digital inputs *low*—and the current measured at OUT 2, with all digital inputs *high*.

Least-Significant Bit (LSB). In a system in which a numerical magnitude is represented by a series of binary (i.e. two-valued) digits, the *least-significant bit* is that digit (or "bit") that carries the smallest value, or weight. For example, in the natural binary number 1101 (decimal 13, or $(1 \times 2^3) + (1 \times 2^2) + (0 \times 2^1) + (2^0)$), the rightmost digit is the LSB. Its analog weight, in relation to full scale (see *Full-Scale Range*), is 2^{-n}, where n is the number of binary digits. It represents the smallest analog change that can be resolved by an n-bit converter.

In converter nomenclature (viz., fractional binary), the LSB is bit n; in bus nomenclature (integer binary), it is Data Bit 0.

Linearity. (See also *Nonlinearity*.) Linearity error of a converter (also, *integral nonlinearity*—see *Linearity, Differential*), expressed in % or parts per million of full-scale range, or (sub)multiples of 1 LSB, is a deviation of the analog values, in a plot of the measured conversion relationship, from a straight line. The straight line can be either a "best straight line," determined empirically by manipulation of the gain and/or offset to equalize maximum positive and negative deviations of the actual transfer characteristic from this straight line; or, it can be a straight line passing through the end points of the transfer characteristic after they have been calibrated, sometimes referred to as "end-point" linearity (Figure 11.5). "End-point" nonlinearity is similar to relative accuracy error (see *Accuracy, Relative*). It provides an easier method for users to calibrate a device, and it is a more conservative way to specify linearity.

For multiplying D/A converters, the *analog* linearity error, at a specified analog gain (digital code), is defined in the same way as for analog multipliers, i.e., by deviation from a "best straight line" through the plot of the analog output-input response.

Linearity, Differential. In a d/a converter, any two adjacent digital codes should result in measured output values that are exactly 1 LSB apart (2^{-n} of full scale for an n-bit converter). Any positive or negative deviation of the measured "step" from the ideal difference is called *differential nonlinearity*, expressed in (sub)multiples of 1 LSB. It is an important specification, because a differential linearity error more negative than −1 LSB can lead to non-monotonic response in a d/a converter and missed codes in an a/d converter using that DAC.

Similarly, in an a/d converter, midpoints between code transitions should be 1 LSB apart. Differential nonlinearity is the deviation between the actual difference between midpoints and 1 LSB, for adjacent codes. If this deviation is equal to or more negative than −1 LSB, a code will be missed, as shown in Figure 7.2.

Often, instead of a maximum differential nonlinearity specification, there will be a simple specification of "monotonicity" or "no missing codes", which implies that the differential nonlinearity cannot be more negative than −1 for any adjacent pair of codes. However, the differential linearity error may still be somewhere more positive than +1 LSB.

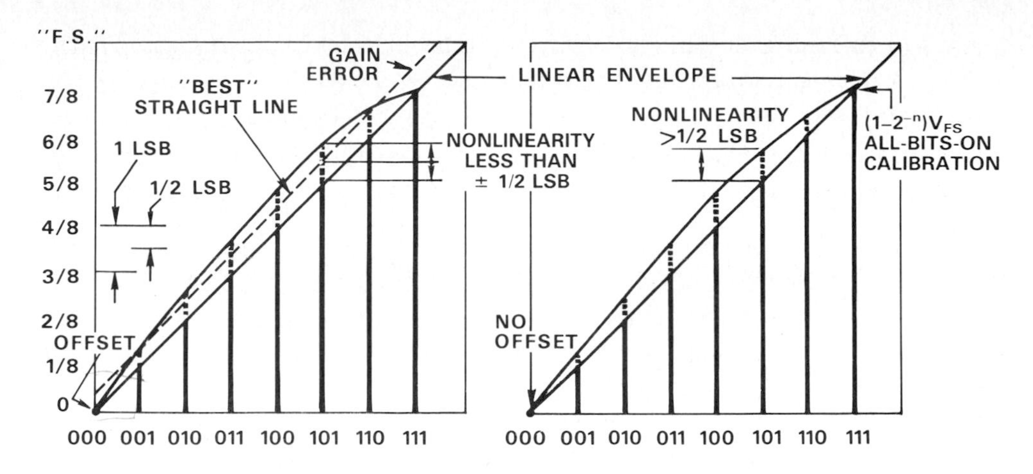

a. ½-LSB nonlinearity achieved by arbitrary location of "best straight line."

b. Nonlinearity reference is straight line through end points. ½-LSB nonlinearity error is half that of (a).

Figure 11.5. Comparison of linearity criteria for 3-bit d/a converter. Straight line through end points is easier to measure and provides a more-conservative specification.

Linearity, Integral. See *Linearity*. While *differential linearity* deals with errors in step size, *integral linearity* has to do with deviations of the overall shape of the conversion response. Even converters that are not subject to differential-linearity errors (e.g., integrating types) have integral-linearity (sometimes just "linearity") errors.

Missing Codes. An a/d converter is said to have missing codes when a transition from one quantum of the analog range to the adjacent one does not result in the adjacent digital code, but in a code removed by one or more counts. Missing codes can be caused by large negative differential-linearity errors, noise, or changing inputs during conversion. A converter's proclivity towards missing codes is also a function of temperature.

Monotonicity. A DAC is said to be *monotonic* if its output either increases or remains constant as the digital input increases, with the result that the output will always be a single-valued function of the input. The condition "monotonic" requires that the derivative of the transfer function never change sign. Monotonic behavior requires that the differential nonlinearity be more positive than -1 LSB.

Most Significant Bit (MSB). In a system in which a numerical magnitude is represented by a series of binary (i.e., two-valued) digits, the *most-significant bit* is that digit (or "bit") that carries the greatest value or weight. For example, in the natural binary number 1101 (decimal 13, or $(1\times 2^3) + (1\times 2^2) + (0\times 2^1) + (1\times 2^0)$), the leftmost "1" is the MSB, with a weight of ½ nominal peak-to-peak full scale (full-scale range). In bipolar devices, the sign bit is the MSB. In A/D converters having overrange bits, the MSB is the most-significant "overrange" bit.

In converter nomenclature (viz., fractional binary), the MSB is bit 1; in bus nomenclature (integer binary), it is Data Bit (n − 1).

Multiplying DAC. A multiplying DAC differs from the conventional fixed-reference DAC in being designed to operate with varying (or ac) reference signals. The output signal of such a DAC is proportional to the product of the "reference" (i.e., analog input) voltage and the fractional equivalent of the digital input number. See also *Four-Quadrant*.

Noise, Peak and RMS. Internally generated random noise is not a major factor in d/a converters, except at extreme resolutions and dynamic ranges. Random noise is characterized by rms specifications for a given bandwidth, or as a spectral density (current or voltage per root hertz); if the distribution is Gaussian, the probability of peak-to-peak values exceeding 7 × the rms value is less than 0.1%.

Of much greater importance in DACs is interference, in the form of high-amplitude, low-energy (hence low-rms) spikes appearing at a DAC's output, caused by coupling of digital signals in a surprising variety of ways; they include coupling via stray capacitance, via power supplies, via inadequate ground systems, via feedthrough, and by glitch-generation (see *Glitch*). Their presence underscores the necessity for maximum application of the designer's art, including layout, shielding, guarding, grounding, bypassing, and deglitching.

Noise in a/d converters in effect narrows the region between transitions and can cause missing codes in a converter with marginal differential nonlinearity. Sources of noise include the comparator, the reference, the analog signal itself, and pickup in infinite variety.

Nonlinearity—or "gain nonlinearity"—in an instrumentation or isolation amplifier is defined as the deviation from a straight line on the plot of output vs. input. The magnitude of linearity error is the maximum deviation from a "best straight line," with the output swinging through its full-scale range. Nonlinearity is usually specified in percent of full-scale output range.

Normal Mode. For an amplifier used in instrumentation, the *normal-mode* signal is the actual difference signal being measured. This signal often has noise associated with it. Signal-conditioning systems and digital panel instruments usually contain input filtering to remove high-frequency—and line-frequency—noise components. *Normal-mode rejection* (NMR), is a logarithmic measure of the attenuation of normal-mode noise components at specified frequencies in dB.

Offset, Bipolar. For the great majority of bipolar converters (e.g., ±10-volt output), negative currents are not actually generated to correspond to negative numbers; instead, a unipolar DAC is used, and the output is offset by half full scale (1 MSB). For best results, this offset voltage or current is derived from the same reference supply that determines the gain of the converter.

Because of nonlinearity, a device with perfectly calibrated end points may have offset error at analog zero.

Offset Step. See *Pedestal*.

Output Voltage Tolerance. For a reference, the maximum deviation from the nominal output voltage at 25°C and specified input voltage, as measured by a device traceable to a recognized fundamental voltage standard.

Overload (Digital Panel Instruments). An input voltage exceeding the instrument's full-scale range produces an overload condition, usually indicated by conspicuous manipulation of the display, such as all dashes, flashing zeros, etc. On a 3½-digit DPM with a range of 199.9 mV, a signal exceeding 200 mV will produce an overload condition.

Overrange (Digital Panel Instruments). An input signal that exceeds all nines on a DPM, but is less than an overload. On a 3 ½-digit DPM with a full-scale range of 199.9 mV, the all-nines range is 0 to 99.9 mV, and signals from 100 to 199.9 mV are said to fall in the 100% overrange region. Some panel meters have higher overrange capability; a 3 3/4-digit meter has a full-scale range of 3.999, or 300% overrange.

Pedestal, or Sample-to-Hold Offset Step. In sample/track-hold amplifiers, a shift in level between the last value in *sample* and the value settled-to in *hold*; in devices having fixed internal capacitors, it includes *charge transfer*, or *offset step*. However, for devices that may use external capacitors, it is often defined as the residual step error after the *charge transfer* is accounted for and/or cancelled. Since it is unpredictable in magnitude and may be a function of the signal, it is also known as *offset nonlinearity*.

Power-Supply Rejection Ratio (PSRR). The ratio of a change in dc power-supply voltage to the resulting change in the specified device error, expressed in percentage, parts per million, or fractions of 1 LSB. It may also be expressed logarithmically, in dB (PSR = $20 \log_{10}$(PSRR)).

Power-Supply Sensitivity. The sensitivity of a converter to changes in the power-supply voltages is normally expressed in terms of percent-of-full-scale change in analog value—or fractions of 1 LSB – (D/A output, A/D input) for a 1% dc change in the power supply, e.g., 0.05%/% ΔV_S. Power-supply sensitivity may also be expressed in relation to a specified maximum dc shift of power-supply voltage. A converter may be considered "good" if the change in reading at full scale does not exceed ±½–LSB for a 3% change in power-supply voltage. Even better specs are necessary for converters designed for battery operation.

Propagation Delay. In CMOS DACs, a measure of the internal delays of the circuit, propagation delay is defined as the time from a digital input change to the time at which the analog output current reaches 90% of its final value.

Quad-Slope Converter. This is an integrating analog-to-digital converter that goes through two cycles of *dual-slope* conversion, once with zero input and once with the analog input being measured. The errors determined during the first cycle are subtracted digitally from the result in the second cycle. The scheme can result in high-accuracy conversion.

Quantizing Uncertainty (or "Error"). The analog continuum is partitioned into 2^n discrete ranges for n-bit conversion and processing. All analog values within a given quantum are represented by the same digital code, usually assigned to the nominal mid-range value. There is, therefore, an inherent quantization uncertainty of ± ½ LSB, in addition to the actual conversion errors. In integrating a/d converters, this "error" is often expressed as "±1 count." Depending on the system context, it may be interpreted as a truncation (round-off) error or as noise.

Ratiometric. The output of an a/d converter is a digital number proportional to the *ratio* of (some measure of) the input to a reference voltage. Most requirements for conversions call for an absolute measurement, i.e., against a fixed reference; but this presumes that the signal applied to the converter is either reference-independent or in some way derived from another fixed reference. However, real references are not truly fixed; the references for both the converter and the signal source vary with time, temperature, loading, etc. Therefore, if the converter is used with signal sources that also rely on references (for example, strain-gage bridges, RTDs, thermistors), it makes sense to replace this multiplicity of references by a single system reference; reference-caused errors will tend to cancel out. This can be done by using the converter's internal reference (if it has one) as the system reference. Another way is to use a separate external system reference, which also becomes the reference for a *ratiometric* converter.

Over limited ranges, ratiometric conversion can also serve as a substitute for analog or digital signal division (where the denominator changes by less than ½ LSB during the conversion). The signal input is the numerator; the reference input is the denominator.

Resolution. An n-bit binary converter should be able to provide 2^n distinct and different analog output values corresponding to the set of n-bit binary words. A converter that satisfies this criterion is said to have a *resolution* of n bits.

The smallest output change that can be resolved by a linear DAC is 2^{-n} of the full-scale span. Thus, for example, the resolution of an 8-bit DAC would be 2^{-8}, or 1/256. On the other hand, a nonlinear device, such as the AD7111 LOG-DAC™,* can ideally achieve a dynamic range of 89.625 dB, or 30,000:1, in 0.375-dB steps, using only 8 bits of digital resolution.

For digital panel instruments, *resolution* is the smallest voltage increment that can be measured. It is a function of full-scale range and the number of digits. For example, if a 3 ½-digit DPM has a resolution of 1 part in 2,000 (0.05%) over a full-scale range of 199.9 mV, the DPM can resolve 0.1 mV.

Sample-to-Hold Offset. See *Pedestal*.

Settling Time—Amplifier. Settling time is defined as the time elapsed from the application of a perfect step input to the time when the amplifier output has entered and remained within a specified error band symmetrical about the final value. Settling time, therefore, includes the time required: for the signal to propagate through the amplifier, for the amplifier to slew from the initial value, recover from the slew-rate-limited overload (if it occurs), and settle to a given error in the linear range. It may also include a "long tail" due to the time required to reach thermal equilibrium, or the settling time of compensation circuits. Settling time is usually specified for the condition of unity gain, relatively low impedance levels, no (or a specified value of) capacitive loading, and any specified compensation. A full-scale unipolar step is used, and both polarities are tested.

Although settling time can generally be grossly inferred from the other amplifier specifications (i.e., an amplifier that has extra-wide small-signal bandwidth, extra-fast slewing, and excellent full-power response may reasonably—but not always—be expected to have fast settling), settling time cannot usually be predicted from the other dynamic specifications.

Settling Time—DAC. The time required, following a prescribed data change, for the output of a DAC to reach and remain within a given fraction (usually ± ½ LSB) of the final value. Typical prescribed changes are full-scale, 1 MSB, and 1 LSB at a major carry. Settling time of current-output DACs is quite fast. The major share of settling time of a voltage-output DAC is usually contributed by the settling time of an output op-amp chip.

Single-Slope Conversion. In the single-slope converter, a reference voltage is integrated until the output of the integrator is equal to the input voltage. The time period required for the integrator to go from zero to the level of the input is proportional to the magnitude of the input voltage and is measured by an internal clock. Measurement accuracy is sensitive to clock speed and integrating capacitance, as well as the reference accuracy.

*LOGDAC is a trade mark of Analog Devices, Inc.

Slew(ing) Rate. A limitation in the rate of change of output voltage, usually imposed by some basic circuit consideration, such as limited current to charge a capacitor. Amplifiers with slew rate greater than about 75 volts/microsecond are usually seen only in more sophisticated (and expensive) devices. The output slewing speed of a voltage-output d/a converter is usually limited by the slew rate of the amplifier used at its output.

Stability. In a well-designed, intelligently applied converter, *dynamic stability* is not an important question. The term stability usually applies to the insensitivity of the converter's characteristics to time, temperature, etc. All measurements of stability are difficult and time-consuming, but stability vs. temperature is sufficiently critical in most applications to warrant universal inclusion in tables of specifications (see *Temperature Coefficient*).

Staircase. A voltage or current, increasing in equal increments as a function of time and having the appearance of a staircase (in a time plot); it is generated by applying a pulse train to a counter, and the output of the counter to the input of a DAC.

A very simple a/d converter can be built by comparing a staircase from a DAC with the unknown analog input. When the DAC output exceeds the analog input by a fraction of 1 LSB, the count is stopped, and the code corresponding to the count is the digital input. If the counter is an up/down counter, where up/down is controlled by the comparator state, with correct polarity, the converter will tend to track the input signal.

Subranging A/D Converters. In this type of converter, an extremely fast—i.e., *flash*—conversion produces the most-significant portion of the output word. This portion is stored in a holding register and also converted back to analog with a fast, high-accuracy d/a converter. The analog result is subtracted from the input, and the resulting residue is amplified, converted to digital at high speed, and combined with the results of the earlier conversion to form the output word. In *digitally corrected subranging* (DCS), the two bytes are combined in a manner that corrects for the error of the LSB of the most-significant byte. For example, using 8-bit and 5-bit conversion, plus this technique and a great deal of video-speed converter expertise, a full-accuracy high-speed 12-bit converter can be built.

Successive Approximations. Successive approximations is a high-speed method of conversion by comparing an unknown against a group of weighted references. The operation of a successive-approximation a/d converter is generally similar to the orderly weighing of an unknown quantity on a precision chemical balance, using a set of weights, such as 1 gram ½ gram, ¼ gram, etc. The weights are tried in order, starting with the largest. Any weight that tips the scale is removed. At the end of the process, the sum of the weights remaining on the scale will be within 1 LSB of the actual weight (± ½ LSB, if the scale is properly biased—see Zero).

Switching Time. In a d/a converter, the switching time is the time taken for an analog switch to change to a new state from the previous one. It includes delay time, and rise time from 10% to 90%, but does not include settling time.

Temperature Coefficient. In general, temperature instabilities are expressed as %/°C, ppm/°C, fractions of 1 LSB per degree C., or as a change in a parameter over a specified temperature range. Measurements are usually made at room temperature and at the extremes of the specified range, and the temperature coefficient (tempco, T.C.) is defined as the change in the parameter, divided by the corresponding temperature change. Parameters of interest include gain, linearity, offset (bipolar), and zero.

a. *Gain Tempco*: Two factors principally affect converter gain stability with temperature. In fixed-reference converters, the reference voltage will vary with temperature (generally less than 5 ppm/°C for the AD581L). The reference circuitry and switches (and comparator in a/d converters) will add a few more ppm/°C.

b. *Linearity Tempco*: Sensitivity of linearity (integral and/or differential linearity) to temperature, in % FSR/°C or ppm FSR/°C, over the specified range. Monotonic behavior in DACs is achieved if the differential nonlinearity is less than 1 LSB at any temperature in the range of interest. The *differential nonlinearity temperature coefficient* may be expressed as a ratio, as a maximum change over a temperature range, and/or implied by a statement that the device is monotonic over the specified temperature range. To avoid missing codes in noiseless a/d converters, it is sufficient that the differential nonlinearity error magnitude be less than 1 LSB at any temperature in the range of interest. The temperature coefficient is often implied by the statement that there are no missed codes when operating within a specified temperature range.

c. *Zero TC (unipolar converters)*: The temperature stability of a unipolar fixed-reference DAC, measured in % FSR/°C or ppm FSR/°C, is principally affected by current leakage (current-output DAC), and offset voltage and bias current of the output op amp (voltage-output DAC). The zero stability of an ADC is dependent on the zero stability of the DAC or integrator and/or the input buffer and the comparator. It is expressed in μV/°C or in percent or ppm of full-scale range (FSR) per degree C.

d. *Offset Tempco*: The temperature coefficient of the all-DAC-switches-off (minus full-scale) point of a bipolar converter (in % FSR/°C or ppm FSR/°C) depends on three major factors—the tempco of the reference source, the voltage zero-stability of the output amplifier, and the tracking capability of the bipolar-offset resistors and the gain resistors. In an a/d converter, the corresponding tempco of the negative full-scale point depends on similar quantities—the tempco of the reference source, the voltage stability of the input buffer and the comparator, and the tracking capabilities of the bipolar offset resistors and the gain resistors.

Thermal Tail. The slow drift of an amplifier having a thermally induced offset due to self-heating as it settles to a final electrical equilibrium value corresponding to internal thermal equilibrium.

Total Unadjusted Error. A comprehensive specification on some devices which includes full-scale error, relative-accuracy and zero-code errors, under a specified set of conditions.

Zero- and Gain-Adjustment Principles. (This is not a substitute for the manufacturer's instructions, which should be followed in detail.)

a. *DACs*: The output of a unipolar DAC is set to zero volts in the all-bits-off condition. The gain is set for F.S. $(1 - 2^{-n})$ with all bits on. The "zero" of an offset-binary bipolar DAC is set to $-$F.S. with all bits off, and the gain is set for $+$F.S. $(1 - 2^{-(n-1)})$ with all bits on.

b. *ADCs*: The zero adjustment of a unipolar ADC is set so that the transition from all-bits-off to LSB-on occurs at $\frac{1}{2}(2^{-n})$ F.S. The gain is set for the final transition to all-bits-on at F.S. $(1 - 3/2\ 2^{-n})$. The "zero" of an offset-binary bipolar ADC is set so that the first transition occurs at $-$F.S. $(1 - 2^{-n})$ and the last transition at $+$F.S. $(1 - 3 \times 2^{-n})$.

11.4 SYSTEM-COMPONENT SELECTION PROCESS

The most natural process for selection of appropriate off-the-shelf components to meet a system requirement involves a method of "successive approximations:" Choose the least costly device that meets the most significant requirements, and perform an error analysis to check its adequacy.

If its performance seems far better than that needed for the other requirements (at possibly excessive cost), or inadequate in some respects, inspect the discrepancies for possible design tradeoffs, make a new choice (if necessary) and repeat the analysis. Remember, though, that in a maturing industry, costs can be expected to decline. *It is often less costly, in the long run, to go for better performance (rather than lowest possible cost) in the initial stages of a design.* Also, efforts aimed at reducing the cost of any element of a system should bear in mind the criticality of that element and the relationship of its cost relative to that of the entire project.

Where new designs are concerned and early results are essential, unless one is an experienced system designer with plenty of component and manufacturing experience (if quantities are involved), it is usually good judgement to ignore initial cost (within the limits of good sense) and go for performance, convenience, and the highest level of system integration that the budget will allow (see Chapter 4).

11.5 EXAMPLE OF A SELECTION AND VERIFICATION PROCESS*

A computer data-acquisition system is to be assembled to process data from a number of strain gages. Signal-conditioning hardware, to be purchased with the gages, delivers ± 10V full-scale signals with 10-ohm source impedance. The signal channels must be sequentially scanned in no more than 50 microseconds per channel. Maximum allowable error of the system is about 0.1% of full scale. System logic is to be TTL, and hardware may work in either binary or 2's complement code. Parallel data readout will be used.

Probable temperature range in the equipment cabinets (including equipment temperature rise) is +25°C to +55°C. Sufficient power at both ± 15V and +5V is available, but the regulation of the ± 15V supply is 150mV.

The objective: specify a set of conversion components having appropriate accuracy and speed.

11.5.1 FIRST APPROXIMATION

A useful rule of thumb that usually provides satisfactory results is this: For the critical specs of a multi-component system, choose each component to perform roughly 10 times better than the overall desired performance. Thus, for a system that needs 0.1%-grade performance, use a 0.01% converter (12 bits) with a compatible multiplexer and sample-hold amplifier.

Reviewing the available a/d converter ICs, we find AD574A to be a possible choice; it completes a conversion in 25 μs. For a sample-hold, the compatible AD585 is chosen, adding 0.5 μs of settling time. Thus, the combination appears to be amply capable of meeting the 50 μs/channel scanning requirement. Since the multiplexer will scan sequentially, its settling time is unimportant. The multiplexer can be switched to the next address as soon as the SHA goes into *hold* on data from the current address. Thus it has at least 25 μs to settle before a measurement is called for. For convenience and small size, one should consider the AD7501 CMOS 8-channel multiplexer.

11.5.2 ERROR ANALYSIS

It's clear that the AD7501, the AD585, and the AD574A generally meet the problem's requirements. Now we must look further into the details of errors, to determine if the worst-case situation is within the allowable 0.1% system error.

*For maximum tutorial benefit, to avoid clutter, and to fit the available space, some of the known but less-salient sources of error have been intentionally omitted. For any that readers are concerned about for particular applications but don't see treated here, it is worth noting that most converter manufacturers, including Analog Devices, make available competent applications engineers who can be reached by mail or by telephone.

Multiplexer

The switches of the AD7501, being MOSFET's, with variable-resistance channels, are not subject to voltage offset errors. Errors here will be due on two factors:

1. Leakage current into the *on* channel from the *off* channels develops an offset voltage across the source impedance.

Leakage current @ 25°C 10 nA
Source impedance 10 ohms
Error voltage $= 10 \times 10 \times 10^{-9} = 10^{-7}$ V (0.01 ppm, negligible)

2. Transfer error due to voltage division across the MOSFET *on* resistance and input impedance of the sample-hold amplifier (AD585):

ON resistance 300 ohms maximum
AD585 R_{in} 10^{12} ohms
Divider ratio attenuation error: 3×10^{-10} (0.0003 ppm, negligible)

Sample-Hold

1. Nonlinearity is 2mV over the 20V range, or 0.01% . . . . . **100 ppm**

2. Gain error of 0.01% maximum (and other similarly small initial gain errors in the system) may be compensated for overall performance when calibrating

Resolution	Family	Output Type	Bus Interface	Reference	Conversion Time	Comments
8 Bits	AD570	3-state buffer	TTL-compatible	Internal	20μs	
	AD670	3-state buffer	TTL-compatible	Internal	10μ s	Instr. amp. input
	AD673	3-state buffer	TTL-compatible	Internal	20μs	
	AD7574	3-state buffer	TTL-compatible	External	15μs	CMOS
	AD7576	3-state buffer	TTL-compatible	External	10μs	CMOS
	AD7581	3-state buffer	TTL-compatible	External	Continuous	8-Channel memory
	AD7820	3-state buffer	TTL-compatible	External	1.5μs	CMOS on-board S/H
10 Bits	AD579	Binary	TTL-compatible	Internal	1.8μs	Hybrid
	AD573	Binary	TTL-compatible	Internal	20μs	
	AD7571	3-state buffer	CMOS-compatible	External	80μs	CMOS
12 Bits	AD574A	3-state buffer	TTL-compatible	Internal	25μs	
	AD578	Binary	TTL-compatible	Internal	3μs	Hybrid
	AD ADC80	Ser/Par bin.	TTL-compatible	Internal	25μs	Hybrid
	AD7582	4-8 latch 3-state buffer	TTL-compatible	External	100μs	CMOS 4-channel MUX
16 Bits	AD ADC71	Ser/Par bin.	TTL-compatible	Internal	45μs	Hybrid

Table 11.1. General-Purpose Analog-to-Digital Conversion ICs

the system by setting the scale constant of the ADC. It is not considered as part of the error budget.

3. Input bias current of 2 nA (max) causes an offset error voltage in the source resistance.

Source resistance = 10 ohms (source) + 300 ohms (MPX switch)
Offset error = 2 nA × 300 ohms = 600 nV = 0.6 μV (0.06 ppm, negligible)

4. Offset vs. temperature = 3 mV/110°C = 27.3 μV/°C
Since the temperature of the chip may change by as much as 28°C, the total change over the range will be

27.3 μV × 28°C = 756 μV, or . . . . **75.6 ppm**

of ± 10 V. An offset adjustment is provided for initial trimming.

Resolution	Family	Output	Digital Input	Reference	Multiply Capability	Comments
8 Bits	AD558	Voltage	8-bit latched	Internal	1-quadrant	Single supply
	AD7224	Voltage	8-bit latched Double-buffered	External	1-quadrant	CMOS
	AD7226	Voltage	8-bit latched	External	1-quadrant	4-channel CMOS
	AD7524	Current	8-bit latched	External	4-quadrant	CMOS
	AD7528	Current	8-bit latched	External	4-quadrant	Dual CMOS
	AD7523	Current	8-bit direct	External	4-quadrant	CMOS
9 Bits	AD7115	Current	9-bit latched	External	2-quadrant	LOGDAC™ CMOS
10 Bits	AD7533	Current	10-bit direct	External	4-quadrant	CMOS
12 Bits	AD7542	Current	4-4-4 latched Double-buffered	External	4-quadrant	CMOS
	AD7545	Current	12-bit latched	External	4-quadrant	CMOS
	AD7548	Current	8-4 or 4-8 latched Double-buffered	External	4-quadrant	CMOS
	AD7549	Current	4-4-4 latched Double-buffered	External	4-quadrant	2-channel CMOS
	AD7541	Current	12-bit direct	External	4-quadrant	CMOS
	AD567	Current	4-4-4 latched Double-buffered	Internal	2-quadrant	
	AD667	Voltage	4-4-4 latched Double-buffered	Internal	2-quadrant	
	AD390	Voltage	12-bit latched Double-buffered	Internal		4-channel Hybrid
	AD565A	Current	12-bit direct	Internal	2-quadrant	
	AD566A	Current	12-bit direct	External	2-quadrant	
	AD DAC80	Optional	12-bit direct	Internal	2-quadrant	
	AD7240	Voltage	12-bit direct	External	1-quadrant	CMOS
14 Bits	AD7534	Current	6-8 latched Double buffered	External	4-quadrant	CMOS
	AD7535	Current	14 or 6-8 latched Double buffered	External	4-quadrant	CMOS
16 Bits	AD569	Voltage	8-8 latched Double-buffered	External	2-quadrant	
	ADDAC71/72	Optional	16-bit direct	Internal	2-quadrant	Hybrid
	AD7546	Voltage	16-bit latched	External	2-quadrant	CMOS

Table 11.2 General-Purpose Digital-to-Analog Conversion ICs

5. Offset vs. power supply is equal to -70 dB, or 0.316×10^{-3} V/V. Since the supply may vary by 150mV, or 1% of 15V, the error contribution is 47.4μV, or . . . . **4.7 ppm**

By an analysis comparable to the above, we would normally also prepare a system timing diagram, and assign operate-and settling-time allowances. However, the components selected for this example have more than adequate settling time, even for 0.01% operation; consequently, we can overlook the need for a formal timing analysis to determine whether settling times are adequate.

Converter

1. Specified linearity error (relative accuracy) of AD574AL: ½ LSB, or . . . . **122 ppm**

2. Quantizing uncertainty: ½ LSB, or 0.0125%. This is a resolution limitation, not normally considered in the error budget.

3. Temperature errors
   a. Gain temperature coefficient: 10 ppm/°C for 30°C
      10 ppm/°C × 30°C = 0.03%, or . . . . **300 ppm**
   b. Zero temperature coefficient: 10 ppm/°C for 30°C
      10 ppm/°C × 30°C = 0.03%, or . . . . **300 ppm**

4. Power supply sensitivity error: 2 ½ LSB for worst-case 10% change in all three supplies. For a 1% shift, the error is 0.25 LSB = 0.000061, or . . . . **61 ppm**

5. Differential nonlinearity temperature coefficient: guaranteed 12-bit resolution with no missing codes over temperature.

In this example, the worst-case arithmetic sum of these errors is 0.096%, and the rms sum is 0.046%. Because they are based on conservative assumptions, these error levels in the conversion stage are consistent with a specified error of 0.1%. The designer can go on to the other hardware, software, interface, and wiring problems.

CONCLUSION

In this chapter, we have sought to make the designer's process of choosing a converter easier and more effective by providing checklists of relevant questions in making a choice, definitions of specifications and related features, converter charts, and an example of selection and evaluation. We now go on to some considerations for what must be done to make the system work as expected.

Chapter Twelve

Applying Converters Successfully

In Chapter 11, we pointed out that selecting the most economical converter for an application is not a simple task, considering the many different types of converters on the market, the complex manner in which converter specifications relate to a specific system application, and the fact that prices of converters range from less than $10 to several hundreds of dollars.

In this chapter, we will discuss system aspects of selecting converters, a continuation of the discussion in the last chapter.

To make the most appropriate converter choice requires that the user consider a number of questions: What are the real objectives of the conversion process, and how do they relate to the converter's specifications? How may the system be configured to relax the converter's performance (and price) requirements, and at what overall cost? How will the other system components limit and degrade the converter's performance? What tradeoffs are available in the system error budget? Is it more economical to make a long-term choice of one "general-purpose" converter, which will meet the needs of a large number of system designs with a single standard configuration, or to go through an optimum selection process for each individual application?

After selecting the appropriate converter, the user should be fully aware that the thorough preliminary analysis and economic component choice usually involved are not by themselves sufficient to ensure that the system performance needs will be met. The system designer must take into account the physical surroundings, interconnections, grounding and power supplies, protection circuitry, and all the other details that constitute good engineering practice.

While these few pages cannot (and are not intended to) be a primer on engineering practice, it is essential for the converter user to become aware of those elements of practice that are of particular relevance to converter-system design.[1]

12.1 MAKING THE PROPER SYSTEM CHOICES

A general rule of thumb used by some designers may be expressed as follows: "As the converter performance requirements approach state-of-the-art converter capabilities in *both speed and accuracy*, the price of the converter will increase exponentially." If substantial cost savings are desired, the user must relax one of these parameters.

12.1.1 DATA-ACQUISITION SYSTEMS

An example will serve to illustrate the process of elimination and winnowing that can be profitably employed to determine a converter's minimum performance requirements. Figure 12.1 shows, in the simplest terms, a block diagram of an analog data-acquisition system, the primary application for which A/D converters are used.

The data-acquisition system, under the direction of the control unit, selects the multiplexer input points, one at a time, and directs the signal appearing on each point to the analog input of the A/D converter via the associated multiplex channel. The signal level is encoded by the converter and outputted to storage. The storage unit retains each piece of data in a predetermined format, and holds it for further processing.

12.1.2 THREE CLASSES OF CONVERTER SPECIFICATIONS

In attacking the problem of determining the converter performance requirements, it is useful to divide the converter specifications into three classes: Those that determine accuracy under optimum conditions, those that are dependent on time (or speed of response), and those that are substantially affected by the environment.

In the first class are included resolution, relative accuracy, differential linearity, noise, quantization uncertainty, monotonicity, and differential-linearity

[1]Useful (and strongly recommended) references include:

1. *Analog Devices 1984 Databook* (or subsequent editions). Integrated Circuits, Vol. I. (Norwood: Analog Devices, Inc.) Application Notes:
   "An I.C. Amplifier User's Guide to Decoupling, Grounding, and Making Things Go Right for a Change," pp. **20**-13 to **20**-20.
   "Gain Error and Gain Temperature Coefficient of CMOS Multiplying DACs," pp. **20**-37 to **20**-40.
   "Shielding and Guarding," pp. **20**-85 to **20**-90
   "Understanding Interference-Type Noise," pp. **20**-81 to **20**-84.
2. Brokaw, A. Paul, "Analog Signal-Handling for High Speed and Accuracy," *Analog Dialogue* 11-2 (1977), pp. 10-16.
3. Burton, Phil, *CMOS DAC Application Guide*. (Norwood: Analog Devices, Inc.), 1984.
4. Morrison, Ralph. *Grounding and Shielding Techniques in Instrumentation* Second Edition. (New York: John Wiley & Sons, 1977).
5. Ott, Henry W. *Noise Reduction Technique in Electronic Systems*. (New York: John Wiley & Sons, 1976).

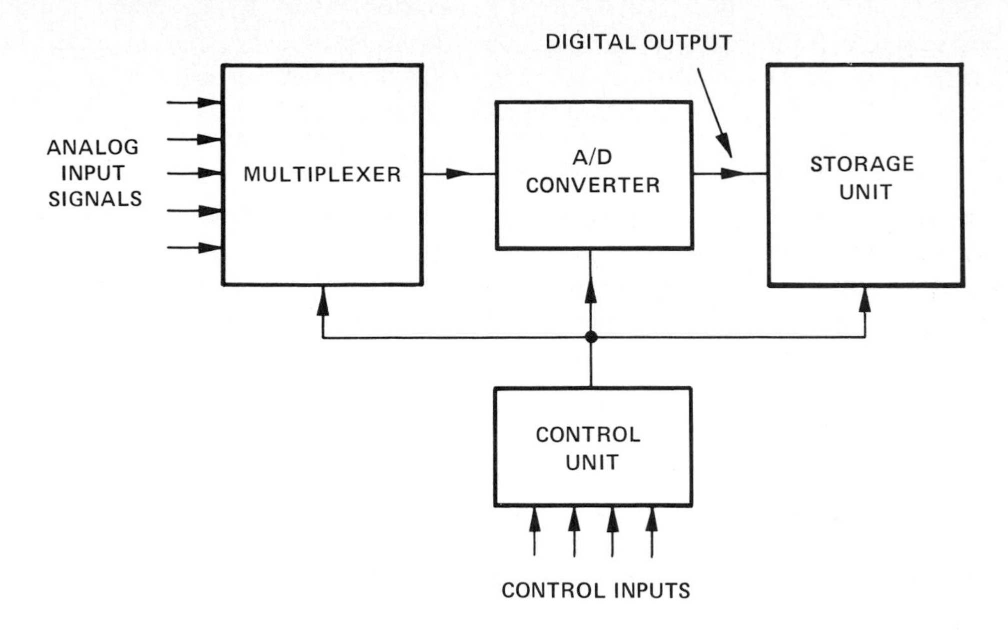

Figure 12.1. Data-acquisition system.

temperature coefficient. The reason this last term is included, though it would appear to be an environmental specification, may be somewhat unexpected: Although the ambient temperature may be in the steady state, it can be elevated (e.g., 25°C above normal room temperature) by virtue of enclosure in a cabinet. Although calibration *in situ* can correct for errors produced by variation of gain and offset with temperature, no correction can normally be effected for errors characterized by the differential-linearity T.C., due to individual bit variations with temperature. For this reason, the differential-linearity error at 25°C is augmented by the product of the steady-state temperature rise and the differential-nonlinearity temperature coefficient.

Speed-dependent specifications, in the second category, include conversion time, bandwidth, settling time of the input circuitry, etc. Environment-related specifications in the third category include gain (i.e., scale-factor) tempco, offset tempco, limits of the operating temperature range, etc.

12.1.3 APPROACHES TO RELAXING THE SPECIFICATIONS

Relaxation of the specifications in the first class may be effected through the use of signal conditioners. Choice of the specific form signal conditioning may take is based on our knowledge of the input signals to be encoded and the information to be extracted from the encoded data. Known or unwanted signal components may be removed from the input signals, and the peak-to-peak variation of the remaining signals may be scaled to equal the input voltage range of the a/d converter, using an analog subtractor having adjustable gain.

For example, if the signal conditioner is a differential instrumentation amplifier, such as the Analog Devices AD524 or AD624, it may be used to bias out dc offsets, and to scale the input appropriately (Figure 12.2).

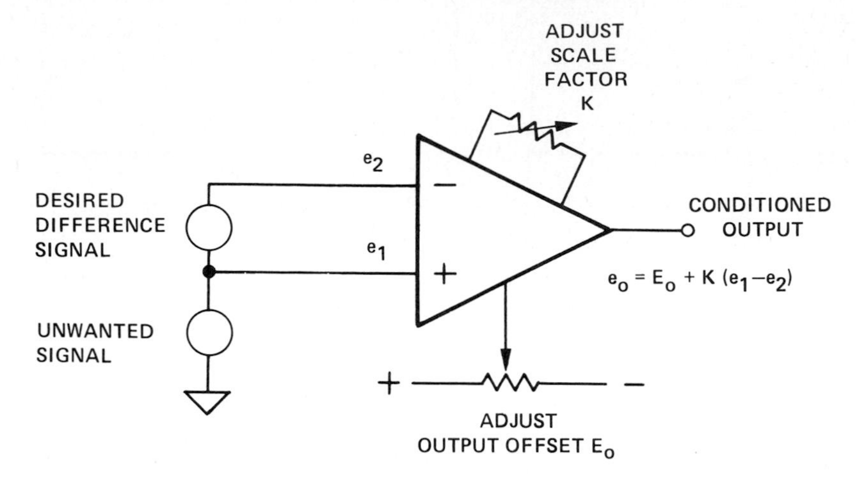

Figure 12.2. Differential instrumentation amplifier as a signal-conditioner for data acquisition.

The level-shifting-and-scaling operation can be used to obtain efficient use of the converter's input range. Scaling may allow voltage increments in the original signals, that were less than 1 LSB of the converter's input voltage range, to be measured.

12.1.4 LOGARITHMIC COMPRESSION

In applications calling for wide dynamic signal range but capable of tolerating constant fractional error (e.g., 1% of actual value), rather dramatic efficiency can be realized through the use of logarithmic amplifiers for data compression, as shown in Figure 12.3.

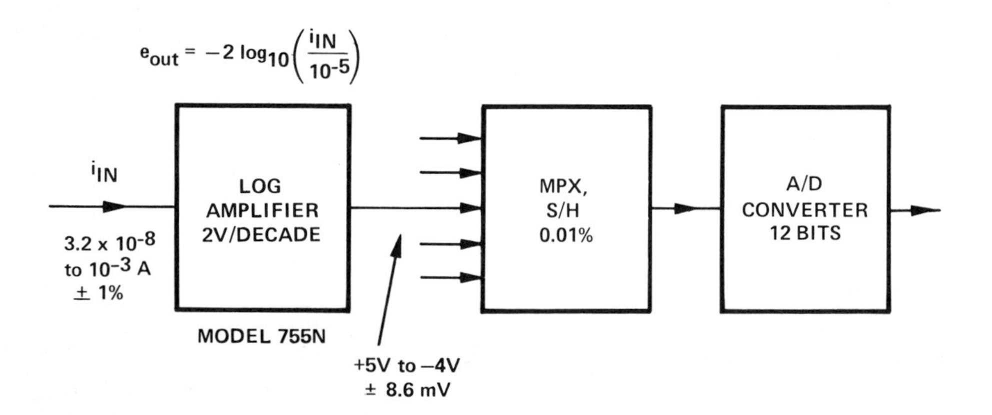

Figure 12.3. Using a log amplifier for range compression in a data-acquisition system.

Here, a logarithmic amplifier allows encoding of signals, that would ordinarily require a minimum of 20-bit conversion to handle the dynamic range, with a far-less-costly 12-bit converter. Modest accuracy in a fixed ratio to (e.g., % of) actual value is substituted for extreme accuracy in relation to the entire full-scale range, at considerably less cost. For many applications, this is an ideal performance mode; an exception is the set of applications for which extremely small errors are required at all points in the range (e.g., measuring long-term stability of voltage sources).

The logarithmic data can be dealt with easily if the data is to be processed digitally; or it can be recovered in linear analog form (if it is simply to be stored or transmitted digitally) by the use of another log amplifier, in the antilog connection, with a D/A converter; or in some cases it can be recovered by a LOG-DAC™ with appropriate connections.

12.1.5 FILTERING

Another commonly used signal-conditioning unit is the filter. Low-pass filters are used to extract carrier, signal, and noise components, above the frequencies of interest, from the input signals. These components appear as aliased low-frequency noise if they are above one-half the sampling frequency. A/D converters often incorporate (or require) follower circuits for impedance buffering. With a modicum of external wiring, they can be connected as active low-pass filters (Figure 12.4).

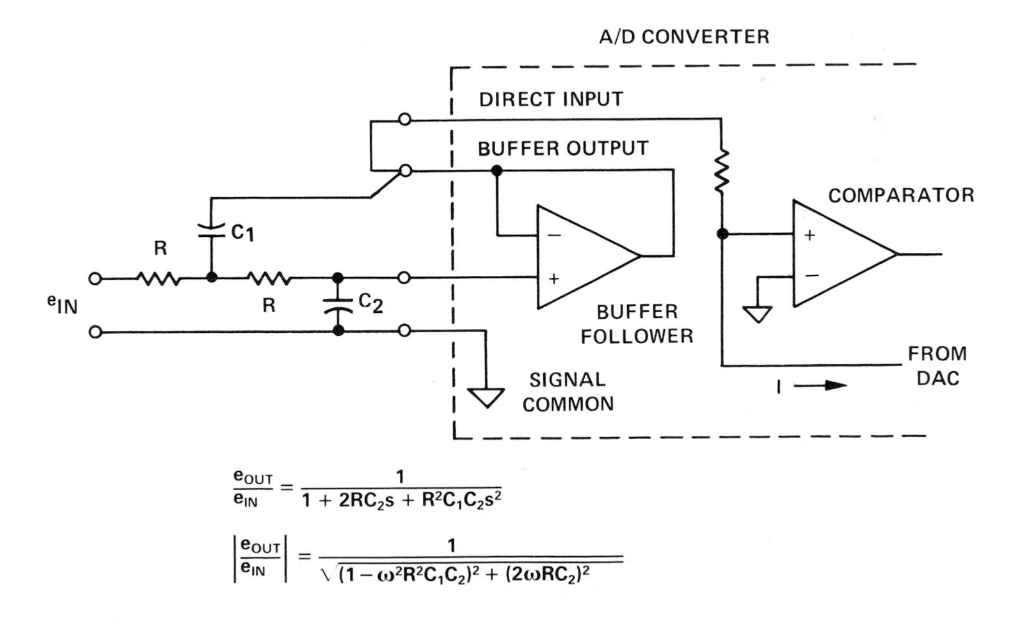

$$\frac{e_{OUT}}{e_{IN}} = \frac{1}{1 + 2RC_2s + R^2C_1C_2s^2}$$

$$\left|\frac{e_{OUT}}{e_{IN}}\right| = \frac{1}{\sqrt{(1-\omega^2R^2C_1C_2)^2 + (2\omega RC_2)^2}}$$

Figure 12.4. Input buffer of an a/d converter connected as an active low-pass filter.

12.1.6 SAMPLE-HOLD

A relaxation of the second class of specifications can also be effected by adding a sample-hold amplifier to the system configuration, as depicted in Figure 12.5.

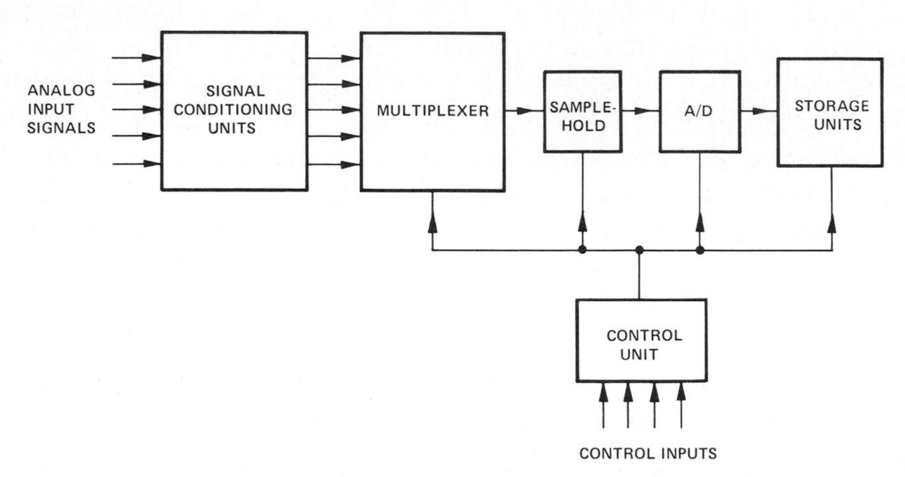

Figure 12.5. Data-acquisition system with sample-hold and input pre-conditioning.

The use of a sample-hold amplifier can substantially increase the highest signal frequency, of a given amplitude, that may be encoded within the resolution of the converter, as well as increasing the system throughput rate.

There are two sources of limitation on the maximum analog frequency that can be handled. One is the Nyquist criterion (the highest component of frequency in the analog signal cannot exceed one-half the sampling frequency); this limits analog signals to one-half the throughput rate. The other has to do with the timing of the sample. The following example shows what the problem is and indicates the improvement possible with a sample-hold: If the input is a sine wave, $E_p \sin (2 \pi f t)$, with $E_p = \text{F.S.}$, the maximum rate of change occurs at zero and, as can be found by differentiating with respect to t, is equal to $2 \pi f E_p$. If the change in the input to the converter is to be less than 1 LSB during the conversion, then

$$\text{Max. Rate} = 2\pi f E_p = \frac{1\,\text{LSB}}{\Delta t} = \frac{2^{-n} \cdot 2\,E_P}{\Delta t} \tag{12.1}$$

and the highest frequency that can be applied is

$$f_{MAX} = \frac{2^{-n}}{\pi \Delta t} \tag{12.2}$$

where $2\,E_p$ is the peak-to-peak signal span, and Δt is the time uncertainty as to when the conversion took place (equal to the conversion time for successive-approximation converters). For a 12-bit converter with a 20-μs conversion time, f_{max} is about 4Hz! Using a sample-hold, one can reduce the uncertainty in the time of measurement from the ADC's conversion time to the aperture-time uncertainty of the sample-hold, thus effecting a possible improvement in f_{max} by the ratio of the conversion time to the aperture-time uncertainty.

Since aperture jitter of the order of nanoseconds is routinely available in sample-holds designed for operation with 12-bit converters (the AD585 has an aperture-jitter specification of 0.5 ns), an improvement greater than 10,000:1 is quite feasible, assuming that the S/H has adequate bandwidth. Thus, the limitation of 4 Hz would become 40 kHZ; but at a sampling rate of 50 kHz 1/(20 μs), the highest signal component would be limited to 25 kHz by the Nyquist criterion. The sample-hold's phase shift for 20-kHz signals is only 0.6°. Thus, this sample-hold application would be quite conservative with regard to timing.

The sample-hold can also increase the system throughput rate. The system throughput rate, without the sample-hold, is determined primarily by the multiplexer's settling time, plus the A/D converter's conversion time.

The multiplexer settling time is the time required for an analog signal to settle to within its share of the system error budget, as measured at the input to the converter, after selection by the multiplexer. For a 12-bit conversion system, with a ± 10V range, multiplexers typically settle within 1 microsecond, and typical conversion times are 20 microseconds. The sample-hold can be used to hold the last channel's signal level for conversion, while the next channel is selected and settles. Since sample-hold amplifiers with acquisition times of less than 5 μs to within 0.01% are readily available (for example, the AD585 has a maximum acquisition time of 3 us to 0.01% for 10-volt changes), the throughput period can be reduced to approach the conversion time. Pairs of sample-holds and A/D converters can be used for alternate conversions to increase throughput rate even further, though at somewhat higher cost.

If the speed of a/d conversion is significantly limited by the settling time of an ADC's input buffer-follower, the sample-hold may be connected to bypass it, providing an even greater increase in throughput rate.

Relaxation of the third class of errors, those due to environment-related specifications, may be abetted by allotting one multiplexer channel to carry a ground-level signal, and another to carry a precision reference-voltage level that is near-full scale. Data obtained from these channels may be used by a processor to correct gain and offset variations common to all channels, generated in the sample-hold, the A/D converter, and the associated wiring.

12.1.7 DRIVING THE ANALOG-TO-DIGITAL CONVERTER

Since the sample-hold is used to create an effectively "dc" input signal during the conversion interval, a sample-hold will be unnecessary if the signal is already varying slowly enough to be, in effect, a dc signal.

However, there may be a trap! Many designers overlook the fact that, besides holding the input signal constant during the conversion, the sample-hold amplifier also has a quite low output impedance (in most cases), which is a requirement for driving the analog input of many a/d converters.

To understand the need for low dynamic output impedance, consider this: for a 12-bit converter, such as the AD574A, with a conversion time of 25 μs, the clock rate of the successive-approximation register is about (12 comparisons)/(25 μs) = 480 kHz. At each trial, the programmed current output from the d/a converter flows through the input resistor, developing a voltage drop, in opposite sense to the analog input voltage. The net voltage is sensed by the comparator, which then makes its decision (Figure 12.6a).

Since this set of 480-kHz current pulses has to flow through the output circuit of the device that furnishes the input signal, it will cause transient changes in the input (glitches)—unless the analog input is driven by an ideal voltage source having zero dynamic impedance; if these transients are sufficiently large or long-lasting, they may lead to erroneous conversions and missed codes.

Ideal voltage sources do not exist, but as long as the impedance is sufficiently low at the switching frequency of the successive-approximations register (480 kHz, in this example), the a/d converter should be able to convert accurately without encountering this kind of dynamic errors.

The way to get an amplifier with low output impedance is to configure a wideband high-slew-rate amplifier as a voltage follower. The closed-loop output impedance of an op amp is equal to the open-loop output impedance (usually a few hundred ohms) divided by the loop gain *at the frequency of interest*.

It is often assumed that the loop gain of a voltage follower is sufficiently high to reduce the closed-loop output impedance to a negligibly small value, especially if the signal is at low frequency. However, the amplifier driving the ADC must have either sufficient loop gain at 480 kHz to reduce the closed-loop output impedance to a low value, low open-loop output impedance, or both.

This can be accomplished either by using a wideband op amp (or sample-hold), or by connecting a discrete-transistor or integrated buffer inside the amplifier's feedback loop (Figure 12.6b) The voltage follower's output should be physically within one or two centimeters of the a/d converter; this minimizes inductance—and the problems that inductive reactance introduces.

Finally, when the system is laid out, unshielded analog signal lines should never run in channels with either digital signal lines or power lines. In applications employing printed-circuit boards, where possible, analog leads should be guarded by paralleled common leads and ground planes on the reverse side.

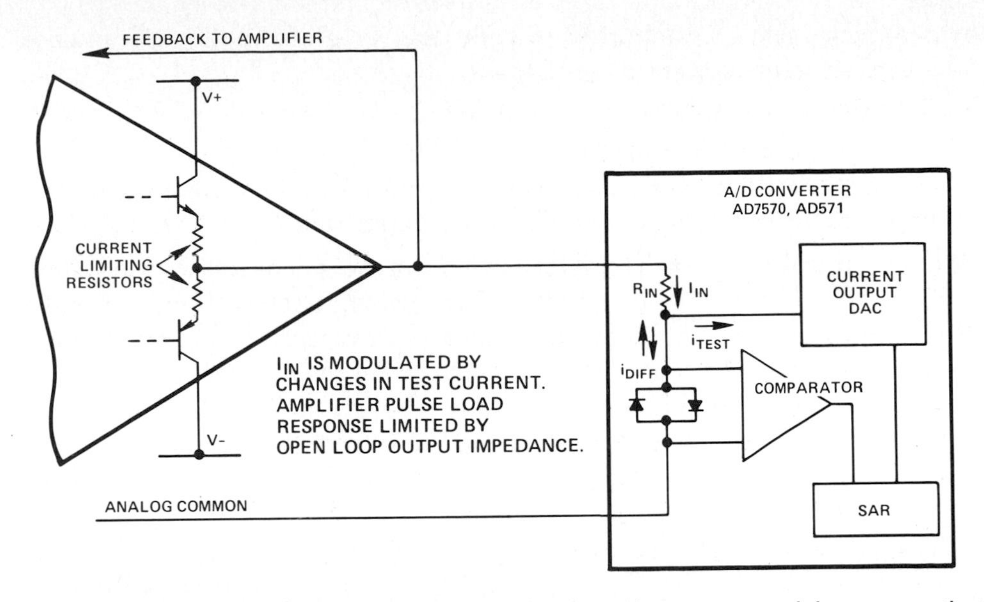

a. Relationship between successive-approximation a/d converter and the op amp that is the source of the input signal.

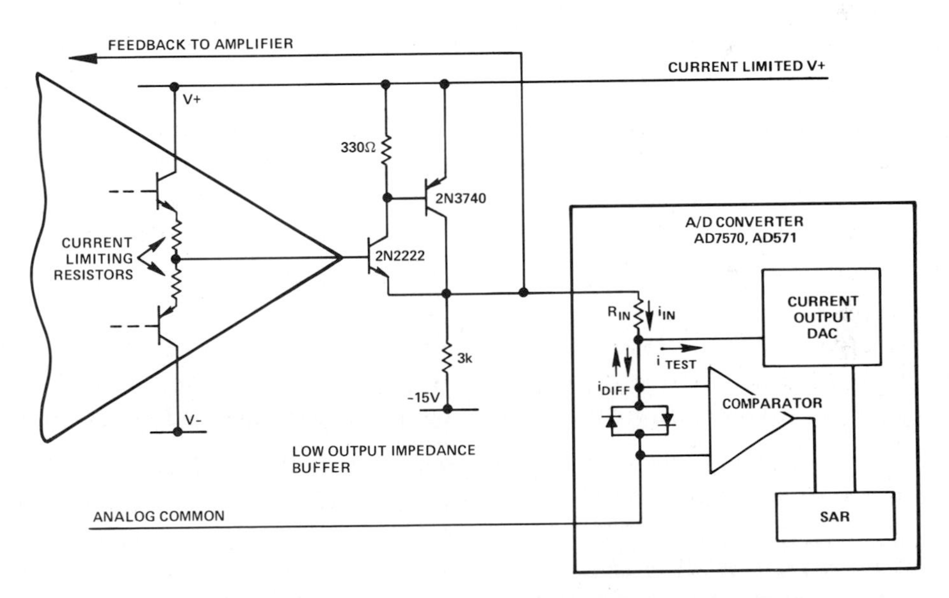

b. Inside-the-loop buffer provides stiff drive for unipolar ADC.

Figure 12.6. Driving a/d converters.

12.1.8 CONTRIBUTIONS TO ERROR

The decision to seek means of relaxing the required specifications is based on the availability and cost of devices that meet the original specifications, as compared to the cost of alternatives and any additional problems engendered by departing from a straightforward approach. To evaluate the performance tradeoffs, an error budget is a useful tool.

Three classes of errors should be considered:

Errors due to the non-ideal nature of each component
Errors due to the physical interconnections of the system components
Errors due to the interaction of system components.

The first group of errors can be determined from the spec sheets for the system components. The second group result from parasitic interactions that are a function of the way the interconnections are managed, e.g., grounding, shielding, contact resistance, etc. The third group result from specific interactions between components in the system; though they are not specifically called out in spec sheets, they can be predicted from careful reading of the specifications of the individual devices, or from the user's knowledge of how they are designed.

An example of this class of error sources might be the offsets created by series impedances in the signal path (signal-source impedance, multiplexer-switch *on* impedance) and the bias and leakage currents of the stages following these impedances, to which they are connected. A second example might be disturbances caused at the signal source as the multiplexer switches it into the circuit.

By showing where the important contributions to error are, the error budget is used as a tool for establishing tradeoffs to set the final performance requirements for the system. The error budget can be used as a tool in predicting the overall expected error, whether by worst-case summation, by root-sum-of-the-squares summation, or by combinations of the above using specific knowledge of possible compensations and common sense.

12.2 INSTALLATION AND GROUNDING

The current popularity of modular converters, in the form of modules, hybrids, and ICs, makes it worthwhile to consider some elements of their design.

For one thing, many types are "customer-programmable." This means that the user may select one of several possible signal voltage ranges by choosing the appropriate jumper-wiring configuration at the device's terminals. It goes without saying that all terminals used to determine the signal voltage range involve analog signals; to protect their low resolution levels, they should be kept away from circuit-card etch runs that carry logic signals.

User-programmable inputs, in addition to their jumpering possibilities, also permit modification by the connection of external resistors. Care should be exercised in doing this, for the reasons mentioned above. In addition, it should be noted that the excellent gain and offset T.C.'s of these devices are achieved by depending, not on absolute stability with temperature, but rather on the close tracking with temperature of key resistors within the device.

Therefore, even if 0 ppm/°C TCR resistors are used externally, the overall gain and offset performance vs. temperature may be appreciably degraded. Since there may be ways of avoiding excessive errors, the manufacturer should be consulted before external resistors are "frozen" into the design.

In the design of the converter module, great care is taken to separate the analog and digital signal lines. This procedure should also be followed with the external layout of the board on which the converter is mounted. Etch runs of digital signal lines should not run parallel in close proximity with etch runs of analog signal lines. If these lines must cross, they should do so at right angles.

Particular care should be taken with sensitive low-level points, e.g., the comparator input on a/d converters and the summing junction of the output amplifier on d/a converters. Etch runs to these points should be as short as possible. Analog-ground guard runs may also help reduce interference.

12.2.1 GROUNDING

Converter modules (actually, most data-acquisition components) have a number of ground terminals, which are generally not connected together within the module. These "grounds" are usually referred to as Digital Common, Analog Common (Analog Power Return), and Analog Signal Ground (or Sense). These grounds must be tied together at one point, the system star point, or "Mecca," usually at the system power-supply ground. Ideally, a single solid ground would be desirable. However, since current flows through the ground wires and etch stripes of the circuit cards, and since these paths have resistance and inductance, hundreds of millivolts could be generated between the system star point and the ground terminal of the module. Separate ground returns are provided to minimize the current flow in the path from sensitive points to the system star point. In this way, supply currents do not flow in the same return path with analog signals, and logic-gate return currents are not summed into the return from a precision reference-zener diode. (Figure 12.7)

In any event, the connections between the system star point and the ground terminals should be as short as possible and should have the lowest feasible impedance.

Each of the device's supply terminals should be capacitively decoupled to the appropriate ground point (see Section 12.4.2), as close to the device as possible. A large-value capacitor with a high resonant frequency should be used. A 15 μF solid-tantalum capacitor is usually sufficient. Analog supply terminals are bypassed to the appropriate power return terminal, and the logic-supply terminal is bypassed to the logic power return terminal.

When gain and offset adjustments are available and are intended to be used, the potentiometers for performing the adjustments should be mounted (with

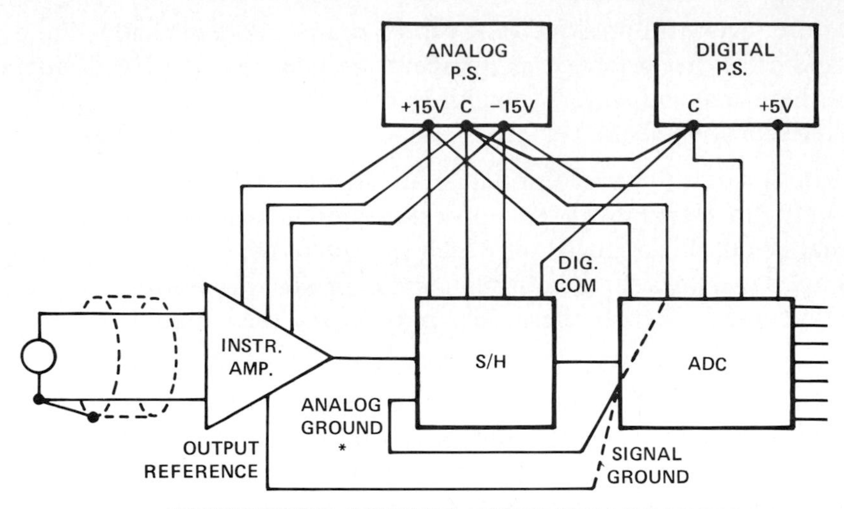

Figure 12.7. A popular (but by no means universal) approach to grounding. Grounding practice must be individually tailored to the structure of a particular system and the characteristics of the components employed, taking into account where the various signal and foreign currents actually appear at the device terminals.

short leads) in such a position that they will be accessible when the mounting board is installed in the system.

The same care should be taken to locate a conversion subsystem properly within a system as is taken to mount a conversion device on its circuit board. A converter should never be located near a transformer or fan blower motor. Using mu-metal shielding to protect against electromagnetic and RFI pickup is an expensive and not-always-successful proposition.

D/A converters should be located at their loads. This may require long cable runs for the digital control signals; however, the reduction in noise pickup and ground-potential differences between the D/A's output and the load can easily justify the expense. An alternative where this is simply not possible, as in component testing, is to use force-sense connections, in which the voltage at the load is sensed, compared with the desired value, and feedback circuitry applies whatever forcing voltage is required to minimize the error (see Figure 6.3).

For more on grounding and bypassing, see Section 12.4.2.

12.2.2 REDUCING COMMON-MODE ERRORS

As we have indicated, a differential amplifier may be used to eliminate the effects of ground-potential differences in various parts of the system in which the converter is used. In Figure 12.8, the signal source is a remotely located transducer, and the differential amplifier is located near the A/D converter.

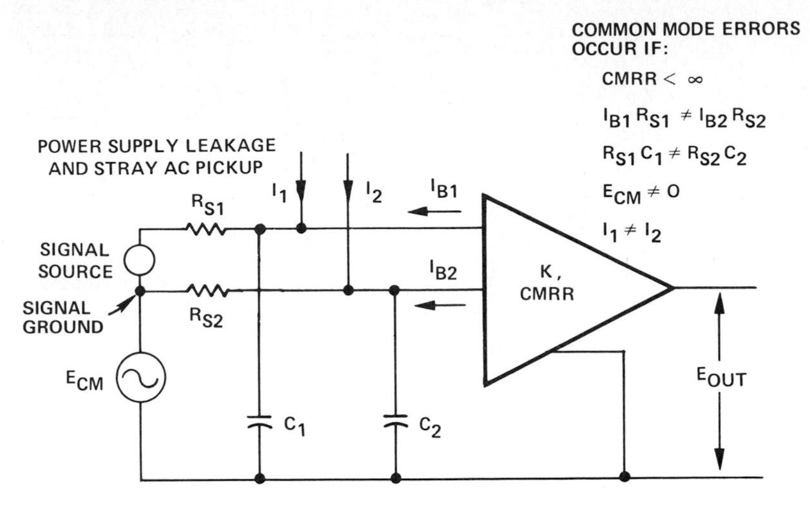

Figure 12.8. Common mode and the difference signals due to line unbalance.

The common-mode signal is the potential difference between the ground signal at the converter and the ground signal at the transducer, plus any undesirable common-mode signal produced by the transducer, and any voltages developed across the unbalanced impedances of the two lines.

If the signal source is the output of the system's d/a converter, the differential amplifier would be located near the remote load. The common-mode signal is developed by the differences in ground potential at the two locations.

The amount of dc common-mode offset that is rejected depends on the dc common-mode rejection (CMR) of the amplifier. However, bias currents flowing through the signal source leads can cause offsets, if either the bias currents or the source impedances are unbalanced. DC CMR specifications generally include a specified amount of source unbalance (e.g., 1 kΩ). Such specifications also indicate a top frequency for which the dc spec is valid, usually the line frequency (50-60 Hz), but sometimes 100 Hz. At higher frequencies, unbalanced RC time constants (balanced or unbalanced series resistance and shunt capacitance to common, plus the amplifier's internal unbalances) reduce the common-mode rejection, by producing a normal-mode signal. This source of error can be greatly reduced by proper use of a guard shield, as shown in Figure 12.9.

Here, no part of the common-mode signal appears across the capacitors C_A and C_B, since the shield is driven by the source of the common-mode signal. The shield also provides electrostatic shielding to minimize coupling to other signal lines in close proximity to the input leads.

When installing a guard shield, it is important that the guard shield connect only at one point, to the source of the common-mode signal, and that the shield be continuous, i.e., through multiplexers, connectors, patch panels, etc. Since the shield is carrying a common-mode signal, it should be properly

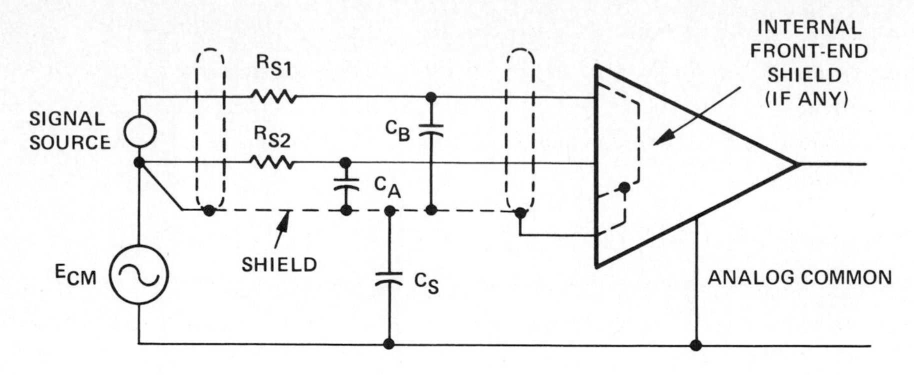

Figure 12.9. Use of guard shield to improve common-mode rejection at higher frequencies.

insulated to prevent it from shorting to other shields or the earth ground. A final precaution that should be taken is to make sure that a conductive return path exists for the bias and leakage currents of the differential amplifier (unless it is a true isolator with transformer- or optically isolated, floating inputs).

It is helpful, in reducing noise and improving common-mode rejection, to connect the largest tolerable capacitance *between* the input leads. It will provide some filtering, and will reduce the capacitive unbalance by more than its ratio to the stray capacitance. (Figure 12.10)

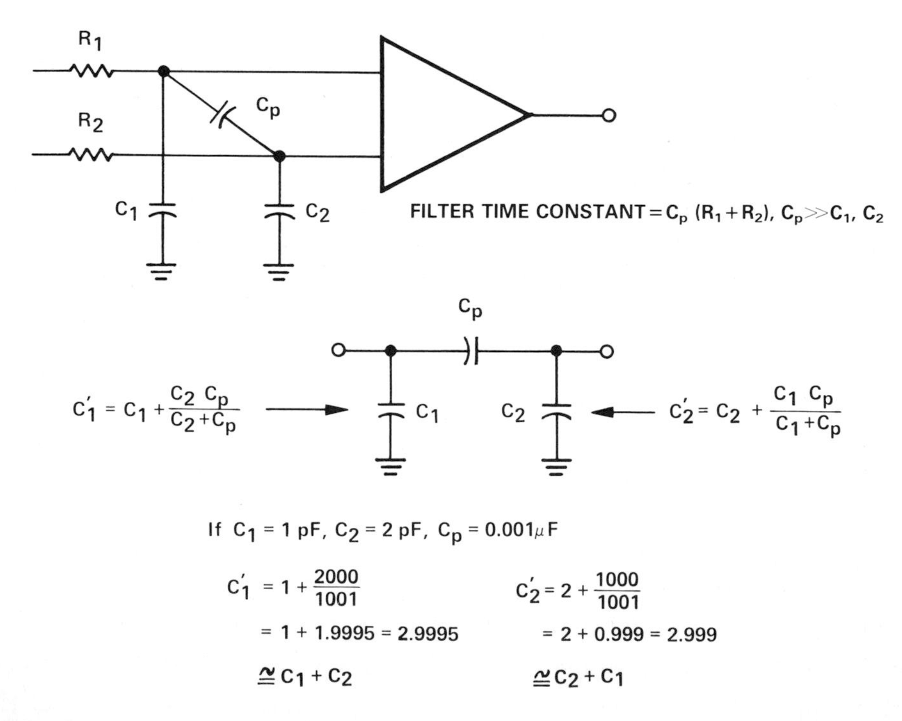

Figure 12.10. Capacitance between the input leads to reduce unbalance and provide filtering.

In portions of a system where differential amplifiers are not used, i.e., where signals are all treated as single-ended signals having a common ground, sufficient precautions should be taken to insure that significant voltages are not induced in ground return leads to the single-point ground, and that the system is free from ground loops.

12.3 HOW TO ADJUST ZERO AND GAIN OF CONVERTERS

Many converter types are pretrimmed and require no adjustments as purchased to meet specified accuracy. However, there are cases where (1) adjustments are called for in the specifications, (2) cost-savings can be effected by trimming inexpensive pre-trimmed devices, with modest specs, for use over narrow ranges for better-than-specified accuracy, or (3) long-term corrections are necessary during years of operation. If adjustments are desired or required, for any of the above reasons, these general principles and guidelines may be helpful. Naturally, the user should follow the specific instructions published in product data sheets, especially where there are conflicts.

Proper adjustment of zero and gain in DACs and ADCs is a procedure that requires great care and extremely sensitive reference instruments. The voltmeter used to read the output of a DAC, or the voltage source used as a driving signal for the ADC, must be capable of stable and clear resolution of 1/10 LSB at both ends of the range of the converter; e.g., at zero and full scale.

Most DACs and successive approximation ADCs manufactured by Analog Devices are provided with Zero and Gain adjustments *which are completely independent of each other*, as long as the adjustment of Zero is attempted *only* when the input code is calling for Zero, and as long as the Zero (or Offset) adjustment is accurately completed before proceeding to adjustment of Gain (at Full Scale − 1 LSB). Of course, it is possible to make Zero and Gain adjustments in reverse order and at other points on the transfer function—but it must be expected that the adjustments will no longer be independent, and the procedure will require a series of successive approximations.

12.3.1 ADJUSTMENT PROCESS

Particularly for bipolar converters, fast and successful adjustment requires knowledge of the technique used in the circuit to configure the inherently unipolar DAC or ADC for bipolar operation.

1. Sign & Magnitude Codes are generally obtained by use of a unipolar converter with separate means of reversing polarity. The Zero adjustment is always made by calling for a zero from the converter. (Logic zero into a DAC produces zero volts output, or zero volts into an ADC produces data-zero output.)

2. Bipolar binary converters utilizing offset binary or twos complement coding usually employ analog offsetting to convert a unipolar design into bipolar.

For instance, a 0 to +10V DAC may have its output amplifier offset by −5V, resulting in an output of −5 volts corresponding to 00 . . . 0 input and +5 volts (minus 1 LSB) corresponding to a 11 . . . 1 input. *Such a converter should have its "Zero" adjusted at* −5V (100 . . . 0 in 2's complement).

Stated another way: with positive-true logic, converter Zero controls should always be set at the "All Bits Off" condition, and then Gain should be set at the "All Bits On" condition.

12.3.2 ADJUSTMENT FOR DACS

ZERO: set the input code so that all bits are "off", then adjust the pot until the output signal is within 1/10 LSB of proper reading, or zero.

GAIN: set the input code so that all bits are "on", then adjust the pot until the output signal reads within 1/10 LSB of *Full Scale less 1 LSB*.

12.3.3 ADJUSTMENT FOR ADCS

ZERO: set the input voltage precisely at ½ LSB above the "all bits off" specified input. The zero control should be adjusted so that the converter just switches in its LSB.

GAIN: set the input voltage precisely at ½ LSB less than "all bits on" input. Note that this is 1½ LSBs less than the nominal full scale value: i.e., all-1's value of a 0 to +10V 12-bit ADC is actually +9.9976V. The gain adjustment should be made with an input ½ LSB less, or +9.9963 volts. With the input voltage set as described, the GAIN control is rotated to the point where the last bit just comes on. For instance, in a 12 bit binary converter, a reading of 1111 1111 1110 would change to 1111 1111 1111.

It is important to note that this discussion is relevant for *offset-binary, positive-true* coding. For 2's complement, the "all-bits-off," positive-true condition is 100 . . . 00 and "all bits on" is 011 . . . 11. For negative-true devices, the "all-bits-off" condition is 111 . . . 11 in offset binary, 011 . . . 11 in two's complement. When in doubt (or to avoid doubt), consult the data sheet.

Resolution	½ LSB	1 LSB	FSR − 1½ LSB	FSR − 1 LSB	FSR
8 bits	39 mV	78 mV	19.88 V	19.92 V	20.00 V
10 bits	9.8 mV	19.5 mV	19.971 V	19.980 V	20.000 V
12 bits	2.4 mV	4.88 mV	19.9927 V	19.9951 V	20.0000 V
14 bits	610 μV	1.22 mV	19.9982 V	19.9988 V	20.0000 V
16 bits	152 μV	305 μV	19.99954 V	19.99969 V	20.00000 V

Table 12.1. Checkpoints for unipolar d/a and a/d Converters with *20-V* Full-Scale Range; Subtract 10 V for ± 10 V range.

Resolution	½ LSB	1 LSB	FSR – 1 ½ LSB	FSR – 1 LSB	FSR
8 bit	19.5 mV	39 mV	9.941 V	9.961 V	10.000 V
10 bits	4.9 mV	9.8 mV	9.985 V	9.990 V	10.000 V
12 bits	1.2 mV	2.4 mV	9.9963 V	9.9976 V	10.0000 V
14 bits	305 μV	610 μV	9.99908 V	9.99939 V	10.00000 V
16 bits	76 μV	153 μV	9.999771 V	9.999847 V	10.00000 V

Table 12.2. Checkpoints for d/a and a/d Converters with 10-V Full-Scale Range (0 to 10 V); subtract 5 volts for ± 5 V.

Resolution	½ LSB	1 LSB	FSR – 1 ½ LSB	FSR – 1 LSB	FSR
8 bits	9.8 mV	19.5 mV	4.971 V	4.980 V	5.000 V
10 bits,	2.4 mV	4.9 mV	4.9927 V	4.9951 V	5.0000 V
12 bits	610 μV	1.2 mV	4.9982 V	4.9988 V	5.0000 V
14 bits	153 μV	305 μV	4.99954 V	4.99969 V	5.00000 V
16 bits	38 μV	76 μV	4.999885 V	4.999923 V	5.000000 V

Table 12.3. Checkpoints for d/a and a/d Converters with 5-V Full-Scale Range (0 to 5 V); subtract 2.5 V for ± 2.5 V.

12.4 OTHER WAYS TO IMPROVE PERFORMANCE

12.4.1 PRESERVING DAC ANALOG OUTPUT ACCURACY.

In all too many applications, a DAC's accuracy is subverted by the associated op-amp circuitry.

Dynamic Problems. The output impedance of a current-mode DAC can generally be treated as a parallel combination of resistance and capacitance. But capacitance at the output of a DAC (Figure 12.11) can produce undesirable results. The capacitance and the feedback resistance combine to add a pole to the open-loop response, resulting in reduced phase margin, which will hurt the closed-loop response.

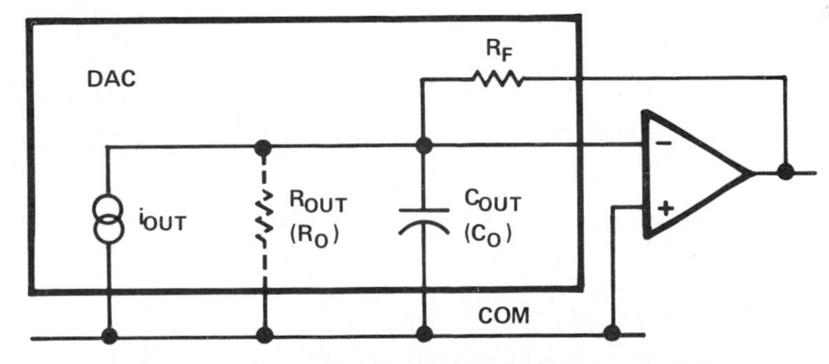

Figure 12.11. Equivalent circuit of current-output ADC.

Figure 12.12 shows how the open-loop amplitude and phase response might appear if the spurious pole due to C_o is below the crossover frequency of the undisturbed system. Not only will the closed-loop bandwidth be reduced, but—more seriously—excess phase shift will be introduced. The extra phase shift reduces the system frequency stability margins and may cause ringing (and perhaps even oscillation).

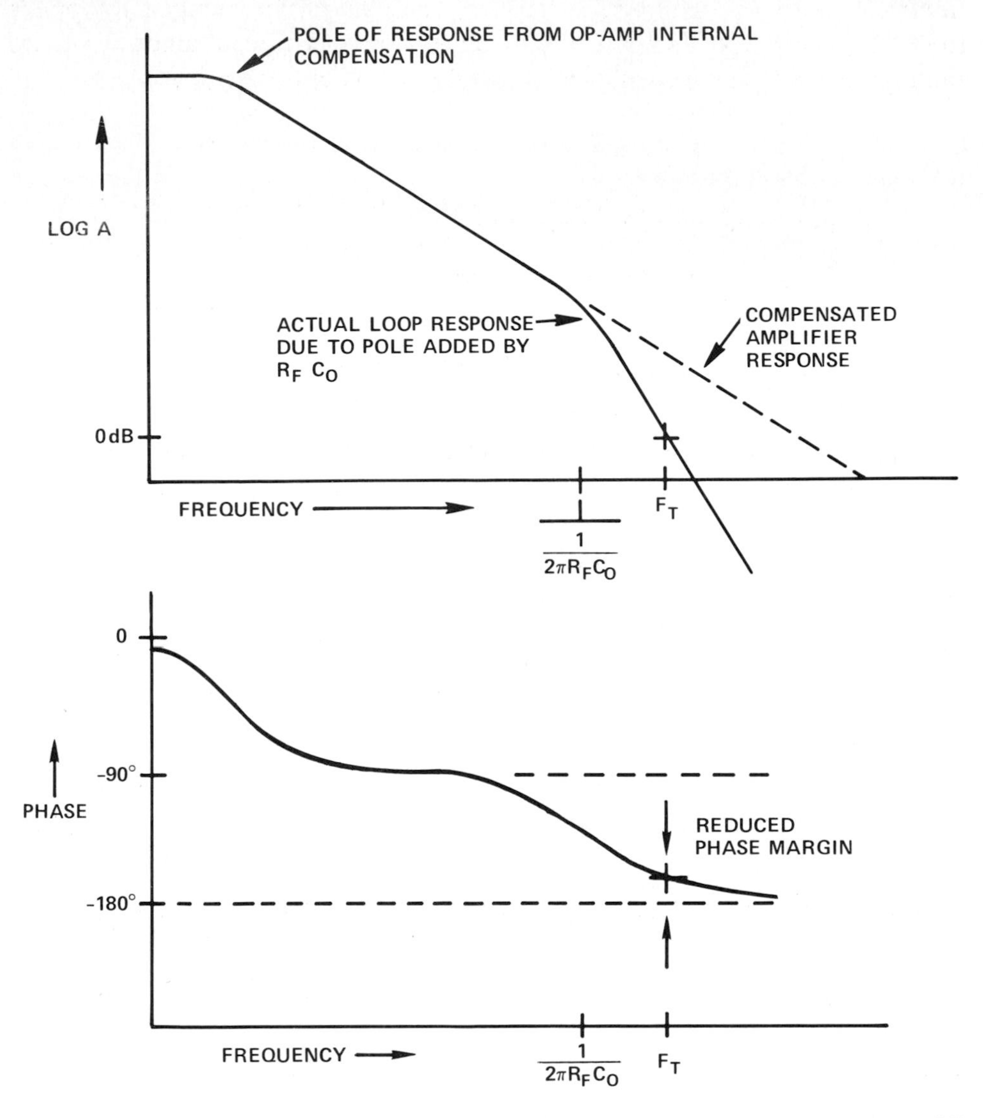

Figure 12.12. Amplitude and phase response of the circuit of Figure 12.11. The additional pole increases settling time by reducing bandwidth and increasing both overshoot and ringing.

As Figure 12.13a shows, the loop-stability margins can be restored by connecting a feedback capacitor, C_F, in parallel with the feedback resistor. This capacitance creates a zero in the open-loop transfer function, which can be adjusted to correct the phase margin. However, if R_o is quite large (as is often the case with current-output DACs), a large pole-zero mismatch may remain (Figure 12.13b).

Even with finite values of R_o. a small residual pole-zero mismatch (Figure 12.13c) may result. In addition, pole-zero mismatch within the amplifier itself can lead to long-settling "tails": the DAC output voltage may appear to settle quickly, but then it slowly changes—by a significant amount—to its final value, over the course of tens of microseconds, or even milliseconds.

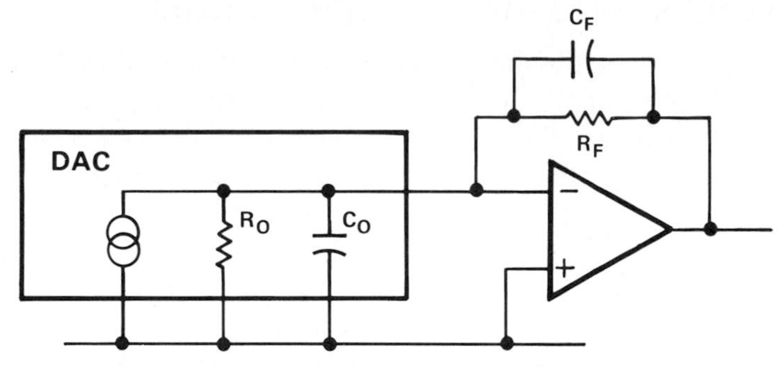

a. Improving loop stability by the use of feedback capacitance, C_F.

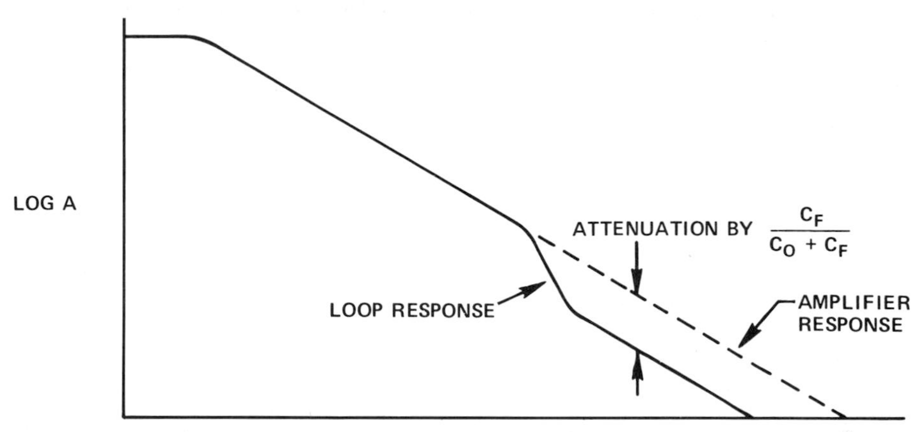

b. Response of circuit (a), neglecting R_o. Pole-zero mismatch may yield poor transient response.

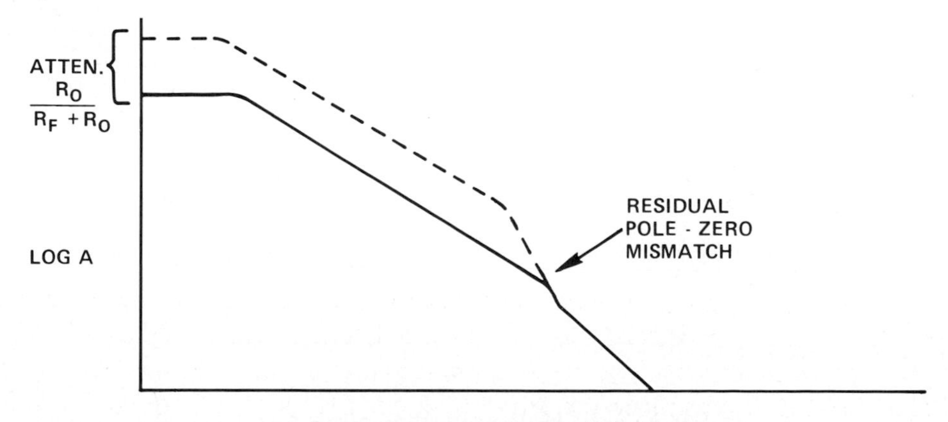

c. Response of circuit (a) with finite R_o.

Figure 12.13. Use of feedback capacitor.

The residual external mismatch will be eliminated when the DAC's output circuit and the feedback network form a frequency-compensated voltage divider, i.e., when $R_o C_o = R_F C_F$. This condition can usually be satisfied, but sometimes it requires large values of C_F. Unfortunately, C_F—which is used to introduce an open-loop zero—also produces a closed-loop pole, which reduces the overall bandwidth and results in increased settling time.

R_F is generally fixed by the desired DAC gain; the minimum value of C_o is a property of the converter that is not under the system designer's control. Therefore, C_F and R_o are the only two parameters that can be manipulated to improve performance (by shunting). As R_o' (the effective value of R_o) is reduced by shunting the DAC output with a resistor, the required value of C_F is reduced, and the closed-loop bandwidth is increased (Figure 12.14a). The unity-gain bandwidth of the op amp, b, limits the open-loop system bandwidth, which—in turn—limits the realization of closed-loop bandwidth. As R_o' is reduced, the *open-loop* bandwidth obtainable for a fixed op-amp bandwidth, b, is also reduced (b).

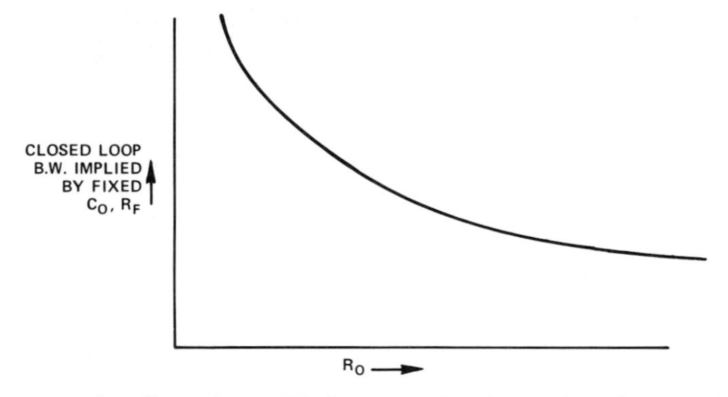

a. Smaller values of R_o increase the closed-loop bandwidth.

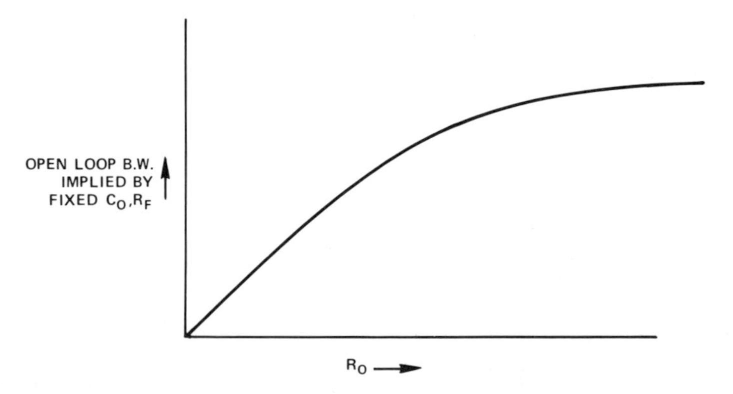

b. Smaller values of R_o decrease the open-loop bandwidth.

Figure 12.14. Effect of varying R_o (R_o') on open-loop and closed-loop bandwidth.

An empirical compromise can be reached by adjusting R_o' to provide the same open- and closed-loop bandwidth. For a fixed C_o and R_F, the values of R_o' and C_F can be determined from

$$R_o' = \frac{1 + \sqrt{1 + 8 b \pi R_F C_o}}{4 b \pi C_o}$$

$$C_F = \frac{1 + \sqrt{1 + 8 b \pi R_F C_o}}{4 b \pi R_F} \qquad (12.3)$$

Offset Problems in CMOS DACs.

Perhaps the best way to control V_{os} in an op amp used with a DAC is at the source—choose an op amp with sufficiently low offset and bias current over the temperature range. The next best way is to null the op-amp's offset by the manufacturer's standard recommended V_{os} trim, taking pains to connect the pot wiper to the appropriate supply terminals *at the device*.

The amplifier's offset-trim adjustment should be used *only* for V_{os} nulling; if it is used to compensate for offsets caused by the flow of bias current through the feedback resistor, as well as for offsets occurring in external circuitry, the amplifier input stage will have to be unbalanced, which will cause its V_{os} tempco to be degraded.

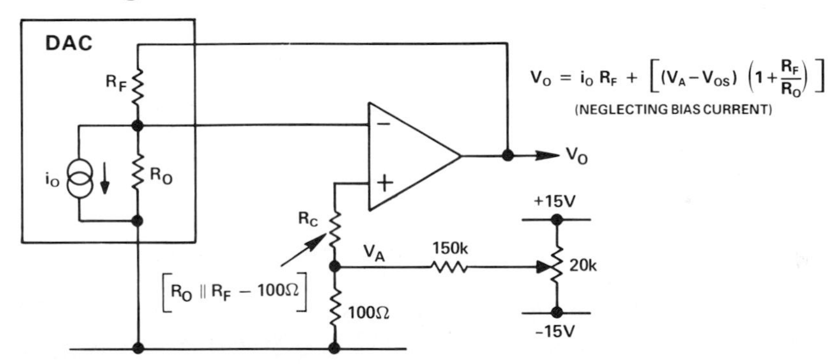

a. Nulling offset with voltage applied to op-amp reference input.

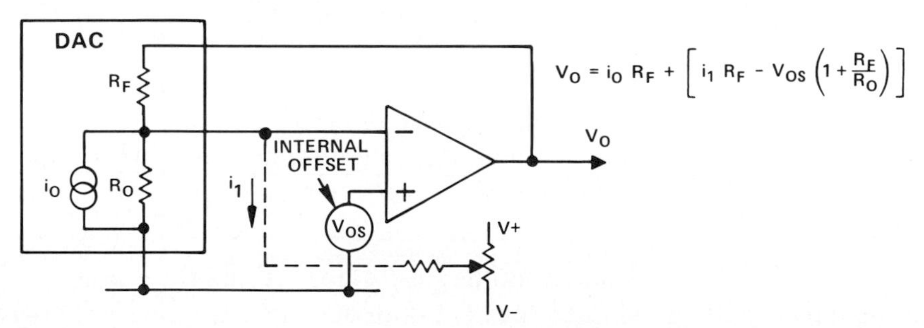

b. Nulling offset with current added at op-amp summing point (not recommended).

Figure 12.15. External offset-null methods.

If the amplifier lacks offset terminals, there are two commonly used ways of providing the trim; they are shown in Figure 12.15. The more-desirable approach is shown in (a); the correction is applied to the amplifier's positive input terminal, as a voltage. Since it is effectively in series with V_{os}, the V_{os} correction is unaffected by changes of R_o'; if the amplifier has low bias current, the output offset will be independent of changes of R_o'.

The less-effective way (popular, but not recommended) is to introduce a current at the summing point, as shown in (b). If the resistances in the circuit (including R_o) are constant, there is no problem. However, if R_o' can vary, the output offset will change, even if bias current is negligible. If the change of R_o' is a function of the applied *digital code*, the result can be increased differential nonlinearity.

For example, if the DAC is an inverted R-2R-ladder type, as shown in Figure 12.16, the output resistance, R_o, is a function of the number of switches that are closed. If all switches are open, R_o is infinite; if all (or many) switches are closed, R_o will approach R; and if only a single switch is closed, R_o will be about 3 R.

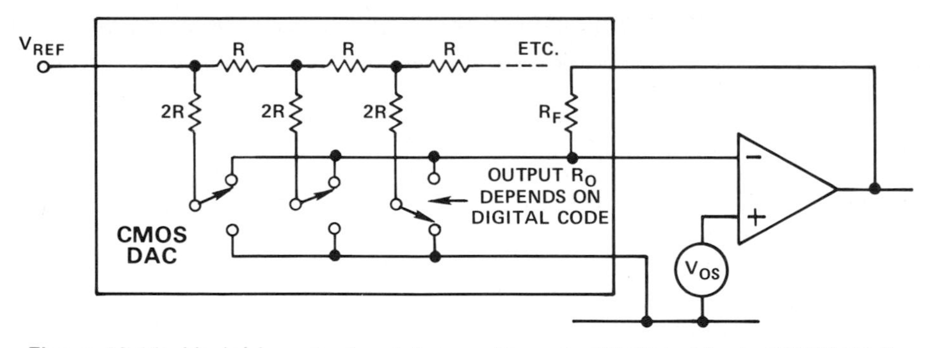

Figure 12.16. Variable output resistance of inverted R-2R ladder in CMOS DACs.

If R = 10 kΩ, the resistance looking back into the network is about 10kΩ for codes containing more than about four 1's, and 30 kΩ for a single 1. Thus, for the one-bit transition from 0011 1111 1111 to 0100 0000 0000, the error voltage, V_{os} $(1 + R_F/R_o)$, changes from 2 V_{os} to (4/3) V_{os}. If the offset had been nulled by current summation at all-zeros ($1 + R_F/R_o = 1$, since $R_o \longrightarrow \infty$), the offset error will be $+V_{os}$ at the first code and $+(1/3)$ V_{os} at the second code; the incremental change of error will be $(-2/3)$ V_{os}. If V_{os} is not much smaller than the voltage equivalent of the least-significant bit (see Table 12.1), differential nonlinearity errors can result in nonmonotonic behavior.

The solution to this dilemma is simple: use (a) for zeroing the amplifier. Better yet, use an amplifier having extremely low offset and negligible bias current.

"Foreign" currents in common ground and power lines can introduce offset, noise, and other errors that will be amplified in the same way as V_{os} errors.

It is important to refer the amplifier circuit (and its external V_{os} trim), the load across which the output voltage is being developed, and the DAC's reference input—all of these—to the DAC terminals, in the manner shown in Figure 12.17.

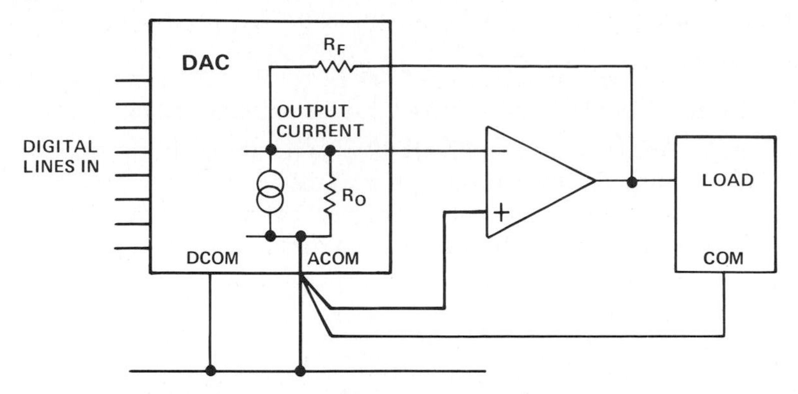

Figure 12.17. Referring buffer amplifier and load circuits to analog common.

12.4.2 MORE ON BYPASSING AND GROUNDING

In "virtual-ground" systems, such as an op amp, driven by a current-output DAC, the DAC output current doesn't actually return to ground, but to one of the power supplies, by way of the op amp's output stage (Figure 12.18). To reduce the impedance in the high-frequency current path, the bypass capacitor should be connected so as to return the currents from one (or both) power terminals to ground *at the DAC*. If the DAC output is active, it may require bypassing of its own supplies for the same reason.

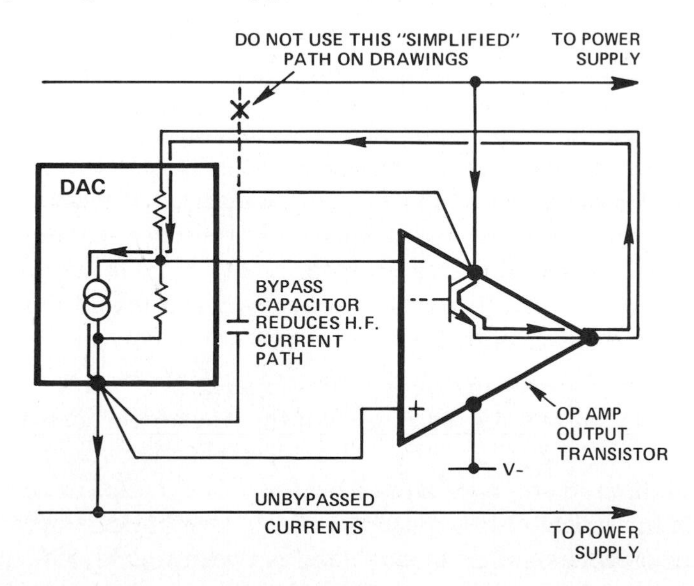

Figure 12.18. Bypassing power supplies for virtual-ground applications. Arrows show unbypassed current flow.

WARNING: You and your drafting department may have conflicting objectives. Your objective is to design circuits that work and communicate the important details to whoever assembles them. Your drafting department (or so it may often seem) has the objective of drawing nice, neat, squared-off diagrams, in which the lines representing conductors are remarkably equipotential. You may have noticed that, where it counts in Figures 12.17 and 12.18, these niceties have been avoided. The lines are configured to resemble closely the job that the wires perform, converging at the common analog connection.

For example, the bypass lead, in Figure 12.18, though artistically squared off, wends its way purposefully in the direction of the op-amp's power-supply terminal, rather than shooting straight up to meet the power-supply line (a sure recipe for costly debugging). If you think your drafting department may have a mind of its own, you may want to include a special message for the person who builds the circuit, to be sure that it gets built the way you want it built. If you have a computer-aided-engineering setup that allows you to create drawings that directly reflect your wishes, this is one less matter for you to worry about.

Figure 12.19a shows an example of ineffective decoupling. Here the op amp drives a load, which connects to a long ground line (returning to the power-supply terminal), and the supply decoupling for the amplifier returns to the power supply through another long line. The return path for the load current is as long as, or longer than, the supply lines powering the op amp. The "local" decoupling is not only ineffective; it may actually contribute to noise on the power-ground bus.

The cardinal rule of decoupling is:

Make it easy for the current to get back by the shortest path.

Figure 12.19b shows a more effective scheme, in which the decoupling capacitor connects by the shortest path between the load return and the load-voltage control element. Here an op amp, swinging a resistive load-circuit negative, drives the load from an internal PNP transistor, connected to V−. Decoupling the V− pin of the op amp to the low side of the load provides the most direct return path for high-frequency currents and bypasses them around ground and power buses.

Well-designed data sheets will offer specific suggestions for grounding and/or decoupling. For example, the data sheet for the Analog Devices AD390 quad 12-bit d/a converter suggests: "The power supplies used with the AD390 should be well filtered and regulated. Local supply decoupling, consisting of a 10-μF tantalum capacitor in parallel with 0.1-μF ceramic, is suggested. The decoupling capacitors should be connected between the AD390 supply pins and the load ground (ideally the AGND pin). If an output booster is used, its supplies should also be decoupled to the load ground."

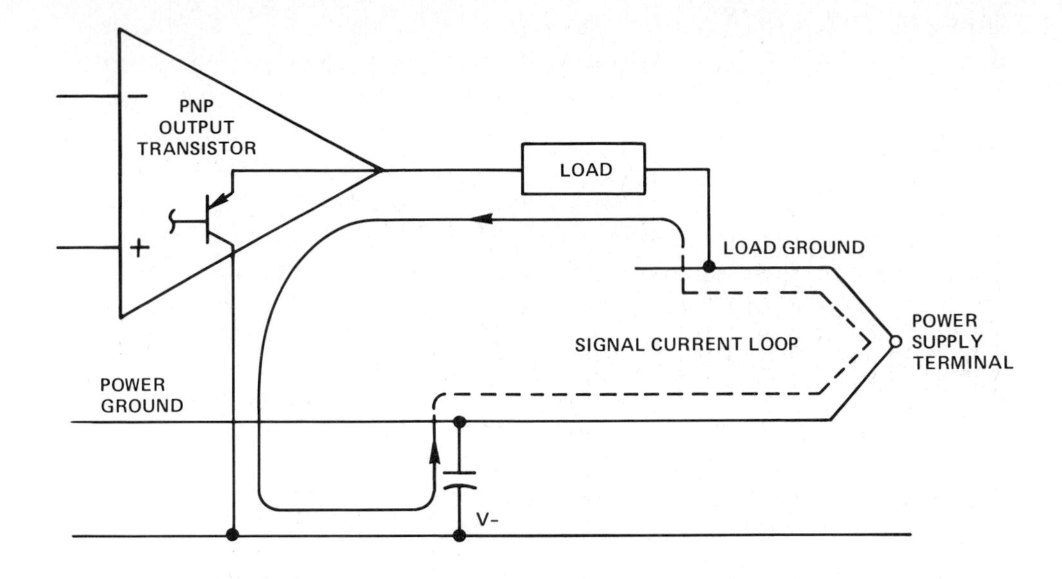

a. Decoupling for negative supply is ineffective.

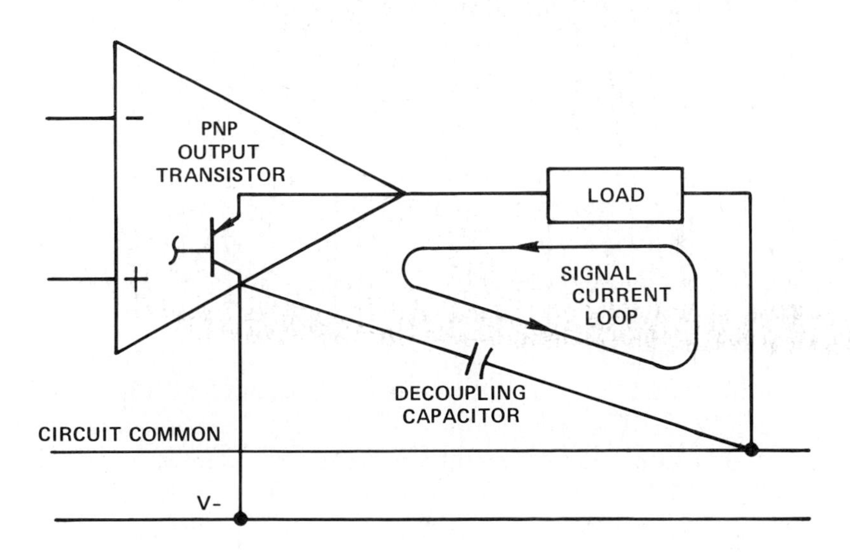

b. Decoupling of negative supply is optimized for "grounded" load.

Figure 12.19. Effective and ineffective decoupling.

Substitutes for Ground

In large systems, it is often impractical to rely on a single common point for all analog signals. In these cases, some form of differential (or even *isolation*) amplifier is required to translate signals between ground systems. For the inveterate op-amp user, a simple subtractor, or "dynamic bridge" circuit may come to mind. These circuits translate a signal which is referred to one ground system into a similar or amplified signal, referred to a different ground signal (Figure 12.20). The common-mode rejection of the amplifier and a resistance-ratio match are used to eliminate the effects of voltage differences between the two grounds, or common points.

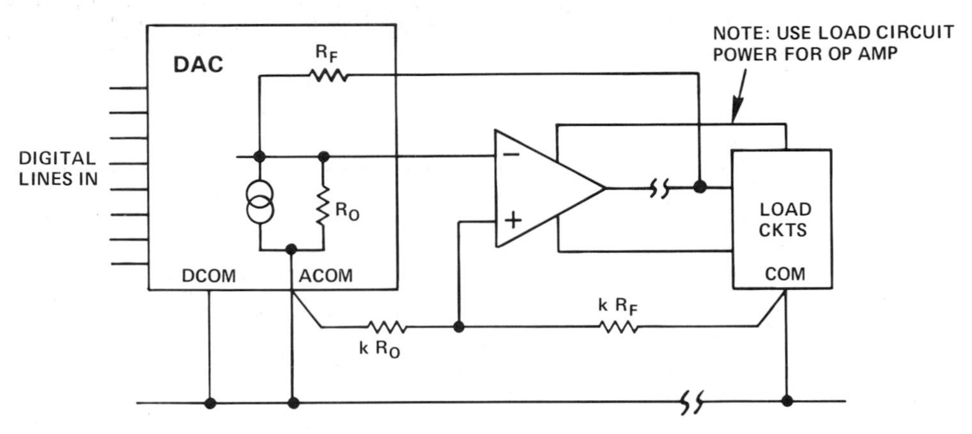

Figure 12.20. Use of differential amplifier to eliminate the effects of common-mode voltage.

It is generally wise to power the op amp from the power available at the *load* side of the circuit, and/or to decouple it with respect to the *load* common. The reason for this can be deduced from the circuit architecture of the most-common types of op amps (Figure 12.21).

An op amp converts a differential input signal to a single-ended output signal. In many popular op amps, the differential-to-single-ended conversion is done with respect to V− (but some use V+), and the resulting signal drives an integrator[2]. The integrator characteristic is used to frequency-compensate the amplifier, and the integrator input is referred to the single-ended output, at V−. The integrator acts as a unity-gain follower for fast signals applied to its non-inverting (or reference) input. As a result, signals applied to the V− terminal have their high-frequency components conveyed directly to the output. Signals having frequency components *above* the amplifier's *closed-loop* bandwidth will be transmitted from V− to the output with little or no attenuation.

[2]The Application Note mentioned in footnote (1), "An I.C. Amplifier User's Guide to Decoupling, Grounding, and Making Things Go Right for a Change," provides considerable detail regarding the integrator-reference and compensation schemes of more than 40 device families.

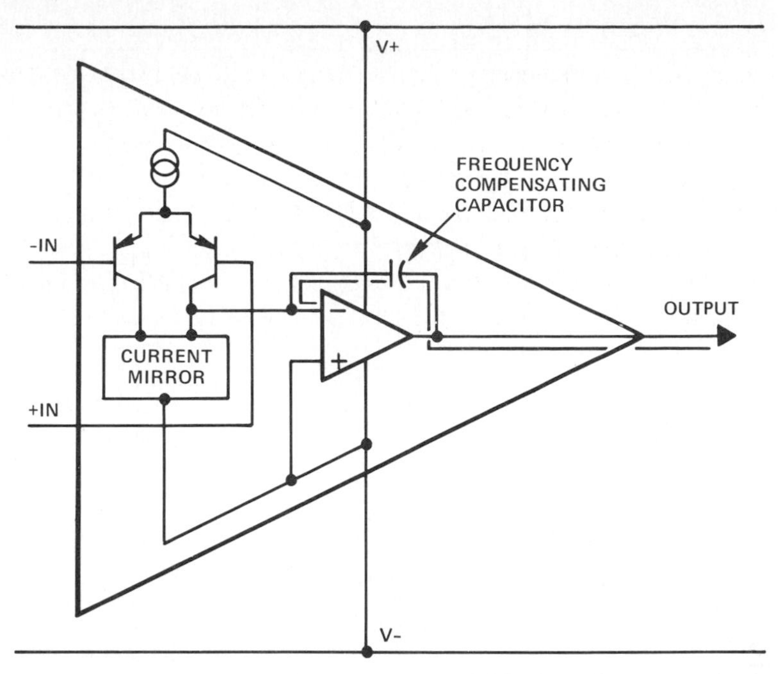

Figure 12.21. Typical op-amp circuit architecture. Reference for output integrator is V−.

As Figure 12.22a shows, if the op amp used as a subtractor amplifier is powered from or bypassed to the same common line as the input signal, any high-frequency signals associated with the common will appear as part of the output signal. If the ground noise includes appreciable high-frequency noise (such as are produced by logic currents), the common-mode rejection will be defeated.

If, on the other hand (Figure 12.22b), the op-amp supply terminals are referred to the *output* signal common, no extraneous signals are coupled into the integrator. Any ground noise appears as a common-mode input signal and is reduced by the common-mode rejection of the amplifier (which is typically very much better than the negative-supply rejection at high frequencies).

Since noise-rejection performance of the subtractor depends on carefully matched source and feedback resistance ratios, it cannot be used in all situations. Whenever the source impedance cannot be controlled, or is exceptionally high, the subtractor (or dynamic bridge) becomes impractical. In this situation, ground noise and other remote-grounding difficulties can often be avoided by the use of an *instrumentation amplifier*.

IC instrumentation amplifiers, such as the AD524, accept differential input signals at high impedance, provide a fixed gain (which can be selected without introducing overall feedback that joins the input and output circuitry), and develop the output voltage with respect to a reference terminal, which may be connected to the input common of a remote load circuit (Figure 12.23).

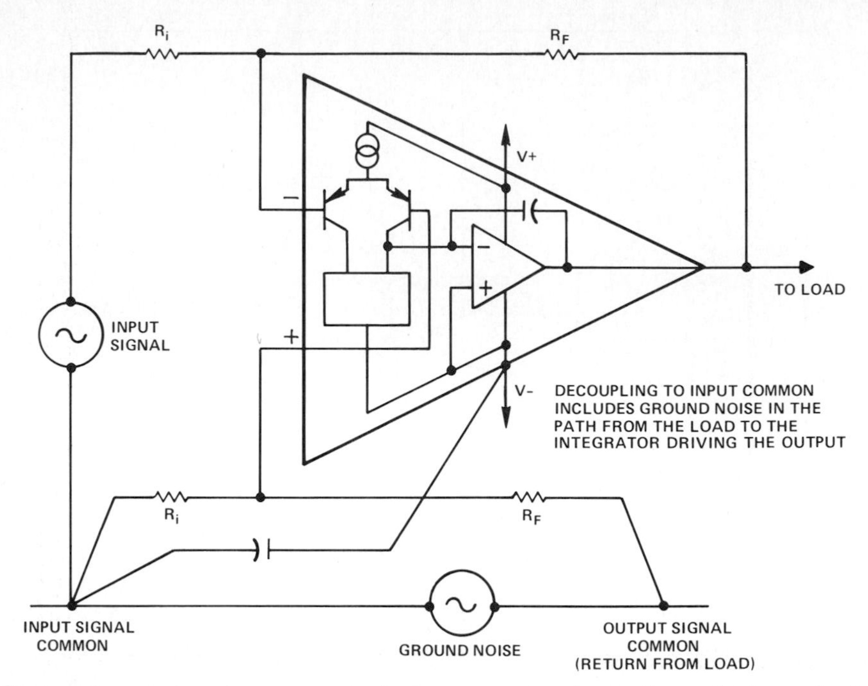

a. Decoupling to *input* common includes ground noise in the path from the load to the integrator driving the output.

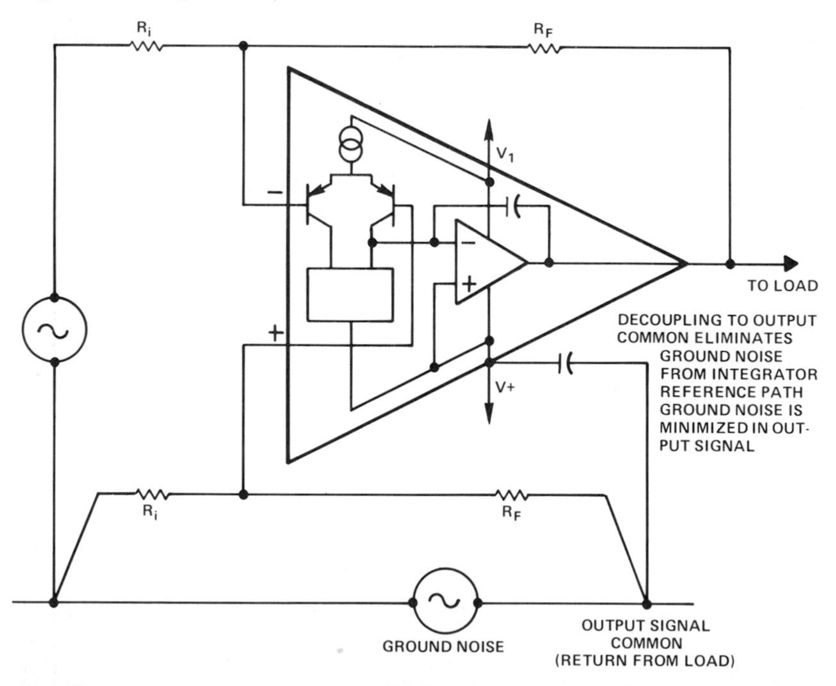

b. Decoupling to *output* common eliminates ground noise from the integrator reference path. Ground noise is minimized in the output signal.

Figure 12.22. Proper and improper decoupling of subtractors using op amp with integrator referred to V−.

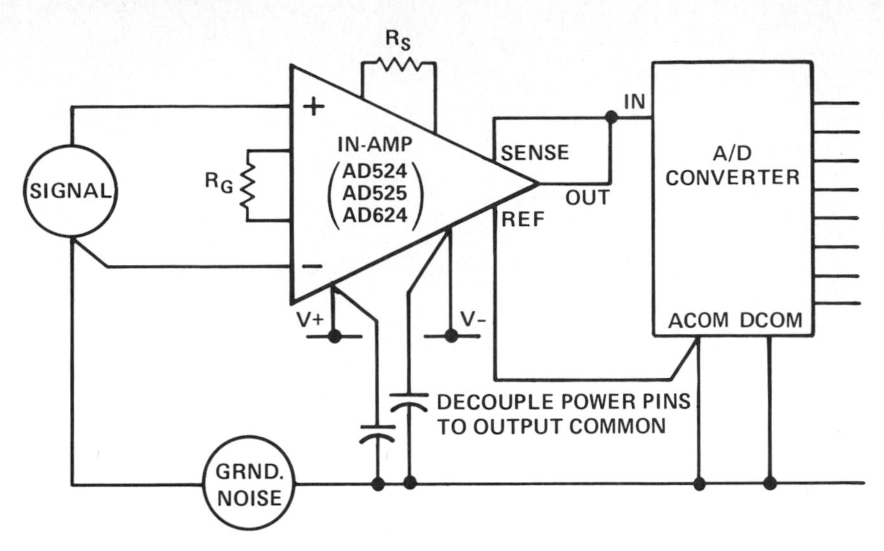

Figure 12.23. Instrumentation amplifier interfaces separate ground systems.

Some instrumentation amplifiers are quite versatile and can provide additional functions, while isolating the common returns. For example, the output-reference terrminal can be used to add a fixed or variable bias voltage to the output.

If the common-mode voltages are very large, or if galvanic isolation is essential for safety, isolation amplifiers—or amplifiers powered by isolated dc-to-dc converters—may be highly desirable.

PART III

ANALOG-DIGITAL CONVERTERS FOR SPECIAL APPLICATIONS

Chapter Thirteen

Video Converters

Video-speed a/d and d/a converters, with conversion rates in excess of one million words per second, present a unique challenge to the designer and user. Converter architecture, performance specifications, application rules, and testing procedures all differ significantly from those for lower-speed devices.

In the case of d/a converters, for instance, the accuracies at low speeds depend upon the design of the resistor network, the stability of the reference voltage, and nonlinearities within the converter's op amp. But as speed and resolution increase, full-scale settling time and the magnitude and duration of "glitches" take on overriding importance.

In the case of a/d converters, as conversion rates pass through the 1 MHz mark, numerous application-dependent performance parameters begin to deteriorate. Spurious in-band harmonic products—created by missing codes, differential nonlinearities, and other nonlinear frequency-dependent effects—may cause signal-to-noise levels and total harmonic distortion to degrade rapidly in improperly designed applications.

This chapter is intended to provide appreciation for and guidance in the design, specification, testing, and application of high-speed data converters.

13.1 APPLICATIONS OVERVIEW

High-speed a/d and d/a converters find use in a wide range of video, digital signal-processing, radar, transient-detection, and communications applications—where the analog bandwidths require digital word rates in excess of 1 million words per second. For these applications, converters must be characterized, not only by specifications common with lower-speed converters (such as linearity, temperature coefficients, etc.), but further, in terms that are heavily application oriented. Table 13.1 lists typical examples of specifications, that in general have not been emphasized elsewhere in this book, and the applications where they are of interest.

We will briefly illustrate the range of high-speed data-converter applications, then examine converter specifications in greater depth.

SPECIFICATION	APPLICATION
Signal-to-noise ratio (S/N)	Radar, communications spectrum analysis
AC linearity	Radar, spectrum analysis
Noise power ratio (NPR)	Communications
Two-tone intermodulation distortion	Radar spectrum analysis
Transient response	Transient analysis, radar
Overvoltage recovery	Radar
Aperture uncertainty	All
Differential phase	Television
Differential gain	Television

Table 13.1. Application-specific specifications.

13.1.1 DISPLAY SYSTEMS

Driven to ever-higher scanning speeds to obtain increased resolution without flicker, new-generation raster display systems require extremely fast information transfer. This topic has been discussed at length in Chapter 6. It will suffice to reiterate here why high-speed digital-to-analog converters are needed.

The most popular forms of display systems include raster-scan alphanumerics, raster-scan graphics, vector-scan graphics, and storage tubes. Although there are differences in the way they operate, as well as certain characteristics that might suggest one type or another for various applications, display systems for digital information generally require some form of high-speed digital-to-analog converter for beam positioning or intensity modulation.

The raster-scan system, shown in Figure 13.1, is a good example for the purpose of illustrating the role of d/a conversion. Figure 13.1 depicts a typical computer-controlled raster-scan graphic display, using a single d/a converter to control beam intensity. In a color display, three such converters are required (and are available as a triad in a single package), one for each color gun of the CRT (red, green, blue). The system consists of a MOS random-access memory buffer for storing display data in digital form, one or more memory controllers for managing the updating of the display and controlling the refresh cycle of the CRT, and a programmable microprocessor for generating display graphics and manipulating the image. The entire system operates as an intelligent peripheral to a host computer; most of the processing associated with image and graphic display is down-loaded to the graphic subsystem.

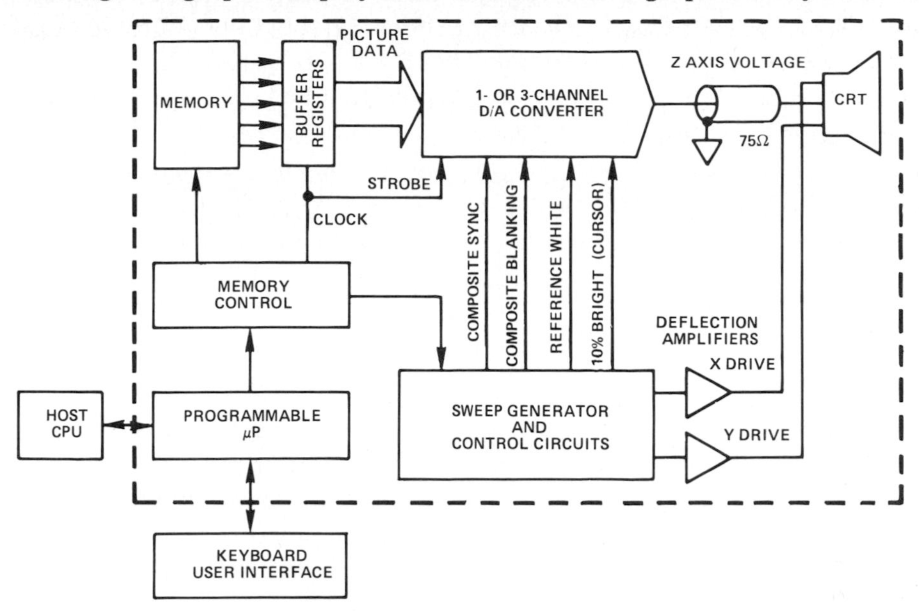

Figure 13.1. Digital display system.

The high-speed data conversion challenge becomes apparent when data rates for such displays are examined. If the picture resolution is specified as $1{,}024 \times 1{,}024$, there are 1,024 horizontal lines; each line has 1,024 independent dots, and each dot has its own programmable intensity level. Thus, there are 1,048,576 picture elements (pixels). If the picture on the CRT is to be refreshed 60 times per second, then a new pixel must appear on the screen at least every 15.9 nanoseconds, even without considering "overhead time," associated with vertical and horizontal blanking during the sweep-retrace portion of the cycle).

The d/a converter controls the Z axis of the CRT, to modulate the brightness of the raster-scan beam. For the example given above, the d/a converter must be capable of being updated somewhat faster than a pixel rate corresponding to 15.9 nanoseconds; it should be capable of settling to a new value in less than 10 nanoseconds. If the display plots a white dot on a black background, the d/a converter output must make a full scale transition between adjacent pixels. For sharp, clean raster-type displays, it is evidently of vital importance that the output d/a converter have fast settling times with imperceptible transient "glitches."

13.1.2 DIGITAL SIGNAL PROCESSING

Digital signal processing (DSP) is a broad term which encompass a wide variety of digital techniques for the solution, manipulation, enhancement, and presentation of information in instrumentation and systems applications.

One of the more dramatic uses of DSP, which probably has the most familiarity to engineers not otherwise engaged in it, is *image processing*. Although many were unaware of it, the startling degree of detail to be found in images of the moon and Mars, displayed on home TV sets during the 1970's, and thereafter, was the result of digital techniques designed to extract a maximum amount of information for each image.

Both optical and digital techniques are used in image processing; optical techniques are simpler, but digital techniques are more flexible and have benefited by the increasing speed and availability of devices for digital data-handling technology—including analog-to-digital converters—and their decreasing cost. Because of the large amount of information typically handled in digital signal-processing, converters having both wide bandwidth and high resolution are generally required for such systems.

DSP, and especially in its applications to image processing, has been of major benefit to many fields; perhaps one of the most important is medicine. Noninvasive tools have become available that permit physicians to observe interior detail and physical functioning of body organs; these tools are helpful in both diagnosis and treatment. The growing list includes computerized axial tomography (CAT scanners), digital fluorography, and phased-array ultrasound. CAT scanners, which employ high-resolution a/d converters, have been discussed in Chapter 6. Here, we will touch on the last two techniques.

Digital Fluorography

Digital fluorography, often called digital radiography, is a computerized x-ray technology that is replacing conventional radiography using photographic film. The technique is relatively inexpensive; it can be incorporated with existing equipment; and it lowers the risks, discomfort, and costs for patients. It also provides more image information because film is limited in its ability to capture the dynamic range of x-ray signals.

X-ray signals consist of high-frequency radiation with the ability to pass through human tissue. The amount of penetration, and the resulting image character, depend on the densities of bone and tissue structure under examination. While a conventional x-ray machine captures these signals directly on a film that is sensitive to x-rays, a digital radiography machine captures x-ray signals with a TV camera, using a high-speed information-processing system to analyze and display the diagnostic image.

The major components of a digital radiography system are shown in Figure 13.2. Basically, the camera and film used in conventional x-ray examinations are replaced by the TV camera, shown in the illustration, along with the associated digital processing elements and display system.

The patient is positioned between the x-ray source and a photomultiplier fluoroscope with an image intensifier, which converts the x-rays into photons

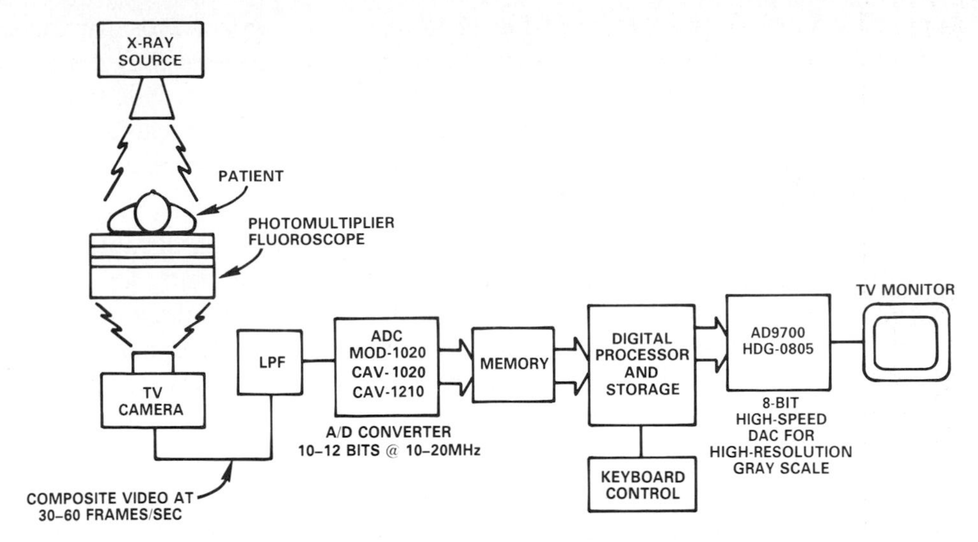

Figure 13.2. Digital fluorography.

(light). The low-level light signals are scanned by a high-quality TV camera, that changes the light into analog electrical signals. The resulting composite-video signal is updated at from 30 to 60 frames per second to eliminate flicker, low-pass filtered, and applied to an a/d converter. The digital data from the converter are stored in memory; and the results of the subsequent image processing can then be stored and/or displayed (essentially in real time, as perceived by the physician) for examination and analysis.

The need for good resolution and wide dynamic range imposes stringent requirements on the system a/d converter, to permit as much information as possible to be captured for use in processing the image. Typically, the ADC must have 10- to 12-bit resolution and be capable of 10-20 MHz word rates.

At the other end of the system, the d/a converter used in the raster-type display monitor must have high gray-scale resolution (for example, 8 bits), so that subtle gradations in tissue density can be observed; if the processor produces an image in color, a triple DAC with good resolution will be needed to display the red-green-blue color pixels.

Compared to conventional x-ray techniques, digital radiography offers improved diagnostic images with less radiation exposure, permitting a safer, less invasive diagnostic procedure. Also, the system's memory and sophisticated information-processing capabilities provide faster image generation and increased flexibility in image display and analysis, enabling the physician to arrive at a diagnosis, and appropriate course of treatment, more quickly.

Digital radiography can be subdivided into categories based on system characteristics, such as types of x-ray beams and detectors, number of pictures taken per second, the type of digital processing used, and—of course—the ultimate clinical application.

A typical example of a major use of digital radiography is *angiography*, an x-ray technique for examining blood vessels and the major organs they supply, especially the heart, head, and neck. Typically, a contrast dye, such as radioisotope iodine, is injected into the patient, and an x-ray scan is taken. The resulting image is then compared with earlier images (or *masks*) of the same area, obtained prior to the injection.

The comparison takes place in the processor by means of a digital subtraction algorithm, which isolates the iodine-infiltrated vessels and organs by eliminating all unaffected tissues from the image. Besides subtraction, the processor also uses contrast enhancement, averaging, edge enhancement, magnification, and other image-processing techniques to help present maximum information to the medical diagnostician.

Digital subtraction angiography (DSA) has emerged as a relatively safe, cost-effective way to examine blood vessels. Patient safety is enhanced, because the technique reduces the need for invasive arterial catheterization. Lower doses of the contrast dye are required for DSA, because image-processing subtraction and enhancement can amplify the image data; this, in turn, improves patient comfort and reduces motion during the procedure. The cost savings are also an important consideration. Elimination of the need for catheterization in many cases can replace a two-day hospitalization by an outpatient procedure. Since the data can be saved on inexpensive magnetic disks or tape, there is also a considerable saving of the cost of film and its storage.

Phased-Array Ultrasound

Phased-array ultrasound is most aptly described as a very-high-frequency sonar system. A typical arrangement is shown in Figure 13.3. An array of from 16 to 32 piezoelectric (crystal) transducers is arranged in a hand-held pickup head, which is placed against the patient's body in the area to be examined. On command from the operator, the transducers are excited with an electrical signal which causes each transducer to transmit a high-frequency (supersonic) pulse.

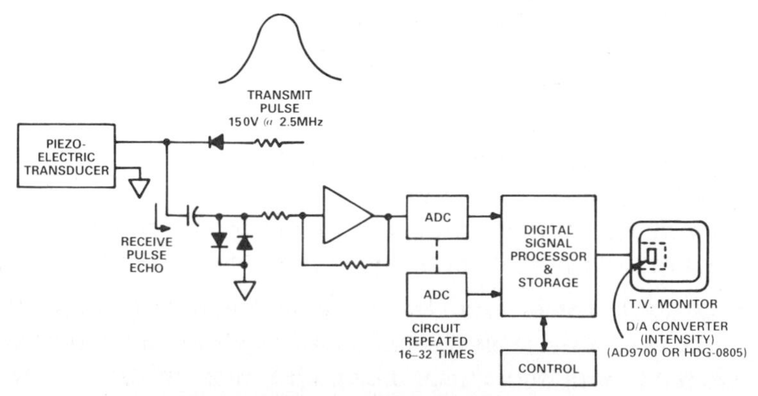

Figure 13.3. Phased array ultrasound.

The excitation is applied to the transducers in some sequence, rather than simultaneously, resulting in a spatially swept acoustic signal; the pattern can be varied by computer control for optimal results in various kinds of examinations.

As the sonic signals impinge on bone and tissue within the patient's body, echo signals will be returned to receiver elements in the array, whose focus is swept in synchronization with the range of returning echoes. These sonic echoes are transformed into electrical signals by the transducers, individually amplified, and applied to each individual channel's a/d converter. The ADC may have a resolution between 6 and 8 bits, and must be capable of word rates between 15 and 20 MHz (and more), for digitizing return signals with bandwidths of from 5 to 10 MHz.

After processing, the signal can be stored and/or displayed on a TV monitor, using a high-speed wide-dynamic-range (8-bit gray scale) d/a converter. Good image reproduction quality is required, because ultrasound systems are real-time diagnostic tools.

The high degree of mobility of the hand-held pickup head used in phased-array systems has made this technique particularly attractive for the study of the heart, because ultrasound is non-intrusive and lessens the physical dangers for the patient. Large strides have been made in recent years in examination techniques and interpretation of the results in this segment of medicine, called echocardiography.

Phased array is just one of several sub-categories in ultrasonic medicine. Others include linear array and mechanical-sector scanning. All are highly demanding on converter performance.

13.1.3 RADAR

In theory, all radars operate in essentially the same way, i.e., a pulse is transmitted from an antenna, and its echo is received from a target. The measurement of the echo's return-time (*range*) and other target information contained in the return pulse are extracted via signal processing.

In *monopulse radars*, the direction from the radar to the target is determined by the direction in which the antenna is pointed. In *tracking radars*, the antenna is continually re-positioned to follow the flight of the target. *Acquisition radars*, designed to show the presence of targets in the zone of interest, are not designed to track the target; instead, they scan periodically to find and indicate that one or more targets are in the vicinity. Performance is limited by the speed with which the familiar parabolic *dish* antenna can be moved.

For most modern applications, the monopulse radar has been supplanted by the *phased-array radar* system, in which the beam is steered electronically. A phased-array antenna comprises a multitude of transmit/receive antenna elements mounted on a flat plane, which is usually fixed in place. The number of elements ranges from 20, or less, for airborne radar, to hundreds for large ground-based radars, such as might be used in missile-tracking systems.

Phased-array systems use the same antenna as both a search radar and a tracking radar by varying the phase- and time delays among the many transmit/receive elements via computers, which are an integral part of the system. In the search mode, the beam is focussed broadly and transmitted at right angles to the plane of the arrays; in tracking, one or more beams are focussed more narrowly and electronically steered as necessary to pinpoint targets and provide accurate tracking data on their locations.

Switching among the many search and track conditions is accomplished at very high rates; this multiplexing achieves the effect of using a single antenna as some number of antennas. Large phased-array systems can outperform parabolic antenna systems of equal size, and they can operate at lower frequencies, of the order of 500 MHz (vs. frequencies in the GHz range).

Processing a radar return signal, or echo, to obtain the maximum amount of information, is a complex procedure, but the technique is not markedly affected by the type of antenna system. Figure 13.4 shows the technique for processing signals in "I and Q" (*in-phase* and *quadrature*) radars. In these systems, the return pulse is separated into two channels of information with a 90° phase difference.

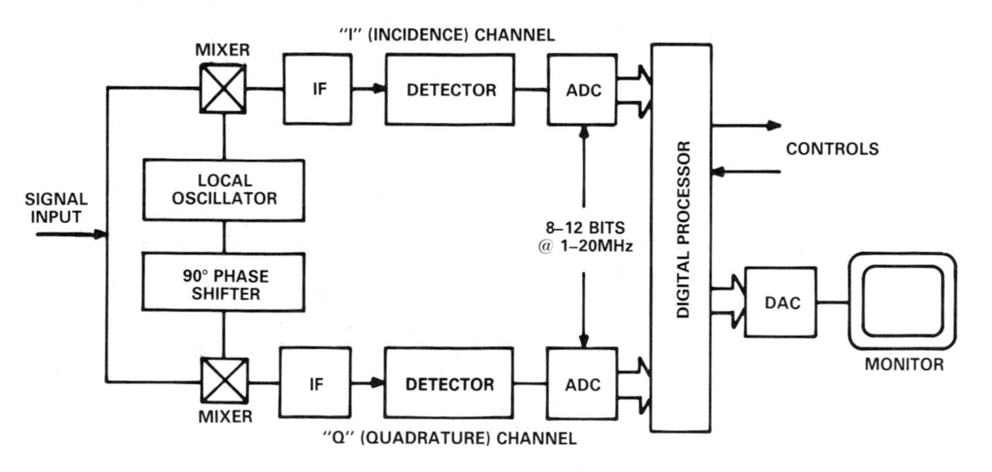

Figure 13.4. I & Q radar processing.

Direction and ranging information is rather straightforward; it is just a matter of knowing the direction in which the antenna is pointed and measuring the time interval between the transmission of the pulse and the receipt of its echo. However, the I and Q technique makes it possible to obtain more information about the target more quickly. Processing a set of returns with I and Q information allows the system to determine the size, speed, direction of travel, target type, and other parameters of the target.

The reason for this is that targets have distinctive radar "signatures," which cause echo signals to differ slightly from one another, depending on the type of target. This kind of data can be extracted more easily with signal-processing algorithms when the return pulse has been split into its in-phase and quadrature components. Most modern radars use this kind of processing, regardless of how the antenna is positioned.

Radar systems call for fast, moderate-to-high-resolution a/d converters, with word rates from 1 to 20 MHz or more, and 8 to 12-bit resolutions, depending on the function of the radar, the location of the converter within the processing chain, and its expected role in the signal-processing function. The higher the speed and resolution, the faster and smaller the targets that can be followed; this is especially important where high signal-to-noise ratios are required.

High-speed, high-resolution converters (10 MHz/10 bits or better), in association with fast processing systems, make it possible, in special cases, to process the signals at intermediate frequencies (IF). This eliminates the need for physical detection (demodulation) circuitry, since any desired detection function can be achieved with high accuracy and repeatability—and low noise—in the course of processing.

For return to the analog world for display and/or recording, fast deglitched or low-glitch d/a converters are used in both vector and raster-type displays, typically with 12-bit and 8-bit resolutions.

13.1.4 TRANSIENT DETECTION

Transient recorders (Figure 13.5) are instruments which perform the function described by their name—they capture, for later analysis, unique "one-shot" events, such as those encountered in nuclear or seismic applications.

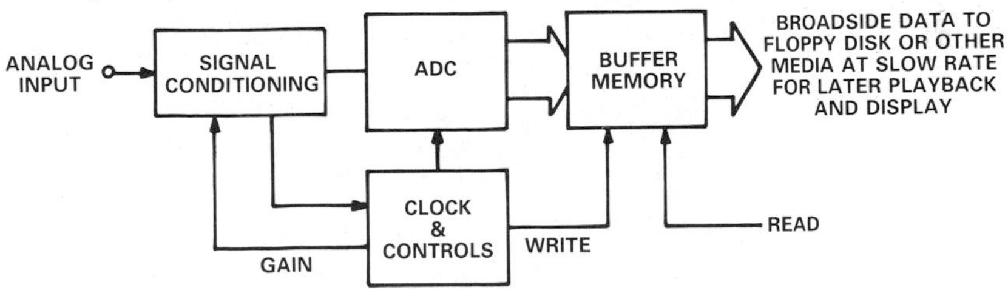

Figure 13.5. Transient recorder.

The input signal-conditioning circuits, the a/d converter, and the memory are all under the command of clock and timing circuits, which control the system's front-end gain, operate the memory's write circuits, and dictate the application of encode commands to the converter.

In one mode of operation, new conversions are triggered at a constant rate, and data flows continuously through a first in-first out buffer memory until an input threshold is exceeded, indicating the presence of an event; then all data within a predetermined time before and after the occurrence of the event may be preserved.

Once stored in buffer memory, the user's program determines whether the data is to be read out into longer-term memory, displayed on a CRT, or processed.

A/D converters used in transient detectors range from low resolution and very high speed (e.g., 6 bits at 50 to 100 MHz word rates) to high resolution at lower speeds (e.g., 12 bits at 20 MHz).

13.1.5 COMMUNICATIONS

Baseband Characteristics

The *baseband* is the band of frequencies occupied by the signal before it modulates a carrier or subcarrier frequency to form the transmitted signal. An important objective of communications receivers in dealing with a received signal is to find out the baseband, if unknown, discard the carrier, and process the signal to determine what information it contains. This is easily accomplished with "friendly" signals; it is much more difficult with intercepted signals—a great deal of processing, analysis, decrypting, etc., may be required. Converters having high resolutions and the highest speeds are essential in order to minimize loss of data.

Fibre Optics and Satellite Communication

An increasingly common communications technique in recent years has been the use of a/d and d/a converters in digitizing, transmitting, and recovering baseband information via fibre optics and satellites. Fiberoptic links are less costly than coaxial cable, easier to maintain, and more secure from noise and interception. In these applications, the baseband frequencies are generally dc to 5 MHz, which is sufficient bandwidth for television or frequency-division multiplex (FDM) voice signals (Figure 13.6).

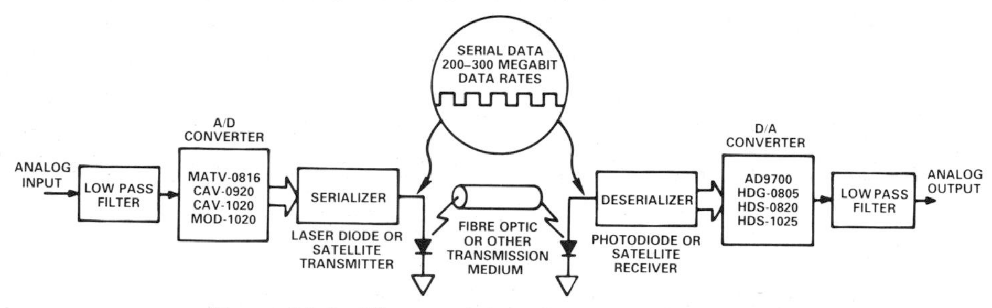

Figure 13.6. Fibre optic/satellite communications.

In typical systems, the analog input is filtered to prevent aliasing, then converted to parallel digital data with medium resolution (8 to 10 bits). To maintain the integrity of signals containing frequencies as high as 5 MHz, the word rates for sampling and conversion must be at least 10 MHz, to satisfy the Nyquist criterion, but are greater in practice to avoid degradation of data.

The parallel data are changed to a serial stream of data at rates of 200 megabaud or more, using a high-speed clock. In a fiber-optic link, the serial data stream modulates a laser diode, which emits pulses of light; this pulse train is transmitted via the link to a photodiode, which is sensitive to light at the same (light) frequency as that emitted by the laser diode.

The electrical pulse-train output of the photodiode is returned to parallel by the deserializer, and then converted back to analog by a d/a converter of appropriate speed and resolution.

Typically, a fiberoptic link can carry signals at 300-MHz data rates, which are adequate for up to three multiplexed digital television signals, or some combination of TV and FDM voice signals.

Satellite communication systems are conceptually similar to the system of Figure 13.6, but the laser diodes are replaced by microwave transmitters and receivers operating at gigahertz (GHz) frequencies.

FDM/TDM Transmultiplexers

In telephony, there are two standard formats for multiplexing voice signals Frequency division multiplex (FDM) has been used throughout the world for many years to transmit long-distance telephone calls; time-division multiplex (TDM) principles have been known for some time but have been widely applied only recently.

FDM stacks voiceband signals in adjacent 4-kHz channels into 12-channel groups and 60-channel supergroups, using single-sideband amplitude modulation; in TDM, each voice signal is digitized, using pulse-code modulation (PCM), at a sampling rate of 8 kHz; The pulse streams which result are then interleaved in time and transmitted.

Inevitably, systems employing FDM and those employing TDM must communicate. To make this possible, digital signal processing provides a convenient means of translation, the FDM/TDM transmultiplexer. However, FDM signals must be digitized before they can be translated to TDM; this calls for an a/d converter. On the other hand, after TDM signals are translated digitally, they must be converted to analog. Figure 13.7 shows a translation process from FDM to TDM for handling signals from the outside within a given office—and ultimately back to FDM again when communicating with the outside system.

The special-purpose matched ADC and DAC used for this purpose, strictly speaking, are not video converters. Nevertheless, their most important

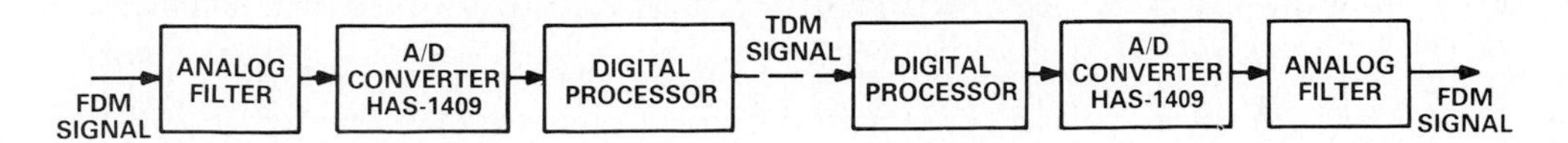

Figure 13.7. Digital FDM/TDM translation.

specifications are based on their ability to deal with high-frequency ac signals, a feature shared with video converters when used for signal-processing applications. In these applications, such characteristics as noise-power ratio (NPR) and signal-to-noise ratio (SNR) are important. The devices labeled in the block diagram are high-performance 14-bit types designed for high resolution and good mid-scale linearity; they operate at a word rate of 112 kHz.

13.2 ABOUT VIDEO CONVERTERS

13.2.1 D/A CONVERTERS

The accuracies of d/a converters at low speeds are primarily a function of the design of the resistor network and switches, the stability of the reference voltage, and the nonlinearities of the op amp (if used). At low speed and resolution, most modern d/a converters exhibit good linearity and monotonicity. However, at video speeds, full-scale settling time and the magnitude and duration of "glitches" become crucial.

DAC Settling Time

For a d/a converter, full-scale settling time is the worst-case time elapsed from the application of a digital input, representing a full-scale change in the analog variable, until the time when the output has entered (and remains within) a specified error band around its final value (usually given as a percentage of full scale or ± ½ LSB). This figure takes into account all internal factors affecting settling time, i.e. turning the switches on and off, current changes within the resistor network, and time required by op-amp or buffer outputs to settle to within their error bands.

DAC Glitches

Glitches, or output spikes caused by *skew* (differences in turn-off and turn-on times of switches or logic), represent another error source in very high speed d/a converters. All DACs produce output glitches that occur during digital input transitions. The worst spikes occur at the major transitions, at half full scale (when the MSB changes state), and at ¼ and ¾ scale, when the next-most-significant bit change state. Figure 13.8 illustrates the glitches at these transitions (and there are glitches of lesser magnitude at other transitions, not shown) for increasing input—in the case where the lower-order bits turn off faster than a higher-order bit can turn on.

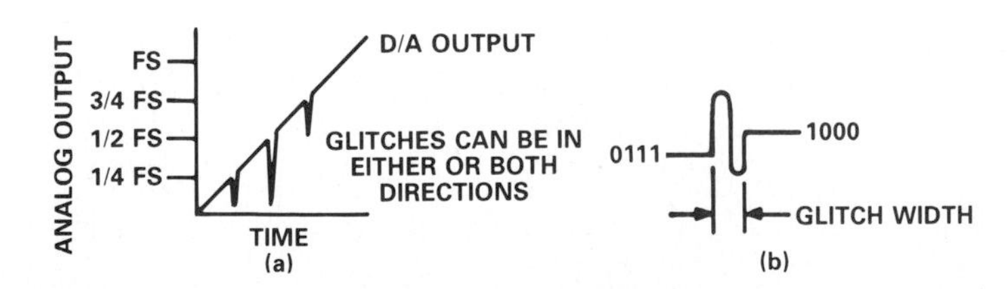

Figure 13.8. Glitches in d/a converter outputs.

DAC glitches cause significant distortion and create real problems in high-speed automatic test equipment and digitally controlled CRT displays. In CRTs the glitches actually appear on the screen as either shaded stripes, discontinuities or varying intensities in straight lines. In other applications, glitches can produce in-band harmonics, and generally higher noise levels.

Glitches can be characterized by measuring the "glitch impulse" (sometimes called glitch energy, or glitch charge), as Figure 13.9 illustrates for a glitch having a doublet shape. The glitch impulse is the net area under the voltage-versus-time curve and is expressed in picovolt-seconds, or millivolt-nanoseconds. It can be estimated by approximating the waveforms by triangles, computing the areas, and subtracting the negative area from the positive area, as illustrated in Figure 13.9. In a well designed DAC, the net glitch impulse can be made much less than 100 picovolt-seconds.

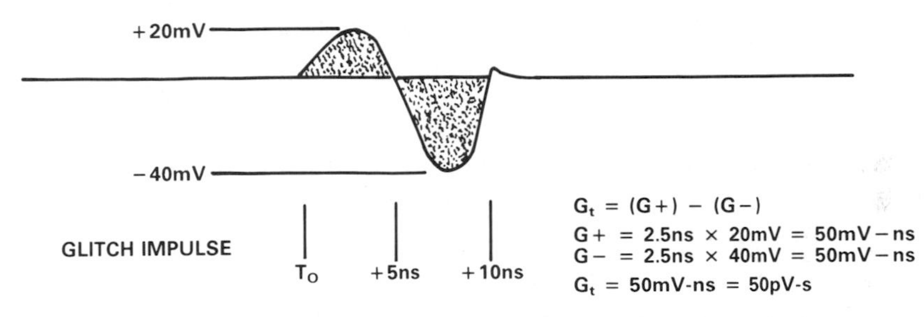

Figure 13.9. Minimal-area (glitch-impulse) doublet glitch.

Glitches in high speed d/a converters result from input bit timing differences, asymmetries in on-to-off and off-to-on times of d/a current switches, and circuit layout. Input bit timing differences, for instance, can cause severe glitches. Time skew in the parallel digital input will increase the width of the glitch by an amount equal to the skew. This time skew can be greatly reduced by adding a set of input registers either in the d/a or very near to it.

TTL signals, by nature, contribute significantly to the time skew. TTL logic utilizes saturated voltage-switching devices and has significant differences in delays and rise times between the positive-going and negative-going changes in logic levels. These differences contribute directly to the time skew at the front end of the DAC. For these reasons, a TTL-compatible d/a converter usually will have significant glitches and in most cases needs to be "deglitched" as will be illustrated later.

Emitter-coupled logic (ECL), on the other hand, utilizes non-saturated current switching, and the delay times for positive and negative-going transitions are almost identical. Acceptable 8-bit performance can usually be obtained from an ECL d/a converter by adding small delay-equalizing deskewing capacitors on the first few most significant bits that drive the DAC, as shown in Figure 13.10. Careful board layout and selection of input registers help to decrease glitches.

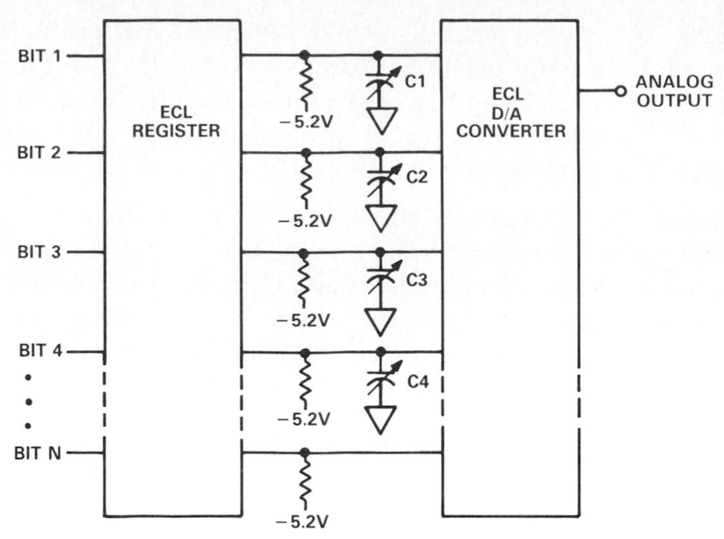

Figure 13.10. Trim Capacitors to neutralize skew.

Since the glitch in d/a converters is a nonlinear and code-dependent phenomenon, it cannot be simply filtered. One of the most effective methods to reduce the effects of glitches associated with video DACs is to use a track-and-hold amplifier (T/H). In Figure 13.11, a track-and-hold is connected to operate on the DAC's analog output. The T/H tracks the DAC output until just before the glitch occurs; then the T/H switches into the "hold" mode and remains there until the glitch has settled to an acceptable level, after which it returns to the track mode.

As Figure 13.11 illustrates, the Hold command begins just prior to and ends just after the worst-case glitch. This effectively masks the widely differing glitches associated with the D/A and instead introduces nearly identical T/H-related pulses at a frequency equal to the update rate. The track-and-hold may also introduce a pedestal (dynamic offset) error, but both the T/H pulse and pedestal errors can be significantly reduced through careful component selection and circuit design. Since the track-and-hold related pulses are consistent in magnitude and frequency, they can be readily low-pass filtered. The T/H technique loses effectiveness for update rates greater than about 20 MHz.

A viable alternative in very fast converters is to minimize the overall magnitude of the glitch by careful balancing of both the logic and the switching skews. If effectively performed, this can further reduce the glitch's net area by forcing it to be a fast doublet, which then becomes more susceptible to filtering, because it is at a doubled fundamental frequency and has no large one-sided excursion requiring a long settling time in linear filters. Though this is difficult to accomplish in TTL circuits because of the inherent asymmetry of saturated logic, nonsaturating ECL is essentially independent of the direction of the logic transition and can be deskewed (as noted above) by the addition of small capacitors.

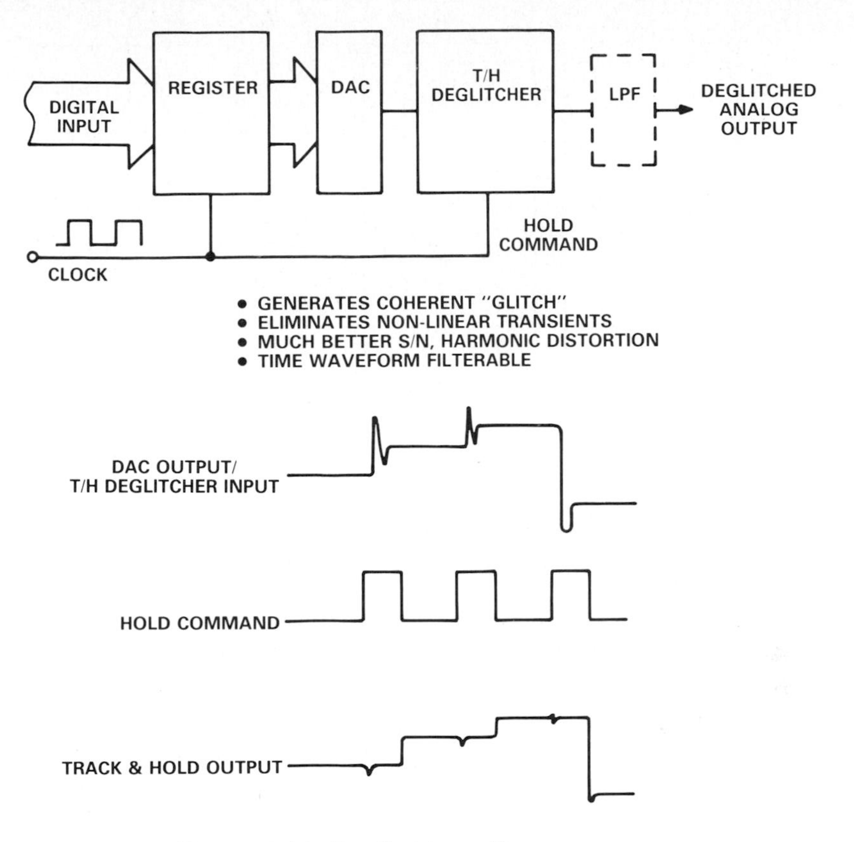

Figure 13.11. Deglitching a d/a converter.

12-Bit Deglitched DAC

Figure 13.12 is the block diagram of a high speed 12-bit DAC, which embodies—within a single thick-film 32-pin hybrid package—the deglitching techniques discussed above. The package includes the 12-bit current-output DAC network and switches, a current reference, an output amplifier with track-and-hold switching, timing circuits, and the associated electronics.

As a first step in minimizing the glitch, an input register removes the effects of differences in the arrival times of the individual bits. All bits are simultaneously latched and arrive at the input of the current-output DAC at the same time. Nevertheless, despite the best efforts at balanced design for the converter, there still exists a difference between on and off times for the switches, causing a glitch.

The block labeled "timing generator" includes a one-shot multivibrator and gating circuitry. This circuit provides a fixed hold period, starting at the time the strobe arrives and continuing beyond the expected glitch duration, irrespective of the strobe frequency. At the conclusion of the hold interval, the output amplifier is returned to the track mode, slews to the new value established by the digital input and settles quickly.

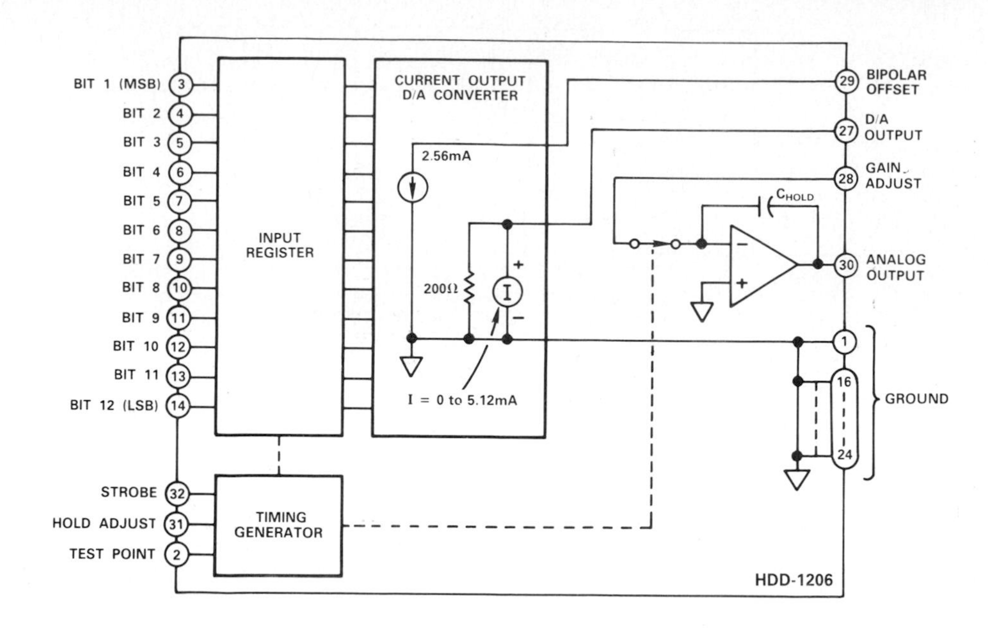

Figure 13.12. Functional block diagram of 12-bit deglitched DAC.

The role of the FET switches used in the track-and-hold is critical, because they affect the dissipation and complexity of the drive circuit, and the smoothness of the converter's output waveform. Chosen for low input capacitance, low gate cutoff voltage, and low drain-to-source "on" resistance, they make possible 6-MHz update rates with low power dissipation. Additional benefits are low charge transfer within the device—hence small residual spikes and simplified drive circuitry, resulting in efficient use of space and low cost.

DACs for Raster Scan

High speed DACs find growing use in raster-scan video-display applications. Devices especially designed for these applications not only meet the speed requirements, but also simplify the generation of composite video waveforms by providing digitally controlled auxiliary output currents of appropriate magnitude and polarity for blanking, synchronization, and 10% brightness levels.

Figure 13.13 shows the industry-standard video composite intensity waveform over 1 ½ cycles of the horizontal sweep. The controlled range of the D/A converter's full scale (0 to −643mV) is from reference white (−71mV) to reference black (−714mV). In the illustration, the intensity is varying from full white to full black. At the beginning of the sync portion, the intensity signal drops to the blacker-than-black "front porch" (−785mV), and then to the extreme black level (−1071mV) during the horizontal retrace. As the next sweep starts, the intensity returns to the "back porch" (−785mV), and, as the first element of the picture is triggered, to the

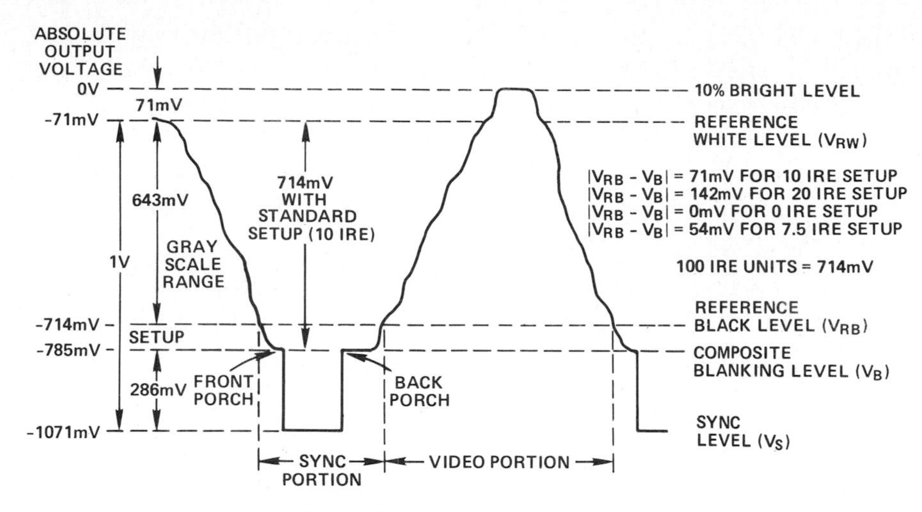

Figure 13.13. Composite DAC output waveform.

controlled range of the D/A converter. During this scan, the 10% "brighter than white" level (0mV) has been activated briefly to display the cursor.

Typical DACs for this purpose, as exemplified by the 8-bit HDG-0805 DAC diagrammed in Figure 13.14, are specifically designed to meet the needs of raster-scan systems. Housed in 24-pin metal hybrid packages with 0.6" double-DIP spacing, they require only a single −5.2V power supply. Such a DAC provides, in a single package, all the circuitry required for 256-level intensity modulation in raster-scan displays at video dot rates up to 100 MHz.

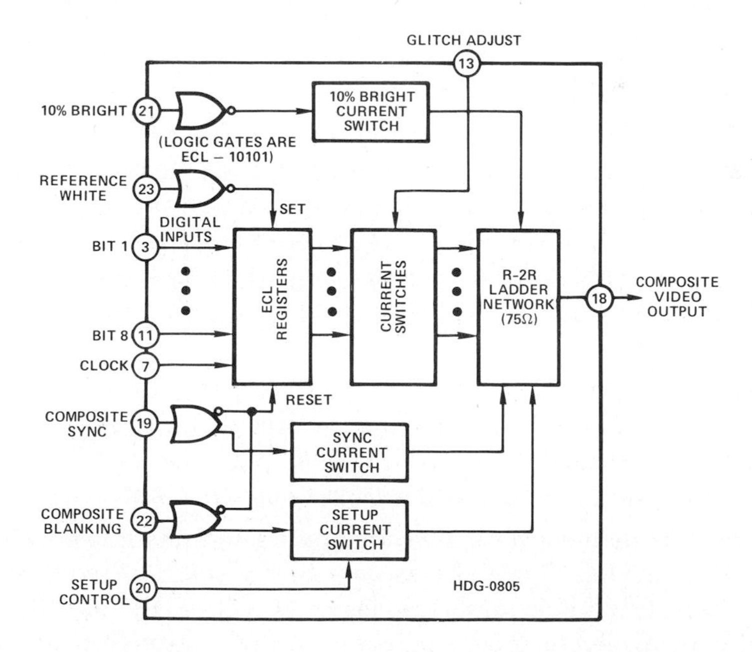

Figure 13.14. Block diagram of HDG-0805 D/A Converter.

For the HDG-0805, resolution is 8 bits, and full-scale settling to within 1 LSB is typically 7 nanoseconds. The output impedance is 75 ohms, thus simplifying impedance matching to video coaxial cable. Full-scale output current develops 1 V p-p video across a 75-ohm load. In order to minimize the glitch, a set of internal ECL registers provides minimum time skew between bits. Typical net glitch impulse is 50 picovolt-seconds. A Glitch-Adjust input lets the system designer optimize glitch performance.

As the block diagram shows, such devices provide self-contained digitally controlled sync and blanking capability, compatible with EIA Standards RS-170, RS-330, and RS-343A, to produce composite video waveforms like the one in Figure 13.13. In addition to the 256 levels of gray scale provided by the 8-bit digital input, three additional digitally controlled switches independently provide auxiliary output currents of appropriate magnitude and polarity for blanking, sync, and 10% bright levels.

Because of their small physical size, display DACs in the form of hybrids or ICs can be located quite close to the CRT's intensity input, eliminating degradation caused by coaxial-cable-related losses. The grounded metal package effectively screens out the effects of the high noise environment usually associated with CRT drives.

13.2.2 A/D CONVERTERS

The challenges of high speed a/d conversion are being increasingly met by a wide range of components from a variety of manufacturers. Unfortunately, in many cases, these devices do not achieve their full potential because of improper application.

Flash Converters

All-parallel, or so-called *flash* converters offer the fastest throughput of available ADC designs. Figure 13.15 illustrates a typical block diagram. The converter employs $2^n - 1$ latched analog comparators in parallel, where n is the number of bits. A resistive voltage divider provides the reference voltages for the comparators. The reference voltage for each comparator is one least-significant bit (LSB) higher than the reference voltage for the comparator immediately below it.

When an analog input signal is present at the input of the comparator bank, all comparators which have a reference voltage below the level of the input signal will assume a logic "1" output. The comparators which have their reference voltage above the input signal will have a logic "0" output.

Decoding logic then converts the comparators' outputs into a binary digital output. Because flash encoders having 8-bit resolution require 255 comparators, and comparable amounts of logic, it can be easily seen that these converters are relatively impractical to construct from discrete comparators and logic elements due to power, size, wiring, and cost considerations. Advances

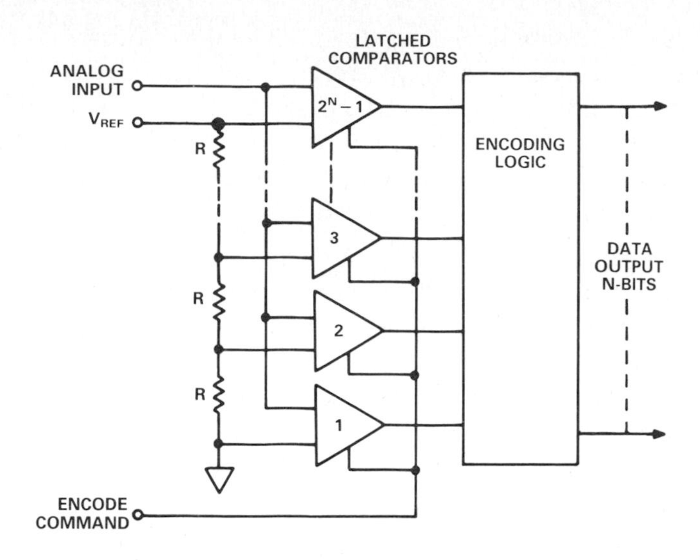

Figure 13.15. All-parallel a/d (flash) encoder.

in large scale integration techniques, however, have resulted in the development of commercially available monolithic devices ranging from 6 to 10 bits of resolution.

Even though the latches essentially perform a track/hold function, high-speed, high-resolution systems require an external track/hold for best performance. However, in low-resolution applications, if the relative aperture delay match between each comparator and the total system aperture uncertainty is better than, for example, about 100 picoseconds, no track-and-hold circuit is required for video signals having bandwidths up to 4 MHz.

Figure 13.16 is a functional block diagram of a 6-bit flash converter, the AD9000 a/d converter, packaged in a standard ceramic 16-pin DIP. It achieves 75 MHz word rates over a temperature range extending from −55°C to +125°C, making it useful for a variety of applications in a wide diversity of environments. As the block diagram shows, 64 parallel comparators are employed to digitize fast-moving analog input signals. An overflow bit makes it possible easily to connect multiple units in a cascade arrangement to obtain up to eight bits of digital data at word rates comparable to those achieved by the devices operating independently. Wired-OR logic circuits within the device encode the comparator outputs into a binary format of six bits of parallel data, along with the overflow bit.

The outputs of the comparators are applied to latches controlled by the ENCODE input. When the encode command is low (digital "0"), the latches are transparent creating the "track" mode. When the ENCODE input changes to high (digital "1"), the latches go into a "hold" or latched condition, thus freezing the most recent digital outputs of the comparators and applying them to the encoding circuits.

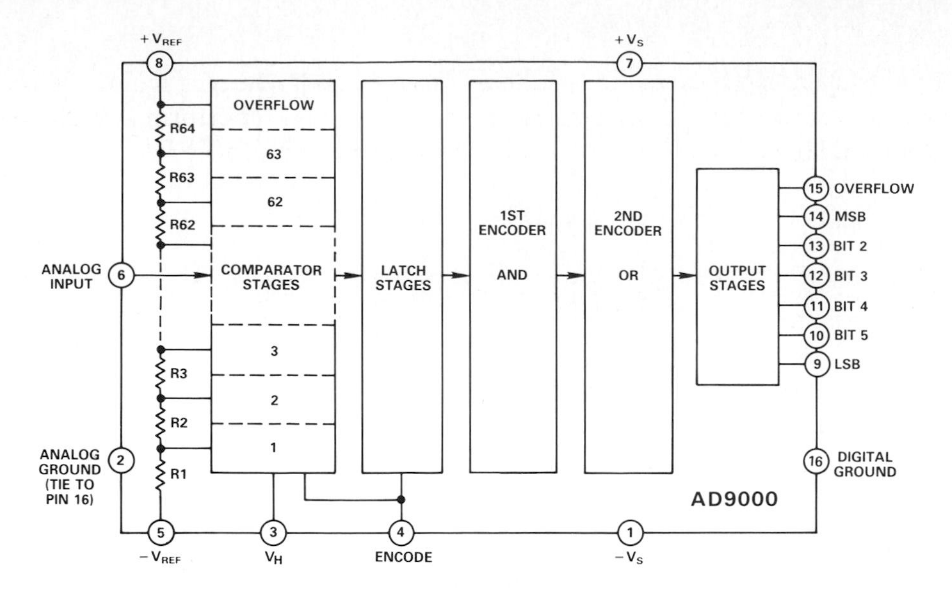

Figure 13.16. AD9000 block diagram.

The signal held in the latches is converted to binary form by the encoders and applied to the output stages as a 6-bit digital representation of the analog signal which was present at the comparator inputs at the instant the ENCODE command made the change to the "hold" mode.

All-parallel flash a/d converters tend to have fairly random linearity errors. The overall linearity in monolithic flash converters is determined primarily by comparator offset-voltage matching and tolerance of the resistors comprising the voltage divider. Codes can be missed if adjacent comparators have offset voltages of opposite polarity and sufficient magnitude. Figure 13.17 shows a typical error characteristic (the difference between a linear unit slope and the typical ADC staircase of Figure 7.2) for a monolithic flash converter.

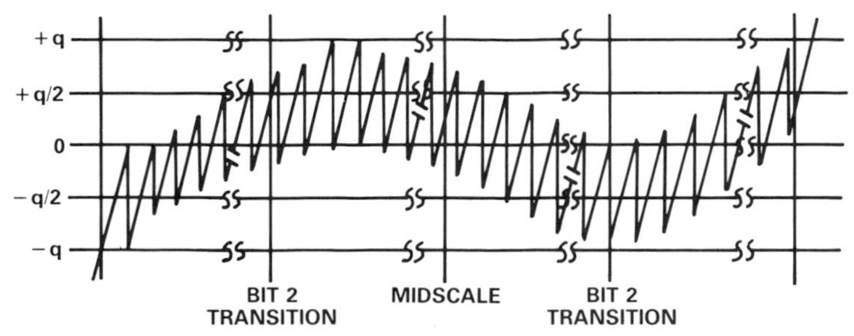

Figure 13.17. Error Characteristic (linear analog response − quantized staircase) for a/d converter. Only a few codes are shown.

Due to the high input capacitance associated with the large number of parallel comparators, the input to the encoder portion of such a converter must be driven with a video operational amplifier which is stable when driving a large capacitive load.

Using Track-Holds

Traditionally, designers have tended to use flash converters without track-and-holds, because the internal latches of the converter perform a track/hold function. However, there are distinct advantages to using a T/H.[1]

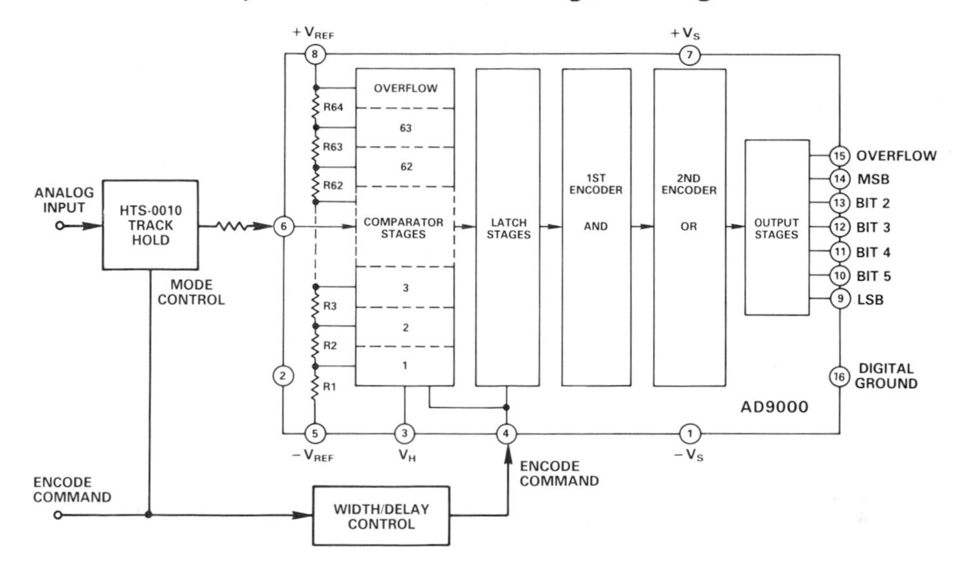

Figure 13.18. Typical flash ADC-track/hold subsystem.

Figure 13.18 illustrates the use of a T/H with a flash ADC. The T/H "holds" the analog input constant while the comparators are settling. Then the comparators are latched and, while the data is being encoded to complete the conversion, the T/H returns to the "track" mode to acquire the new analog value for the next conversion. When the T/H is switched to hold, the comparators are unlatched and switch to the new state. Thus, the comparators are always converting a new "dc" value, rather one which is continually changing. This process provides a substantial improvement, for reasons discussed below.

As indicated earlier, all flash converters have inherent capacitance characteristics which that affect their ability to operate on high-frequency analog signals. There are sets of problems that are specific to each of the inputs.

The analog input capacitance, C_{ai}, of each comparator (Figure 13.19) is determined primarily by the base-emitter junction capacitance of the transistor on the analog-input side. The value of this capacitance is affected by junction bias; forward-bias capacitance is higher than reverse-bias capacitance. Since the inputs of all comparators are in parallel, the total capacitance at the flash converter's analog input is the sum of the individual comparators' base capacitances.

[1]See "Flash Converters Work Better with Track/Holds," by Jerry Neal and Jim Surber, *Analog Dialogue* 18-2 (1984) 10-15.

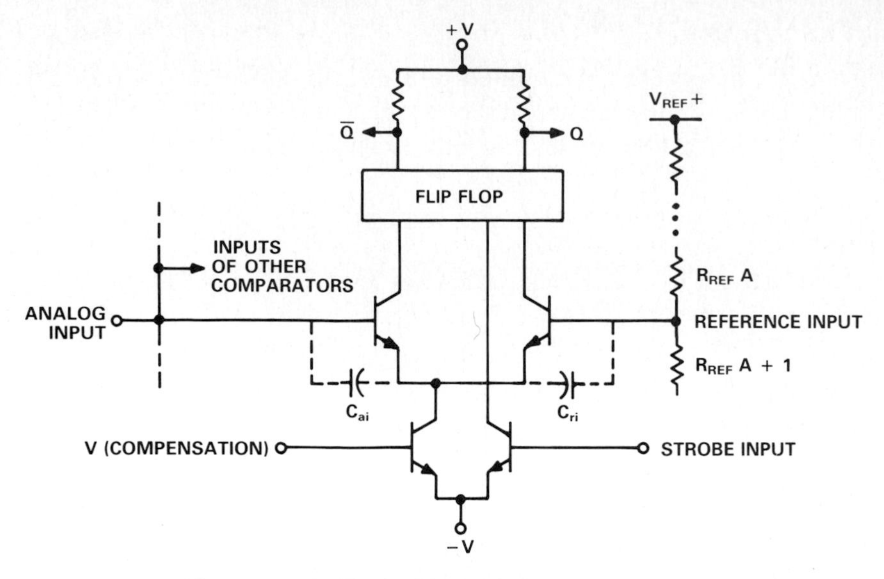

Figure 13.19. Typical flash ADC comparator cell.

Unfortunately, the total capacitance is essentially random, because it depends on the amplitude of the analog input signal; the signal amplitude and the on-off state of each individual comparator dictate its base-emitter voltage, hence its contribution to the total capacitance. The total capacitance can vary from 30 to 120 pF. Since the comparator can be driven by an analog buffer, this is not a serious problem (but it *must* be dealt with). However, the buffer doesn't solve capacitance problems associated with the reference input.

Since the comparator has a differential input, the reference input is similarly subject to capacitance variations introduced by its base-emitter capacitance, C_{ri}. But the problems they raise are not as easily dealt with; see Figure 13.20.

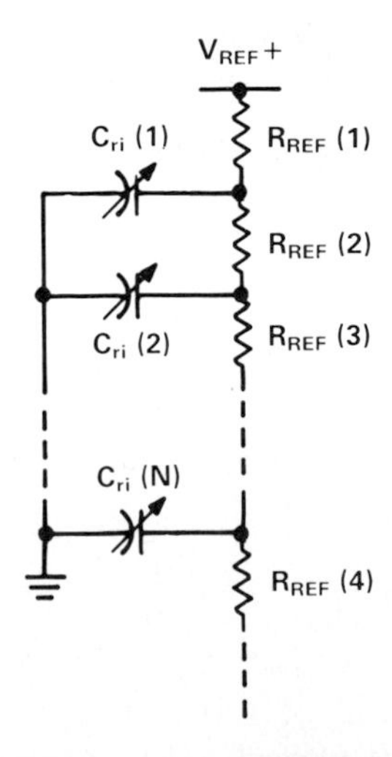

Figure 13.20. Equivalent circuit of the reference resistor network.

The charging path for each individual C_{ri} is through the reference ladder. This path constitutes an R-C time constant at the reference input; and the time required to charge the capacitances through the resistors tends to cause the reference voltages to lag fast changes in the current flowing through the reference ladder, dynamically distorting the voltage division of the reference ladder.

This capacitive reactance effect depends on the differential input voltage to the comparator, the speed at which it varies (frequency), and the comparator's position within the ladder. When signals having high slew rates are applied, the ac integral-linearity error of the comparator array will increase because of the reactance variations among the many comparators.

In addition, errors also occur due to noise introduced by parasitic coupling from the strobe circuit to the reference ladder via C_{ri}. Although the strobe is equally coupled to both sides, the lower source impedance at the analog side tends to reduce its amplitude, resulting in an imbalance between the comparator inputs.

One of the most effective ways to equalize the differential input impedances of the comparators is to use an external resistor in series with the analog input of the converter, as Figure 13.18 shows, to produce better common-mode rejection of unwanted noise voltages and improve the integral linearity of the comparator array. The resistance value should be about one-fourth of the total resistance of the reference ladder (i.e., the parallel resistance of the upper and lower halves of the ladder resistance).

Another potential source of dynamic conversion error in flash ADC designs is the variation of effective sample delays. Individual comparators within an array can be visualized as having variable delay lines in series with their latch inputs. The magnitude of delay for each comparator is determined by comparator inconsistencies, chip layout, and the strobe frequency.

For low-frequency analog inputs, these sample-delay variations among adjacent comparators are not a significant problem. But as the input frequency is increased, latch-time disparities among comparators can result in missing codes and excessive differential nonlinearities. The errors differ with slew-rate magnitude and direction. As a practical matter, it is realistic to expect sample-delay variations as large as 200 to 300 picoseconds, rather than to calculate them based on just the published aperture-jitter specifications.

Considerable improvement can be realized if we can replace the sample-delay variations of a flash converter by the much faster switching of a track/hold. For example, the track/hold shown in Figure 13.18 has a specified aperture uncertainty of only 5 picoseconds. Since the track-and-hold amplifier ahead of the flash converter freezes the input signal, it maintains a constant input while the comparators are latching, and it can also be timed to allow the comparator input capacitances to charge and settle before the conversion takes place.

It is important to remember that a track/hold amplifier used ahead of a flash converter has no effect on the conversion time; it simply allows the converter to digitize higher-frequency (faster slew rate) analog signals. The total time required for conversion, taking into account the delays of both the track/hold amplifier and the flash converter, can be calculated and compensated for by the system timing.

Multi-Stage Digitally Correcting Subranging A/D Converters

When higher resolutions than those obtainable with flash converters are needed, a multi-stage converter is required. Multi-stage converters, also referred to as subranging converters, provide high resolution with markedly fewer comparators and simpler logic than single-stage converters, but give up some of the inherently higher-speed capabilities of the single-stage types. Subranging converters are used instead of parallel-, or flash-type, converters whenever higher resolution and/or less complexity are required for the conversion function.

Figure 13.21 illustrates a 12-bit subranging a/d converter constructed with two encoders having resolutions that add up to 13 bits (6 and 7, in this case). The analog signal from the track-and-hold is applied through the two buffer amplifiers to a 6-bit flash encoder and a video delay line simultaneously. The 6-bit encoder converts the analog signal to binary, producing the six most significant bits. These bits are stored in a register and also applied to a 6-bit d/a converter having an accuracy of at least 12 bits. The output of the d/a converter is inverted and subtracted from the delayed track-and-hold output in a

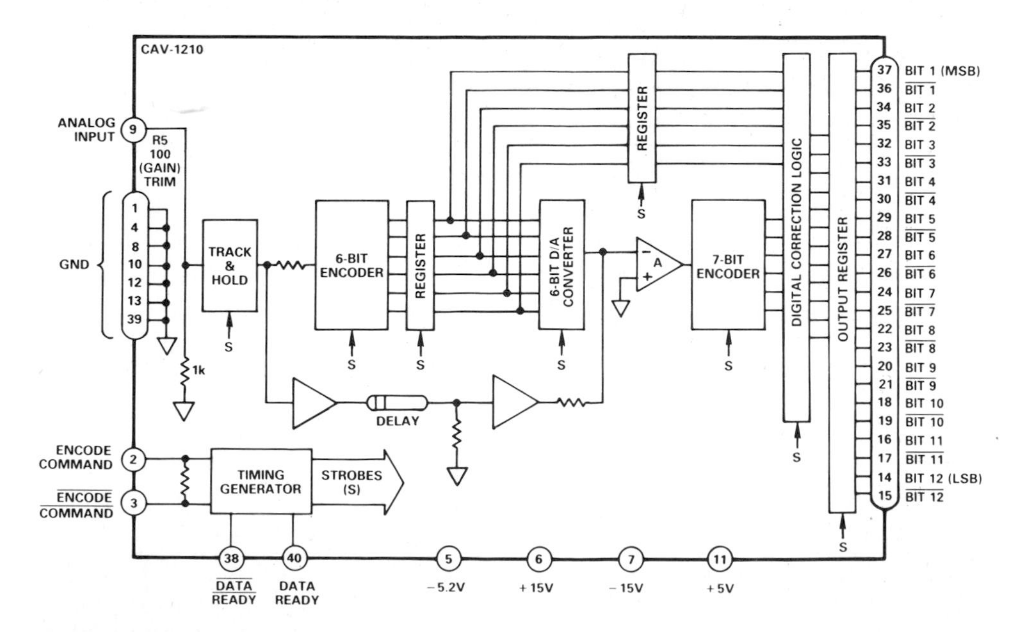

Figure 13.21. Functional diagram of two-stage flash converter with digitally corrected subranging.

summation network, and the resulting "residue" is amplified to the 7-bit converter's full-scale span.

The "residue" is then converted to digital form by a 7-bit flash encoder and represents the less-significant portion of information. The outputs of the 6-bit holding register and the 7-bit flash are then combined in an output register to yield a 12-bit parallel binary output.

Timing is of extreme importance in this type of a/d, as each element in the conversion process must settle to its optimum point before the required strobe signals are applied.

It would not be at all unusual for converters of this type to exhibit differential-linearity discontinuities around the bit-6 transition points, due to mismatch between the front-end and residue encoding circuits. However, in converters employing digitally corrected subranging (DCS), these discontinuities are eliminated by the use of digital correction logic, since the information to accurately characterize the bit-6 transition is already present in digital form because of the accurate d/a conversion and summing, and the 7th bit in the second conversion.

Here's a simplified explanation of how it works: The result of the first conversion—in the holding register—is equal (digitally) to: (Input − Residue + 6-bit Error). It is converted to analog and subtracted from Input. Assuming a perfectly accurate DAC and subtraction, the result is (Residue − 6-bit Error). That is,

$$I - (I - R + E6) = R - E6 \qquad (13.1)$$

This difference is scaled up and converted, giving a digital quantity equal to (Residue − 6-bit Error + 12-bit Error). Finally, the two digital words are added, giving: Input + 12-bit Error. That is,

$$(I - R + E6) + (R - E6 + E12) = I + E12 \qquad (13.2)$$

The extra bit is necessary because (R − E6) can exceed 1 LSB of 6 bits.

Although the explanation is simple, the execution is not. Fast 6-bit DACs with better than 12-bit accuracy are not easy to make, nor are fast subtracting amplifiers with adequate dynamic linearity. Remember that both the DAC and the output amplifier must settle before the second conversion can be completed.

However, this technique, incorporating fast emitter-coupled logic, has resulted in 10-bit converters capable of 40-MHz sampling rates and 12-bit converters that can handle 20-MHz sampling, with similar architecture to the one described by Figure 13.21 and depicted in Figure 13.22. The technique seems promising for even faster and higher-resolution converters.

[1]See "Flash Converters Work Better with Track/Holds," by Jerry Neal and Jim Surber, *Analog Dialogue* 18-2 (1984) 10-15.

Figure 13.22. Physical construction of the converter of Figure 13.21.

13.3 PRACTICAL DESIGN INSIGHTS

Much of the challenge in using high-speed converters lies, not in the electronic design, but rather, in the physical design and layout of the printed circuit boards. It is the old problem, reiterated many times in these pages, of the need for high-resolution analog signals to maintain their integrity in the presence of fast, pervasive digital edges. Grounding problems, cross-coupling, and parasitic effects associated with the physical circuit layout can degrade the performance of even the best designs.

13.3.1 BOARD-LEVEL GROUNDS

In general, a very low impedance ground is a must. Every portion of a printed circuit board which doesn't contain circuits or conducting runs should be ground plane. Sometimes circuit density must be purposely reduced in order to create more room for ground plane. All breadboards should be constructed on double sided copper clad boards. Avoid using general digital (purely insulating) breadboards and thin hookup wires.

Another basic rule for working with high-speed (high-frequency) printed-circuit (PC) board designs is to connect analog ground and digital ground together within the PC board. Board designs that employ separate digital and analog grounds often end up with ground loop problems that are very hard to solve. But there is a dilemma: although connecting the two grounds together at the board enhances the performance of converters at the board or subsystem level, it can create other problems at the system level. This problem will be discussed below.

As another practical rule in maintaining low ground impedance, use as many of the PC board's connector pins as possible (wired in parallel) for ground connections to the system. This lowers the contact resistance, thus avoiding IR-drop related noise. For perspective, consider that, for a 10-volt input range on a 12-bit a/d converter, the least significant bit (LSB) of the converter will have a value of only 2.5 millivolts (high-speed systems generally use even lower voltage ranges). Assume that a single pin of the PC connector, to be used for ground, has a resistance of 0.05 ohm—and that the PC card draws a total of 1.5 amperes (not unusual for cards carrying high-speed logic circuitry).

The subsequent voltage drop at the ground pin could be as much as 75 millivolts. If only digital logic were used, this voltage drop would be minuscule and hardly worth considering. However, if the circuit involves both analog and digital signals, and if a fraction of that voltage drop is coupled into the analog circuit, the effects could be devastating for converter accuracy.

Consider, for instance, the case of TTL logic. Since TTL is a saturated logic, ground currents vary widely, and varying current flowing through the ground often produces noise signals which modulate the ground plane. This noise, created by digital switching, can couple into the analog portion of the circuit. If only 10% of the 75 millivolt IR drop cited above couples into the analog signal, that would represent 3 LSBs of lost converter resolution. Connecting multiple pins in parallel to reduce effective contact resistance helps solve the problem.

13.3.2 LAYOUT

Component layout is just as important a consideration as grounding. A massive low-impedance ground will simply not help a poor basic layout. Signal cross-coupling generally represents the biggest problem. Digital signals should not couple to analog signals, and the analog channels must remain isolated from each other. Coupling between analog channels often leads to oscillation. Any subsystem or circuit layout operating at high speeds with both analog and digital signals needs to have those signals physically separated as much as possible to prevent crosstalk.

Digital signals leaving or entering the layout should use runs that have minimum length. The shorter the digital runs, the less the likelihood of coupling to the analog circuits.

Analog signals should be routed as far from digital signals as size constraints allow; and the two ideally should never closely parallel one another's paths. If they must cross, they should do so at right angles to minimize interference. Coaxial cables may be necessary for analog inputs or outputs—a demanding condition mechanically, but sometimes the only solution electrically.

If analog and digital signals must run parallel, design a ground path run between them. For signals entering and leaving a board, interpose ground pins between signal pins for added isolation.

When combining track-and-hold and a/d converter hybrids or modules on the same board, they should be mounted as near one another as practicable. All grounds need to be connected to the single, low-impedance ground plane; and the connections should be made right at the components themselves.

13.3.3 POWER SUPPLIES

Improper selection and application of power supplies can also impede high speed converter performance. Every power-supply line leading into a high speed PC card or data acquisition circuit must be carefully bypassed to its ground return to reduce noise entering the card. Ceramic capacitors, ranging in value from 0.01 to 0.1 microfarads should be used generously in the layout, mounted as closely as possible to the device or circuit being bypassed. In addition, at least one high-quality tantalum capacitor of 3 to 20 microfarads should be assigned to each power supply voltage, mounted as near as possible to the incoming power pins to keep potentially high levels of low frequency ripple off the card.

Low-noise, low-ripple, temperature-stable linear regulated power supplies are the preferred choices for high speed circuits. Switching power supplies often seem to meet those criteria. But because their ripple specifications are often expressed in terms of rms levels, the actual spikes generated in switchers (as an unavoidable by-product of the technique) may often produce hard-to-filter, uncontrollable noise peaks with amplitudes of several hundred millivolts. Their high frequency components may be extremely hard to keep out of the ground system.

If switchers cannot be avoided for high speed designs, then they should be carefully shielded and located as far away from sensitive circuits as possible. And of course, their outputs should be heavily filtered.

13.3.4 SYSTEM CONSIDERATIONS

Although analog and digital grounds should be tied together at the board, this can create problems for the system-level designer. At the system level, data converters should be considered as analog, not digital, components. The system design must be assigned to capable analog engineers who are experienced at defending millivolt signals against interference and wideband signals against degradation.

ADCs and DACs should be placed near other parts of the analog section in order to avoid transmission-line effects associated with long runs carrying high speed signals. Signal reflections and dispersion on long runs, for instance, can significantly reduce bandwidth and amplitude. In addition, if not adequately isolated from sensitive analog boards, digital subsystems can couple noise into analog circuits via the ground plane or power supplies, or via direct radiation to nearby analog components.

Each card in the system should be returned directly to the power supply common, using heavy-gauge wire. Where system considerations impose separate analog and digital board-level grounds, each should be returned to the power supply individually.

13.4 TESTING VIDEO CONVERTERS

In addition to being characterized by specifications common to general-purpose converters, such as linearity, temperature coefficients, etc., video converters require further characterization in terms that are heavily application oriented. Since the devices must be used at very high bandwidths in widely disparate applications, a diverse range of application-oriented specifications and testing methods have evolved.

Many tests of high speed converter circuits fail, due to improper use of test equipment at high speeds. To work with an oscilloscope at video speeds, for instance, requires special care, particularly when measuring settling time, troubleshooting a noise problem, or making certain other critical measurements. Oscilloscope ground clip-leads can lead to erroneous readings due to capacitive/inductive pick-up. Bayonet-type adapters, or BNC connectors should be used on or instead of probe tips. Some measurements may even require a special test fixture that plugs in directly to the scope without the use of cables. Component layout and ground considerations should also be carefully examined, as discussed in the previous section.

13.4.1 ADC TESTING

At low speeds and resolutions, a/d converters can be tested by operating in cascade ("back-to-back") with a high-speed, high-resolution d/a converter and using the reconstructed analog signal for the measurement. For testing high-resolution converters at high speeds, however, this introduces significant d/a-converter errors into the test loop. Most dynamic high speed a/d converter testing is today done using computer techniques. In typical tests, an analog signal is applied to the input of the a/d converter, while the converter's digital output is stored in a buffer memory. The digital samples are then time-weighted and a Fast Fourier Transform (FFT) is computed. From this basic data, various performance parameters can be calculated.

ADC Signal-to-Noise Ratio

As the name implies, this parameter measures the ratio of signal to noise at the output of the a/d converter. An a/d does generate some internal noise. This noise floor can create difficulties, particularly in high-resolution ADCs, where it can reduce resolution and dynamic range.

Figure 13.23 illustrates a typical test for signal-to-noise. A single frequency full-scale sinewave is applied to the a/d converter. The results are loaded into buffer memory, the samples are time weighted, an FFT is computed, and the sinewave power is computed. Then a band-stop filter at the sinewave frequency is switched in, removing the fundamental signal; and the total remaining power is computed. The ratio is then computed and converted to dB; the result is the S/N ratio. In an ideal a/d converter with a full-scale sinewave input, the theoretical rms signal-to-noise ratio is (6n + 1.8) dB, where n is the number of bits.

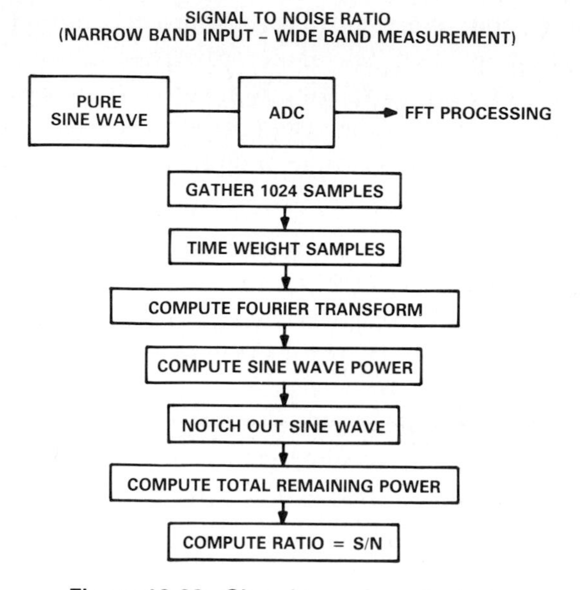

Figure 13.23. Signal-to-noise ratio test.

AC Linearity

In an ideal a/d converter, a pure sine wave on the analog input appears at the digital output as a pure (sampled) sine wave. In real-world ADCs, however, spurious signals due to nonlinear distortion within the ADC appear in the output. These spurious signals generally are combinations of the harmonics of the fundamental—and intermodulation products. The intermodulation products are produced when the fundamental and its harmonics beat with the sampling frequency. Generally, only spurious signals that fall within the video bandwidth (½ the sampling rate) are considered important. For example, in a system with sample rate of 10.74 MHz (video bandwidth less than 5.37 MHz), the third harmonic of a 3 MHz sine wave appears at 9 MHz. Spurious signals are generated at 1.74 MHz (sampling rate minus third harmonic) and at 19.74 MHz (sampling rate plus third harmonic). Note that, even though the harmonics may be out-of-band, the intermodulation products, such as the 1.74 MHz signal here, may fall within the band. It is desirable for such intermodulation products to be so small as to be negligible.

AC linearity can most easily be measured by applying a single frequency sine wave to the ADC input and observing the reconstructed d/a converter output with a spectrum analyzer. This of course may introduce additional errors due to the DAC—which is why it should be chosen to have high speed, resolution, and linearity. AC linearity specifications generally include all in-band spurious signals and are specified in dB below full scale.

Two-Tone Intermodulation Distortion

Two-tone testing is a carryover from r-f testing techniques. With inputs consisting of sine waves at two frequencies, f_a and f_b, any active device with non-linearities will create distortion products, of order $(m + n)$, at sum and difference frequencies of $m\,f_a \pm n\,f_b$, where $m, n = 0, 1, 2, \ldots$. Intermodulation terms are those for which m or n is not equal to zero. For example, the second-order terms include $(f_a + f_b)$ and $f_a - f_b)$, and the third-order terms include $(2\,f_a + f_b), (2\,f_a - f_b), (f_a + 2f_b)$, and $(f_a - 2\,f_b)$.

Traditionally, in r-f work, the two third-order difference terms are the ones principally considered, since their frequencies are the closest to the input frequencies (hence, in-band components); the others are filtered out (see Figure 13.24). However, in baseband applications, the second-order terms may well be more significant.

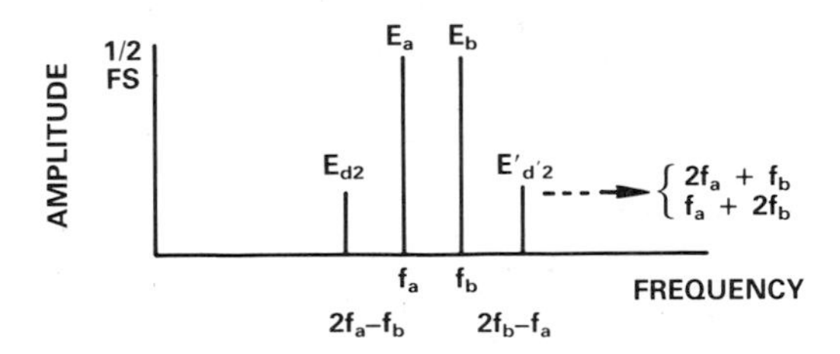

Figure 13.24. Two-tone test spectrum.

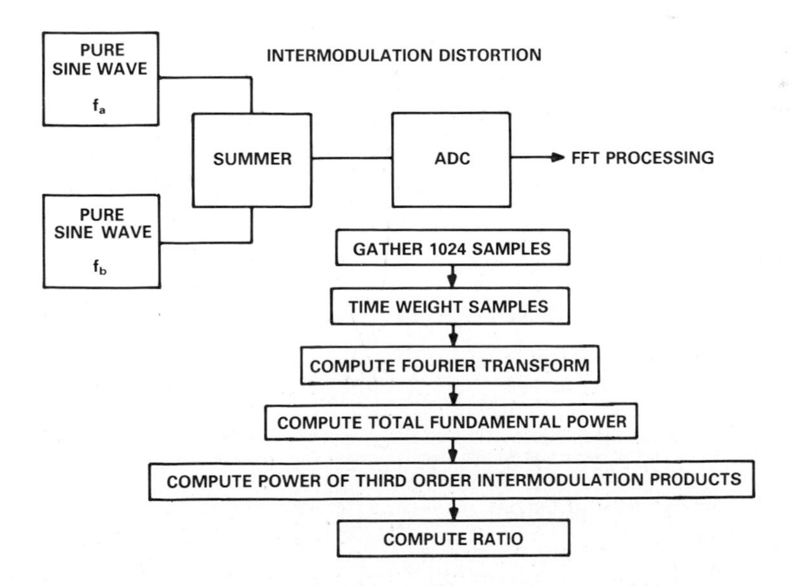

Figure 13.25. Intermodulation-distortion test.

Figure 13.25 illustrates the test procedure. Two half-scale sine waves of slightly different frequencies are summed, and the resulting signal is applied to the ADC under test. The resulting data loads into memory and is time weighted. An FFT is computed and the total fundamental power is calculated. Then the power of the third order intermodulation products, and their ratios compared to the total fundamental signal power, are computed.

Noise Power

Noise power (NPR) is the measure of the spectral power of all contributed errors, such as intermodulation and harmonic distortion, in a narrow frequency-slot within the baseband of the composite signal being processed. An example of such a communication system, including the test set-up that may be used, is shown in Figure 13.26.

In such a system, the baseband signal ranges from dc to about 8 MHz, and the a/d and d/a converters typically encode at rates of about 20 megawords/second. The NPR test consists of encoding a limited band of white noise, created by a noise generator, and examining this signal at the output of the DAC, using a noise receiver. The noise generator is equipped with band-stop filters, which eliminate very narrow slots (typically 3kHz) from the transmitted frequency spectrum (13.26b). At the receiving end, the noise receiver is equipped with complementary filters to allow the receiver to examine power spectral density of the noise contributed by the transmission medium (including the ADC and DAC) within these ideally noiseless slots (13.26c).

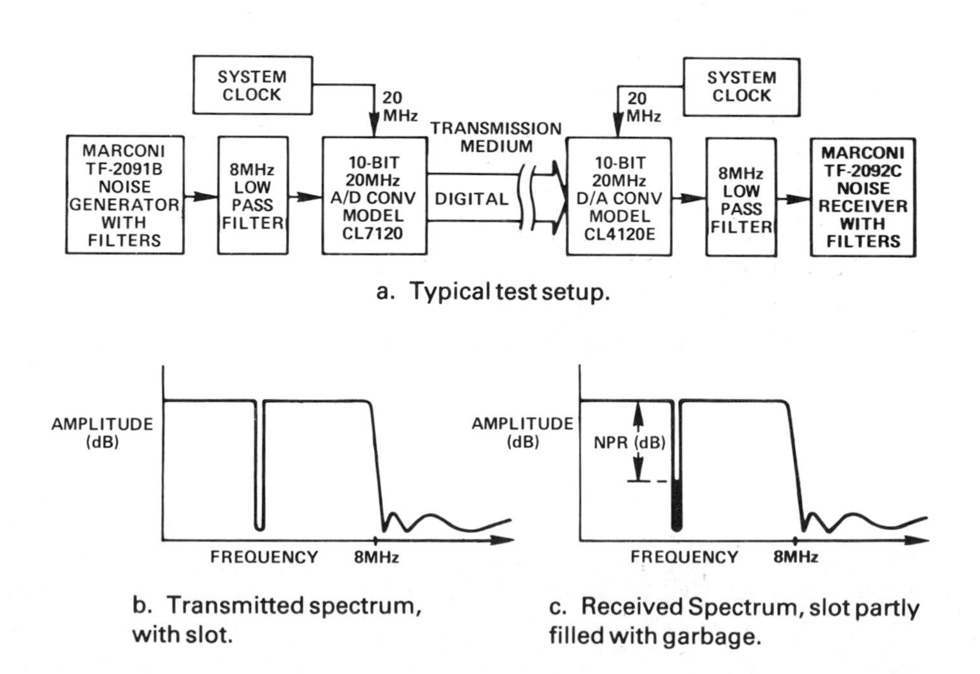

b. Transmitted spectrum, with slot.

c. Received Spectrum, slot partly filled with garbage.

Figure 13.26. Noise power ratio test setup.

This noise, the total cumulative effect of all transmission and encoding errors, such as intermodulation and harmonic-distortion products, aperture errors, and the like, is displayed as a weighted ratio of the output noise found in the slot to the power of the total transmitted noise spectrum. This number, expressed in dB, is called noise-power ratio—the larger its magnitude, the better. Noise-power ratio may also be computed in digital form, using fast Fourier transforms and/or down-conversion. Figure 13.27 diagrams a typical noise power ratio test procedure using FFTs.

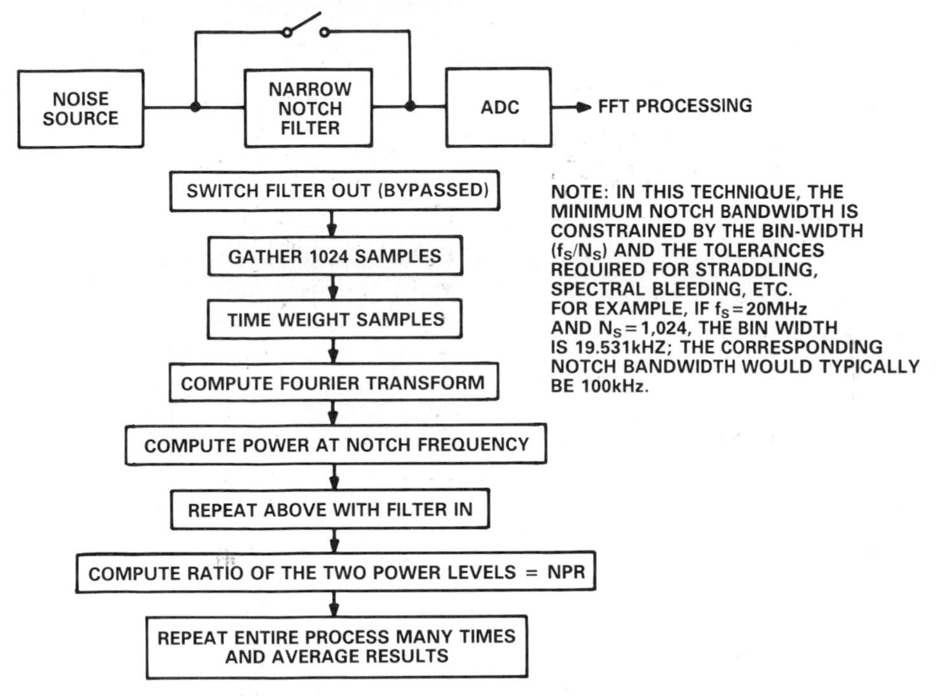

Figure 13.27. Noise-power ratio test.

Transient Response

As in the case of low-frequency converters, transient response is defined in terms of the time necessary for an ADC to achieve its rated accuracy after a full scale step function is applied to its input. Computerized testing does not readily lend itself to this test, so it is usually done manually, as illustrated by Figure 13.28.

In Figure 13.28, an encode generator is synchronized to a flat pulse generator. The flat pulse generator is adjusted for an output of just under full scale. A good high-frequency dual-trace scope is connected as shown and synchronized to the generators. The encode pulse is moved in time in relation to the flat pulse. At the setting for which the bit display indicates that the output is within 1 LSB of the final value, the amount of time from transition of the step function to the leading edge of the encode command is the transient-response settling time.

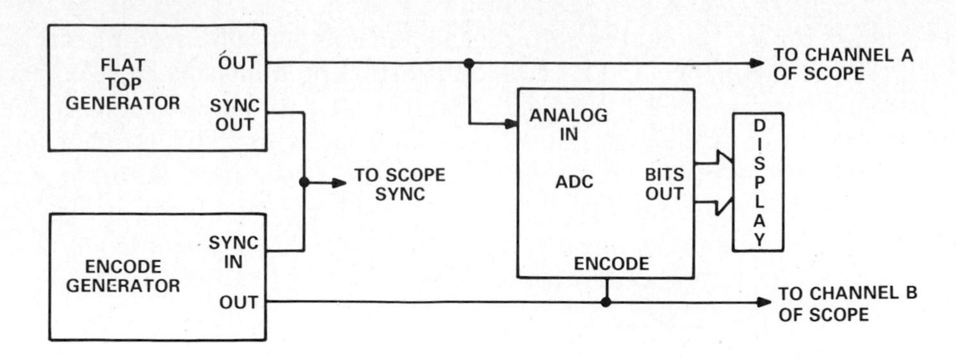

Figure 13.28. Transient-response test setup.

Differential Gain and Phase

This specification is performed specifically for video and television applications. Differential gain is defined as the percentage difference between the output amplitudes of a small high-frequency sine wave at two stated levels of a low frequency signal on which it is superimposed. Differential phase is the difference in the output phase of a small high frequency sine wave for two stated levels of a low frequency signal on which it is superimposed.

Figure 13.29 shows a typical test set-up. Specialized test equipment is used for these tests and a color video monitor is used for a final visual check of the performance of the ADC.

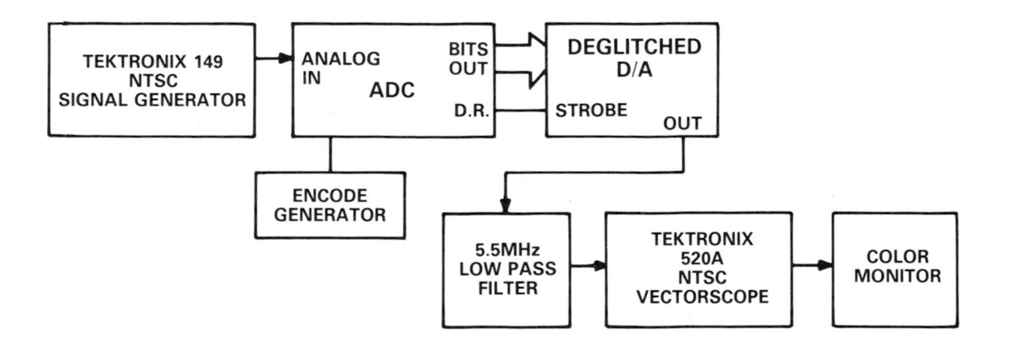

Figure 13.29. Differential-gain and differential-phase test.

13.4.2 DAC TESTING

Many of the testing methods that apply to low-speed DACs also apply to high-speed components. Measuring settling time of high-speed DACs, however, presents an especial challenge.

The high vertical gain required to determine accurately the *full-scale settling time* of fast d/a converters by observations during during the final settling period tends to overload the oscilloscope during the initial phase of the response, driving it into a state that will give erroneous readings. This results in serious degradation of measurement accuracy. Therefore the preferred, although less direct, approach is to measure rise time and then calculate the settling time, or to determine settling time by observing the transition time between two voltage-comparator windows. In both cases, the oscilloscope is teamed with a digital driver (to drive the inputs of the d/a converter) and a flat-pulse generator. A window comparator circuit may also be necessary.

Calculating Settling Time

Because of the high vertical gain required to measure the full-scale settling time of an 8-nanosecond 8-bit d/a converter to within 0.4% of full scale accuracy, overloading develops and the results obtained from the oscilloscope are incorrect. However, the gain can be reduced, and the settling time can then (in many cases) be calculated from a rise-time measurement.

The rise time (10-90% points) of the all-ZEROs to all-ONEs transition, t_r, is measured by means of a real-time or sampling oscilloscope that has a bandwidth of at least 500 MHz. In a single-pole system, the associated time constant, T, is related to t_r by $t_r = 2.2\ T$. If an exponential function is assumed, then the d/a converter output, V, is calculated from the equation,

$$V = (1 - e^{-t/T})\,V_o \tag{13.3}$$

where V_o is the final, settled output.

Table 13-2 lists the settling time required to reach a percentage of the final output, V_o, for various rise times, t_r.

Settling times to reach ± 1/2LSB of output

Resolution (± 1/2LSB)	Settling time (t_s)
12 bits	$3.8\,t_r$*
11 bits	$3.5\,t_r$
10 bits	$3.1\,t_r$
9 bits	$2.8\,t_r$
8 bits	$2.5\,t_r$
7 bits	$2.2\,t_r$
6 bits	$1.9\,t_r$
5 bits	$1.5\,t_r$
4 bits	$1.2\,t_r$

*t_r = rise time for input step from all-ZEROS to all-ONES (10% to 90%)

Table 13.2. Settling time vs. resolution for single time-constant exponentials.

Accurate *midscale bit-transition settling time* (as opposed to full-scale) is relatively easy to measure using direct oscilloscopic techniques. Since an interval of only 1 LSB is being monitored, the response time is generally quite a bit shorter, and the glitch at the midscale carry point (011 . . 11 to 100 . . 00) usually has an amplitude of less than 100 mV—maintained very briefly—little danger exists of excessively overdriving the input amplifier of the oscilloscope.

Window Measurement of DAC Settling Time

An alternative way to measure full-scale settling time (from the start of the transition) is to employ fast window comparators to establish T_1, when the response has started—i.e., exceeds 1 LSB—and when it is within 1 LSB of the final value (T_2). The oscilloscope is used only to observe the binary switching of logic levels rather than a graded analog response. Figure 13.30 shows a typical measurement setup.

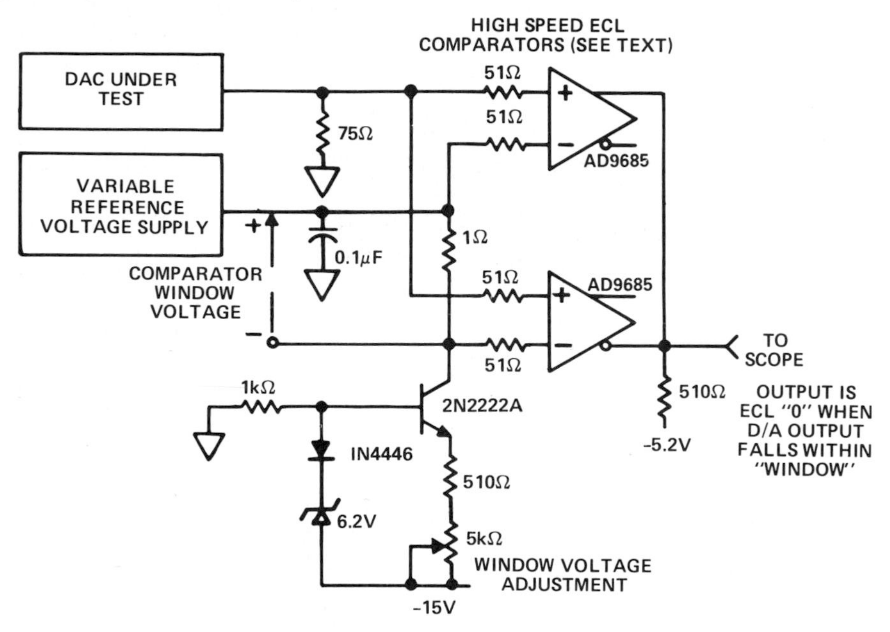

Figure 13.30. Settling-time test setup employing comparators.

To measure full scale settling time, T_1 and T_2 must be measured. The window is set for 1 LSB (for instance, 2.5 mV for the HDG-0805 8-bit display DAC). For T_1, the reference is set at the most positive value that will still leave the comparator output at logic 0 during the time immediately preceding the output transition of the DAC. T_1 is defined as the time the fast comparator output reaches 50% of its transition following the start of the DAC output transition (Figure 13.31).

For T_2, the comparator reference is shifted positively until the comparator output returns to logic 0 during the time following the DAC's output transition, and the 50% point of the 1 to 0 transition is the shortest possible time after T_1. $T_2 - T_1$ represents the settling time, as defined above, independently of the fixed delay through the DAC input registers.

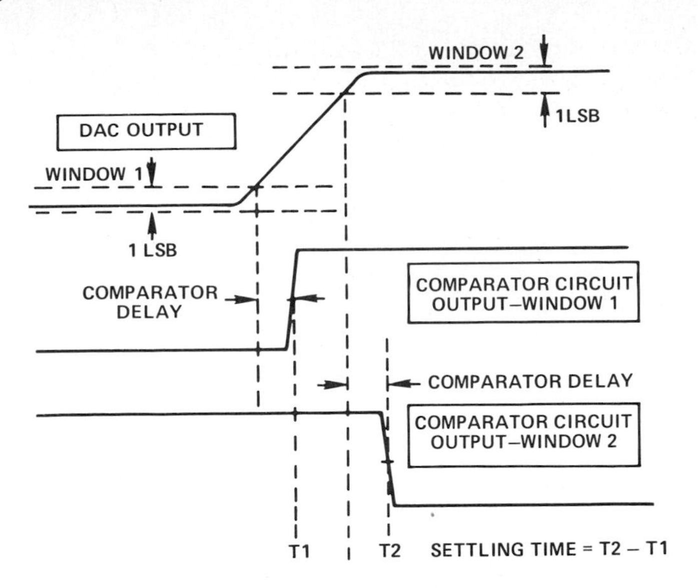

Figure 13.31. Response plots in settling-time test.

Since T_1 is established with both comparators in a state of large overdrive, and T_2 with the comparators in small overdrive, it is desirable to consult the specifications for the specific comparators being used to see if the difference in comparator response is sufficiently significant to require a correction, and to establish the size of the correction. Figure 13.32 shows the comparator waveforms for T_1 and T_2 superimposed in a double exposure. The settling time measures about 7.5 nanoseconds.

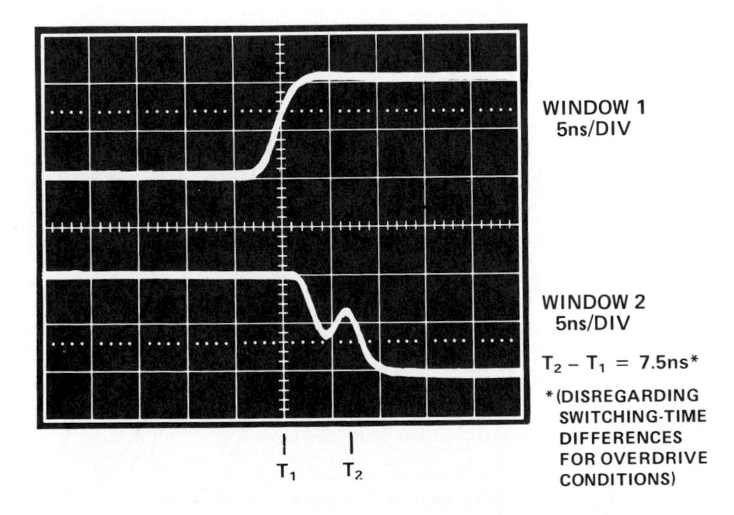

Figure 13.32. Waveforms in settling-time test.

Chapter Fourteen

Converters for Resolvers and Related Devices

The conventional approach to signal conditioning and data acquisition (Chapters 2 and 4) is useful in the many cases where a variety of transducers, chosen for a variety of reasons, must have their outputs interpreted or their inputs furnished digitally. However, there are a great many applications in position measurement where a simple subsystem, consisting of a specialized electromagnetic sensor having standardized analog data format and a special-purpose microprocessor-compatible data converter, provides accurate, reliable, cost-effective measurement and control. This chapter is about some subsystems of this kind.

The boom in robotics, computer-aided manufacturing, and factory automation has generated increasing demands for linear and rotational position-measuring transducers that are rugged, reliable, and easily automated. Resolvers, synchros, Inductosyns™* (flat-transformer-like position transducers), and LVDTs (linear variable differential transformers) have been available to meet these needs for years. Until recently, however, they have found only limited use in digital systems, primarily because the interfacing was somewhat complicated and expensive.

The situation has now changed considerably. Integrated-circuit manufacturers, such as Analog Devices, offer compact digital interface circuits, such as resolver-to-digital and synchro-to-digital converters (Figure 14.1a), in the form of hybrid integrated circuits. These devices make it significantly easier to use resolvers, synchros, Inductosyns, and—more recently—LVDTs in automated position-measurement applications. Data-acquisition systems that

*TM–Inductosyn is a registered trademark of Farrand Controls, Inc.

team these converters up with the respective transducers achieve superior reliability, noise performance, accuracy, and survivability in harsh environments when compared with other commonly used position-transducer alternatives, such as optical encoders and potentiometers.

(Courtesy of Moore-Reed, Ltd.)

Figure 14.1. Resolvers and resolver-to-digital converters.

14.1 POSITION MEASUREMENT IN PERSPECTIVE

Modern machine-tool and robot systems increasingly rely on precision position transducers as the primary feedback element in their closed-loop control systems. The robots of Figure 14.2, for instance, have nineteen angular movements, each of which must be measured and controlled. Not only do resolvers, in combination with resolver-to-digital converters (RDCs), provide absolute digital representations of angular position; they also offer the option of an analog voltage output proportional to angular velocity. Designers of robot and machine-tool controllers find this signal indispensable for control-loop stabilization, or speed-profile control.

Resolvers, synchros, Inductosyns, and their associated digital converters have already found use in a wide range of applications, and the list is growing. A brief listing might include:

Avionics:

- airborne radar scanners
- aircraft control surface feedback (flap control, for instance)
- acquisition of pitch, roll, heading data for aircraft navigational equipment
- sonars

Industrial measurement and control, including:

- robots
- machine-tool controllers
- speed controllers
- control of printing machinery

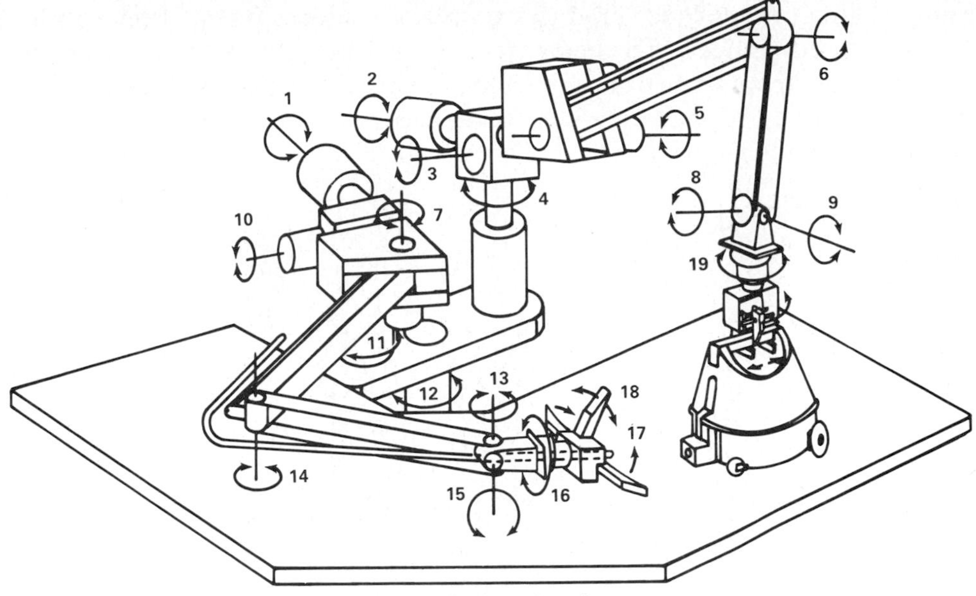

Figure 14.2. Multi-axis robot arms.

Military:

- missile gyros and launchers
- naval fire control systems, data retransmission units, and compass repeaters
- radars
- torpedos and torpedo launchers
- electronic countermeasures equipment
- tank gun sights and gun controls
- ground-based missile launchers

Of available position sensors, most—such as resolvers, synchros, rotary inductosyns, potentiometers, and optical encoders—are inherently *rotary* measuring devices. Through the use of lead screws, however, such rotary transducers are readily converted into *linear-position* transducers. In contrast, linear Inductosyns, linear variable differential transformers (LVDTs), and straight wire potentiometers measure linear positions directly. Position sensors can be further categorized as providing either *absolute* or *incremental* measurements.

14.1.1 OPTICAL ENCODERS

Among the most-popular position measuring transducers, optical encoders find use mainly in relatively low-reliability and low-resolution applications. An *incremental* optical encoder (Figure 14.3a) is a disc divided into sectors that are alternately transparent and opaque. A light source is positioned at one side of the disc, and a light detector at the other side. As the disc rotates, the output from the detector switches alternately on and off depending on whether the sector appearing between the light source and the detector is opaque or transparent. Thus, the encoder produces a stream of square wave pulses which, when counted, indicate the angular position of the shaft. Available encoder resolutions (the number of opaque and transparent sectors per disc) range from 100 to 65,000, with absolute accuracies approaching 30 arc seconds (1/43,200 per rotation).

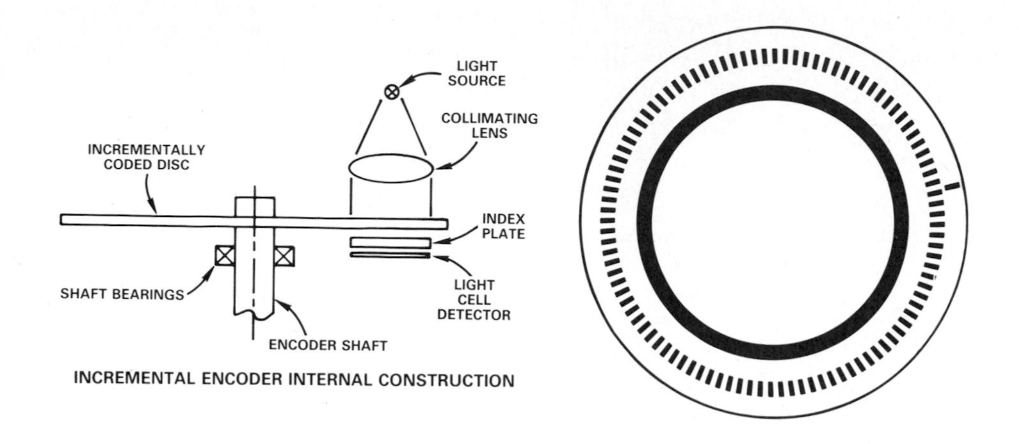

a. Incremental encoder.

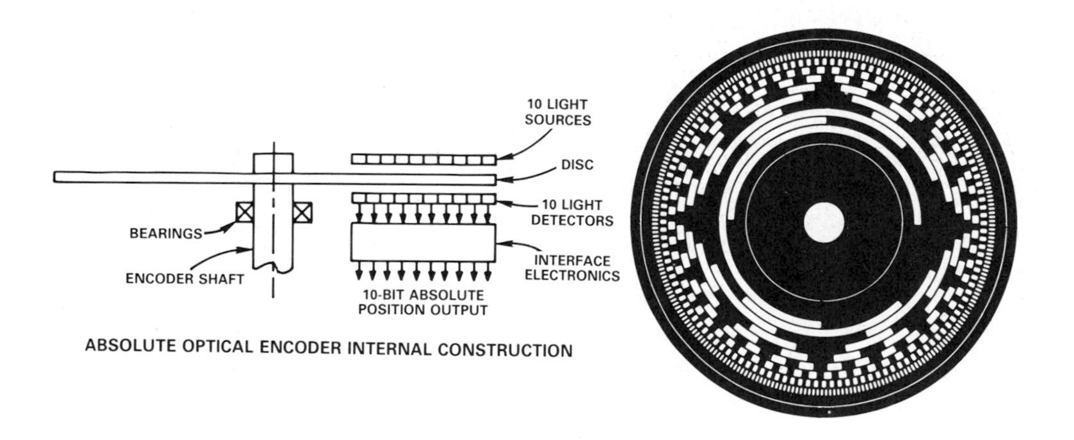

b. Absolute encoder.

Figure 14.3. Optical encoders.

Most incremental encoders feature a second light source and detector, atan angle to the main source and detector, to indicate direction of rotation. Many encoders also feature a third light source and detector to sense a once-per-revolution marker. Without some form of revolution marker, absolute angles are difficult to determine.

As a potentially serious disadvantage, incremental encoders require external counters to determine absolute angles within a given rotation. If the power is momentarily shut off, or if the encoder misses a pulse due to induced noise or a dirty encoder disk, the resulting absolute angular indications will be in error. For robotics applications, in particular, reliable determination of absolute angular position is increasingly required. The absolute position of a robot's arm is critical in most applications, for instance, welding or assembly.

Absolute optical encoders overcome these disadvantages but typically cost more than twice as much as their incremental counterparts. An absolute optical encoder's disc is divided up into 2^N sectors (Figure 14.3b). But each sector is further divided radially along its length into opaque and transparent sections, forming a unique N-bit digital word with a maximum count of 2^N. The digital word formed radially by each sector increments in value from one sector to the next, usually employing Gray code (see Table 7.6 and Figure 7.5).

A set of N light detectors, arranged radially on one side of the disc, responds to N light sources positioned on the other side. The detectors' output forms an N-bit digital word corresponding to the disc's absolute angular position. Industrial absolute optical encoders achieve up to 16-bit resolutions, with absolute accuracies that approach the resolution (20 arc seconds).

Both absolute and incremental optical encoders, however, experience difficulties, particularly in harsh factory environments. When subjected to vibration or sudden shocks, they may fall out of optical alignment. Light-source lifetimes can be seriously temperature limited, and incandescent light sources can experience damage through vibration. The gallium-arsenide light sources used in many optical encoders may suffer degraded performance due to aging and extremes of temperature. For this reason, designers often hesitate to mount optical encoders where they are needed most, such as directly on drive motors that operate in harsh environments. For low-resolution (10-bits or less) applications not requiring the utmost in reliability, however, optical encoders are easy to use and low in cost.

14.1.2 POTENTIOMETERS, CONTACT ENCODERS, RVDTS, LVDTS, AND GLASS SCALES

As common position-measuring devices, *potentiometers* exhibit exceptionally low cost but suffer from significantly poorer reliability when compared to other transducers. Since typical pots employ a brush or wiper contacting a resistive track, they often experience wear-related electrical and mechanical noise problems. Their performance depends upon the stability of their reference voltage supply, and the accuracy of the signal conditioning circuitry and associated analog-to-digital converter. Designers occasionally use potentiometers for coarse position measurement, in conjunction with other transducers. Available potentiometers approach 12-bit resolutions, with absolute accuracies approaching 7 arc-minutes (about $1/3000$ per rotation).

So-called *brush* or *contact encoders* work on a principle similar to that of optical encoders. However, instead of using optics to generate the angular signal, contact encoders employ conducting and insulating surfaces in conjunction with brush contacts. This type of encoder cannot sustain high rotational speeds and remains limited to applications requiring small and infrequent shaft movements. Contact encoders with resolutions of up to 10 bits and accuracies of 26 arc-minutes (about 1/830 per rotation) are available.

Rotary (or linear) variable differential transformers (RVDTs, LVDTs) are in common use. LVDTs, for instance, find use in metrology, gauging, and even cash dispensers. They feature contactless operation, small size, and high resolution and repeatability. Some devices do suffer from drift as high as 0.08%/°C. RVDTs find use extensively in avionic applications, such as vane-position indication in turbines, aircraft angle-of-attack indicators, and control-surface feedback. RVDTs suffer from a limited rotational range (85 degrees) and limited accuracies (1 to 2%).

Direct interfacing of LVDTs and RVDTs to digital systems has recently become practical through the use of LVDT-to-digital and RVDT-to-digital converters, such as those in the Analog Devices 2S56 series. They have many features in common with resolver-to-digital converters.

Glass-scale or Moire-fringe transducers represent still another transducer technology. Glass scales are widely used for high-accuracy linear and rotational measurement. A linear scale for example will give a basic resolution of 10 microns, which can often be improved by a 4- or 5-bit interpolation method. Glass scales are significantly less expensive than some alternative transducer technologies, but lack ruggedness.

14.1.3 RESOLVERS AND SYNCHROS

Because of the serious disadvantages of competing transducer technologies, machine-tool and robotics manufacturers increasingly turn to resolvers and Inductosyns as the position transducers of choice. These components particularly excel in demanding factory applications requiring small size, long-term reliability, absolute position measurement, high accuracy, and low-noise operation.

Figure 14.4 is a diagram of a typical brushless resolver. In appearance, synchros and resolvers resemble small cylindrical AC motors. They vary in diameter from 0.8 inches to 3.7 inches. One end of their bodies has an insulated terminal block, and the other end a mounting flange. Rotor shafts are normally threaded and splined.

In their simplest forms both synchros and resolvers employ single-winding rotors that revolve inside fixed stators. In the case of a simple *synchro*, the stator has three windings oriented 120 degrees apart and electrically connected in a Y-configuration . Resolvers differ from synchros in that their

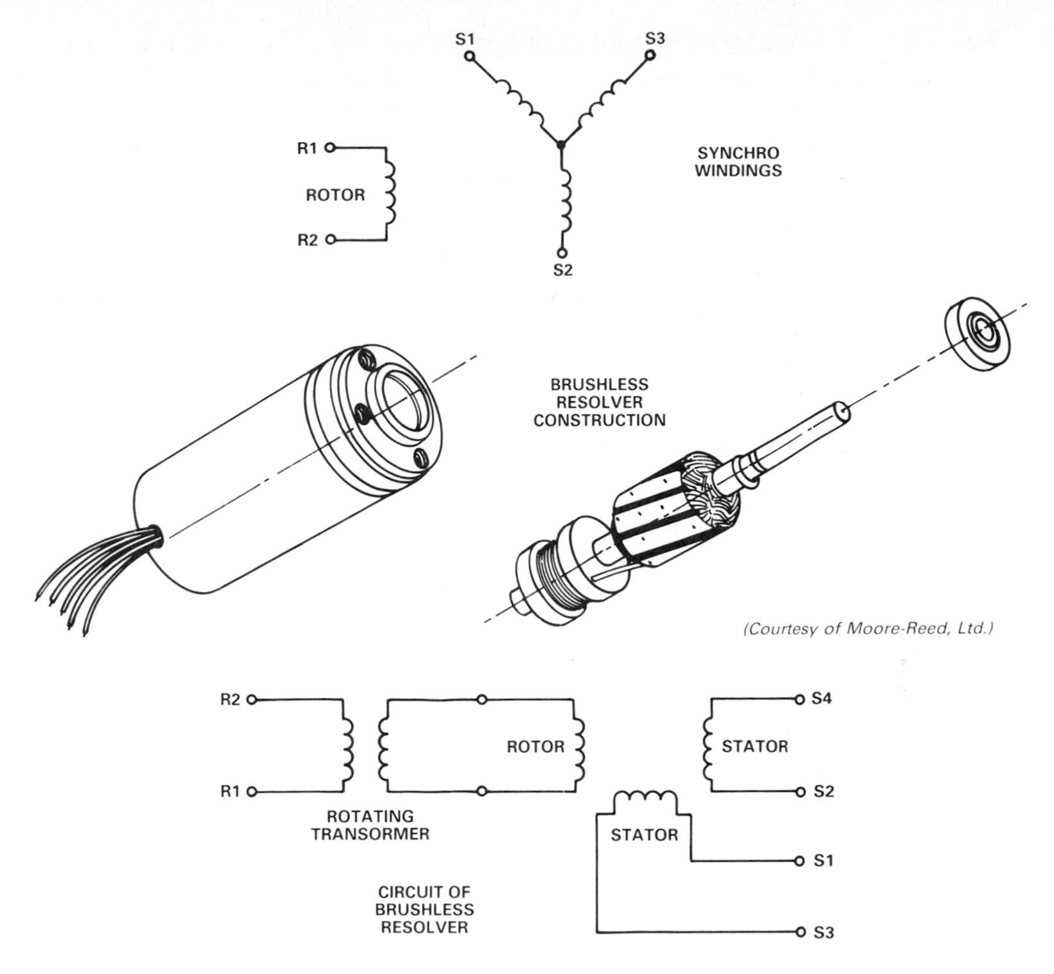

Figure 14.4. Synchro and resolver windings. Brushless resolver construction.

stators have only two windings oriented at 90 degrees.

Because synchros have three stator coils in a 120° orientation, they are more difficult than resolvers to manufacture and are therefore more costly. Today, synchros find decreasing use, except in certain military and avionic retrofit applications.

Modern *resolvers*, in contrast, are available in a brushless form that employs a rotating transformer on the rotor to couple the rotor signals. Because brushless resolvers have no slip rings or brushes, and because there are only two stator coils wound at right angles, they are cheaper and easier to make compared to synchros. They are also more rugged than synchros because there are no brushes to break or dislodge. In fact, the life of a brushless resolver is limited only by its bearings. Furthermore, most brushless resolvers are specified to work over 2-40 volts and at frequencies from 400 to 10,000 Hz. This makes them easier to tailor to the user's needs.

Typical resolvers achieve the following accuracies:

Standard Size 11:	± 5 arc-minutes
Selected Size 11:	± 3 arc-minutes
High Accuracy Slab Type:	± 1 arc-minutes
Two-speed electrically geared type:	0.5 arc-minutes

In operation, synchros and resolvers resemble rotating transformers. The rotor winding is excited by an AC reference voltage, at frequencies up to a few kilohertz. The magnitude of the voltage induced in any stator winding is proportional to the cosine of the angle, θ, between the rotor coil axis and the stator coil axis. In the case of a synchro (Figure 14.4), the voltage induced across any pair of stator terminals will be the vector sum of the voltages across the two connected coils.

For example, if the rotor of a synchro is excited with a reference voltage, V sin ωt, across its terminals, R1 and R2, then the stator's terminals will see voltages of the form

$$
\begin{aligned}
V(S1 \text{ to } S3) &= V \sin \omega t \sin \theta \\
V(S3 \text{ to } S2) &= V \sin \omega t \sin (\theta + 120°) \\
V(S2 \text{ to } S1) &= V \sin \omega t \sin (\theta + 240°)
\end{aligned} \tag{14.1}
$$

where θ is the shaft angle.

In the case of a resolver, with a rotor a-c reference voltage of V sin ωt, the stator's terminal voltages will be

$$
\begin{aligned}
V(S1 \text{ to } S3) &= V \sin \omega t \sin \theta \\
V(S4 \text{ to } S2) &= V \sin \omega t \sin (\theta + 90°) \\
&= V \sin \omega t \cos \theta
\end{aligned} \tag{14.2}
$$

Resolver-to-digital converters (RDCs) transform these voltage relationships into digital representations of the actual angle, θ.

When teamed up with such converters, synchros and resolvers easily achieve 12-bit resolutions, corresponding to an angular resolution of 5.3 arc-minutes. Absolute accuracies can approach 20 arc seconds, or better, for models having higher resolution. If the resolver, connected to a 0.1-inch-pitch leadscrew, is used with a 12-bit converter, linear resolution becomes 25 microinches.

Resolvers and synchros offer the advantages of proven long-term reliability, absolute position measurement, high accuracy and resolution, small size, and moderate cost. In fact, a standard resolver in combination with an RDC costs less than many optical encoders, especially in the case of optical encoders specifically designed for use in harsh industrial environments. In addition, users report that typical resolvers and synchros have a mean time before failure (MTBF) four or more times longer than that for optical encoders.

In the past, resolvers and synchros were handicapped by the need for complicated, expensive interface circuitry. But with the availability of easy-to-implement RDCs and SDCs in hybrid form, this roadblock no longer exists.

14.1.4 INDUCTOSYNS

Synchros and resolvers inherently measure rotary position, but they can make linear position measurements when used with lead screws. An alternative, the Inductosyn* (Figure 14.5) measures linear position directly. In addition, Inductosyns are accurate and rugged, well-suited to severe industrial environments, and do not require ohmic contact.

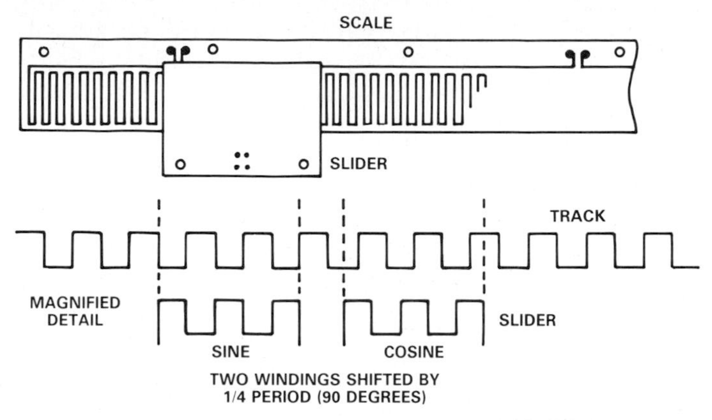

Figure 14.5. Inductosyn, showing detail of slider pattern.

The linear Inductosyn consists of two magnetically coupled parts; it resembles a multipole resolver in its operation. One part, the scale, is fixed (e.g. with epoxy) to one axis, such as a machine tool bed. The other part, the slider, moves along the scale in conjunction with the device to be positioned (for example, the machine tool carrier).

The scale is constructed of a base material such as steel, stainless steel, aluminum, or a tape of spring steel, covered by an insulating layer. Bonded to this is a printed-circuit track, in the form of a continuous rectangular waveform pattern. The pattern typically has a cyclic pitch of 0.1 inch, 0.2 inch, or 2 millimeters. The slider, about 4 inches long, has two separate but identical printed circuit tracks bonded to the surface that faces the scale. These two tracks have a waveform pattern with exactly the same cyclic pitch as the waveform on the scale, but one track is shifted one-quarter of a cycle relative to the other. The slider and scale remain separated by a small air gap of about 0.007 inch.

Inductosyn operation resembles that of a resolver. When the scale is energized with a sine wave, this voltage couples to the two slider windings, inducing voltages proportional to the sine and cosine of the slider's spacing within a cyclic pitch. If S is the distance between pitches, X is the slider displacement within a pitch, and the scale is energized by the voltage, $V \sin \omega t$, then the slider windings will see terminal voltages of

*Inductosyns are manufactured by Farrand Controls, Inc., Valhalla, NY, and their licensees. Inductosyn® is a registered trademark of Farrand Controls, Inc.

$$V(\text{sine output}) = V \sin \omega t \sin \frac{2\pi X}{S}$$

$$V(\text{cosine output}) = V \sin \omega t \cos \frac{2\pi X}{S} \quad (14.3)$$

As the slider moves a distance equivalent to one pitch, the voltages produced by the two slider windings are similar to those produced by a resolver rotating through 360 degrees. However, in contrast to a resolver's highly efficient transformation ratio of 1:1 or 2:1, typical Inductosyns operate with transformation ratios of 100:1. This results in a pair of sinusoidal output signals in the millivolt range which generally require amplification. Dual preamplifiers are available to amplify both signals in identical fashion (see Figure 14.14).

Since the slider output signals are derived from an average of several spatial cycles, small residual errors in conductor spacing have minimal effects. This is an important reason for the Inductosyn's very high accuracy. In combination with 12-bit Inductosyn-to-digital converters (IDCs), 0.1-inch-pitch linear Inductosyns readily achieve 25 microinch resolutions.

Rotary Inductosyns can be created by printing the scale on a circular rotor and the slider's track pattern on a circular stator. Such rotary devices can achieve very high resolutions. For instance, a typical rotary Inductosyn may have 360 cyclic pitches per rotation, and might use a 12-bit Inductosyn-to-digital converter. The converter effectively divides each pitch into 2^{12}, or 4,096, sectors. Multiplying by 360 pitches, the rotary Inductosyn divides the circle into a total of 1,474,560 sectors. This corresponds to an angular resolution of less than 0.9 arc seconds. Higher-resolution models can achieve resolutions to 0.05 arc seconds (about 25-bit resolution), with absolute accuracies approaching 1.5 arc-seconds.

In the above example, the absolute orientation of the Inductosyn is determined by counting successive pitches in either direction from an established starting point. Although the basic Inductosyn is a single-track incremental device, absolute distance measurements can also be made by combining two tracks. The most common device is often called an "N and N – 1" Inductosyn. The device employs two tracks with, say, 255 and 256 pitches. A separate converter on each track and a small amount of digital logic results in absolute position measurements at all times *including turn-on*. Schemes such as this provide the most cost-effective high-resolution system available, typically achieving resolutions comparable to those of absolute optical encoders (for instance) at less than one fourth the cost.

14.1.5 COMPARING ALTERNATIVES

Figure 14.6 compares in a general way the absolute accuracies and costs of the competing position-measuring transducers. In addition to these parameters, resolution is an important consideration. The resolution necessary for a given application, however, largely depends upon the smoothness required for the finished workpiece and/or the number of bits of digital resolution specified for input to the computing system. Today, 10-bit resolutions are most common (1,024 increments per revolution), but 12-bit (4,096 increments) and 14-bit (16,384 increments) resolutions are becoming more popular, particularly in advanced machine-tool controllers and drive systems. The designer must carefully match the system resolution to the absolute accuracy of the transducer used. It makes no sense to use a 1%-accuracy potentiometer with a 14-bit control system.

Repeatability often has greater significance than absolute accuracy, particularly in robot or tracer applications. Repeatability should at least match the resolution of the control system. Repeatability is particularly important in "teaching-type" robots where it is crucially important that the robot return to the same instructed positions each time.

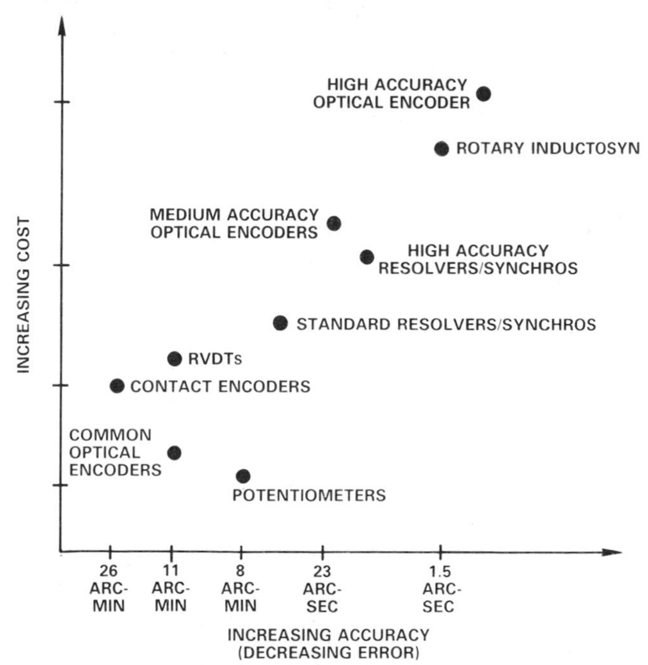

Figure 14.6. Relative cost vs. accuracy of implementing common position transducers digitally.

14.2 RESOLVER-TO-DIGITAL CONVERTERS

14.2.1 TRACKING CONVERTERS

The attractiveness of resolvers, synchros and Inductosyns is significantly enhanced by the availability of easy-to-implement interface circuits, for example, tracking resolver-to-digital converters (RDCs), such as those pictured in Figure 14.1.

Figure 14.7 is a functional diagram of a typical tracking resolver to digital converter. To begin with, it is worth noting that RDCs and Inductosyn-to-digital

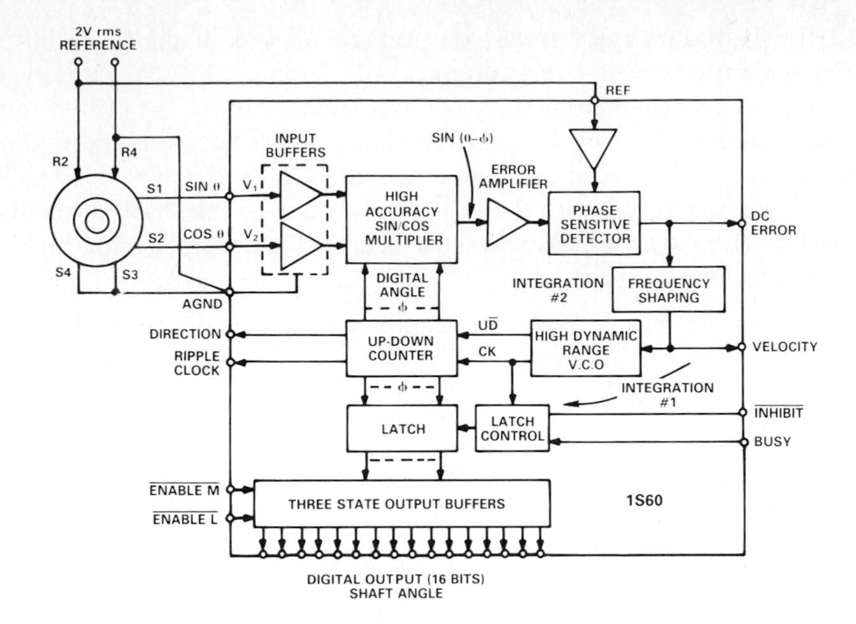

BIT WEIGHT TABLE

Bit Number	Weight in Degrees
1 (MSB)	180.0000
2	90.0000
3	45.0000
4	22.5000
5	11.2500
6	5.6250
7	2.8125
8	1.4063
9	0.7031
10	0.3516
11	0.1758
12	0.0879
13	0.0440
14	0.0220
15	0.0110
16	0.0055

Figure 14.7. Functional diagram of tracking resolver-to-digital converter.

converters work in the identical way, but synchro-to-digital converters require specialized input circuitry to convert the 3-wire input into resolver format. This circuitry can be either a Scott-T connected transformer or its solid state equivalent.

In operation, the sine and cosine multipliers of Figure 14.7 are in fact multiplying digital-to-analog converters, which incorporate sine and cosine functions. Begin by assuming that the current state of the up-down counter is a digital number representing a trial angle, ϕ. The converter seeks to adjust digital angle, ϕ, continuously to become equal to, and to track θ, the analog angle being measured. The resolver's stator output voltages, from equation 14.2, are written as

$$V_1 = V \sin \omega t \sin \theta$$
$$V_2 = V \sin \omega t \cos \theta \tag{14.4}$$

where θ is the angle of the resolver's rotor. The digital angle ϕ is applied to the cosine multiplier, and its cosine is multiplied by V1 to produce the term,

$$V \sin \omega t \sin \theta \cos \phi \tag{14.5}$$

Digital angle ϕ is also applied to the sine multiplier and multiplied by V2 to produce

$$V \sin \omega t \cos \theta \sin \phi \tag{14.6}$$

These two signals are subtracted by the error amplifier to yield an AC error signal of the form

$$V \sin \omega t (\sin \theta \cos \phi - \cos \theta \sin \phi) \tag{14.7}$$

This, in turn, through a simple trigonometric identity, reduces to

$$V \sin \omega t \sin (\theta - \phi) \tag{14.8}$$

The phase-sensitive detector demodulates this AC error signal, using the resolver's rotor voltage as a reference. This results in a DC error signal proportional to $\sin(\theta - \phi)$.

The DC error signal feeds an integrator, the output of which drives a voltage controlled oscillator (VCO). The VCO, in turn, causes the up-down counter to count in the proper direction to cause

$$\sin(\theta - \phi) \rightarrow 0 \tag{14.9}$$

When this is achieved

$$\theta - \phi \rightarrow 0 \tag{14.10}$$

Therefore,

$$\phi = \theta \tag{14.11}$$

to within one count. Hence the counter's digital output, ϕ, represents the angle θ. The latches enable this data to be transferred without interrupting the loop's tracking.

The circuit of Figure 14.7 is equivalent to a so-called type-2 servo loop, because it has, in effect, two integrators. One is the counter, which accumulates pulses; the other is the frequency-shaping filter at the output of the phase-sensitive detector. In a type-2 servo loop, with a constant rotational velocity input, the output digital word continuously follows, or tracks the input, without needing externally derived convert commands, and with no steady state phase lag between the digital output word and actual shaft angle. An error signal appears only during periods of acceleration or deceleration.

In contrast, since a type-1 servo loop has only has one stage of integration, its error signal must remain non-zero for constant rotational velocities. In a type-1 loop, the digital angle output therefore lags the resolver's actual angular position by an amount proportional to the required error signal. As a consequence, type-1 loops are hardly ever used.

Here is a heuristic explanation of how this works: In Figure 14.7, if the resolver shaft is rotating at a constant rate, then the output digital word must follow at a constant rate, the counter must be counting pulses at a constant rate, and they can be produced at a constant rate only if the voltage at the input to the VCO (voltage-controlled oscillator) is constant and nonzero.

Therefore, for a constant rotational velocity, the second integrator (the frequency-shaping network) must maintain a constant nonzero dc output voltage to drive the VCO. This can only occur when its input, the loop error signal, remains at zero volts (since if it were not zero, the output of the integrator would have to be changing). Consequently, a type-2 servo loop enforces zero steady-state error signal (i.e., $\phi = \theta$) when the shaft is rotating at constant velocity. In other words, the digital output signal follows the resolver's actual shaft angle.

As an added bonus, the tracking RDC of Figure 14.7 provides an analog dc output voltage directly proportional to velocity. This is a useful feature if velocity is to be measured or used as a stabilization term in a servo system—it makes tachometers unnecessary.

Since the operation of these converters depends only on the ratio between input signal amplitudes, attenuation in the lines connecting them to resolvers doesn't substantially affect performance. For similar reasons, these converters are not greatly susceptible to waveform distortion. In fact, they can operate with as much as 10% harmonic distortion on the input signals; some applications actually use *square-wave* references with little additional error.

Because the tracking converter doubly integrates its error signal, the device offers a high degree of noise immunity (40 dB-per-decade rolloff). The net area under any given noise spike produces an error. But typical inductively coupled noise spikes have equal positive and negative going waveforms. When integrated, this results in a zero net error signal. The resulting noise immunity, combined with the converter's insensitivity to voltage drops, lets the user locate the converter at a considerable distance from the resolver. Noise rejection is further enhanced by the phase detector's rejection of any signal not at the reference frequency, such as wideband noise.

The family of converters exemplified in Figure 14.7 will handle resolver or Inductosyn signals at input frequencies in the range 400-10,000 Hz. The converter's inputs are designed to accept voltages of 2 volts rms, which is the lowest reference voltage likely to be encountered. Higher voltages can be scaled down as needed. Consequently the converters can interface to just about any resolver or Inductosyn - and can produce digital output words with resolutions (depending on which family member) of 10, 12, 14, or 16 bits.

14.2.2 PHASE-ANALOG APPROACH

Some users of resolvers, synchros and Inductosyns choose not to use proprietary converters, such as the one illustrated in Figure 14.7. Instead, they seek economies by designing their own interface circuitry. The so-called phase-analog method represents the most common approach.

In contrast with the tracking converter method, the phase-analog converter excites the resolver's stator (instead of rotor) with reference signals $V \sin \omega t$ and $V \cos \omega t$. The rotor winding then sees an induced voltage of the form $V \cos (\omega t + \theta)$, where θ is the shaft angle. Controlled by zero-crossing detectors, a pulse generator and counter together time the phase delay between rotor and stator signals. This delay is, of course, directly proportional to the rotor shaft angle θ.

While this approach is less costly for large numbers of channels, it is noise-sensitive and not capable, in practice, of the accuracy with which theory blesses it. Here are some points to consider, when comparing the phase analog approach with commercially available tracking-type converters. Tracking converters have:

- High tracking speed and no velocity error or lag.
- Higher accuracy. Whereas phase analog converters are capable of high resolution, depending upon clock rate, the technique has difficulty achieving accuracies beyond 10 or 11 bits in practice.
- Better immunity to variations in resolver signal and noise. The tracking converter concept virtually eliminates problems with signal level variations and noise. Phase-analog converters, on the other hand, produce large errors when the reference source is unstable, and require filters on the resolver rotor when using non-sine wave references.
- Cost. Here, the phase analog approach may have an advantage today in multichannel systems (more than about 5 channels) if accuracy and noise considerations are of less importance. Even so, new low-cost hybrid converters (and the monolithics that are likely to appear in the future) will extend their cost effectiveness to increasing numbers of channels.
- *Perhaps most important*, the accuracy of the phase analog approach is highly dependent upon the ratio of the sine and cosine drive-signal amplitudes, as well as the actual phase difference between those signals. Figure 14.8 plots the number of bits of achievable accuracy as a function of the drive signal amplitude and phase relationships. If, for instance, the drive signals have a 1% amplitude mismatch (an amplitude ratio of 1.01) and deviate from a 90° phase relationship by only 0.5° (corresponding to the $\beta = 89.5°$ curve in Figure 14.8), then maximum achievable accuracy is less than 10 bits.

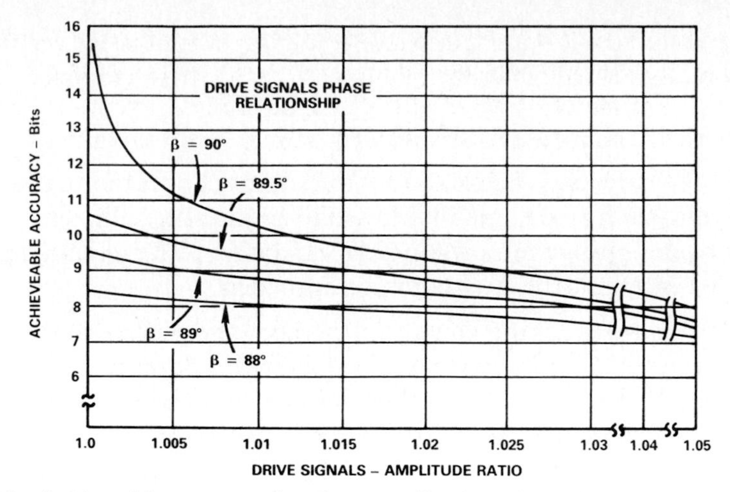

Figure 14.8. Achievable accuracy level vs. amplitude ratio and drive-signal phase error ($= |\beta - 90°|$) in a phase-analog conversion system.

This underscores the need for analog design expertise in implementing the phase-analog approach.

14.2.3 SUCCESSIVE-APPROXIMATION APPROACH

The successive-approximation conversion technique has been discussed in Chapter 7 and elsewhere. Although widely used for general-purpose a/d conversion and once popular for synchro-to-digital conversion, this technique is not often used in commercial SDC designs. Nonetheless, it warrants discussion, since elements of the technique are often adopted for in-house designs, especially in multi-channel applications.

In one approach, the successive approximation converter accepts sine and cosine inputs (via Scott-T transformers if the information is in synchro form), converts them to digital, and finds the angle via a tangent lookup or an algorithm. Thus, the loop does not act in a continuous tracking mode; each conversion is independent and requires a convert command signal to initiate it. Noise represents a serious problem for this approach. Any transient appearing on the sine or cosine lines as they are being converted will cause large errors. For this reason, many designs employing this technique take several readings and produce an average.

The inputs to the converter are two dc voltages representing $V \sin \theta$ and $V \cos \theta$, where θ is the input angle. The converter itself cannot accept ac voltages in the synchro or resolver format; the angular information requires signal conditioning to translate it to dc.

What then is the advantage of the successive-approximation converter? It is sometimes cheaper, when a large number of channels of synchro or resolver information have to be converted into digital form, along with other analog data, such as temperature, pressure, etc.

Suppose for example there are 16 channels of slowly changing data to be processed, including eight channels of synchro data. Clearly, the best angle-measuring performance could be obtained by using 8 tracking converters, from which fresh angular data from any channel could be taken at any time by digitally multiplexing the outputs on a bus.

However, the cost of 8 tracking converters (in recent years, but not necessarily at the time you read this or in the future) might appear to make this method undesirable. As an alternative lower-cost approach, a system of Scott-T-connected transformers, peak detectors, and sample/hold amplifiers, together with a multiplexed successive approximation converter, could serve as a solution to this problem. Figure 14.9 illustrates such an approach, showing three channels.

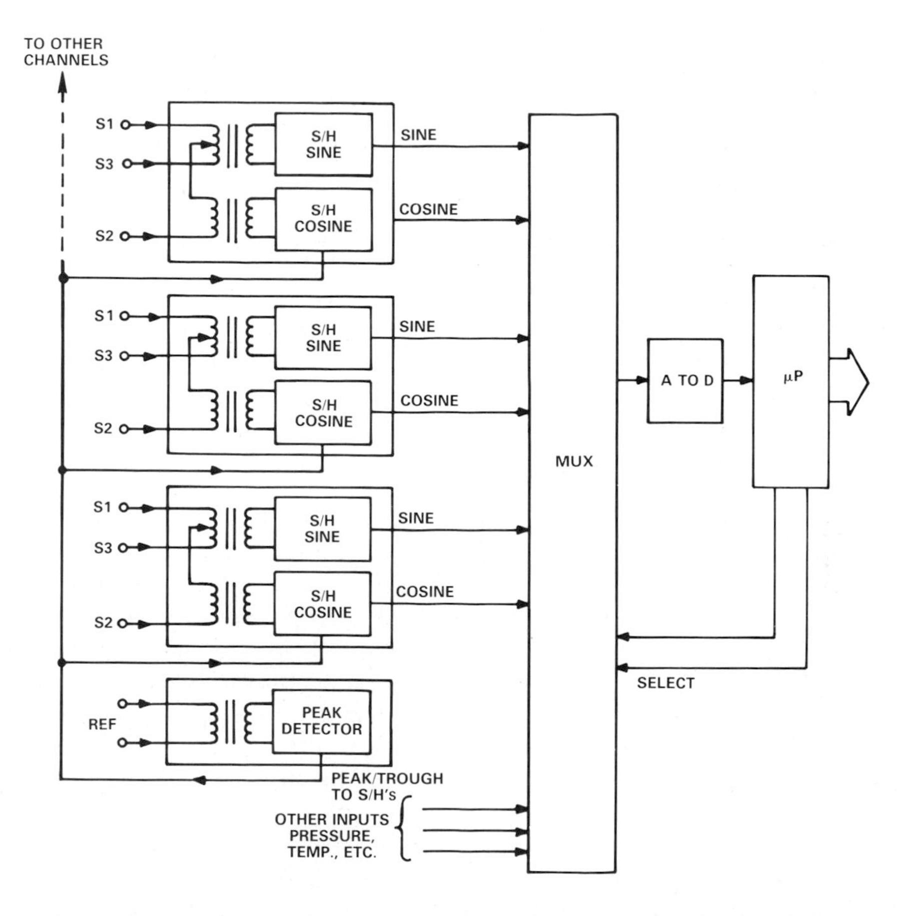

Figure 14.9. Multichannel conversion system using successive-approximation conversion.

The transformers convert from synchro to resolver format, if necessary; the peak detectors determine the amplitudes of the sine and cosine components of each angle in synchronism with the peaks of the reference waveform, and the sample/holds store the values until they are read out via the multiplexer and converted.

In an ideal system, where no phase shift exists between the reference and signal waveforms, the resolver-format signals would be sampled at the peaks of their waveforms. In practice, small phase shifts which may be present have no effect on the accuracy of the system. When a channel is to be converted, the sine and cosine outputs of its sample/hold are applied in sequence at the inputs of the a/d converter, and conversions are initiated.

Note that sampling converters, such as this one, are more sensitive to noise on the input than are tracking converters. This is because the input is usually obtained from a single sample of the voltage, while the tracking converter's frequency-shaping circuits effectively average the waveform over many periods of the reference. For applications where the inputs are varying slowly, it is possible to use precision phase-sensitive rectification and smoothing instead of sample/hold modules, in which case the smoothing will give some immunity to noise.

14.2.4 TACHOMETER OUTPUTS

Most servo-loop designs, particularly in systems where position measurement is critical, employ some form of tachometer signal for stabilization or speed control. For this reason, it is important to note that many modern converters offer the user a tachometer output with specifications comparable to what can be achieved with top-grade electromechanical tachometers.

Therefore, if a resolver is used as a feedback element, and the RDC provides a tachometer output, the designer can eliminate an electromechanical tachometer from the system design.

14.3 DIGITAL-TO-RESOLVER CONVERTERS

Digital-to-synchro and digital-to-resolver converters (DSCs and DRCs) interface digital systems to synchro or resolver angular control systems, and to other forms of application, such as PPI radar displays. The converters take binary digital inputs, representing angles, and produce outputs in either synchro or resolver format at standard frequencies and voltages. Practical converter outputs normally include a 2 volt-ampere amplification stage, which lets the units drive many electromechanical loads directly.

The most common loads include:

Control Transformers (CT)
Torque Receivers (TR)
Control Differential Transmitters (CDX)
Torque Differential Transmitters (TDX)

Synchro or resolver control transformers, used as feedback elements in electromechanical servo control loops, represent the most common loads. Figure

14.10 shows such a control loop being driven from a digital-to-synchro converter.

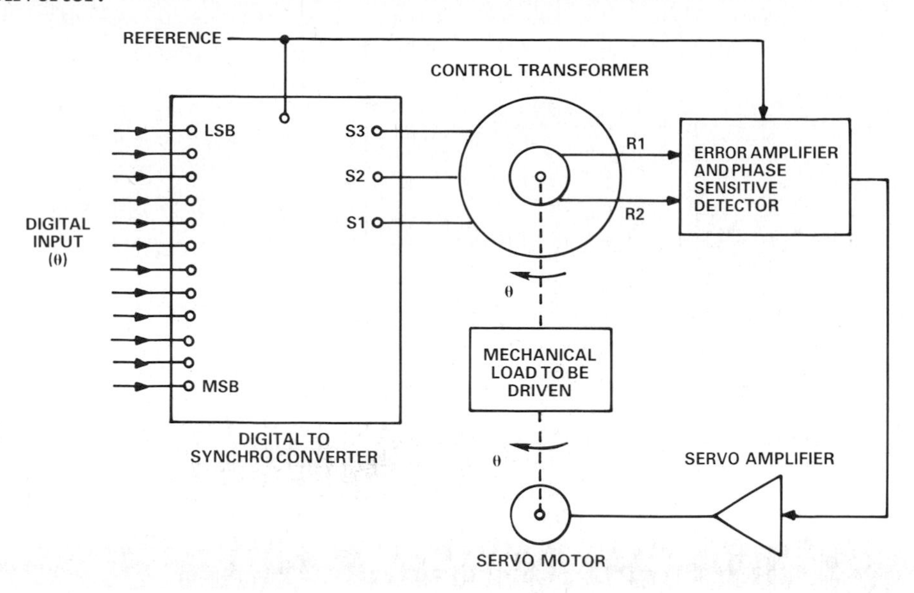

Figure 14.10. Digital-to-synchro converter in control-loop setpoint application.

Figure 14.11a illustrates the design of a digital-to-synchro converter and Figure 14.11b a digital-to-resolver converter. Differences in the secondary connections of the output transformers signify the only difference between the two designs. In the DSC, the output transformers are connected in a Scott configuration, with the outputs 120 degrees apart; the DRC's transformers are connected to provide outputs 90 degrees apart.

Like the tracking-type SDCs and RDCs, the converters of Figures 14.11 use multiplying digital to analog converters incorporating sine and cosine laws. The converters multiply the reference input voltage, V sin ωt by sin θ and cos θ to produce the resolver-form voltages:

$$V(\text{sine}) = E \sin \omega t \sin \theta$$
$$V(\text{cosine}) = E \sin \omega t \cos \theta \qquad (14.12)$$

where θ is the desired shaft angle applied in digital form to the sine and cosine multipliers.

For digital-to-resolver converters, these voltages are fed directly into the power amplifiers and isolation transformers. In the case of DSCs, the voltages feed into the power amplifiers and the Scott-T-connected transformers to produce the required synchro-format voltages:

$$V_{S1-S3} = E \sin \omega t \sin \theta$$
$$V_{S3-S2} = E \sin \omega t \sin (\theta + 120°)$$
$$V_{S2-S1} = E \sin \omega t \sin (\theta + 240°) \qquad (14.13)$$

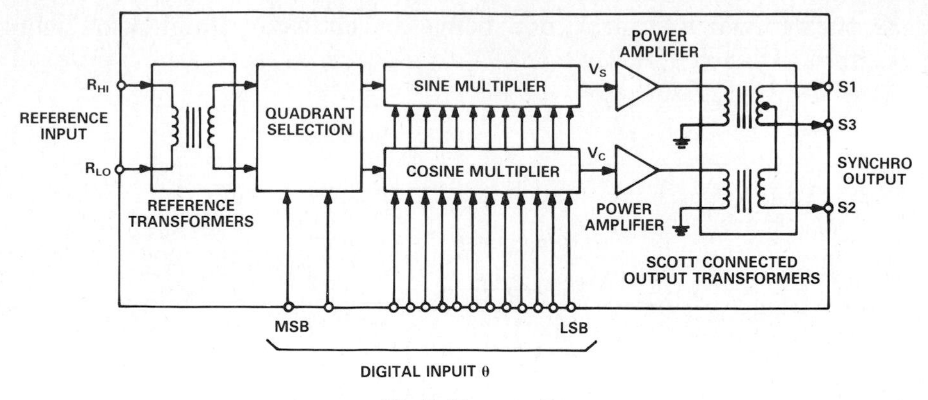

a. Digital to synchro.

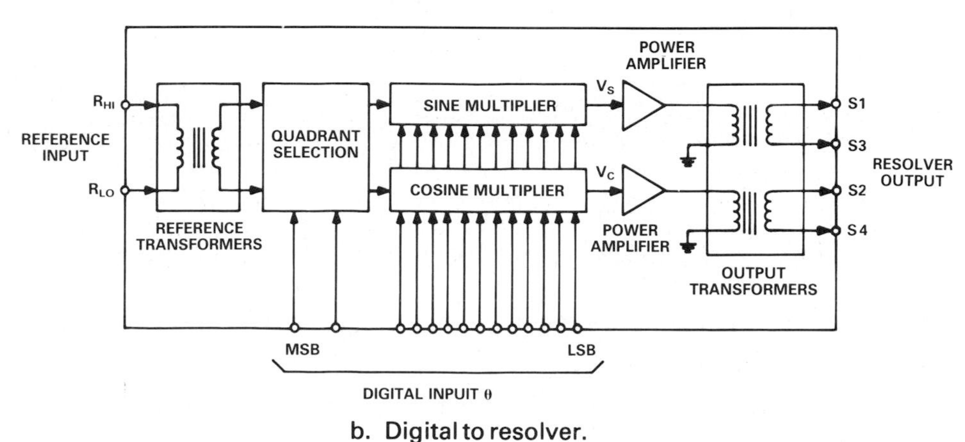

b. Digital to resolver.

Figure 14.11. Digital-to-synchro and digital-to-resolver converters.

In certain converter designs, the sine and cosine multipliers do not follow the sine and cosine laws exactly, but vary from them by up to ± 7%. Although the sine and cosine are not accurate *per se*, in such devices, their ratio forms a very close approximation to the tangent, and they are useful in establishing shaft angles accurately. However, the value of E, and hence the *magnitude* of the resulting vector in Equations 14.12 and 14.13, is a function of the angle θ; the phenomenon is called *radius-vector variation*. If a digital-to-resolver converter of this type were used to drive the X and Y plates of an oscilloscope or radar display, the resulting trace—which is expected to be circular—would appear distorted, as shown in Figure 14.12, with accurate angles but erroneous amplitudes.

Modern converters (for example, the Analog Devices DRC1745) achieve low radius-vector variations, typically ± 0.1%, and consequently experience few problems of this sort, whether driving an electromechanical servo system, an oscilloscope, or a radar display. Some older converter designs, with significant radius-vector errors, can still be found driving certain electromechanical loads that depend on the angle of the vector and are tolerant of magnitude errors. In those cases, it is still important that the *ratio* of sine and cosine, i.e., the tangent, retain sufficient accuracy.

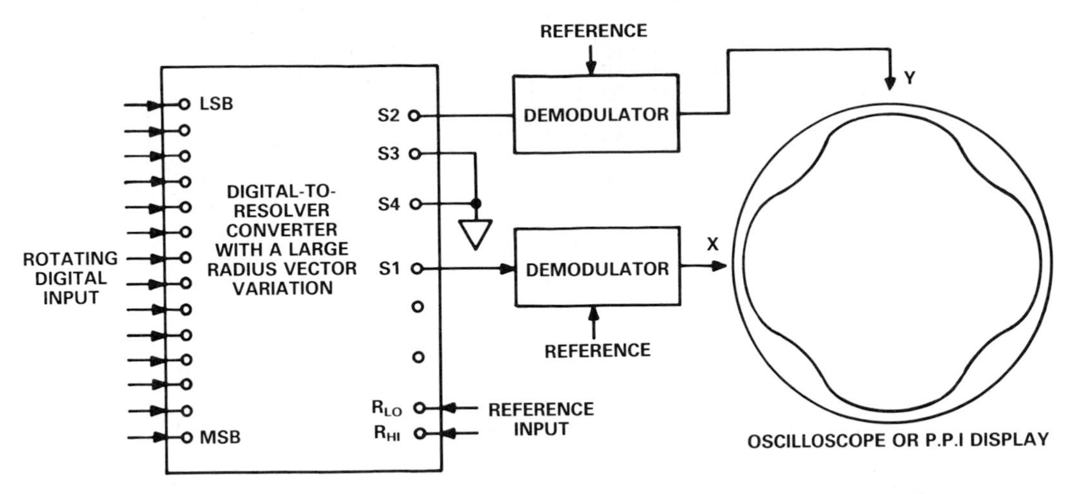

Figure 14.12. Effect of radius-vector variation on oscilloscope circle plot.

14.4 DESIGNING WITH RESOLVERS, SYNCHROS, INDUCTOSYNS, AND CONVERTERS

Selecting the right resolver, synchro, Inductosyn, and converter for a particular application has been considerably simplified by tables, such as those digested to form Figure 14.13. Before employing such tables, however, the designer should begin by asking a few fundamental questions.

Which measurement technology fits the needs of the application most closely? For instance, does the application call for angular resolutions of 10 bits or more, with absolute errors of less than 30 arc-minutes? Does the application need absolute measurement of angles (or will incremental representation suffice)? Must the transducer operate within a mechanically rugged or stressful environment? If the designer answers yes to any or all of these questions, then the use of resolvers, Inductosyns, or perhaps synchros is clearly called for.

Next, the designer should determine the amount of space available for the transducer, the required range of motion, amount and type of power available, and any unusual system interfacing requirements.

With this data in hand, the designer can select an appropriate transducer and converter from the tables. Figure 14.13 is a sampling of data appearing in tables available from Analog Devices. For each listed transducer, the tables list the range of compatible converters and the subsequent combined system performance.

RESOLVER TYPE	BASIC SPECIFICATIONS		ANALOG DEVICES CONVERTER PART NUMBER				
			1732/510 12 Bits	1S20/510 12 Bits	1S40/510 14 Bits	1S60/510 16 Bits	1S61/510 16 Bits
Singer Kearfott CR01093103 (Size 11) (400Hz)	System Resolution	(arc min)	5.27	5.27	1.32	0.33	0.33
	System Resolution	(bits)	12	12	14	16	16
	Converter Accuracy	(arc min)	21.00	8.50	5.30	4.00	10.00
	Resolver Accuracy	(arc min)	3.00	3.00	3.00	3.00	3.00
	System Accuracy	(arc min)	21.21	9.01	6.09	5.00	10.44
	Worst-Case Accuracy	(arc min)	24.00	11.50	8.30	7.00	13.00
	Tracking Rate	(rev/sec)	50.00	50.00	12.50	3.00	3.00
			1732/560	1S20/510	1S40/510	1S60/510	1S61/510
Harowe 11BRCX-30-C (Size 11) (1000Hz) Intrinsicially Safe	System Resolution	(arc min)	5.27	5.27	1.32	0.33	0.33
	System Resolution	(bits)	12	12	14	16	16.
	Converter Accuracy	(arc min)	21.00	8.50	5.30	4.00	10.00
	Resolver Accuracy	(arc min)	7.00	7.00	7.00	7.00	7.00
	System Accuracy	(arc min)	22.14	11.01	8.78	8.06	12.20
	Worst-Case Accuracy	(arc min)	28.00	15.50	12.30	11.00	17.00
	Tracking Rate	(rev/sec)	100.00	50.00	12.50	3.00	3.00
			1732/560	1S20/560	1S40/560	1S60/560	1S61/560
Thomson CSF 12T 11 RX4a (Size 11) (10kHz)	System Resolution	(arc min)	5.27	5.27	1.32	0.33	0.33
	System Resolution	(bits)	12	12	14	16	16
	Converter Accuracy	(arc min)	21.00	8.50	5.30	4.00	10.00
	Resolver Accuracy	(arc min)	5.00	5.00	5.00	5.00	5.00
	System Accuracy	(arc min)	21.58	9.86	7.29	6.41	11.18
	Worst-Case Accuracy	(arc min)	26.00	13.50	10.30	9.00	15.00
	Tracking Rate	(rev/sec)	100.00	170.00	42.50	10.50	10.50
			1732/560	1S20/510	1S40/510	1S60/510	1S61/510
Moore Reed 11RS236 (Size 11) (2kHz)	System Resolution	(arc min)	5.27	5.27	1.30	0.33	0.33
	System Resolution	(bits)	12	12	14	16	16
	Converter Accuracy	(arc min)	21.00	8.50	5.30	4.00	10.00
	Resolver Accuracy	(arc min)	1.00	1.00	1.00	1.00	1.00
	System Accuracy	(arc min)	21.02	8.56	5.39	4.12	10.05
	Worst-Case Accuracy	(arc min)	22.00	9.50	6.30	5.00	11.00
	Tracking Rate	(rev/sec)	100.00	50.00	12.50	3.00	3.00
			1732/560	1S20/510	1S40/510	1S60/510	1S61/510
Clifton Precision HSBJ-20-C-1 Multispeed 1:1 and 32:1 Size 20 (1200Hz)	System Resolution	(arc min)	0.16	0.16	0.04	0.01	0.01
	System Resolution	(bits)	17	17	19	21	21
	Converter Accuracy	(arc min)*	0.66	0.27	0.17	0.13	0.31
	Resolver Accuracy	(arc min)*	0.50	0.50	0.50	0.50	0.50
	System Accuracy	(arc min)	0.82	0.57	0.53	0.52	0.59
	Worst-Case Accuracy	(arc min)	1.16	0.77	0.67	0.63	0.81
	Tracking Rate	(rev/sec)†	3.12	1.56	0.39	0.09	0.09

*Refers to 32:1 Channel
†Refers to 1:1 Channel (i.e., Resolver Shaft)

a. Resolver to digital.

Figure 14.13. Excerpts from Converter Selection Charts.

14.4.1 DESIGNING WITH RESOLVERS AND SYNCHROS

Transducer Accuracy and Size

The first column in Figure 14.13a lists a variety of resolvers, their accuracies, and their sizes. For instance, the first section illustrates data for a resolver offered by Singer Kearfott, along with combined characteristics when employing various matching resolver-to-digital converters from Analog Devices. A model CR01093103 resolver has a body diameter of 1.1-inches (denoted as "size 11"), and an absolute accuracy of 3 arc minutes. This means that the resolver's analog electrical output, if measured ideally, will represent the resolver shaft angle to within ± 3 arc minutes.

System Resolution

The numbers in each subsequent column measure overall system performance when the resolver in the first column is mated with the RDCs noted in

BASIC SPECIFICATIONS		ANALOG DEVICES CONVERTER PART NUMBER					
		1732/560 12 Bits	1S10/560 10 Bits	1S20/560 12 Bits	1S40/560 14 Bits	1S60/560 16 Bits	1S61/560 16 Bits
Inductosyn Size	4″						
Number of Pitches	256						
Inductosyn Frequency	10kHz						
System Resolution	(arc secs)	1.24	4.94	1.24	0.31	0.08	0.08
System Resolution	(bits)	20	18	20	22	24	24
Converter Accuracy	(arc secs)*	4.92	5.39	1.99	1.24	0.94	2.34
Inductosyn Accuracy	(arc secs)*	3.00	3.00	3.00	3.00	3.00	3.00
System Accuracy	(arc secs)	5.76	6.17	3.60	3.25	3.14	3.81
Worst-Case Accuracy	(arc secs)	7.92	8.39	4.99	4.24	3.94	5.34
Tracking Rate	(revs/sec)†	0.39	2.65	0.66	0.17	0.04	0.04

*Refers to 256 Pitch Channel
†Refers to Inductosyn Itself

b. Rotary Inductosyn to digital.

BASIC SPECIFICATIONS		ANALOG DEVICES CONVERTER PART NUMBER					
		1732/560 12 Bits	1S10/560 10 Bits	1S20/560 12 Bits	1S40/560 14 Bits	1S60/560 16 Bits	1S61/560 16 Bits
Inductosys Size	0.1″						
Inductosyn Accuracy	0.0001						
Length	9.99″						
System Resolution	(microinches)	24.41	97.66	24.41	6.10	1.53	1.53
Converter Accuracy	(microinches)	97.22	106.48	39.35	24.54	18.52	46.30
System Accuracy	(microinches)	139.47	146.08	107.46	102.97	101.70	110.20
Worst-Case Accuracy	(microinches)	197.22	206.48	139.35	124.54	118.52	146.30
Velocity	(inches/sec)	10.00	68.00	17.00	4.25	1.05	1.05

c. Linear Inductosyn to digital.

the top row. For instance, when the CR01093103 resolver is connected with a 1S60/510 (16-bit) converter, *system resolution* becomes 0.33 arc minutes, or 16 bits. This is calculated from:

$$\frac{360^\circ \times 60\,\text{arc-minutes}/^\circ}{2^{16}\,\text{increments}} = 0.33\,\text{arc-minutes/increment} \tag{14.14}$$

Converter Accuracy

The specification for *converter accuracy* measures the worst case error of the RDC itself, specified in arc-minutes. For instance, the 1S60/510 converter's digital output will always be within ± 4.00 arc minutes of its analog input.

System Accuracy

This entry measures the overall rms error of the resolver in combination with the converter. System rms accuracy, in arc minutes, is calculated as the square root of the sum of the squares of the resolver accuracy specification and the converter accuracy specification. For the case of the CR01093103 resolver and 1S60/510 converter shown in the Figure 14.13a, for instance, the system rms accuracy is calculated to be 5.00 arc minutes.

Worst-Case Accuracy

This entry, also measured in arc minutes, represents the algebraic sum of the resolver accuracy specification and converter accuracy specification. For the example considered here, the converter's digital output will always indicate the resolver's shaft angle to within ± 7.00 arc minutes of the resolver's actual shaft angle.

Tracking Rate

Converter *tracking rate* is a measure of the fastest shaft rotational speed, in revolutions per second, that the converter can measure without losing lock. The figures shown in the tables of Figure 14.13 are guaranteed minimums.

Multiple-Speed Resolvers

Multi-speed resolvers have sets of both coarse and fine windings on the stator. When the shaft is turned through 360 degrees, the output from the coarse winding is equivalent to that obtained by turning the shaft of a single-speed resolver through 360 degrees. The fine output, however, completes a number of electrical cycles that depends on the quoted resolver ratio. Typically, the ratio will be a binary number, such as 8:1, 16:1, or 32:1, in order to simplify the combinational logic.

Multi-speed resolvers permit much higher accuracies and resolutions than would be possible with single-speed systems. The resolvers need two converters, one on the fine output and one on the coarse, although it is the performance of the fine converter which determines the overall accuracy and resolution. A "shift and add" circuit combines the outputs of the two converters to give a single digital word representing the shaft angle, which is absolute at all times including switch-on.

Consider, for example, the Clifton Precision Multi-speed resolver, type HSBJ-20-C-1, the last section of Figure 14.13a. The ratio in this case is 32:1. If we use a 14-bit converter on the fine winding output, the resolution will be:

$$14 + \log_2(32) = 19 \text{ bits.}$$

The worst-case accuracy (error) will be:

$$0.5 + 5.3/32 = 0.67 \text{ arc-minutes.}$$

In other words, the coarse winding does not contribute to the error and converter accuracy error is reduced to $1/32$ of its specified value. In the limit, the accuracy can never be better than 0.5 arc-minutes, no matter what converter is used, but the *resolution* could be increased by using a higher resolution converter on the fine output. For accuracies better than this, Inductosyns, which are really high ratio multi-speed resolvers, can be used very effectively.

Frequency

Finally, note that each example in Figure 14.13 specifies nominal frequency of operation. Frequency primarily influences allowable tracking rates. Higher frequencies lead to higher allowable tracking rates, as can be seen if the tracking rate specs in Figure 14.13a for 400 Hz and 10 kHz resolvers are compared.

14.4.2 DESIGNING WITH ROTARY INDUCTOSYNS

Figure 14.13b illustrates system performance of Farrand rotary Inductosyns in combination with Inductosyn-to-digital converters from Analog Devices. In analyzing system performance, it is useful to think of multi-pole rotary Inductosyns as multi-speed resolvers. In a rotary Inductosyn, there are 2 poles/pitch, and the number of pitches is equivalent to the number of speeds of an equivalent multi-speed resolver.

In the case of the 256-pitch rotary Inductosyn and IRDC1732/560 (12-bit) converter of Figure 14.13b, for instance, resolution is calculated as:

$$\frac{360° \times 60\,\text{arc-min/°} \times 60\,\text{arc-sec/arc-min}}{256\,\text{pitches} \times 2^{12}\,\text{increments/pitch}} = 1.24\,\text{arc-sec/increment} \quad (14.15)$$

Converter Accuracy

In the case of the rotary Inductosyn, *converter accuracy* in the table of Figure 14.13b is calculated from the inherent converter error, 21.00 arc-minutes for the IRDC 1732/560, divided by the number of Inductosyn pitches, as follows:

$$\frac{21\,\text{arc-min} \times 60\,\text{arc-sec/arc-min}}{256\,\text{pitches}} = 4.92\,\text{arc-sec} \quad (14.16)$$

System RMS Accuracy and Worst-Case Accuracy

In the case of rotary Inductosyns, *system rms accuracy* and *worst-case accuracy* are calculated in the same way as for resolvers, as discussed above. In addition, Inductosyns will experience a further error, of the order of fractions of an arc second, due to gain-mismatch in the preamplifier.

Tracking Rate

At comparable operating frequencies, the *tracking rate* of a rotary Inductosyn-and-converter combination can be calculated from the basic tracking rate of the converter divided by the number of pitches. In the case of the IRDC1732/560, TRACKING RATE is calculated as:

$$\frac{100\,\text{revolutions/second} \times 360°/\text{revolution}}{256\,\text{pitches}} = 140.6°/\text{second} \quad (14.17)$$

where 100 pitches per second is the tracking speed of the IRDC 1732/560 at 10 kHz.

14.4.3 DESIGNING WITH LINEAR INDUCTOSYNS

Figure 14.13c lists the overall performance of a typical linear Inductosyn from Farrand Controls when used with Inductosyn-to-digital converters from Analog Devices.

The Inductosyn is characterized by the length of its pitch, its overall length, and its inherent accuracy. Overall system performance figures are listed in the right hand columns.

Resolution

For instance, in the case of the Inductosyn shown, with 0.1-inch pitch, operating with an IRDC1732/560 Inductosyn-to-digital converter, *system resolution* (in microinches) is calculated from the Inductosyn *pitch* (in microinches) divided by 2^N, where N is the converter's rated number of bits. In this case:

$$\begin{aligned}\text{Resolution} &= 100{,}000\ \text{microinches}/2^{12}\\ &= 24.41\ \text{microinches}\end{aligned} \tag{14.18}$$

Converter Accuracy

The *converter accuracy* of Figure 14.13c is calculated by multiplying the Inductosyn *pitch*, in microinches, by the *converter accuracy* number in arc-minutes, divided by a conversion factor. For example, consider again the case of the IRDC1732/560. *Converter accuracy* is calculated by:

$$\begin{aligned}\underset{\text{(error)}}{\text{Converter accuracy}} &= 21\ \text{arc-minutes} \times \frac{100{,}000\ \text{microinches/pitch}}{360^\circ \times 60\ \text{arc-min}/^\circ}\\ &= 97.22\ \text{microinches}\end{aligned} \tag{14.19}$$

System Accuracy

As in previous cases, *system rms accuracy* is calculated from the square root of the sum of the squares of the *converter accuracy* and *Inductosyn accuracy*.

Worst-Case Accuracy

This term is calculated by simply adding the *Inductosyn accuracy* to the *converter accuracy*.

Velocity

The velocity specification represents the maximum Inductosyn velocity in inches per second for which the IDC doesn't lose lock. Velocity is calculated from the *tracking rate* of a comparable converter at comparable operating frequency, multiplied by inches per pitch. For instance, in the case of the IRDC1732/560 converter of Figure 14.13c, and an 0.1-inch Inductosyn, the converter's *tracking rate* is 100 revolutions per second. Since one revolution of a resolver is equivalent to one pitch of an Inductosyn, 100 revolutions per second is equivalent to 100 pitches per second. Maximum *velocity* for this case is then calculated by:

$$\begin{aligned}\text{Velocity} &= 100\ \text{rev/second} \times 1\ \text{pitch/rev} \times 0.1\ \text{inches/pitch}\\ &= 10\ \text{inches/second}\end{aligned} \tag{14.20}$$

14.4.4 DESIGNING WITH DIGITAL-TO-RESOLVER/SYNCHRO CONVERTERS

When designing with digital-to-resolver, or digital-to-synchro converters, the accuracy of the converter has an important influence on the overall performance of the system. Such systems often are in the form of a feedback control loop, which also includes a resolver and RDC for position sensing.

To deal with an important concern when designing with digital-to-resolver converters, the designer should be certain to provide sufficently large amounts of power to ensure proper performance of the resolver itself.

14.5 APPLICATIONS OF RESOLVERS, SYNCHROS, INDUCTOSYNS, AND CONVERTERS

The increasing industry attention to factory automation, robotics, and computer-controlled machine tools is resulting in expanded use of resolvers and Inductosyns.

14.5.1 MACHINE-TOOL APPLICATIONS

Figure 14.14 illustrates a typical central computer-based position-control system where an Inductosyn and an Inductosyn-to-digital converter constitute the primary feedback elements. As seen in the tables of Figure 14.13, the IDC produces a 12-bit digital word proportional to the Inductosyn slider's position within one pitch of the scale. The IDC's output also includes a direction signal (DIR), indicating the direction of motion, and a ripple-clock signal (RC), which indicates when the scale moves from one cyclic pitch to the next. A similar circuit configuration can be used in applications where a resolver and converter are used with a lead screw.

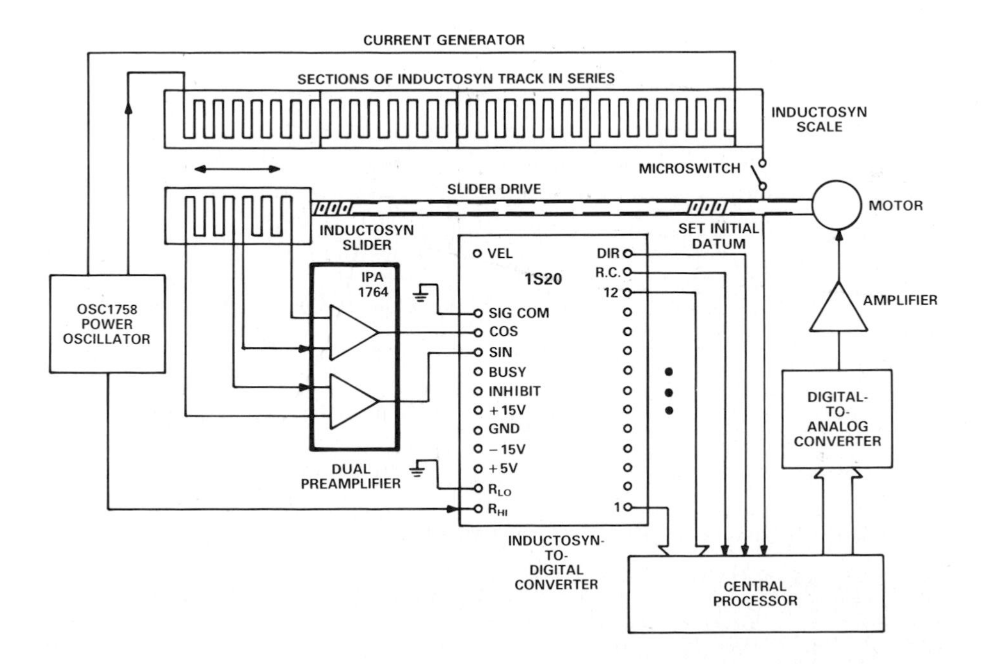

Figure 14.14. Inductosyn and IDC in a typical application.

Designers of systems employing resolver/leadscrew or Inductosyns may prefer a serial output to a parallel output. This is often the case where a resolver is replacing an incremental optical encoder. Such a system is shown in Figure 14.15. The IRDC 1731 has a serial output of 4000 counts per pitch, as well as a zero-crossing pulse which indicates when the traveler has moved from one pitch to the adjacent one. External counters convert the serial output to a parallel binary or BCD word. In this example, it is a parallel BCD word, which could drive a display to facilitate readout of position by the machine operator. Also, when used with Inductosyns, the choice of 4000 counts per pitch (instead of 4096, or 12 bits) makes it easier to divide the 2 mm pitch or the 0.1 inch pitch of typical Inductosyns into more-easily processed increments.

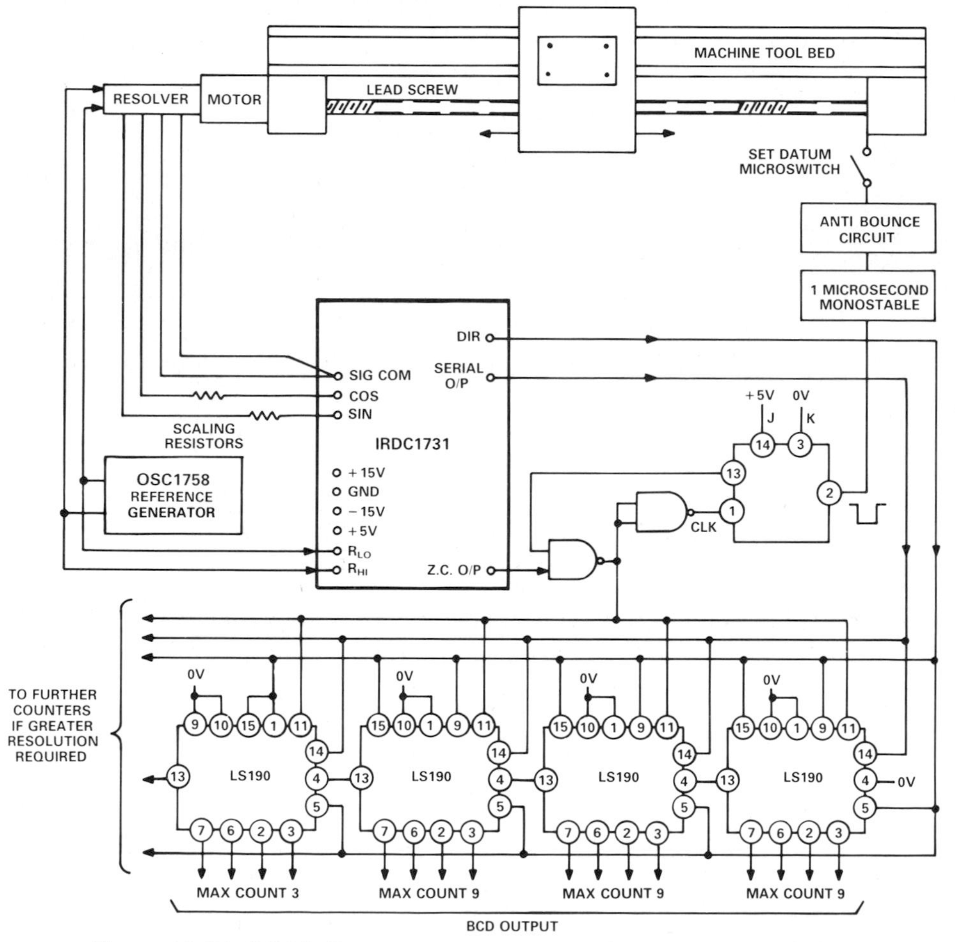

Figure 14.15. RDC in linear measurement application, with leadscrew.

14.5.2 GUN CONTROL

Military fire-control systems, such as that outlined in Figure 14.16, often use synchros to measure angular position, employ a processing system which accepts signals in synchro format and produces outputs, usually in coarse/fine synchro format, to point a gun or missile launcher. Tracking synchro-to-digital converters can be used to process the synchro inputs, and a digital-to-resolver converter, in combination with control transformers, provides the output control signal.

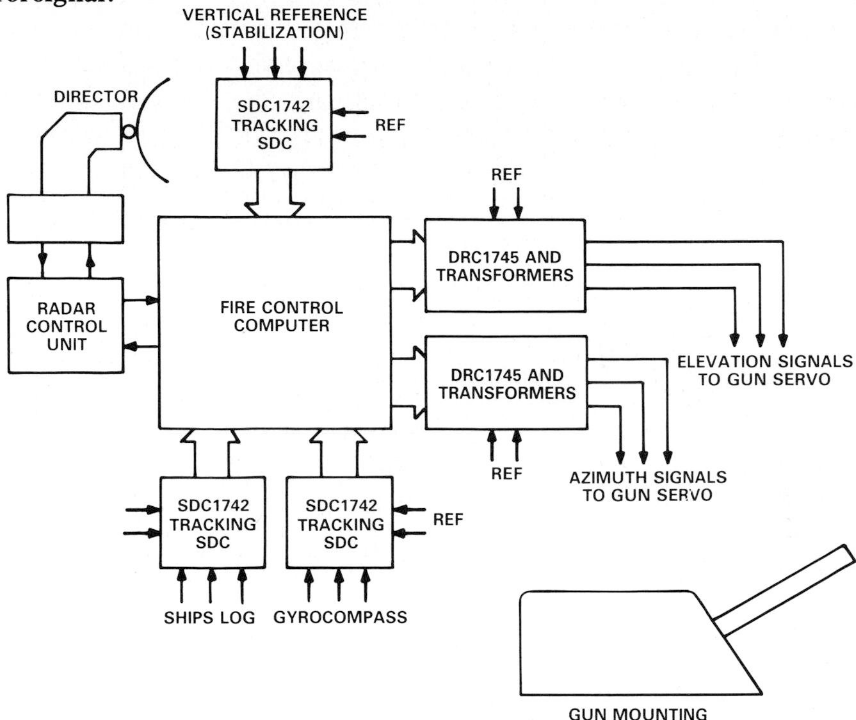

Figure 14.16. Fire-control application.

The fire-control computer of Figure 14.16 employs inputs from the ship's tracking radar in addition to various other ship sensors.

Figure 14.17 illustrates an alternative closed-loop mechanical positioning system that makes use of the velocity output voltage of the 1S24 RDC—a device having a high-grade tachometer output.

14.5.3 PLAN-POSITION INDICATOR

Figure 14.18 shows how a hybrid digital-to-resolver converter can be used in conjunction with a synchro-to-digital converter to generate the waveforms required for radar PPI displays.

The synchro signal from a control transformer representing the radar antenna angle is converted to a binary word by the synchro to digital converter. This digital angle is applied to the digital input of the DRC. A dc voltage is applied to the analog input to control the radius of the displayed raster. The outputs of the DRC, which are proportional to the sine and cosine of the antenna angle, provide the voltages required by the X and Y time base of the PPI display.

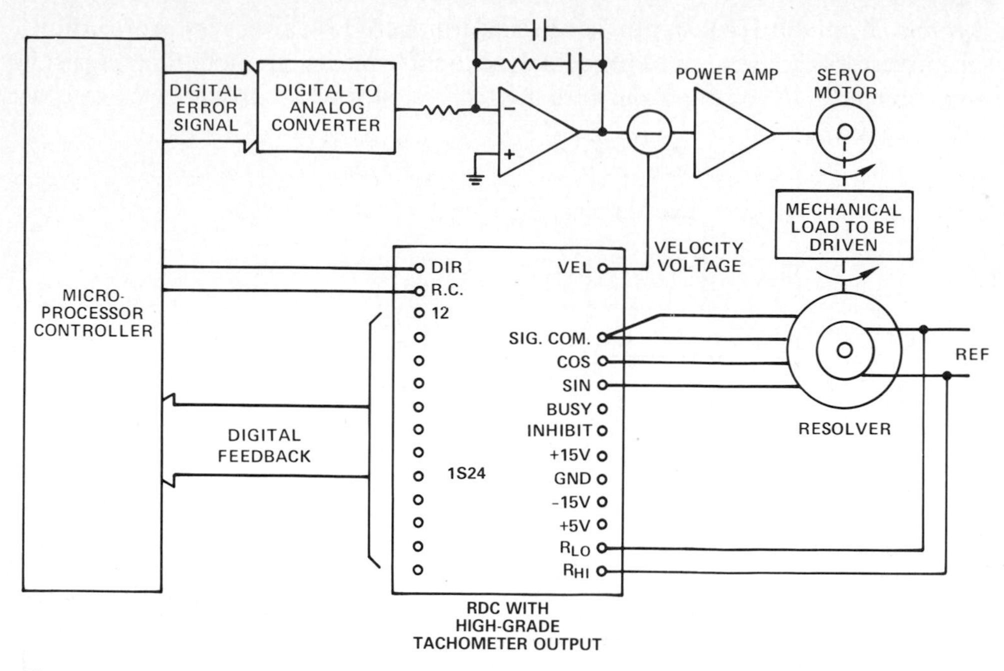

Figure 14.17. Resolver-to-digital converter in servo-loop application, showing use of velocity output for tachometric feedback.

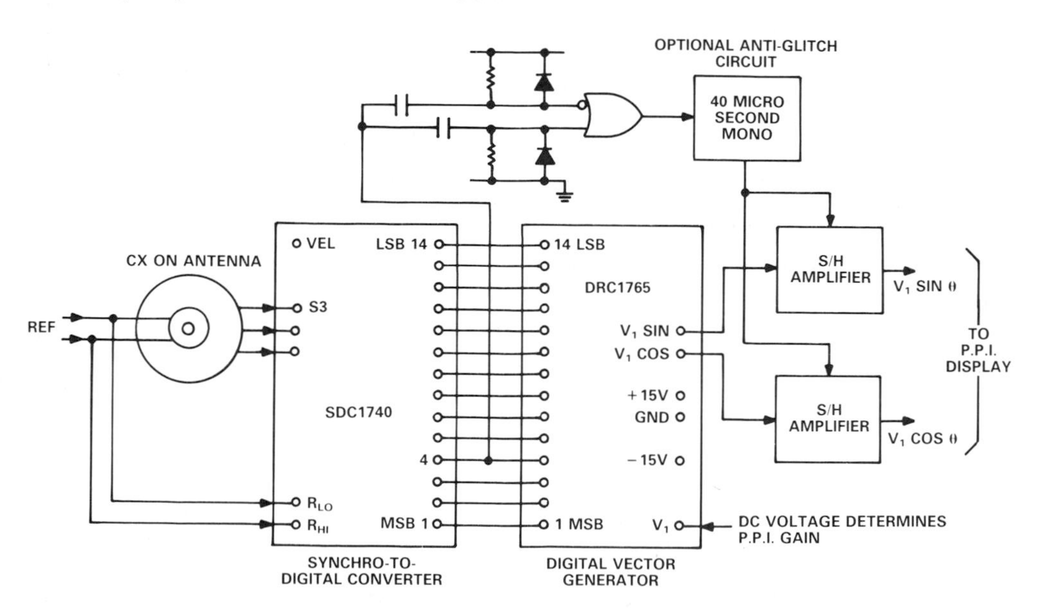

Figure 14.18. Converters in PPI application.

Radar antenna direction information is sometimes transmitted in serial digital form from the antenna to the processing equipment. This method is used, for example, in the case of microwave transmission of position data. At the receiving end, it may be required to convert these azimuth change pulses, or ACPs, which usually number 4096 per revolution, into dc sine and cosine voltages to drive the PPI display. The ACPs are usually accompanied by an azimuth rotation pulse, or ARP, which occurs once per revolution of the antenna.

Figure 14.19 illustrates a circuit designed to accomplish the conversion. Note that the azimuth rotation pulse resets the counter to zero every revolution in case any spurious pulses have occurred during the the previous revolution.

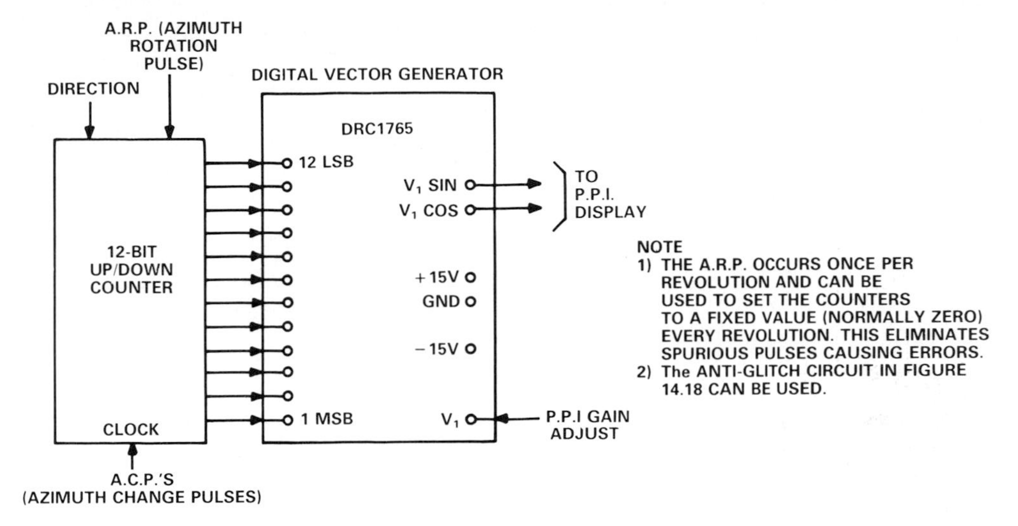

Figure 14.19. PPI application with serial input.

Chapter Fifteen

Voltage-to-Frequency Converters

15.1 INTRODUCTION

A voltage-to-frequency converter (VFC) accepts an analog voltage or current signal and provides at its output a train of pulses or square waves with periods inversely proportional to the average value of the input over each cycle of output. For a given input value, the number of pulses per second, or *frequency*, is proportional to that input value. Thus, the output digital word from a counter clocked by a VFC and read out at regular intervals will be proportional to the analog input. When used in an appropriate feedback loop, a VFC circuit will function as a *frequency-to-voltage* converter (FVC).

The VFC is a versatile building block that is quite useful in modern data-acquisition systems. In its most elementary application (measuring the frequency resulting from an applied voltage), an analog-to-digital data conversion is achieved in a very simple and economical manner. An integrating device, the VFC voltmeter has guaranteed monotonicity, very high resolution, and rejection of hum and noise. The time required to convert an analog voltage into a digital number is related to the full-scale frequency of the VFC and the required resolution of the measurement.

In general, a VFC as an a/d converter is slower than successive-approximation and "flash" types (See Chapter 7) but comparable in speed to other types of integrating converters, such as dual-slope types. A VFC voltmeter provides a kind of flexibility that none of the other data converters offers: conversion speed is easily traded for measurement resolution, when a VFC is used in conjunction with a computer-controlled frequency counter. In addition, a VFC voltmeter can provide measurement resolution far greater than can be achieved with any other integrated-circuit data converter. Resolution can, in principle, be increased almost indefinitely by simply waiting long enough to accumulate a sufficient number of pulses to resolve the output frequency to the degree desired.

For example, an Analog Devices AD650 VFC, operating with a full scale frequency of 1 MHz, will achieve a resolution of 18 bits,—or four parts in a million—by allowing a counter to accumulate pulses for little more than a quarter of a second. The ultimate resolution of a VFC voltmeter is limited by noise and drift, and not by architectural constraints (i.e., length of a resistance ladder, number of comparators, or width of an internal counter). A VFC voltmeter offers a cost advantage when high resolution (rather than high speed) is of primary importance. And the cost savings can be quite dramatic if many signals are to be monitored simultaneously with high resolution.

A voltage-to-frequency converter can also perform the function of a digital-to-analog data converter (DAC), again with very high resolution. To achieve this, the VFC is configured as a frequency-to-voltage converter (FVC), and the digital information is presented as a train of pulses or square waves at a given frequency. In the very simplest DAC application—configured as a "tachometer"—using converters such as the ADVFC32 or AD650, each cycle of the input frequency causes a fixed quantity of charge to be dumped into a lossy integrator (this will be discussed in relation to the circuit of Figure 15.13). This kind of circuit is useful in low-speed applications, such as driving a panel meter or controlling power applied to a motor or a heater. At constant frequency, the output voltage is not purely dc; it has some ripple at the input frequency, but its average value is strictly proportional to frequency.

It is possible use filtering to reduce the amount of ripple on the voltage output, but only at the expense of settling time. An FVC with a much more favorable tradeoff between ripple and settling time can be achieved by using a VFC in a phase-locked loop (PLL). In the PLL (Sec. 15.4.3), the frequency of the input is compared with the frequency produced by a VFC, and the signal applied to this VFC is adjusted until the two frequencies are equal. This signal, which is proportional to frequency, becomes the voltage output. This circuit achieves very fast settling time of the voltage output with low ripple—at the expense of increased circuit complexity. When an AD650 v/f converter is used in a PLL application, the circuit can easily handle signal bandwidths of 70 kHz with a dynamic range (a measure of noise and ripple) of 80 dB.

Other uses of the VFC are simple variations and combinations of the two fundamental circuits described above. For example, an integrator can easily be built from the same components as a VFC voltmeter. A counter is used to accumulate pulses of the output frequency, and the total number of pulses over the gate time is interpreted as an integral (i.e., total charge delivered to the VFC), rather than as an average frequency. Very long integration times can be achieved using this method. For example, to evaluate the total energy in a signal, an analog multiplier continuously calculates the power delivered to a load (V × I), and the VFC can integrate the power signal to keep track of the total watt-hours used.

Another important application of a VFC is to transmit a high-accuracy analog signal through a noisy environment without interference. To accomplish this, the VFC converts the analog signal to a pulse train at a frequency proportional to the input, and that frequency signal is then transmitted. The frequency signal, consisting of large-amplitude digital pulses, is much less susceptible to interference than a high-resolution analog signal. At the receiving end, the signal is converted back into a voltage by one of the methods mentioned earlier.

The selection of the medium of transmission of the frequency signal is primarily determined by the nature of the task at hand. For example, galvanic isolation is achieved with a simple transformer or opto-isolator; extremely high-voltage isolation or transmission through environments having severe radio-frequency noise levels can be accomplished with a fiber-optic link; and telemetry can be accomplished with a radio link. In a sense, time can be considered as a noisy environment: a common need is to transmit information from the past into the future; this can be easily fulfilled by recording the frequency signal on magnetic tape and then replaying the tape when needed in the future.

15.2 DEFINITIONS AND SPECIFICATIONS

A voltage-to-frequency converter is a device that produces an output frequency in response to an input signal. The input signal may be presented as either a voltage or a current (in the discussions and examples that follow, voltage will be used as the input variable, but the reader should bear in mind that it could just as well be current). In the ideal case, the output frequency is directly proportional to the input. Thus,

$$\frac{f_{out}}{f_{FS}} = \frac{V_{in}}{V_{FS}} \tag{15.1a}$$

where f_{out} is the frequency of the output signal, V_{in} is the input voltage, and f_{FS} and V_{FS} are the respective full-scale quantities. Since they will be used as constants, they may be combined in a transfer constant,

$$f_{out} = G\,V_{in} \tag{15.1b}$$

where G is the gain of the VFC in events-per-second (Hz) per volt.

A real device will not have this perfectly linear transfer function—there will be offsets, gain errors, and some warping or *nonlinearity*. Nonlinearity is the primary specification of a VFC device; it warrants careful consideration since it determines the irreducible error of a system which uses the device (other errors can be reduced at a fixed temperature by trimming and calibration, either statically or dynamically).

In a VFC with *offset*, zero frequency does not occur at zero voltage. This means that, for example, the VFC ceases to oscillate when the input is at a small but non-zero voltage, or that it will not cease to oscillate with an input of zero volts. The offset of a VFC transfer function at constant temperature is typically not a problem, since it may be adjusted to be zero with very simple circuit trimming. If the offset should change, as a function of temperature, however, the resolution of the VFC will be compromised, because it is not possible to distinguish between a small change in the input and a change in the offset.

The *dynamic range* of the system will also be compromised by any offset drift. A dynamic range of a million to one means that a VFC which accepts a 10-volt full scale signal should also accept a 10-microvolt signal and produce an output frequency one millionth as great, with any difference due only to nonlinearity. However, if the offset of the VFC transfer function should drift away from its trimmed value of zero, then a larger error in the ratio of frequencies produced by the 10-microvolt and 10-volt signals will result. If the offset drifts by $+5\mu V$, for example, it would effectively reduce the $+10\text{-}\mu V$ input to $+5\mu V$, and the frequency ratio would be two million. However, if the drift were $-5\mu V$, the effective minimum input would be $15\mu V$, and the ratio would be two thirds of a million. This example of distortion at the low end dramatically shows the need for an "autozero" circuit in any wide-dynamic-range system. An example of a circuit to provide compensation of offset errors will be found in the Applications section (15.4).

The linearity error of a VFC is usually specified by the *end-point* method. That is, the error is expressed in terms of the deviation from the ideal voltage-to-frequency transfer relation after calibrating the converter at full scale and "zero". To be able to verify the linearity specification, one must have available a switchable voltage source (or a DAC) that has a linearity error less than 20ppm, and one must use very long measurement intervals to minimize counting uncertainty. Fortunately, reputable manufacturers submit every VFC to automatic testing for linearity, and it will not usually be necessary to perform this verification, which is both tedious and time consuming. However, if a nonlinearity test is required, either as part of an incoming quality screening or in final product evaluation, an automated "benchtop" tester would prove useful.[1]

Figure 15.1 shows voltage-to-frequency transfer relationships, with the nonlinearity exaggerated for clarity. The first step in determining nonlinearity is to connect the measured end points of the operating dynamic range (for example, 10mV and 10V) with a straight line; this straight line is the ideal relationship, which is desired from the circuit. The second step is to find the differ-

[1]See "V-F Converters Demand Accurate Linearity Testing," by L. DeVito, *Electronic Design*, March 4, 1982, for an example of implementation of such a system, based on the Analog Devices' LTS-2010 benchtop tester.

ence between points on this line and the actual response of the circuit at a few frequencies between the end points—typically, ten intermediate points will suffice. The difference between the actual and the ideal response is a frequency error, measured in Hz. Finally, these frequency errors are normalized to the full-scale frequency, and expressed either as parts per million (ppm) or parts per hundred (percent) of full scale.

For example, if 100 kHz is full scale and the maximum frequency error is 5Hz, the nonlinearity would be specified as 50 ppm, or 0.005%. The gain of the VFC is simply the slope of the "end-point" line which was used in the determination of the nonlinearity as described above.

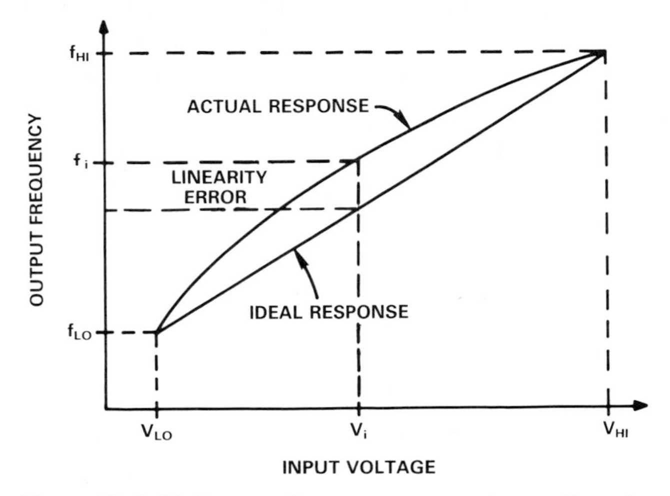

Figure 15.1. Voltage-to-frequency converter nonlinearity.

The specification of *gain*, (or *full-scale error*) is usually quite loose—especially in monolithic devices. This specification is not very critical, since the gain can always be trimmed at a given temperature with simple circuit adjustments. Digital systems can be designed for periodic gain calibration to remove the effects of temperature drifts.

The offset of the voltage-to-frequency transfer relation is simply the intercept of the extrapolated "end-point" ideal straight line, through the measured end points, with the frequency axis, as shown in Figure 15.2. It is equally valid, in concept, to refer the offset to the input, i.e., express the offset as a voltage, by finding the intercept of the "end-point" line with the voltage axis. However, this is would be inconvenient for computations of nonlinearity using the "end-point" line, since it would mix dependent and independent variables; also, offset may stem from more than one source.

The offset to be expected from a VFC must be calculated using the specifications of both input voltage offset and input bias current. The offset is primarily due to offset voltage in the operational amplifiers used as pre-amps, integrators, or voltage-to-current converters—see Section 15.3. However, in the charge-balance type of VFC (e.g., the AD650), the input bias current of the integrator's operational amplifier also contributes an offset. And in multivibrator-type VFCs, such as the Analog Devices AD537, the input bias current of the op amp used as a voltage-to-current converter will also contribute an offset. For this reason it is more helpful to express the combined offset effects in terms of frequency.

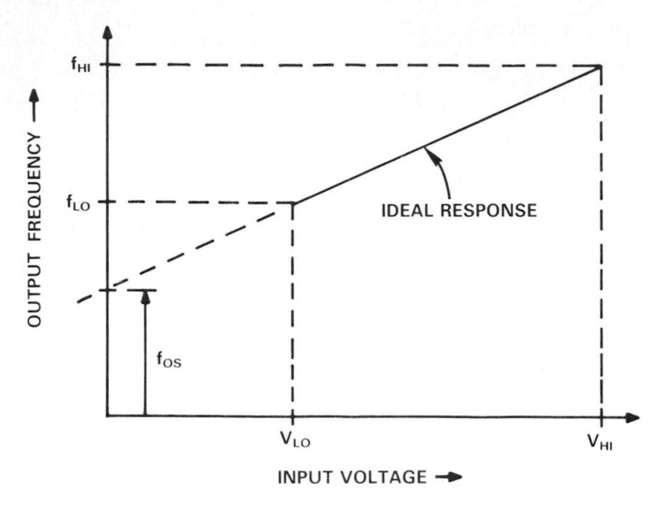

Figure 15.2. Voltage-to-frequency converter offset.

Gain tempco, or full-scale error versus temperature, is simply a measure of the change in the slope of the "end-point" line with a change in temperature. This parameter of a VFC must be carefully considered, because the ultimate accuracy of a system may be limited by temperature changes and not by the linearity of the converter. For example, consider an Analog Devices Model 460 VFC, with a 1-MHz full-scale frequency: the linearity error is ±150ppm, and the gain tempco of the L grade is ±15ppm/°C. With 10 volts full-scale, the 150-ppm nonlinearity will allow an absolute measurement accuracy of 1.5 millivolts; however, the gain TC will contribute an error of equal magnitude if the temperature should change by 10°C after calibration.

In general, the linearity specifications of VFC devices are so good that they cannot be fully exploited unless deliberate steps are taken to minimize offset and gain changes with temperature. It is usually a simple matter to install the device (especially if it is monolithic) in a constant-temperature oven, to stabilize its operating temperature against a changing ambient. Another approach to gain stabilization is to use an "auto-gain" circuit, which can adjust the scale factor of the VFC in order to produce a constant output in response to a known reference input. Such a system is described in the Applications section (15.4).

The *power-supply rejection ratio* (PSRR as a ratio, PSR in logarithmic form—dB) is a specification of the change in gain of the VFC as the power supply voltage is changed. The PSRR is usually expressed in units of parts-per-million change of the gain per percent change of the power supply—ppm/%. For example, consider a VFC with a 10-volt input applied and an output frequency of exactly 100 kHz when the power supply potential is ±15 volts. Changing the power supply to ±12.5 volts is a 5 volt change out of 30 volts, or 16.7%. If the output frequency changes to 99.9 kHz, the gain has changed 0.1% or 1000ppm. The PSRR is thus 1000ppm divided by 16.7% which equals 60ppm/%.

Alternatively, the PSRR may be expressed in units of percent change in gain per volt change of the power supply—i.e., %/volt. For the example given, the equivalent specification would be 0.1%/5 V = 0.02 %/volt. The user must beware when comparing specifications of various devices: units are not always commensurate. For example, PSRR of the AD537 is specified as ±0.1% per volt max. However, to convert this into ppm/% we must first assume a nominal operating power supply voltage. If we take the usual ±15 volt supplies, then a one percent change of supplies is seen to be 0.3 volt, which yields a PSRR of 1000ppm/3.3% = 300 ppm/%; if, on the other hand, the device is operated on a single 5-volt supply, the sensitivity will be 1/6 as much, i.e., 50 ppm/%.

15.3 SURVEY OF VFC DEVICES

Today's system designer has a choice among a variety of voltage-to-frequency converter types, including both monolithic circuits and modules; each has strengths and weaknesses which must be understood to make an appropriate choice for the task at hand. There are two basic voltage-to-frequency converter architectures; one uses a multivibrator, the other is a charge-balancing technique.

15.3.1 MULTIVIBRATOR TYPES

Figure 15.3 shows an example of a monolithic VFC design using a current-controlled multivibrator as the primary timing element. An operational amplifier converts the input voltage into a proportional current; this current determines the charging rate of external capacitor, C, which in turn determines the frequency of operation of the multivibrator. The device has an internal bandgap reference (Chapter 20), which provides both a constant 1-volt output and a thermometer output scaled to 1mV per kelvin.

The frequency output is delivered via an open-collector transistor, and only the emitter of this output device is returned to digital ground. The dedicated digital ground allows the complete separation of noisy digital spikes from the high-accuracy analog circuits.

Note that signal current is relayed to the multivibrator from the input amplifier through n-p-n transistors; this means that current of only one polarity can be delivered to the timing capacitor. Thus, the net input voltage must always be positive.

In the case of a small input signal that has a large amount of noise, the instantaneous input voltage may be negative, which would call for a current of opposite polarity. However, this circuit cannot provide a bipolar current, so the input amplifier will saturate—and continuous integration of the input signal is lost. This means that positive peaks of noise will increase the charging current, but negative noise peaks cannot decrease the current below zero in order to compensate. The result is that the noise-rejection characteristics of the multivibrator VFC are not as good as those of the charge-balance architecture.

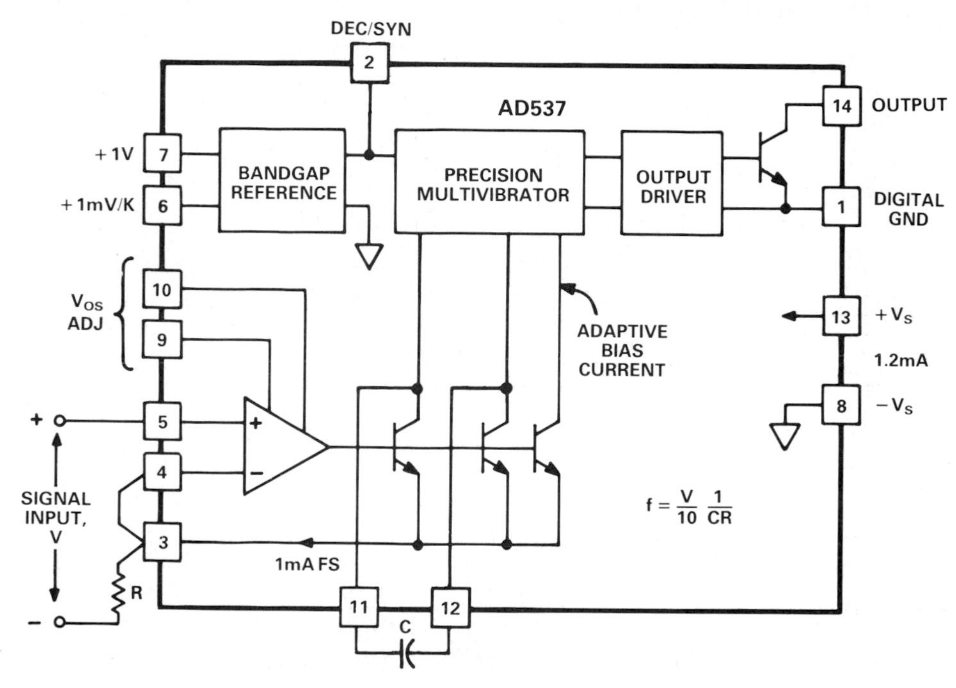

Figure 15.3. Multivibrator-type VFC architecture.

However, apart from this disadvantage, there are a number of advantages. The power consumption of a multivibrator VFC can be very low; for example, the AD537—a representative example of the type—can operate on a single supply voltage as low as +4.5 volts and consume a maximum of 2.5mA quiescent current. The device can operate with full-scale output frequencies up to 100 kHz and voltage-to-frequency nonlinearity of 0.1% (or 1000 ppm), referred to full scale. In addition, its output is a square wave, which allows it to be a-c coupled without introducing dc level shifts when the frequency is changed.

This combination of characteristics makes such a device an ideal choice for telemetry applications, where measurements at a remote location must be relayed to a central data collection area. The flexibility of single-supply operation and low power consumption allows battery-powered operation with long life, or convenient two-wire operation, in which a single wire pair will deliver power to the remote device and relay the frequency back to the central location. The two wire telemetry application is shown in Figure 15.4. The thermometer output of the device is especially convenient for remote thermometry.

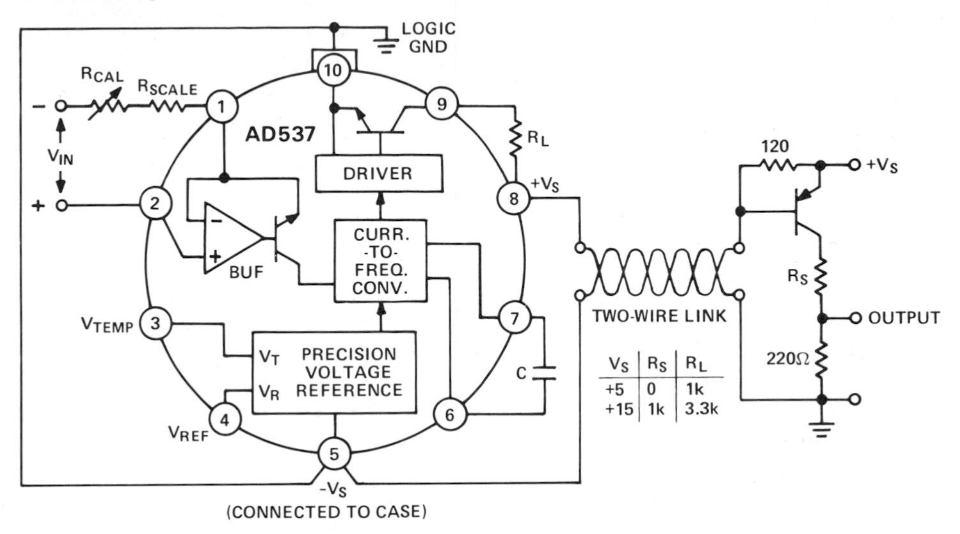

Figure 15.4. VFC in two-wire single-supply operation. Both input power and output signal share the interconnecting line.

As suggested earlier, it is important that the signal appearing at the device inputs not be contaminated by hum or noise. If the signal is tainted, then the dynamic range of the system will be reduced. To preserve the continuous integration of the signal and thereby avoid the consequences of the device's susceptibility to noise, it is desirable to provide filtering if possible, and then to constrain the minimum d.c. level of the input voltage signal to be greater than the expected zero-to-peak value of the residual noise. This is often not a problem, especially when using the thermometer output. In remote measurements, care should be taken to locate the VFC as close as possible to the signal source in order to minimize noise pick up, since noise is usually introduced on long lengths of cable; this mode of operation is of course the forte of a VFC: the measurement circuitry is placed near the signal source, and only a digital frequency signal need traverse any length of cable.

15.3.2 CHARGE-BALANCE TYPES

If flexibility and low power consumption are the hallmarks of the multivibrator type of VFC (as exemplified by the AD537), then high speed, high linearity, and high noise rejection are the salient features of the *charge-balance* VFC architecture, employed by devices like the ADVFC32 and the high-performance AD650. The latter type is used as our example, because it provides some features that are especially useful in conversion systems—not found on most general-purpose VFCs, e.g., separate digital ground, offset nulling for the integrator op amp, and a bipolar offset current for convenient half-scale offset when converting bipolar signals. Key specifications of such a device include typical nonlinearity of 50ppm for 100-kHz full-scale frequency and maximum nonlinearity specifications of 1000 ppm for 1-MHz full-scale frequency. Power requirements are conventional split supplies, at ±9 to ±18 volts, with maximum quiescent current of 8 mA.

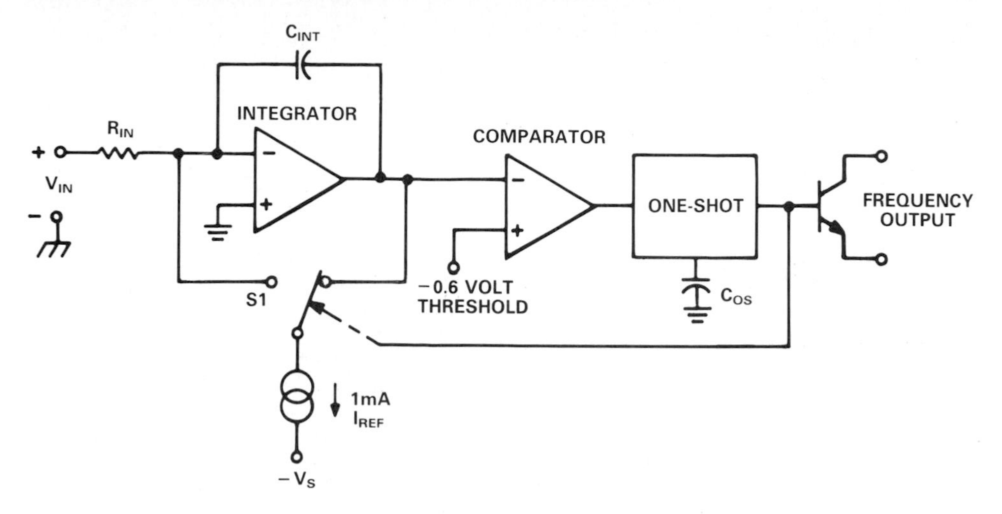

Figure 15.5. Charge-balance converter architecture.

Figure 15.5 shows a block diagram of such a device (AD650), connected to operate as a voltage-to-frequency converter. It comprises a summing integrator, a current source and steering switch, a comparator, a one-shot, and an output transistor. The integrator, in a feedback loop, forces an internal feedback current to exactly balance (over time) the input signal current, I_{IN}, which is either furnished directly to the op amp's summing point, as a current, or developed across R_{IN} by the input voltage. The current from the precision current source is applied as a feedback via S1 as short, accurately timed bursts of current—of opposite polarity to the input—in effect, as precisely defined packets of charge ($\Delta Q = I\,\Delta t$). The net charge is accumulated in the integrating capacitor, and the output of the inverting integrator is compared against a threshold; when the output falls below the threshold, indicating that the accumulated input charge is larger than the accumulated feedback, the comparator changes state and stimulates the one-shot to emit another pulse of current; at the same time an output pulse is emitted.

Thus, the number of charge packets required to keep the feedback loop balanced (each accompanied by one pulse of the output transistor), depends upon the magnitude of the input signal. Since the input signal is being balanced by a proportional number of discrete, equal charge packets, a linear voltage-to-frequency transformation is accomplished. The frequency output is furnished via an open-collector transistor.

Here is a more detailed description and analysis of the circuit's behavior: When the output of the one-shot is low, the current steering switch, S1, diverts all the current to the output terminal of the op amp, where it has no effect on the integrator's rate of charge-accumulation; this is called the Integration Period (Figure 15.6a). When the one-shot has been triggered, and its output is high, the switch, S1, diverts the current to the summing junction of the op amp; this is called the Reset Period (Figure 15.6b). The figure shows the various branch currents and the integrator output voltage in both states. It should be noted that the output current from the op amp is the same for either state (in the integrating period, the current from the current source flows directly from the output; in the reset period, it flows through the capacitor), thus minimizing transients.

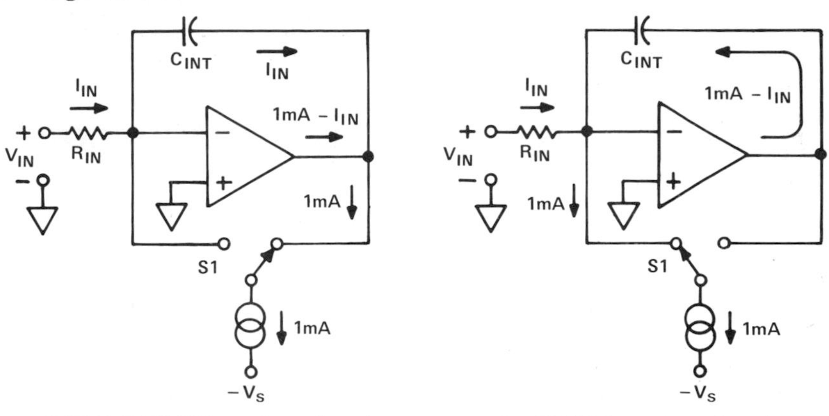

a. INTEGRATION portion of cycle. b. RESET portion of cycle.

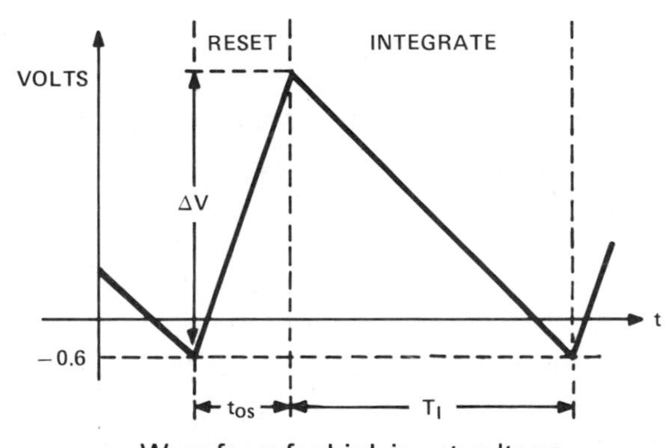

c. Waveform for high input voltage.

Figure 15.6. Charge-balance converter operation.

The positive input voltage develops a current ($I_{IN} = V_{IN}/R_{IN}$), which charges the integrating capacitor C_{INT}, the output voltage of the integrator ramps downward towards ground. When the integrator output voltage crosses the comparator threshold (-0.6 V), initiating the Reset Period, the comparator triggers the one-shot; its time period, t_{os}, establishing the Reset Period, is determined by the one-shot capacitor, C_{os} (Figure 15.5). The integrator now ramps upward during the interval, t_{os}, by an amount:

$$\Delta V = t_{os}\frac{dV}{dt} = t_{os}\frac{1\,mA - I_{IN}}{C_{INT}} \tag{15.2}$$

After the Reset Period has ended, the device starts another Integration Period, as shown in Figure 15.6, and starts ramping downward again. The amount of time, T_1, required to reach the comparator threshold is:

$$T_1 = \frac{\Delta V}{dV/dt} = t_{os}\frac{1\,mA - I_{IN}}{C_{INT}}\frac{1}{I_{IN}/C_{INT}}$$

$$= t_{os}\frac{1\,mA - I_{IN}}{I_{IN}} \tag{15.3}$$

Thus, the output frequency is:

$$f_{OUT} = \frac{1}{t_{os} + T_1} = \frac{I_{IN}}{t_{os} \times 1\,mA}$$

$$= \frac{V_{IN}}{R_{IN}}\frac{1}{t_{os} \times 1\,mA} \tag{15.4}$$

Note that C_{INT}, the integration capacitor, has no effect on the transfer relation; its principal effect is to establish the amplitude of the sawtooth signal out of the integrator.

A key factor of the analysis is t_{os}, the one-shot time period. This time period can be divided into two time segments, approximately 300ns of propagation delay, and an interval which depends linearly on timing capacitor C_{OS}. When the one-shot is triggered, a voltage switch, that clamps the capacitor at analog ground, is opened, allowing the capacitor voltage to change. An internal 0.5mA-current source connected to the capacitor terminal draws current through C_{OS}, causing its voltage to decrease linearly. At approximately -3.4V, the one-shot resets itself, ending the timed period and starting the V/F conversion cycle over again. The one-shot time period is:

$$t_{os} = \Delta V\,\frac{C_{os}}{I} + T_{delay} \tag{15.5a}$$

substituting the above values,

$$t_{os} = 3.4\,V\,\frac{C_{os}}{0.5\,mA} + 300\,ns \tag{15.5b}$$

Therefore,

$$f_{OUT} = \frac{V_{IN}}{R_{IN}} \frac{1}{2(C_{OS} \times 3.4V + 300\,ns \times 0.5\,mA)} \tag{15.6}$$

Component selection is not difficult. Since R_{IN} and C_{os} are the only two parameters available to set the full-scale frequency and accommodate any given input voltage range, all of the necessary design information can be presented in graphical form. The selection guide of Figure 15.7 allows the user to quickly and easily pick a combination of R_{IN} and C_{os} to suit almost any operating requirements.

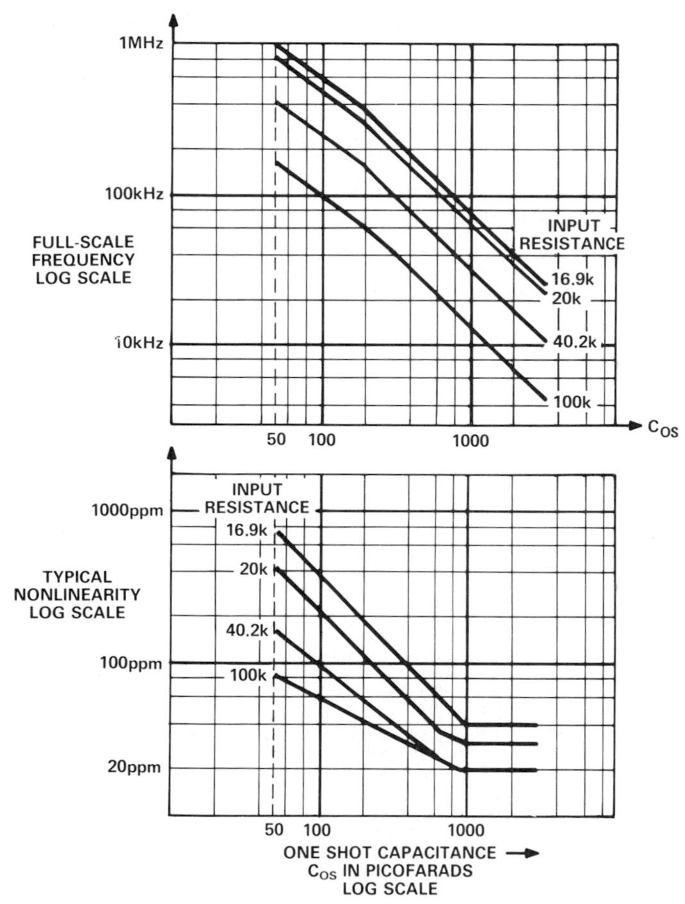

Figure 15.7. VFC parameter selection guide.

The linearity is also affected by the choice of R_{IN} and C_{os}; this information has also been provided in Figure 15.7. In general, larger values of C_{os} and lower full scale input currents (larger values of R_{IN}) result in better linearity. Although the selection guide is based on a 0-to-10-volt input signal range, the results are easily extended by simply scaling the input resistor in proportion to the desired input voltage span.

Selection of the integrating capacitor, C_{INT}, depends more on the particular application of the VFC than upon the input voltage span and output frequency range. If the input voltage is a steady signal with no interference in the form of hum or noise pulses (for example, in an application as a clock source, where the input signal is derived from a clean voltage reference), C_{INT} is most easily determined by the equation:

$$C_{INT} = \frac{100\mu F/s}{f_{MAX}\,Hz} \text{ microfarads} \tag{15.7}$$

The minimum value of C_{INT} is 1000 pF.

A major advantage of integrating-type converters, such as the charge-balance VFC, is the ability to reject large amounts of additive noise by integrating both signal and noise. If the noise has a mean value of zero, then its effect on the measurement is reduced or even eliminated. This kind of integrating analog-to-digital converter uses a counter to accumulate pulses of the VFC output frequency for a fixed gate time. An integrating converter behaves like a filter having a transfer function of the form:

$$H(f) = \frac{\sin \pi fT}{\pi fT} \tag{15.8}$$

where f is the frequency of the input signal and T is the gate time of the counter. For very low frequencies the transfer function is unity; this means that the dc and low-frequency components are measured correctly by the integrating converter. However, for a frequency which corresponds to the gate period, and for its harmonics, the transfer function is zero. This means that additive noise is completely rejected at certain frequencies; the gate time is usually selected to provide rejection of power line hum.

Figure 15.8 is a logarithmic plot of the magnitude of the transfer function of Equation 15.8, normalized to 60 Hz, as a function of frequency. Note that, even for frequencies not integrally related to the reciprocal of the gate time, the envelope of the function provides some rejection of additive noise. Since the pulses of the output frequency of the VFC are totalized over a gate time, T, to give an average frequency over that time interval, samples of the input signal are generated at a rate of:

$$f_s = 1/T \tag{15.9}$$

where f_s is the sampling rate and T is the gate time of the counter. As mentioned elsewhere in this volume, the Sampling Theorem explains that it is not possible to reconstruct from the samples any component of the input signal having a frequency greater than half the sampling frequency. Furthermore, if there are components present (noise or hum) at frequencies higher than half the sampling rate, those components will be aliased into the baseband. The amplitude of these aliased signals is determined by Figure 15.8; however, the apparent frequencies of the interfering components will all lie below half the sampling rate. Therefore, additional filtering may be needed if normal-mode noise components unrelated to line frequency are significant.

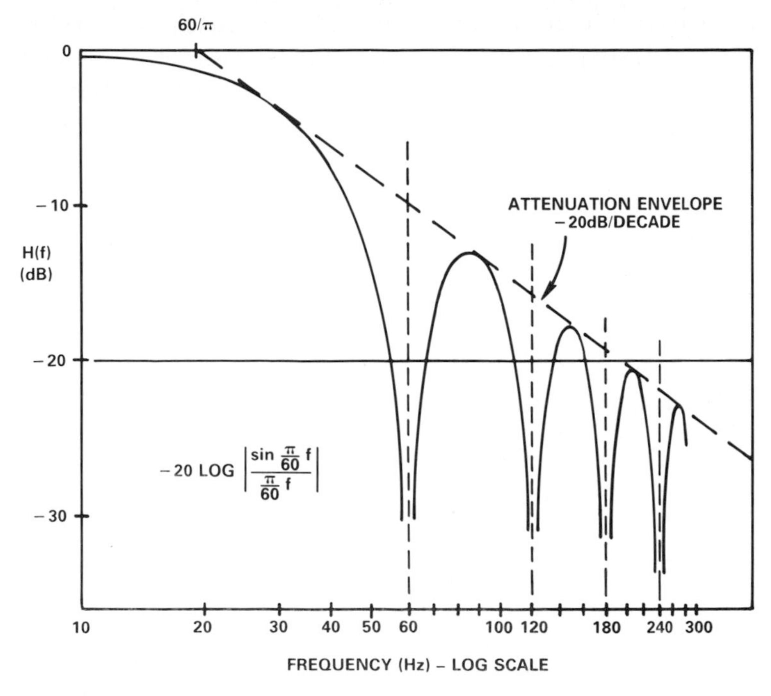

Figure 15.8. Magnitude response of sampling filter as a function of frequency.

These properties of noise rejection at certain frequencies depend upon continuous integration of the input signal; fortunately, the charge-balance architecture of the AD650 VFC depicted in Figure 15.5 performs continuous integration. Consider the Integration Period of the conversion cycle, as shown in Figure 15.6. When the input voltage is low, the downward ramp of the integrator output will be slow. If there is a noise pulse on the input signal which causes the instantaneous input voltage to be negative, then the integrator output will actually tend to move upward, away from the comparator threshold, but only during the duration of the noise impulse. Clearly, this is a transient situation; the instantaneous voltage does not remain negative for very long, and eventually the finite d.c. value of the input voltage causes the integrator output to ramp down and trigger the comparator. If a positive noise peak causes the integrator output to move down and trigger the comparator prematurely, no error is committed, since the period of the next VFC cycle will be longer, and the average frequency will still be correct.

The only situation which will allow an error to be made in the charge-balance process is if the continuous integration of the input signal is interrupted. For example, if the negative noise peak is too large in amplitude or too long in duration, the integrator output can drift all the way up to the plus rail and saturate. When the input again becomes positive, and the integrator returns to normal operation, there will be a lapse in the memory of the signal, leading to errors.

This situation is most likely to arise with low power supply voltages and long, unshielded input leads, which are subject to hum and interference pick up. The amount of noise which would be required to interrupt the continuous integration depends upon the frequency of the interference and the values of the input resistance, R_{IN}, and the integrating capacitance, C_{INT}. The voltage waveform on the output of the integrator, in the presence of interfering noise, is simply the sum of the sawtooth—as derived in Figure 15.6—and the response of the integrator to the noise alone. For example, consider a value of 1000 pF for C_{INT}, corresponding to a full-scale frequency of 100 kHz. Let us use a 40.2-kΩ input resistor and a 330-pF one-shot capacitor. Suppose that the interference is hum at 60Hz; the output magnitude of the voltage from the integrator at this frequency is:

$$V_{OUT} = \frac{V_{IN}}{2\pi \times 60\ \text{Hz} \times 1000\ \text{pF} \times 40.2\ \text{k}\Omega} \tag{15.10}$$

Thus, there is a gain of 66 for the interfering hum. Noise at higher frequencies (for example, impulse noise from a spark plug) will be amplified less, and therefore constitute less of a problem than hum.

The highest voltage normally seen at the output of the integrator (see Figure 15.6) is the peak of the sawtooth signal at the end of the Reset Period. From Equation 15.2 we see that ΔV is most positive when the input signal (I_{IN}) is smallest. Using Equation 6 to calculate t_{os} (2.5μs), we find from Equation 15.2 that the maximum value of ΔV is 2.5 volts. Since the rising edge of the sawtooth starts from the −0.6 volt comparator threshold, the maximum value of the waveform is 1.9 volts.

Let us now assume that the device is operating on ±9-volt power supplies; with an allowance of 3 volts as headroom for the op amp output stage, the integrator output voltage cannot exceed 6 volts. Since the sawtooth signal would go to 1.9 volts in the absence of any interference, we now see that there are 4.1 volts of noise margin before the hum causes the op amp to saturate and sawtooth-plus-hum waveform to form a "flat top". Since there is a gain of 66 for the 60-Hz hum (see Equation 15.10), the noise may have any amplitude up to 62mV.

To provide greater noise rejection, the value of the integrating capacitance can be increased. If C_{INT} were increased to 5,100 pF in this example, the allowable 60-Hz noise amplitude would be 470mV. This factor-of-7.6 improvement is due in part to a 5.1× reduction in gain seen by the hum, as calculated in Equation 15.10, and partly to a reduction in the ΔV of the sawtooth signal, as calculated using Equation 15.2. Another way to increase noise margin is to increase the input resistance. If R_{IN} of the example were 100 kΩ, then the value of C_{os}

would have to be 100 pF in order to maintain the full-scale frequency at 100 kHz. In this example, we could allow 210 mV of hum pick-up amplitude with a 1000-pF C_{INT}. Finally, combining approaches, if the integrating capacitance were 5,100pF, then 1.2-volt noise-pickup amplitude at 60 Hz could be tolerated at the op-amp input with no significant errors.

Synchronous VFCs

A variation of the charge-balance type of VFC is the *synchronous voltage-to-frequency converter* (SVFC). Devices of this kind require an external clock signal; the period of the clock signal is used as the length of the Reset period (Figure 15.6), making the scale factor proportional to the external clock frequency (In Equation 15.4, let $t_{os} = 1/f_{clock}$). In this way, the temperature drifts associated with the one-shot and its capacitor are eliminated, since the clock can be derived from a crystal oscillator for great temperature stability.

In an ordinary charge-balance VFC, both the dc reference (the 1-mA current source in Figure 15.6) and the one-shot can cause the gain to drift with temperature. Even if the on-chip portion of the one-shot were perfect, the temperature coefficients of the one-shot capacitor would still introduce errors. In the synchronous VFC, since the one-shot is completely eliminated, only the dc errors remain, and they can be made very small.

In general, the linearity and offset errors of SVFCs are comparable with those of other types of charge-balance devices, but the gain drifts are much better. In the SVFC, the pulses of the output frequency are generated in phase with the input clock; that is, the Reset period is initiated and terminated by a rising (or falling) edge of the input clock. This characteristic can be exploited in system applications (see section 15.4.5).

15.3.3 ASSEMBLED MODULAR DEVICES

Monolithic charge-balance voltage-to-frequency converters, such as the AD650 and ADVFC32, offer high-speed operation and outstanding linearity. However, for the most demanding applications, these devices cannot match the temperature stability of the best high-performance modular VFC devices. For example, the best grade of the Analog Devices Model 458 provides a temperature coefficient of gain equal to ±5ppm/°C at a 100-kHz full-scale, compared to the monolithic AD650's 150ppm/°C gain-TC for the same frequency scale. However, both devices have approximately equal nonlinearity, in the several hundred ppm range. As a further example, the Model 460 modular VFC's gain-tempco is ±15ppm/°C at 1-MHz full-scale, and nonlinearity is 150 ppm at the same frequency scale; this can be compared with the AD650's gain tempco of of −400, +200ppm/°C and nonlinearity of 1000ppm on the 1-MHz scale.

However, a price must be paid for the superior performance of the modular devices, in size and power consumption. They are substantially more expensive, by a factor of about 10. Table 15.1 compares the salient performance specifications of the VFC devices discussed here.

	INTEGRATED CIRCUITS			DISCRETE MODULES	
	AD537	**ADVFC32**	**AD650**	**458**	**460**
Frequency (Max. Full Scale):	**150 kHz**	**500 kHz**	**1 MHz**	**100 kHz(fixed)**	**1 MHz(fixed)**
Linearity Error (PPM)					
10 kHz Full Scale	700	100	50	–	–
100 kHz Full Scale	1000	500	200	100	–
500 kHz Full Scale	–	2000	500	–	–
1 MHz Full Scale	–	–	1000	–	150
Gain Drift (PPM/°C)	50	75	75	5	15
Offset Voltage (mV)	5	4	4	10	10
Bias Current (nA)	100	100	20	–	–
Supply Voltage Range (Volts)	4.5 to 36	±9 to ±18	±9 to ±18	±13 to ±18	±13 to ±18
Quiescent Current (mA)	2.5	8	8	+25, – 8	+25, – 8
Minimum Dissipation (mW)	12	144	144	429	429

TABLE 15.1 Specifications of Typical Devices

15.4 APPLICATIONS

V/F converters, as indicated earlier, can serve in a wide variety of applications. In this section, we will discuss in depth a few of the more popular and useful applications, including VFC voltmeters, signal isolation, and two possibilities for phase-locked-loop f/v conversion.

15.4.1 VFC VOLTMETER

A voltage-to-frequency converter can be used as the heart of a high-resolution voltmeter. In an instrument where the ultimate consumer of the data is human, or where extremely high resolution is necessary, a conversion rate of one or two per second is adequate. In this case, the VFC provides accuracy and linearity effectively, economically, and compactly. In a system where speed requirements are more aggressive, the VFC can still be used, but with somewhat reduced resolution; in these applications, the VFC provides the benefits of guaranteed monotonicity and no missing codes. This is a very important consideration in closed-loop control systems, where any non-monotonic behavior can lead to instability, with possibly dire consequences.

The conversion speed of a VFC voltmeter is related to full-scale frequency of the VFC and the required resolution of the measurement. For example, a measurement with 14 bits of resolution requires that changes as small as one part in 16,384 of full scale be detected. If we use a VFC with a 1-MHz full scale, and accumulate pulses of the output frequency for 1/60 of a second, then the full-scale reading of 16,666 would exceed 14 bits. Note that the gate time will provide infinite rejection of 60-Hz power line hum and its harmonics if continuous integration of the signal is preserved, as detailed in Section 15.3.

The accuracy of a VFC voltmeter is bounded by the linearity of the VFC. For example, consider a Model 460 modular VFC, with nonlinearity of $\pm$150ppm and a full scale of 1 MHz. Since a 14-bit measurement implies resolution of 61 ppm of full scale, it is seen that the Model 460 will provide a relative accuracy to within 3 LSB. If resolution greater than 14 bits is required, the gate time can be increased to accumulate more pulses of the output frequency. In principle, there is no limit to the resolution obtainable with a VFC-based voltmeter, provided one is willing to wait. In practice, however, noise will limit useful resolution.

In order to determine the maximum usable resolution of a VFC-based voltmeter, it is necessary to measure a constant, highly stable voltage repeatedly and observe the variation in successive answers. As the gate period of the counter is increased, a point will be reached where the variation between successive measurements will equal the resolution of the measurement. In other words, there will be jitter in a supposedly constant frequency; there is no point in extending the resolution any further.

In order to get full accuracy from a VFC as a voltmeter, it is necessary either to adjust gain and the offset of the device by itself or to calibrate the measurement chain, starting with preamplifiers, signal conditioners, multiplexers, etc. These adjustments are easy to accomplish with trim potentiometers; however, this type of calibration only compensates for channel-to-channel or unit-to-unit variations. Since the errors caused by temperature drifts and component shifts after the initial calibration are not addressed by this technique, especially in system applications, it is often necessary and desirable to use an automatic gain and zero adjustment.

Such auto-cal adjustments are easy to do in software; simply measure the output frequency produced in response to two known input voltages—for example 10 volts and 1.2 volts. Then, by assuming a linear voltage-to-frequency transfer relation (always a safe bet with a VFC) the actual input voltage can be calculated for any output frequency. Also, if several intermediate voltage-reference points are provided, it is possible to make some first order corrections for nonlinearity.

These techniques are valuable if computational ability is already in place in a system (and what instrument these days does not have a microprocessor?) However, it may be more desirable to have a hardware auto-zero and auto-gain system if, for example, the necessary computations would overburden the computer, or if the results are required at a rate which is too fast to allow time for the arithmetic. Another system where a hardware gain and offset calibration would be necessary is in a frequency-to-voltage application, for example, as a tachometer, where a calibrated output voltage is produced in response to an input frequency.

As an example of the approach, a system will be described that has automatic adjustment of both gain and offset for an AD650 monolithic VFC. The *offset autozero loop* uses a sample-and-hold amplifier (SHA), as shown in Figure 15.9. The input voltage to the VFC is switched between the signal to be meas-

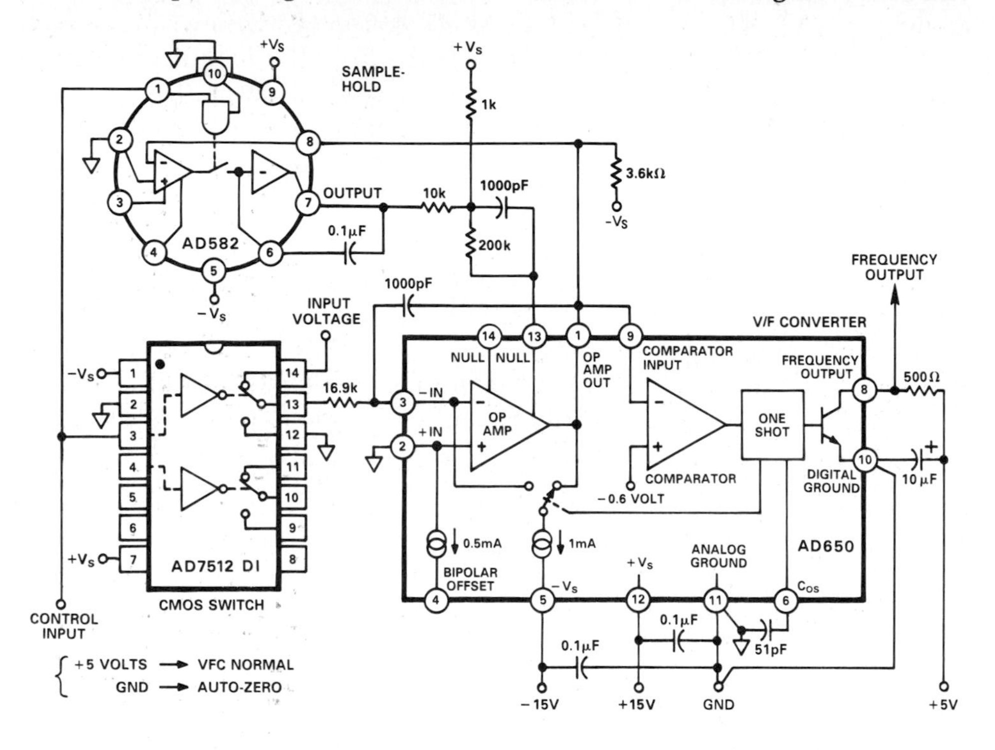

Figure 15.9. Autozero circuit for v/f converter.

ured and a ground reference, via an analog switch. In the *sample* mode, while the VFC's voltage input is grounded, the SHA measures and establishes the required correction for the VFC's offset; in the *hold* mode, the SHA maintains the zero adjustment, while the VFC is performing its measurement.

The sample-hold adjusts the offset of the AD650 by forcing a current at pin 13, one of the offset null pins. During the *sample* mode (VFC input grounded), the sample-hold and the integrating amplifier form a negative feedback loop; since the sample-hold's input is grounded, the signal returned to its feedback input must also be at zero; the sample-hold drives just enough current into the nulling input of the integrator to bring the output of the integrator to zero and keep it there. Since the output of the integrator is constant, its input current must be zero; thus the combined effects of amplifier offset voltage and bias current have been forced to zero. When the auto-zero cycle is complete, the SHA simply holds its output voltage to maintain the AD650 at effectively zero offset.

Note that the output of the DUT could have been forced to any convenient voltage other than ground, just by the choice of constant input voltage to the SHA. The 1000-pF capacitor shunting the 200-kΩ resistor is dynamic compensation for the two-amplifier servo loop; two integrators in a loop require a single zero for compensation. The 3.6-kΩ resistor from pin 1 of the AD650 to the negative supply is not part of the auto-zero circuit; it is required for VFC operation at 1 MHz.

The *auto-gain* loop, shown in Figure 15.10, uses a multiplying d/a converter to adjust the effective input resistance to the VFC. The signal reaches the

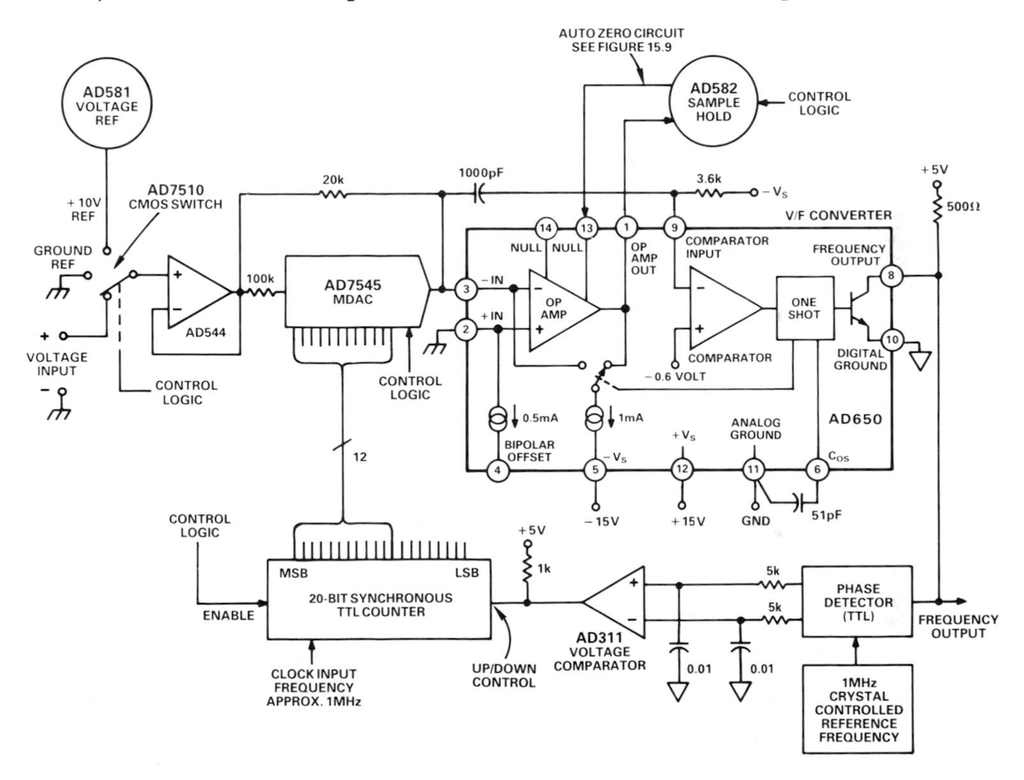

Figure 15.10. Automatic gain control for VFC.

VFC's summing junction via two parallel paths: about 90% of the input signal is applied directly to the VFC through a fixed 20-kΩ resistor; the remaining 10% of the input signal is diverted through the MDAC, which acts like a digitally adjustable input resistor. Since the full 12 bits of the MDAC is being applied to only a tenth of the scale, the gain can be adjusted with a resolution of 16 bits!

In order to set the gain of the VFC, an external 1-MHz reference signal is required, preferably derived from a crystal oscillator. A phase detector is used to compare with a reference frequency the output of the VFC when connected to a 10-volt reference. The phase detector's voltage output drives a voltage comparator, which decides which frequency is higher; this output drives the UP/DOWN control of a TTL counter, which adjusts the gain of the DAC in the appropriate direction.

If, for example, the VFC's output frequency is higher than the reference frequency, the comparator's output will cause the counter to count down, reducing the gain of the VFC. The counter will continue to count down until the comparator changes sign, indicating that the signal frequency has been adjusted to be within 1 LSB of the reference frequency. When the auto-gain operation is completed, the data is latched into the DAC, which will hold the gain setting of the VFC until the next calibration cycle. The auto-gain loop requires the use of synchronous counters (TTL ...S169 are used in this circuit) to avoid the effects of ripple carry errors, which would cause the MDAC to assume erroneous values before settling to the proper result, leading to instability in the loop.

15.4.2 SIGNAL ISOLATION

A very important application of a VFC is analog signal isolation: transmission of an analog signal across a barrier with high accuracy. The analog signal is converted into a pulse train at a proportional frequency; this pulse train is relayed across a barrier and then converted back into an analog signal. The signals can thus be transmitted across a barrier accurately and without interference (assuming that the pulses are significantly higher in amplitude than any interfering signals), since they have been converted into a digital form.

The barrier may be physical (e.g., distance); it may be electrical, as in transmitting data across a large potential difference; or it may be a noisy environment such as a factory floor. The barrier may even be time; one may wish to save data by recording it on magnetic tape (for example) and reproduce it faithfully at some future time. The signal, once transformed into a pulse train with digital properties, may be transmitted in any fashion suitable to the application at hand. The analog signal may be converted into a frequency by straightforward application of the VFC; it is also possible to improve the accuracy of the transmitted signal by adding the auto-gain and auto-zero circuits described above.

The high operating frequency and excellent linearity of a 1-MHz VFC permits very accurate transmission of dynamic signals. However, the operation of a VFC at high frequency is not without concerns; remember that 1 MHz is a frequency usually classified as RF (it is central to the AM broadcast band). Most opto-isolators are not fast enough to relay the very narrow pulses produced by VFCs with high-frequency outputs. Figure 15.11 is an example of a circuit, using a 6N137 opto-isolator, which has fast-enough response to keep pace with the VFC.

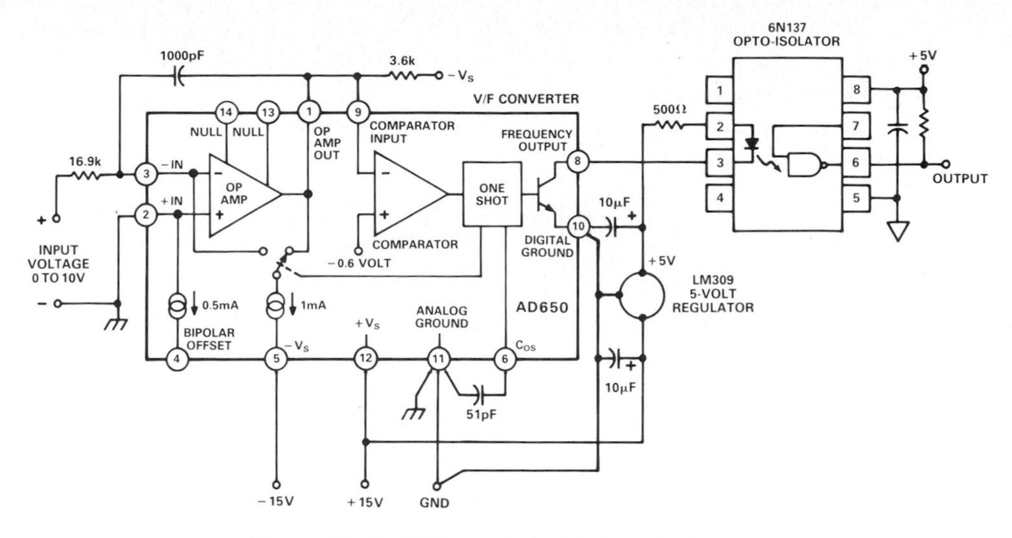

Figure 15.11. VFC applied with opto-isolator.

Another concern is the difficulty of relaying digital pulses over a long length of cable. Simply connecting a TTL output to a very long cable without regard for proper termination will not work. The logic signal cannot properly drive the cable capacitance, and also the pulses may be distorted by reflections caused by improper termination.

Indeed, it is possible that a single pulse launched from the transmitter may be reflected at the receiver and return to the transmitter. At the transmitter, it may be reflected again and then return to the receiver where it will register as a pulse of the signal frequency for a second time. This echo effect may cause considerable errors in the transmission of the signal frequency.

A circuit is shown in Figure 15.12 to drive a 100-foot length of shielded twisted-pair cable, using the open-collector output of a VFC. The circuit operates in a current mode, and no voltage pulses are transmitted over the cable. The LED of the opto-isolator presents a load which is essentially a very low impedance in series with a 1.5 volt DC source. That is, there can be large changes in current through the LED, with only very small voltage changes, but it needs 1.5 volts to turn on.

Since there are no large voltage signals on the cable, stray capacitance is not a problem. With VFCs like the AD650, the cable can be conveniently driven directly, with no extra components. And, as an added bonus, a frequency-output voltage signal is available at the output-transistor collector for local use. At the receiving end of the 100-foot cable, only a local ground is required to power the opto-isolator's detector circuit. Thus this circuit provides complete galvanic isolation, in addition to transmission of the signal over a great distance. Ground loops and power-line faults can all be avoided with this simple and effective circuit.

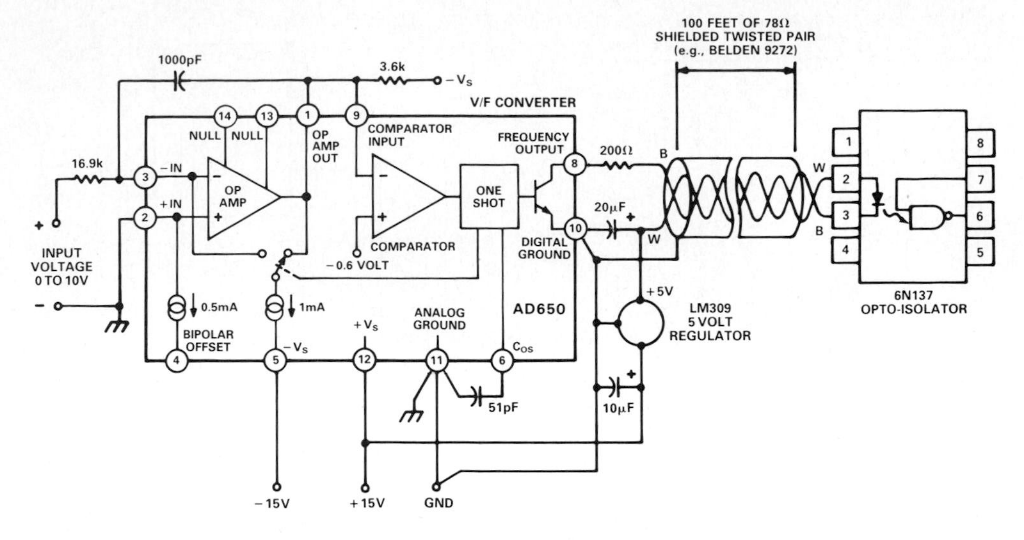

Figure 15.12. VFC in current-mode application with opto-isolator.

Signal Reconstruction

Once a frequency signal has been received, it is necessary to reconstruct the original analog signal by using a frequency-to-voltage converter—FVC. A simple approach is the tachometer circuit shown in Figure 15.13. Each time the input signal crosses the comparator threshold going negative, the one-shot is activated and switches 1 mA into the integrator input for a measured time period (determined by C_{OS}). As the frequency increases, the total amount of charge injected into the integrator summing junction increases proportion-

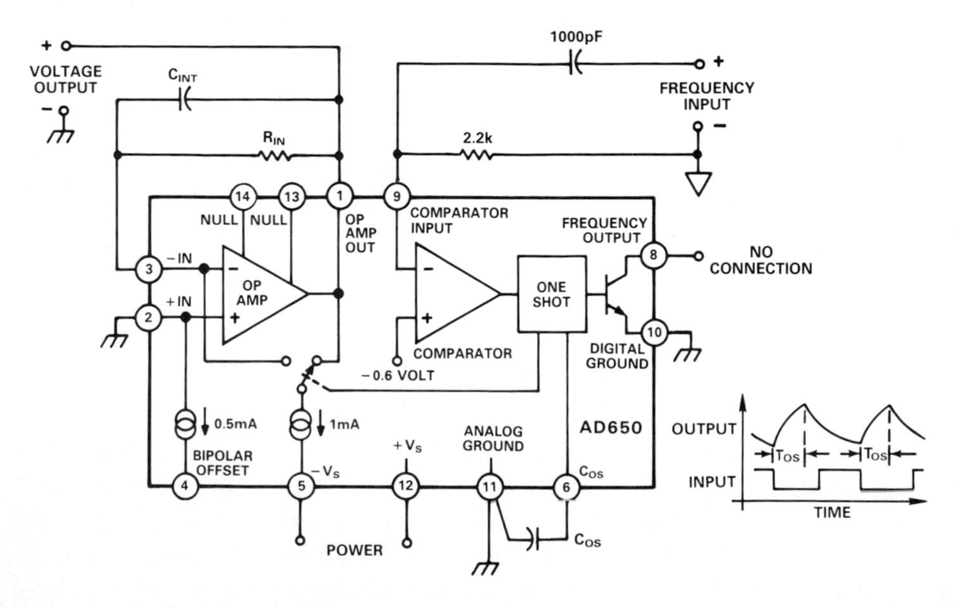

Figure 15.13. VFC in simple frequency-to-voltage conversion application.

ately. The voltage across the integrating capacitor is stabilized when the feedback leakage current through R_{IN} equals the average current being switched into the integrator. The net result is that the average output voltage is proportional to the input frequency.

The output voltage waveform is a series of exponential sections; at high frequencies, it will look somewhat like the inset of Figure 15.13, but at very low input frequencies it will look a lot more like a series of "spikes," i.e., a small d.c. level with high ripple content. The ripple can be reduced by increasing the value of the integrator capacitor, but at the expense of reducing the speed of response of the tachometer circuit. The ripple can also be reduced by filtering the output with a lowpass filter; however, care must be taken not to contaminate the signal by introducing offsets or gain errors in the filter.

The circuit shown can accommodate almost any input signal waveform. With the 1nF coupling capacitor and the 2.2-kΩ resistor, a TTL input creates a clean negative going spike that triggers the one-shot on each falling edge. For input signals with slower edges, higher capacitance or resistance may be used, as long as the comparator is never exposed to a voltage lower than −0.6V for longer than the one-shot's time period. If this happens, the one-shot will trigger itself more than once per cycle, creating discontinuities in the F/V transfer function.

The tachometer has the advantage of being a very simple circuit, but it has difficulty handling rapidly changing information. Where the bandwidth required of the voltage output signal is low, such as might be expected from thermocouple signals, the economy provided by this circuit can be attractive. Analog signals at higher frequency, however, are more efficiently reconstructed with a phase-locked-loop circuit.

15.4.3 PHASE-LOCKED LOOP F/V CONVERSION

A phase-locked loop (PLL) provides a much more favorable compromise between speed of response and output ripple than the simple tachometer circuit, at the price of added complexity. In a phase-locked servo loop, a local VFC is driven to produce an output frequency which exactly matches the frequency and phase of an input signal. Since the voltage that drives the VFC (derived from the phase or frequency error signal) is proportional to the input frequency, it is taken as the output voltage of the frequency-to-voltage converter circuit.

The linearity of the frequency-to-voltage conversion depends almost entirely on the linearity of the VFC used in the loop. In certain circumstances, it is possible to achieve cancellation of linearity errors, if the nonlinearity of the VFC used in the receiver is almost identical to that of the VFC used in the transmitter. In order to do this effectively, though, the user must select matched pairs of converters, since random selection will be inadequate.

In some PLL applications, linearity is not an issue; for example, in a frequency synthesizer. In such applications, the oscillator output frequency is first processed through a programmable "divide by N" before being applied to the phase detector as feedback. Here the oscillator frequency is forced to be equal to "N times" the reference frequency, and it is this frequency output from the VFC that is the desired output signal, not the input voltage.

A very simple PLL circuit is shown in Figure 15.14. The phase detector consists of two D-type flip-flops and a NAND gate, while the charge pump and loop filter consists of an exclusive-or gate, a DMOS field-effect transistor, and an op amp. The charge pump relies upon power supply voltages and logic levels to establish the phase detector gain: among designers, this is usually considered a heinous crime—but the spirit of this circuit is simplicity.

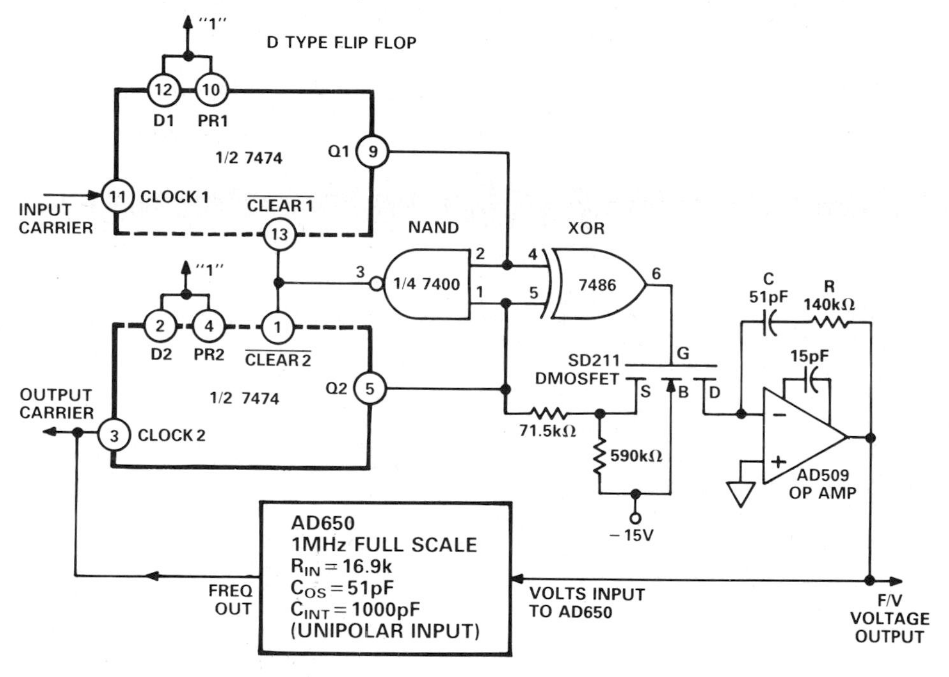

Figure 15.14. Phase-locked-loop in frequency-to-voltage conversion.

The phase detector is actually a so-called phase-frequency detector (PFD), which locks on edges. It provides proper feedback in the event of unequal frequencies; this means that the loop can never loose lock (unlike the situation when an analog multiplier is used as a phase detector; despite its superior noise rejection, a multiplier has only a limited range over which the output is a measure of the relative phase of the two inputs; if the phase error between the two inputs exceeds this range, the signal applied as feedback will not be correct and the loop will thrash about).[2]

[2]For more information on phase detectors and a full discussion of phase-locked-loop circuits, see F. M. Gardner, *Phase Lock Techniques*, 2nd ed. (New York: John Wiley and Sons, 1979).

For an analysis of the loop of Figure 15.14, start with the 7474 dual-D-type flip flop. When the input carrier matches the output carrier in both phase and frequency, the Q outputs of the flip flops will rise at exactly the same time. With the inputs of the exclusive-or (XOR) gate first at two zero's, then at two one's, the output will remain low, keeping the DMOS FET switched off. Also, the NAND gate will go low, resetting the flip-flops to zero.

Throughout the entire cycle just described, the DMOS integrator gate remains off, allowing the voltage at the output of the op-amp-connected-as-integrator to remain unchanged from the previous cycle. However, if the input carrier leads the output carrier by just a few degrees, the XOR gate will be turned on for the small time span that the two signals are mismatched. Since Q2 will be low during the mismatch time, a negative current will be fed into the integrator, causing its output voltage to rise. This will, in turn, increase the frequency of the VFC slightly, driving the system towards synchronization. In like manner, if the input carrier lags the output carrier, the integrator output will be forced down slightly to synchronize the two signals.

The ± 25-μA pulses from the phase detector are incorporated into the phase-detector gain expression (for K_d):

$$K_d = \frac{25\ \mu A}{2\pi\ \text{rad}} \simeq 4 \times 10^{-6}\ \frac{\text{amperes}}{\text{radian}} \tag{15.11}$$

Also, the V/F converter is configured to produce 1 MHz in response to a 10-volt input, so its gain, K_o, is:

$$K_o = \frac{2\pi \times 1 \times 10^6\ \text{Hz}}{10\ \text{V}} = 6.3 \times 10^5\ \frac{\text{radians}}{\text{volt-sec}} \tag{15.12}$$

The dynamics of the phase relationship between the input and output signals can be characterized as a second order system with natural frequency, ω_n:

$$\omega_n = \sqrt{\frac{K_o K_d}{C}} \tag{15.13}$$

and damping factor

$$\zeta = \frac{R}{2}\sqrt{C K_o K_d} \tag{15.14}$$

For the values shown in Figure 15.14, these relations simplify to a natural frequency of 35 kHz, with a damping factor of 0.8.

As a simple approach to determining component values for other PLL frequencies and VFC full-scale voltage, the following cookbook steps can be used:

1. Determine K_o (in radians per volt second) from the maximum input carrier frequency, f_{max} (in hertz), and the maximum output voltage, V_{max}:

$$K_o = \frac{2\pi f_{max}}{V_{max}} \qquad (15.15)$$

2. Calculate a value for C based upon the desired loop bandwidth, f_n. This is the desired frequency range of the output signal. The loop bandwidth (f_n) is not the maximum carrier frequency (f_{max}): the signal may be very narrowband, even though it is transmitted over a 1-MHz carrier.

$$C = \frac{K_o}{f_n^2} \times 10^{-7} \frac{\text{VF}}{\text{rad-s}} \text{ farads} \qquad (15.16)$$

where C is in farads, f_n is in Hz, and K_o is in radians/(volt-second).

3. Calculate R to yield a damping factor of approximately 0.8 using this equation:

$$R = \frac{f_n}{K_o} \times 2.5 \times 10^6 \frac{\text{rad}\Omega}{\text{V}} \text{ ohms} \qquad (15.17)$$

where R is in ohms, f_n in hertz, and K_o in radians/volt-second.

If, in actual operation, the PLL overshoots or hunts excessively before reaching a final value, the damping factor may be raised by increasing the value of R. Conversely, if the PLL is overdamped, a smaller value of R should be used.

Phase-Locked-Loop Performance

The performance of the PLL circuit is demonstrated by the system shown in Figure 15.15. An analog signal is converted into a frequency, and then this

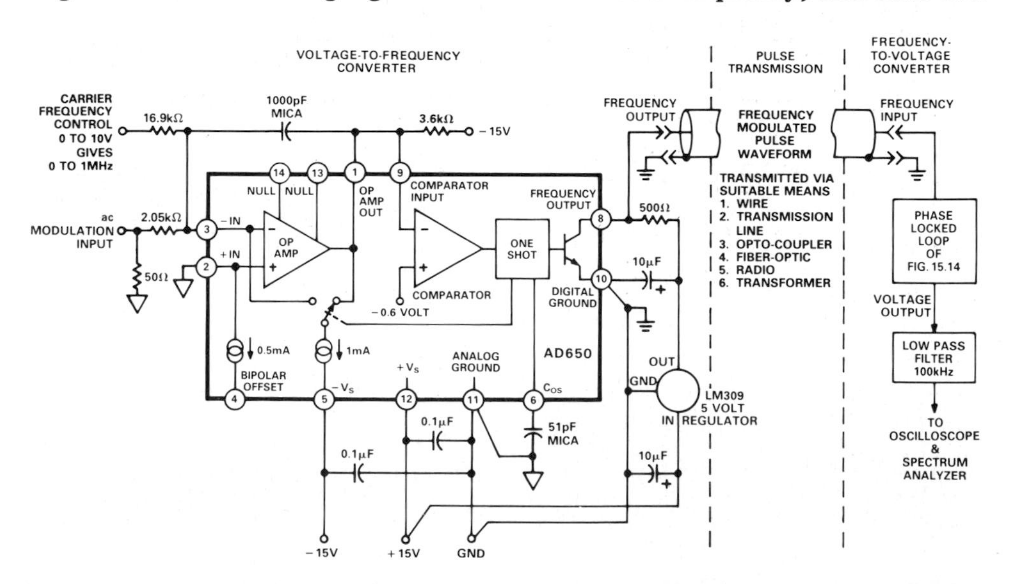

Figure 15.15. Circuit to demonstrate performance of PLL in frequency-to-voltage conversion.

frequency is converted back into an analog voltage by the PLL. The signal to be converted to analog is produced by a VFC with two additive inputs—one at dc, to set the carrier frequency, the other to establish an ac modulation signal. For this purpose, it is useful for the VFC's summing junction to be available.

The output frequency is then relayed to the PLL via a jumper cable (the signal at this point is a 5-volt digital pulse train, which may be transmitted in any fashion suitable to the application at hand). The filter on the output signal attenuates carrier feedthrough to allow easy interpretation of the signal with an oscilloscope and spectrum analyzer.

The response of the system to a step change of modulating frequency is shown in Figure 15.16a. The signal output is swinging between 5 volts and 10 volts, for an input step of 500 kHz to 1 MHz. Note that, despite the overshoot to 1.1 MHz, the response remains well-controlled. Note also the slight irregularity during the transition: it is caused by cycle-slipping during slewing, when feedback is lost temporarily and the PLL actually loses phase lock.

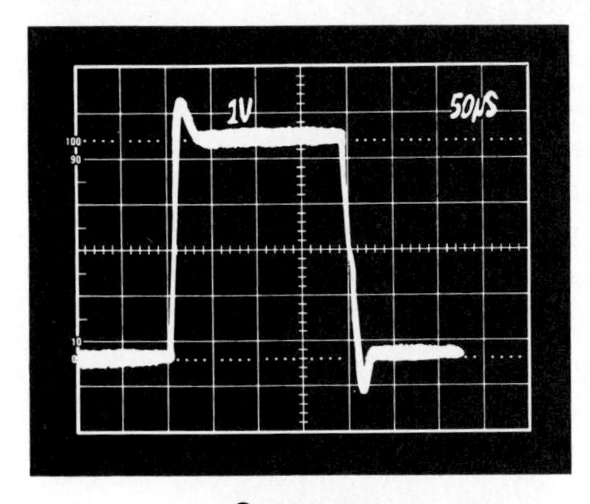

a. Step response.

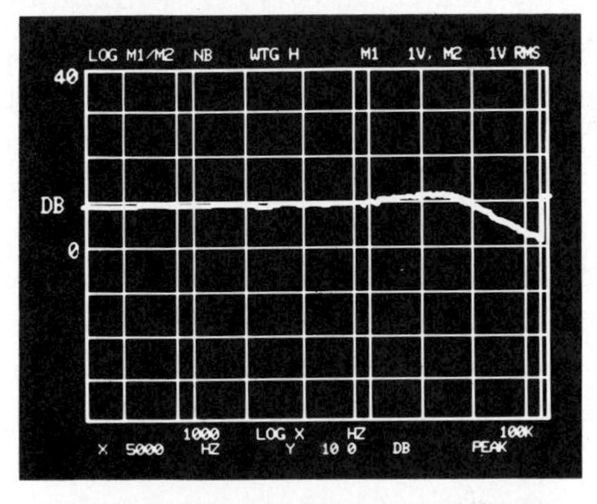

b. Frequency amplitude response.

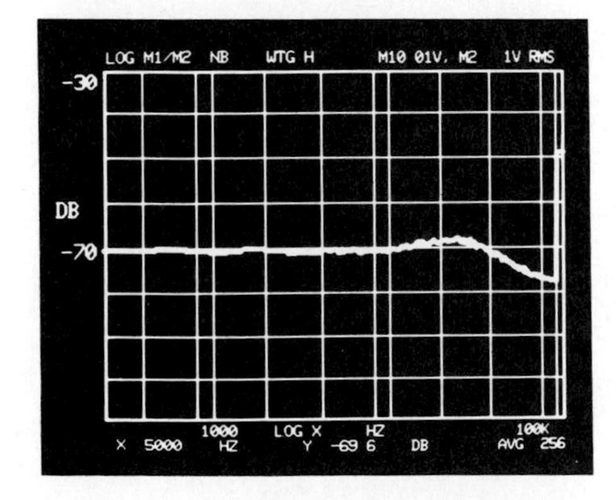

c. Zero-signal noise.

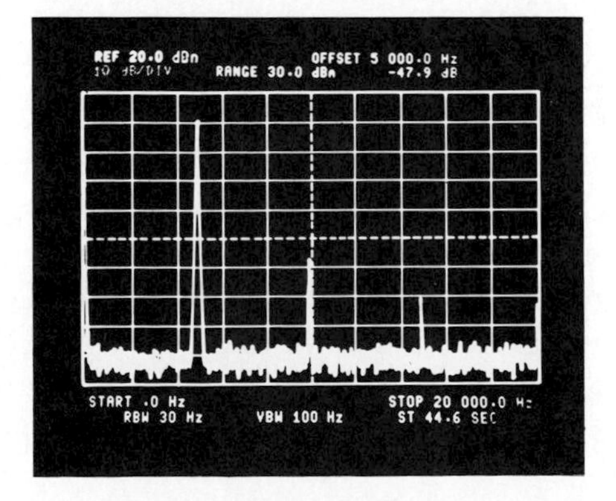

d. Harmonic distortion.

Figure 15.16. Performance of PLL as f/v converter.

The frequency response of the system, when driven with sine-wave excitation, is shown in Figure 15.16b. Here the output level is set to 2 volts peak-to-peak, and the carrier is 800 kHz. Note that the − 3dB bandwidth is about 70 kHz, which is consistent with a damping factor of 0.8 and a natural frequency of 35 kHz*.

When an unmodulated carrier is applied to the PLL, the noise that appears at the output signifies the lower limit of the dynamic range of the system. The spectrum of the noise at the output of the PLL is shown in Figure 15.16c. By comparing this with the output levels of Figure 15.16b, the dynamic range of the system is seen to be about 80dB. The harmonic distortion of the system is shown in Figure 15.16d for a 2-volt peak-to-peak sinewave at 5 kHz; the amplitude of the first harmonic is seen to be 48dB below the fundamental. The harmonic distortion can be improved to the level of 60dB if the amplitude of the modulation is reduced, but this is at the expense of dynamic range, since the the noise floor remains unchanged.

15.4.4 SOPHISTICATED PHASE-LOCKED LOOP

In some circumstances it is imperative that there be absolutely no feedthrough of the carrier into the output voltage of a PLL used in frequency-to-voltage conversion. Although it is possible to filter the voltage produced by the PLL to remove any residual carrier, it is not a trivial matter to design a filter with absolutely no d.c. offset voltage and no gain error. The post-filter may actually add errors worse than those to be eliminated.

A better strategy is to include the filter inside the phase-locked loop, where any offset or gain error will be compensated for by the feedback action of the loop. The voltage delivered to the VFC will always be forced to be the value required to produce an output signal of proper frequency and phase; hence, small gain and offset errors in the filter will not affect the output. Furthermore, since the voltage delivered to the VFC is a pure d-c level with no a-c components, there will be no phase noise, or incidental frequency modulation of the output frequency signal. This can be very important in such applications as frequency synthesis.[3]

Figure 15.17 shows a PLL circuit with a 5-pole Chebyshev active filter inside the loop. The phase detector is similar to the one used in the circuit of Figure 15.14. The charge pump operates in a differential mode and uses a separate reference voltage to establish the phase detector scale factor; also a differential integrator is used in the loop. The differential structure of the charge pump and integrator to some extent rejects digital noise.

*See page 13 of Gardner.[2]

[3]More details on a low-phase-noise synthesizer can be found in "Linear V-F Converter Chip Invades Module Territory," by L. DeVito, *Electronic Design*, February 23, 1984.

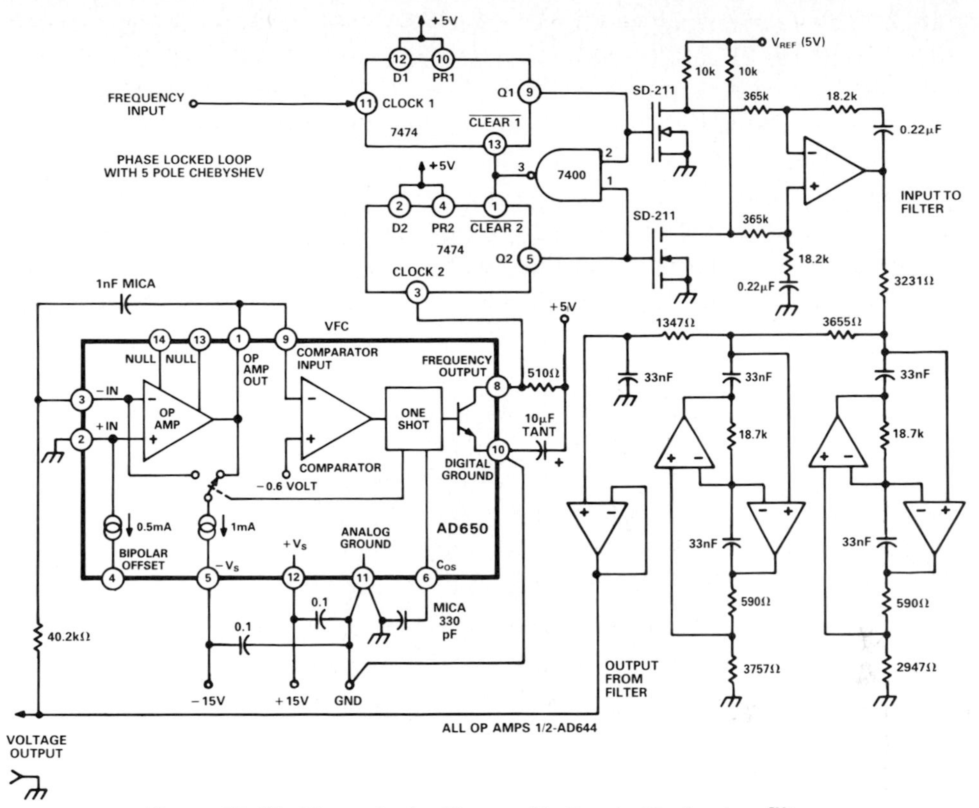

Figure 15.17. Phase-locked loop with 5-pole Chebyshev filter.

The low-pass filter has a 2.3-kHz cutoff frequency, based upon a frequency dependent negative resistor (FDNR) simulation of a doubly terminated ladder filter. The VFC used in this loop has a gain of 10^4 hertz per volt, or 100 kHz full scale for a 10-volt input. The performance of this PLL as a high quality F to V converter can be judged from the frequency spectra of Figure 15.18,

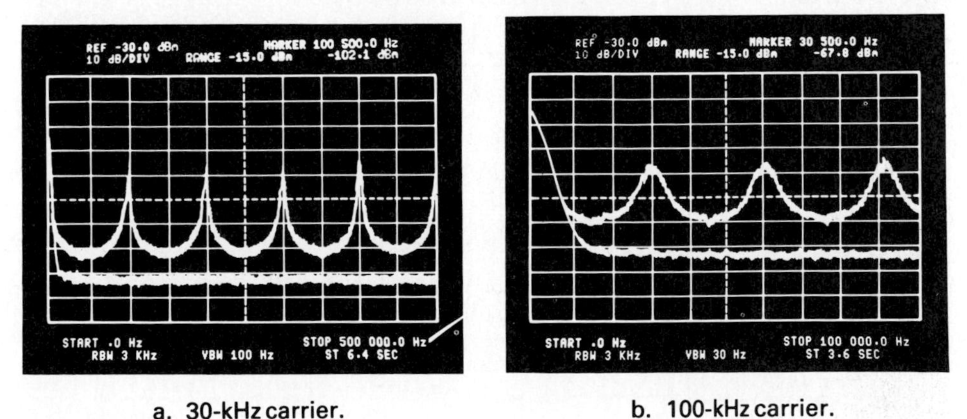

a. 30-kHz carrier.

b. 100-kHz carrier.

Figure 15.18. Output noise spectra of phase-locked loop of Figure 15.17, with and without filter.

which shows the output voltage in response to two constant-input carriers, 30 kHz and 100 kHz. The spectra in a and b show the performance of the circuit with and without the Chebyshev filter inside the loop. Note that the attenuation of carrier feedthrough and the concomitant improvement in dynamic range is as much as 45dB. The response shown is for a static carrier (the output voltage is constant); in the dynamic situation, the carrier feedthrough is more pronounced, and the improvement afforded by this circuit will be even more dramatic. As Figure 15.19 shows, the noise output voltage of the loop is less than 1 millivolt, peak to peak, and full-scale can be as much as 10 volts p-p; thus the dynamic range is 80dB.

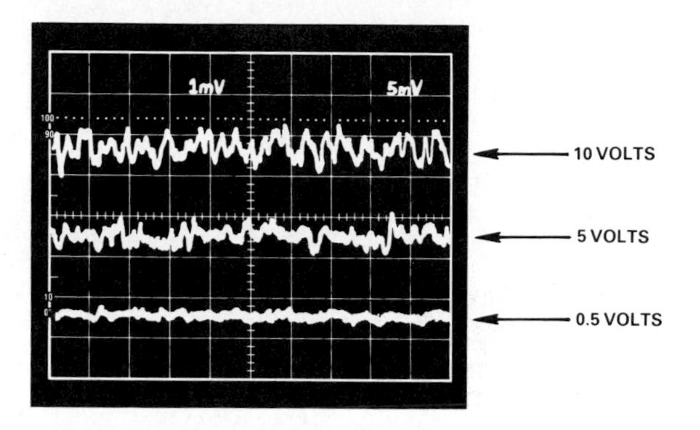

Figure 15.19. Output noise of 5-pole-filtered PLL: waveforms at various input levels.

For analysis of the circuit, the block diagram of the PLL is shown in Figure 15.20. The loop transmission of the PLL is:

$$L(s) = \frac{5\,V}{2\pi\,\text{rad}} \frac{sR_zC + 1}{sCR_i} K_f \frac{2\pi\,10^5\,\text{Hz}}{s\,10\,V} \tag{15.18}$$

The transfer function of the Chebyshev filter is designated as K_f; it will be set equal to unity for now. The loop transmission then simplifies to:

$$L(s) = \frac{4.004 \times 10^{-3}s + 1}{1.6 \times 10^{-6}s^2} \tag{15.19}$$

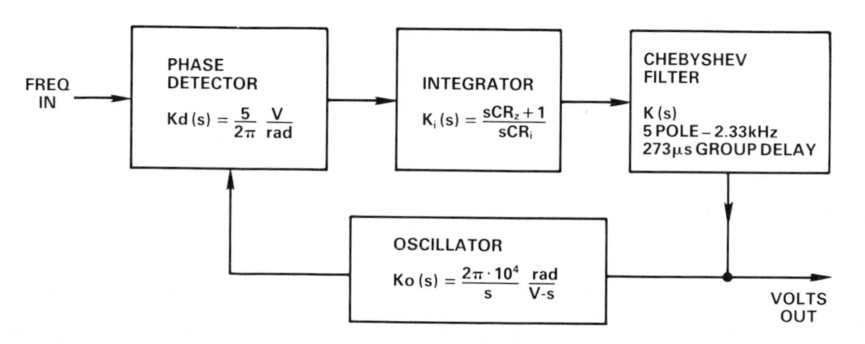

Figure 15.20. Block diagram of PLL for transfer-function analysis.

A Bode plot of the loop transmission is shown in Figure 15.21. Note that the loop crossover is at 400Hz and the frequency of the zero is 40Hz; this would provide a phase margin of 84°. The effects of the Chebyshev filter must now be considered. The corner frequency of the filter is 2.33 kHz; in the region of the 400Hz crossover, the magnitude of the filter response is unity, hence the loop crossover frequency is unaffected by the filter. The group delay of the Chebyshev filter at frequencies well below the corner frequency is 273μs, which translates into a phase shift of 39° at 400Hz. Thus the phase margin of the loop is reduced to 45° by the presence of the filter.

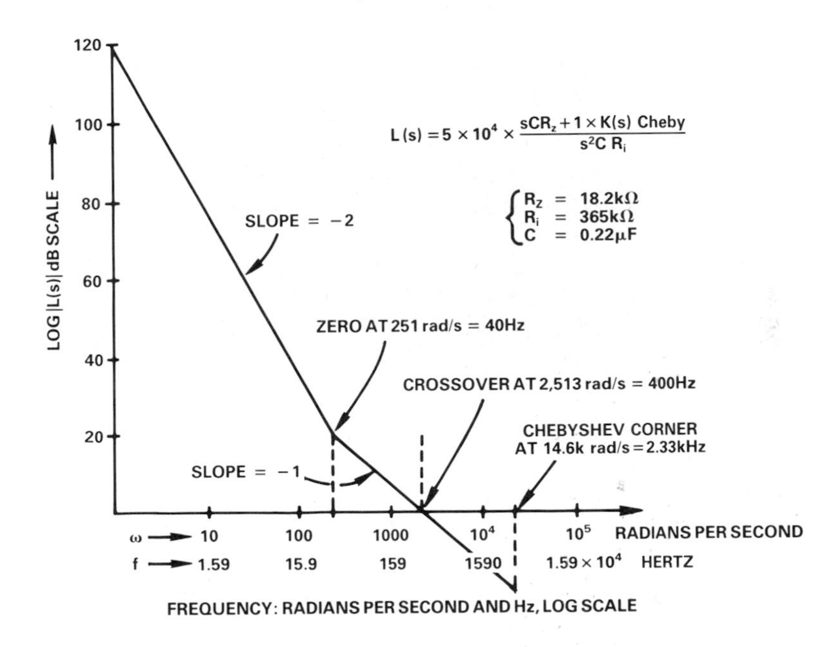

Figure 15.21. Bode amplitude plot of PLL.

The dynamics of the PLL circuit are shown in Figure 15.22. The top photo shows the small-signal step response, with its rise time (10% to 90%) of 400μs, which is consistent with a 400-Hz crossover. The bottom photo shows the response to a 1-volt step. Note that most of the response time is used for slewing, at about 30 V/μs, to the vicinity of the final voltage level. While the integrator is capable of slewing faster, the error signal from the phase detector has only an average duty cycle of about 50% under this transient condition.

15.4.5 SYNCHRONOUS VFC APPLICATIONS

Two examples of synchronous v/f converter (SVFC) designs are the monolithic Analog Devices AD651 and the hybrid AD379. The AD651 operates at clock frequencies of up to 2 MHz, with linearity within 250 ppm; at lower clock frequencies, linearity approaches 20 ppm. With a 100-kHz clock,

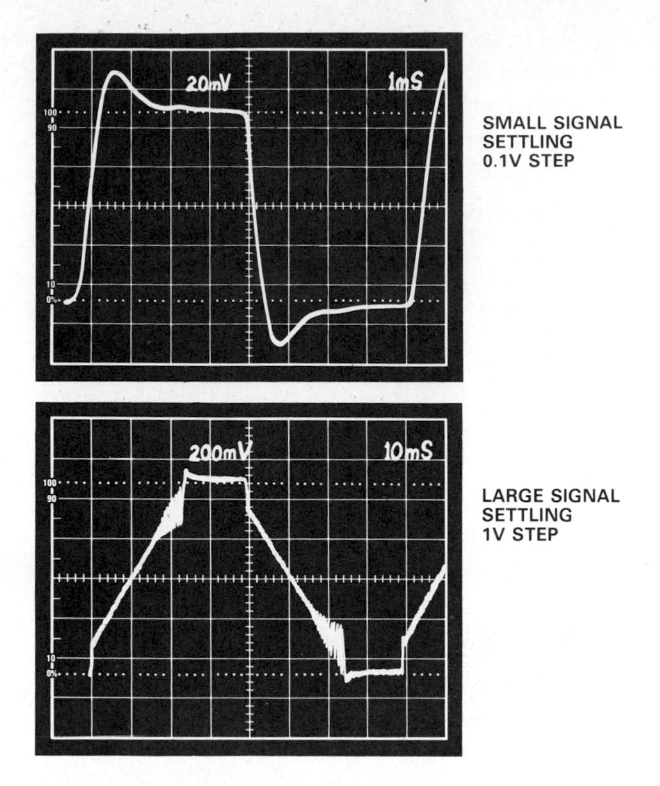

Figure 15.22. Transient response of PLL.

its gain drift can be as low as 40 ppm per °C. It can operate with either a single supply—or in the more-traditional dual-supply mode, with supply voltages as low as 12 volts, or ±6 volts. The hybrid AD379's gain and offset errors are about ten times smaller; it is also capable of handling bipolar input signals—it has a precision rectifier preceding the SVFC section and a sign bit, which indicates the polarity of the input signal.

Because the AD651 is a synchronous VFC, it doesn't require a one-shot in the conversion process. However, there is an on-chip one-shot to control the width of the frequency-output pulses; it drives the open-collector output. This is a useful feature in applications where the output pulse must be shorter than the clock period. For example, opto isolators require significant amounts of current; a very short pulse may be required to save power.

In the most elementary application of the SVFC, a stable clock signal is used, and output pulses are counted over a gated period, to measure frequency. However, if the gate period used to count the output pulses from the SVFC is derived from the clock signal by frequency division, a temperature-stable clock is not necessary. An example of such an application is shown in Figure 15.23.

Since the gain of the SVFC is proportional to the clock frequency, and the gate time is inversely proportional to the same clock frequency, the total

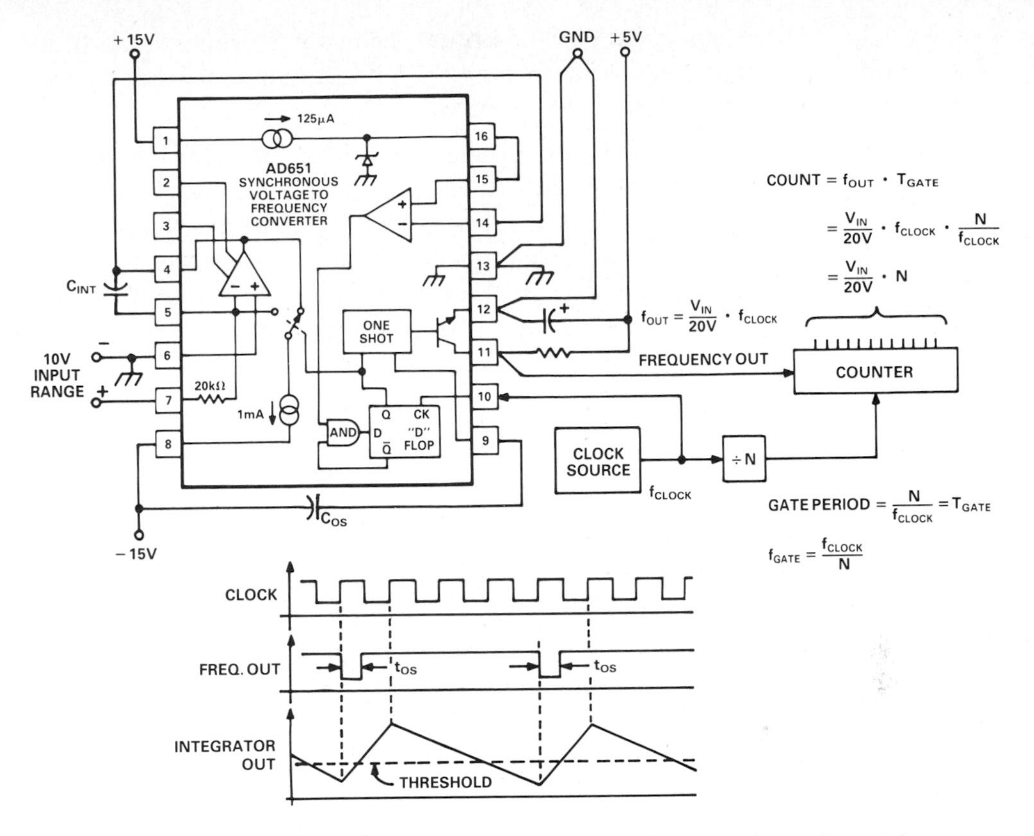

Figure 15.23. Synchronous VFC application. Accumulated count is a ratiometric measure of input voltage, independent of clock frequency, when counter clock and gate drive are derived from the same frequency source.

number of pulses during the gate time (for a given input voltage) will remain constant when the clock frequency changes.

Another application that exploits the proportionality between the SVFC gain and the clock frequency is the simple multiplier circuit of Figure 15.24. In this circuit, the clock frequency is established by the output of another VFC,

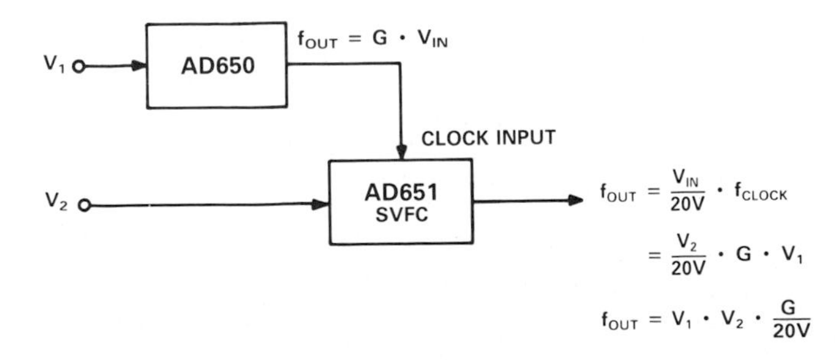

Figure 15.24. Multiplier application employing a free-running VFC to establish the scale factor of a synchronous VFC.

such as the AD650; since the VFC's output frequency is proportional to its input voltage, the frequency output of the SVFC is proportional to the product of the two voltages.

In an interesting system application, which benefits by the AD651's controllable-period one-shot, several data channels are multiplexed onto one digital line, as shown in Figure 15.25. Several SVFC devices, with very short output pulse-widths, are driven from different phases of a multi-phase clock, and all the open collectors are simply wire-or'd together. Thus, the presence or absence of an output pulse at each phase of the clock is identified with one of

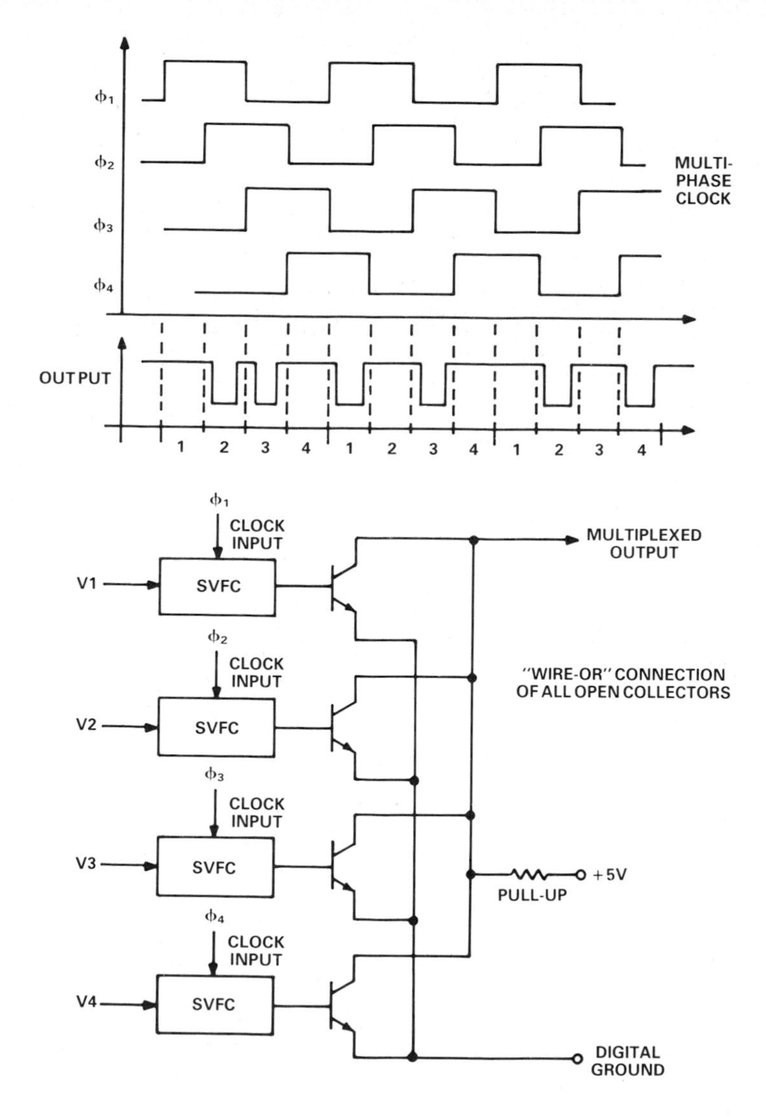

Figure 15.25. Multiplexed application. Each input channel is synchronized to a different clock phase.

the inputs. This arrangement can be quite cost-effective in a data-transmission type of application, such as that shown in Figure 15.12, since a single transmission line and opto-isolator can be shared among many data-channel outputs.

15.5 PRACTICAL MATTERS

15.5.1 DECOUPLING AND GROUNDING

It is good engineering practice to use bypass capacitors directly at the device's supply-voltage pins and to insert small-valued resistors (10 to 100 ohms) in the supply lines to provide a measure of decoupling between the various circuits in a system. Ceramic capacitors (0.1 μF to 1.0 μF) should be connected between the supply-voltage pins and analog signal ground for proper bypassing of the VFC. In addition, a higher-capacitance board-level decoupling capacitor (1 μF to 10 μF) should be located relatively close to the VFC on each power supply line.

Such precautions are imperative in high-resolution data acquisition applications, where one expects to exploit the full linearity and dynamic range of the VFC. Although some types of circuits may operate satisfactorily with power-supply decoupling at only one location on each circuit board, such practice should be strongly discouraged in high accuracy analog design. Devices such as the AD650 and the AD537 have separate digital and analog ground terminals. The emitter of the open-collector frequency-output transistor is the only signal node that should be returned to the digital ground. All other signals are referred to analog ground.

The purpose of the two separate grounds is to allow the high-precision analog signals to be isolated from the digital section of the circuitry; substantial amounts of noise on the digital ground can be tolerated without affecting the accuracy of the VFC. Such ground noise is inevitable when switching the large currents associated with the frequency-output signal.

For instance, at 1 MHz full-scale, it is necessary to use about 500 ohms of pull-up resistance in order to get a rise time fast enough to provide well-defined output pulses. This means that, with a 5-volt logic supply, the open-collector output will draw 10mA. The switching of this much current will surely cause ringing on long ground runs due to self inductance of the wires. For instance, #20 gauge wire has an inductance of about 20nH per inch; a current of 10 mA, switched in 50 ns, will produce a voltage spike of 50mV at the end of 12 inches of 20 gauge wire. With a separate digital ground, the VFC will easily handle these types of switching transients.

A remaining problem will be interference caused by radiation of electromagnetic energy by these fast transients. Typically, voltage spikes produced by inductive switching transients can capacitively couple into other sections of the circuit. Another problem is ringing of ground lines and power supply lines, due to the distributed capacitance and inductance of the wires. Such ringing can also couple interference into sensitive analog circuits.

The best solution to these problems is proper bypassing of the logic supply at the VFC. A $1\mu F$ to $10\mu F$ tantalum capacitor should be connected directly between the supply side of the pull-up resistor and the digital ground pin. The pull-up resistor should be connected directly to the frequency output pin. The lead lengths of the bypass capacitor and the pull up resistor should be as short as possible. The capacitor will supply (or absorb) the current transients, and large ac signals will flow in a physically small loop through the capacitor, pull-up resistor, and frequency-output transistor. It is important that the loop be physically small for two reasons: first, short wires have less self-inductance, and second, smaller loops will not radiate radio-frequency interference as efficiently.

The digital ground should be separately connected to the power-supply ground. Since the leads to the digital power supply are only carrying dc current, they will not produce RFI.[4]

15.5.2 COMPONENT SELECTION

When applying typical monolithic VFCs, the user must select external parameter-setting and support components. Since the input resistor directly affects the scale factor, a high quality resistor must be used, especially if a low gain-temperature coefficient is desired. Do not use a carbon composition resistor here under any circumstances. Resistors of this type have a voltage-dependent nonlinearity which can be greater than the nonlinearity of the VFC, especially a high performance device such as the AD650.

The one-shot capacitor of the AD650 and the multivibrator capacitor of the AD537 also directly affect the scale factor and must be carefully chosen when low temperature drift is required. In general, a mica capacitor is good enough and quite cheap. However, in extreme cases, a ceramic C0G (formerly NP0) capacitor is necessary. The C0G ceramics also provide a bonus; they generally have low dielectric absorption compared to less-expensive ceramics. Teflon, polypropylene, and polystyrene capacitors also have low dielectric absorption, but their temperature coefficients are usually significant.

Dielectric absorption in the one-shot capacitor of charge-balance VFC types can cause difficulties. The problem occurs when the output frequency has been relatively constant for a long period of time and then changes. For some time after the change the capacitor will "remember" the previous average voltage it has seen and will have an apparent leakage current while it soaks

[4]More information on proper grounding and reduction of interference can be found in Ott, H. W., *Noise-Reduction Techniques in Electronic Systems* (New York, John Wiley and Sons, 1976).

at the new average voltage. This leakage current, not used to charge the main timing capacitor, produces an error. The result is that, after a change, the VFC will not immediately go to the proper frequency; it will slowly creep to the final value after a quick step to a nearby value. These effects can also cause excess nonlinearity of the voltage-to-frequency transfer relation, especially where the intrinsic linearity of the device is very good, such as at the low-frequency scales of the AD650.

Dielectric absorption can also create inaccuracies when a poor-quality capacitor is used for the integrating capacitor; however, the effect on linearity is not as severe as the degradation which can be caused by defects in the one-shot capacitor. The reason is that the average voltage seen by the integrating capacitor does not change as much, as a function of frequency, as the average voltage seen by the one-shot capacitor.

However, defects in the integrating capacitor can still create problems for the system designer. For example, consider the situation where one wishes to "squelch" an AD650 (i.e., stop it from oscillating) by applying a negative input voltage, which causes the integrator output to rise up to the positive supply level and saturate. The integrating capacitor will then soak at a voltage equal to the supply potential; when the VFC is taken out of this "standby" mode, the integrating capacitor will remember its previous situation and slowly release absorbed charge as it soaks at its new average voltage. The charge being given up by the integrator capacitor is like a leakage and contributes directly as an input error. As a result the output frequency will not immediately be stable; it will slowly creep to a final value as the integrating capacitor forgets its past.[5]

[5]More information about dielectric absorption can be found in "Understand Capacitor Soakage to Optimize Analog Systems," by R. Pease, *EDN*, Oct. 13, 1982.

Chapter Sixteen

Intentionally Nonlinear Converters

D/A converters map the set of n-bit binary numbers into an equivalent set of 2^n voltages, currents, or gains. In most converters, for a given binary integer—of value, N—the nominal equivalent voltage is proportional (*linear*): In DACs, $V_{OUT} = V_{FS} N/2^n$, where V_{FS} is the nominal full-scale voltage. For example, if $n = 8$ and $N = 200$ (binary value: 11001000), then the nominal voltage corresponds to the value of the ratio, 200/256 of full-scale.

For such converters, the linear relationship is the key specification. It is in wide use because most measurements are linear measurements (we use linear measures for length, weight, voltage, current, power, etc.); but there do exist circumstances for which a nonlinear relationship between a physical quantity and a digital number is necessary or desirable. Some of these will be discussed below.

A nonlinear relationship can be achieved in three basic ways—through nonlinear analog signal conditioning, through lookup tables or nonlinear digital processing, or by means of a nonlinear conversion. Some aspects of nonlinear *analog* signal conditioning were discussed briefly in Chapter 2; A venerable but still useful reference on this subject is the *Nonlinear Circuits Handbook*.[1] Nonlinear *digital* signal processing can be achieved through software and/or hardware operations (see Chapters 5 and 21). The present chapter discusses aspects of the third approach, nonlinear *conversion*, with emphasis on logarithmic multiplying d/a converters. Nonlinear conversions involving transducer variables in trigonometric form i.e, resolvers, are discussed in Chapter 14.

[1] D. H. Sheingold, ed., *Nonlinear Circuits Handbook* (Norwood: Analog Devices, Inc., 1974).

16.1 WHY INTENTIONALLY NONLINEAR CONVERTERS?

Many important applications involve the handling of analog signals over a wide dynamic range. Two important classes of signals requiring such handling are: (1) signals requiring accurate correspondence between physical quantities and numbers, and (2) signals calling for wide dynamic range with a specified accuracy at any level. The difference can be seen by a discussion of two examples.

In numerical machine control (NMC) applications, a typical requirement might be to move a workbed over a total distance of 1 foot, with a minimum increment distance of 0.001″. This implies that the controller has a resolution of 1 part in 12,000 over the working distance. Any point in the range must be accurately located. Thus, a linear-coded D/A converter, with at least 14-bit resolution (1 part in 16,384), would be needed to meet this requirement. A discussion of high-resolution conversion will be found in Chapter 17.

On the other hand, a DAC might be required to set a system gain, with a gain tolerance no better than ±12% (1 dB)*, but over a range of 10,000:1, or 80 dB. For this purpose, an attenuator with (say) 128 2/3-dB steps would seem suitable to map the entire gain function. Since only 128 steps are needed, the digital input to a multiplying d/a converter with an appropriate conversion relationship would need a resolution of only 7 bits ($2^7 = 128$).

In this case, the appropriate conversion relationship is exponential; and the digital input is proportional to the logarithm of the desired gain ratio. Note that, if the conversion were linear, the accuracy would have to be (1/8)/10,000, or 1:18,000, i.e., 17 bits—and considerable accuracy would be wasted at the higher gains.

The ability to perform the desired function at a much lower resolution level provides savings in cost, package size, and system complexity; however, there may be a need for software or a lookup table to provide the digital data in the proper form for conversion, if it does not already exist in that form.

16.2 UNDERSTANDING LOGDACS

A LOGDAC™ is a multiplying d/a converter with gain proportional to the exponential of the digital input. Equal changes of digital input produce equal ratios of analog gain change. When used in the forward path of an op amp (its specified mode of operation), it produces attenuation. Used in the feedback path, it provides gain. Figure 16.1 is the block diagram of an attenuator using a typical LOGDAC; it has an 8-bit μP-bus-compatible digital data input, and it employs decoding logic and a high-resolution DAC to provide nominal 0.375-dB gain steps for 1-bit input changes.

*Using the relationship, $dB = 20 \log_{10}$ (ratio), if the ratio is 1.12, the corresponding logarithmic quantity is 0.98 dB.

LOGDAC™ is a trademark of Analog Devices, Inc.

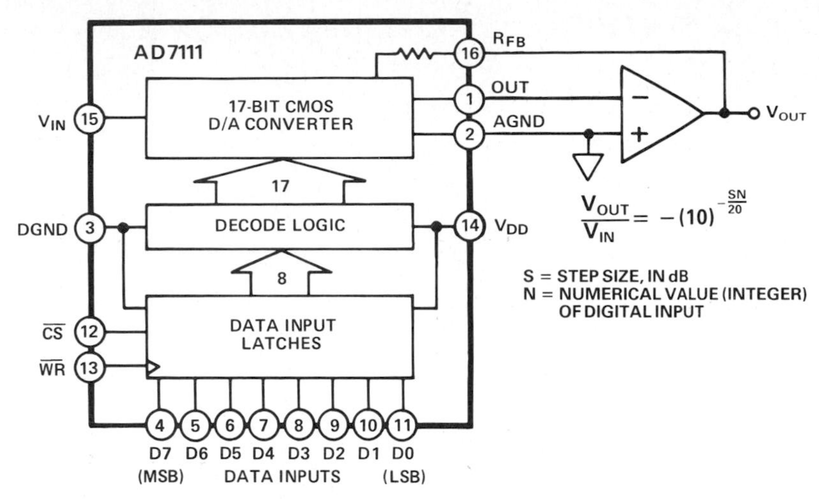

Figure 16.1. Block diagram of 88.5-dB μP-compatible LOGDAC.

To understand the operation of a LOGDAC™, consider first the operation of a linear-coded CMOS D/A converter as an attenuator. The simple block diagram shown in Figure 16.2a treats the D/A converter as a black box with analog input V_{IN}, digital input N and analog output, V_{OUT}. The transfer function is simply

$$V_{OUT} = \alpha V_{IN} \tag{16.1}$$

where α is the attenuation factor.

In a linear DAC, the attenuation factor's relationship to the value, N, of the digital input word is

$$\alpha = \frac{N}{2^n} \tag{16.2}$$

where N is the base-10 integer equivalent to the digital input code and n is the resolution (in bits) of the linear converter. In a LOGDAC, α is an exponential function of the digital input code. This nonlinear relationship affects only the *gain*; the analog input and output are always linearly related, with low distortion.

Figure 16.2b shows the basic current-switching circuitry employed in CMOS DACs (see Chapter 7). Because the input resistance of the R-2R ladder is constant and equal to the ladder resistance, R, regardless of digital input code or signal level, we can define an input signal current, $I_{IN} = V_{IN}/R$. If the lower end of each shunt arm is assumed to be at zero volts, regardless of switch position (in normal operation this assumption is valid), the input current, I_{IN}, will be subdivided equally at each ladder node, as indicated in the diagram. Thus, the current in any shunt arm is one half the current in the next left shunt arm; the input current is divided equally between the first shunt arm and the series arm leading to the rest of the ladder. Each attenuation of one-half is equivalent to -6 dB (actually, -6.02 dB), or $20 \log_{10}(\frac{1}{2})$.

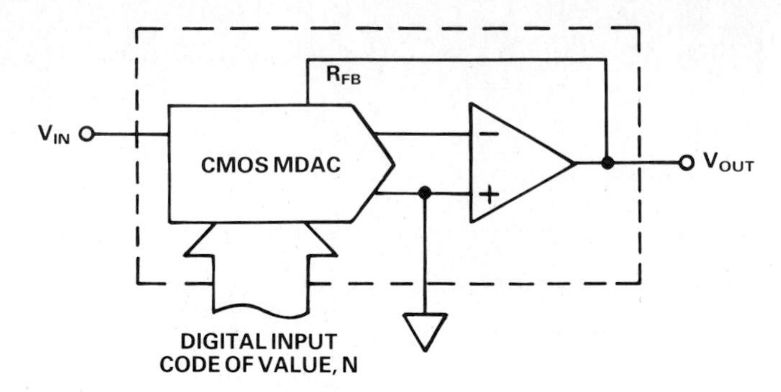

a. Basic application.

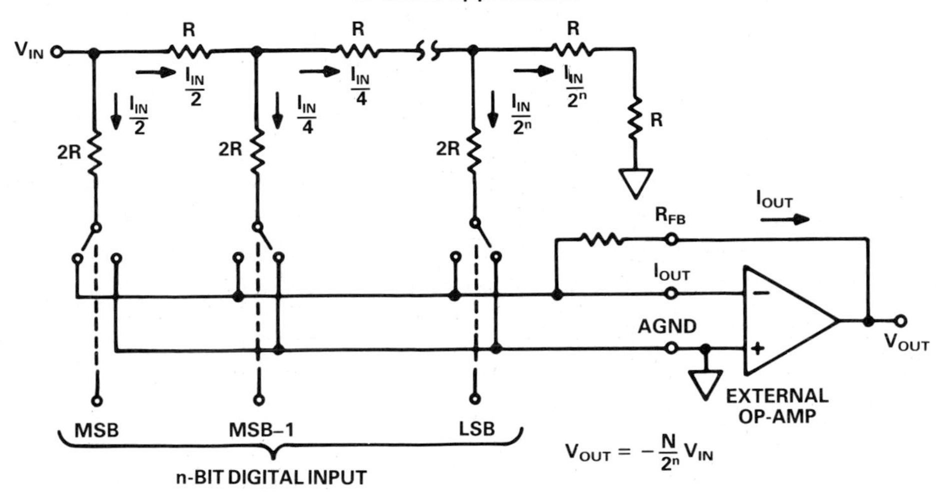

$$V_{OUT} = -\frac{N}{2^n} V_{IN}$$

b. Current division in R-2R ladder.

Figure 16.2. Review of linear multiplying DACs.

A linear converter has a logarithmic relationship requiring no computation or code manipulation, if one considers just the individual bits; since each bit is a power of two, its bit number, starting with the MSB, is simply the negative of the log of its attenuation. Thus, the logarithm of the MSB (ratio to full scale = ½), to base 2, is −1, $\log_2$ of the next bit is −2, etc. Log_2 of the LSB is −n.*

For any linear-coded converter with n-bit resolution, there are 2^n possible input codes. If the all-zeros, or all-bits-off, input code is excluded,† there remain $2^n - 1$ input codes, corresponding to a choice from among $2^n - 1$ output attenuation values. These levels of attenuation are linearly related to the input voltage; each level is separated from the adjoining level by one least significant

*To transform $\log_2$ to $\log_{10}$, multiply by $\log_{10}(2)$, or 0.3010. Thus, $\log_{10}$ (MSB) = −0.3010, $\log_{10}$ (Bit 2) = −.6020, etc. Using the definition, dB = 20 $\log_{10}$ (ratio), the dB measure of the MSB is −6.02 (approximately 6), of Bit 2, −12 dB, etc., down to the LSB, with an approximate dB measure of −6n.

†In a logarithmic device—which is the goal of this transitional discussion—a gain of zero (0-V output), or perfect muting, is not a step in the logarithmic sequence; however, it is often available in actual devices.

bit (LSB), which is a constant percentage of input full-scale range (FSR), since

$$1\,\text{LSB} = \frac{\text{FSR}}{2^n} = \frac{V_{IN}}{2^n} \tag{16.3}$$

For an n-bit converter of this type, the maximum digitally controlled attenuation ratio, from all ones to 1 LSB is

$$20 \log_{10}(2^n - 1) \tag{16.4}$$

For a hypothetical 6-bit device (Figure 16.3), chosen as an example, the range is approximately 36 dB. Each 1-bit increase in the resolution of the converter increases the attenuation range by 6 dB. A 10-bit device has a maximum attenuation range of $10 \times 6\,\text{dB} = 60\,\text{dB}$.

Thus, it is possible to use just the individual bits of a linear DAC for logarithmic operations, without further errors or decoding, if one is willing to put up with the rather coarse 6-dB spacing. However, as Figure 16.3 shows, there are a great many codes available for interpolation, if properly chosen—but they occur mainly at the lower attenuation values.

To achieve constant dB step sizes, i.e., ouput gain changes which are equal percentages of reading, it is necessary to select the closest available individual codes which will yield the proper values of gain.

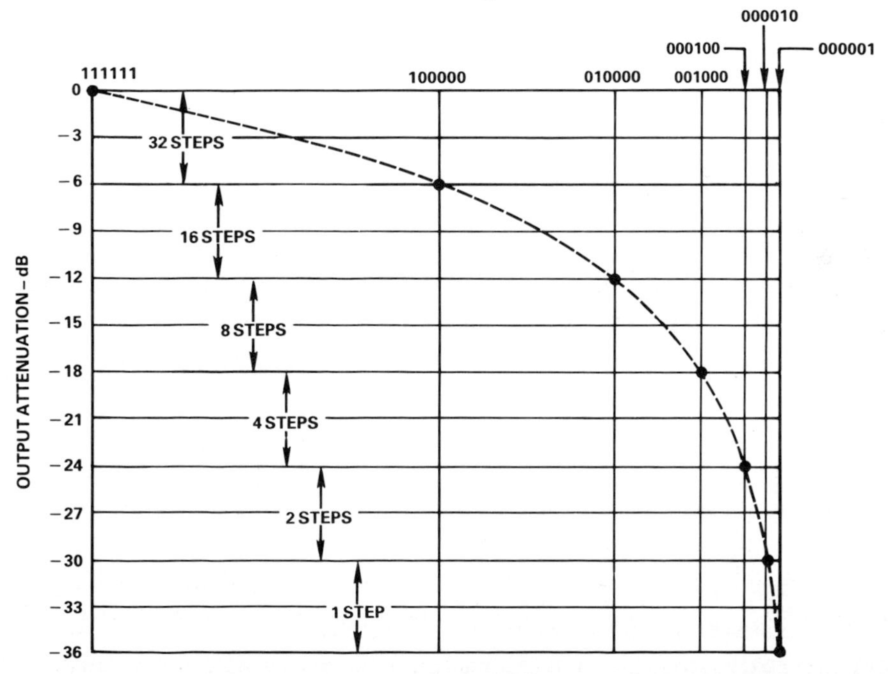

Figure 16.3. Log-scale plot of attenuation at individual bit codes for a linear 6-bit DAC.

For example, suppose we want 3-dB spacing, and are willing to choose the nearest code that interpolates between each pair of accurate 6-dB steps.

Since -3dB corresponds to a ratio of 0.707, the 6-bit code corresponding to -3dB would be about 45/63 (= 0.714), or 101101. The difference in the actual ratios, which is an error due to limited resolution, must be added to errors in the accuracy of the DAC (assume that they are within ± ½ LSB) to establish the expected overall accuracy.

As Table 16.1 suggests, and Figure 16.4 illustrates, the errors, which are quite reasonable at the upper end of the range, and conservative for the 3-dB resolution, become rather large at the lower end of the range.

Required Attenuation dB	Possible Input Code MSB LSB	Actual Attenuation dB	Accuracy Actual Step dB	Possible Output Attenuation Range in dB with Respect to V_{IN}
0	11 11 11	−0.137	+0.069 −0.069	−0.068 to −0.206
−3	10 11 01	−3.06	+0.097 −0.096	−2.963 to −3.156
−6	10 00 00	−6.021	+0.135 −0.137	−5.886 to −6.158
−9	01 01 11	−8.889	+0.187 −0.191	−8.7 to −9.08
−12	01 00 00	−12.041	+0.267 −0.276	−11.774 to −12.317
−15	00 10 11	−15.296	+0.386 −0.404	−14.91 to −15.7
−18	00 10 00	−18.062	+0.527 −0.561	−17.535 to −18.623
−21	00 01 01	−22.144	+0.828 −0.915	−21.316 to −23.059
−24	00 01 00	−24.082	+1.023 −1.16	−23.059 to −25.242
−27	00 00 11	−26.581	+1.339 −1.584	−25.242 to −28.165
−30	00 00 10	−30.103	+1.938 −2.499	−28.165 to −32.602
−33	00 00 01	−36.124	+3.522 −6.021	−32.602 to −42.145

Table 16.1. Output attenuation and errors when 6-bit linear DAC is employed as logarithmic DAC with 3-dB gain steps.

Some interesting observations can be drawn from these results:

1) Highest step accuracy occurs at the smallest value of attenuation; accuracy decreases, expressed in increasing dB of error, with increasing attenuation. At the 6-bit converter's maximum required attenuation level of -33 dB, the actual output can range from -32.6 dB to -42.1 dB, a spread of almost 10 dB.

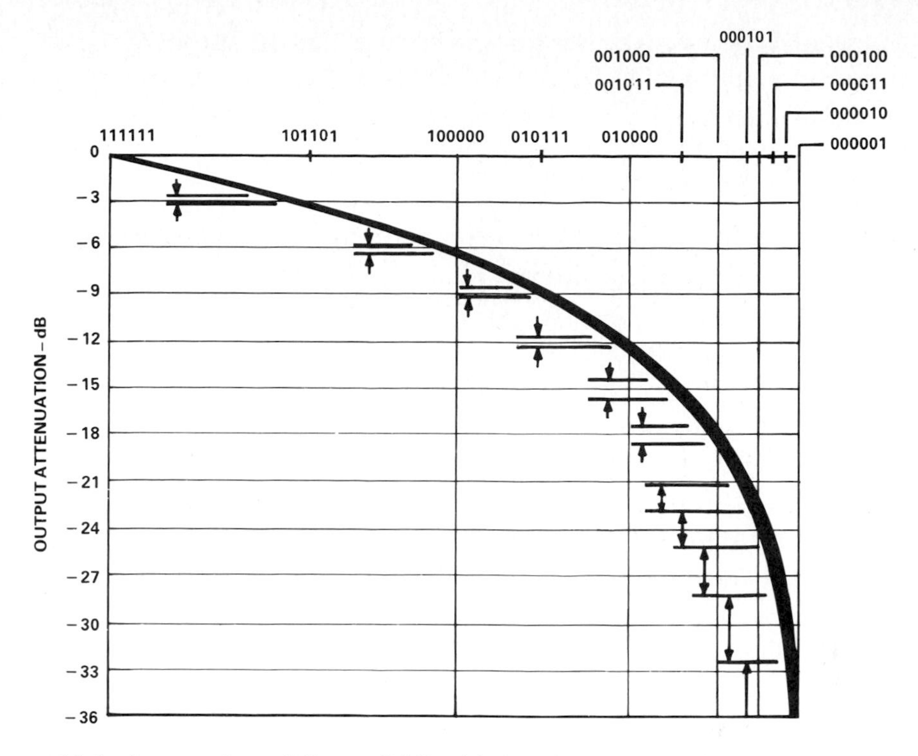

Figure 16.4. Attenuation of linear DAC with key input codes interpolated at 3-dB intervals. Error bands are shown.

2) From attenuation levels of −21 dB and lower, the bottom of one accuracy band aligns with the top of the next and so on. This is because, for steps of −21 dB, −24 dB, −27 dB, −30 dB and −33 dB, so little resolution is available that the digital input code to the DAC can decrease by only one LSB per 3 dB step. Although the accuracy bands align they do not overlap, hence the attenuator, though inaccurate at the low end, is monotonic over the entire range in 3 dB steps.

This should not be entirely unexpected, since we are dealing with a DAC having 6-bit resolution *and* 6-bit accuracy. Such a DAC is inherently monotonic, whether the output is expressed in LSBs or dB. Note that if the DAC were only 5-bit accurate, the error bands—which align in Figure 16.4—could overlap, allowing possible non-monotonic operation at the high attenuation levels. Under those conditions, the attenuator would only be monotonic in 3 dB steps down to a maximum of −21 dB.

3) From Table 16.1, at the 0-dB setting, the range of possible output attenuations is from approximately −0.07 dB to −0.2 dB. This corresponds to the output signal level being (worst case) approximately 2.3% below the input signal level. Note also that the errors are all negative, which means that the circuit is always acting as an attenuator. The reason for this is that at maximum gain (minimum attenuation), in the all-1's condition, the nom-

inal gain is $1 - 2^{-n}$, which is less than unity. The dB output of the attenuator, using a linear 6-Bit DAC, is (where N is the integer value of the binary number)

$$20 \log_{10} \frac{V_{OUT}}{V_{IN}} = 20 \log_{10} \frac{N}{2^n} \tag{16.5}$$

Since $N/2^n$ will always be less than unity, the output signal level will always be less than the input signal level.

4) Only 12 codes out of a possible 63 (again neglecting the all-zeros code) are required for 3-dB spacing, using a 6-bit DAC. The small number of codes employed suggests the possibility of reducing the number of bits in the digital input word from 6 to 4 and using an internal lookup table to select the codes for the R-2R ladder.

These observations for a linear 6-bit DAC performing an attenuation function with equal dB step-sizes are central to an understanding of LOGDAC behavior. The design philosophy behind the LOGDAC is to use an R-2R ladder network of sufficient resolution, concomitant with the required dynamic range, and sufficient accuracy to allow monotonic operation over a desired dynamic range, using defined step sizes. It will be instructive to look at the main features of three different commercially available LOGDACs and to look more closely at the specifications of a representative one, the Analog Devices AD7111, for an interpretation of the important d-c specifications published in the data sheets.

	AD7111	**AD7115**	**AD7118**
Dynamic Range	0 to −88.5 dB	0 to −19.9 dB	0 to −85.5 dB
Step Size	0.375 dB	0.1 dB	1.5 dB
Gain Error	±0.1 dB	±0.1 dB	Not separately specified; included in accuracy specs.
Digital Input	8-bit parallel	2 ½ BCD Coding	6-bit parallel

Both the AD7111 and the AD7118, based upon a 17-bit-resolution R-2R ladder network, have dynamic ranges comparable with 16-bit-linear DACs, but they require only 8- and 6- bits, respectively, of digital input control. To simplify the internal code-conversion logic, which drives the R-2R ladder switches, a scheme is used that differs slightly from the basic one-to-one lookup table mentioned in the 6-bit linear DAC example.[2] However, from the user's point of view, the internal logic design of the devices is likely to be of little consequence.

The AD7115 high gain-resolution attenuator accepts a 2½-digit binary-coded-decimal (BCD) word to provide a dynamic range of 0 to −19.9 dB, in

[2]A full explanation of the actual scheme can be found in Hynes, Michael J. and Burton, D. Philip, "A CMOS Digitally Controlled Audio Attenuator for Hi-Fi Systems," *IEEE Journal of Solid-State Circuits*, Vol. SC-16, Feb. 1981, pp. 15-20.

0.1-dB steps. The AD7115's attenuation is produced by a 12-bit-resolution R-2R ladder network, which is accurate to 12 bits. This device uses a lookup-table technique to convert its BCD input to a 12-bit-wide binary word, which controls the gain.

16.2.1 WHAT THE SPECIFICATIONS MEAN

As mentioned earlier, the AD7111 provides a dynamic range of 0 to −88.5 dB, controlled by an 8-bit-wide digital input word. Nominal gain resolution, or step size, for this device is 0.375 dB. A tabulation of ideal attenuation, in dB, as a function of input code, is shown in Table 16.2. The four more-significant bits (D7 through D4, corresponding to the MSB through bit 4) select the rows; the four less-significant bits choose the columns. The all-zeros condition represents *0 dB of attenuation, i.e., unity gain.** The input codes selected by the last row, 1111XXXX (FX_H) represent a mute condition: no throughput, or infinite attenuation; this is equivalent to an all-zero code in a conventional linear DAC, such as the 6-bit device in the earlier example.

D7–D4 \ D3–D0	0000	0001	0010	0011	0100	0101	0110	0111	1000	1001	1010	1011	1100	1101	1110	1111
0000	0.0	0.375	0.75	1.125	1.5	1.875	2.25	2.625	3.0	3.375	3.75	4.125	4.5	4.875	5.25	5.625
0001	6.0	6.375	6.75	7.125	7.5	7.875	8.25	8.625	9.0	9.375	9.75	10.125	10.5	10.875	11.25	11.625
0010	12.0	12.375	12.75	13.125	13.5	13.875	14.25	14.625	15.0	15.375	15.75	16.125	16.5	16.875	17.25	17.625
0011	18.0	18.375	18.75	19.125	19.5	19.875	20.25	20.625	21.0	21.375	21.75	22.125	22.5	22.875	23.25	23.625
0100	24.0	24.375	24.75	25.125	25.5	25.875	26.25	26.625	27.0	27.375	27.75	28.125	28.5	28.875	29.75	29.625
0101	30.0	30.375	30.75	31.125	31.5	31.875	32.25	32.625	33.0	33.375	33.75	34.125	34.5	34.875	35.25	35.625
0110	36.0	36.375	36.75	37.125	37.5	37.875	38.25	38.625	39.0	39.375	39.75	40.125	40.5	40.875	41.25	41.625
0111	42.0	42.375	42.75	43.125	43.5	43.875	44.25	44.625	45.0	45.375	45.75	46.125	46.5	46.875	47.25	47.625
1000	48.0	48.375	48.75	49.125	49.5	49.875	50.25	50.625	51.0	51.375	51.75	52.125	52.5	52.875	53.25	53.625
1001	54.0	54.375	54.75	55.125	55.5	55.875	56.25	56.625	57.0	57.375	57.75	58.125	58.5	58.875	59.25	59.625
1010	60.0	60.375	60.75	61.125	61.5	61.875	62.25	62.625	63.0	63.375	63.75	64.125	64.5	64.875	65.25	65.625
1011	66.0	66.375	66.75	67.125	67.5	67.875	68.25	68.625	69.0	69.375	69.75	70.125	70.5	70.875	71.25	71.625
1100	72.0	72.375	72.75	73.125	73.5	73.875	74.25	74.625	75.0	75.375	75.75	76.125	76.5	76.875	77.25	77.625
1101	78.0	78.375	78.75	79.125	79.5	79.875	80.25	80.625	81.0	81.375	81.75	82.125	82.5	82.875	83.25	83.625
1110	84.0	84.375	84.75	85.125	85.5	85.875	86.25	86.625	87.0	87.375	87.75	88.125	88.5	88.875	89.25	89.625
1111	MUTE	MUTE	MUTE	MUTE	MUTE	MUTE	MUTE	MUTE	MUTE	MUTE	MUTE	MUTE	MUTE	MUTE	MUTE	MUTE

Table 16.2. Theoretical attenuation in dB vs. digital code for AD7111 LOGDAC.

There are some significant differences between the ideal attenuator, of Table 16.2, and real attenuators. For example, the last three input codes of the AD7111 (1110 1101, 1110 1110, and 1110 111), representing attenuations from 88.875 dB to 89.625 dB, are not recommended for use because of possible nonmonotonic behavior of the associated output levels. The code corresponding to the greatest usable attenuation, short of muting, is 1110 1100 (EC_H) corresponding to an attenuation level of 88.5 dB. Other differences will be found in a comparison between the ideal performance of Table 16.2 and the device specifications, in Table 16.3.

Before performing the comparisons, it may be instructive to consider what resolution and accuracy would be required for an R-2R ladder network to

*The all-zeros condition for LOGDACS corresponds essentially to all-ones for linear DACS, since minimum attenuation means maximum gain.

meet the ideal performance figures in practice. Assume that the ideal accuracy of any step (or level) in Table 16.2 is to within one-half the step size, i.e., ±0.17 dB. It can be shown that an R-2R ladder with 17-bit resolution and 22-bit accuracy is required. Such DACs are difficult to procure, especially if one seeks them in the form of 16-pin plastic DIPs! It should therefore come as no surprise to find that the actual performance of practical, low-cost logarithmic DACs falls somewhat short of ideal.

Parameter	AD7111L/C/U GRADES $T_A = +25°C$	AD7111L/C/U GRADES $T_A = T_{min}, T_{max}$	AD7111K/B/T GRADES $T_A = +25°C$	AD7111K/B/T GRADES $T_A = T_{min}, T_{max}$	Units	Conditions/Comments
NOMINAL RESOLUTION	0.375	0.375	0.375	0.375	dB	
ACCURACY RELATIVE TO 0dB ATTENUATION						
0.375dB Steps:						
Accuracy ≤ ±0.17dB	0 to 36	0 to 36	0 to 30	0 to 30	dB min	Guaranteed attenuation ranges
Monotonic	0 to 54	0 to 54	0 to 48	0 to 48	dB min	for specified step sizes
0.75dB Steps:						
Accuracy ≤ ±0.35dB	0 to 48	0 to 42	0 to 42	0 to 36	dB min	
Monotonic	0 to 72	0 to 66	0 to 72	0 to 60	dB min	
1.5dB Steps:						
Accuracy ≤ ±0.7dB	0 to 54	0 to 48	0 to 48	0 to 42	dB min	
Monotonic	Full Range	0 to 78	0 to 85.5	0 to 72	dB min	Full Range is from 0 to 88.5dB
3.0dB Steps:						
Accuracy ≤ ±1.4dB	0 to 66	0 to 54	0 to 60	0 to 48	dB min	
Monotonic	Full Range	Full Range	Full Range	Full Range		
6.0dB Steps:						
Accuracy ≤ ±2.7dB	0 to 72	0 to 60	0 to 60	0 to 48	dB min	
Monotonic	Full Range	Full Range	Full Range	Full Range		
GAIN ERROR	±0.1	±0.15	±0.15	±0.20	dB max	
V_{IN} INPUT RESISTANCE (PIN 15)	9/11/15	9/11/15	7/11/18	7/11/18	kΩ min/typ/max	
R_{FB} INPUT RESISTANCE (PIN 16)	9.3/11.5/15.7	9.3/11.5/15.7	7.3/11.5/18.8	7.3/11.5/18.8	kΩ min/typ/max	

Table 16.3. Accuracy specifications of a LOGDAC (AD7111).

Table 16.3 is the section of the Specifications page from the AD7111 Data Sheet pertaining to device accuracy. Other specifications, such as logic input levels and logic input timing for the data-input latches, are omitted here.

Note that the accuracy is specified as relative to 0 dB attenuation and not with respect to V_{IN}. The accuracy of the 0 dB setting is established by the gain-error specification of ±0.1 dB at +25°C and ±0.15 dB over the temperature range (L/C/U grades). This is similar to a linear DAC specification, in which accuracy and gain error are specified separately. For both LOGDACs and linear DACs, gain error is measured with full-scale (FS) output, i.e., at 0 dB attenuation for a LOGDAC and with all 1's on the digital inputs for a unipolar binary-coded DAC.

Gain error results from a mismatch between R_{FB} (the feedback resistance) and the R-2R ladder resistance; its effect in a LOGDAC is to produce a constant additive attenuation error in dB over the whole range. Since the gain error of CMOS multiplying DACs is normally less than 1%, the accuracy error contribution due to gain-error effects is typically less than 0.09 dB.

For the AD7111 and the AD7115, which have step sizes comparable to the gain-error contribution, it is especially desirable to separate the specifications of gain error at 0 dB—and device accuracy, relative to 0 dB attenuation. On the other hand, for the AD7118, which has a nominal step size of 1.5 dB, much greater than any gain-error contribution, the accuracy is specified relative to V_{IN}, and gain-error is included in the accuracy specification. This is similar to a specification of total unadjusted error for a d/a converter.

For simplicity in the earlier discussion of the 6-bit DAC, gain error was not considered; the DAC was assumed to have negligible gain error (not an unwarranted assumption, at the 6-bit level). Thus, the attenuation data of Table 16.1 (and its graphical representation in Figure 16.4) is specified with respect to V_{IN}.

In Table 16.3, the accuracy specification over the full 0 to −88.5-dB attenuation range is divided into various subgroups. Each subgroup specifies—for a given step size—two ranges of attenuation:

the range for which the error is always less than the specified amount
the range for which the device is always monotonic.

To illustrate the point, consider the first subgroup for the AD7111L/C/U grades at +25°C; from 0 dBto 36 dB, the attenuation can be increased in 0.375-dB steps with a step accuracy, relative to 0 dB, of ±0.17 dB. Increasing the attenuation beyond 36 dB, in 0.375 dB steps, can result in attenuation steps with accuracy bands wider than ±0.17 dB relative to 0 dB. The associated monotonic range of 0 dB to 54 dB suggests that, from 36 dB to 54 dB, the accuracy of each increasing step is deteriorating, yet the output level will always respond in the proper direction to a 0.375 dB step increase or decrease. Within each subgroup, the step accuracy specification is chosen to be less than one half of the relevant step size.

If the attenuation is increased beyond 54 dB, in 0.375-dB steps, the accuracy bands of adjoining 0.375 dB steps can overlap each other, causing non-monotonic performance.

The difference between the V_{IN} and the R_{FB} input resistance specifications is worth noting. For a linear-coded DAC the feedback resistance, R_{FB}, and the R-2R ladder resistance, R, are made equal—or as nearly equal as possible. The equality was assumed for the linear 6-Bit DAC example and its importance can be easily seen from Figure 16.2 where:

$$
\begin{aligned}
V_{OUT} &= -I_{OUT} R_{FB} \\
&= -\alpha I_{IN} R_{FB} \\
&= -\alpha V_{IN} \frac{R_{FB}}{R} \qquad (16.6)
\end{aligned}
$$

Equation (16.1) assumes that $R = R_{FB}$. It was noted earlier that, since α is always less than unity, because the gain at all-1's is not equal to full scale, an ideal ($R_{FB} = R$) 6-Bit linear DAC will produce an actual attenuation at its "0 dB" setting that can range from approximately −0.07 dB to −0.2 dB.

Equation (16.6) suggests that, by a deliberate increase in the ratio, R_{FB}/R, the gain can be increased, shifting the entire DAC transfer function towards 0 dB. With the right ratio, this corrects the negative skew of the 0-dB code and results in a distribution of the "0 dB" error band around 0 dB. The typical ratio of R_{FB} to R for Analog Devices LOGDACs is:

AD7111;	1.044 : 1
AD7115;	1.04 : 1
AD7118;	1.05 : 1

Since the all-1's code is nominally within 1 LSB of full scale, and the AD7111 and the AD7118 employ a 17-bit-resolution R-2R ladder, while the AD7115 is based upon a 12-bit-resolution R-2R ladder, one would expect the R_{FB}-to-R ratios for the AD7111 and the AD7118 to be much less than the equivalent ratio for the AD7115 if a similar internal decoding technique were used in all three. The ratios are actually greater because differing decoding techniques are used for the 17-bit and the 12-bit R-2R ladders.

16.3 APPLICATIONS OF LOGDACS

The LOGDAC is a multiplying d/a converter with gain that is exponentially related to the digital input. The analog input can be voltage or current, ac or dc, positive or negative in polarity. This flexibility suggests a variety of applications, including dB-programmable attenuators and amplifiers, logarithmic a/d converters, level-independent automatic gain control, wide-range programmable state-variable filters, digitally programmable oscillators, distortion generators, audio panners, etc. Dynamic range, at a given accuracy level, can be extended by using fixed blocks of attenuation with the programmable attenuation provided by the LOGDAC.

In this section, we will consider some basic configurations of LOGDACs, from which can be derived many circuits with wider ramifications. Treated here will be exponential attenuators and amplifiers, logarithmic a/d converters, and range extension for attenuators. Considerable additional information is available from manufacturers in the form of data sheets, application notes, and applications-engineering advice.

16.3.1 BASIC ATTENUATOR OR DIGITAL POTENTIOMETER

Figure 16.5a shows the basic connection for using the LOGDAC as a digitally controlled attenuator. The nominal response of this circuit is

$$V_{OUT} = -(10)^{-\frac{NS}{20}} V_{IN} \qquad (16.7)$$

where S is the step size, in dB, and N is the integer value of the binary or BCD digital input, over the specified range for which device performance is valid.

Expressed as a digitally controllable gain for an analog signal, V_{IN},

$$\text{GAIN} = \frac{V_{OUT}}{V_{IN}} = -(10)^{-\frac{NS}{20}} \tag{16.8}$$

For example, if the step size is 0.1 dB, as in the AD7115, and N = 100, then $-V_{OUT}/V_{IN} = 10^{-1/2} = 0.316$. In general, the response is of the form shown in Figure 16.5b; equal increases of N result in equal (increasing attenuation or decreasing gain) ratios of output.

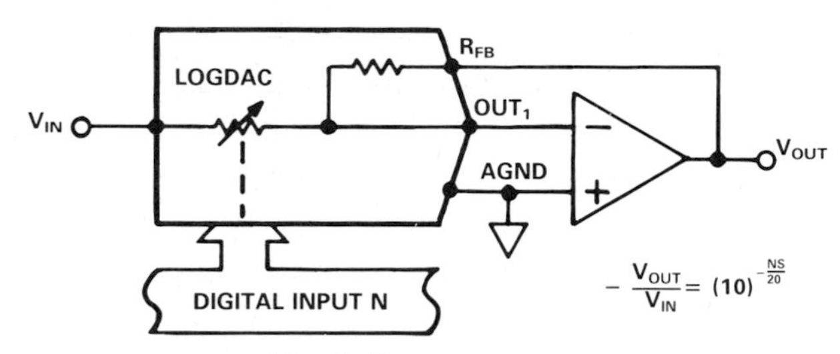

a. Block diagram of circuit.

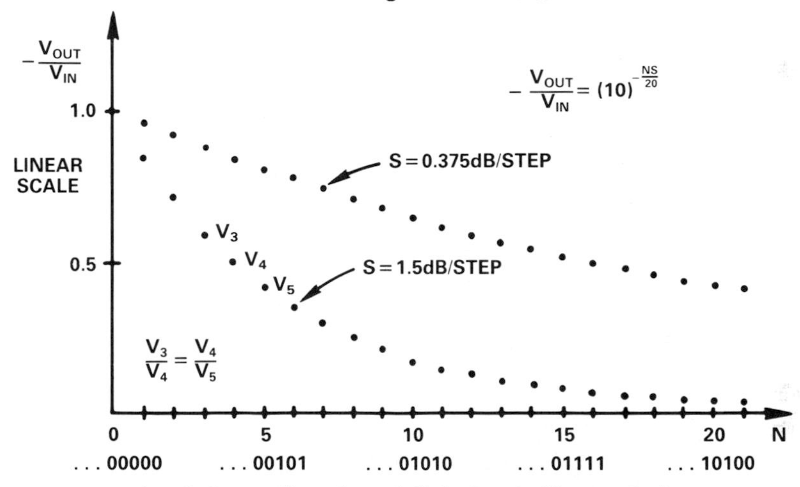

b. Gain as a function of digital code (linear plot).

Figure 16.5. Application of LOGDAC as attenuator.

16.3.2 EXPONENTIAL AMPLIFIER OR POT WITH FEEDBACK

If the LOGDAC is used as the feedback element, as in Figure 16.6a, the output and input change sides of the equation; now the output must generate whatever voltage is necessary such that, when attenuated by NS dB, it produces a current equal to the input current. The result is that the polarity of the exponent becomes positive, and each step of N produces an increment of gain, starting from 1.00 (i.e., 0 dB) at $N = 0$:

$$V_{OUT} = -(10)^{+\frac{NS}{20}} V_{IN} \tag{16.9}$$

Using the same example as in the case of the attenuator, with a 0.1-dB step size and $N = 100$, then $-V_{OUT}/V_{IN} = 10^{1/2} = 3.16$. A typical response function is shown in Figure 16.6b. Again, equal changes in N result in equal (but increasing) ratios of output gain.

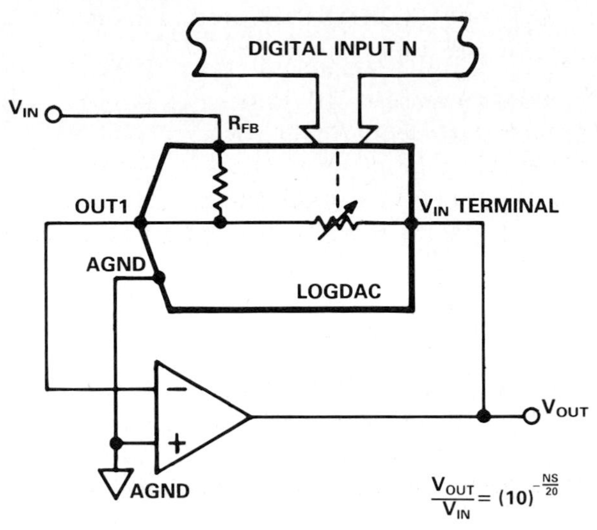

a. Block diagram.

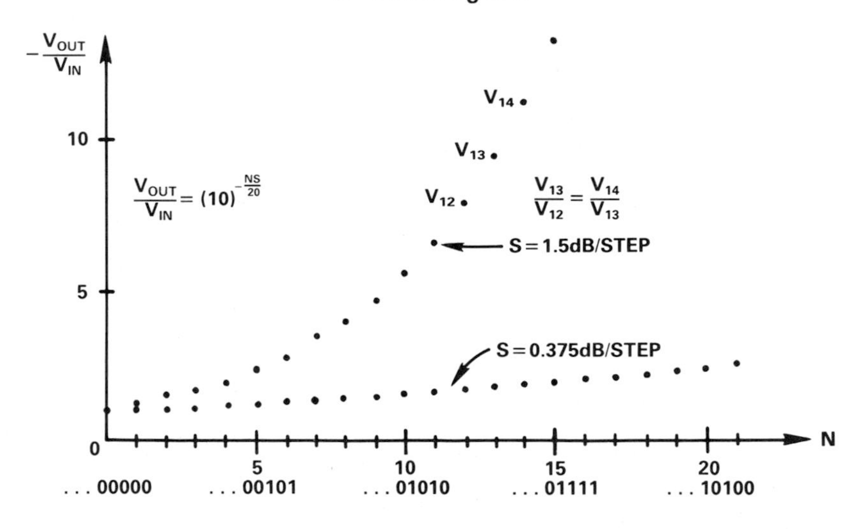

b. Gain as a function of digital code (linear plot).

Figure 16.6. Application of LOGDAC as amplifier with digitally controlled gain in dB steps.

By using fixed gain or attenuation in conjunction with the circuits of Figures 16.5 and 16.6, the overall attenuation or gain can be programmed to start from a level different from 0 dB. For example, in the case of Figure 16.5, an associated gain (say) of 10 volts/volt in cascade with the LOGDAC and its amplifier will result in a maximum gain of 20 dB, occurring at the 0-dB code, with the normal complement of S-dB steps. Thus, the dynamic range of gain remains the same, but its starting point is 20 dB higher. In such applications, it may be tempting to consider economizing on amplifiers by using series or parallel resistance with R_{FB}, but with an extra amplifier stage, you can avoid the consideration of matching and temperature-tracking problems that is necessary when the on-chip resistors are connected with external resistors.

Naturally, in all cases, with or without additional gain or attenuation, the limits to dynamic range presented by amplifier output bounds and input noise levels must be considered. Within these ranges, however, the LOGDACs, like all CMOS multiplying DACs, are able to handle a wide range of positive and negative analog signals with wide bandwidth and low distortion and noise—once the gain is digitally set.

16.3.3 LOGARITHMIC A/D CONVERTER

The LOGDAC can be used to replace the linear DAC in a successive-approximation or tracking-type converter. In each case, the conversion loop chooses a code that reduces the unbalance to within 1 LSB of the analog value; when that state is reached, the digital input to the LOGDAC is proportional to the log of the input, that is

$$V_{IN} = -(10)^{-\frac{SN}{20}} V_{REF}$$

Therefore, the code represents the digital value,

$$N = -\frac{20}{S} \log_{10}\left(-\frac{V_{IN}}{V_{REF}}\right) \quad (16.10)$$

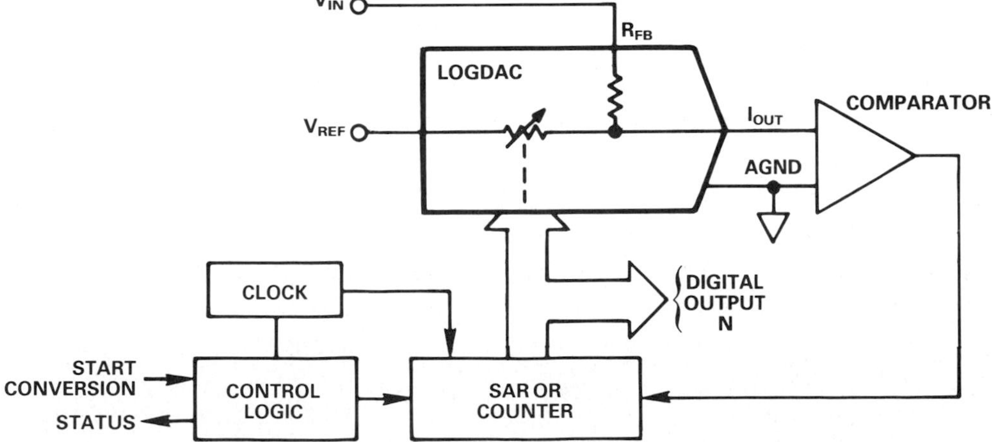

Figure 16.7. Application of LOGDAC as logarithmic (log-ratio) a/d converter.

Again, within the system limitations of dynamic range and accuracy, this type of a/d conversion allows a wide dynamic range of input to be compressed into a relatively small digital resolution. Advantages might include single-precision 8-bit bus operation and fast floating-point conversion without the need for programmable-gain amplifiers and repeated conversions (logarithmic conversion in inherently a limited floating-point operation).

Logarithmic DACs can also be used in automatic gain-control loops, which function essentially as tracking a/d converters.[3]

16.3.4 RANGE EXTENSION

A popular requirement is to extend the attenuation range. The purpose may be to extend a limited range, as in the case of the AD7115—perhaps to double its inherent range of 20 dB to 40 dB, in 0.1-dB steps. Or one may wish to extend the most-accurate range—for example, to double the 0 to 36-dB range of the AD7111 to 0 to 72 dB, while maintaining the same basic accuracy specification for 0.375-dB steps.

To achieve this, one or more stages of switched fixed-gain attenuation or programmable-gain amplification are made available in cascade with the LOGDAC to provide the range of total attenuation required, when combined with the attenuation range of the LOGDAC. At low levels of system attenuation (within the basic LOGDAC's accurate attenuation range), the precision attenuator is switched out of the circuit and contributes 0 dB attenuation to the signal path, all attenuation being controlled by the LOGDAC. At some user-defined input code, the precision attenuator switches in an attenuation equal to that defined by the input code, while the LOGDAC code is reset to all 0's to give 0 dB attenuation. As input codes increase further, to call for increased attenuation, the LOGDAC again controls the incremental attenuation as required.

This technique extends the required range of accurate performance by the amount introduced by the precision attenuator, assuming that it introduces no additional errors itself. For example, the AD7118 (L/C/U grades), with a step size of 1.5 dB, has a step accuracy relative to V_{IN} of ± 0.35 dB, from 0 dB to -30 dB, increasing to ± 0.7 dB from -31.5 dB to -48 dB. A table of ideal attenuation vs. input code for the AD7118 is shown in Table 16.4.

To extend the specified dynamic range by, say, 24 dB, the input code change from $N = 15$ to $N = 16$, i.e., 001111 ($0F_H$) to 010000 (10_H), corresponding to an attenuation level change from 22.5 dB to 24 dB, provides an opportunity to implement the scheme. If the AD7118 is driven only by the 4 less significant bits, and a range switch (switching in a 24-dB precision attenuator on "1") is driven by the next largest bit, the desired performance will be obtained for

[3]John Wynne, "Level-Independent Automatic Gain Control," *Analog Dialogue* 17-1 (1983), pp. 16-17.

N	Digital Input D5 D0	Attenuation dB	N	Digital Input	Attenuation
0	00 00 00	0.0	31	01 11 11	46.5
1	00 00 01	1.5	32	10 00 00	48.0
2	00 00 10	3.0	33	10 00 01	49.5
3	00 00 11	4.5	34	10 00 10	51.0
4	00 01 00	6.0	35	10 00 11	52.5
5	00 01 01	7.5	36	10 01 00	54.0
6	00 01 10	9.0	37	10 01 01	55.5
7	00 01 11	10.5	38	10 01 10	57.0
8	00 10 00	12.0	39	10 01 11	58.5
9	00 10 01	13.5	40	10 10 00	60.0
10	00 10 10	15.0	41	10 10 01	61.5
11	00 10 11	16.5	42	10 10 10	63.0
12	00 11 00	18.0	43	10 10 11	64.5
13	00 11 01	19.5	44	10 11 00	66.0
14	00 11 10	21.0	45	10 11 01	67.5
15	00 11 11	22.5	46	10 11 10	69.0
16	01 00 00	24.0	47	10 11 11	70.5
17	01 00 01	25.5	48	11 00 00	72.0
18	01 00 10	27.0	49	11 00 01	73.5
19	01 00 11	28.5	50	11 00 10	75.0
20	01 01 00	30.0	51	11 00 11	76.5
21	01 01 01	31.5	52	11 01 00	78.0
22	01 01 10	33.0	53	11 01 01	79.5
23	01 01 11	34.5	54	11 01 10	81.0
24	01 10 00	36.0	55	11 01 11	82.5
25	01 10 01	37.5	56	11 10 00	84.0
26	01 10 10	39.0	57	11 10 01	85.5
27	01 10 11	40.5	58	11 10 10	87.0
28	01 11 00	42.0	59	11 10 11	88.5
29	01 11 01	43.5	60	11 11 XX	∞
30	01 11 10	45.0			

NOTES
X = 1 or 0. Output is fully muted for $N \geqslant 60$
Monotonic operation is not guaranteed for N = 58, 59

Table 16.4. Ideal attenuation vs. input code for the AD7118.

precision attenuations of up to a maximum possible of 46.5 dB, since—at the next code change—the input code to the AD7118 is reset to all 0's and the precision 24 dB attenuator is switched out of the signal path. Maximum attenuation range has been traded for increased accuracy. Figure 16.8a illustrates the basic scheme and (b) illustrates a circuit that will implement the switched attenuator.

Differing circuit configurations can be used, depending upon the performance required (i.e., accuracy and total range), employing more or less of the attenuation range of the LOGDAC, one or more steps of switched precision attenuation, and magnitude of the external attenuation. For example, if a wider dynamic range desired, a dedicated control bit for the precision attenu-

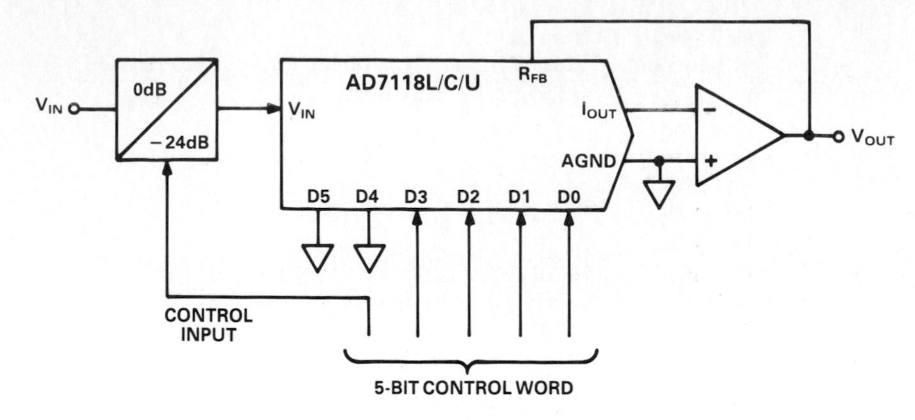

a. Block diagram.

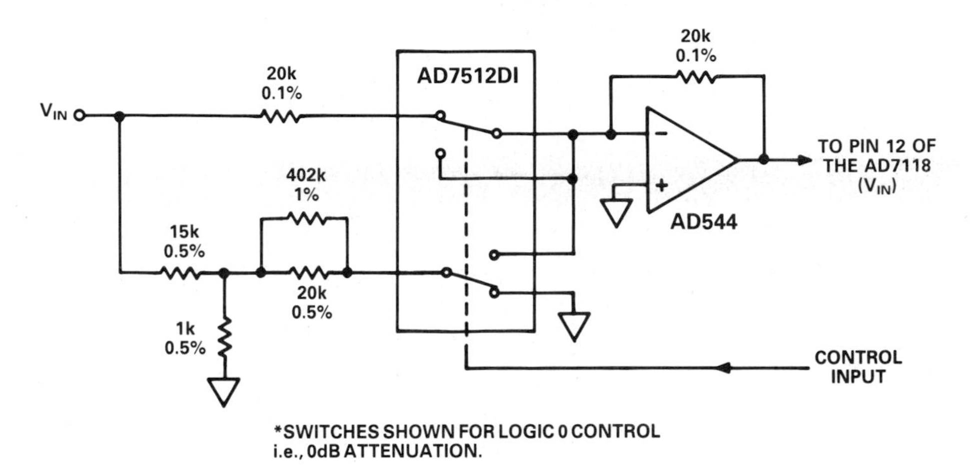

b. Fixed attenuator.

Figure 16.8. Application of LOGDAC with switched fixed attenuation to extend range of accurately set gain.

ator can be added to the AD7118's entire 6-bit control word to provide a (theoretical) dynamic range from 0 dB to − 109.5 dB.

Whatever configuration is used, the step accuracy over the extended range will be a function of the errors of both the LOGDAC and the precision attenuator. A fixed attenuation step, when switched in, should be large enough in relation to that for the prior code, so that the transition (for example, from 01111 to 10000, in Figure 16.8), will be monotonic.

The idea of augmenting the LOGDAC's range can be extended by replacing the fixed-value attenuator with a variable attenuator, which might even be a d/a converter. A DAC as an attenuator could either provide large programmable chunks of attenuation for range extension, or interpolate a fine trim between the LOGDAC's coarse (e.g., 1.5-dB) steps, for range intensification.

Chapter Seventeen

High-Resolution Data Conversion

This chapter deals with the problems, challenges, and opportunities afforded by conversion devices and techniques that result in high resolutions and accuracies. High-resolution converters are a special breed, and—with the possible exception of very high-speed converters—they have been the most likely candidates for custom design to meet unusually demanding system requirements. However, converter manufacturers are rising to the challenge, and a growing number of high-resolution-and-accuracy converters are becoming available at competitive prices.

Converters meeting this definition guarantee resolutions beyond 17 bits or at least 16 bits of resolution with comparable accuracy. The principal requirement of a 16-bit-accurate converter is an integral nonlinearity specification of $\pm \frac{1}{2}$ LSB, or 7.6 parts per million (ppm). Data converters with resolutions of 17 bits and beyond, even without a commensurate improvement in accuracy, can still offer the wide dynamic range required for many signal processing applications.

The definition of "high resolution" has been shifting upward as data-converter design and IC process technology improve. High resolution, a dozen years ago, was considered to be 12-bits or greater. Sixteen bits of resolution was a laboratory curiosity for designers with undeniable need to brave the world of microvolts. Today, 12-bit resolution and accuracy are available in low-cost ICs (despite gloomy predictions by some hardshelled converter design experts), and 16-bit converters have inherited the mantle of "high resolution".

At this writing, although 16-bit monolithic, hybrid and modular converters fill pages of manufacturers' data books, few offer true 16-bit performance. Most are "16/14" bit converters with only fourteen bits of accuracy. The non-monotonic (or missed codes) behavior of many of these devices falls so far short of 16 bits of real resolution that one might wonder why users want the extra pins. One reason may be the hope that—some day—a device with real 16-bit performance will evolve that can be plugged into the same socket; another is that the low cost of many of these devices has led to their use in applications where designers, not needing the sometimes illusory dynamic range that "16-bit" devices seem to offer, use them to get *true 12-bit* performance with less need to trim the converter.

17.1 APPLICATIONS

Applications for high resolution thus fall into two basic categories: high accuracy and wide dynamic range. High-accuracy applications include: automatic test equipment (ATE), calibration standards, instrumentation, process control, and precision positioning. High-accuracy applications usually involve low bandwidth; the key specifications are linearity, offset, and gain error. Designs calling for converters with wide dynamic range, without especial regard for dc or gain accuracy, include such applications as digital audio and waveform reconstruction; they require low total harmonic distortion (THD) and a high signal-to-noise ratio (S/N).

A major use for high-resolution converters is in testing lower-resolution converters. To test an a/d converter or d/a converter accurately requires another converter with at least two additional bits of resolution and accuracy. For example, in DAC testing, a digital code is fed to both the the reference DAC and the converter being tested, and the outputs from both converters are connected to a differential amplifier. Its amplified voltage output, the difference between the output of the reference DAC and the converter being tested, corresponds to the error of the DAC being tested, since the reference converter is "known" to be accurate. As a rule, accurate 16-bit DACs are used to test 12-bit DACs, and accurate (or calibrated) 18-bit-or-better DACs are used to test 16-bit DACs.

Semiconductor equipment manufacturers use high-resolution-converters in computer-based IC processing machines. An example of this is a beam steering application in electron-beam lithography. To shrink device feature sizes, say from 3 to 1 microns, calls for a threefold improvement in the accuracy of the components used in the processing equipment. An electron beam is used, either to make the mask for standard photolithographic processing or—directly steered across the surface of the wafer—to write the appropriate patterns on the die.

A computer controls the shape, intensity, focus and deflection of the electron beam as it runs the gauntlet of lenses and electrostatic and electromagnetic deflection stages (Figure 17.1). High-resolution d/a converters provide analog signals to control the beam's deflection, in the same way as in a CRT. To resolve 1-micron (1-μm) features to 0.25 microns, for a die size of 5 millimeters, the beam must be steered with a resolution and accuracy of better than 1:20,000. Since there will be additional error sources, the resolution of a 16-bit (or better) converter, 1:65,536, is needed. The wafer is positioned by an X-Y positioning table.

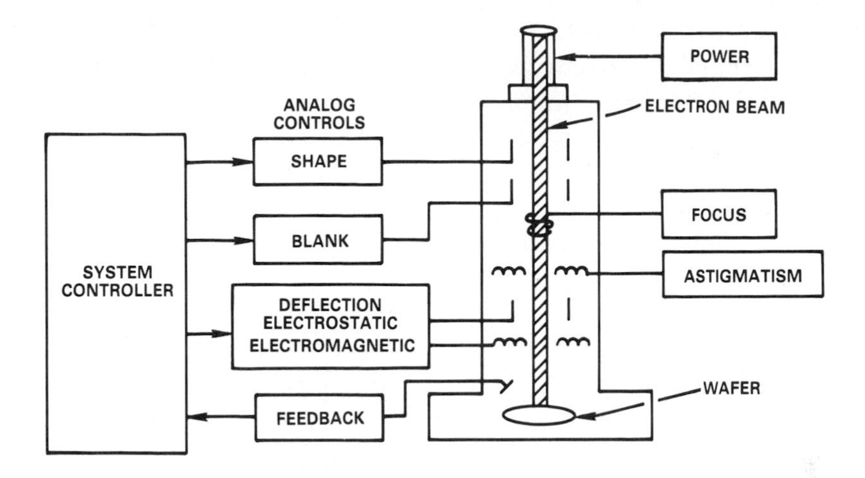

Figure 17.1. Electron-beam lithography uses a shaped electron beam to write patterns on integrated-circuit wafers. High-resolution d/a converters control the deflection of the electron beam.

Similarly, in numerical control, resolution of the a/d and d/a converters used in positioning determine the tolerance of machined parts. Using a DAC with 16 bits of accuracy and resolution means that a machine tool can mill a three-foot (about one-meter) precision steel part to within about 0.6 mils (15 micrometers). To get a feeling for how precise this is, it's worth noting that the same metal part will expand about 13 micrometers for each 1°C increase in temperature, which calls for great effort to control the temperature of the part while it is being machined (or at least while the measurement is made), as well as high accuracy for the other electronic components in the system.

17.2 HIGH-RESOLUTION D/A CONVERTERS

17.2.1 TESTING DAC INTEGRAL AND DIFFERENTIAL NONLINEARITY

High-resolution converters do not readily lend themselves to high-speed testing using automatic test equipment. Testing can be semiautomated or automated, but a custom test fixture is usually necessary; it will provide an electrically quiet, thermally stable environment, as well as a test-instrumentation front end comprising components having the required degree of accuracy, stability, and timeliness of calibration.

Differential nonlinearity, as defined in Chapter 11, is the difference between the actual analog output change when the digital input is increased by 1 bit, and the theoretical value of 1 LSB for the device being tested, i.e., FSR/2^n, where FSR is the full-scale voltage range (output reference) and n is the resolution in bits.

A simple way to measure differential nonlinearity is to connect a parallel-load down counter to the device under test (Figure 17.2). The inputs to the counter are first preset to the initial digital value (for example, 1000 . . 00). When the clock goes low, the digital input is asynchronously loaded into the counter, and impressed on the converter. Then, on the rising edge of the clock pulse, the counter counts down by one, and thereby decrements the code presented to the converter (e.g., to 0111 . . . 11). On the next cycle, the counter is preset to the initial value again, and the cycle repeats.

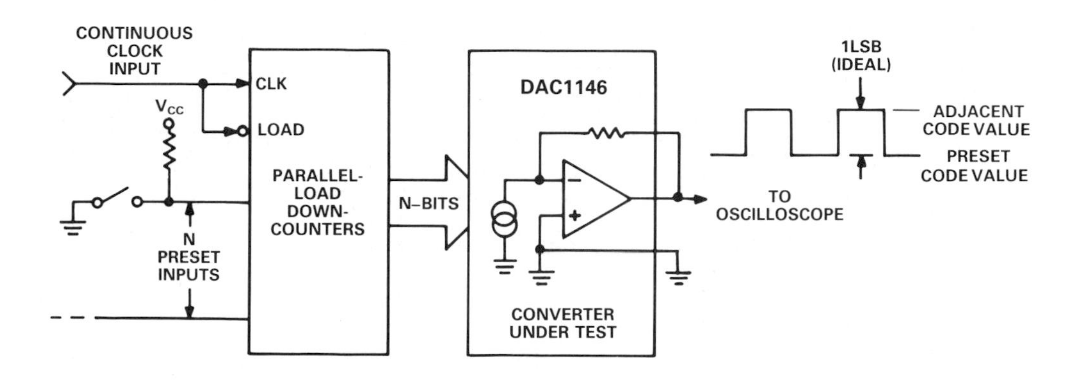

Figure 17.2. A simple scheme for checking differential linearity of a DAC without high-precision equipment.

Ideally, the resulting analog output will be a square wave with an amplitude of 1 LSB at the clock frequency, biased at a dc level equal to the initial value. The deviation of the square-wave's amplitude from the expected value of 1 LSB is the differential nonlinearity for the device at that particular digital input transition. Since a 1-LSB change is small (10 V/65,536 = 153 μV), the signal will have to be amplified with the dc level removed. The dc level can be removed by a-c coupling or by taking the difference between the DAC's output and the output of a reference DAC, set at the same initial code; the former technique is simpler, but the latter has the advantage of also indicating the absolute error of the initial code.

For 18-bit converters, there are 2^n—or 262,144—codes. Such a large number of codes makes it impractical to test each individual combination, unless a semi-automated test setup—and plenty of time—is available. Even with an automated tester that required as little as 10 ms per test, the entire sequence would take almost one hour. Fortunately, it is not necessary to test all input combinations to characterize a high resolution device with no summation errors*—a total of only n tests is required. If summation errors are small, then a total of 2 n tests may be sufficient. These test are usually performed at the major carries, such transitions as 1000 . . . 00 to 0111 . . . 11; 0100 . . . 00 to 0011 . . . 11; 0000 . . . 100 to 0000 . . . 011, etc., and, if necessary, at second major carries, such as 1100 . . . 00 to 1011 . . . 11.

Integral nonlinearity, INL, also referred to as *relative accuracy*, is the deviation of the actual converter output from a straight line drawn between the end points of the converter's output-vs.-input function. INL includes both bit errors and summation errors; it is difficult to measure, because the wide dynamic range of the signal requires finding the difference between two large numbers. A 6 ½-digit meter, accurate to 0.0002% (i.e., 2 ppm), would be necessary to measure the integral nonlinearity of a 16-bit converter.

For nonlinearity measurements, a better approach is to compare the output of the converter under test to a voltage having a known accuracy. The errors are then read with a null meter. This test requires a high-precision 18-bit d/a converter to measure a 16-bit-accurate converter. A precision divider, traceable to the National Bureau of Standards, must be used to test an 18-bit converter.

17.2.2 MAINTAINING HIGH RESOLUTION AND ACCURACY

The high price paid for 18 bits of resolution makes it economical, and sometimes mandatory, to perform fine-tuning to improve or maintain the accuracy by adding correction circuitry. To improve the linearity of an 18-bit converter from 16 to 18 bits of integral nonlinearity, for example, the integral linearity error can be measured at all codes. Then a corresponding correction signal for each code is added using a converter having lower resolution and a full scale value equal to only a few LSBs (viz., the maximum error) of the device to be corrected.

In such a scheme, a programmable read-only memory (PROM), addressed by a common input bus, stores the digital input to the correction converter. The drawback to this approach is the large amount of PROM memory required: 256K × 8-bits (assuming 8 bits, or 256 levels, of correction). However, the memory requirements can be reduced, with essentially the same results, by correcting for only the major error sources, rather than every single code.

For a converter with no summation errors, the worst-case integral linearity error will be less than or equal to one-half the of the worst case differential

**Summation errors*: For a linear device, superposition holds. This means that, for (say) eighteen terms with individual values equal to either b_i or zero, the output of a DAC with no offset and exactly unity gain—which sums those terms—will ideally be equal to their algebraic sum, for any combination of b_i's and zeros. The difference between the actual output and the mathematical sum is a *summation error*. It is due to a nonlinear summation, such as might be caused by an amplifier with distortion or a nonlinear feedback resistor. In defining summation error, it doesn't matter what the individual terms, b_i, are; they are not necessarily exactly equal to $V_{FSR}/2^i$; if they aren't, their deviations from the ideal are called *bit errors*.

linearity error. Therefore, if the summation errors and differential linearity errors of a converter are corrected, this will result in correction for integral linearity errors as well. Information about the size and location of summation errors for any given converter must be determined experimentally.

Many high resolution converters, instead of having a single ladder network, comprise several independent internal d/a converter stages of 4, 8 or 12-bits. For example, the Analog Devices DAC1146, an 18-bit d/a converter, has three stages, with independent resolutions of 4, 12, and 2 bits. The internal architecture can be designed to guarantee that these stages do not interact to produce interstage summation errors. Summation errors in the less-significant stages are suppressed with respect to full scale and so should be negligible.

The only summation errors that are greater than ¼ LSB (at 18 bits) occur in the four MSBs. Since any value of DAC output must include one of the sixteen bit-combinations, to correct for the summation errors in these four MSBs requires only 16 bytes of memory. To ensure correction for interstage errors, the number of bytes can be increased to 32 or 64—still many orders of magnitude less than a full correction scheme. The only remaining task is to correct for the differential nonlinearity errors in the less-significant bits, if necessary. The net result is equivalent to an 18-bit-accurate converter.

Figure 17.3 shows the principles of a semiautomated calibration scheme to make a true 18-bit d/a converter. In this example, the major-carry and summation-error corrections for the first six bit combinations are established in RAM, and the corrections for the lower-order bits are set by means of adjustable trim resistors. The correction codes are different for each DAC used, and they should be expected to change over time and temperature. For this reason, the long-term correction memory should be EEPROM or non-volatile RAM.

The correction circuitry, shown at the lower left, consists of an 8-bit DAC, driven by the correction RAM, and the trim potentiometers for the lower bits. The circuitry employed to determine the errors and implement the correction, shown above and to the right, includes an 8-bit counter, driven by a clock and gated by a differential comparator with a sampling capacitor on one input, a precision instrumentation amplifier with a gain of 1000, to amplify the error, and a 12-bit DAC, to hold the bit level while two adjacent bits are compared.

Here's the principle: For each of the 64 major carries involving the first 6 bits—for example, from **11 0100** 0000 0000 0000 to **11 0011** 1111 1111 1111—the DAC's output difference should be exactly 1 LSB. To check and adjust it, a 12-bit DAC is latched at the upper value, and its output is subtracted from that of the DAC under test (DUT), first at the upper value, then at the lower value, with the difference amplified by 1,000. The first of the differences is stored on a capacitor; the second is added to a current equal to 1,000 LSBs

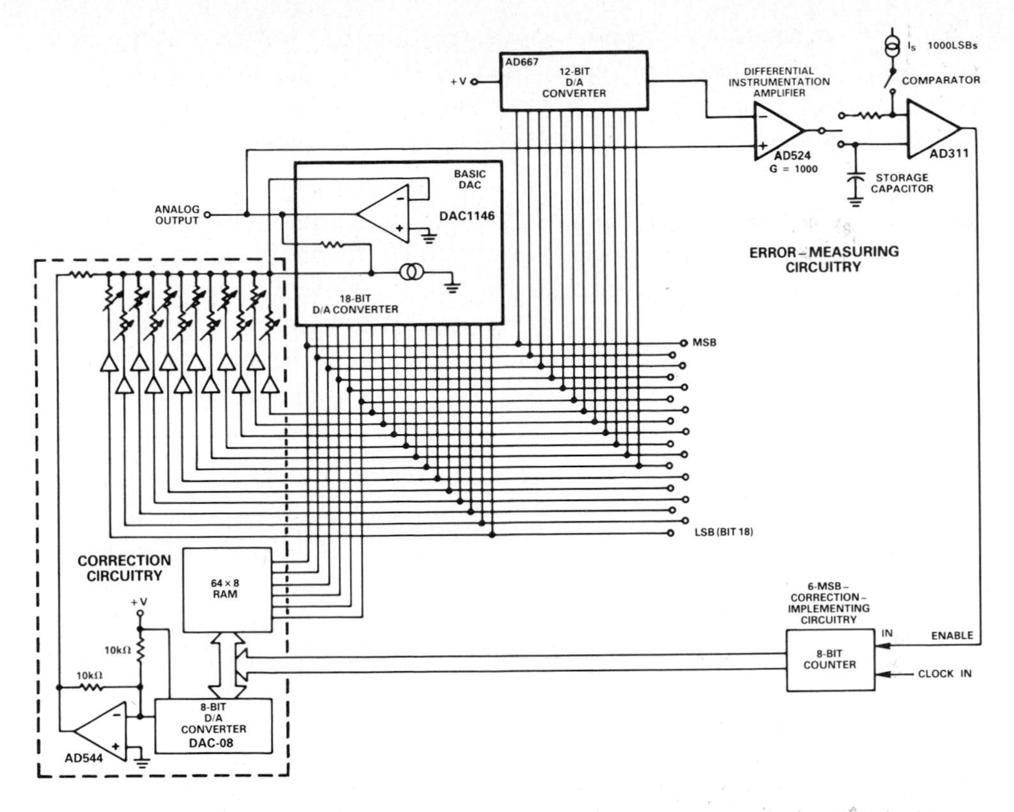

Figure 17.3. Scheme for trimming the accuracy of a high-resolution DAC, using an auxiliary DAC and memory.

and compared with the stored value. The 12-bit DAC removes the d-c common-mode value of the DUT's output, so the state of the comparator will reflect only the difference between the DUT's outputs for two adjacent codes, plus 1 LSB. Any difference from zero is an error. The counter is clocked up or down, and the measurement is repeated until the comparator changes state. The value of the correction furnished by the 8-bit DAC is stored in RAM.

For each code value, the output is thus compared to the output for one bit less, plus one LSB, and the correction converter's input is incremented until the analog output of the DUT is correct. This value is stored in RAM. The less significant bits are corrected in a similar manner, for major transitions of the lower twelve bits (for example, from 00 0000 1000 0000 0000 to 00 0000 0111 1111 1111), with trimming potentiometers substituted for the correction converter and RAM.

If the DAC in this example has no summation error, the trimming potentiometers and some memory could be eliminated, by using a modicum of computation. With the deviations between the ideal and actual outputs for each of the 18 bits measured and stored, the 8-bit DAC can correct all codes with only 18 bytes of memory. The μP adds up the correction factors for each of the *on* bits to generate the appropriate correction code for the d/a converter. For example, with the desired output code at 3/4 full scale, i.e., the two MSBs *on* (11 0000 00 . . . . 00), the processor adds the correction terms for bits 1 and 2 and loads the sum into the 8-bit correction DAC, which adds the analog value of the correction to the output.

Because the LSB is so small in high-resolution converters, few potential contributions to error can be neglected. Once a converter has been calibrated, the principal source of error is the variation of parameters with temperature. When a high-resolution-and-accuracy converter is used in an environment having wide ranges of temperature, many of the specifications may suffer serious degradation. This is easy to understand when you recognize that, for an 18-bit converter, 1 part-per-million is a large change.

For example, a DAC's worst-case monotonic temperature range or an ADC's worst-case no-missing-codes range is computed by subtracting the specified differential nonlinearity (DNL) from 1 LSB and dividing the result by the DNL temperature coefficient. This gives the minimum temperature deviation around the specified temperature (usually 25°C) at which the converter will remain monotonic or exhibit no missing codes. A 16-bit converter with ½-LSB DNL (7.5 ppm) and DNL tempco of 1 ppm/°C offers a monotonic temperature range of ±7.5°C around the initial 25°C operating temperature. An 18-bit d/a converter with DNL of ½ LSB (1.9 ppm)and a drift tempco of 0.4ppm/°C will remain 18-bits monotonic for a range of only ±4.7°C around 25°C or from 20.3°C to 29.7°C. These are minimum monotonic temperature ranges; the conservative assumption is that worst-case trims and worst-case tempcos occur at the same code. However, this is not necessarily the case; typical performance can be considerably better.

Differential nonlinearity is only one component of a converter's accuracy. Absolute accuracy error, at any output value, consists of gain error, zero error and integral linearity error. Gain and zero errors can be trimmed to zero at a specific ambient temperature. As the converter temperature deviates from the calibration temperature, errors accumulate and degrade the accuracy of the converter. For example, a unipolar 16-bit d/a converter, trimmed to initial accuracy of ½ LSB, with drift specifications such as ±10ppm/°C gain drift, and ±0.5ppm/°C offset drift, will have a maximum full-scale deviation from the initial accuracy of ±10.5ppm/°C. (If the drift of the reference is not included in the gain drift specification, it must be added on.) The total ±10.5ppm/°C drift means that the converter is 16-bits accurate (±1 LSB) for approximately ±1.4°C around the specified operating range.

However, one must retain the system perspective. The converter is not used alone, and a precision measurement system will usually require that the ambient temperature variation be kept small, in order that the performance of all analog system elements, including the converter, suffer as little degradation as possible. When necessary, analog and digital temperature-compensation techniques may be used to reduce temperature-related drifts to some degree. In addition, one should consider that, in a great many cases, high-resolution converters are purchased in order to get improved resolution and accuracy for lower resolutions over wider temperature ranges; for these applications, *full* accuracy-and-resolution are not needed over temperature.

Analog compensation involves sensing the changes with temperature in the major error sources and injecting compensating currents or voltages into the circuit. Sources of errors, within the internal circuitry, include the reference (which has a major impact on determining the full-scale output, and thereby the gain), the offset (zero), and the bit current weights. Since the MSBs have the largest current weights, they are the primary determinants of errors in the resistor ladder network.

Digital correction involves measuring the deviation of the converter's output, using a known reference source, and either adding an analog current or voltage to provide the correct value, or summing appropriate corrections digitally so that the code which determines the output incorporates the correction. There are a number of ways to provide this. For traceable standards a system can provide an autocalibration cycle on a periodic basis.

The 16-bit d/a converter shown in Figure 17.4 is connected to two 8-bit d/a converters, which can be programmed to provide automatic calibration of offset and gain errors. During a calibration cycle, the output of the precision DAC is compared with the correct value for each calibration point, and comparator measures the errors. Offset error is measured as the difference between the the converter output at zero and "ground". After zeroing, gain error is measured as the the difference between actual output with an all-1's input code and (FS – 1 LSB) output. Once measured, the correction codes

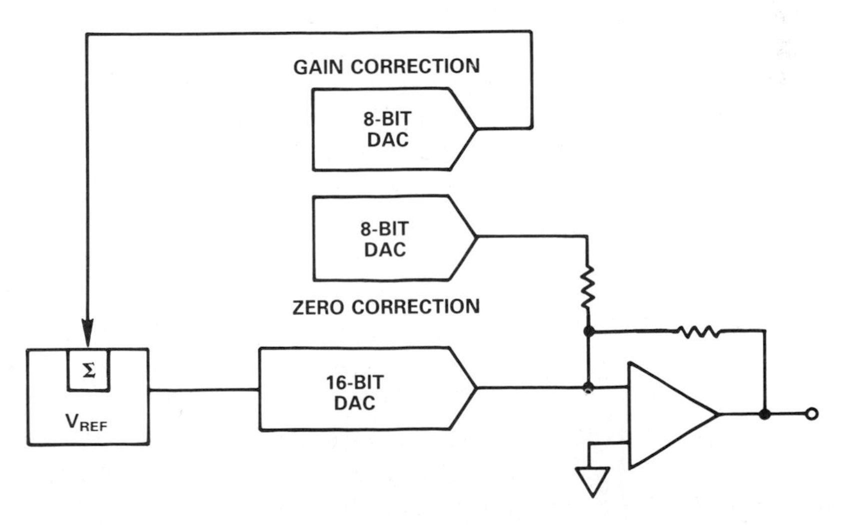

Figure 17.4. Use of auxiliary DACs for offset and gain correction of a high-resolution DAC.

are applied to the two correction DACs to compensate for the errors. The offset DAC injects the required current into the summing node of the output current-to-voltage amplifier. The gain DAC adjusts the reference voltage.

With a dedicated microprocessor, linearity—as well as gain and offset—can be corrected automatically. A typical scheme, which can be implemented as a system or a complex module, is shown in Figure 17.5. The elements of the scheme include a 16-bit d/a converter, a stable, temperature-compensated voltage reference, an error-measuring circuit, and a microcomputer to calculate correction factors for offset, gain and linearity. To trim the converter (initially within a few LSBs) it is adjusted, using the reference, offset, gain, and manual adjustments of (say) the four most-significant bits. The accuracy of the reference over temperature and time ultimately determines the accuracy of the d/a converter after calibration, since the reference is used as the "standard" to compensate the converter.

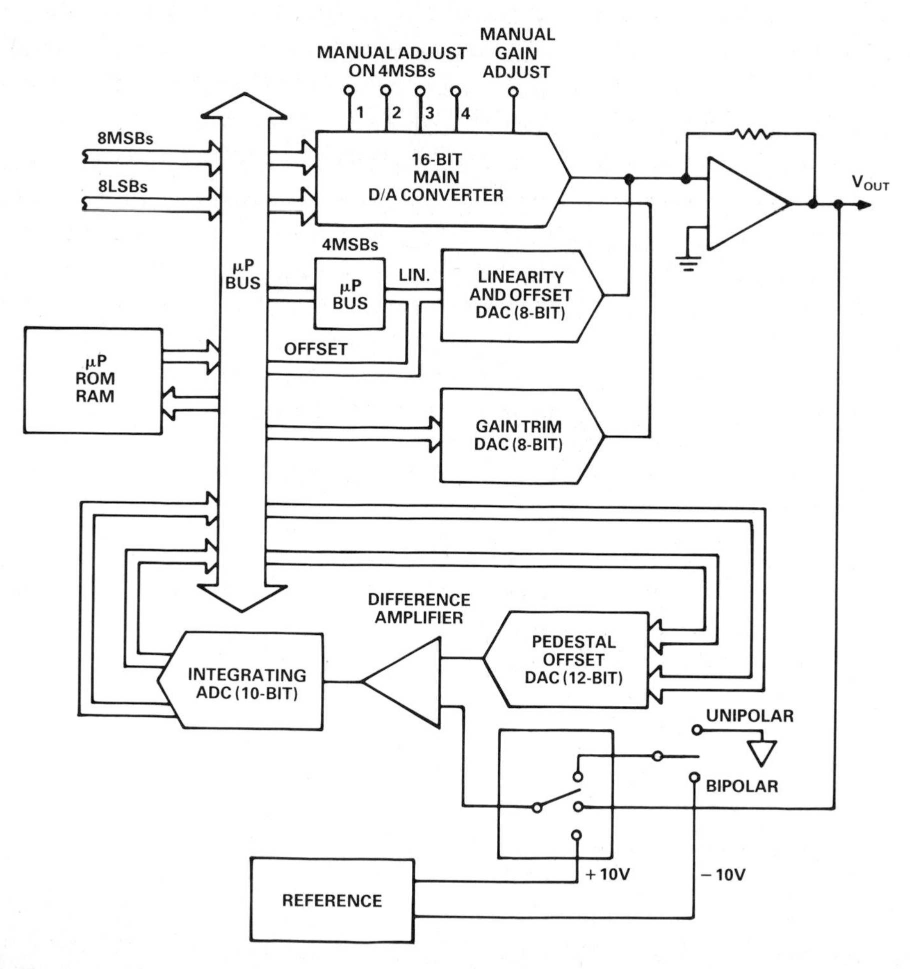

Figure 17.5. D/A converter with gain, offset, and linearity correction, using a dedicated microprocessor.

The error-measurement circuit compares a series of voltage pairs to derive the compensation factors. For example, the circuit compensates offset errors by measuring the zero output of the converter against ground, gain errors by comparing the DAC's full-range output (full scale—1 LSB) and a similar fraction of the reference, and nonlinearity errors by conducting a linearity test on the four MSBs, in much the same manner as described in Figure 17.3, except that an integrating ADC may be used to determine the error digitally, instead of the tracking measurement and conversion scheme. The correction factors are then stored in RAM and used as inputs to the linearity/offset-trim, and gain-trim d/a converters.

Some applications use high-resolution converters as programmable voltage or current sources. Since many high resolution d/a converters produce a current output, which can be applied to the inverting input terminal of an external op amp, with feedback via an on-chip application resistor—matched to other resistors in the device—the user can select from a universe of operational amplifiers to match a particular application's requirements. For example, a low-drift amplifier, with an inside-the-loop booster follower, can provide a large output drive for a programmable power supply.

It is also possible for some current-output DAC types to drive load resistance directly, with the voltage developed buffered by a follower op amp, in order to avoid the inherent sign inversion of inverting-amplifier configurations. However, a high-level inverting amplifier, used as an I-to-V converter, will make it possible to avoid voltage-compliance and summation errors. Voltage compliance is the maximum voltage that can appear at the current output terminal while maintaining a specified linearity. While some converters, such as the 10-bit AD561, have large compliance voltage ranges (e.g., -2 V to $+10$V), others have low compliance voltage ranges; for example, the typical compliance for the DAC1146 is ± 500 mV. Exceeding that specification leads to linearity errors.

Summation (integral-linearity) errors may occur in a voltage-output converter as the analog output increases from zero to full scale. The power dissipated by the feedback resistor increases, and the resistor heats up. This causes a change in the resistance value—and a corresponding change along the transfer function. In addition, if an external resistor is used with the internal feedback resistor for a non-standard gain, the external resistor will not track with internal resistors, resulting in larger than expected gain tempcos.

17.2.3 DYNAMIC APPLICATIONS OF DACs

Important Specifications

Dynamic applications take advantage of the wide dynamic range inherent in a high resolution converter. A 16-bit DAC, for example, offers 96 dB of dynamic range, versus the 72-dB range of a 12-bit converter. Although a number thic ICs do offer 16-bit resolution, the lower accuracy, 14-bits, reduced signal-to-noise ratio, since differential linearity errors, especially around zero, reduce the signal-to-noise ratio of the converter.

In waveform reconstruction, the basic goal is to reconstruct the waveform as closely as possible to its original form. Specifications that impact on this accuracy include such results-oriented specs as total harmonic distortion, intermodulation distortion, noise, limited dynamic range, settling time, and aliasing. Users of high-resolution data converters in waveform reconstruction applications are generally not concerned with differential nonlinearity, as such, or such traditional precision-dc-measurement specifications as dc offset and gain.

Dynamic range is the ratio of the maximum output signal to the smallest output signal the converter can produce (1 LSB), expressed logarithmically, in decibels (dB $= 20 \log_{10}$ (ratio)). For an N-bit converter, the ratio is theoretically very nearly equal to 2^N (in dB, $20 \text{ N} \log_{10}(2) = 6.02$ N). However, this theoretical value is degraded by converter noise and inaccuracies in the LSB weight.

The *signal-to-noise ratio* of a perfect digital waveform-reconstruction system is given by the ratio of the full-scale rms input to the rms quantization error.

For sine waves, the rms value is ½ the peak-to-peak value, divided by $\sqrt{2}$; for an N-bit system, the full-scale rms value is thus $2^{(N-1)}/\sqrt{2}$. The quantization error, an expression of the fact that each quantized level represents an uncertainty as to the actual value of points in its neighborhood, is determined by the difference between a linear response (the expected output) and the characteristic analog-to-digital staircase function (the actual ideal digital output); the shape is a sawtooth oscillating once per LSB (or quantization interval, Q) between ±½ LSB (i.e., ±Q/2). The rms value of this sawtooth wave is the peak output divided by $\sqrt{3}$, or: $(Q/2)/\sqrt{3}$. Thus, the signal-to-noise ratio is

$$\text{S/N} = \frac{2^{(N-1)}}{\sqrt{2}} \frac{\sqrt{3}}{1/2} = 2^N \frac{\sqrt{3}}{\sqrt{2}} = 1.225 \times 2^N \tag{17.1}$$

Expressed in dB,

$$\text{SNR} = 20\,(\log 1.225 + \text{N} \log 2) = 1.76 + 6.02\,\text{N dB} \tag{17.2}$$

Thus, the resolution and quantization level establish a noise floor; system noise and other sources of random noise will decrease the signal to-noise ratio.

17.2.4 TESTING HARMONIC DISTORTION OF DACs

Total harmonic distortion (THD) is the difference between an ideal sine wave and its reconstructed version using an a/d and a d/a converter. THD is the ratio of the square-root-of-the-sum-of-the-squares of the RMS values of the harmonics to the RMS value of the fundamental. For a converter with a finite number of digital inputs, N, and associated output voltages, THD is customarily calculated from the formula:

$$\mathrm{THD} = \frac{\text{RMS error}}{\text{RMS signal}} = \frac{\sqrt{\frac{1}{N}\sum_{i=1}^{N}\left[E_L(i) + E_Q(i)\right]^2}}{E_{RMS}} \tag{17.3}$$

Where $E_L(i)$ is the linearity error and E_Q (i) is the quantization error of the converter at the sampling point i. Intermodulation distortion is caused by the additional error produced when the ideal output is composed of two sine waves of different frequency.

A computer-generated sine wave stored in a programmable read-only memory (PROM) can be used to demonstrate and test harmonic distortion. The PROM in Figure 17.6 contains one cycle of a computer-generated sine wave. A complete cycle is generated by counting through all address values; the frequency depends on the clock rate. If there are N values stored in the PROM to form a complete cycle, the sine-wave frequency will be C/N, where C is the clock rate. If a constant clock frequency is used, N must be reduced by skipping data points in order to increase frequency.

Thus, in the scheme of Figure 17.6, if the adder is set to increase the PROM address by one on each count, 4,096 inputs will be presented to the converter

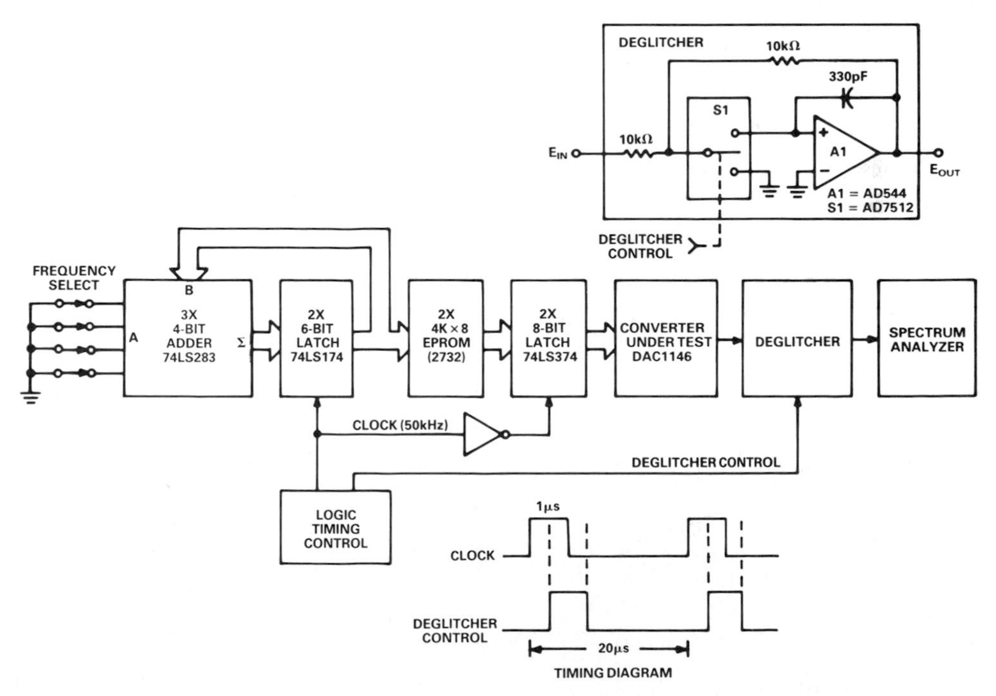

Figure 17.6. Testing harmonic distortion as a function of frequency.

on each cycle of the PROM, for a very close approximation to the sine wave. If the adder is set to increase the PROM address by 1,024 on each count, four inputs will be presented on each cycle (an adequate number of samples). Frequency-select switches program the adder with the number of codes that the converter should skip on each count. In this way, any of 2,048 discrete frequencies between 12 Hz and 25 kHz (with corresponding count densities) can be generated by a constant 50-kHz sampling rate.

This series of quasi-sinusoidal digital codes is fed to the converter to generate staircase approximations of an analog sine wave. The DAC output is deglitched (see below) and displayed on a spectrum analyzer. Total harmonic distortion can be computed for each set of samples by comparing the amplitude of the fundamental frequency with the amplitudes of the harmonics.

17.2.5 IMPROVING DYNAMIC PERFORMANCE OF HIGH-RESOLUTION D/A CONVERTERS

Glitches and Deglitching

Because they distort the waveform, glitches in a d/a converter's output increase total harmonic distortion. Their spiky nature produces large distortion components at frequencies considerably greater than the fundamental, and their asymmetry introduces dc components, all of which makes glitches extremely difficult to filter with linear filters.

Glitches have two principal components, one introduced by charge coupling from the digital logic signals to the analog output, the other caused by asymmetry in timing as codes change. The former is less serious, because the pulses it produces, though variable, tend to change less in amplitude than the latter.

The asymmetry in timing, caused by skewed response in the up- and down-directions in both digital logic and analog switches, produces false intermediate codes, which can vary tremendously in amplitude.

For example, if the code is changing by 1 LSB at mid-scale, from 0111 . . . 11 to 1000 . . . 00, and the switches turn off (or receive their *off* drive signals) faster than they turn on (or receive their *on* drive signals), there will be a brief interval during which the output is seeking to respond to 0000 . . . 00; this will result in a large negative-going spike; on the other hand, the change to the next code, from 1000 . . . 00 to 1000 . . . 01, will only result in a less-than-1-LSB negative-going spike. Changes at other code transitions will result in spikes of differing amplitudes—largest at major carries—depending on which codes are turning off and which are turning on.

Converters using CMOS technology have additional glitch impulse introduced by the charge stored on the gate-to-source and gate-to-drain capacitance of CMOS switches during switching. The output capacitance, associated with the relatively large N-channel devices used for the DAC switches, is highest when all the switches are on, and lowest when all the switches are off.

A deglitcher is essentially a sample-hold, which holds the signal from just prior to conversion until just after the signal settles. Although the deglitcher itself introduces a small glitch, it is small relative to the DAC's glitch, of constant amplitude, and independent of the digital code. Since it injects energy at the sampling frequency rather than the signal frequency, it introduces no harmonic distortion within the signal passband. The deglitcher can be band-limited with a linear filter to suppress distortion caused by slew-rate limiting of the output amplifier.

Most digital audio applications are stereo and require a right and left channel to be fed from a single fast, high-resolution 16-bit DAC, updated at 50 kHz. The L and R channel signals update the DAC alternately. A popular scheme for simultaneously updating both audio output channels—while eliminating the DAC glitch—employs a track-hold to store the right channel, while the left channel signal updates the DAC. Then two sample-holds ("deglitchers"), one on the DAC output, the other on the R track-hold, update the two output channels and hold the signal until the next joint update, while the R-channel track-hold is once again updated to start the new cycle.

The deglitcher response is band limited to eliminate distortion of the amplifier. A time constant of 3.4 μs ensures a full 20-kHz sine wave response without distortion. A 16-bit converter, track and hold, and deglitchers can produce an audio signal with distortion of less than 0.005%.

17.2.6 BIPOLAR-OUTPUT DAC PERFORMANCE

Dynamic signals, especially audio, are inherently bipolar. Performance around zero is important, since zero represents "silence" in a digital recording; many audio recordings consist of silences punctuated by sound. Performance around zero also determines the signal-to-noise ratio of the converter.

Signal polarity in d/a conversion is commonly handled in two basic ways—by offsetting (offset binary and twos complement) and by absolute value (sign-magnitude). Offsetting techniques employ a basic unipolar binary converter with an offset of ½ scale. A sign-magnitude converter senses the polarity of the signal and switches the output circuit of the basic unipolar converter between the inverting and noninverting modes.

An offset-type bipolar converter exhibits its worst temperature performance around zero volts because it requires the MSB to track with the sum of all the less-significant bits, as well as with the bipolar offset resistor; it also suffers dynamically, because the offset zero is the point at which the major-carry glitch occurs. Large signals tend to mask error-induced noise, while noise buries small signals. A sign-magnitude converter exhibits its best performance around zero volts because all the bits are OFF, and transitions involve only the least-significant bit.

A simple unipolar binary converter is converted to sign-magnitude by the addition of a fast, linear, low-drift inverting amplifier with accurately set unity gain, a SPDT CMOS switch and an output buffer amplifier. The result is a maximum DNL drift of ±½ ppm/°C for ±⅛-full-scale range and ±1 or 2 ppm/°C over the full range.

17.2.7 EXTENDED RESOLUTION

A substantial number of high-resolution converters are converters with extended resolutions without commensurate increase in accuracy. These designs aim strictly at high (monotonic) resolution to obtain the wide dynamic range required in applications such as sonar, radar, optical signal processing, where the actual value of any data point is not required to great accuracy, but where the range of values is quite large, typically extending well beyond 80 dB. For example, high-quality photodiodes used in optical measurements produce linear current outputs over a 100-dB range, requiring at least a 17-bit a/d converter.

17.3 HIGH-RESOLUTION A/D CONVERTERS

High-resolution analog-to-digital conversion techniques are used to obtain high accuracy, improve system resolution, or boost the dynamic range. Goals for a system may specifically include lower noise and improved temperature stability.

17.3.1 FLOATING-POINT CONVERTERS

A floating-point converter improves dynamic range by acquiring data in two parts, usually employing a two-step approach. In a typical approach, a binary programmable-gain amplifier (PGA) scales the signal to the proper range, typically between 1 MSB and full scale; the input code to the PGA controls the exponent of 2 (i.e., $G = 2^X$). The scaled signal is then converted by a an ADC whose output code forms the mantissa, while the digital signal used to set the gain prefaces it with the exponent of 2, viz., X. For example, a 12-bit converter with a 4-bit preface and the code, **0100** 1100 0000 0000 would have an overall gain of 8×0.75.

The dynamic range of a converter is the ratio of the full-scale input range to the smallest signal the converter can detect. With a floating-point converter, the smallest value of the LSB corresponds the the value of the LSB of the converter when the PGA is programmed for its highest gain.

$$\text{LSB} = \frac{\text{FSR}}{\text{GAIN} \times 2^N} \tag{17.4}$$

For a floating-point converter with full-scale range of 10 volts, 256 V/V maximum gain and a 12-bit ADC, this corresponds to:

$$\text{LSB} = 10\text{V}/(256 \times 4096)$$

or 9.5μV. The dynamic range in decibels is based on the logarithm of the product of the amplifier gain and the converter's dynamic range, i.e.,

$$\text{dB} = 20 \log (2^{12} \times 2^{8})$$

$$= 20 \log 2^{20} = 120 \text{ dB}$$

This corresponds to a 12-bit resolution, with a dynamic range of 20 bits.

There are a number of different approaches to designing a floating-point converter. In some systems, software is used to set the gain of the the PGA. The signal is converted, and if the MSB is not equal to 1, the gain is increased by a number of binary steps (up to a specified maximum), essentially equal to the number of leading zeros. This maximizes the full-scale range of the conversion process and insures a wide dynamic range. Another approach is to use a logarithmic converter; it is fast, conceptually simple, maintains essentially constant percentage error over a major portion of the dynamic range, and uses a minimal bus width.

Another approach to floating-point converter design, with improved throughput rate, involves setting the gain of the PGA with a flash autoranger. The 20-bit converter diagrammed in Figure 17.7 consists of a pair of track-and-hold amplifiers, flash octave a/d converter (reference levels are at octave intervals, instead of equally spaced), a nine-range programmable gain amplifier, and a fast 12-bit a/d converter. The first track-and-hold amplifier holds the signal for the flash autoranger, which determines which binary quantum the input falls in, relative to full scale (e.g., 1/2 to 1/4, 1/64 to 1/128, etc.), and encodes the information as a 4-bit digital byte. Responding to it, the PGA

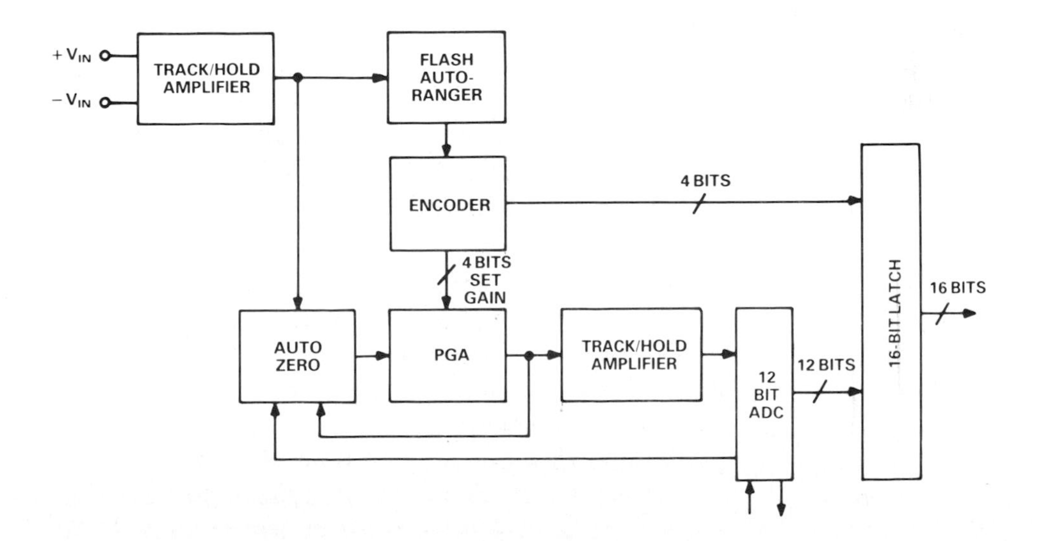

Figure 17.7. Floating-point a/d converter, employing a 12-bit ADC, with 20 bits of dynamic range represented by a 16-bit output word.

adjusts its gain to the appropriate level, and the track-and hold amplifier holds the amplified signal while the two-step a/d converter translates it to a binary number. A sixteen-bit latch holds the 4-bit output from the encoder and the 12-bit output from the converter.

There also exist modified floating-point converters, in which the prefix digits produce a selection of gains that are not in strict binary relation, for example, 1, 10, 100, or 1, 2, 5, 10, 20. An interesting example is a 12-bit a/d converter, with an amplifier having a maximum gain of 256, expressed by a prefix code based on the rule, 2^{2x} (i.e., 1, 4, 16, 256); here, a wide range of gains can be controlled with a 16-bit word.

17.3.2 EXTENDED RESOLUTION ADCs

Stochastic techniques

Resolution can be improved beyond the actual resolution of the converter by means of multiple conversions. With stochastic conversion techniques, the resolution of an a/d converter (for example, a successive-approximation type) can be extended by several bits. The stochastic a/d technique (Figure 17.8) adds a pseudorandom dither with an average value of zero to the input signal, makes several conversions on the signal, and then computes the average. The time period of the conversion is set to be an integer number of power-line cycles to maximize the normal-mode noise rejection.

When pseudorandom noise is added to the signal input, the output of the converter will vary by some LSBs from a nominal value. As each conversion is completed, its output is summed with the previous ones and stored in an accumulator. When the desired number of conversions have been completed, the sum of all the data samples is averaged. In the example of Figure 17.8, an input of 0.3 is added to 8 samples of digitally generated dither and converted each time with 2-bit resolution. Without the dither, the digital value of the converter output is 0.25 (code 01); when dithered and averaged, using the eight terms and arbitrary precision, the result is 0.3125, corresponding to the code, 0101, representing the binary fraction, 0.0101 (instead of the lower-resolution 0.01).

As the simple 2-bit example of Figure 17.8 shows, the result will depend upon where the main analog input signal lies within a particular code width of the main a/d converter. For example, if the input lies precisely in the center of the code, the data samples will have equal positive and negative distributions about the nominal code value. If the analog input lies elsewhere between the limits of the quantum, the computed average will show either a positive or negative bias from the center of the code, and—since the sum is obtained digitally to arbitrary precision—the digital result is, in effect, a meaningful extension of resolution.

This approach requires time and software, but little or no additional hard-

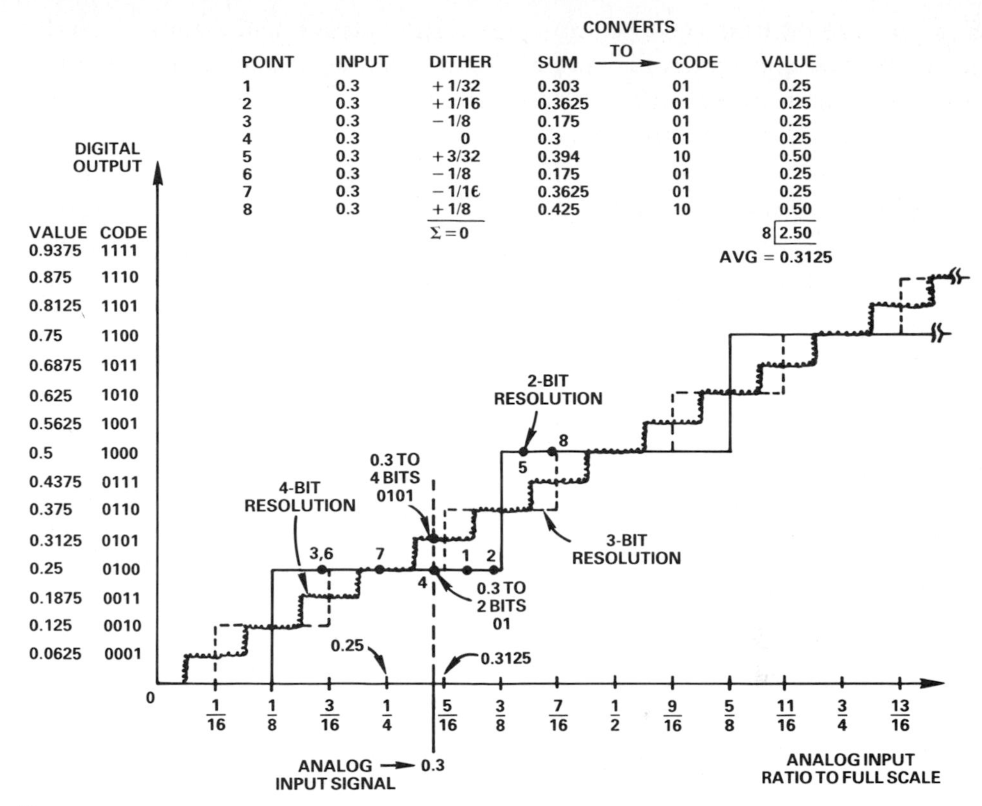

Figure 17.8. Random dither improves effective converter resolution. Getting four-bit resolution from a 2-bit converter.

ware. The noise signal can be furnished by a low-resolution dither DAC, using noise numbers that are computed or retained in memory. The approach provides the optional tradeoff of either high-speed, medium-resolution conversion or slower, high-resolution conversion.

17.3.3 CHALLENGES IN DESIGNING A HIGH-RESOLUTION A/D CONVERTER

In a high-resolution, fast, successive-approximation a/d converter, all components—and in particular, the reference DAC and the comparator, must be accurate to considerably better than N bits (for example, 18 bits in an 18-bit converter) in order for the converter's overall accuracy to be better than N bits. We have already had some discussion of the difficulties of designing 18-bit DACs; since ½ LSB of 18 bits is 1.9 ppm of full scale, a true 18-bit DAC would drift less than 1.9 ppm over the operating temperature range; for ordinary environments, this requirement implies resistor matching too close to contemplate. In addition, for an ADC, the comparator is an additional crucial component; it must respond quickly to an overdrive of ½ LSB, or 19 microvolts out of 10 volts. It must also have input noise significantly smaller than this value.

No complete "off the shelf" comparator exhibits these specifications, so the designer must construct one from discrete components, consuming a lot of power to lower noise and increase speed, but also with very high gain to achieve the desired sensitivity. As the speed of the converter increases, however, the design starts to approach some quantum limitations.

For example, the charge involved in measuring one LSB of an 18-bit ADC in 10 μs is only about 24,000 electrons. Determining the charge of a single electron, and the current flow of a single LSB demonstrates this. The charge on an electron is 1.6×10^{-19} coulombs, and thus 1 coulomb $= 6.2 \times 10^{18}$ electrons. Since, by definition, 1 ampere = 1 coulomb/second, the current flowing from the LSB is 7.6 nA, or $= 2.4 \times 10^{10}$ e-/second. A 10-μs bit-conversion time, therefore, corresponds to about ½ μs/bit ≈ 24,000 electrons.

The above discussion highlights a few of the problems associated with the design of a high resolution data converter. Although the design problems of the manufacturer are not the particular concerns of the user, many of the issues that face the designer also face the user. A key factor is the effect of noise on converter performance.

Once a converter has been selected, installed and tested, system designers often encounter the perplexing problem of the digital codes deviating from their expected values. This deviation is manifested as a number of unexpected codes over many conversions or over time with a steady analog input or output. These errors are caused by noise in the converter circuitry, in external circuitry or inherent in the analog signal itself. Noise produces code errors in the output code of an ADC or analog output errors in a DAC if the peak value of the noise, after filtering or integration exceeds one LSB. The percentage of time for which a given noise level will exceed the 1-bit threshold may be calculated.

17.3.4 A LITTLE NOISE THEORY

Noise in a/d converters arises from two principal sources, the "ideal" no-noise *quantization (or roundoff) error* that is inherent in the data conversion process, and *input noise*—which includes noise generated in the converter, noise arriving with the signal, and noise coupled in from the environment. If the analog signal is recovered in a d/a converter and compared with the analog input, there will be an error that depends on a combination of the error due to input noise and the quantization error.

In the same way that op-amp noise is referred to the op-amp input, all input noise in an ADC can be referred to the comparator input (where the decision is made as to which quantum an input belongs to), whether it is the comparator's own noise, noise in the summing resistors at the input to the comparator, noise in the reference DAC output, noise that arrives with the signal, or noise that is coupled in from the power supply or the environment. All these noise sources can be summed to provide the total error signal the comparator will see and that will determine the maximum realizable conversion error.

If the comparator were perfect, the arriving signal "clean," and all interference eliminated, there would still be unavoidable noise due to thermal resistance noise (Johnson noise) of the signal-source input resistance and the DAC's resistance ladder. Johnson noise voltage is given by

$$E_n = \sqrt{4kTRB} \tag{17.5}$$

Where E_n is the rms value of voltage generated in the effective source resistance, R (ohms); k is Boltzman's constant (1.381×10^{-23} J/K); T is the temperature in kelvins (°C + 273.2°); and B is the effective bandwidth ($f_2 - f_1$), "brick wall," in hertz. In more practical units,

$$E_n = 0.129\sqrt{R \times B}\ \text{microvolts rms} \tag{17.6}$$

where $R \times B$ is in MΩ-Hz, kΩ-kHz, or Ω-MHz. For example, if R = 1kΩ and B = (5 kHz − 4 kHz) = 1 kHz, then E_n is equal to 0.129μV. If R = 100 ohms and B = 1 kHz, then E_n = 41 nanovolts.

E_n is the theoretical minimum; in practical conversion systems, noise will be several times as large due to the other factors mentioned. Other additional noise sources will add to this minimum to further degrade the resolution of the converter. For noise sources that are independent of each other, a designer can calculate the overall noise from all known or estimated random noise sources as

$$E_n = \sqrt{(E_{n1})^2 + (E_{n2})^2 + \ldots + (I_{n1} R_1)^2} \tag{17.7}$$

Where E_n is the total noise, E_{ni} are the various voltage noise sources, and I_{ni} are the various uncorrelated current-noise sources. In actual practice, the contributions of these noise sources can be as much as (or more than) 10 times the theoretical minimum Johnson noise.

The significance of the total noise in a high-resolution conversion system depends on the magnitude of the quantum step (1 LSB). If the a/d converter under consideration is assumed to be linear, the LSB will equal

$$LSB = \frac{FSR}{2^N} \tag{17.8a}$$

where FSR is the full-scale range of the converter and N is the resolution in bits. The equivalent input noise can be compared to this value (ratio = K) to determine its digital significance:

$$K = \frac{E_n}{LSB} = \frac{2^N E_n}{FSR} \tag{17.8b}$$

For example, in a 16-bit ADC, with 10-volt full-scale range, if the effective total input noise has an rms value of 25μV, the value of K, in terms of the number of least-significant bits, is $65{,}536 \times 25 \times 10^{-6}/(10\ V) = 0.16$.

If the noise is Gaussian, the rms value corresponds to the standard deviation, σ, of the distribution (Figure 17.9). Since the noise is presumed to be random, it contains all amplitudes, some very large, but the probability of large values decreases quite rapidly with amplitude. For example, the probability of a peak greater than 3 σ is 0.27%, while the probability of a peak greater than 7 σ is 2.6×10^{-12}.

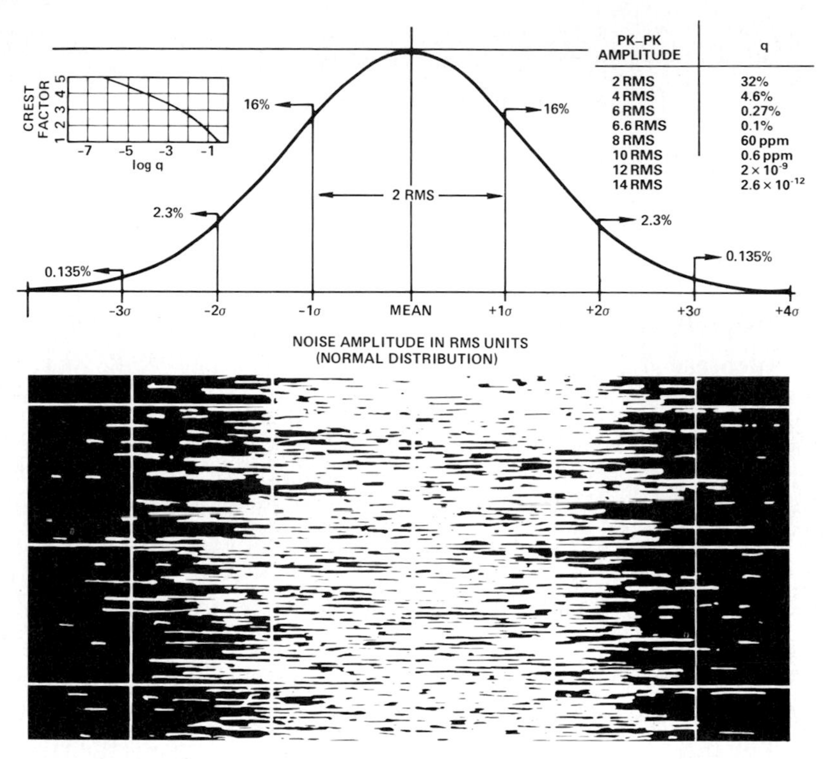

Figure 17.9. Relationship between rms value of Gaussian noise and probabilities of various peak amplitudes.

If we are concerned about the probability of missing codes (with error larger than ½ LSB), we can relate that to the probability of a peak greater than 0.5/K, since K corresponds to bit value of σ. For the above example, the probability of a peak larger than ½ LSB will correspond to the probability of a peak greater than 3 σ, or 0.27%.

Figure 17.9 provides information from which the probability of missing codes due to noise can be computed for any assumed value of input noise, converter resolution, and full-scale range, based on the above discussion and a Gaussian distribution.

It is important to note, however, that this discussion is addressed to *random* noise arising from natural phenomena. If the input noise is dominated by large spikes picked up from the power supply or the electrical environment, the fact that their rms value is minuscule will be of small comfort.

The size of the error, in LSBs, depends on the noise factor, K, and where the signal is located in the quantization interval. This value will be at a minimum when the actual code is centered in the interval and increase as the value deviates from the center value. For an expected random distribution of signals over time, the rms value of quantization noise can be treated as one of the inputs to equation 17.7.

Noise leads to random errors and missing codes in an ADC or nonmonotonic behavior in a DAC. Since these errors can lead to a closed-loop system making inappropriate responses to specific false data points, high-resolution systems, where random noise tends to limit resolution and to cause errors, can benefit by digital or analog filtering to smooth the data.

17.3.5 BOARD LAYOUT CONSIDERATIONS FOR HIGH-RES ADCs

Printed wiring board layout is extremely critical when using high-resolution analog-to-digital converters. High-speed logic level signals are present on the same board as low-level analog signals with microvolts of resolution. If signal conditioning or high-gain amplification is also included, the problems are compounded.

Figure 17.10 shows what can happen in an inadequate design. An amplifier with a gain of 1,000 is used to amplify a low-level 0-to-10 mV signal and present the resulting 0-to-10 V signal to the ADC. Suppose that, due to board strays, there are 1,000 megohms of resistance and 0.1 pF of capacitance between the summing node of A1 and one of the digital logic lines. The amount of dc pickup, relative to the analog input, is 10^{-6} (1,000 ohms/1,000 megohms), or 5 microvolts—0.05% of full scale. On the other hand, assuming feedback capacitance of 10 pF, a 5-volt logic edge would be attenuated by 0.1

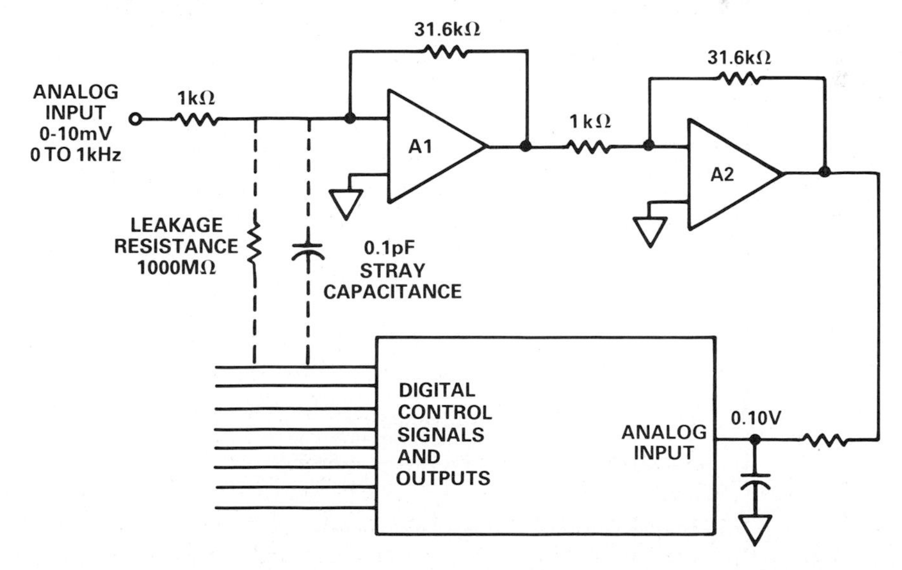

Figure 17.10. Example of effects of stray capacitance and leakage resistance.

pF/10 pF, in the first stage, while the analog signal experienced a gain of 31.6, so 5 volts of logic would insert a 1.6 mV spike, referred to the input, or 16% of full-scale. However, it would be damped out within 2 μs.

Effective solutions to this problem involve *distance*—keeping high-level and digital lines as far as possible from low-level analog lines, *isolation*—using shielding and guarding to isolate low-level signals, and *orientation*—where leads must cross, doing so at right angles, using twisted pairs to balance pick-up, etc.

If grounded guards are placed around the summing nodes of the amplifiers, stray capacitance from digital signal leads is to ground, rather than to the sensitive nodes. Not all guards are grounded guards; In order to be fully effective for low-frequency and dc common-mode pickup, as well as AC strays, guarding must be done at the same potential as the signal to be guarded.

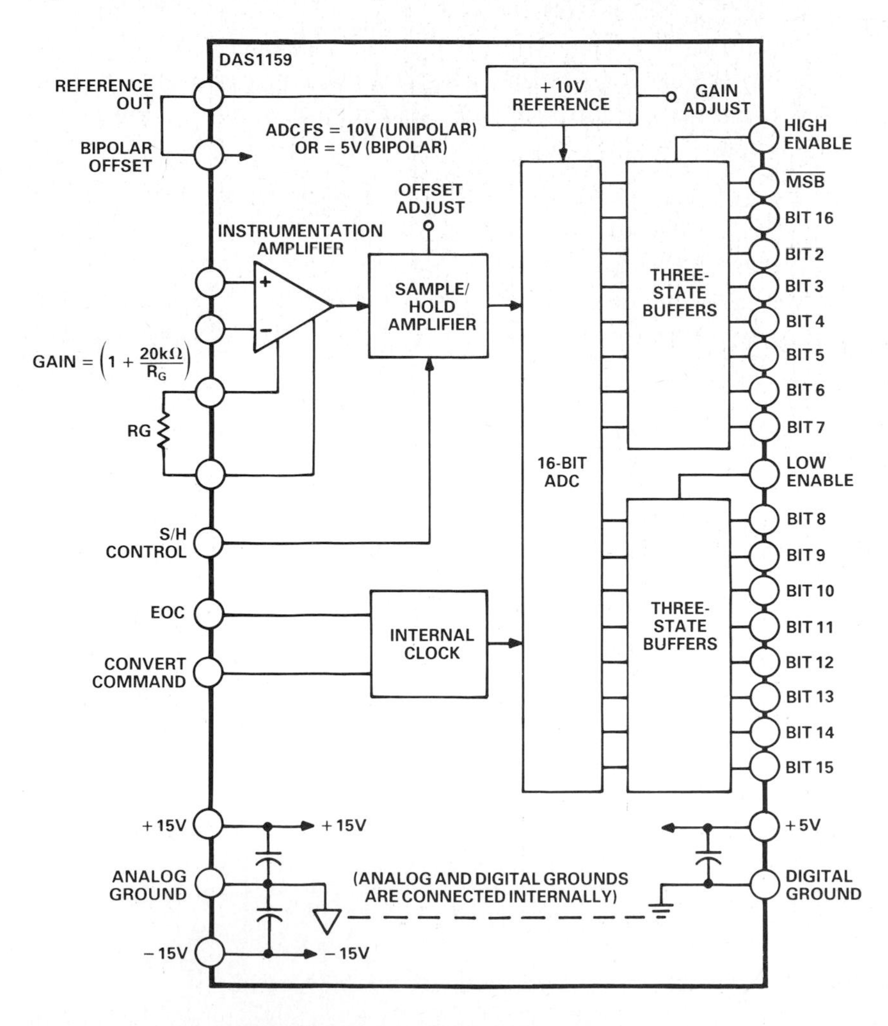

Figure 17.11. A 16-bit sampling a/d converter.

Unfortunately, due to space limitations, optimum guarding and grounding practice in the neighborhood of high-resolution ADCs is sometimes difficult to achieve, and converter noise is the likely result. With the analog input grounded in the data-acquisition system of figure 17.11, noise could come from many sources. It is helpful to have a workable procedure for tracking down interference-noise problems. The internal architecture of a converter or data-acquisition system can be helpful in a rational search for noise sources.

17.4 HIGH-RESOLUTION SAMPLE/HOLDS

In high-resolution conversion, the dynamic characteristics, even of slowly varying signals must be considered. In order to faithfully digitize a signal, of frequency, f, and resolution, n, to 1 LSB, the conversion time (aperture) uncertainty, τ_a, must be less than:

$$\tau_a = \frac{2^{-n}}{\pi f} \tag{17.9}$$

If, for example, a 16-bit successive-approximation converter can complete a conversion within 35 μs, the highest frequency that can be converted with 16-bit resolution is *0.14 Hz*. In order to convert at a sampling rate of 25 kHz, to handle, say, 10-kHz input signals, a sample-hold must be used ahead of the converter, with an aperture uncertainty better than 0.5 ns.

Sampling a 20-kHz signal to 16 bits requires the following specifications (actually, they should be considerably better when considering worst-case performance, but these are generally considered acceptable in the industry):

Aperture Jitter	0.25 ns
Slew Rate (20V pk-pk)	1.26V/μs
Feedthrough (½LSB Max)	−102dB
Droop Rate (½ LSB Max in 15μs)	5.1 μV/μs
Acquisition time (±½LSB max (with 20-kHz signal and 15/μs ADC)	10μs
Pedestal shift (max)	−96.3dB
Gain Tempco (±10°C ambient)	1.5ppm/°C
Thermal tail	0.3mV
Linearity error (max)	±.0015%FSR

Aperture jitter is the uncertainty about when the sample is taken; it must be considered, even though the T-H control line is driven by a precise clock. All high-speed sampled-data systems depend on low aperture jitter for digitizing high-frequency signals for spectrum analysis and accurate signal reconstruction.

The T-H amplifier's slew rate determines the maximum switching rate when following changes between multiplexed input signals. The feedthrough from input to output while in the *hold* mode should be less than 1 LSB. The *hold-*

mode droop rate rate should be less than 1 LSB of droop in the output during the conversion time of the a/d converter. For a 16-bit ADC with a 15-μs conversion time, for example, the maximum droop rate, as noted above, is ½LSB per 15μs; since 1 LSB $= 10/2^{16}$ V $=$ 152.6μV, the maximum droop rate is 5.1 μV/μs.

The linearity error should be less than 1 LSB over the transfer function, as set by the relative accuracy of the a/d converter. The track-hold's acquisition time and settling time (t_{a+s}), along with the conversion time of the of the a/d converter (t_c), determine the highest sampling rate, f_s.

$$f_s = \frac{1}{t_{a+s} + t_c} \tag{17.10}$$

This, in turn, will determine the highest input signal frequency that can be sampled at a minimum of twice per cycle, according to Nyquist sampling theory.

The pedestal shift due to input signal changes should either be linear, to be seen as gain error, or negligible. Feedthrough should also be negligible. The temperature coefficients for drift should be low enough so that the full accuracy is maintained over some minimum temperature range. The droop rate and pedestal will shift more over temperature for temperature ranges above +70°C, generally a considerably higher temperature range than these devices will experience for most of the applications in which they are used. An additional factor to consider with high-resolution converters is the noise in the T-H during the *track* mode, since it will affect the value that is sampled when the T-H is switched into *hold*. This noise must be added (root sum-of-squares) to converter noise when calculating the actual noise error in an ADC.

Minimal thermal-tail effects are another requirement for high-resolution applications. The self-heating-induced transient errors due to transient thermal imbalances resulting from the changing current levels in the output stages of T-H amps may cause more than 1 LSB of error because of the time required for the converter to settle to equilibrium temperature.

The profusion of T-H specifications and the need to carefully monitor the design might be better solved by turning to a high-resolution data acquisition subsystem or sampling a/d converter, which has been engineered to perform an overall task. The sampling a/d converter contains both a track-and-hold and an a/d converter. An example of a high-resolution sampling converter's structure can be seen in Figure 17.11.

PART IV

RELATED CIRCUITS AND DEVICES (DATA-ACQUISITION-PERIPHERAL COMPONENTS)

Chapter Eighteen

Sample-Hold Circuits

A sample-hold amplifier (SHA) is a device that has a signal input, output, and a control input. As its name implies, a SHA has two steady-state operating modes. In the *sample* (or *track*) mode, the output tracks the input as faithfully as possible until the hold command is applied at the control input. In the *hold* mode, the output retains the last value of the input signal that it had at the time the Hold command was applied.

In data-acquisition systems, sample-hold amplifiers (SHA) are frequently required to "freeze" fast-moving signals prior to their processing by the system. Accurately holding the amplitude of the signal for an appropriate length of time is critical in measurement systems involving analog-digital conversion, peak detection, multiplexing, and other functions where precise timing is of the essence.

The control input is usually TTL- or ECL-logic-compatible; one logic state initiates and maintains the Sample or Track mode; the other does the same for the Hold condition. Most track-holds and sample-holds are identical in both function and circuit implementation. The only distinction between them is how they are used in the system. A *sample-and-hold* implies that the device samples the input for a short time and stays in the hold mode for the duration of the duty cycle. A *track-and-hold*, on the other hand, spends most of the time tracking the input and is switched into the hold mode for only brief intervals.

In data-acquisition systems operating at high update rates (greater than 1 MHz), the terms track-hold and sample-hold lose their distinction; for the purposes of this discussion, all such devices will be referred to as SHAs. The circuitry, characteristics, and applications of various types of SHAs will be discussed.

18.1 SAMPLE-HOLD OPERATION

Regardless of the circuit details or type of SHA in question, all such devices have four major components. The input amplifier, energy storage device (hold capacitor), output buffer, and switching circuits are common to all SHAs, as shown in the typical configuration of Figure 18.1.

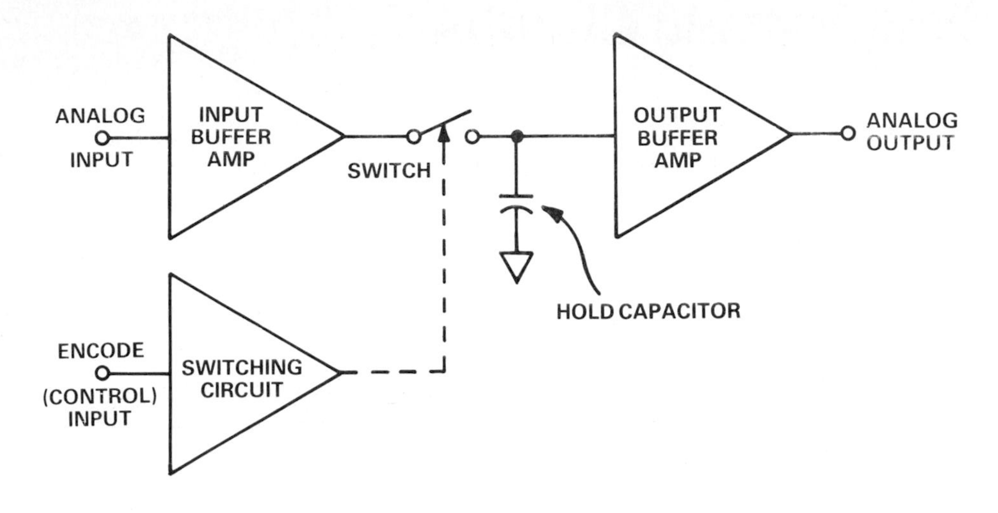

Figure 18.1. Basic sample-hold-amplifier structure.

The energy-storage device, the heart of the SHA, is almost always a capacitor. The input amplifier buffers the input by presenting a high impedance to the signal source and providing current gain to charge the hold capacitor. In the track mode, the hold capacitor usually determines the frequency response of the device; in the hold mode, the capacitor retains the voltage existing before it was disconnected from the input buffer. The output buffer offers a high impedance to the hold capacitor to keep the held voltage from discharging prematurely. The switching circuit and its driver form the mechanism by which the hold capacitor is alternately switched between Track and Hold.

18.2 SPECIFICATIONS

There are four groups of specifications that properly describe SHA operation. They are the *static* and *dynamic* characteristics that describe operation in and between the *track* and *hold* modes. Unique to SHAs are the dynamic specifications that describe the transitions from Track to Hold, and Hold to Track. An understanding of the terminology used to describe these devices is of key importance to the proper selection and use of SHAs. Figure 18.2 shows errors (exaggerated) during a complete cycle from Track to Hold and back.

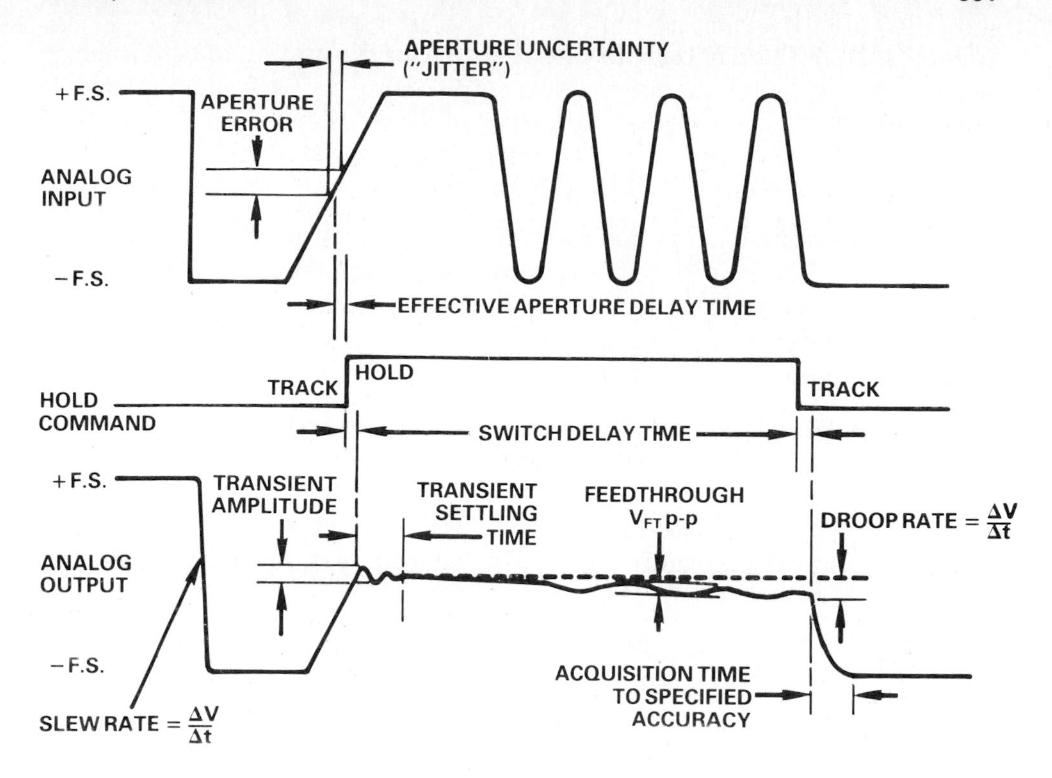

Figure 18.2. Sources of error (exaggerated) in a sample-hold.

18.2.1 TRACK MODE

While in the track—or sample—mode of operation, the SHA is simply a limited-bandwidth amplifier that may or may not provide gain. Operation in this mode is described by the same specifications that are used to characterize any analog amplifier. Indeed, many SHAs are little more than an operational amplifier with a capacitor and a switch. The principal specifications used to describe sample-mode operation are:

Offset: For zero input, the extent to which the output deviates from zero, over time and temperature.

Nonlinearity: The amount by which the plot of output vs. input deviates from a "best straight line." It is usually specified as a percentage of full-scale.

Gain: The multiplication factor describing the input to output "dc" transfer function.

Settling Time: The time required for the output to attain its final value within a specified fraction of full scale when a full-scale analog input step is applied, either as an input step or by a hold-to-sample transition (0 to ±FS, −FS to +FS, or +FS to −FS).

Bandwidth: Describes the frequency response in terms of output attenuation over frequency; it is usually characterized by the −3-dB value.

Slew(ing) Rate: The maximum rate of change of the output voltage when an analog step is applied, either as an input step or by a hold-to-sample transition.

18.2.2 TRACK-TO-HOLD TRANSITION

The specifications used to describe the transition from Track to Hold and from Hold to Track seem to be the most confusing to users of SHAs. These terms are unique to SHA devices and deserve special attention. Perhaps the most misunderstood and misused specifications are those that include the word "aperture" in their definition.

Aperture Time: The most essential dynamic property of a SHA is its ability to disconnect quickly the hold capacitor from the input buffer amplifier. The short—but non-zero—interval required for this action is called *aperture time*. The actual value of voltage that gets held at the end of this interval is a function of both the input signal and the errors introduced by the switching operation itself.

Sample-to-Hold Offset or Pedestal: There is a step error, which causes the held value to differ from the last value in *sample*; it is caused by charge dumped onto the hold capacitor via stray capacitance from the switch-control circuit. A design objective is to keep this error, resulting from a non-ideal switch, independent of the input signal level; the degree to which it in fact deviates from a constant over the input signal range can result in nonlinearities in the output with respect to the input signal. The constant offset portion of the error, called *charge transfer*, or *offset step*, can be compensated by coupling a signal of opposite phase onto the hold capacitor through an auxiliary switching circuit and compensation capacitor.

Thus, the *sample-to-hold-offset*, or *pedestal*, specification depends on the actual device configuration. For SHAs having fixed internal hold capacitance, it includes the residual uncorrected step and the *offset nonlinearity*. For integrated-circuit SHAs, requiring external capacitors, it is the residual step error after the *charge transfer* is accounted for and/or cancelled. In a device for which the capacitance can be chosen by the user, these effects can be reduced by increasing the capacitance in proportion, but at the cost of increased acquisition time.

Aperture Delay (Or, more descriptively, *Effective Aperture Delay Time*): This specification is important because it helps the SHA user know when to strobe the device with respect to the input-signal timing. Figure 18.3 shows the sequence of what happens when the hold command is applied with an input signal of arbitrary slope (for clarity, the sample-to-hold offset error and switching transients are ignored). The value that finally gets held is a delayed version of the input signal, averaged over the aperture time of the switch. Effective Aperture Delay Time is defined as the interval between the leading edge of the hold command and the instant when the *input* signal was equal to the held value. This is a more useful specification than aperture time alone, because it includes the effects of the analog and digital propagation delays, as well as the aperture time.

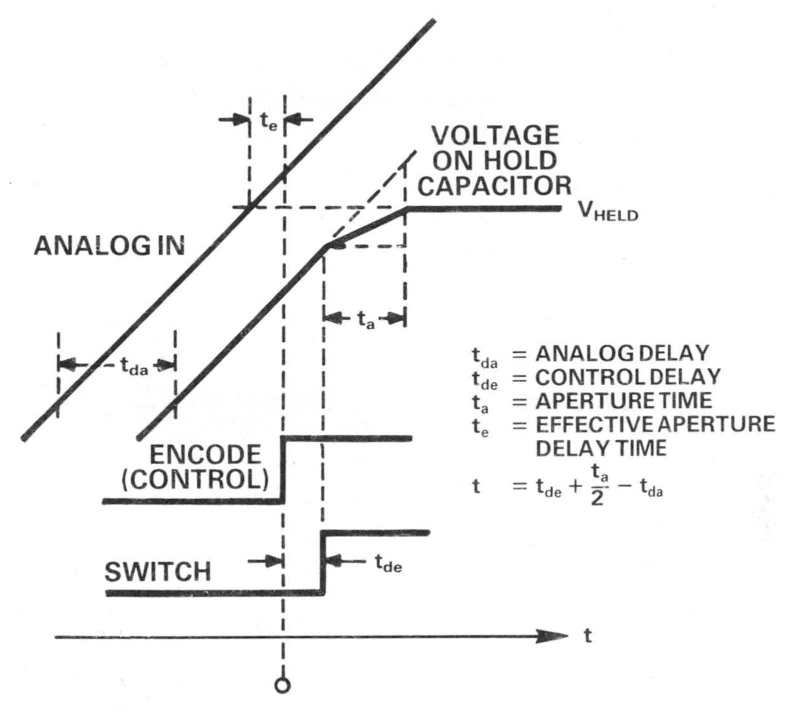

Figure 18.3. Internal sample-hold-amplifier timing.

The formula in Figure 18.3 is based on the assumption that the value of voltage on the hold capacitor is approximately equal to the average value of the signal applied to the switch over the interval during which the switch changes from a low to high impedance (aperture time).

Aperture Uncertainty (Jitter): Aperture uncertainty, or "jitter", is the result of noise which modulates the phase of the hold command. This jitter shows up as a sample-to-sample variation in the value of the analog signal which is being "frozen".

Aperture uncertainty manifests itself as an aperture error, as shown in Figure 18.2. The amplitude of the error is related to the rate-of-change of the analog input. For any given value of aperture uncertainty, aperture error will increase as the input dV/dt increases.

Switching Transient: As shown in Figure 18.2, most SHA specifications include the maximum amplitude and duration of the transient that appears at the output as a result of the sample-to-hold transition. A similar transient appears during the hold-to-sample transition but is usually not noticeable because of the dominant effect of acquisition time.

Switch Delay Time: The interval between the edges of the hold command and the beginning of change of state at the analog output. This delay, as shown in Figure 18.2, occurs at both the sample-to-hold and hold-to-sample transition.

18.2.3 HOLD MODE

During the hold mode there are errors due to imperfections in the switch, the output amplifier, and the hold capacitor.

Droop: A constant drift of the output voltage due to charge leakage from the hold capacitor through the switch, output buffer, circuit board or substrate—or within the capacitor itself (dV/dt = I/C). This error can be reduced by increasing the hold capacitance (at the cost of increased acquisition time) and/or reducing leakage currents by component choice, component placement, and shielding or guarding.

Feedthrough: The fraction of input signal that appears at the output in Hold, caused primarily by capacitance across the switch. Usually measured by applying a full-scale sinusoidal input at a fixed frequency (e.g., 20 V peak-to-peak at 10 kHz) and observing the output during Hold.

Dielectric Absorption: The tendency of charges within a capacitor to redistribute themselves over a period of time, resulting in "creep" to a new level when allowed to rest after large, fast changes. This effect, less than 0.01% for good polystyrene and teflon capacitors, can be as large as several percent for ceramic and mylar capacitors.

18.2.4 HOLD-TO-SAMPLE TRANSITION

Acquisition Time: The length of time during which the SHA must remain in the sample mode in order for the hold capacitor to acquire a full-scale step input; adequate acquisition time makes it possible for the subsequent hold-mode output to be within a specified error band of the final value.

Acquisition time is a key SHA dynamic specification. The maximum sample rate of any SHA is limited by the sum of the time intervals required for the sample and the hold modes. The interval spent in the hold mode (after transients have settled) is primarily determined by the system in which the SHA is used. The minimum time spent in the Sample mode, however, is established by the sample-hold's acquisition time for a given degree of accuracy.

When acquisition time is measured as the interval from the hold-to-sample transition to the instant when the output buffer has settled, the resulting value of acquisition time is generally pessimistic. As defined, acquisition time measures the time required to acquire the signal *at the hold capacitor, not necessarily at the output of the buffer* (unless they are the same). For devices in which the hold capacitor is buffered by a follower, rather than used as an integrator (see the next Section), a measurement at the output includes output buffer settling time as well as switch delay time. In practice, the voltage has already settled at the hold capacitor, and a Hold command can be applied before the output buffer completely settles.

Figure 18.4 is a sketch of waveforms in a method of determining acquisition time that is independent of output-buffer effects. The input analog square wave is sampled and converted at twice the analog square-wave frequency with relatively narrow Sample pulses. The Figure indicates that, on the transition to Sample, the output starts to change, in order to follow the step input. Initially set wide enough for the signal to be accurately acquired, the pulsewidth, t_{aq}, is reduced until the output waveform, measured digitally, begins to collapse to some defined percentage of full scale. When that occurs, the width of the Sample pulse is equal to the acquisition time, because the hold capacitor can no longer acquire the signal accurately with further reduction in t_{aq}.

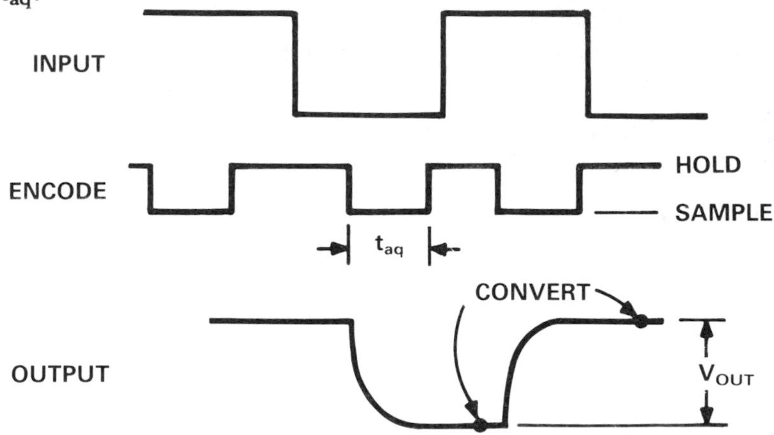

Figure 18.4. Acquisition-time measurement waveforms.

18.3 SAMPLE-HOLD CIRCUITS

Most SHA designs fall into one of two categories, open- or closed-loop circuits. Closed-loop SHAs exploit the accuracy, low drift, and gain flexibility available with operational amplifiers. Open-loop designs take advantage of the high-speed characteristics of unity-gain buffer amplifiers.

18.3.1 OPEN-LOOP CIRCUITS

Figure 18.5 shows the conceptually simplest SHA circuit. When the switch is closed, the capacitor charges exponentially to the input voltage, and the amplifier's output follows the capacitor's voltage. When the switch is opened, the charge remains on the capacitor. The capacitor's acquisition time depends on the series resistance and the current available to charge the capacitor. Once the charge is acquired with the appropriate accuracy, the switch can be opened, even though the amplifier has not yet settled, without affecting the final output value.

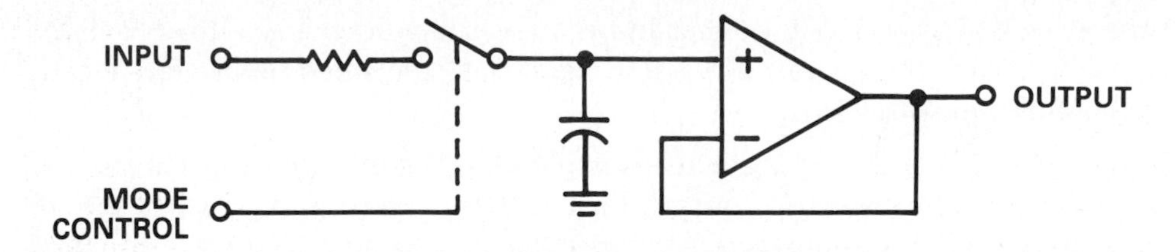

Figure 18.5. SHA structure – simple follower.

A disadvantage of this circuit is that the switched capacitor dynamically loads the input source, which may not have low-enough output impedance and sufficient current-drive capacity. The circuit of Figure 18.6 is similar but includes an input buffer amplifier to isolate the source. The Analog Devices HTS-0025 and HTS-0010 SHAs use this scheme to achieve acquisition times as low as 10 ns. These designs utilize a high speed diode switching bridge for sampling wide-bandwidth signals at update rates of up to 50 MHz. The high speeds are achieved by employing buffer amplifiers that do not use voltage feedback; the nonlinearities of these amplifiers limit their use to systems requiring resolutions of 12 bits or less.

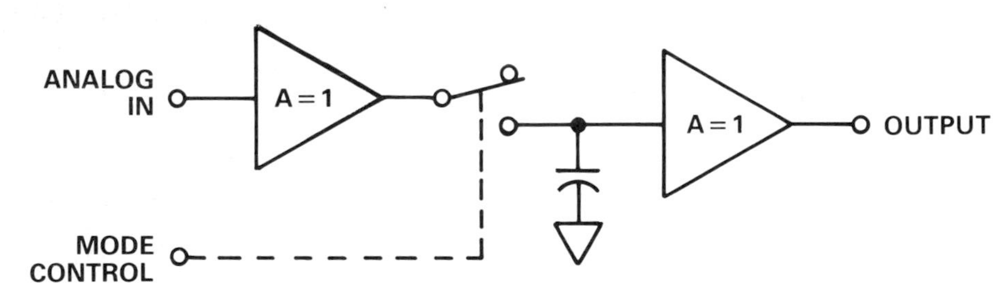

Figure 18.6. SHA structure – follower with input buffer.

The HTS-0010 is an example of a SHA that has the hold capacitor's active terminal brought out so that additional capacitance can be employed to improve the droop rate and sample-to-hold offset (at the expense of bandwidth and acquisition time, as noted earlier).

SHA specifications and terminology are generally applicable to both open-loop and closed-loop SHAs. Applications requiring update rates higher than 10 MHz and acquisition times less than 100 ns usually require open-loop circuits. At these speeds, subtle distinctions in terminology can mislead the system designer. Before selecting a high-speed SHA, the user should have a clear understanding of the intent and meaning of the manufacturer's usage of such specifications as aperture time and acquisition time.

18.3.2 CLOSED-LOOP CIRCUITS

The circuits of Figures 18.5 and 18.6 have the essential advantage of potentially fast acquisition and settling time because they are open-loop devices.

If low-frequency tracking accuracy is more important than speed, this can be accomplished by closing the loop around the storage capacitor and using high loop gain to enforce tracking accuracy.

Figure 18.7 shows a configuration in which the input follower of Figure 18.6 is replaced by a high-gain difference amplifier. Now, when the switch is closed, the output (which represents the charge on the capacitor) is forced to track the input, within the capability of input amplifier's gain, bandwidth, common-mode error, and current-driving capabilities. SHAs using this circuit configuration to achieve higher accuracy than that available in open loop circuits include the Analog Devices AD582, AD583, and ADSHC-85.

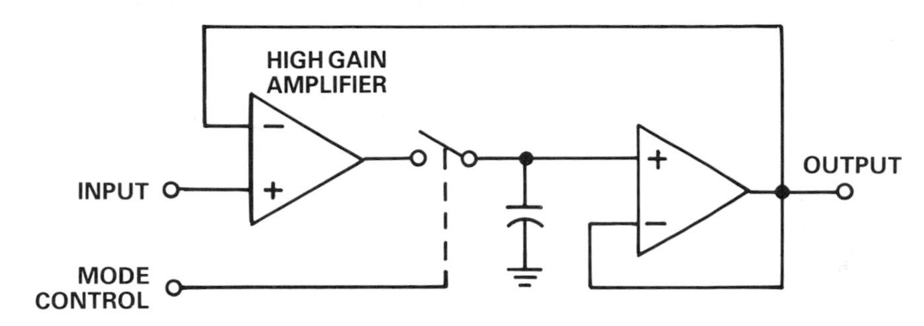

Figure 18.7. SHA structure – feedback circuit with output buffer follower.

In Figure 18.8 (which illustrates the architecture of the closed-loop AD585 and SHA1144), an integrator configuration is used to charge the capacitor, permitting the switch to operate at ground potential; this simplifies leakage problems.

Because both the output and the input affect the charge on the capacitor, the acquisition time and the settling time are identical in the circuits of both Figure 18.7 and Figure 18.8. If the circuit of Figure 18.7 is switched into Hold before the output has settled at its final value, the sample may be in error. In addition, since the loop is open during Hold, the input stage must reacquire the input when the circuit is returned to Sample, even if the input

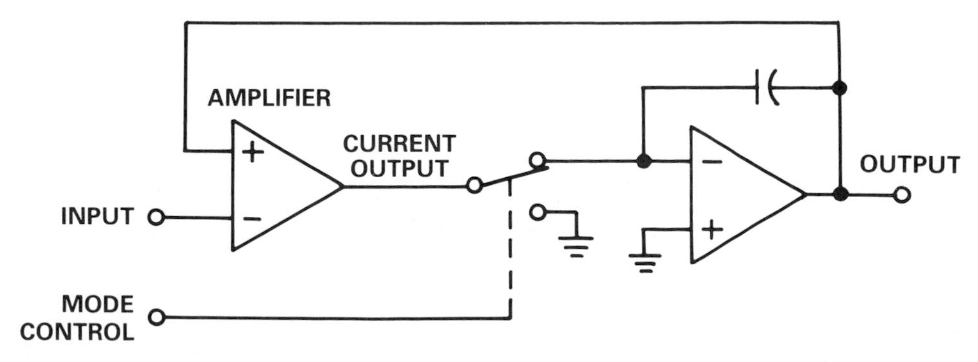

Figure 18.8. SHA structure – feedback circuit with output integrator.

is unchanged. As a rule, this will result in a spike if the input amplifier has high voltage gain.

If input impedance and acquisition time are not critical, and sufficient drive current is available from the source, the circuit of Figure 18.9 may be desirable. Here, only a single operational amplifier is required; and the input impedance and gain are a function of the choice of R_I and R_F. In both Sample and Hold, the input impedance to ground is R_I. In the Sample mode, the input circuit sees a virtual ground through R_I, and the hold capacitor is charged by the amplifier. In the Hold mode, the resistance node is switched to analog ground potential to disconnect the capacitor while minimizing signal feedthrough and maintaining a constant input impedance. The Analog Devices AD346, AD389, HTC-0300, and HTC-0500 all utilize this principle.

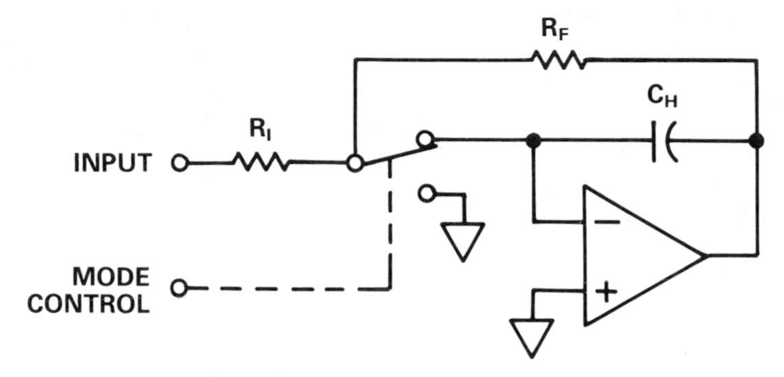

Figure 18.9. SHA structure – inverting integrator switched at summing point.

18.4 APPLICATIONS

Sample-holds are most widely used in data acquisition systems, typically as shown in Figure 18.10. The sample-hold maintains the input to the a/d con-

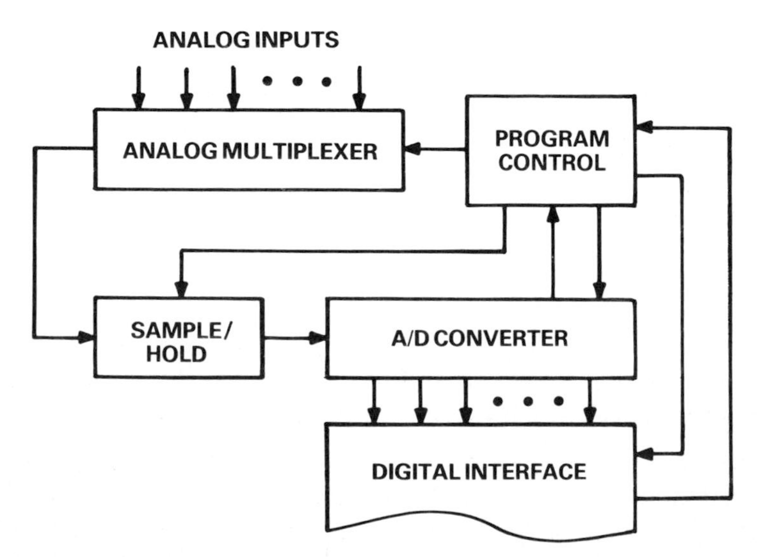

Figure 18.10. Typical data-acquisition system.

verter constant during the conversion interval; meanwhile, the multiplexer is seeking the next channel to be converted, either randomly or sequentially. As soon as conversion is completed, the sample-hold samples the newly established input, and the cycle is repeated.

This mode of operation is known as synchronous sampling; the sample-hold operates in synchronism with the other system elements. In another mode (viz., asynchronous), a large number of sample-holds may be used to acquire and store data at rates pertinent to each individual channel. They are then either interrogated by analog multiplexers, or the signals are individually converted asynchronously, and then multiplexed digitally.

In data distribution, 0.01% sample-holds may be less costly than large numbers of D/A converters having comparable accuracy. A typical data distribution system is shown in Figure 18.11. A fast, accurate D/A converter updates a number of sample-holds at speed and accuracy levels appropriate to the individual channels.

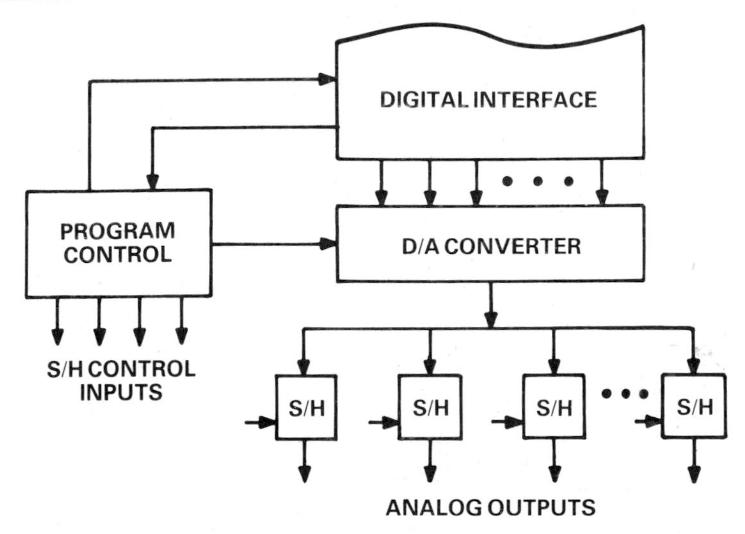

Figure 18.11. Data distribution system with analog storage.

Sample-holds are also used to "deglitch" d/a converters in systems that are sensitive to the D/A glitches that occur at transition points. Figure 18.12 shown the timing relationship used when deglitching DACs. Just prior to latching the new data into the d/a converter, the SHA is put into the hold mode, so that D/A glitches are isolated from the output. SHAs used as deglitchers must have very small sample-to-hold and hold-to-sample transients and pedestal errors, as well as fast acquisition times.

Undoubtedly, the principal usage of SHAs is ahead of analog-to-digital converters. "Aperture time," in an a/d converter system, differs from the SHA specification. In a/d conversion, *aperture time* refers to the period of time over which the analog input must remain stable in order for an accurate conversion

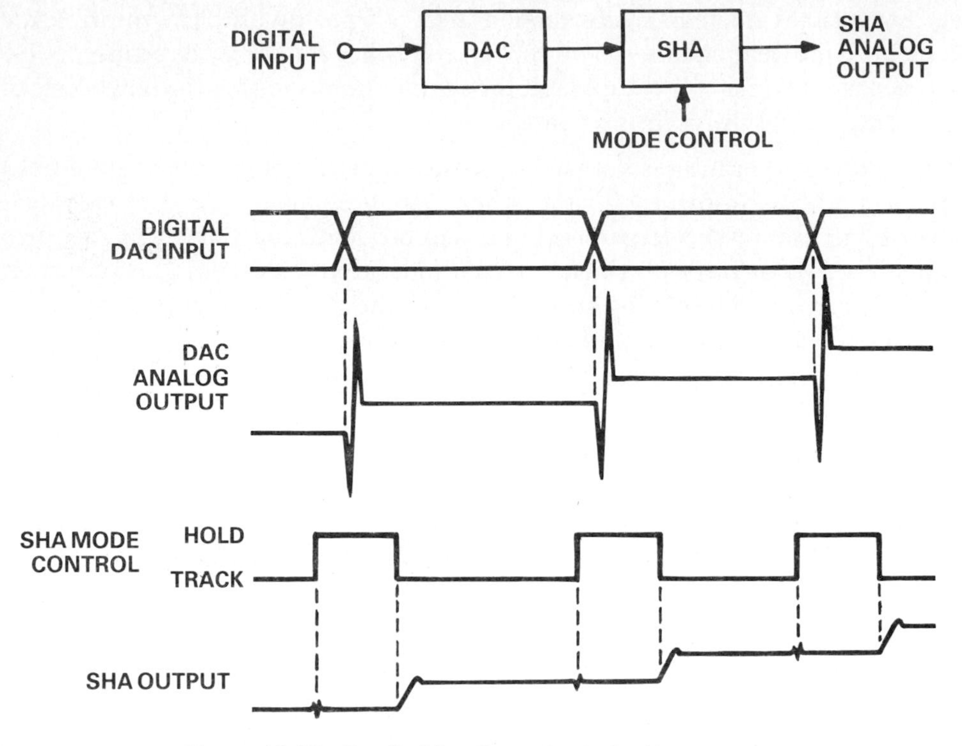

Figure 18.12. Deglitching the output of a d/a converter.

to occur. It is this interval which limits the analog bandwidth (maximum dV/dt) of the A/D converter.

The purpose of the SHA is, in effect, to reduce the aperture time of the a/d converter. By holding the rapidly changing analog input for the period required by the ADC, the analog bandwidth of the system can be increased substantially. In addition, accuracy and linearity—even of low-frequency signals—are enhanced. See Chapters 2 and 13 for discussions of this point.

Most successive-approximation and subranging-type a/d converters require a SHA for all but the lowest analog bandwidths. For this reason, some converters are offered with the SHA incorporated into the design. This makes available to the system designer an ADC-SHA combination in which the timing has already been optimized for the best performance. On the other hand, many state-of-the-art flash type a/d converters offer extremely low aperture times; the inclusion of a SHA with such devices can improve high-frequency performance but may degrade low-frequency performance because of droop associated with the necessarily small capacitors.

Another application of the SHA is as a peak detector. A typical example is shown in Figure 18.13.

When the input is greater than the SHA output, the comparator's positive output causes the SHA to track. When the input decreases and becomes less

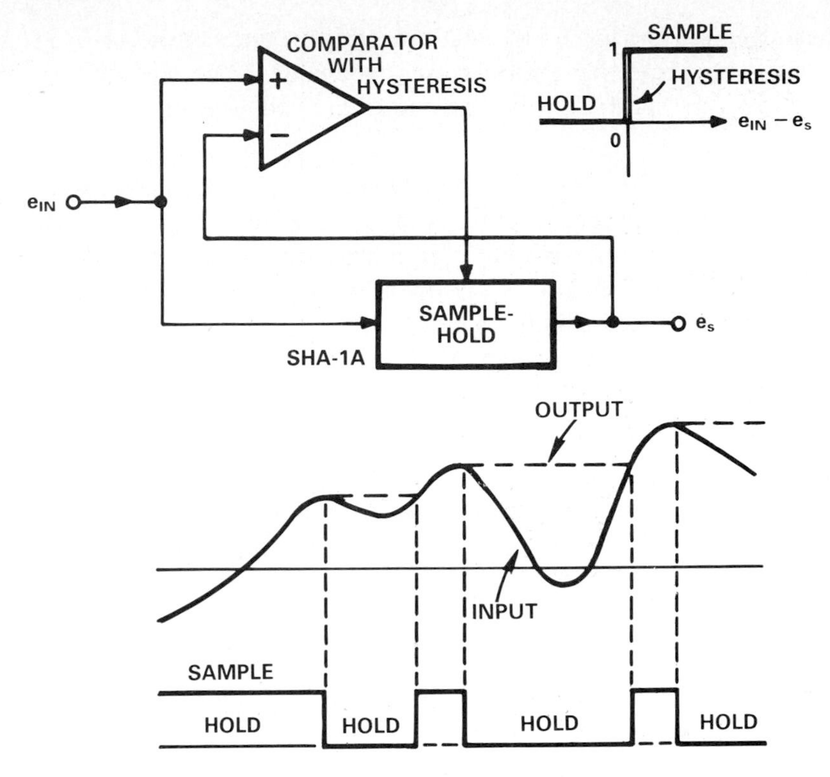

Figure 18.13. Peak follower, using a SHA and a comparator.

than the SHA output, the comparator's "0" output causes the SHA to hold until the input again becomes greater than the output. To reset, the control input is arbitrarily switched into Sample, and the lowest expected level is applied at the input. The sample-hold output (or the comparator input) is biased by a few millivolts of hysteresis to avoid ambiguity during step inputs and minimize false triggering by noise.

18.5 SELECTING AND USING SHAS

In a data-acquisition system, the point where the analog sample is taken—the sample-hold—is often the weakest link, because many system designers overlook the challenging nature of the requirements on both its static and dynamic performance. As an example of the variety of performance characteristics, Table 18.1 lists a number of SHAs available from Analog Devices at the end of 1984 and some of their critical specifications.

In those devices that require an external hold capacitor, the charge transfer is given in picocoulombs (pC). The actual sample-to-hold offset error is given by:

$$V_{error} = \frac{Q}{C} = \frac{\text{Charge Transfer (pC)}}{\text{Hold Capacitance (pF)}}$$

In choosing capacitors for devices with this option, the sample-to-hold (pedestal) error is only one of the specifications to consider. The settling time, droop rate, bandwidth, and acquisition time are all functions of the hold capacitance.

When comparing SHA specifications, be sure to take into account the conditions under which the measurements are made. For example, settling-time measurements made while driving a large capacitive load will rarely be as good as manufacturers' specifications, based on measurements with a simple resistive load. In short, system designs incorporating SHAs require consideration of the associated circuits and layout, in addition to understanding and appreciation of the device's specifications.

With all this for the user to consider, it is little wonder that many manufacturers offer ADCs and SHAs packaged (and specified) together. This is another dimension of choice for the system designer: the advantages of an ADC, complete with SHA; or the flexibility offered by the immense variety of individually available SHAs and ADCs.

	Acquisition Time	Droop Rate or Current	Aperture Uncertainty (ps, rms)	Pedestal Error (Voltage or Charge)	Offset vs. Temp.	Nonlinearity	Technology
HTS-0010	14 ns (0.1%)	100 μV/μs	5	5mV	125μV/°C	0.1%	Hybrid
HTS-0025	25 ns (0.1%)	200 μV/μs	20	5 mV	100 μV/°C	0.1%	Hybrid
HTC-0300A	150 ns (0.1%)	5 μV/μs	100	5 mV	100 μV/°C	0.01%	Hybrid
HTC-0500	700 ns (0.1%)	0.5 μV/μs	60	5 mV	100 μV/°C	0.01%	Hybrid
AD346	1 μs (0.01%)	0.1 mV/ms	400	10 mV	–	–	Hybrid
AD389	1.5 μs (0.01%)	0.1 μV/μs	400	2 mV	–	0.001%	Hybrid
AD585	3 μs (0.01%)	1 mV/ms	500	0.3 pC	–	–	Monolithic
AD583	4 μs (0.1%)	5 pA	5,000	10 pC	–	–	Monolithic
ADSHC-85	4.5 μs (0.01%)	0.2 mV/ms	500	1 mV	25 μV/°C	0.01%	Hybrid
SHA1144	6 μs (0.003%)	1μV/μs	500	1mV	30μV/°C	0.001%	Module
AD582	6 μs (0.1%)	100 pA	15,000	5 pC	–	0.01%	Monolithic

TABLE 18.1. Brief Summary of Sample-Hold Performance

Chapter Nineteen

Analog Switching and Multiplexing

The semiconductor analog switch has become an essential and ubiquitous component in the design of many electronic systems. As an integrated-circuit electronic building block, it has matured greatly since the early days when switches were constructed using discrete semiconductors.

The electromechanical relay continues to be widely used for signal switching. But the many advantages associated with semiconductor analog switches, particularly with regard to their high speed of operation, reliability, and ease of interfacing to microprocessors, have created significant new roles and markets, which are effectively barred to electromechanical devices.

Table 19.1 is a brief comparison of the advantages and disadvantages of relays and analog switches; it serves to indicate the suitability of the two categories to various application areas in analog signal handling.

19.1 POPULAR IC SWITCH PROCESSES

Several circuit and process techniques are used to fabricate analog switches. The most common of these are JFET, PMOS and CMOS. Following a very brief discussion of the three approaches, this chapter will concentrate on the characteristics and applications of *CMOS* switches, which are becoming widely used in multiplexing and general-purpose switching.

19.1.1 JUNCTION FIELD-EFFECT TRANSISTOR (JFET) TYPES.

JFET switches exhibit constant ON resistance with varying signal voltage. This makes JFET devices particularly suitable for low-distortion signal switching.

RELAYS

ADVANTAGES	DISADVANTAGES
• Low ON resistance • High OFF resistance • Galvanic isolation • Switch characteristics are relatively temperature independent	• Limited number of operations • R_{ON} and R_{OFF} deteriorate throughout the usable life of the device • High power dissipation • Large physical size/weight • Slow (1-millisecond t_{ON}) • Prone to "bounce" • High cost per channel • Not compatible with standard logic (TTL, CMOS) • Acoustically noisy

ANALOG SWITCHES

ADVANTAGES	DISADVANTAGES
• Fast (100-ns t_{ON}) • TTL/CMOS compatible • No switch "bounce" • No channel degradation during switch life • Virtually unlimited number of switch operations • Small size and weight • Low power dissipation • Low cost per channel • Rugged construction • Acoustically quiet	• Relatively High R_{ON} • Potentially Lower R_{OFF} than relays • Switch characteristics are temperature dependent • Lack of galvanic isolation

Table 19.1 Electronic analog switches vs. relays

However, in order for an n-channel JFET switch to be fully turned off, it requires a gate voltage that is more negative than the most negative value of signal voltage to be switched, by at least V_p volts (where V_p is the pinchoff voltage of the FET). The negative supply of the JFET switch driver provides this negative voltage; this means that only signals that are more positive than $(-V_s + V_p)$, where V_s is the negative supply voltage, can be switched.

Figure 19.1 illustrates the range of signal voltages that can be handled by a typical JFET analog switch.

19.1.2 P-TYPE METAL-OXIDE-SEMICONDUCTORS (PMOS)

Analog switches employing PMOS technology are relatively simple to fabricate because of the low number of masking stages necessary during fabrication.

However, the PMOS switch channel exhibits a large variation in ON-resistance with changing signal voltage; this can cause unacceptable degradation of circuit performance.

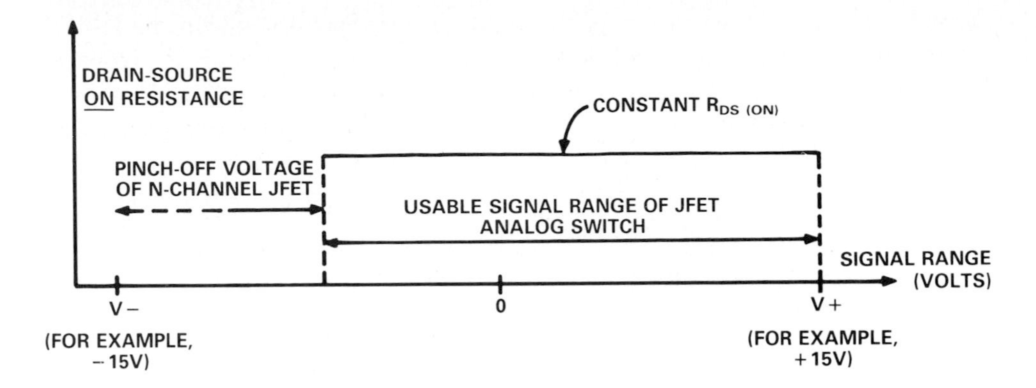

Figure 19.1. Voltage-range limitation of junction FET switches.

For a specified channel ON-resistance, PMOS switches occupy much larger chip areas than either NMOS or CMOS alternatives. Thus, junction capacitances, which are proportional to junction area, are relatively large, making PMOS switches less suitable for high frequency switching.

19.1.3 COMPLEMENTARY METAL-OXIDE-SEMICONDUCTORS

When a PMOS switch is paralleled with an NMOS switch, we arrive at another switch technology—CMOS. CMOS is by far the most popular process used for the fabrication of analog switches. CMOS switches have extremely low quiescent power dissipation, require little drive or supply current while switching, and are low in cost. Their ON resistance is low and varies slightly (typically 10%) with applied voltage. In the OFF condition, leakage is quite small, both across the gate and to the drive and supply circuits. Most types respond to TTL/DTL—as well as CMOS—logic levels.

Like most other processes, CMOS has undergone many and diverse developments over the past decade. Many CMOS process variations have been developed to fulfill the parametric needs of analog switches. These process variations can be conveniently divided into two main types: *junction-isolated CMOS* and *dielectrically isolated CMOS*.

All the various CMOS fabrication technologies make it possible to have low ON-resistance variation with changing signal voltage. Figure 19.2 shows the ON-resistance profiles of separate NMOS and PMOS devices and the effect on R_{ON} when paralleling the two complementary device types to form a CMOS channel.

Junction-Isolated CMOS

This relies on reverse-biased p-n junctions to form the electrical isolation between different devices on the same chip. While junction isolation provides effective isolation under normal conditions, fault conditions can cause junctions to link together to form parasitic bipolar transistors within the CMOS structure; under certain conditions, they can provide the latched low-resistance path that is characteristic of a silicon controlled-rectifier (SCR). While the device behaves normally with specified signal and control voltages, reversed supplies or overvoltage may trigger the parasitic SCR, resulting in high and potentially destructive fault currents in an unprotected device. The phenomenon is known as "latchup."

The reverse-biased isolating junctions exhibit capacitance proportional to junction area; this capacitance tends to degrade high-frequency switch performance. The junction capacitance also induces ac coupling between the switch input/output terminals and the device power supplies, which are effectively at ac ground. This can result in increased insertion loss, reduced OFF-isolation, and susceptibility to high-frequency pickup from the supply leads.

Dielectrically Isolated CMOS

Each DI device is diffused in its own separate pocket of silicon surrounded by an isolating dielectric layer, usually silicon dioxide. The advantages of DI are threefold:

• It allows the IC designer to incorporate devices on-chip which would be difficult, if not impossible, to include using a junction-isolated process. Notable examples of this are current-limiting resistors, which are useful in protecting against input overvoltage.

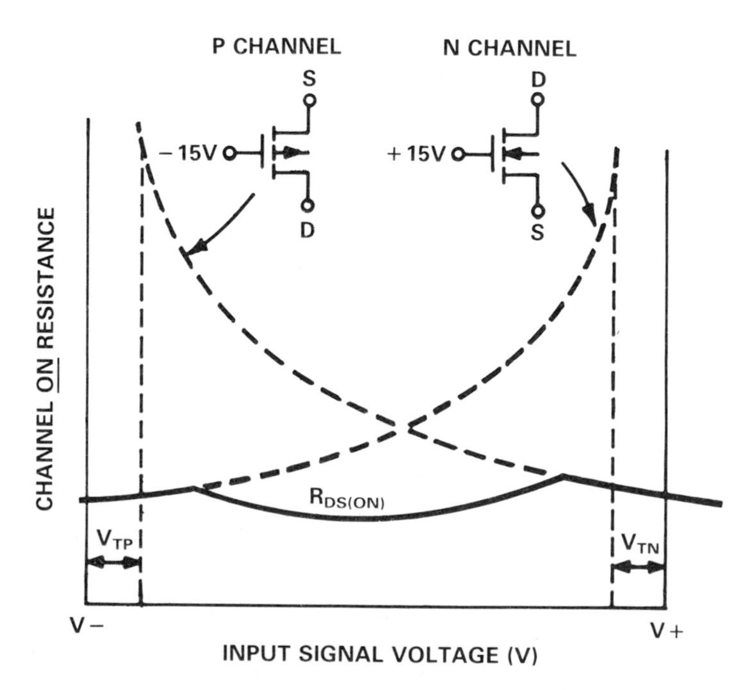

Figure 19.2. ON-resistance profiles of MOS switches.

• It eliminates the parasitic SCRs that can be formed within junction-isolated processes, thus completely avoiding device latchup.

• It reduces circuit parasitic capacitances because of the significant reduction of reverse-biased junction area.

19.1.4 SWITCH FORMS

Analog switches are available in three basic functional forms:

- Single-Pole-Single-Throw—SPST (Figure 19.3a)
- Single-Pole-Double-Throw—SPDT (Figure 19.3b)
- Double-Pole-Single-Throw—DPST (Figure 19.3c)

Note that the DPST function can be made up using two SPST switch channels with their digital control inputs connected together.

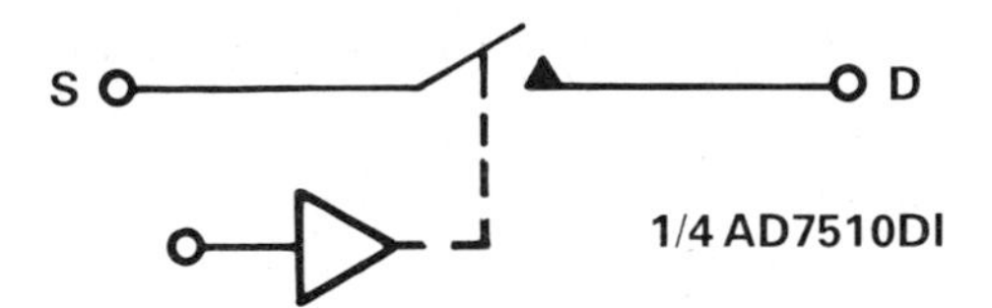

a. Single-pole, single-throw (SPST).

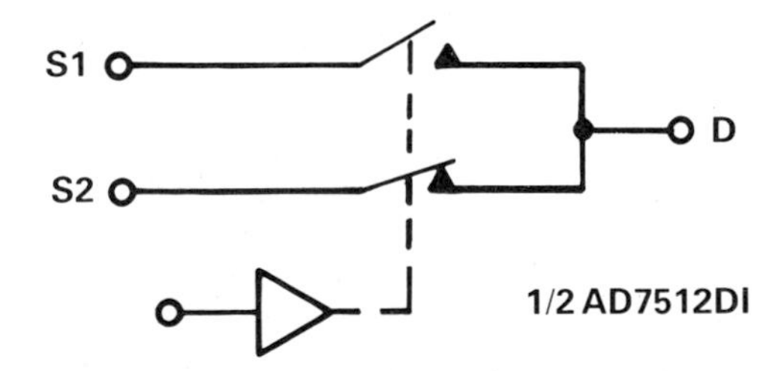

b. Single-pole, double-throw (SPDT).

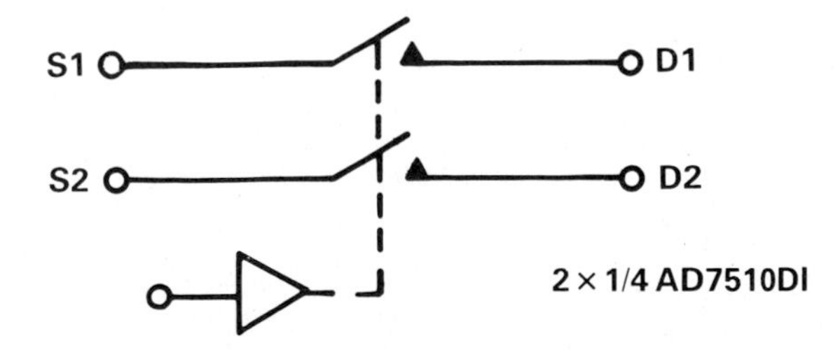

c. Double-pole, single-throw (DPST).

Figure 19.3. CMOS analog switch functions.

Figure 19.4 shows the basic schematic of a typical single-pole-single-throw switch channel. The switch driver includes level shifters, which translate the TTL or CMOS input logic levels into a pair of voltages—which swing between the supply rails—suitable for driving the gates of the two signal-carrying complementary MOSFETS.

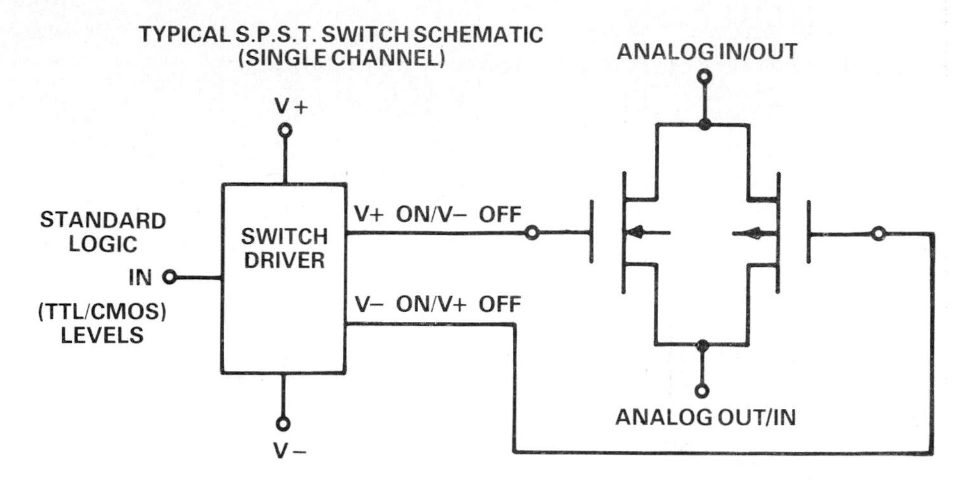

Figure 19.4. Single-pole, single throw CMOS analog switch.

The bodies (back gates) of the n-channel and p-channel FETS are usually connected to the negative and positive supply rails respectively. This ensures that the source-to-body and drain-to-body junctions remain reverse biased for all signal voltages within the supply-rail potentials.

19.2 SWITCH CHARACTERISTICS AND SPECIFICATIONS

There are many switch properties specified on switch and multiplexer data sheets. Key phenomena and specifications will be discussed in the sections that follow. A list of parameters, their abbreviations and definitions, appears in Tables 19.2 (switches) and 19.3 (multiplexers).

19.2.1 R_{ON} VARIATION WITH SIGNAL VOLTAGE

Although the resistance change is not large at higher supply voltages, the ON-resistance of CMOS switches does vary with changes in applied signal voltage. In Figure 19.5, the effect of this phenomenon in the AD7510 is illustrated for various supply voltages.

R_{ON} can introduce attenuation, and its variation with applied voltage can introduce distortion into the signal path. Figure 19.6 shows a typical circuit that exhibits this effect. The nominal 100-ohm resistance of R_{ON} introduces a 1% attenuation, which can be compensated for by a gain trim. However, the $\pm 10\%$ variations of R_{ON} with signal level will modulate the attenuation, introducing a nonlinearity of $\pm 0.1\%$. If the input and feedback resistance are

R_{DS}	Ohmic resistance between terminals D and S.
I_D, I_S	Current at terminals D or S. This is a leakage current when the switch is OFF.
I_{DS}	Current flowing through the closed switch.
$I_D - I_S$	Leakage current that flows from the closed switch into the body. (This leakage will show up as the difference between the current, I_D, going into the switch, and the outgoing current, I_S.
V_D, V_S	Analog voltage on terminal D or S.
C_S, C_D	Capacitance between terminal S or D and ground. (This capacitance is specified for the switch open and closed.)
C_{DS}	Capacitance between terminals D and S. (This will determine the switch OFF isolation as a function of frequency.)
C_{DD}, C_{SS}	Capacitance between terminals D or S of any two switches. (This will determine the cross-coupling between switches as a function of frequency.)
t_{ON}	Delay time between the 50% points of the digital input and switch ON condition.
t_{OFF}	Delay time between the 50% points of the digital input and switch OFF condition.
V_{INL}	Threshold voltage for the low state.
V_{INH}	Threshold voltage for the high state.
I_{INL}, I_{INH}	Input current of the digital input.
C_{IN}	Input capacitance to ground of the digital input.
V_{DD}	Most positive voltage supply.
V_{SS}	Most negative voltage supply.
I_{DD}	Positive supply current.
I_{SS}	Negative supply current.

Table 19.2 Switch Characteristics

sufficiently high, distortion effects can be neglected; for example, 100-kΩ resistors will reduce the nonlinearity to ±0.01%. However, excessively large resistance values can incur high noise levels. An acceptable trade-off between noise performance and distortion may not be possible using this circuit configuration.

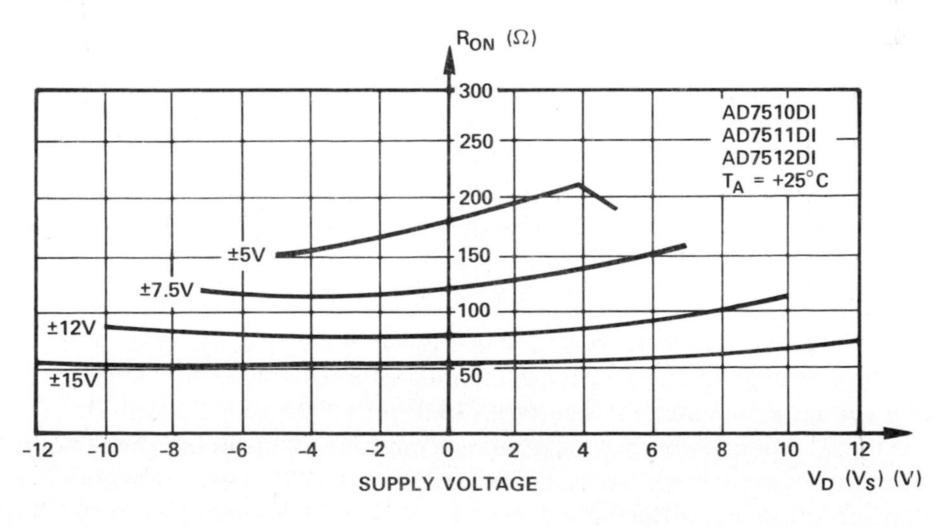

Figure 19.5. Switch ON resistance vs. V_D or V_S, as a function of supply voltage.

R_{ON}	Ohmic resistance between the output and an addressed input.
R_{ON} vs. Temperature	R_{ON} drift over the temperature range.
ΔR_{ON} between switches	Difference between the R_{ON} values of any two switches.
R_{ON} vs. Temperature between switches	Difference between the R_{ON} drifts of any two switches.
I_S	Current at any switch input, S1 through SN. This is a leakage current when the switch is open.
I_{OUT}	Current at the output. This is a leakage current when all switches are open.
$I_{OUT}-I_S$	Difference between the current flowing into terminal S and the current flowing out of terminal "out" when terminal S is addressed.
V_{INL}	Digital threshold voltage for the low state.
V_{INH}	Digital threshold voltage for the high state.
C_S	Capacitance between any open terminal, S, and ground.
C_{OUT}	Capacitance between the output terminal and ground with all switches open.
$C_{S\text{-}OUT}$	Capacitance between any open terminal, S, and the output terminal.
C_{SS}	Capacitance between any two "S" terminals.
$t_{transition}$	Delay time when switching from one address state to another.
t_{open}	"OFF" time of both switches when switching from one address state to another.
$t_{on(En)}$	Delay time between the 50% points of the enable input and the switch "ON" condition.
$t_{off(En)}$	Delay time between the 50% points of the enable input and the switch "OFF" condition.
V_{DD}	Most positive voltage supply.
V_{SS}	Most negative voltage supply.
I_{DD}	Positive supply current.
I_{SS}	Negative supply current.

Table 19.3 Multiplexer Characteristics

An alternative is to make the circuit insensitive to loading by eliminating or buffering the load resistance. In Figure 19.7, the switches are buffered by a non-inverting unity-gain stage. Distortion is substantially eliminated because the modulation of ON-resistance has no effect on the gain of the amplifier. Since the resistors have been eliminated, the noise effects associated with them in Figure 19.6, due to the noise they generate, as well as the voltage developed across them by the input current noise of the amplifier, are also eliminated.

Another way to improve the situation of Figure 19.6 is to connect the switches at the summing point of the unity-gain inverter, as shown in Figure 19.8. The signal voltage across the switch is kept small; thus, ON-resistance modulation effects due to signal-voltage variations are minimized, although the series resistance of the switches must still be accounted for. An interesting advantage of this circuit is that differing values of input resistance may be used to program different gains for the various input signals.

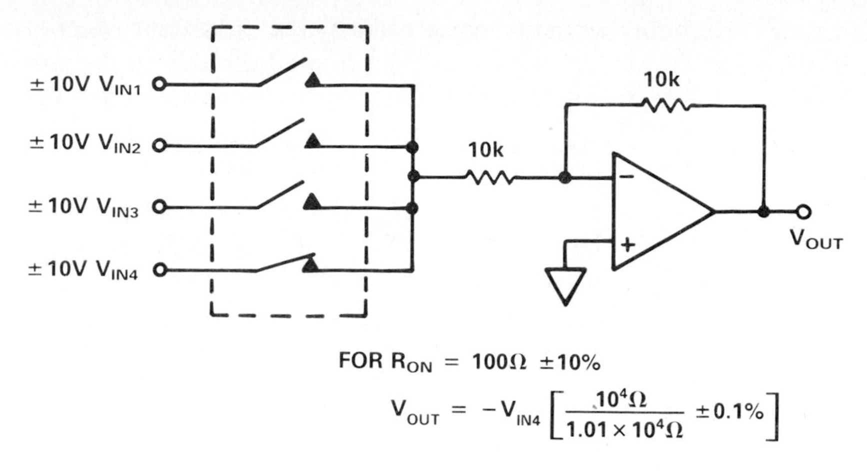

Figure 19.6. Quad switch selects an input for a unity-gain inverter.

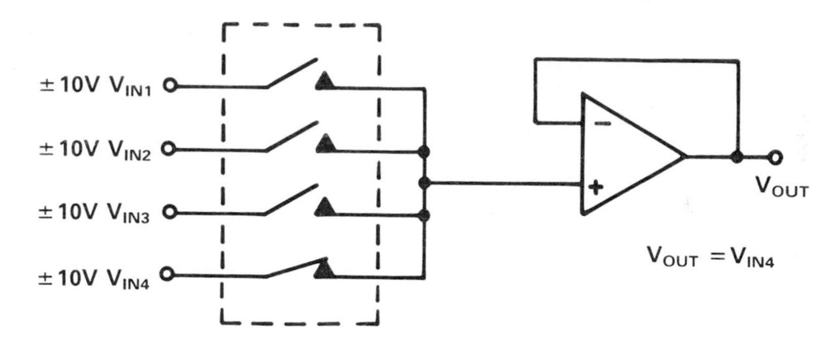

Figure 19.7. Switching the input of a follower-amplifier.

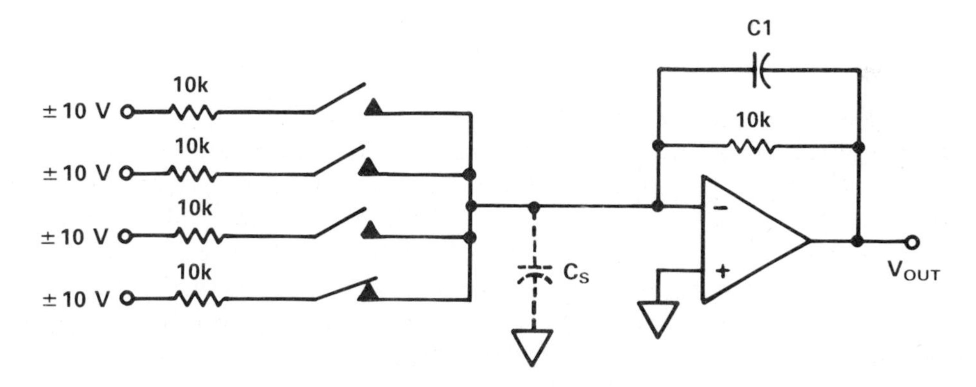

Figure 19.8. Switching at the summing point.

However, this configuration can introduce dynamic problems. First, the switch capacitance, C_S, which appears between the op-amp summing point and ground, can reduce phase margin. To compensate, capacitor C1, with approximately equal capacitance, is used to maintain stability—at the cost of reduced frequency range.

In addition, switching transients, coupled in by the capacitance between the switch drive and the switch points, will be coupled directly to the amplifier summing point, introducing output noise. This is not as much of a problem in Figures 19.6 and 19.7, because the coupled charge is dumped into the (usually low) output impedance of the connected signal source.

Since the switch resistances tend to be approximately equal, a permanently ON switch channel can be inserted in series with the feedback resistor, as shown in Figure 19.9, to reduce the gain error caused by the R_{ON} of the switches in series with the circuit resistors. In addition, since the switch resistances tend to track with temperature, gain drift due to temperature-induced resistor variation can be reduced. However this improvement in amplifier performance is achieved at the cost of one signal channel.

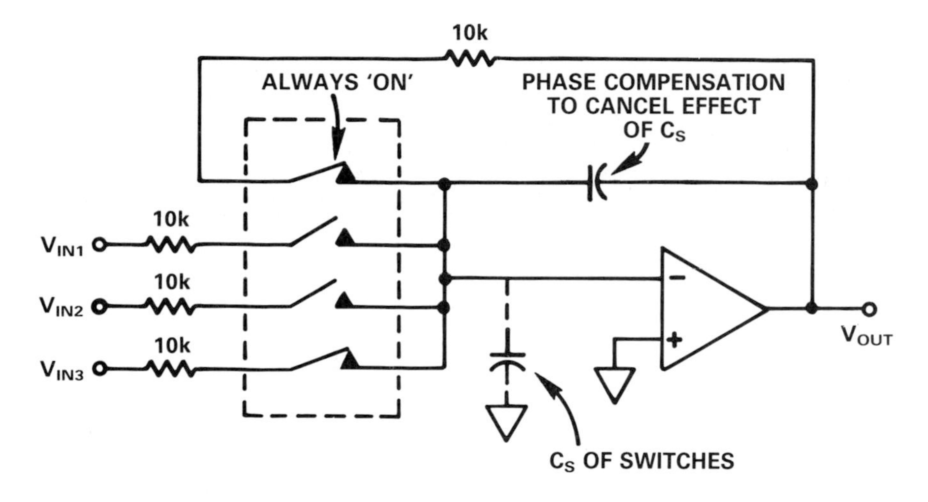

Figure 19.9. Compensating for switch resistance and capacitance.

19.2.2 SWITCH LEAKAGE

The dc OFF-isolation of an analog switch is determined by the leakage current flowing between the source and drain terminals.

However leakage current does not only flow between source and drain. Leakage currents measured at the source or drain connection flow from various parts of the circuit, as well as across the switch channel. The various contributions to switch leakage are difficult to quantify separately. An easier, and often more useful, parametric test is the total leakage at the source or drain. This model current can be used to calculate signal voltage errors due to the flow of leakage currents through channel resistance.

The problem of leakage gets worse at elevated temperatures. Leakage tends to double with every 10°C increase in temperature. ON-resistance also has a positive temperature coefficient.

The cumulative effect is that worst-case ON-resistance-leakage voltage errors occur at the maximum system operating temperature and are normally computed for that temperature.

An illustration of this is shown in Figure 19.10, where the total leakage current is shown as the sum of the ON leakage current for the ON channel, and 3 × the OFF leakage current for the three OFF channels. If $I_{D\ ON}$ is 9 nA max at +25°C and 600 nA max over temperature; and $I_{D\ OFF}$ is 3 nA max at +25°C and 200 nA max over temperature; then for $R_{FB} = 10\,k\Omega$,

$$V_{OUT} = 18 \times 10^{-9}\,A \times 10^{4}\,\Omega = 180\,\mu V, \text{at } +25°C$$

$$V_{OUT} = 1.2 \times 10^{-6}\,A \times 10^{4}\,\Omega = 12\,mV,\ -55°C \text{ to } +125°C$$

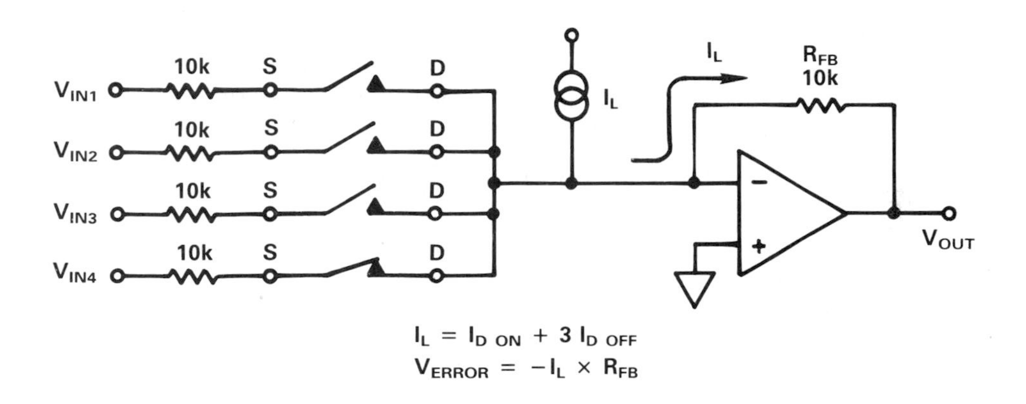

Figure 19.10. Switch leakage effects.

19.2.3 CHARGE INJECTION

As mentioned earlier, charge can be injected into the signal line due to capacitive coupling between the digital control and the source and drain pins.

The coupling is mainly associated with three separate capacitances:

Output switch n-channel gate-to-drain capacitance (C_{GDN}).
Output switch p-channel gate-to-drain capacitance (C_{GDP}).
Digital-control-pin to output-switch capacitance (C_{LIO}).

These capacitances are illustrated in Figure 19.11.

For a positive logic (logic High = Switch ON) SPST analog switch the respective voltage transitions present on these three capacitances when the switch turns OFF are (assuming ±15-volt supplies):

+15 to −15 volts. (n-channel gate capacitance)
−15 to +15 volts. (p-channel gate capacitance)
−5 volts (TTL Logic) or −15 volts (CMOS Logic)

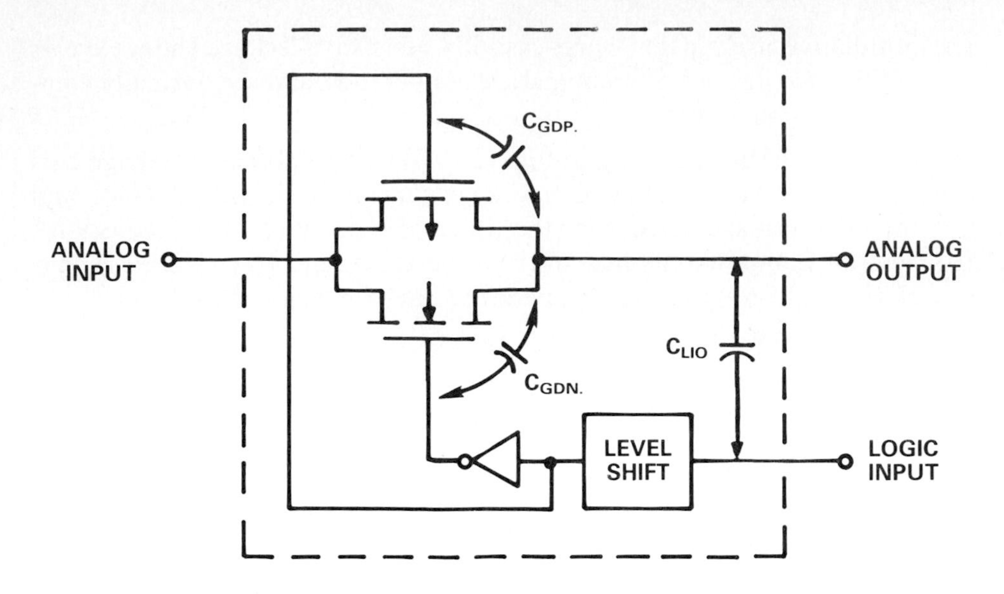

Figure 19.11. Parasitic capacitances.

The charge injections, due to each of these transitions acting on the relevant capacitance, sum algebraically to form a composite charge injection at the switch output.

The p-channel gate capacitance is usually about two to three times greater than the n-channel gate capacitance. The logic-input-to-analog-output capacitance depends largely on the device pin configuration. If these two pins are far enough apart (and the associated wiring maintains the separation), this capacitance can be neglected.

The total charge injection is given by the approximate sum:

$$Q_{INJ}\,(\text{p-channel}) + Q_{INJ}\,(\text{n-channel}) + Q_{INJ}\,(\text{logic input, Off transition})$$

$$Q_{INJ}\,(\text{total}) = C_{GDP}\,\Delta V_{GP} + C_{GDN}\,\Delta V_{GN} + C_{LIO}\,\Delta V_L$$

$$= +C_{GDP}\,30V - C_{GDN}\,30V - C_{LIO}\,5V$$

for $\pm$ 15 volt supplies and TTL-logic, with $V_S = 0$ volts, $R_S = 0$.

This is an oversimplified quantitative approach to computing the charge injection phenomenon. Other parameters which influence the value of charge injection are source voltage, source impedance, and effective load impedance. A typical charge-injection-vs.-source-voltage characteristic is shown in Figure 19.12.

A charge-injection measurement circuit is illustrated in Figure 19.13a, and typical test waveforms that one might see, using this circuit, are shown in Figure 19.13b.

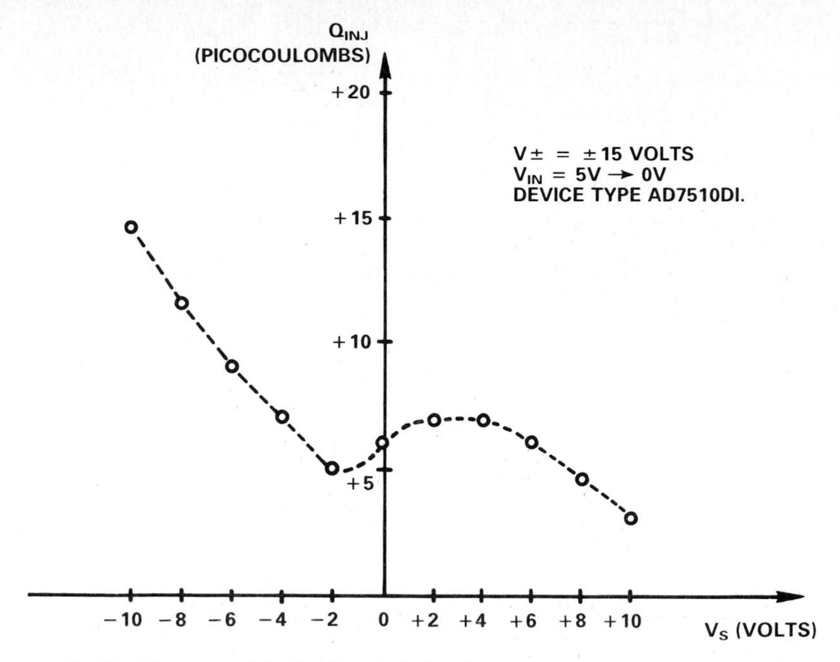

Figure 19.12. Measured drain charge injection as a function of source voltage.

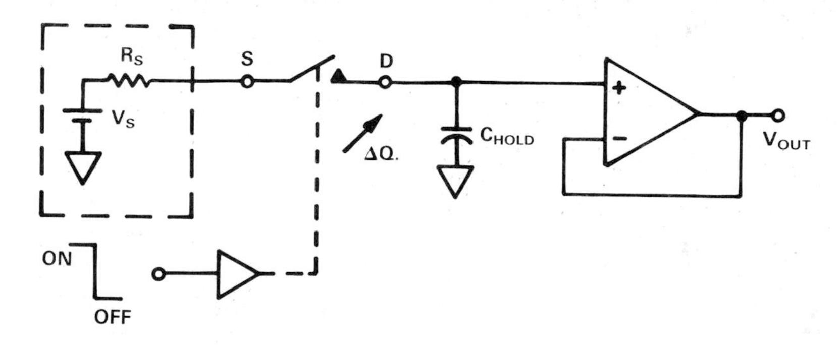

a. Charge-injection measurement circuit.

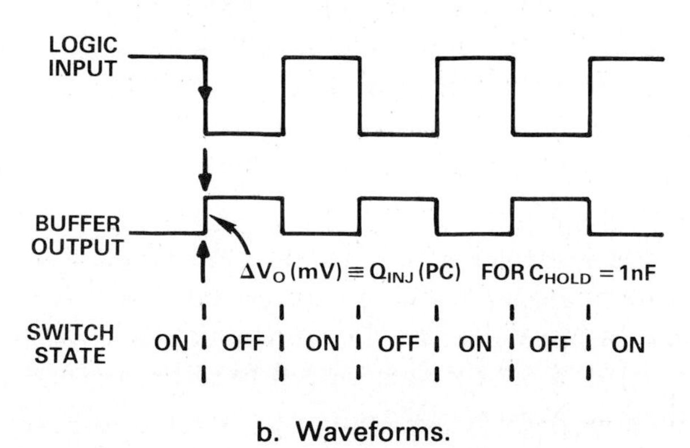

b. Waveforms.

Figure 19.13. Charge-injection measurement circuit and waveforms.

19.2.4 PUMPBACK

Figure 19.14 shows a typical analog switch configuration which can exhibit the phenomenon known as pumpback.

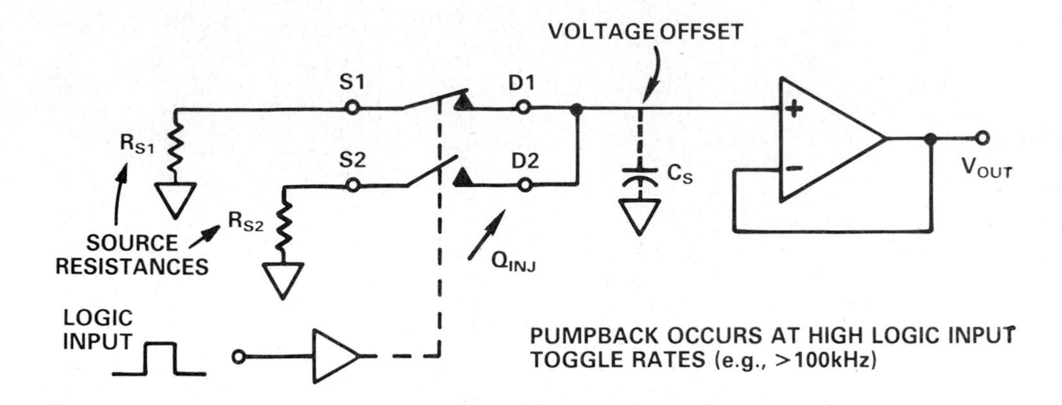

Figure 19.14. Circuit to demonstrate pumpback.

Charge, which is injected into the signal path via D1 when switch 1 turns OFF, normally discharges through the ON-resistance of switch 2, in series with the source impedance at channel 2 (R_{S2}). If the source impedance is low, the injected charge will dissipate quickly. However, if the source impedance is high, discharge will take longer and—if the frequency of changeover between channel 1 and channel 2 is high—the charge may not be sufficiently discharged by the time the next injection of charge, due to the OFF transition of switch 1, takes place. The result of this situation is a standing offset at the switch output. This phenomenon is most readily measured with both switch input voltages at zero. As the switch toggle frequency is increased, this "pumpback" offset voltage will increase.

Pumpback can thus impose an upper limit on the sampling rate of high-speed data-acquisition systems. Reduction of pumpback to increase maximum sampling rates can be effected by selecting low-charge-injection switches and by reducing source impedances.

19.2.5 SWITCHING SPEED

Analog switches and multiplexers are commonly used to sample several input signals for subsequent analog-to-digital conversion. The maximum sampling rate of the switches is determined by the propagation delay of the switch drivers and the time taken for the switch output to settle to within the required error band of the a/d converter.

Propagation delay is quantified by the turn-on and turn-off times of the switch. Turn-on time is the time elapsed from an arbitrary point on the logic input transition (usually 50%) to an arbitrary point (usually 90% of the final value) on the analog output transition. Switching speed and settling time measurements are illustrated in Figure 19.15.

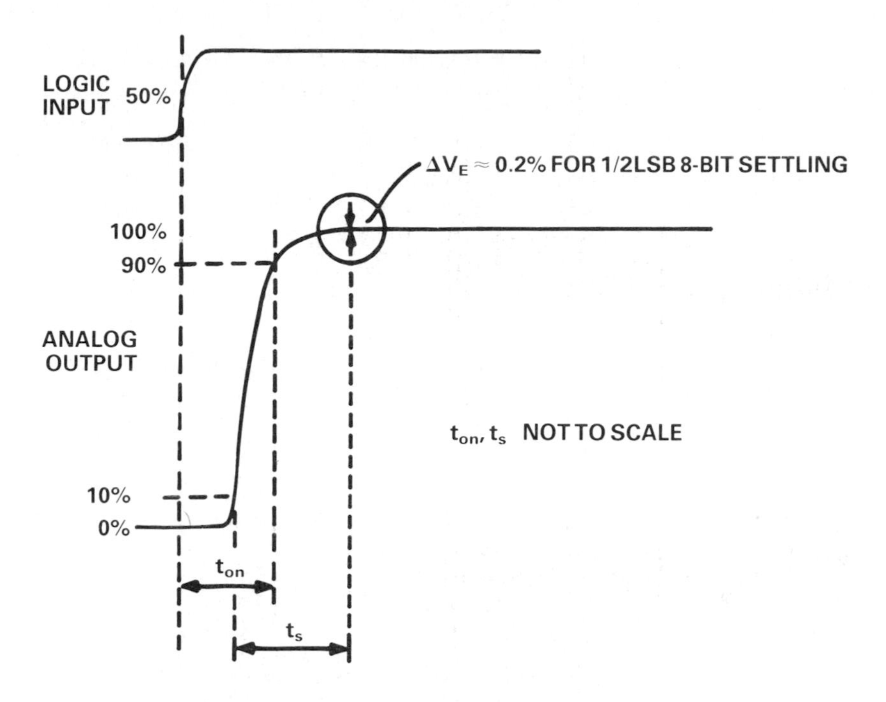

Figure 19.15. Switching speed and settling time.

19.2.6 SETTLING TIME

Settling time, from an arbitrary point on the analog output transition (e.g., 10%) to within a specified percentage of the final value, is a function of the signal source impedance, the switch ON-resistance and the capacitance at the switch output. Measurements of settling time require great care because of the high speed, the role of stray capacitance, and the transient response of the measurement circuitry.

19.2.7 OFF ISOLATION

Analog switches and multiplexers have a limited ability to isolate ac and dc input signals from the switch output. OFF Isolation is limited at dc by the leakage current across the switch. At higher frequencies, the switch capacitances reduce the OFF impedance of the switch. An OFF-Isolation specification must be accompanied by a statement of the frequency of measurement and the load impedance, as both affect the switch isolation. OFF isolation is defined as the ratio, V_{OUT}/V_{IN}, with the switch OFF, and is expressed in dB as $20 \log_{10}(V_{OUT}/V_{IN})$.

19.2.8 INSERTION LOSS

This measure of the ON-channel attenuation in a signal path is defined, at a specific frequency and a given load, as the ratio, V_{OUT}/V_{IN}, with the switch ON. It is expressed in dB as $20 \log_{10} (V_{OUT}/V_{IN})$.

At dc, the insertion loss is approximately $R_L/(R_L + R_{ON})$, where R_L is the load resistance and R_{ON} is the channel ON resistance.

19.2.9 CROSSTALK

This is the amount of spurious signal crossover from the signal input of an OFF-channel to the input of a nearby ON channel.

Crosstalk in analog switches and multiplexers is mainly due to capacitive coupling between channels and is defined as the ratio $V_{IN\ ON}/V_{IN\ OFF}$, where $V_{IN\ OFF}$ is the signal input to the OFF channel and $V_{IN\ ON}$ is the resulting signal at the ON-channel input.

It is important to note that the inherent parameters of the switching device set a *minimum* to the system's expected high-frequency isolation, crosstalk, and insertion loss at high frequencies. These parameters can be further greatly influenced by circuit-board layout.

19.3 MULTIPLEXING AND MULTIPLE SWITCHES

To achieve analog switching of multiple inputs to a common output (or vice versa), using a single IC device, the switching function is designed to make optimum use of each available package pin. The analog multiplexer IC invariably uses binary address-decoding to minimize the number of channel-select control pins.

Features included in the standard multiplexer are an Enable pin, which allows the user to turn off all channels, regardless of the chosen (decoded) channel, and break-before-make switch action, which helps protect the input sources by ensuring that no two channel inputs are connected to one another.

Figure 19.16 shows a typical multiplexed-input data-acquisition system. The 12-bit a/d converter must wait until the output of the multiplexer has settled to within ½ LSB of the decoded input's voltage range before starting a conversion.

The settling time of a multiplexer is directly dependent on the time constant established by the load capacitance and the sum of the input source impedance and the multiplexer ON-resistance.

The time from the application of the multiplexer decode pulse to the point where the a/d converter can start converting is the sum of the switching time (t_{ON}) of the multiplexer and the further time it takes for the buffered multiplexer output voltage to settle to within ½ LSB (12 bits) of the analog input range.

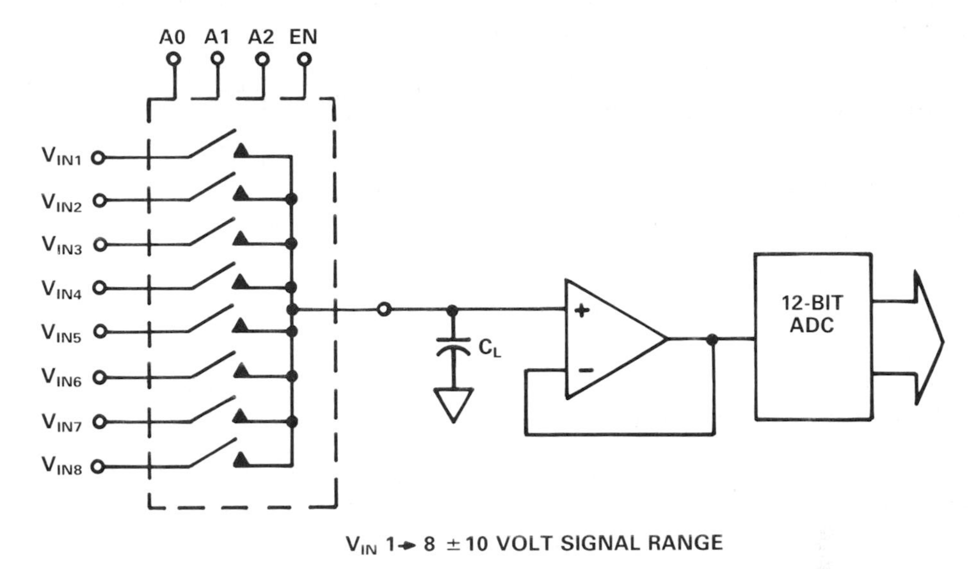

Figure 19.16. Multiplexer application.

19.3.1 LOW-LEVEL SIGNAL MULTIPLEXING

In high-resolution data-acquisition systems, electrically noisy environments can preclude the use of conventional single-ended multiplexing techniques. If a high-CMRR amplifier-per-channel, to eliminate common-mode voltage and amplify the signals substantially before multiplexing, is not feasible, differential multiplexing offers improved common-mode rejection and is particularly useful for switching low-level signals before amplification.

The advantages of differential multiplexing are threefold.

- The data-acquisition system remains differential, from the transducers through to the high CMRR instrumentation amplifier(s).
- Noise induced due to charge injection of the multiplexer switches, being a common-mode effect, tends to cancel if differential multiplexing is used.
- Low-cost twisted-pair wiring can sometimes be used as an alternative to expensive shielded conductors.

Generally, the lower the signal level, the more important the need for differential multiplexing. Figure 19.17 shows a typical differentially multiplexed front end of a low-signal-level data-acquisition system.

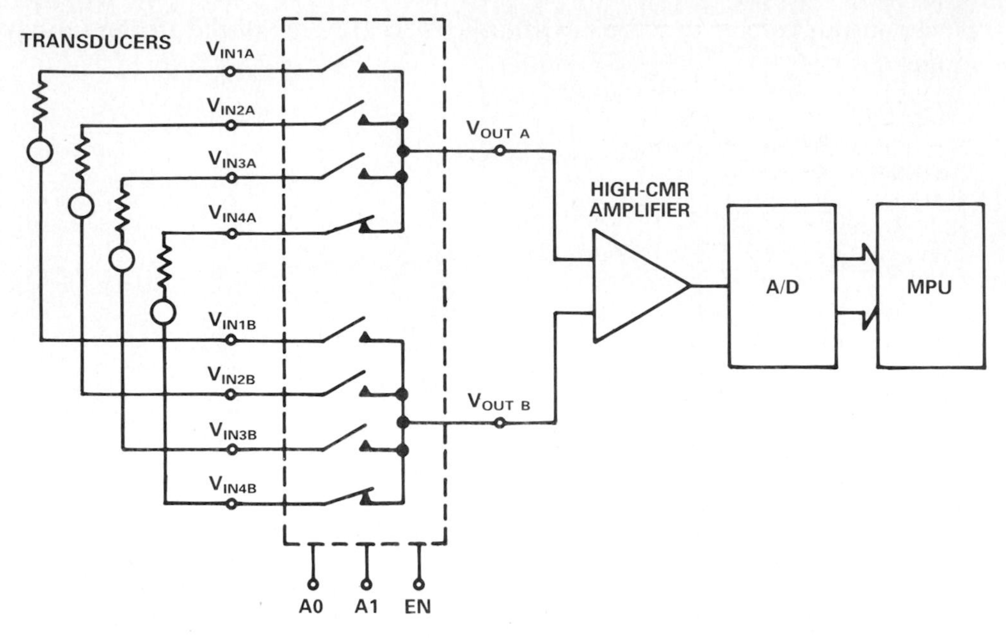

Figure 19.17. Four-channel differentially multiplexed data acquisition.

19.3.2 MULTI-RANK MULTIPLEXING

Figure 19.18 shows a two-level single-ended multiplexing system, used for channel expansion. In this example, eight 8-channel multiplexers are multiplexed into a single 8-channel multiplexer to handle 64 channels of single-ended input. Although it uses an additional multiplexer (the system could have consisted of eight 8-channel multiplexers with their outputs paralleled), it has significant advantages over the single-level n^2-channel system. For example, the n^2-channel system would have n multiplexer outputs connected in parallel, while the 2-rank system only looks at one output at a time, greatly reducing output capacitance and leakage effects.

The error of this circuit due to leakage current is

$$2 \times I_{OUT\ ON} \times (2\,R_{ON} + R_S)$$

where R_S is the analog input source resistance.

The error due to leakage of an equivalent single-level multiplexer system is:

$$(7 \times I_{OUT\ OFF} + I_{OUT\ ON}) \times (R_{ON} + R_S)$$

where $I_{OUT\ ON}$ and $I_{OUT\ OFF}$ are the output leakages of an 8-channel multiplexer (e.g., AD7501) in the Enabled and Disabled modes, respectively.

For source resistances which are significantly greater than R_{ON}, the leakage error of the single-level system approaches $4\times$ that of the two-level multiplexer circuit. This improvement factor is reduced to $2\times$ for source impedances that are low relative to R_{ON}. ($I_{OUT\ OFF} \simeq I_{OUT\ ON}$)

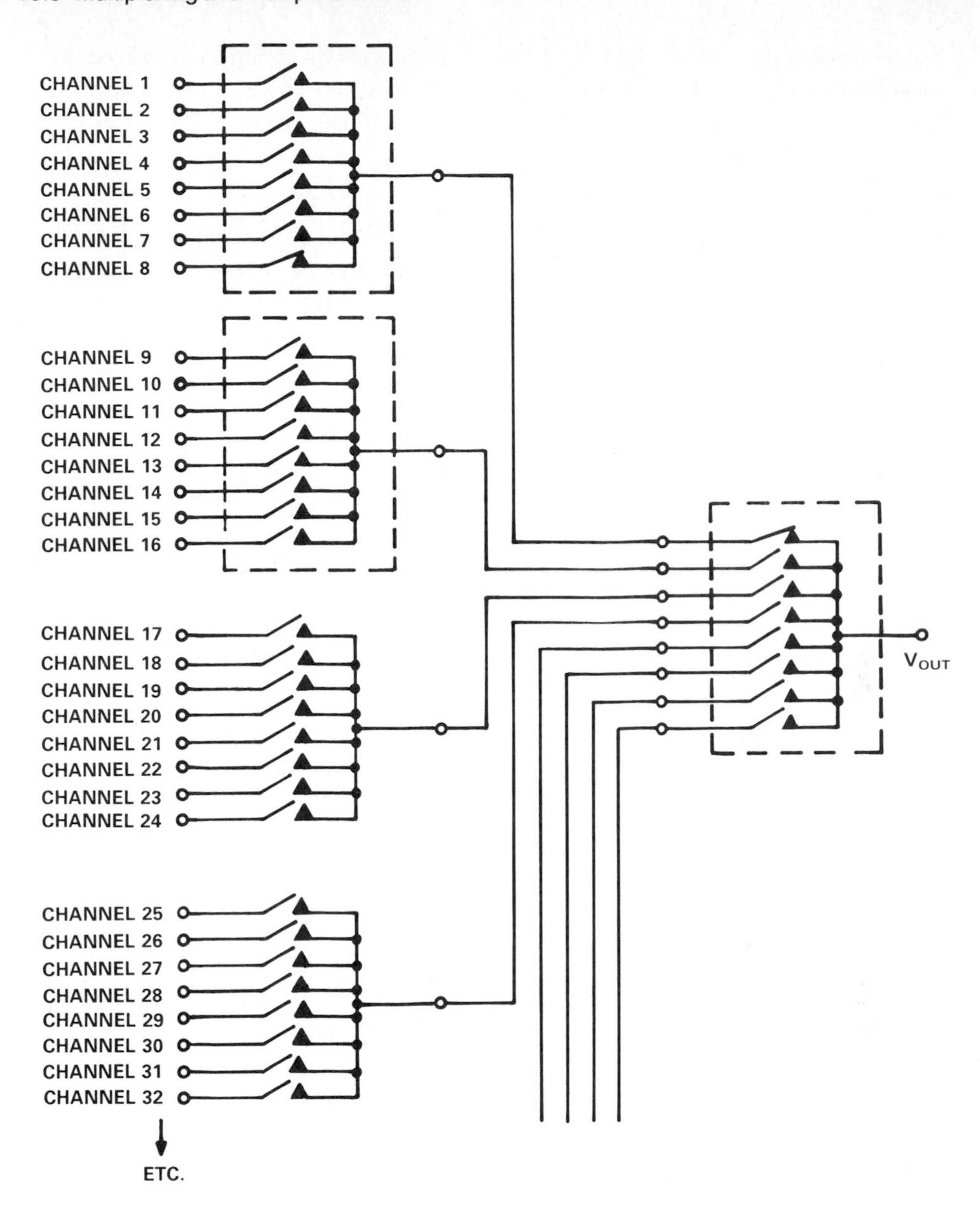

Figure 19.18. Two-level multiplexing. Channel 8 is selected.

A further advantage of two-level multiplexing is potentially faster system switching speed. The maximum toggling rate of the single-level system is limited by the maximum switching speed of each multiplexer. For example, if the multiplexer's minimum t_{ON} and t_{OFF} are 1 microsecond, the maximum toggling rate is 500 kHz. The two-level circuit, however, can be configured using standard 1-μs t_{ON} multiplexers in the first level and fast (e.g., less-than 200-ns t_{ON}) switches in the second level.

Suppose channel 1 of all the first-level multiplexers has been selected; the second-level switches can strobe each multiplexer in sequence. The toggling rate is thus determined by the switching speed of the second-level switches. Approximately half-way through the sequential selection of the eight multiplexers, the first four multiplexers are switched to Channel 2.

By the time that the strobing of channel 1 of the last four multiplexers has been completed, channel 2 of the first four multiplexers is already selected. The maximum system toggling rate has become four times that of the single-level system. This maximum toggling rate is achieved if the second level switches are at least four times as fast as those of the first-level multiplexers.

19.3.3 MULTIPLE-SWITCH CONFIGURATIONS

Combinations of switches can offer improvements in performance over that available from a single switch.

Switches in Parallel

By paralleling two or more switch channels, as shown in Figure 19.19, the ON resistance of the switch is reduced, and the current handling capability is increased.

$$R_{ONT} = \frac{R_{ON}}{N}$$

Where R_{ONT} = ON resistance of switch combination.
R_{ON} = ON resistance of separate switches.
N = No of switches in parallel.

$$I_{DT} = N \times I_D$$

I_{DT} = Maximum drain current of switch combination.
I_D = Maximum drain current of separate switches.

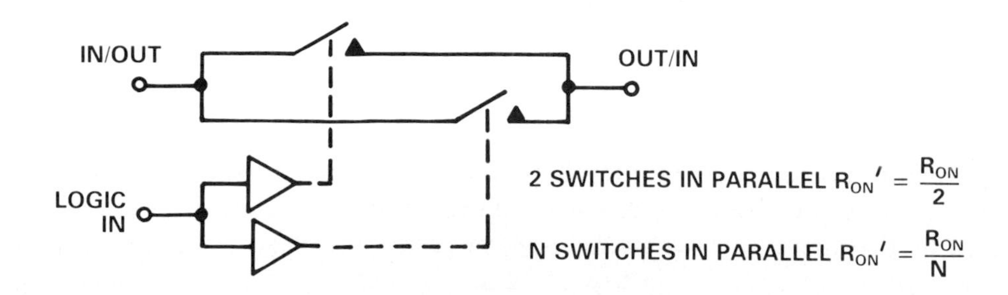

Figure 19.19. Paralleling analog switches.

T-Switch

Figure 19.20 shows three switches configured in T-formation. This configuration provides an extra stage of attenuation, offering superior OFF-isolation performance at both low and high frequencies.

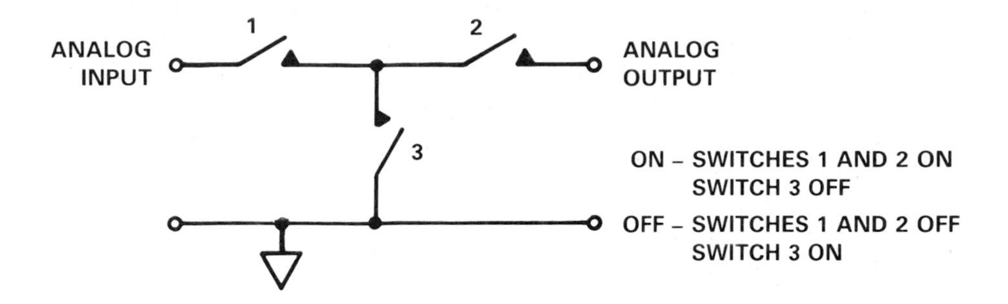

Figure 19.20. T-switch configuration.

19.4 APPLICATIONS

The applications of analog switches and multiplexers can be categorized under the following two broad headings:

Signal Routing
Signal Processing

Signal Routing

Examples of signal routing applications include multiplexed-input data-acquisition systems, automatic test equipment, force and sense selection, and message switching.

Signal Processing

Examples of signal processing applications include programmable-gain amplifiers and programmable filters.

19.4.1 SAMPLE-AND-HOLD

Sample-hold circuits form a significant subset of the applications of analog switches. It is worthwhile to discuss design aspects of such circuits.

Figure 19.21 shows a basic sample-hold circuit suitable for interfacing to 8-bit a/d converters. The time taken for the capacitor to charge to within ½ LSB ($<$ 0.2%) of its final value is $\ln(512) = 6.24$ time constants, or $6.24\times(R_S + R_{ON})\,C_H$. For zero source impedance, $C_H = 10$ nF, and $R_{ON} = 100\Omega$, 8-bit ½-LSB switch settling time will be about 6.2 μs.

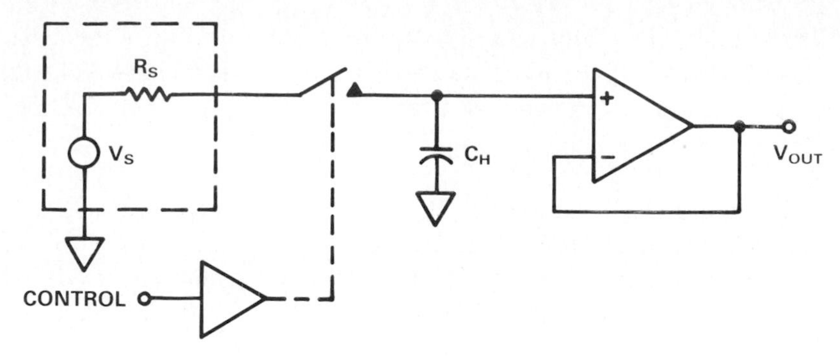

Figure 19.21. Sample-hold circuit.

The *acquisition time* (the time needed to acquire a sample of data), from the time the logic input transition is applied to the switch, is the switch-circuit settling time plus the switch propagation delay. A typical analog switch may have a propagation delay of 300ns; therefore about 6.5 microseconds total sampling time is needed.

Voltage *droop* on the hold capacitor, during HOLD, is caused by buffer amplifier bias-current flow, switch leakage and capacitor leakage. The voltage on the hold capacitor must stay within ½ LSB, at most, of the signal originally applied to the input of the sample/hold, until the a/d converter has completed its conversion.

A design trade off must be made between the following factors:

Magnitude of hold capacitance—C_H.
Settling Time of Sample/Hold—T_S.
Droop Rate of hold-capacitor voltage—$\Delta V_H/\Delta T$.
Charge Injection Error due to switch—ΔV_Q.

Reducing C_H will generally reduce T_S, but lower values of C_H will increase $\Delta V_H/\Delta T$ and ΔV_Q (charge injection is related to the switch capacitance).

The salient parameters affecting the choice of an analog switch for a sample/hold are:

Switch Leakage—I_{DOFF}
Charge Injection—Q_{INJ}
R_{ON}
Switch Propagation Delay—t_{ON} and t_{OFF}

Low charge injection can be achieved by selecting high ON-resistance switches, which exhibit low switch capacitance. Furthermore, switches having higher ON resistance tend to exhibit lower leakage. However, higher ON-resistance increases settling time. The first three listed parameters can be traded off to achieve a suitable system solution.

Propagation delays are generally independent of other analog constraints; the choice is for t_{ON} and t_{OFF} as low as required, based on what is available at reasonable cost. Generally, the shaving of a few hundred nanoseconds off propagation delays will ease the tradeoffs between switch leakage, charge injection and R_{ON}.

19.4.2 DAC DEGLITCHING

Figure 19.22 shows a single analog switch channel being used in a sample-hold circuit to deglitch the output of an 8-bit d/a converter. During the DAC code transition, when the glitch is generated, the switch remains in the HOLD mode, ignoring changes in the DAC output. After the transition is over, the SAMPLE mode is initiated, recharging the hold capacitor to the new value of DAC output voltage.

The resulting performance is determined by the charge injection, glitch energy, and feedthrough of the analog switch.

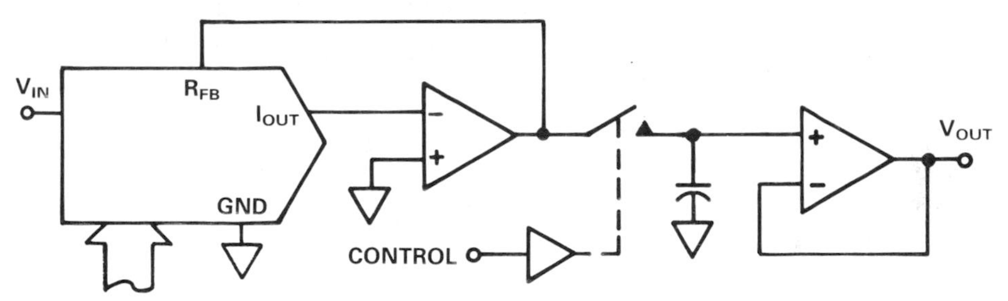

Figure 19.22. DAC deglitching.

19.4.3 DAC OUTPUT DEMULTIPLEXING

An extension of the basic deglitching function is the distributive multiplexing (demultiplexing) of the DAC output to provide a number of independent glitch-free voltage outputs. Figure 19.23 shows a quad analog switch in an arrangement that provides four separate digitally programmable voltages from the same DAC.

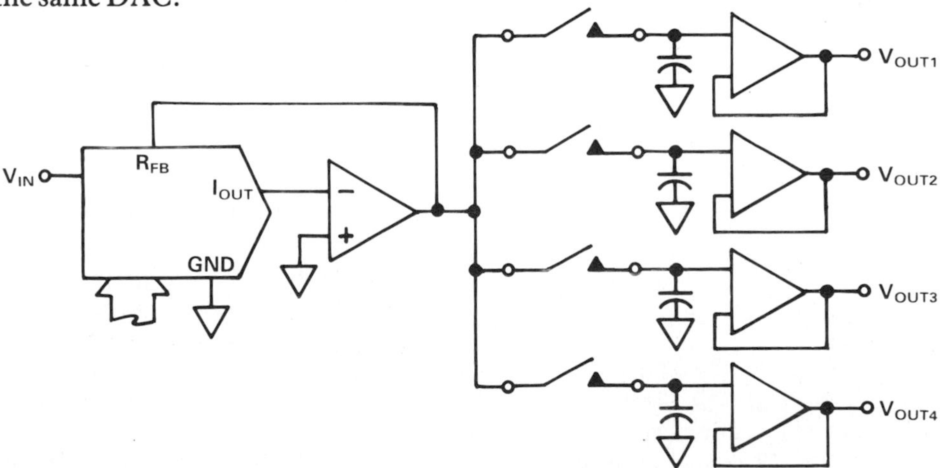

Figure 19.23. DAC output demultiplexing (data distribution).

19.4.4 OTHER SIGNIFICANT APPLICATIONS (Figures 19.24 and 19.25)

Programmable-Gain Amplifier

In Figure 19.24, a quad switch chooses the appropriate output of a tapped feedback voltage divider to set the gain of an operational amplifier. Similar circuitry can be used with some instrumentation amplifiers (e.g., the Analog Devices AD625) to set gains digitally, in such applications as automatic ranging, where dynamic range must be extended.

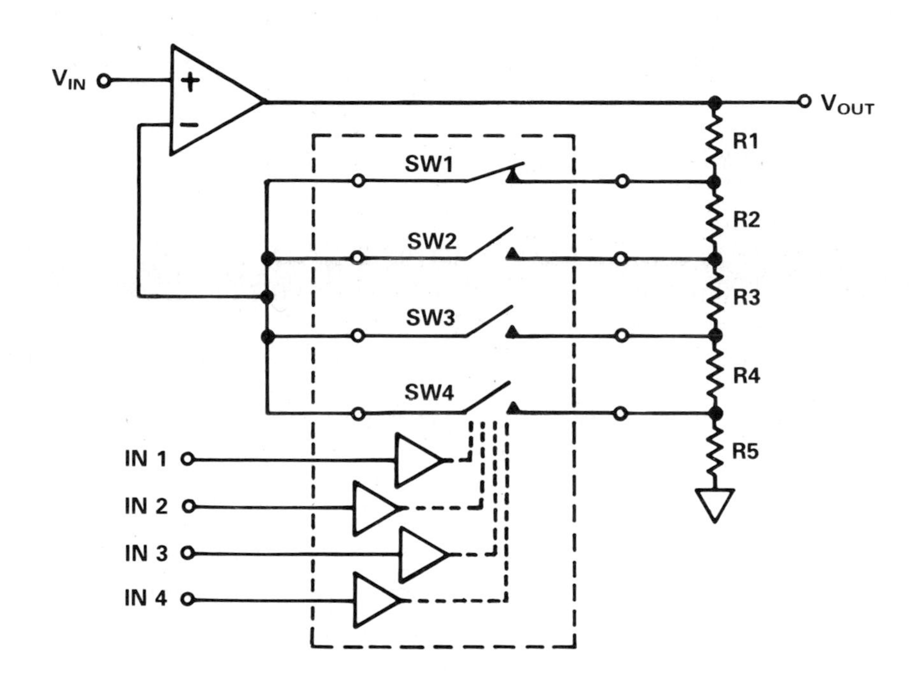

Figure 19.24. Digitally programmable-gain amplifier.

A/D Converter Input Autoranging

In Figure 19.25, a quad switch chooses a digitally determined attenuation level to establish the input range to an a/d converter having a buffered input. The result of the conversion can be used to readjust the attenuation for a smaller or larger input signal, thus achieving a limited degree of floating-point operation.

The above are but a small sampling of the uses for CMOS analog switches, limited to applications related to analog-digital conversion, within the charter of this book. However, there are a great many more potential applications in measurement and control and general electronics.

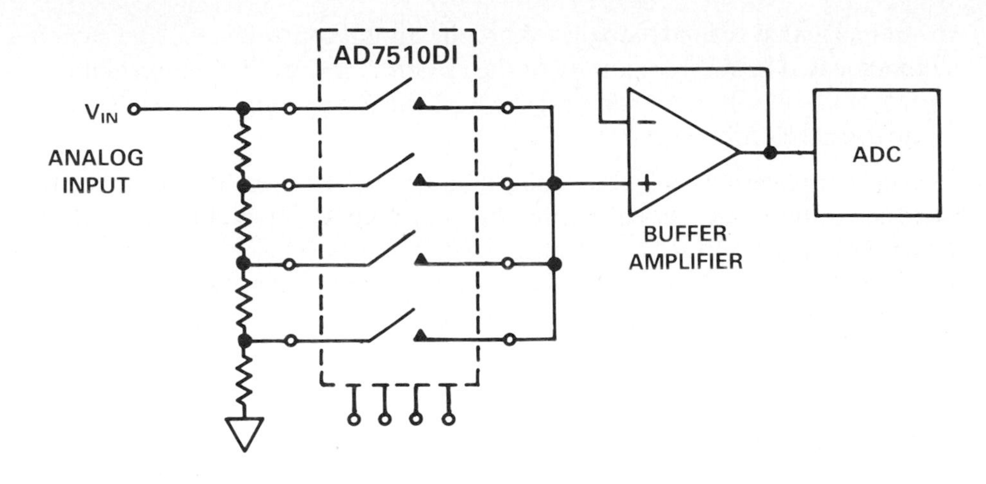

Figure 19.25. Digital scale-factor ranging for converter input.

Just one highly suggestive example: Figure 19.26 shows a circuit exemplifying the use of digitally controlled analog switches in remote control. Here they make possible a remotely programmable cutoff-frequency low-pass filter. The remote source of control can be either automatic digital equipment or a manual switch.

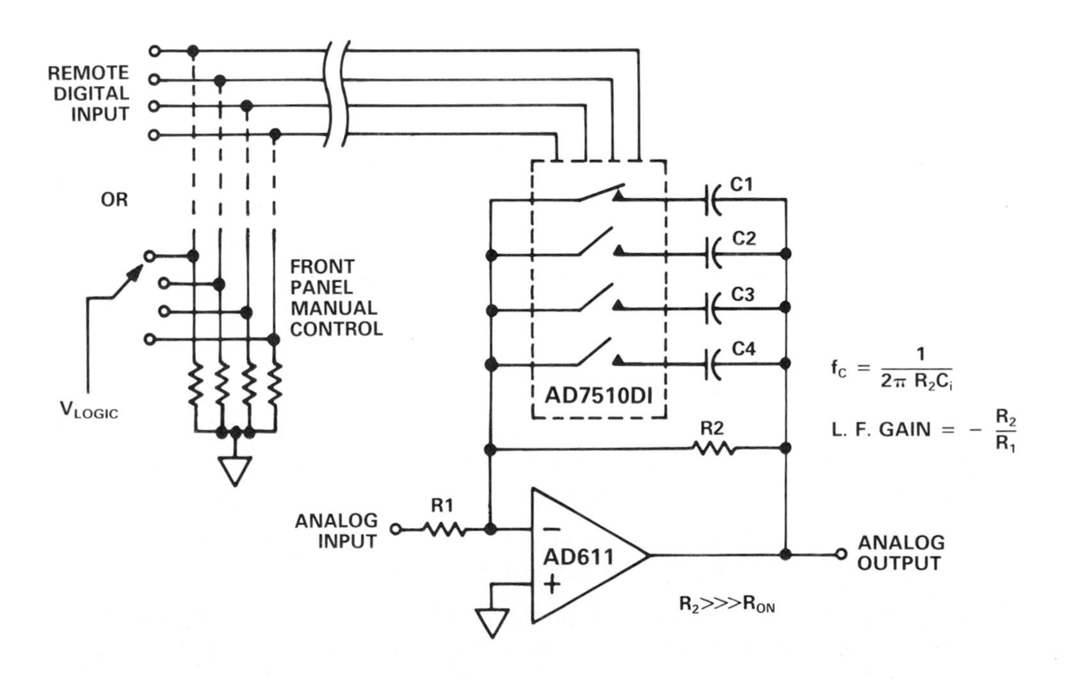

Figure 19.26 Remotely controlled analog filter circuit.

In such circuits, the only long leads are the digital control lines. The switches can be located in close proximity to the circuits they control. optoisolator circuitry can be employed if necessary to deal with large common-mode voltages in the digital circuitry.

Typical applications of the technique range from autocalibration routines in automatic test equipment to "soft keys" in portable laboratory instrumentation.

Chapter Twenty

Voltage References

Voltage references provide accurately known voltage for use in circuits or systems. Measurement systems rely on precision references in order to establish a basis for absolute measurement accuracy. Any reference error undermines the overall system accuracy, thus ideal references are characterized by accurately set constant output voltage, independent of load changes, temperature, input (supply voltage) and time.

20.1 WHERE REFERENCES ARE USED

A digital multimeter, digital communication system, portable instrument for precision measurement or calibration, electronic thermometer, precision switching regulator, or—in fact—any digital system, subsystem, or device with analog inputs or outputs needs at least one accurate reference. Such a reference might be a dc voltage source, with an accurate output somewhere in the range from 1.0 V to 10 V, stable to within a few parts per million per degree C and per month.

For example, in a/d converters, the digital output number depends on the ratio of the quantized input to the "full scale" reference. If the reference is allowed to change in response to a second analog input, the digital output will be proportional to the ratio of the analog input signal to the reference signal (i.e., a ratiometric ADC is equivalent to an analog divider with digital output). Often, the change in response to any "second analog input" may be unwanted, especially if that input is temperature, time, or upstream voltage; in such cases, an insensitive precision reference is essential.

A d/a converter is essentially a digitally controlled potentiometer that produces an analog output (voltage or current) that is a normalized fraction of its "full scale" setting. The full-scale setting is determined by the reference value chosen. Again, if the reference changes, the output will change; if the DAC output must be fixed at any digitally set value, a precision reference with appropriate characteristics must be used.

Desirable voltage reference characteristics include:

- *Accurate output voltage*
- *Low temperature drift*
- *Good load regulation*
- *Good line regulation*
- *Good long-term stability*

20.2 TYPES OF REFERENCES

References come in a variety of shapes and forms, depending on their projected use, ranging from banks of saturated standard cells to Josephson junctions to integrated-circuit voltage regulators. There are a number of ways of classifying references; for example, there are primary standards, secondary standards, transfer standards, etc. For our purpose, we may divide the family of references into two principal categories: *depletable* types—which can be used without a primary power source but cannot deliver power over long periods of time without degradation (i.e., batteries maintaining an accurate voltage with low drift)—and essentially *undepletable* types, which are based on fundamental relationships that give constant or predictable voltage but require a primary power source.

In this chapter, we will consider only the kinds of references commonly employed as components in converter circuits, using available power, with the understanding that their calibration may be traceable to international voltage standards.

20.2.1 ZENER-DIODE REFERENCES

The most widely used reference device is the temperature-compensated Zener* diode. Zeners have constant voltage drop in a circuit when provided with a fairly constant current derived from a higher voltage elsewhere within the circuit. They are often used with operational amplifiers in circuits that stabilize operating points, unload the diode's output impedance, and/or transduce voltage to current.

How The Zener Diode Works

The active portion of a Zener diode is a reverse-biased semiconductor P-N junction. When the diode is forward biased (the P region is made more positive), there is very little resistance to current flow. Actually, in the forward-

*Whatever the reverse breakdown mechanism—avalanche, Zener, or mixed—all diodes used in the breakdown mode have come to be called "Zener diodes," even though many of the diodes purchased for specified breakdown characteristics are in fact purely avalanche diodes. We will follow the custom in this chapter.

biased region, a Zener diode looks very much like a normal high-conductance silicon diode (Figure 20.1). Values of V_F greater than about 0.7V will produce substantial amounts of current.

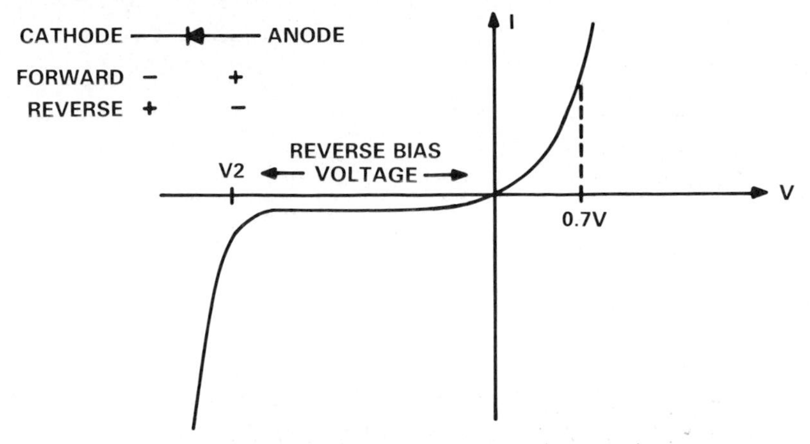

Figure 20.1. Generalized I-V characteristic for Zener diode.

In the reverse-bias region, very little current flows if V is less than V_Z, the *breakdown voltage*. The small amount of leakage current that flows—the reverse saturation current—is relatively insensitive to the actual magnitude of reverse voltage for fixed temperatures.

As the reverse voltage across the diode approaches the breakdown voltage, the reverse current increases more rapidly and will even run away if the applied voltage is sufficient. For this reason, diodes operated in the breakdown region are always used in series with resistsors or current sources. The sharpness of the transition depends on the relative value of the breakdown voltage and the manufacturing process used to make the diode. The most common circuit in which Zener diodes are used employs a series resistor to limit the current, as shown in Figure 20.2.

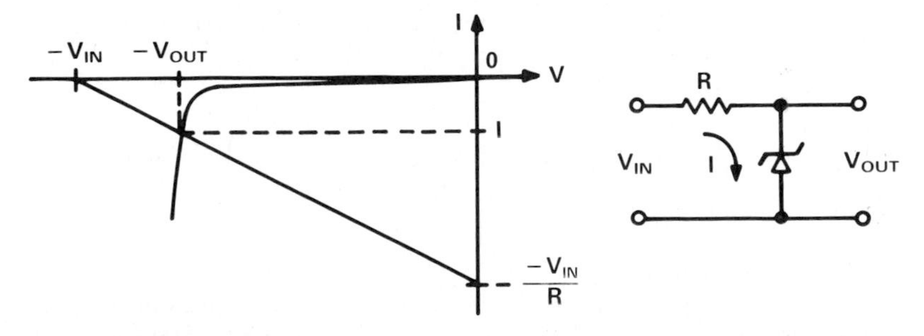

Figure 20.2. Breakdown diode characteristics and application.

It is worth knowing that there are two significantly different reverse-bias voltage breakdown phenomena, Zener breakdown and avalanche breakdown; the difference can be important to both the user and the designer of voltage-reference ICs. In Zener breakdown (Figure 20.3), a low-voltage phenomenon, the breakdown voltage *decreases* as the junction temperature rises. In avalanche breakdown the breakdown voltage *increases* as the junction temperature rises.

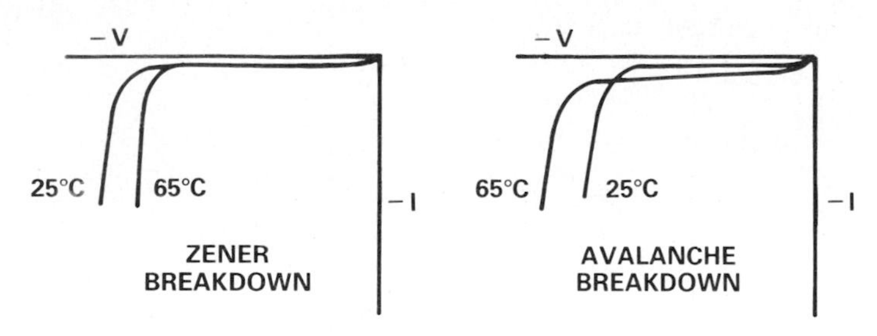

Figure 20.3. Avalanche breakdown vs. Zener breakdown.

The factor determining which manner of breakdown occurs is the relative concentration of impurities in the P- and N-type materials forming the P-N junction. A junction that has a narrow depletion region at a given voltage will develop a high field intensity and will break down by the Zener mechanism at relatively low voltages, as electrons and holes are stripped away from atoms, thus providing the carriers for conduction.

A high electric field supplies the energy required to traverse the "energy gap", i.e., electrons are excited from the valence band to the conduction band. Since conduction is a function of the energy gap (the band gap, or forbidden energy gap), an increase in temperature reduces the band gap and thus the breakdown, or Zener, voltage by increasing the energy of the valence electrons.

In high-resistivity materials, the depletion region is of sufficient width to avoid Zener breakdown; carriers will experience collisions before crossing the region completely at a given temperature. But, as the temperature increases, the kinetic energy of the valence electrons increases to the point that, because of the high energy level, a collision may rip off more carriers from an atom. The newly released carriers now gain sufficient energy from the field to begin collisions of their own, producing the "avalanche" effect.

Because avalanche breakdown is a kinetic process, it is more temperature-sensitive than Zener breakdown. Changes in temperature will cause a change in the mean kinetic energy of a particle and thus affect the avalanche breakdown mechanism. Zener breakdown, on the other hand, is dependent on the strength of the electric field and is theoretically insensitive to changes of temperature.

Zener-Diode Performance

"Zeners" are available with voltages from about 2 to 200V, tolerances of 10% to 20%, and power dissipation from a fraction of a watt to 40 or 50 watts.

Attractive as they might seem for use by themselves as general-purpose voltage references, they have many shortcomings. For use without additional circuitry, it would be necessary to stock a range of values; the voltage tolerance is generally poor, except in high-priced versions; and they are noisy and very sensitive to changes in current and temperature. For example, a 1N5221 (27-volt Zener) has a temperature coefficient of +0.1%/°C and will change by 1% for a variation in current of from 10% to 50% of its maximum rating.

Zener breakdown has a negative temperature coefficient, while avalanche breakdown has a positive temperature coefficient. Both are relatively independent of current, if self-heating effects are ignored. Zener diodes which are in the 6-V range exhibit both avalanche and Zener breakdown and have either positive or negative temperature coefficient, depending on which effect predominates.

For this reason, performance is best for diodes that break down in the neighborhood of 6 volts; they achieve very low temperature coefficients and become relatively stiff against changes in current (Figure 20.4) because the positive and negative temperature coefficients tend to cancel one another.

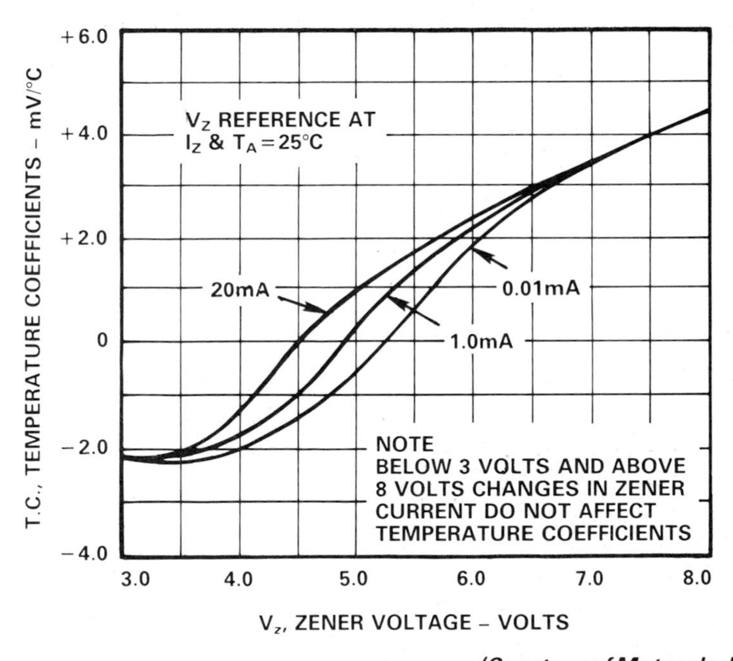

(Courtesy of Motorola, Inc.)

Figure 20.4. Temperature coefficient as a function of Zener voltage and current for a typical device.

Temperature Compensation

For a given Zener voltage, at low current levels, the Zener effect is stronger and the temperature coefficient is negative. At higher current levels, the avalanche effect takes over and the temperature coefficient becomes positive. Because both Zener and avalanche effect are occurring simultaneously and are controlled by the current level, the temperature coefficient varies as the current level changes. At some specific level of current the negative TC of the Zener effect is equal to the positive TC of the avalanche effect, and the net temperature coefficient is theoretically zero. Therefore, as can be seen in Figure 20.4, by proper choice of the diode's reverse current, the temperature coefficient can be adjusted for breakdown at a given voltage. Alternatively, a compensated Zener reference can be built using a Zener diode with a positive temperature coefficient in series with a forward-biased diode. The Zener diode's voltage is chosen to cancel the forward diode's temperature coefficient.

If you need a Zener for use in an application where stability is the prime concern, and you don't care what the exact voltage value is, the best choice would be a compensated Zener reference, made from a 5.6-volt Zener in series with a forward-biased diode.

The temperature coefficient of the forward-biased diode (Figure 20.5) is often more sensitive to current than that of the reverse-biased Zener diode; this can be detrimental to good performance unless the bias current is kept very close to the level specified by the manufacturer.

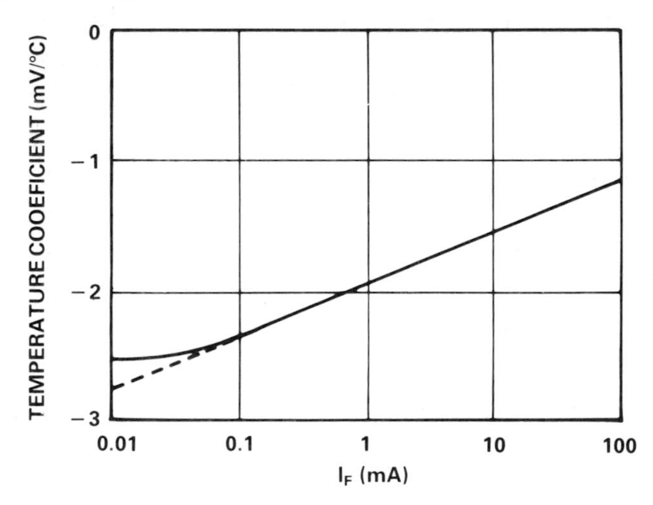

Figure 20.5. Temperature coefficient of a typical forward-biased junction diode as a function of current.

IC Zeners

Successful compensation to achieve near-zero temperature coefficients requires close tracking of the junction temperatures of the Zener and the forward-biased diodes. It is difficult to do this with diodes in different packages—the best arrangement is to place the diodes in direct contact with each other (Figure 20.6).

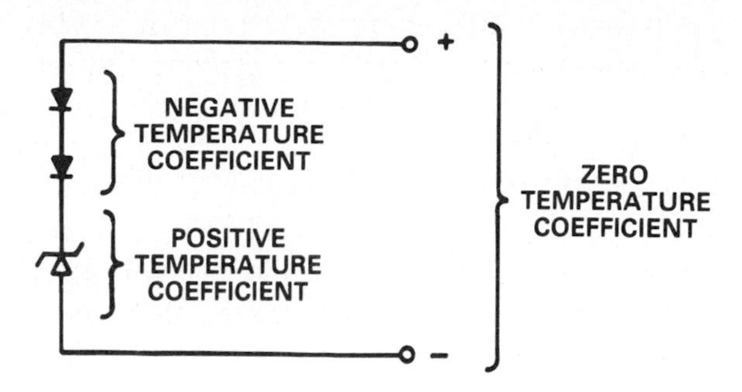

Figure 20.6. Using forward-biased diodes in temperature compensation.

Reference Circuitry Using Zener Diodes

These compensated Zeners can be used as stable voltage references within a circuit, but they must be supplied with constant current. A choice of technologies exists for performing this function; the appropriate choice, as always, depends on the circuit application.

High-performance hybrid references usually employ a temperature-compensated Zener diode and a feedback amplifier. The amplifier serves several purposes, providing constant bias current to the Zener, buffering it, allowing the reference circuit to source and sink current, and providing gain to boost the nominal 6.2-V diode voltage to the desired level (Figure 20.7). A low offset-drift amplifier, such as the AD510, or the AD OP-07, is essential, as the amplifier's offset drift will add to the drift of the Zener diode.

Here's how it works: Assuming that the circuit has started up and is operating properly (R5 provides the diode's startup current), current will flow through R3 and the diode, thus setting the plus input of the op amp at $+V_Z$. However, the minus input must follow, and it can do this only if

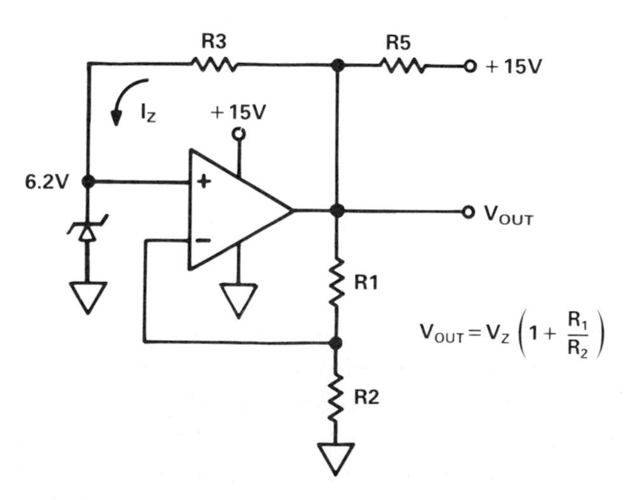

Figure 20.7. Hybrid 10-volt reference, using temperature-compensated Zener.

$$V_{OUT} = \left(1 + \frac{R_1}{R_2}\right) V_z \qquad 20.1$$

The current through R3 must then be equal to $(V_{OUT} - V_Z)/R_3$, and is maintained at that value, independently of the supply voltage, amplifier loading, and—to a first order—temperature.

R3 is chosen to provide the desired value of bias current, which is determined by the stable difference between V_Z and $(1 + R_1/R_2)V_Z$, i.e.,

$$I = \frac{R_1}{R_2 R_3} V_z \qquad 20.2$$

The R_1/R_2 resistance ratio can be trimmed to set the output voltage to the desired degree of accuracy, and R_3 can be trimmed to set the current to the proper value to minimize the drift of the Zener and thus decrease the output temperature coefficient. Hybrid references, with their freedom to mix technology, can usually be obtained with tight initial accuracy, using lasers to trim thin-film resistor networks. An example of a commercially available precision reference using this kind of circuit is the Analog Devices AD2710, which has 1 millivolt of initial error and a temperature coefficient of one ppm/°C.

A major use for references of this kind is with d/a and a/d converter types that require an external reference. Figure 20.8 shows how the reference is used with a high-speed 12-bit DAC. The AD566KD is laser-trimmed for ±¼-LSB maximum nonlinearity, and exhibits a gain temperature coefficient of 3 ppm/°C. When used with the AD2710LN reference, the worst-case total gain temperature coefficient is 4 ppm/°C. After initial calibration of the DAC scale fac-

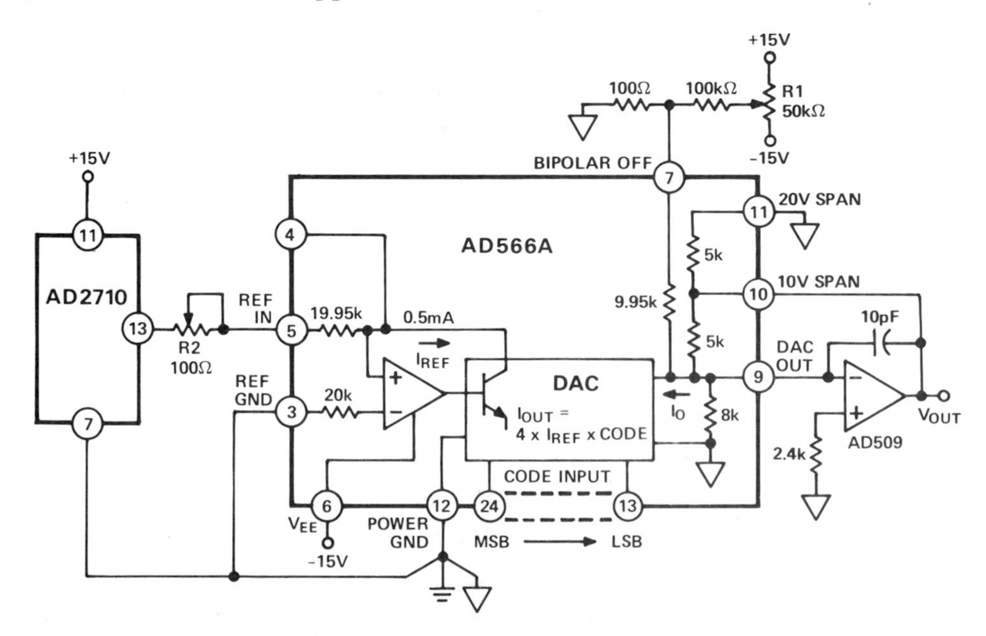

Figure 20.8. Reference for a precision d/a converter.

tor at room temperature, 12-bit accuracy can be maintained over the +15°C to +70°C temperature range. The AD2710's output-current capability makes it possible for it to serve as a reference for up to 10 such converters in a system.

Another approach to providing references for multiple DACs is to use the reference with single-package multiple DACs, such as the Analog Devices AD390. The combination of the AD2710LN and AD390KD (quad 12 bit d/a converter) will yield a compact multiple-DAC system (Figure 20.9) with a maximum full scale drift of ±6ppm/°C and excellent tracking.

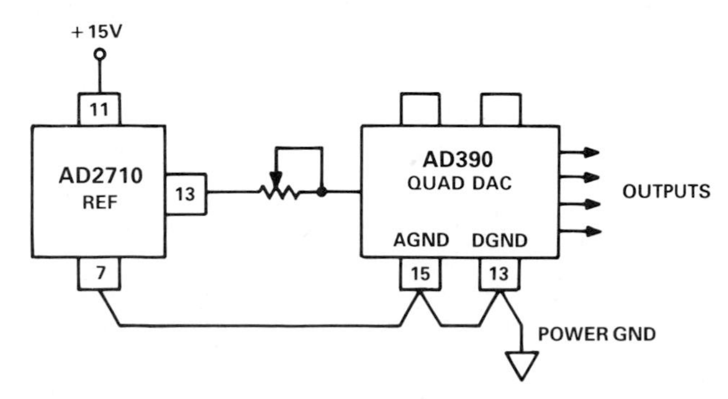

Figure 20.9. Furnishing a reference for four d/a converters at once in a quad DAC.

Buried-Zener References (BZRs)

A major category of precision *monolithic* voltage reference sources (other than band-gap devices) are subsurface, or *buried* Zener diodes. Development of the buried Zener diode was all but essential to the manufacture of accurate, high-resolution complete monolithic DACs and ADCs.

In ICs, Zener references have traditionally been produced using the reverse breakdown of the base-emitter junction of a vertical NPN transistor. This breakdown occurs right at the surface of the device, where the voltage is af-

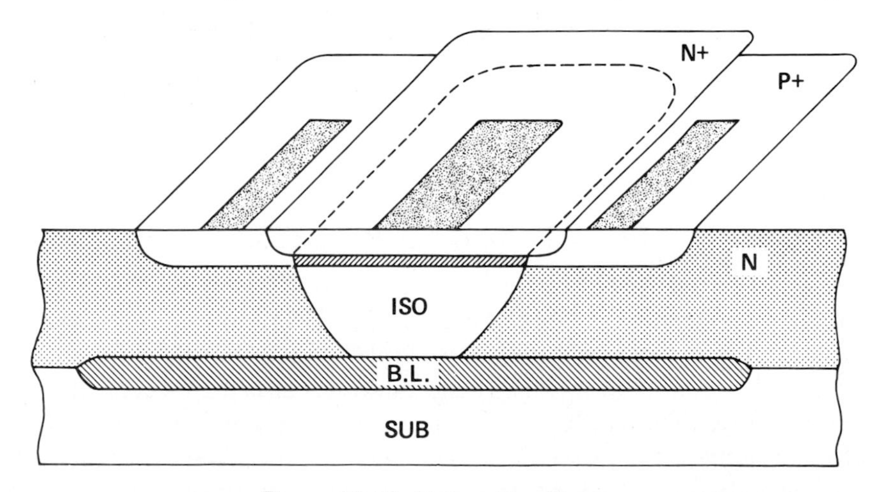

Figure 20.10. Subsurface Zener.

fected by crystal imperfections, mobile charges in the oxide, and other forms of contamination. These effects cause the surface-breakdown diode's noise and long-term stability to be unsuitable for 10- or 12-bit applications. The modified diffusions of the buried Zener (Figure 20.10) cause breakdown to occur well below the surface, thus avoiding surface effects. Long-term stability of 50 ppm/year can be achieved with this IC-process-compatible device.

However, because the diffusion is less controlled below the surface than on the surface, there is greater spread in the absolute values and temperature coefficients of subsurface Zener diodes, often exceeding acceptable tolerances. The circuitry used in conjunction with the diode must be designed to allow trimming of both the absolute value and the temperature coefficient of the whole reference circuit; A typical approach is shown in Figure 20.11. Typically, absolute accuracy is to within 0.1% and temperature coefficient is ± 10 ppm/°C. Buried-Zener reference circuitry is used to great advantage in complete monolithic high-resolution converters, such as the Analog Devices 12-bit AD667 DAC, since the overall *output* voltage and its tempco can be automatically laser-trimmed at the wafer stage—at the same time that other DAC parameters are trimmed.

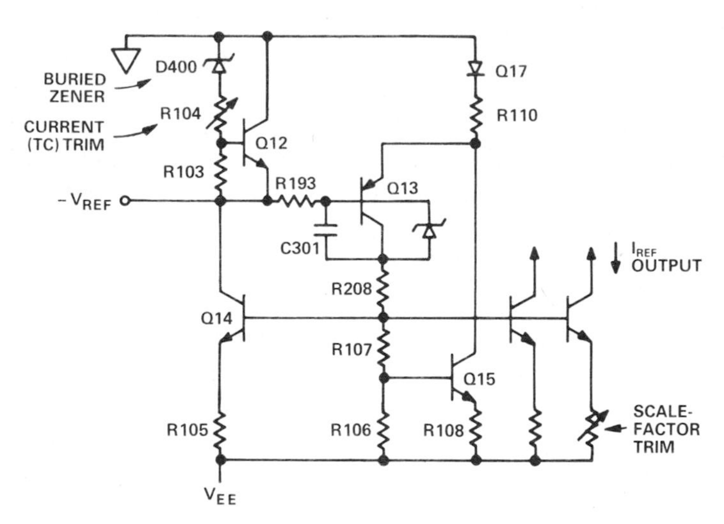

Figure 20.11. Zener-circuitry portion of IC chip.

20.2.2 TEMPERATURE-STABILIZED REFERENCES

Another way to improve temperature stability in IC references is to hold the reference at a constant elevated temperature. Temperature-stabilized (not "compensated") devices include on a single chip a temperature stabilizer (heater) and the reference circuit. The heater maintains the reference at a constant temperature that is independent of ambient temperatures up to the preset stabilizer temperature, but no cooling is available at higher temperatures. For this reason, it is common to see substantially differing temperature coefficients for the two operating ambient temperature ranges (ambients above and below heater temperature) for a device of this type.

For example, an LM199 temperature-stabilized reference is specified as having a temperature coefficient of ±½ ppm/°C from −55°C to +85°C, and 10 ppm/°C from +85°C to +125°C. This type of performance is most suitable for laboratory equipment in controlled environmental conditions where ample power is available. Since the heater can draw nearly 200mA on a cold start at low temperature, equipment using these devices should be designed to furnish enough power for startup; it may be incompatible with low-power systems. It should also be noted that, if this type of device is intended to operate over a wide temperature range—say, +25°C to +125°C—the average *overall* temperature coefficient is

$$\frac{0.5\ \text{ppm/°C}\,(85°\text{C}-25°\text{C}) + 10\ \text{ppm/°C}\,(125°\text{C}-85°\text{C})}{100°\text{C}} = 4.3\ \text{ppm/°C}.$$

20.2.3 BANDGAP REFERENCES

Another popular design technique for voltage references in monolithic circuitry is the "bandgap" approach, based on an underlying physical property of base-emitter voltage in a forward-biased silicon transistor.

Most designers are familiar with the approximate −2 mV/°C temperature coefficient of the base-emittter voltage when biased into the active region. The exact value depends on the value of base-emitter voltage, but it is so repeatable for a given transistor that it can be used for linear temperature sensing—if the emitter current is made proportional to temperature, the temperature coefficient of V_{BE} is nearly constant over a wide temperature range.

If, as in Figure 20.12, the V_{BE} values for several devices are plotted as a function of temperature and extrapolated to absolute zero (−273.2°C), the straight lines would have different slopes—but all would intersect at the same voltage value: 1.205 V. This is the *bandgap voltage of silicon* at 0 kelvin. If one could generate a voltage that *increases* proportionally with temperature at the same rate V_{BE} decreases (for a given transistor) the sum of the two voltages would be a constant 1.205V at any temperature. This voltage can be obtained by amplifying the difference between the V_{BE}'s of similar transistors operating at different values of current density (J_1 and J_2, equal to I_1/A_1 and I_2/A_2, in amperes/meter^2). Then

$$\Delta V_{BE} = \frac{kT}{q}\ \ln\frac{J_1}{J_2} = \frac{I_1}{I_2}\frac{A_2}{A_1} \tag{20.3}$$

k/q is the ratio of Boltzmann's constant to the unit of electronic charge—86.14μV/K. T is absolute temperature, in kelvins, and ΔV_{BE} is the difference of the base emitter voltages, which is proportional to absolute temperature if J_1/J_2 is constant. It is then scaled up to a value that, when summed with V_{BE} at the same temperature, equals 1.205V and is theoretically independent of temperature.

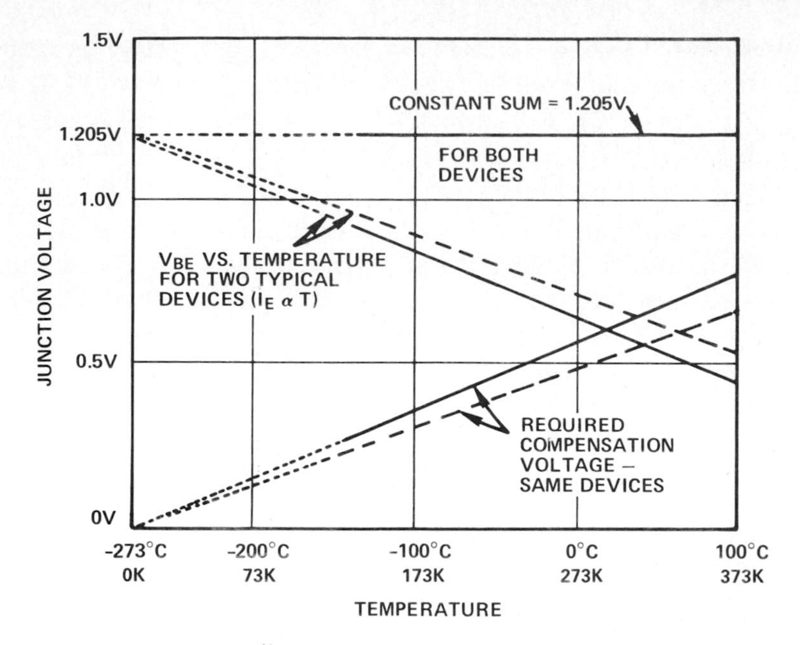

Figure 20.12. Extrapolated variation of base-emitter voltage with temperature (I_E proportional to absolute temperature), and required compensation, shown for two different devices.

Circuit Operation

Bandgap reference circuits are based on the circuit of Figure 20.13. If the amplifier is a high-gain op amp, its two inputs are kept at the same voltage by feedback to the base of Q1 through voltage divider, R4-R5. If R8 and R7 have equal resistances, then equal currents must flow through them and therefore through the collector and emitter circuits of high-β transistors, Q1 and Q2.

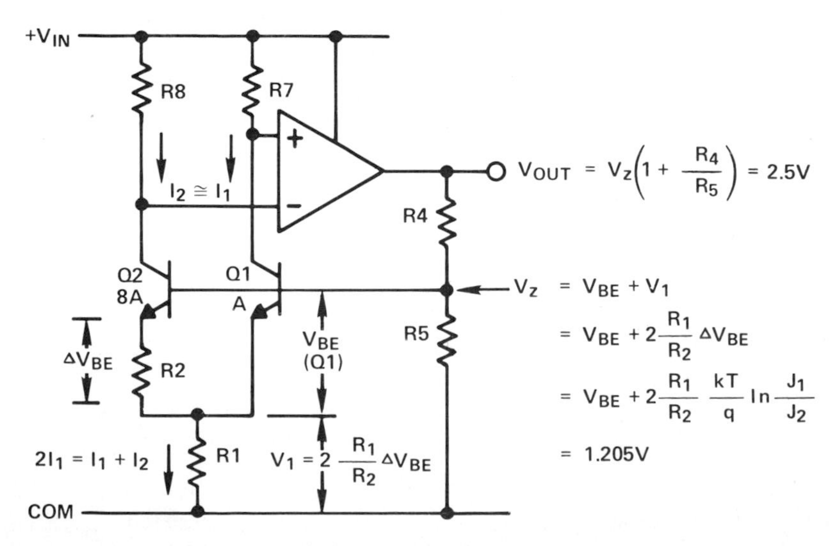

Figure 20.13. Basic bandgap-reference regulator circuit.

The emitter area of Q2, in this example, is eight times as great as the area of Q1, therefore the current density, J_2, is ⅛ J_1, and ΔV_{BE} is (kT/q) ln 8, or 179.2 T microvolts. Since R2 is connected between the emitters, the voltage across R2 is also equal to ΔV_{BE}. The design value of R_2 is determined by the desired level of current through Q1 and Q2; that current, equal to $\Delta V_{BE}/R_2$, also flows through R1, and, since $I_1 = I_2$, the total current is 2 $\Delta V_{BE}/R_2$, and the voltage across R1 is

$$V_1 = 2\frac{R_1}{R_2}\frac{kT}{q}\ln 8 \qquad (20.4)$$

For the proper choice of the ratio, R_1/R_2, the sum of the voltages, $V_{BE1} + V_1$, will be equal to the bandgap voltage, 1.205 V, which in turn will be amplified by the ratio, $R_4/R_5 + 1$, to give the desired value of output voltage.

Integrated-circuit process control makes V_{BE} predictable, so that R_1/R_2 can be predetermined and implemented with stable, low-tracking-tempco, thin-film resistors, deposited on the silicon chip and laser trimmed for increased accuracy.

Monolithic bandgap-reference circuits are available in several common forms: some can be used as three-terminal voltage-output regulator-amplifier circuits, others can be used as two-terminal synthetic Zener diodes, and many types can serve in either application. Because it is an integrated circuit, a bandgap reference may also be integrated on-chip with the converter for which it serves as the reference; an example is the Analog Devices AD558 8-bit DAC.

Using Bandgap References

Bandgap references are often useful substitutes for Zener diodes and circuits employing Zener diodes. Bandgap devices will operate from low voltage supplies (typically V_{out} + 2V), compared to the breakdown voltage-plus-current-generation "headroom" required to set the proper operating current for the Zener. Bandgap references typically have ten times lower output impedance than low-voltage Zener diodes and can be obtained in a variety of nominal output voltages, ranging from 1.2 to 10V.

Integrated-circuit designers are able to incorporate additional features in bandgap circuits to make them even more useful. For example, the feedback circuit of the Analog Devices AD584 has a multi-tap voltage divider, with the terminals brought out, so that external jumpers can be used to program a variety of fixed calibrated voltages (2.5V, 5.0V, 7.5V, and 10V), at currents up to +10 mA at +25°C, or +5 mA over the temperature range, and external precision resistors can be used to set arbitrary voltage values.

The AD584 also has a "strobe" terminal, which permits the device to be turned on or off. When used in this mode, the reference output can be switched to zero with an external signal. When the AD584 is used as a reference for a power supply, the supply can be switched off to conserve energy, via the AD584's strobe terminal.

A typical application of a reference with a multiplying d/a converter is shown in Figure 20.14. The reference output (shown here programmed for +10V)

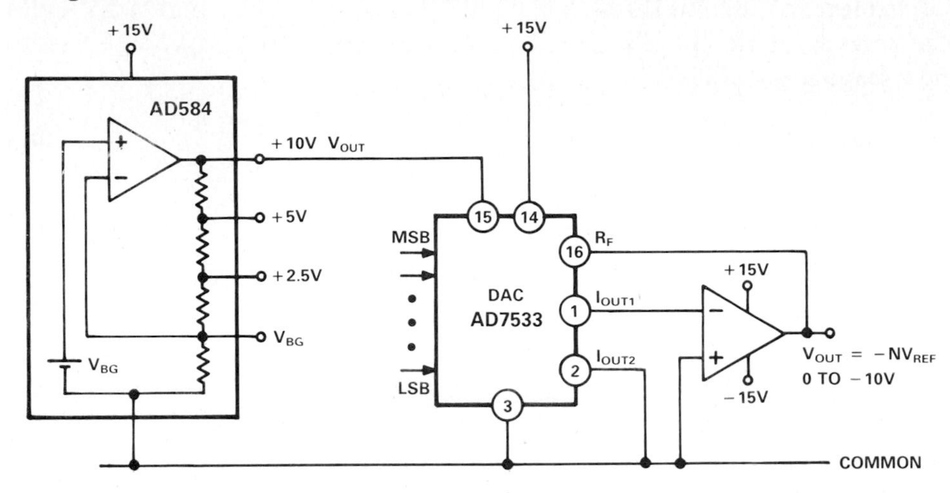

Figure 20.14. Low-power d/a converter, using bandgap reference.

is connected, in the regulator mode, to the reference-input terminal of the DAC. The digital input to the DAC sets the gain, and the op amp inverts the converted output; a +10-volt reference produces a 0 to −10-volt output range. Key specs for the op amp may include speed, dc stability, and/or low power drain.

By using a bandgap reference, such as the AD584, in the two-terminal "Zener" mode, a variety of useful circuits can be derived:

Figure 20.15 shows the device connected as a −5-volt reference. When it is

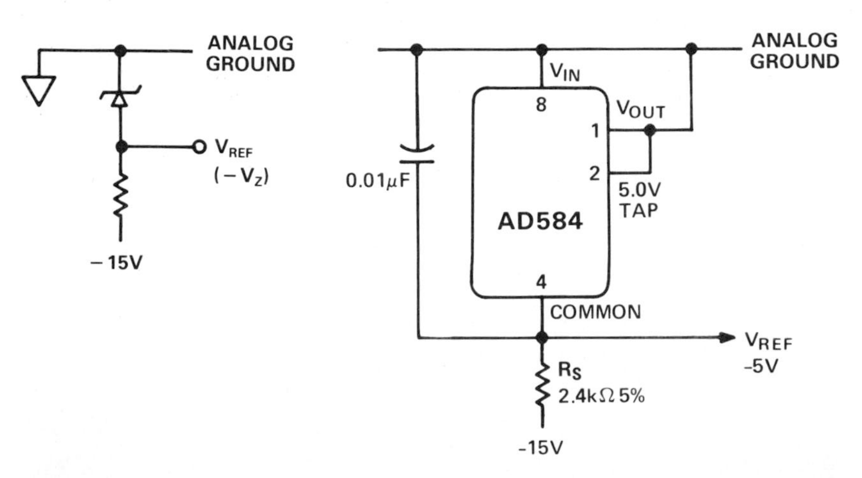

Figure 20.15. Two-terminal −5-volt reference.

used with a multiplying d/a converter and an inverting op amp, the output voltage will be positive. The AD584 has better stability and lower output impedance than a Zener diode.

Figure 20.16 shows how a device of this type can be used as a precision current limiter. The current drawn through the AD584 is equal to the load current (V_{OUT}/R) plus a fixed quiescent current of the order of 0.75 mA (1 mA max). In practice, R_{LOAD} could be adjusted to set the precise value of current.

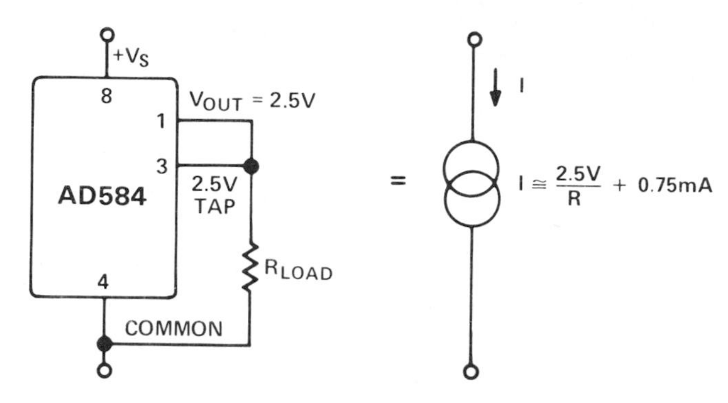

Figure 20.16. Two-component precision current limiter.

It is also possible to use the AD584, with a few external components, to provide a very low-level current source, ignoring the quiescent current. In Figure 20.17, the follower-connected op amp maintains the voltage at its negative input equal to common by driving the common terminal at the load voltage. Since the AD584's output is a constant 2.5V, the current through R_{SENSE} must be equal to $2.5V/R_{SENSE}$, and essentially all of it flows through the load (the FET-input AD547 has very low bias current). The quiescent current

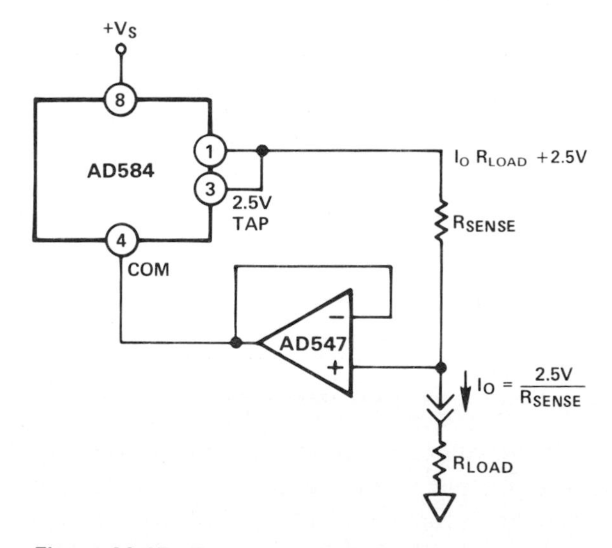

Figure 20.17. Precision low-level current source.

of the reference—whatever its value—is furnished by the op amp's output circuit.

If more than the rated 10-mA (25°C) output current is required, a simple inside-the-loop booster circuit may be added without significantly degrading performance. Figure 20.18 shows such a circuit, capable of furnishing 10 volts

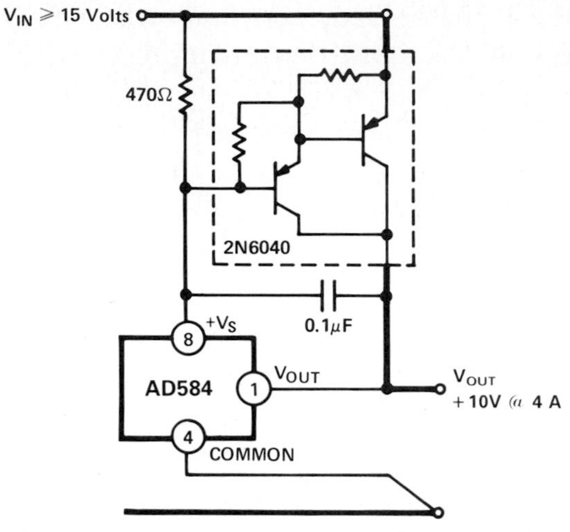

Figure 20.18. High-current precision supply.

at up to 4 amperes. The 2N6040 should be properly heat-sinked, and leads should be short—and of reasonable gauge.

When used with an additional op amp, and no external resistors, a tracking ±5-volt reference circuit can be created, as shown in Figure 20.19. In this circuit, the op amp must keep the +5-volt tap at output common by driving the regulator's COM terminal at −5 V. The 10-volt tap must then be at a voltage equal to +5 volts, with respect to output ground.

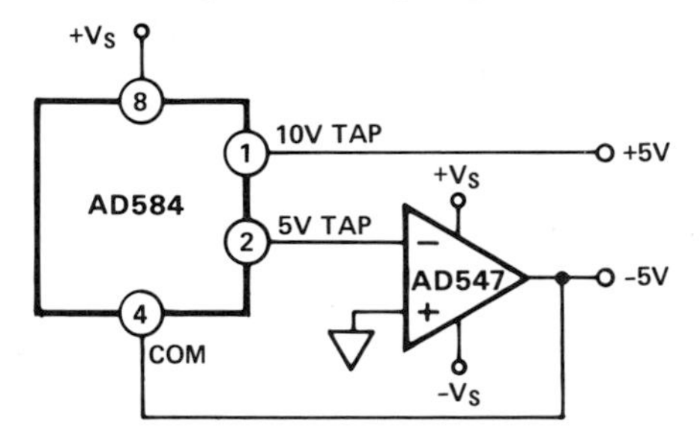

Figure 20.19. ±5-volt tracking reference.

20.3 WHAT THE SPECIFICATIONS MEAN

How do you choose the right reference for your needs? Ideally, a precision reference provides an output voltage that is stable with temperature, input voltage, varying load, and time. If the reference is in error, it will create errors in the device, instrument, or system it is connected to.

In A/D converters, the digital output number depends on the ratio of the quantized input to the "full-scale" reference. If the reference is allowed to (say) increase, the digital output, which is proportional to the ratio of the analog input signal to the reference signal, will decrease by the same percentage.

A D/A converter can be thought of as a digitally controlled potentiometer that produces an analog output (voltage or current) proportional to the product of the reference and the digital input. If the reference increases by 1%, the output will increase by 1%. If the reference is used with a 12-bit converter, that change will create a gain error of about 41 least-significant bits.

Measurement systems of any kind rely on precision references in order to establish a basis for absolute measurement accuracy. Reference errors translate into system errors.

Key sources of reference error include the initial error, and changes of error with line, load, time, and temperature. They lead to the five basic accuracy specifications for any IC or hybrid Zener or bandgap reference used in data-acquisition applications, highlighted in Figure 20.20.

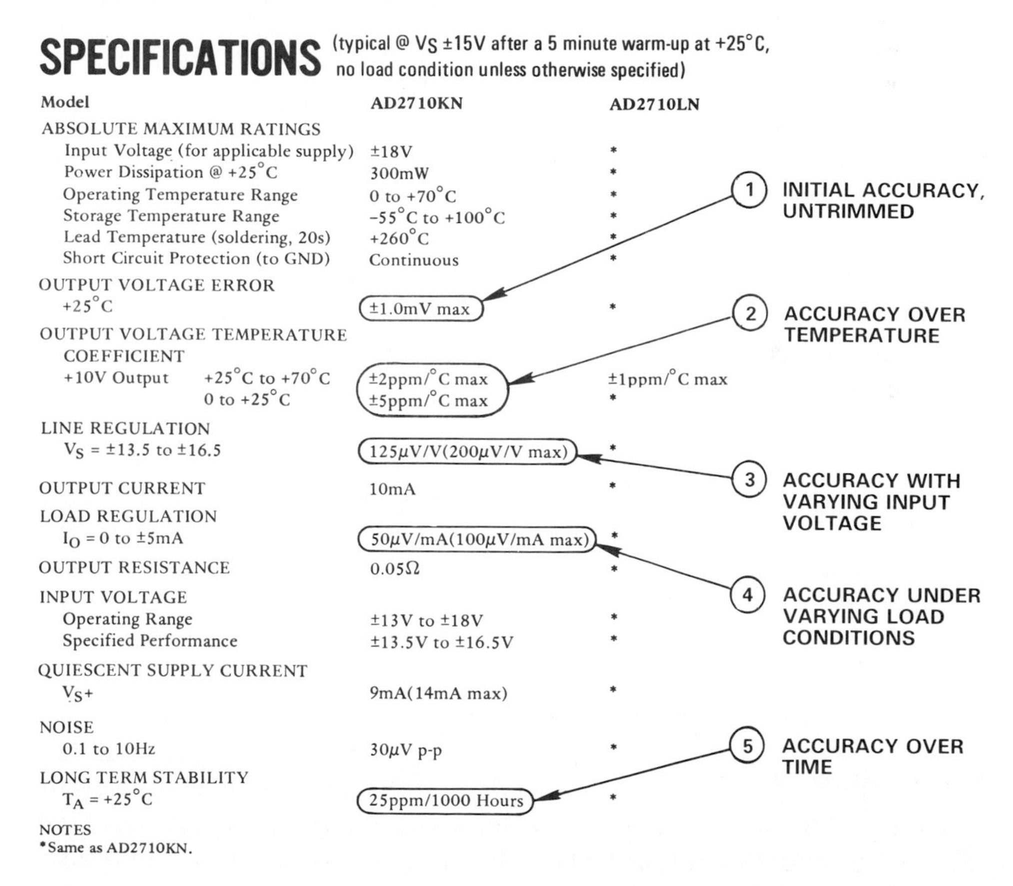

SPECIFICATIONS (typical @ V_S ±15V after a 5 minute warm-up at +25°C, no load condition unless otherwise specified)

Model	AD2710KN	AD2710LN
ABSOLUTE MAXIMUM RATINGS		
Input Voltage (for applicable supply)	±18V	*
Power Dissipation @ +25°C	300mW	*
Operating Temperature Range	0 to +70°C	*
Storage Temperature Range	−55°C to +100°C	*
Lead Temperature (soldering, 20s)	+260°C	*
Short Circuit Protection (to GND)	Continuous	*
OUTPUT VOLTAGE ERROR		
+25°C	±1.0mV max	*
OUTPUT VOLTAGE TEMPERATURE COEFFICIENT		
+10V Output +25°C to +70°C	±2ppm/°C max	±1ppm/°C max
0 to +25°C	±5ppm/°C max	*
LINE REGULATION		
V_S = ±13.5 to ±16.5	125μV/V(200μV/V max)	*
OUTPUT CURRENT	10mA	*
LOAD REGULATION		
I_O = 0 to ±5mA	50μV/mA(100μV/mA max)	*
OUTPUT RESISTANCE	0.05Ω	*
INPUT VOLTAGE		
Operating Range	±13V to ±18V	*
Specified Performance	±13.5V to ±16.5V	*
QUIESCENT SUPPLY CURRENT		
V_S+	9mA(14mA max)	*
NOISE		
0.1 to 10Hz	30μV p-p	*
LONG TERM STABILITY		
T_A = +25°C	25ppm/1000 Hours	*

NOTES
*Same as AD2710KN.

Figure 20.20. Key reference specifications.

20.3.1 INITIAL ACCURACY

Initial accuracy, or output voltage error, or output voltage tolerance, is the deviation from the nominal output voltage at 25°C and specified input voltage. It should be measured by a device traceable to a fundamental voltage standard. Initial accuracy is trimmable on some devices, such as the AD584 and the AD2700 series. Fine-adjust pins are provided for use with an external potentiometer (Figure 20.21). Considerable care must be taken, though, as these external components can affect the temperature coefficient.

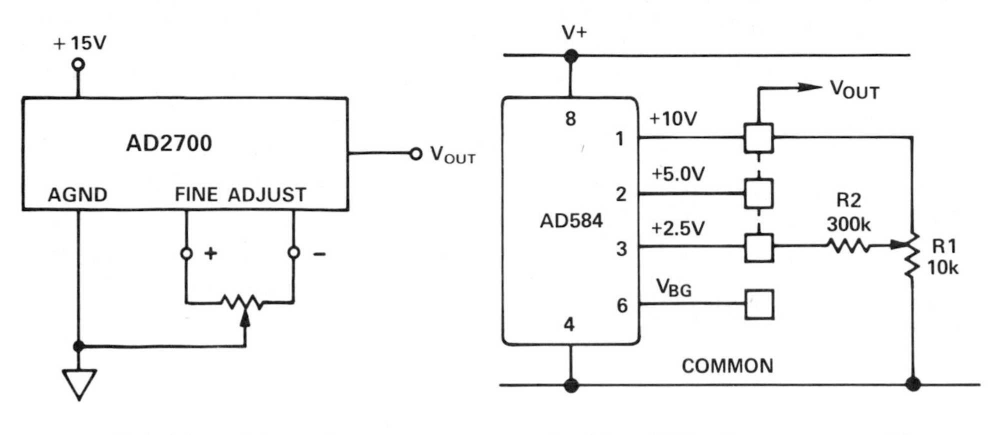

a. Hybrid precision reference.

b. Monolithic pin-programmable precision reference.

Figure 20.21. Trimming the reference output.

20.3.2 OUTPUT VOLTAGE CHANGE WITH TEMPERATURE

Output voltage change with temperature, or output voltage temperature coefficient, or voltage drift. This is the change in output voltage from the value at 25°C ambient. It is independent of variations in other operating conditions. There are two common methods of specifying drift, and the differences between the two must be understood in order to pick the right reference for the job.

The *Box Method* of specifying drift consists of specifying an error band and an equivalent temperature coefficient, usually in ppm/°C. The error band is graphically defined as a box (voltage on the vertical axis, temperature on the horizontal) whose diagonals extend from T_{LOW} to T_{HIGH} through 25°C. The slope of this diagonal is given as the stated temperature coefficient. Thus, the total absolute error for a reference over its specified temperature range is equal to the output voltage tolerance at 25°C plus the error band.

One difficulty with the box method is that manufacturers have defined T_{LOW} and T_{HIGH} in different ways. In some box specifications T_{LOW} is T_{MIN}, T_{HIGH} is T_{MAX} (0°C and 70°C for a "commercial" temperature range, or −55°C and +125°C for the extended "military" range). T_{LOW} has also been defined as 25°C ("ambient") and T_{HIGH} as T_{MAX}. The "box" specification defines the drift by the worst (or biggest) temperature excursion from the initial temperature. It is up to the user to be sure which method is used; for example, a "5-ppm/°C" box from −55°C to +125°C allows 900ppm drift (9 mV, for a +10V reference), while a "5-ppm/°C" box from +25°C to +125°C allows 500ppm (5 mV). If the user is careless, the 5-mV limit could be misinterpreted, in the latter case, to apply for −55°C to +125°C.

A second trap of the box method relates to the location of the voltage "origin"; i.e., is the box measured from $V_{nominal}$ at 25°C or V_{actual} at 25°C? In the above example (5ppm box, 25 to +125°), drift of 5 mV is theoretically allowed. To translate this to an absolute accuracy spec one must know if this drift is measured from $V_{nominal}$ or added to the initial accuracy error. If measured from V_{actual} (at 25°C), the box floats up or down depending on the initial value, and the device meets spec if V_{out} at T_{MIN}, $T_{ambient}$, and T_{MAX} all fall within the box. Figure 20.22 shows an extreme example of this case.

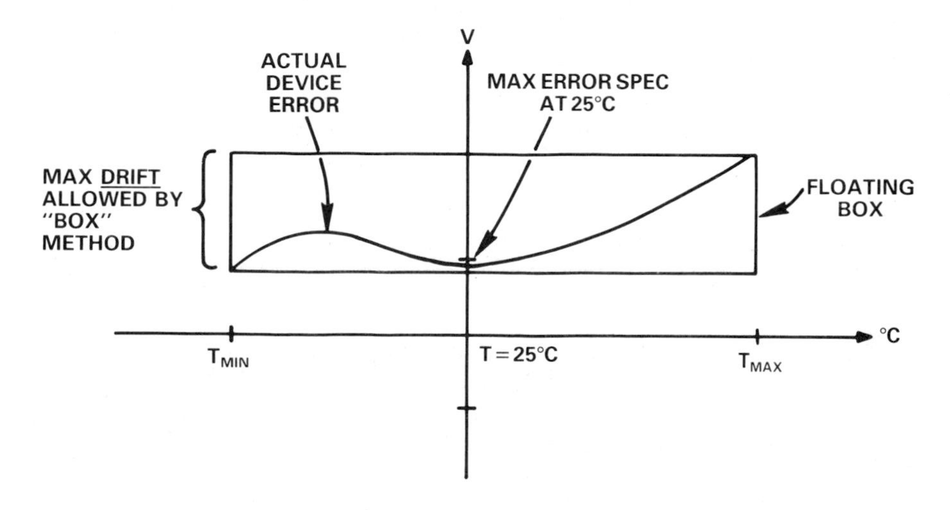

Figure 20.22. Floating box drift specification.

To remove some of the ambiguity of the box method, Analog Devices uses a *"modified" box* method (Figure 20.23) that fixes the box at $V_{nominal}$ and specifies only the absolute error at a temperature rather than initial error plus allowed drift. The AD2700 series is specified in this manner. The "modified" in modified box refers to the notch at 25°C, which tightens the initial error band allowed.

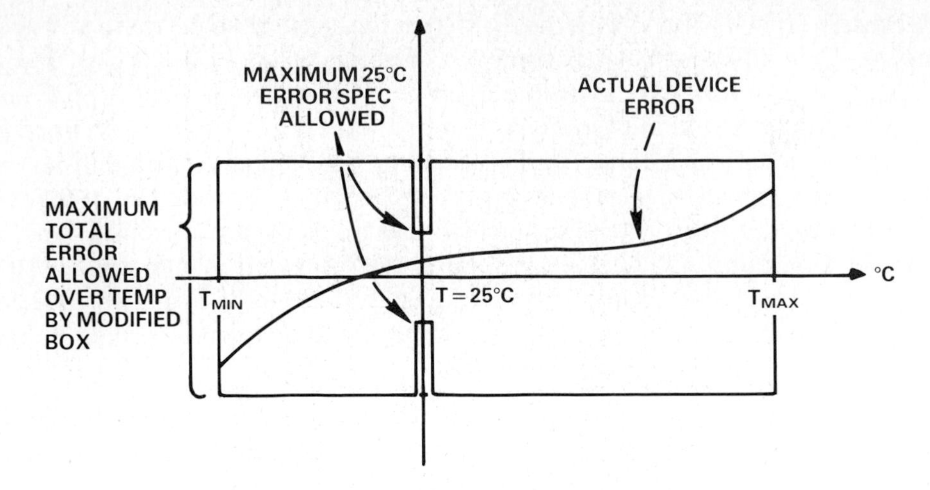

Figure 20.23. Modified box drift specification.

The third method of specifying output voltage change with temperature is the *butterfly* method. This method can be the tightest specification method, as it ties down the maximum excursion or change with temperature and can be extrapolated to any temperature within specification (rather than the end points only). Again, the danger is that the center of the butterfly may be either $V_{nominal}$ or V_{actual} at $T_{ambient}$ (25°C). Figure 20.24 demonstrates the butterfly method, using $V_{nominal}$.

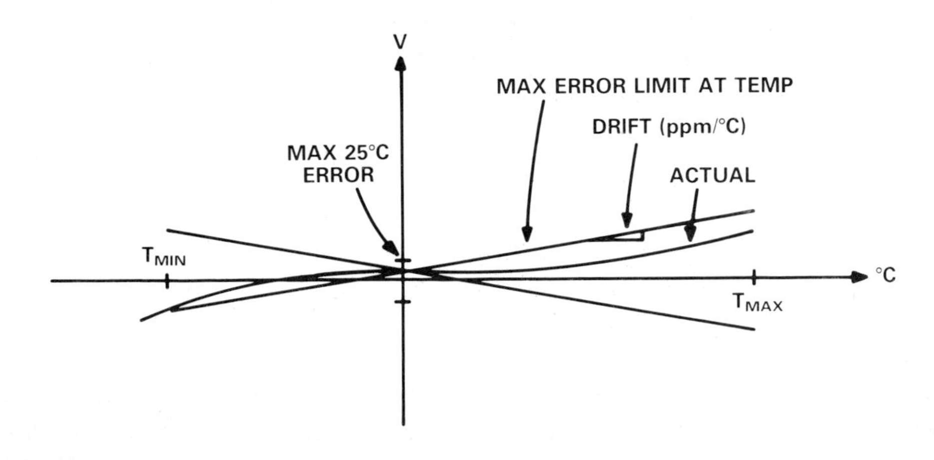

Figure 20.24. Butterfly method of specifying drift.

BOX METHOD	MODIFIED BOX	BUTTERFLY
Box Floats	Box fixed	Can float or be fixed
Limits absolute drift error by end points	Limits total absolute error	Limits absolute drift error over the range
Tends to ignore initial error	Includes initial error	Includes initial error

An example of a device with a tight butterfly drift specification is the AD2710/2712 series of precision references. Top grades are specified for ± 1-millivolt initial error (maximum) and ± 1ppm/°C maximum drift error (25° to 70°). Figure 20.25 shows the typical and maximum drift error of these references.

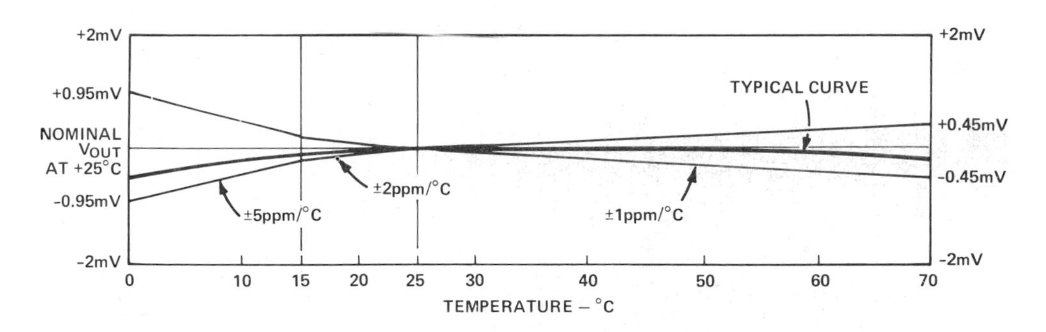

Figure 20.25. Maximum change of +10-volt output from +25°C value vs. temperature.

20.3.3 LINE REGULATION

Line regulation, or accuracy with varying input voltage is the change in output due to a specified change in input voltage, usually specified as % per volt or μV per volt of input change. It is a measure of power-supply rejection and is typically measured at dc. A related specification, ripple rejection, (ratio of ripple components in the output to residual upstream ripple) is rarely specified for IC references; it therefore behooves the careful reference user to choose an upstream power supply with minimal ripple.

20.3.4 LOAD REGULATION

Load regulation, or accuracy under varying load conditions, is the change in output voltage for a specified dc change in load current. It is generally expressed in μV/mA, and sometimes as ohms of dc output resistance. It includes the effect of self-heating due to increased power dissipation at high load currents.

20.3.5 LONG-TERM STABILITY

Long-term stability, or accuracy over time, is usually specified as parts-per-million per 1000 hours at a specified temperature. This is a difficult spec to verify, and is generally given as "typical", based on characterization data.

Zener diodes undergo the major portion of their long-term drift during the first part of their life. They tend to settle down with age, finally reaching a point where only small random variations compatible with 1/f noise occur. Unfortunately, it may take years to arrive at this condition (Figure 20.26). For this reason, most Zeners used in precision references are aged ("burned in") at an elevated temperature to accelerate the process.

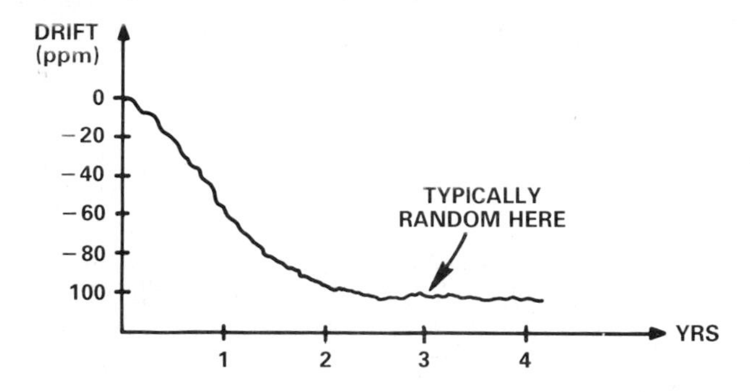

Figure 20.26. Typical 1N829 drift with time.

Chapter Twenty-One

Digital Signal Processing

Digital signal processing (DSP) means the processing of (analog)* signals using digital techniques (i.e., digital hardware and software). While *any* digital processing of signals that originate in the analog world, and have at some point been converted to digital, would qualify under this broad definition, the term has come to be used in a much more specific way—as it will be in this chapter: DSP is the application of fast, specialized hardware, sophisticated algorithms, and the appropriate software for the purpose of manipulating large amounts of data associated with extracting and processing analog-based information in essentially "real time."

The emergence of DSP hardware is changing the role of analog-to-digital conversion in today's signal processing systems. In early days, all processing of a signal, with the goal of obtaining results with sufficient speed to be useful in real time, was of necessity handled by analog components. The principal destinations for analog signals converted to digital format, after substantial analog processing, were off-line computation, data storage, and hard-copy tabulation, rather than real-time instrumentation, computation, and control.

Now, however, system designers have an incentive to perform the signal conversion as early in the loop as possible (see Figure 21.1). The reason for this is that much or all of the required signal processing can be handled by fast, flexible digital components that allow high-performance DSP routines to be

*It should be noted that any electrical signal is by nature an analog signal, even if it represents a digital "1" or "0". This can be understood if one pictures what happens to a chain of 1s and 0s returned from beyond the orbit of Jupiter, buried in cosmic noise: the signal must be received, amplified, converted, and processed (digitally) to reconstruct the original digital information; but until that digital information has been identified, the signal—to all intents and purposes—is purely analog in origin.

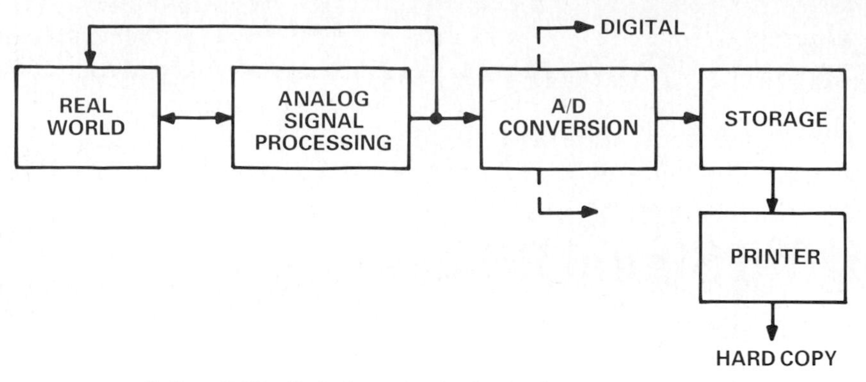

a. Before DSP: digital used principally for record-keeping.

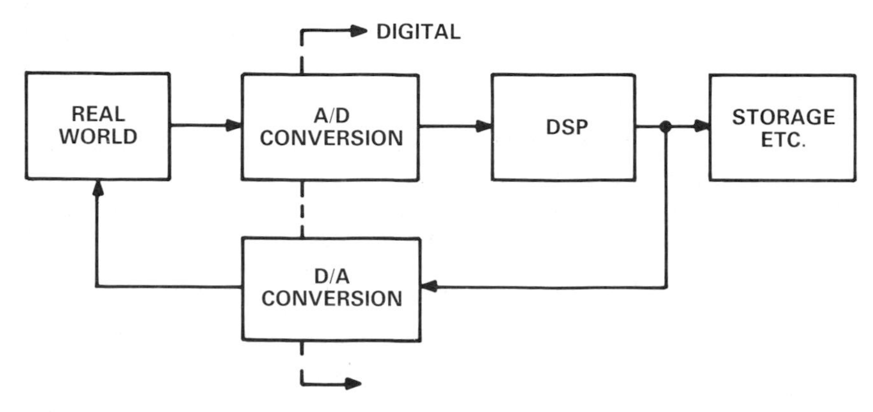

b. DSP era: digital responds to and operates on real-world phenomena.

Figure 21.1. Analog and digital signal processing.

implemented more accurately, reliably, and flexibly than with analog circuitry, yet, in many cases, with sufficient speed to interact in real time.

This chapter first reviews basic signal processing tasks, giving emphasis to the general role played by DSP. Two key DSP algorithms are examined in some detail—digital filters and spectral analysis. The basic hardware required to perform DSP is described. Finally, some applications that exemplify DSP's advantages are reviewed.

21.1. SIGNAL-PROCESSING BASICS

Signal processing revolves around two basic tasks—digital filtering and spectral analysis. *Filtering* smoothes, removes noise from, selects particular signal components from, or predicts future values of an incoming signal. A time-domain signal can be interpreted as a weighted combination of purely sinusoidal spectral components; *spectral analysis* determines the weights corresponding to each frequency in the spectrum.

Signal-processing applications span many areas, including speech analysis and synthesis, telecommunications, instrumentation, radar and sonar, and—using multi-dimensional techniques—graphics and imaging. For example, filtering is used to minimize high-frequency noise and the low-frequency hum in telephone-line transmission. Spectral analysis is used to determine the formant content of incoming speech for recognition. Two-dimensional filtering improves the clarity of a satellite image.

Filtering and spectral analysis have traditionally been implemented with analog components. Filtering is carried out by passing the signal through a circuit consisting of resistors, capacitors, op amps, and/or inductors; the precise configuration of these components and the relationship of the magnitudes of their parameters determine the filter's characteristics. Multiple analog filters—each passing energy in a narrow band—can be cascaded for sharpness and banked together to perform spectrum analysis.

Analog-based signal processing has numerous advantages, including low component cost, the ability to handle wide bandwidths in real time, the availability of pre-packaged modules and ICs, and a large existing base of knowledge. However, analog components introduce noise at each stage; and filter characteristics—requiring effort to tune initially—are sensitive to the effects of temperature and aging. In addition, multi-stage filters pose subtle design challenges. Because coefficients and configurations—once established—tend to be inflexible, signal-processing hardware using analog parts generally is restricted to performing a narrow, dedicated task.

In response to the limitations of analog-based processing, the digital processing of signals has emerged as an alternative. The next section demonstrates how signal-processing tasks—including filtering, spectral analysis, and a host of others—can be carried out with digital arithmetic operating on digitized data. Recent advances in VLSI (very large-scale integration) now make it feasible to perform real-time digital signal processing with just a handful of ICs. The advantages conferred upon a system by such DSP hardware are dramatic—substantially improved performance, stability, and flexibility. Just as digital computers supplanted analog computers two decades ago in general-purpose computing applications, DSP is strongly challenging analog circuit configurations in real-time processing.

Our discussion of spectral analysis and digital filtering will benefit from a brief discussion of DSP nomenclature (there is also a brief glossary at the end of this chapter). Following Figure 21.2, an incoming analog signal is digitized, with the sampled data output points denoted x_i, or x(i). The index, i, corresponds to the discrete sampling time. This sampled data is stored in a buffer and operated on by DSP hardware. The DSP algorithm determines the sequence in which data and coefficients are accessed and how they are processed. In the cases below, the computational outputs are spectral weights or filtered sampled data.

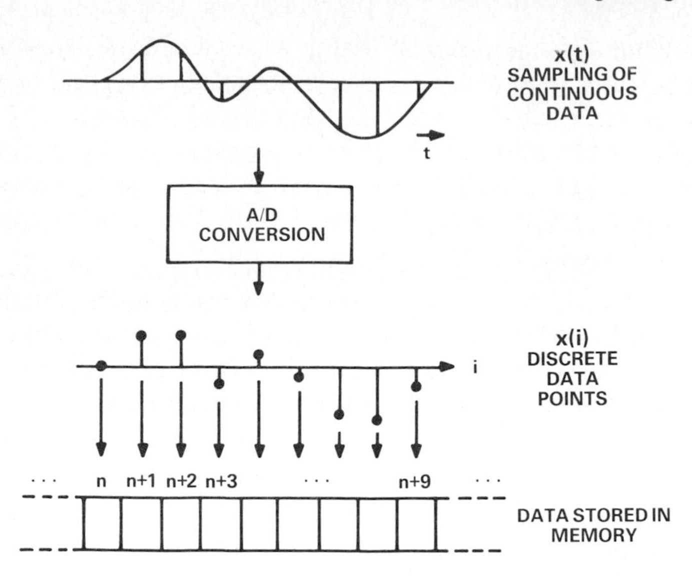

Figure 21.2. In DSP, continuous data is replaced by sampled data and continuous time by discrete time.

21.1.1. SPECTRAL ANALYSIS

The departure point for a discussion of spectral analysis is the Fourier transform equation pair:

$$x(t) = \frac{1}{2\pi}\int_{-\infty}^{\infty} X(\omega)\,e^{j\omega t}\,d\omega \tag{21.1}$$

$$X(\omega) = \int_{-\infty}^{\infty} x(t)\,e^{-j\omega t}\,dt$$

where t is time, ω is angular frequency, (2πf), x(t) is the signal—a function of time—and X(ω) is its counterpart in the frequency (spectral) domain. These equations give us, at least formally, the mechanics for taking a signal's time-domain representation and resolving it into its spectral weights—called Fourier coefficients.

Since the Fourier equations require continuous integrals, they have only indirect bearing on digital processors. However, under certain circumstances, a sampled (digitized) signal can be related faithfully to its Fourier coefficients through the discrete Fourier transform (DFT):

$$x(n) = \frac{1}{N}\sum_{k=0}^{N-1} X(k)\,W_N^{-kn} \tag{21.2}$$

$$X(k) = \sum_{n=0}^{N-1} x(n)\,W_N^{kn}$$

Provided that the signal is sampled frequently enough (at a rate $\geqq$ twice its highest frequency component), and assuming that the signal is periodic, the above DFT equations hold exactly. What is most interesting from the standpoint of DSP is that the DFT equation provides us with a means to estimate spectral content by digitizing an incoming signal and simply performing a series of multiply/accumulate operations.

To see qualitatively why the DFT equation yields spectral information, consider Figure 21.3. A time signal is superposed on a spectral "template" at various frequencies. In the first case, for frequency ω_1, the input signal and the spectral template have little relationship; as a result, the positive products are more or less cancelled out by negative products. The net effect is that the summation in the DFT equation indicates little spectral energy of frequency ω_1 in the input signal. In the second case, however, a reinforcing pattern emerges; the signal and the template tend to be positive or negative concurrently—producing a positive product nearly everywhere. Thus, the sum of the products will be a large positive number, indicating that the incoming signal has significant energy of frequency, ω_2.

Unfortunately, the large number of multiplications required by the DFT limits its use in real-time signal processing. The computational complexity of

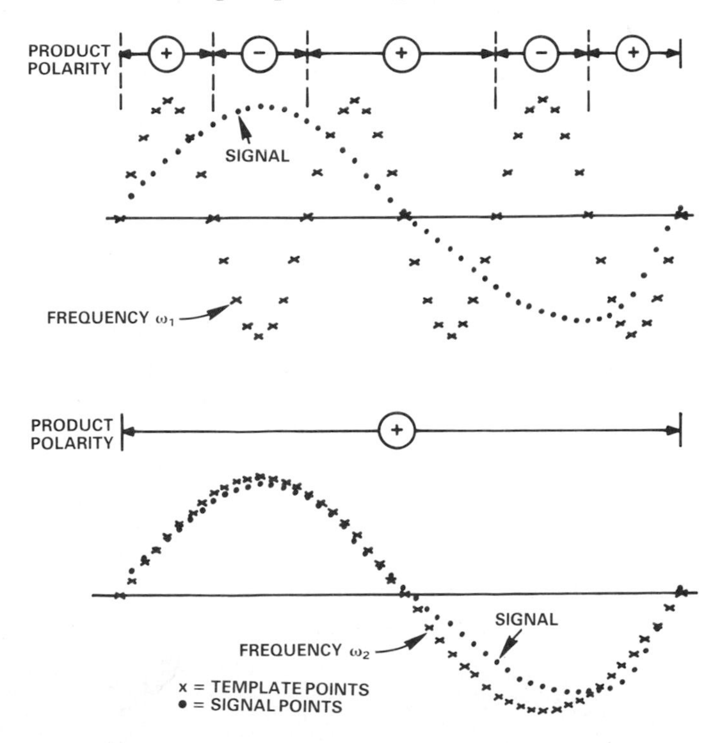

Figure 21.3. Input signal compared with sinusoidal "template" to measure frequency content of input signal. ω_1 is at a much different frequency, ω_2 is at very nearly the same frequency as the signal's fundamental.

the DFT grows with the square of the number of input points; to resolve a signal of length N into N spectral components requires N^2 complex multiplications ($4N^2$ real multiplications*). Given the large number of input points needed to provide acceptable spectral resolution, the computational requirements of the DFT are prohibitive for most applications.

The fast Fourier-transform (FFT) algorithm produces results identical to those of the DFT but reduces computation requirements by several orders of magnitude. The FFT achieves its economies by exploiting computational symmetries and redundancies that exist in computing the DFT. The availability of the FFT makes spectral analysis feasible, at virtually real-time rates.

The basic evolutionary process from DFT to FFT is demonstrated by simply segregating the odd and even terms to break an N-point DFT equation's summation into the sum of two (N/2)-point sub-series:

$$\begin{aligned} X(k) &= \sum_{n \text{ even}} x(n)\, W_N^{nk} + \sum_{n \text{ odd}} x(n)\, W_N^{nk} \\ &= \underbrace{\sum_{r} x(2r)\, W_{N/2}^{rk}}_{\text{N/2–point DFT}} + W_N^{k} \underbrace{\sum_{r} x(2r+1)\, W_{N/2}^{rk}}_{\text{N/2–point DFT}} \end{aligned} \qquad (21.3)$$

In this way, one N-point DFT has been reduced to two N/2-point DFTs. While this decomposition may not seem material, the key is that computational requirements for the DFT grow with the square of the number of points involved. Breaking the DFT in half decreases the number of complex multiplications from N^2 to $2x(N/2)^2$. Each of the above (N/2)-point DFTs can in turn be partitioned into two (N/4)-point DFTs, again reducing computation.

Proceeding in this fashion, an entire N-point DFT can be reduced (for N a power of two) to a series of elementary operations, called butterflies. A butterfly is a two-point DFT, along with a multiplication by a complex rotation factor (see Figure 21.4). The net effect of this systematic decomposition is to

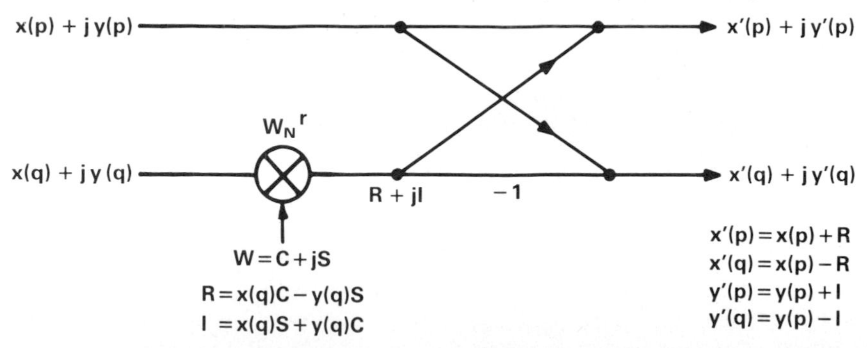

Figure 21.4. Fast Fourier transform radix-2 "butterfly." An N-point transform contains (N/2) $\log_2$ (N) of these operations.

*$(A + jB)(C + jD) = (AC - BD) + j(AD + BC);\ j = \sqrt{-1}$

reduce the total number of complex multiplications from N^2 to $N/2 \log_2 N$ for executing an N-point DFT. For example, a 1,024-point DFT would require more than 1,000,000 complex multiplications, while the corresponding FFT requires only $512 \times 10 = 5{,}120$ complex multiplications.

Figure 21.5 illustrates how an FFT resolves a signal into its spectral components—and the effect of FFT length on spectral resolution. In all three cases, the same input signal is examined. In the first case, we perform a 64-point FFT on the first 64 sample points; the second and third cases perform 256- and 1,024-point FFTs on the first 256 (1,024) data points. The differences observed in spectral resolution underscore a key principle—the longer the time period in which a signal is observed, the sharper the spectral resolution that can be attained.

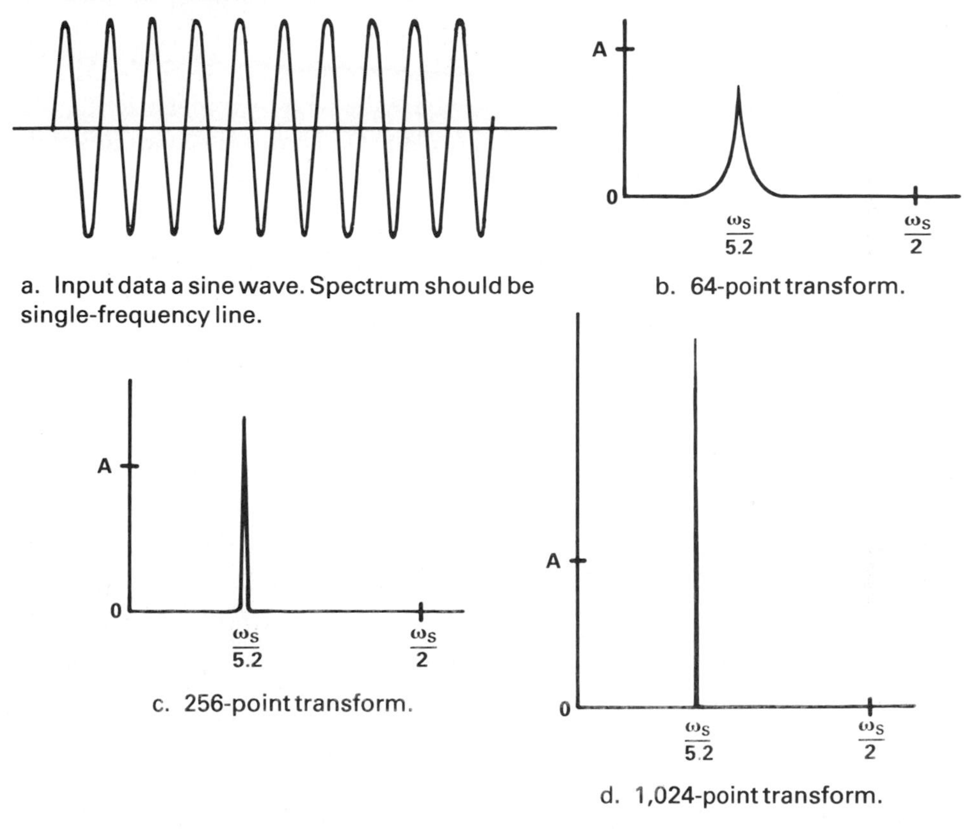

Figure 21.5. Fourier transform: effect of number of points on computed spectrum.

Bringing digital hardware to bear on a spectral-analysis task has numerous advantages. With a long-enough window of data, it can provide very precise spectral resolution. Moreover, the system can be flexibly programmed to vary the FFT size dynamically, according to the spectral resolution needed. Finally, once the data is digitized, it is possible to perform additional DSP tasks, such as spectrum averaging, to further improve FFT performance.

21.1.2. DIGITAL FILTERING

Digital filters have performance attributes similar to those of analog filters—ripple in the passband and attenuation in the stopband. What distinguishes digital filters is their ability to provide arbitrarily high performance. For example, the rolloff slope (i.e., the rate at which the filter makes a transition from the passband to the stopband) can be made virtually as steep as is desired. In general, it is straightforward to design a digital filter that easily out-performs the most complicated analog designs.

The fundamental digital filtering equation is:

$$y(n) = \sum_{i=0}^{N-1} h(i)\,x(n-i) \;+\; \sum_{j=1}^{M} \mathrm{b}(j)\,y(n-j) \qquad (21.4a)$$

The coefficients, $h(i)$, are weighting factors applied to the most recent N sample points; the coefficients $\mathrm{b}(j)$ correspond to terms feeding back the M most recent filtered output points. In the case where the feedback coefficients, b(j), are all equal to zero, the digital filter is termed a Finite Impulse Response (FIR) filter.

$$y(n) = \sum_{i=0}^{N-1} h(i)\,x(n-i) \qquad (21.4b)$$

If feedback terms are used, then the filter belongs to the Infinite Impulse Response (IIR) class. These two types of digital filter manifest important tradeoffs, which warrant further discussion.

Finite Impulse Response Filters

In equation 21.4b, the fundamental FIR filter equation, each output point, y(n), is obtained by convolving the past N input points, x(n – i), with a set of coefficients. An FIR filter can be viewed as a tapped delay line (see Figure 21.6); the parameter, N, corresponds to the number of taps of the FIR filter. The number of taps tells us the number of multiply/accumulate operations required to compute this convolution.

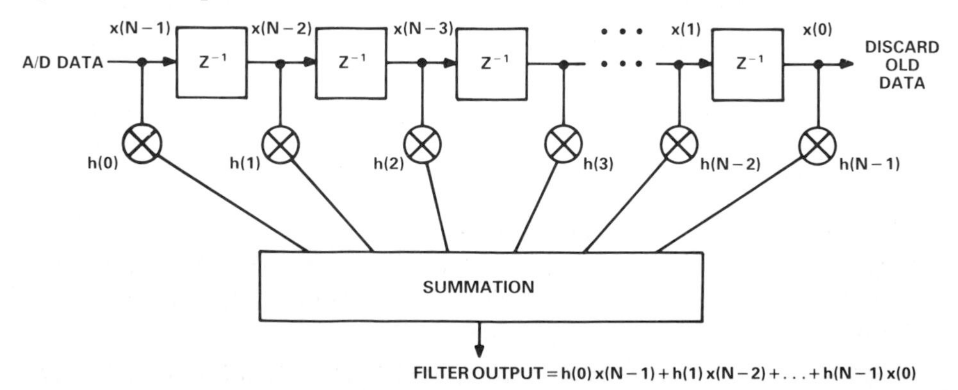

Figure 21.6. FIR filter as a tapped delay line.

The coefficients, h(i), represent the impulse response of the FIR filter. As Figure 21.7 demonstrates, an input of 1 at time 0 (x(0) = 1), and zero at all other times, results in output values equal to h(i) for the periods $i=0,\ldots,N-1$. Note that the h(i) can be non-zero for only a finite number of time periods, hence the term "finite" impulse response. Since they use no feedback, FIR filters are unconditionally stable.

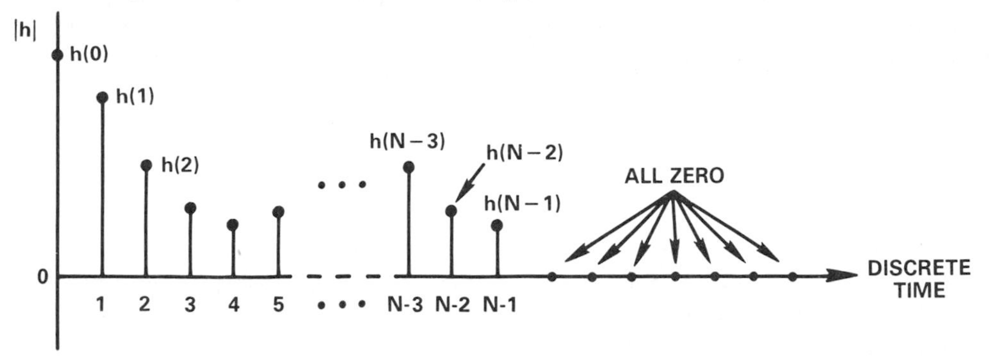

Figure 21.7. FIR-filter impulse response function, h(i).

FIR filters can best be understood in the context of two fundamental relationships. First, a filter's time-domain impulse response, h(i), and its frequency response, H(f), are related via the Fourier transform. Second (a key principle of DSP), multiplication in one domain is equivalent to convolution in the conjugate domain. With respect to FIR filters, this tells us that multiplying the input spectrum by the desired filter transfer function is equivalent to convolving the input time-function with the filter's impulse response in the time domain.

To further amplify the above point, consider Figure 21.8. Figure 21.8(a) illustrates an incoming signal that we wish to low-pass filter. It consists of the sum of two signals at frequencies, f_1 and f_2. Since there are just two frequencies present, its spectrum looks like (b). We'd like to design an FIR filter to filter out f_2, leaving just f_1, as shown vs. time in (e) and frequency in (f).

An ideal low pass filter is suggested in (d); note that multiplying it by the input spectrum in (b) will give the spectral domain representation of a low-pass filtered output, allowing f_1 to pass and completely attenuating f_2. Now, the Fourier transform of (d)'s ideal filter is the sinc function ($\sin x/x$) in (c). Consequently, if the input (a) is convolved with a discretized sinc function, (c), we can directly compute the filtered output signal, as a function of time (e).

More generally, an FIR filter boils down to simply convolving the digitized input signal with the filter's time-domain coefficients, h(i)—an action equivalent to multiplying the frequency representation of the input signal by the filter's transfer function.

Unfortunately, from the perspective of practical implementation, Figure 21.8(c)'s sinc function is infinite in duration. To obtain a filter that can be implemented, we must somehow truncate the number of coefficients used to represent (c); this can be carried out by discarding the tails—or, more effectively—by multiplying the function by some window. This truncation/windowing, however, makes it impossible to realize (d)'s ideal low-pass filter transfer function, and ripple and rolloff are necessarily introduced (see Figure 21.10).

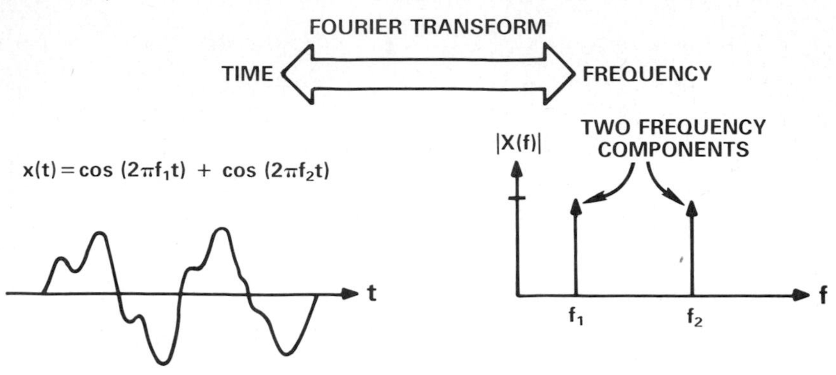

a. Input signal, sum of two sine waves at different frequencies.

b. Spectrum of signal in (a).

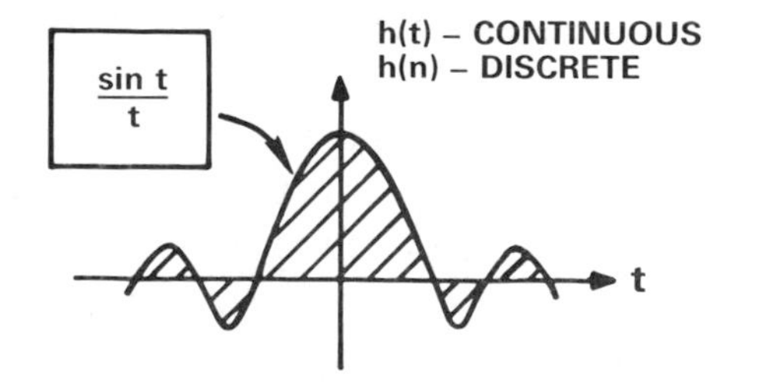

c. Impulse response of desired filter, (sin t)/t function—transform of (d).

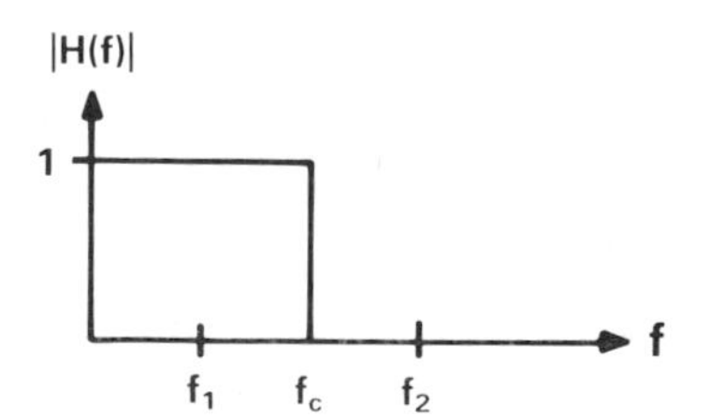

d. Desired low-pass filter transfer function.

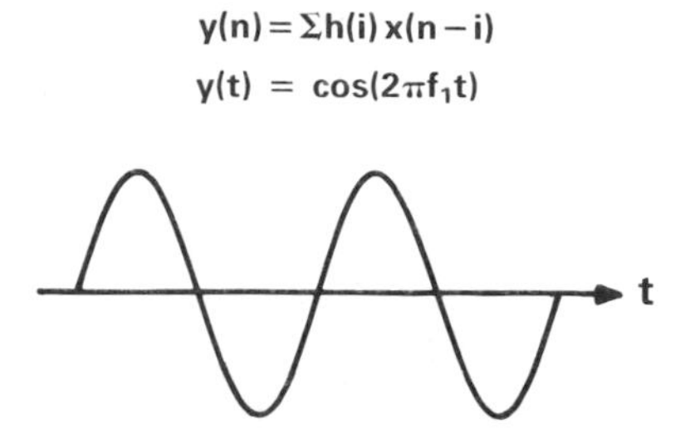

e. Time response of filtered signal, convolution of (a) and (c).

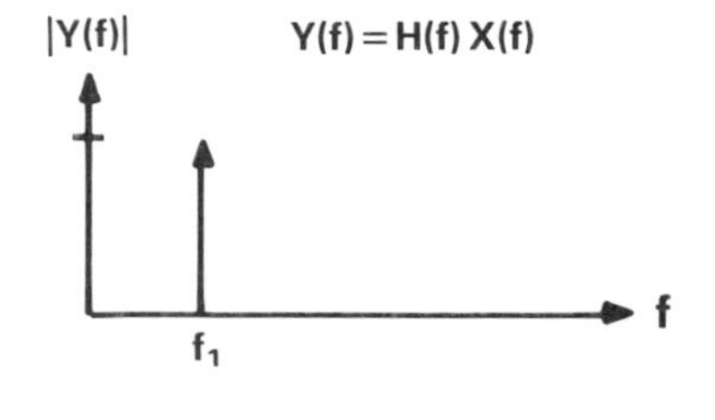

f. Frequency response of filtered signal, product of (b) and (d).

Figure 21.8. Basics of low-pass filter design using Discrete Fourier transform, (a), (c), (e) in time domain, (b), (d), (f) in frequency domain.

The essentials of FIR filters can be illustrated in a very simple example, something many readers may have already used. It is intuitively apparent that noisy data can be smoothed (or filtered) by taking a moving average. For example, noisy laboratory data might be plotted using a moving average of the last five points:

$$y_n = (1/5)\,[x(n) + x(n-1) + x(n-2) + x(n-3) + x(n-4)].$$

This simple scheme is nothing less than a five-tap FIR filter. Its impulse response is constant for five periods and then drops to and stays at zero. In the spectral domain, the Fourier transform of the impulse response, $h_i = 0.2$, $i = 0, \ldots, 4$, and zero elsewhere, is shown in Figure 21.9. This frequency-domain plot tells us, consistent with our intuition, that low-frequency components will largely be passed, while high-frequency noise will be relatively damped, but not uniformly.

The real value of digital filters is not apparent in an elementary example, such as Figure 21.9's rather sloppy transfer function. By taking an adequate

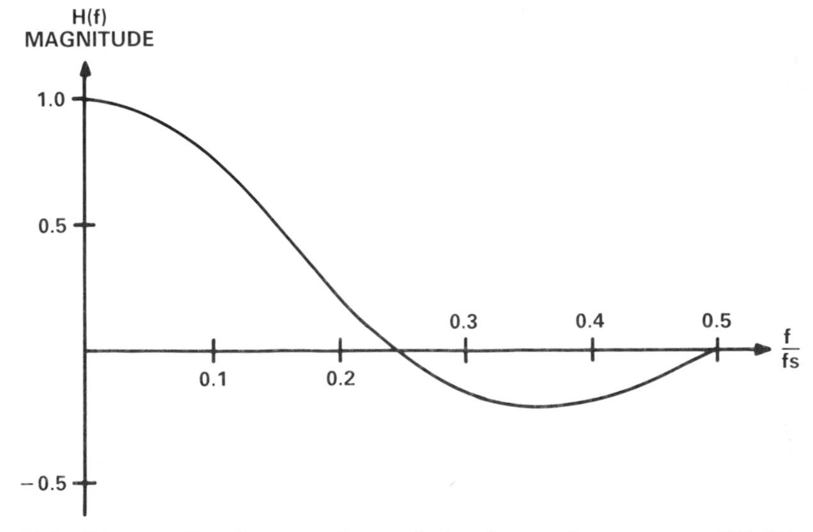

Figure 21.9. Discrete Fourier transform of simple running-average FIR filter with 5 taps.

number of taps and properly choosing the coefficients, an FIR filter can provide excellent discrimination, as the response spectra shown in Figure 21.10 illustrate, for various numbers of taps. In general, the greater the number of taps used in an FIR filter, the better the filter's performance, at the expense of reduced throughput.

In designing FIR filters, tradeoffs must be made among several attributes (i.e., ripple in the passband, ripple in the stopband, width of the transition band, phase distortion, and throughput). These tradeoffs are reflected in the number of coefficients used—and their particular values. This selection can be made directly in the time domain (for example, to implement a pure time

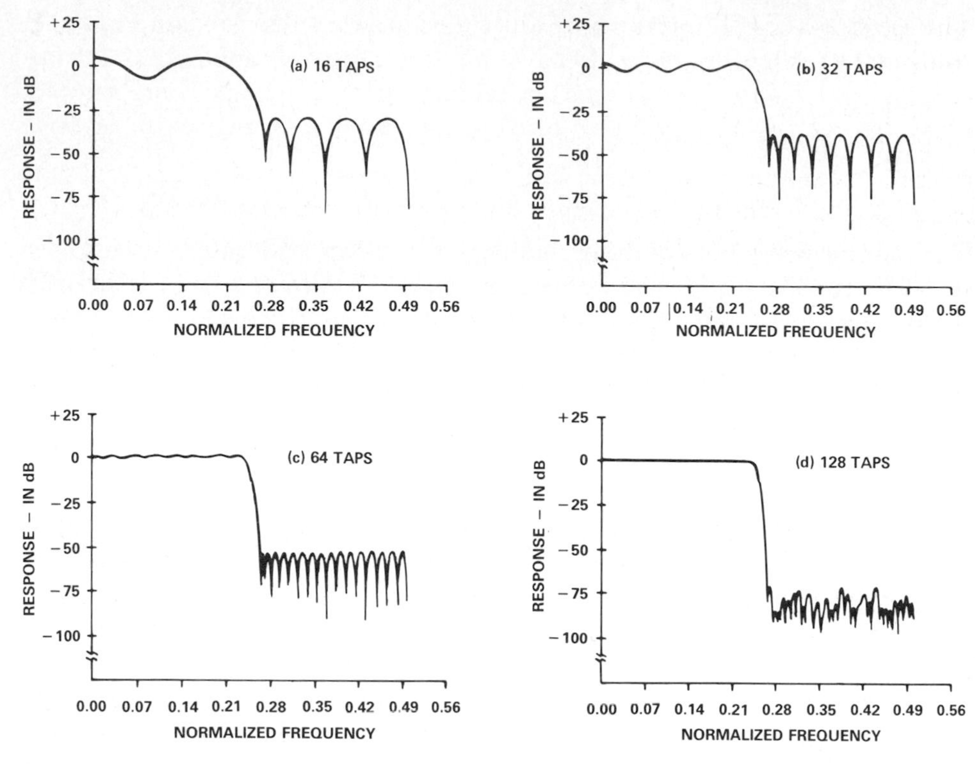

Figure 21.10. Comparison of FIR low-pass filters with (a) 16, (b) 32, (c) 64, and (d) 128 taps. Normalized frequency $= f_{actual}/f_{sampling}$.

delay or an N-point running average), but more commonly is made employing powerful and easy-to-use computer-aided-design (CAD) techniques to determine optimal parameter values for the desired filter performance.[1]

With an FIR filter, there is a direct relationship between the incoming sample rate, the speed of the digital hardware used in the system, and the performance of the FIR filter:

$$\frac{1}{\tau_m} = N \times S \tag{21.5}$$

where τ_m = multiply time
N = number of taps
S = sample rate

For example, a specified performance and incoming sampling rate determine the multiplication rate required of the hardware. Thus, if a filter is to have 100 taps, and the sampling rate is 100 kHz, the multipliers must perform their

[1]While a discussion of these approaches is beyond the scope of this chapter, the interested reader is referred to Rabiner, Lawrence R., and Gold, Bernard, *Theory and Application of Digital Signal Processing*, (Englewood Cliffs, N.J.: Prentice-Hall, 1975).

multiplications within 100 nanoseconds, a speed not difficult to achieve with modern CMOS multipliers.

The highest-performance filters of Figure 21.10 could not be matched by an analog-based implementation. Moreover, these digital filters are straightforward to design and implement in hardware. There are other advantages of digital FIR filters that further increase their desirability. Once designed, they are stable; performance is insensitive to the effects of temperature or aging. In addition, a key consideration is that the filter's performance can be changed simply, just by modifying the number of coefficients used and their values. For instance, a simple software modification would shift a filter's performance from (a)'s to (d)'s—with no change in hardware, except that slightly more memory is used.

Infinite Impulse Response Filters

Infinite Impulse-Response (IIR) filters are the other commonly used digital filter, differing from FIR filters in one fundamental respect: feedback. Because of feedback, the filter's impulse response can continue long after the initial impulse—indeed, for an infinite duration. The use of feedback allows an IIR filter to economize in the number of multiplications required to provide a given filter performance. But this efficiency is not without its costs. As in other recursive systems, input perturbations can ring indefinitely—in some cases causing the filter to be unstable. Also, the accumulated effects of fed-back round-off noise can noticeably degrade the filter's performance.

Equation 21.4a is the basic IIR filter equation, assuming one or more of the coefficients, b(i), is non-zero. Figure 21.11 plots an impulse response function for an IIR filter with a typical set of coefficients. In this case, and in gen-

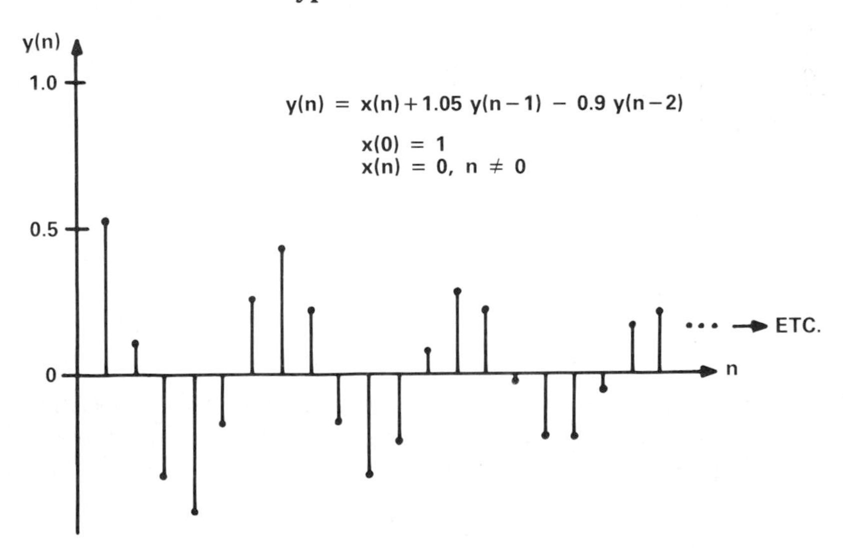

Figure 21.11. A portion of the impulse response of an IIR filter.

eral, the presence of feedback means that the impulse response of an IIR filter never converges to zero—may even diverge—hence *infinite* impulse-response. However, as a practical matter, noise, round-off error, and limited resolution do result in effective convergence when simulating analog filters that settle physically.*

One form of IIR filter that is widely used is the biquadratic, or biquad, form of the generalized IIR equation (equation 21.4):

$$y(n) = a_0 x(n) + a_1 x(n-1) + a_2 x(n-2) - b_1 y(n-1) - b_2 y(n-2) \qquad (21.6)$$

corresponding to the continuous transfer function:

$$\frac{A s^2 + B s + C}{D s^2 + E s + F} \qquad (21.7)$$

The biquad serves as a building block for IIR filter design. Generally, several biquad sections are cascaded together to obtain the desired filter performance.

The goal in IIR filter design, to match a given transfer-function requirement, is to determine the number of biquad sections to be used, and the values of their coefficients. Two principal IIR design techniques exist. The first considers the transfer functions of conventional analog filters, such as the Butterworth, Chebyshev, or Elliptic; a digital filter is then constructed that provides the same impulse response as its analog counterpart. The second relies on computer-aided-design techniques to arrive at an optimal IIR implementation. In this context, "optimal" means that the number of terms needed to meet a specified performance specification is minimized.[2]

An example that demonstrates the efficiency of an IIR implementation is a comparison of an FIR and IIR implementation of a 70-dB stopband attenuation filter. To achieve this performance, an FIR filter would require nearly three times as many multiplications-per-second as an IIR implementation. These performance advantages, however, require tradeoffs to be made in other key respects, as summarized below:

	IIR	FIR
Performance/Throughput	Higher	
Ease of Design		Easier
Filter Stability	Sensitive	Unconditional
Round-off Noise	Sensitive	Insensitive

*Even a simple single-time-constant R-C analog filter will theoretically take an infinite time to reach its asymptotic steady-state condition, but in practice it settles, for example, to 1 LSB of 32 bits within 23 time constants—and so does its IIR-filter equivalent.

[2]A more detailed discussion of IIR filter design can be found in Rabiner and Gold, *op. cit.*[1]

21.1.3 OTHER DSP ALGORITHMS

DSP is not limited to FFTs and digital filters. In fact, one of the prime advantages of DSP is that, once the data is digitized, fast digital hardware can perform a broad range of tasks. Commonly used DSP routines include modulation/demodulation (heterodyning), waveform generation, correlation, estimation, control, power spectrum calculations, and multi-dimensional transforms. While a discussion of these areas would take us far afield of this chapter's focus, their breadth points to an important advantage of DSP—system flexibility. By converting signals early and incorporating fast multiply/accumulate hardware to perform digital filtering and/or spectral analysis, the system can readily offer numerous enhancements.

21.2 DSP HARDWARE

As the previous sections have suggested, DSP algorithms require fast data transfers and, what is most important, a fast multiply/accumulate capability. For example, if a 30-tap FIR filter is applied to a signal sampled at a 100-kHz rate, the system must be able to perform 3 million multiply/accumulates per second (Equation 21.5). As we will see, DSP processors can easily meet this challenge. In contrast, it would take nearly twenty of one of the fastest microprocessors available (the 12.5-MHz Motorola 68000, with its 5.5-microsecond multiply time) in parallel to handle this task.

Although the number of DSP architectures that can be designed is quasi-infinite, they have in common the need for several functional elements—program sequencing, address generation, and number crunching—as well as memories that store program instructions, data, and coefficients.

Program instructions provide the controls that, on a cycle-by-cycle basis, govern the operation of all circuits in the DSP processor. By specifying the addresses to the program memory, the *sequencer* controls the system's instruction flow. It is also responsible for branching, subroutine jumps, interrupt handling, and overall system control.

A digital signal processor's *address-generator* logic determines read/write locations for coefficient and data memory. A fast, flexible addressing element is needed in DSP, since many algorithms have intricate addressing structures and require rapid data transfers to and from the number crunchers. System capabilities are seriously compromised by inadequate addressing logic.

A DSP system's *number crunchers*—multipliers, arithmetic/logic units (ALUs), and barrel shifters—perform the data computation for DSP algorithms. These elements are characterized by three attributes—precision, throughput, and cost. Precision is established by the size of arithmetic word used to handle the digitized signal, ranging from a 4-bit fixed-point number (one part in 16 resolution) to double-precision floating point (dynamic range of better than one part in 10^{300}). Throughput is a function of speed, internal architecture, and port structure. The arithmetic characteristics of these devices must match the system's processing demands.

Two principal approaches are used today: microcoded systems—using building-block ICs—and single-chip signal processors. The advantages of a microcoded system are generally speed and flexibility. Single-chip processors, in contrast, generally benefit from compactness and ease of design.

21.2.1 SINGLE-CHIP PROCESSORS

Single-chip processors have program sequencing, addressing, and arithmetic logic on a single device. In addition, devices available today generally include a modest amount of program and data memory on the device. In the present state of commercial semiconductor technology, it isn't feasible to fabricate high-performance DSP functions and a large amount of memory on a single monolithic device. Consequently, compromises must be made that force today's single-chip processors to be optimized for specific applications (e.g., telecommunications, imaging). They offer adequate performance in such narrowly defined roles—but are precluded from general-purpose high-performance digital signal-processing.

21.2.2 MICROCODED SYSTEMS

A microcoded system employs building-block ICs to construct the digital signal-processor (see Figure 21.12). This gives the designer increased control

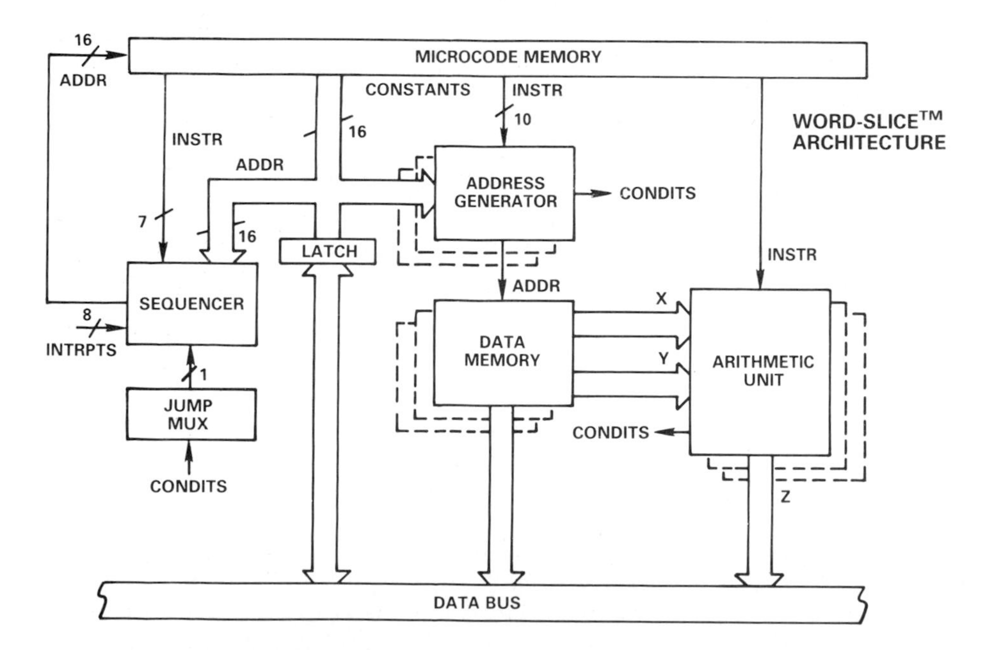

Figure 21.12. Building blocks of a microcoded DSP system. Word Slice™ is a trade mark of Analog Devices, Inc.

over the system's architecture, with sufficient latitude to meet very high throughput specifications. In the past, such designs traditionally were pieced together with numerous power-hungry bit-slice devices, which tended to increase the system's cost and complexity. A second-generation family of microcoded devices, which provide a comprehensive set of building blocks that enable high-performance, relatively compact digital signal processors to be constructed, is shown in figures 21.13 through 21.20.

The power of a microcoded system lies in the fact that, during each clock cycle, each component can execute an instruction, allowing the system to attain high throughput. The operation of each component is governed by the system's microcode memory, which contains the microcode controls (0's and 1's). On a cycle-by-cycle basis, these controls are fed to each component, telling it what instruction to execute. The microcode memory's width is proportional to the number of components in the system; its depth is proportional to the length of the overall microcode program (or microprogram).

The brain of a microcoded system is its program sequencer, which steps the system through the microcode memory. A representative sequencer is the device shown in Figure 21.13, a high-speed, 16-bit microprogram sequencer op-

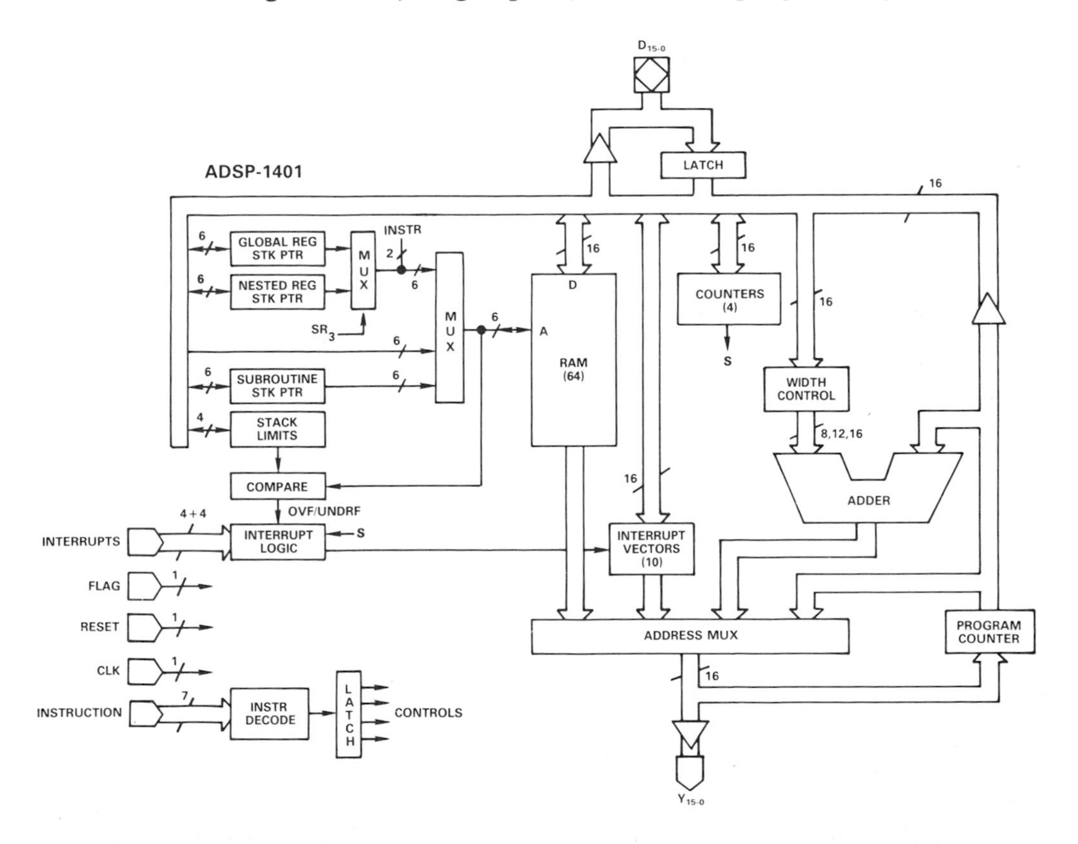

Figure 21.13. Program sequencer for use in microcoded DSP designs.

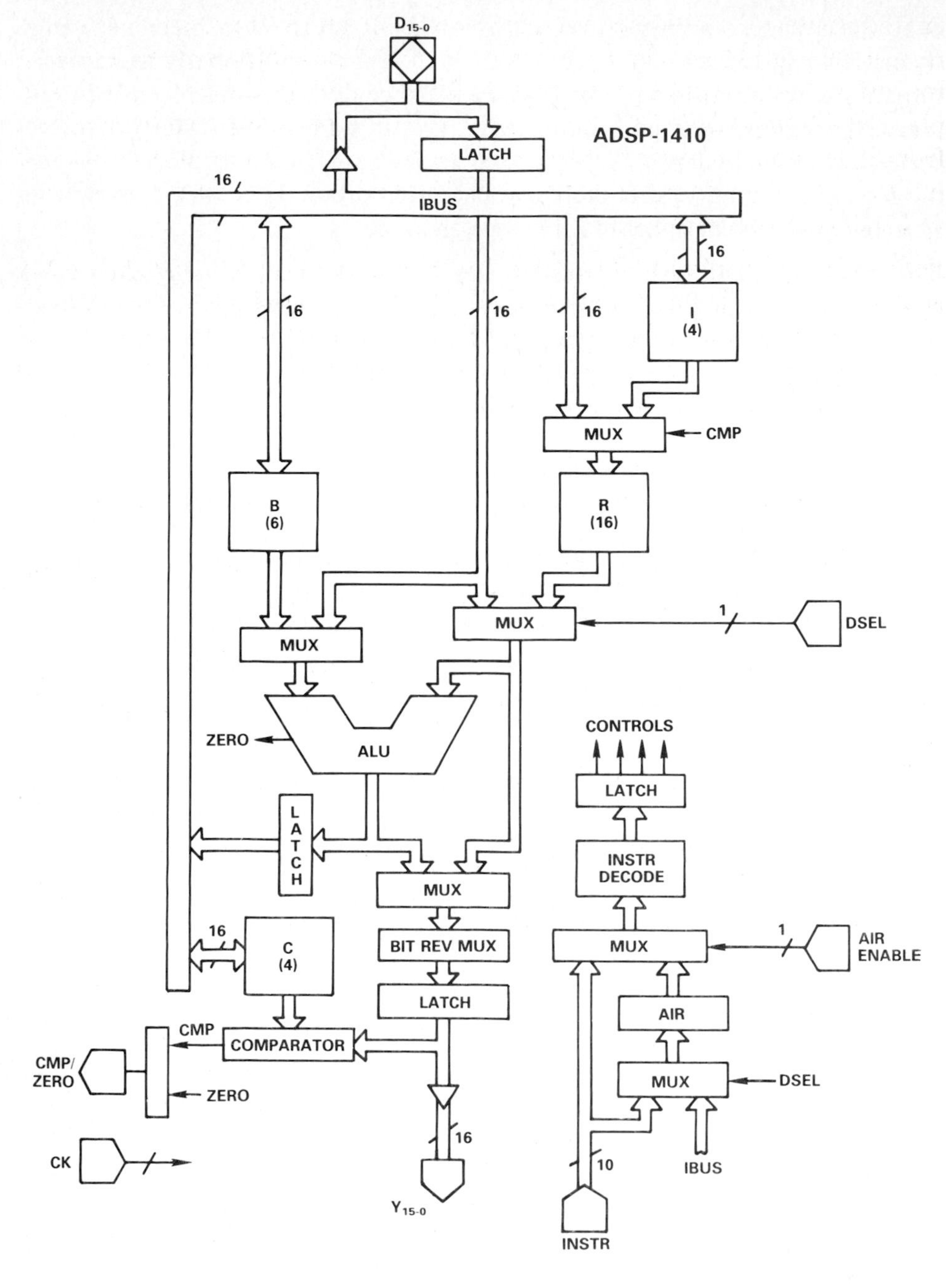

Figure 21.14. Address generator for microcoded DSP designs.

timized for DSP. This device usually performs an elementary function—it stores the current microcode address (instruction number) in the program counter, increments it, and outputs it. This activity corresponds to simply stepping through the instructions of a program. Obviously, though, limiting any computer to executing straightline code is unduly restrictive. For this reason, microprogram sequencers must also have the flexibility to handle subroutine jumps, branches, interrupts, and indirect jump addresses. For example, the 48-pin ADSP-1401 determines the next program memory address from i) a simple increment of the current address; ii) an absolute or relative jump; iii) a jump address from the internal RAM; iv) a subroutine return; or, v) an internal interrupt vector.

For its addressing requirements, a microcoded system can splice together several 4-bit or 8-bit arithmetic and logic units (ALUs). Alternatively, it can use devices such as the high-speed, general-purpose data address generator, shown in Figure 21.14. It performs all addressing tasks required by DSP routines in a single cycle; specifically, it

- Outputs an address pointer to data memory;
- Modifies the pointer by an offset value to determine the next memory read/write address;
- Compares the output pointer to a pre-set value, and, if equal, re-initializes the pointer with a value stored on-chip.

One address generator may be adequate for an application; however, if extremely high throughput is required, several address generators may be needed in the system.

The heart of a DSP microcoded system is its arithmetic units, the devices that perform the DSP number crunching. Until recently, designers of systems demanding high throughput had relatively few options; but numerous recently available chips increase the designer's choices, imparting the flexibility to optimize DSP hardware for the application's specific processing requirements.

21.2.3 FIXED-POINT OPTIONS

Initially introduced in 1976 by TRW, and now available from many manufacturers—Analog Devices among them—industry-standard fixed-point multipliers provide systems with fast multiplication capability. These VLSI devices dedicate a large amount of silicon to a parallel array multiplier that can deliver the product of two numbers at 10-MHz rates. To maximize throughput, these devices have dedicated ports for multiplier inputs and outputs. As a result, device I/O rates match multiply speed. The block diagram of a 16×16-bit example is shown in Figure 21.15.

Since many DSP algorithms require a chain of products to be continuously summed, high-speed multiplier/accumulators (MACs) are also available (Figure 21.16). These devices not only multiply two numbers together rapidly, they can optionally add each product to a value stored in an on-chip accumulator. In this fashion, a single device can perform the high-speed MACs required in DSP.

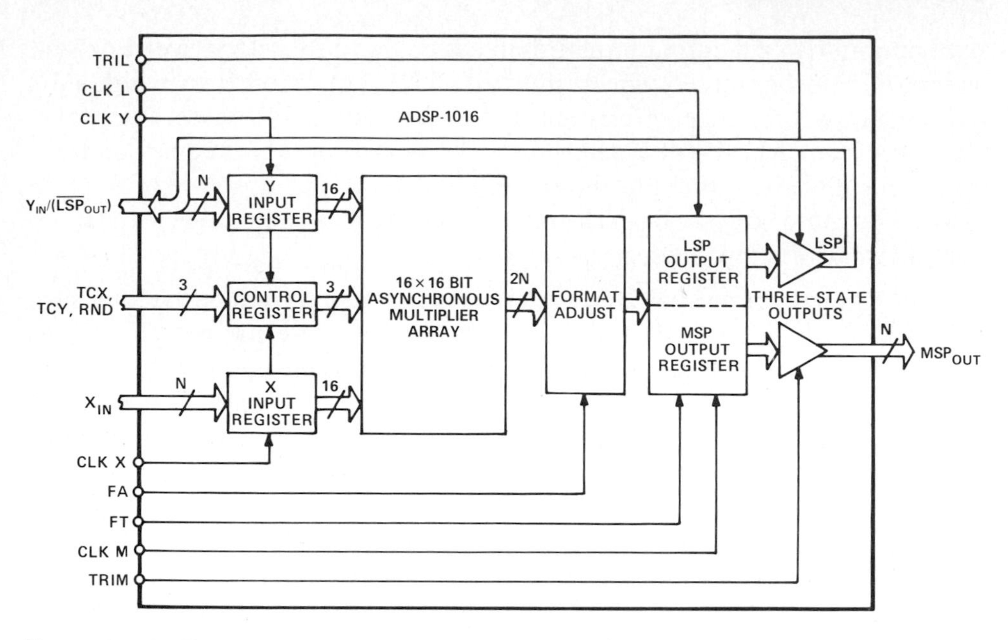

Figure 21.15. Typical 16 × 16 multiplier for DSP. Note that full 32-bit output product is available.

Several different versions of these industry-standard multipliers and MACs are available. Depending on the arithmetic precision required in an application, the designer can choose 8-bit, 12-bit, or 16-bit implementations. Number formats they will support include unsigned magnitude or twos com-

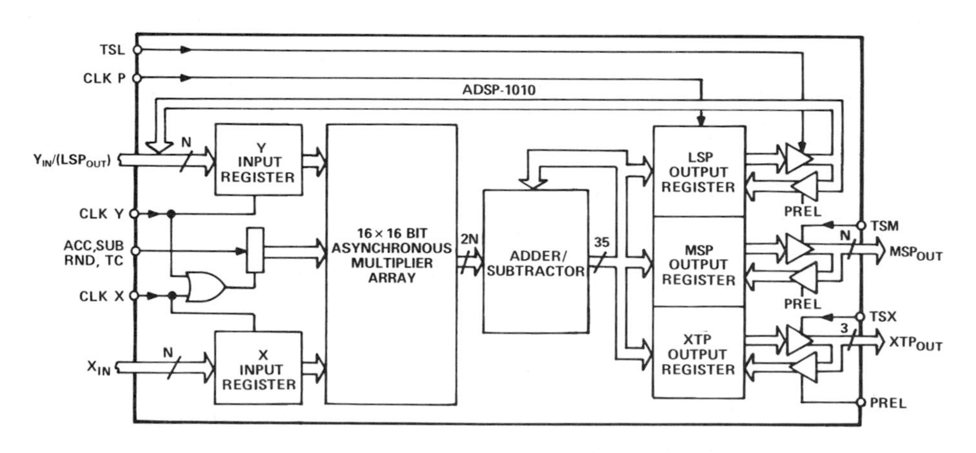

Figure 21.16. Typical 16 × 16 multiplier/accumulator (MAC).

plement. Tradeoffs must also be made among two key attributes of these devices—speed and power—although recent advances in high-speed CMOS are allowing the same device to dominate in both respects. Table 21.1 details some of the available high-speed, low-power CMOS multipliers and MACs.

Designation	Device Type	Resolution (Bits)	Cycle Time Commercial (ns)	Military (ns)
ADSP-1016A	MULTIPLIER	16×16	75	90
ADSP-1010A	MAC	16×16	95	110
ADSP-1012A	MULTIPLIER	12×12	65	75
ADSP-1009A	MAC	12×12	80	90
ADSP-1080A	MULTIPLIER	8× 8	50	60
ADSP-1081	UNSIGNED MULT.	8× 8	100	115
ADSP-1008A	MAC	8× 8	55	85

Table 21.1 CMOS Multipliers and Multiplier/Accumulators

There are also devices that differ from the industry-standard fixed point multipliers and MACs to provide special application conveniences. For example, for applications where cost and board space must be reduced, a single-port MAC (Figure 21.17a) can be used with little sacrifice in throughput. Both inputs, and the output, communicate with the 16-bit computer bus via its single port, but a 6-bit instruction set makes possible a substantial amount of time-saving internal processing. Extra fixed-point precision can be obtained using devices such as a 24 × 24-bit multiplier (b). Extremely high speed can be attained with high-speed multipliers employing *pipelining* (see Glossary); for example, the Analog Devices ADSP-1102 and 1103, 16-bit fixed point devices can, in pipelined mode, deliver products at a 30-MHz rate.

Figure 21.18 shows the architecture of a 16-bit "enhanced" MAC (EMAC), that provides high-speed DSP capabilities, while eliminating considerable external circuitry. A monolithic integrated circuit, packaged in a one-square-inch pin-grid array, the device features a 16 × 16-bit array multiplier, two addressable input registers on each input port, two 40-bit-wide accumulators, and an internal feedback path, which simplifies polynomial expansions.

21.2.4 FLOATING-POINT OPTIONS

Floating-point arithmetic differs from fixed point in that each number has its own exponent, as well as a mantissa. As a result, systems using floating-point arithmetic handle very wide dynamic ranges. Floating point is valuable to a system any time a weak signal must be detected in the presence of noise. Also, if a large number of arithmetic operations are being performed on the data, floating-point largely eliminates the distorting effects of round-off noise. Applications demanding floating point include instrumentation, graphics and image processing, engineering workstations, and general-purpose array processors.

In the past, floating-point capability required considerable LSI, MSI, and SSI logic, configured on one or more circuit boards. Recently, however, monolithic floating-point implementations have become available. Initial offerings included limited-precision multipliers and ALUs—22-bit devices.

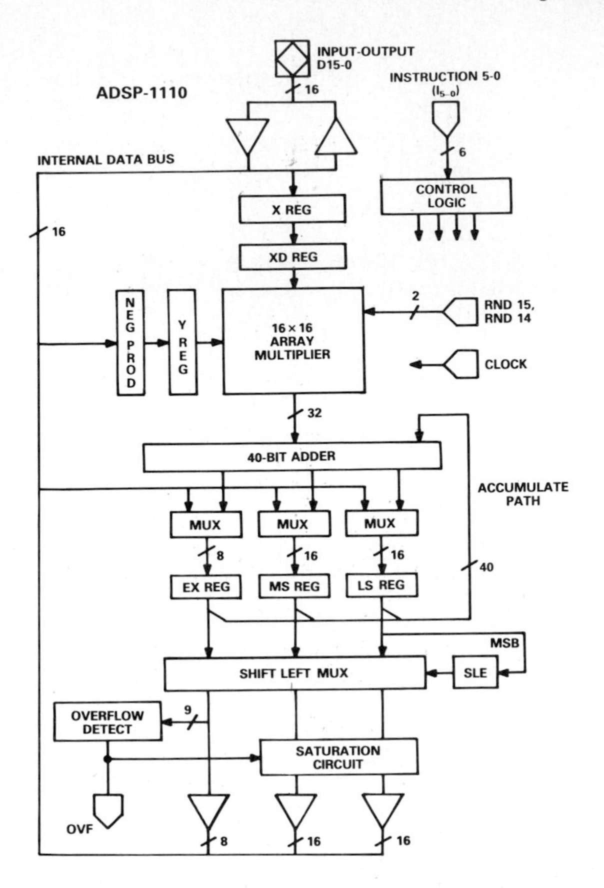

a. Single-port 16 × 16 multiplier/accumulator architecture. 40-bit adder provides extra accumulator capacity, reducing tendency to overflow.

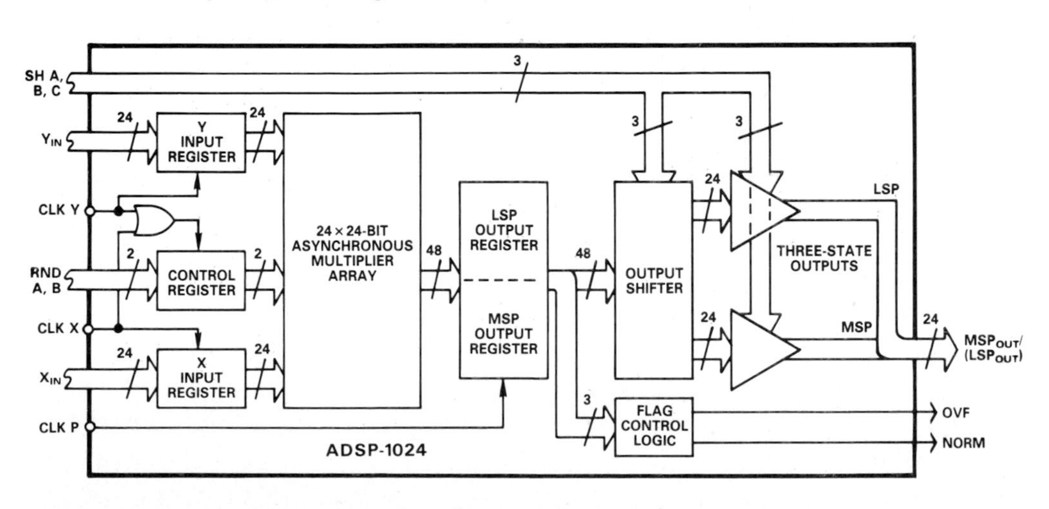

b. 24-bit multiplier with full 48-bit output in two bytes.

Figure 21.17. Special-purpose multipliers.

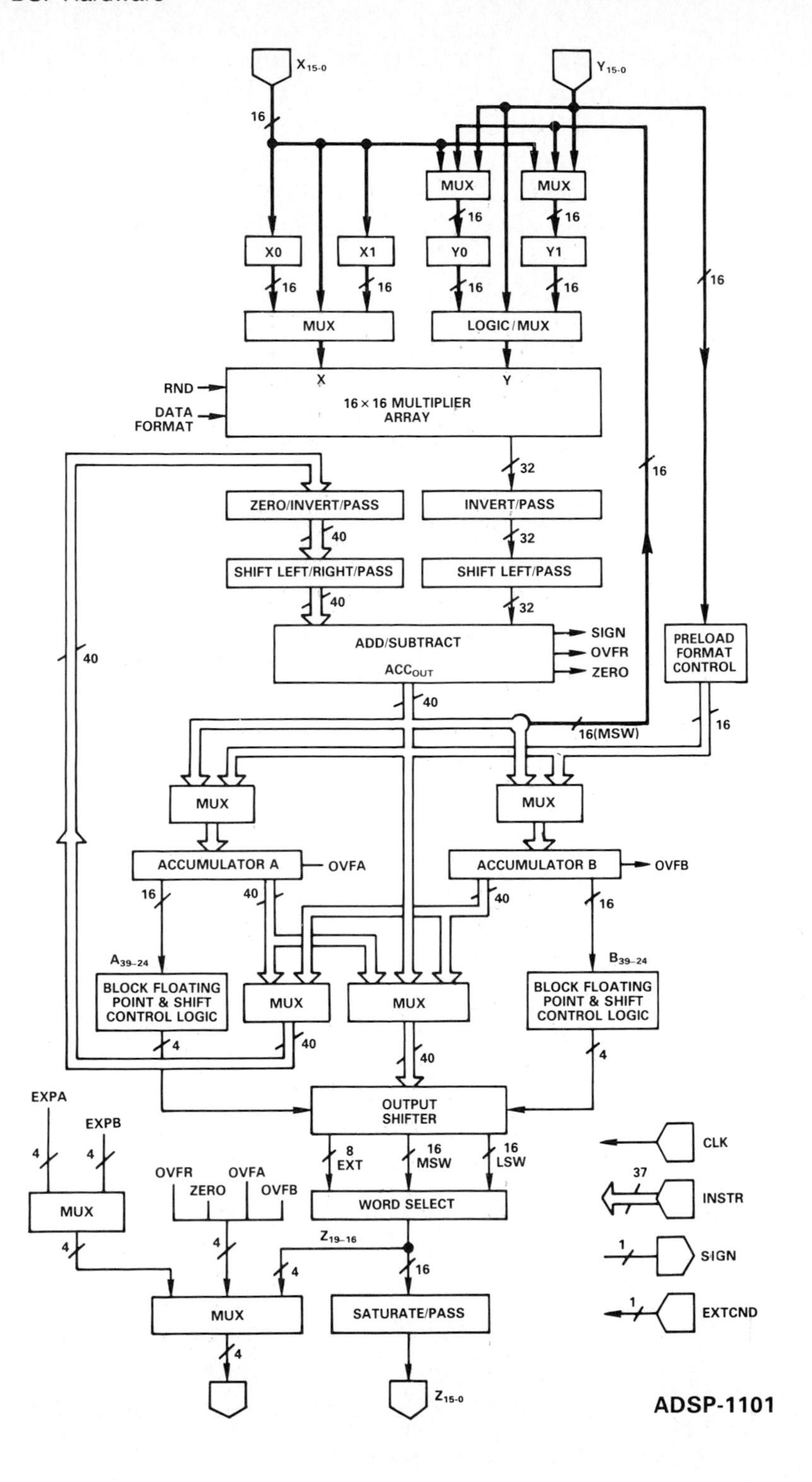

Figure 21.18. Enhanced multiplier/accumulator (EMAC).

The IEEE has issued a 32-bit floating point standard, and monolithic DSP ICs now exist that implement its specifications. Figure 21.19a shows an IEEE-compatible floating-point multiplier, capable of performing single- and double-precision floating-point operations, as well as 32-bit fixed point multiplications. A companion device (b) is an ALU, which handles single- and double-precision IEEE floating-point, as well as 32-bit fixed-point operations.

21.2.5 REPRESENTATIVE SYSTEMS

Figure 21.20 shows several alternative DSP architectures, each with its own characteristic throughput, precision, and cost. These compact, low-power CMOS-based systems can be microprogrammed to carry out any DSP algorithm. A key benchmark in DSP is the 1,024-point complex FFT. This al-

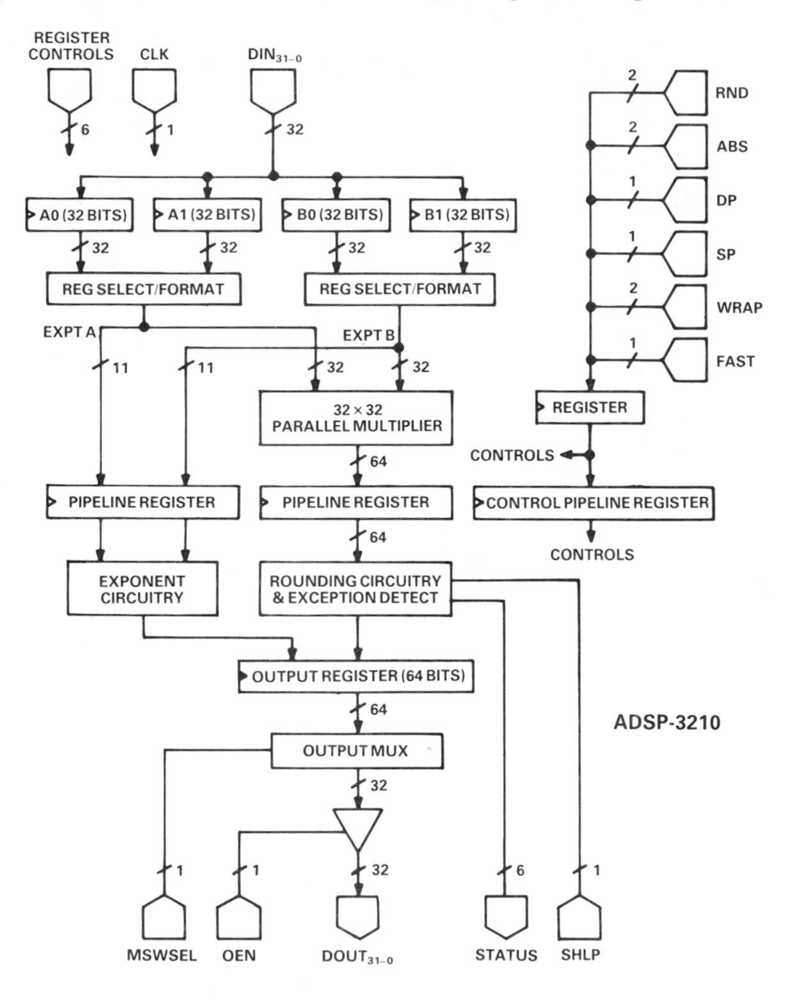

a. Floating-point multiplier.

Figure 21.19. Floating-point DSP units.

gorithm requires highly efficient program-sequencing, address-generating, and number-crunching capabilities. All of Figure 21.20's architectures will execute a 1,024-point complex FFT swiftly. For instance, (a)'s EMAC-based architecture (see Figure 21.18) performs this algorithm in 3 milliseconds. The floating-point architecture of (b) delivers a 1,024-point FFT in 4 milliseconds. Finally, the architecture of (c), featuring the 16 × 16-bit ADSP-1110A single-port MAC, a program sequencer (PS), and an address generator (AG), can execute a 1K complex FFT in 10 milliseconds.

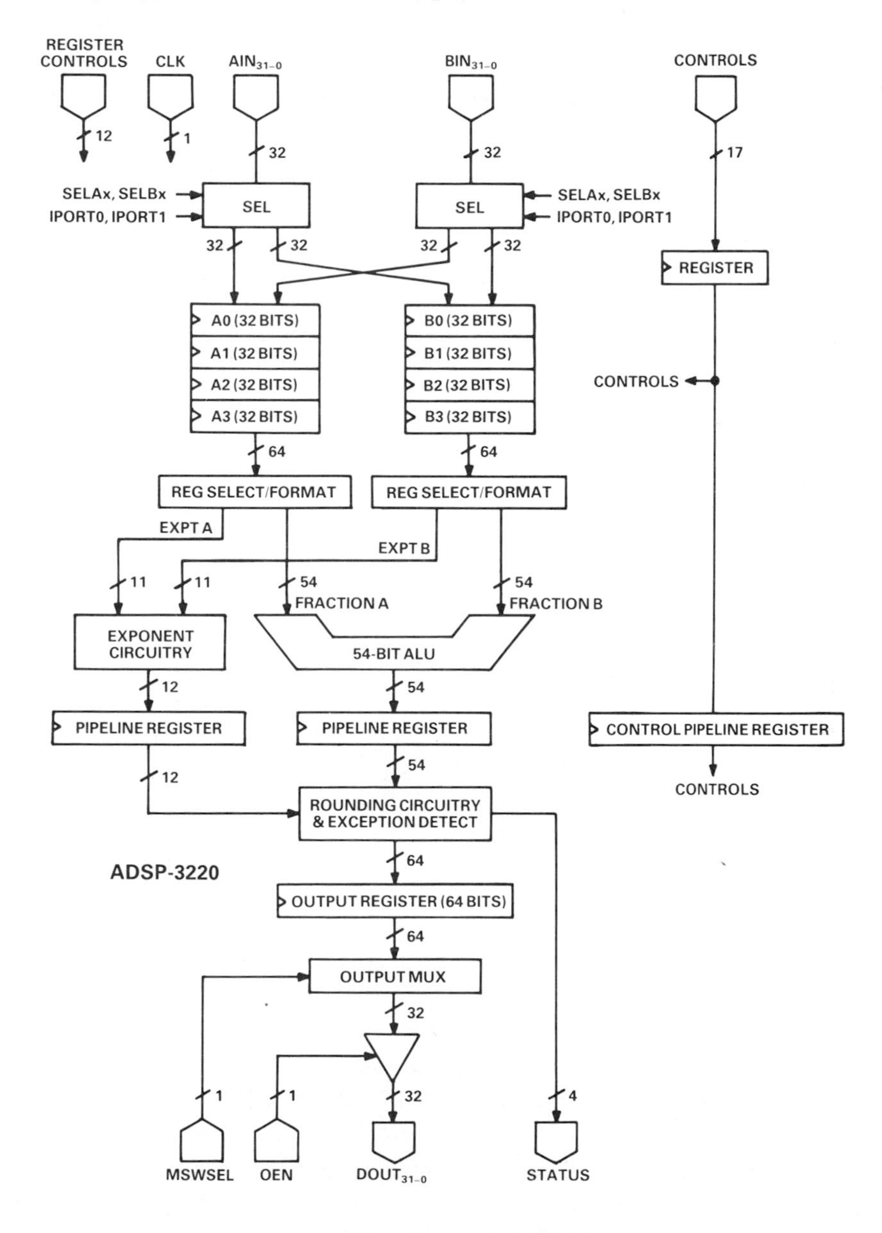

b. Floating-point ALU.

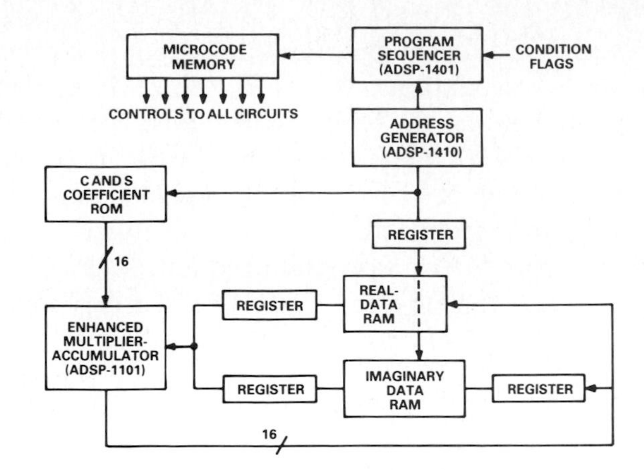

a. Enhanced-multiplier/accumulator (EMAC) based system.

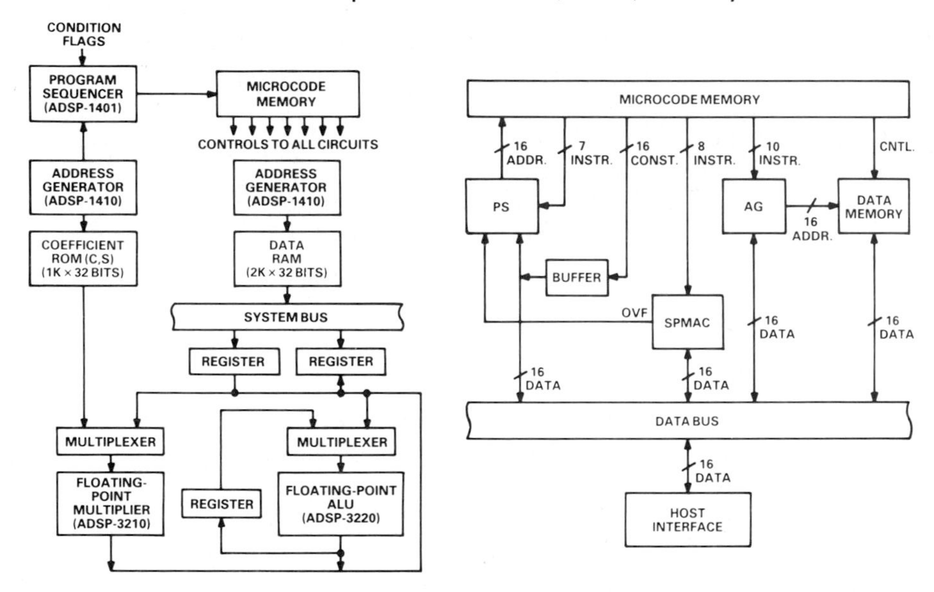

b. Floating-point processor.

c. 16-bit fixed-point FFT processor.

Figure 21.20. Processor architectures.

21.3 DSP APPLICATIONS

DSP began as a specialized technology used in military applications. With government funding, a high-speed integrated-circuit array multiplier was developed and first offered commercially in 1976. This multiplier formed the heart of high-performance radar, sonar, and missile-control systems. Over time, however, the use of DSP has spread from specialized military niches into a broader set of industrial and commercial markets, as the table below confirms. A few important uses—modems, studio recording, ultrasound imaging, and vibration analyzers—are described in detail below, primarily to illustrate how DSP's advantages benefit the application.

PRINCIPAL DSP MARKETS

Instrumentation:	Spectrum analyzers, vibration analyzers, mass spectroscopy, chromatography
Audio:	Studio recording, music synthesis, speech recognition
Communications:	Modems, transmultiplexers, vocoders, satellite transmission, repeaters, voice storage and forwarding systems
Computers & Computer Peripherals:	Arithmetic acceleration, servo controls for disk head positioning, array processors, engineering workstations
Imaging:	Medical, satellite, seismic, bandwidth compression, digital television, machine vision
Graphics:	CAD/CAM, computer animation and special effects, solids modelling, video games, flight simulators
Defense Electronics:	Radar, sonar, missile/torpedo control, secure communications
Control:	Robotics, servo links, skid-eliminator adaptive control, engine control.

21.3.1 MODEMS

A tremendous amount of information is transmitted today over analog communication links, such as telephone lines. With the growing role of computer-based systems, this information is increasingly digital in nature (for example, digital data and digitized voice transmission). The challenge of transmitting digital data over analog links at high speeds, and reconstructing the received data with high noise immunity, thereby reducing communication costs, is met by a modulator-demodulator (modem).

In transmitting digital data over analog communication lines, a digital bit pattern is represented by modulating the phase, frequency, and/or amplitude of an analog signal. Figure 21.21 shows a simple scheme, which involves changing the frequency of the signal to denote a "0" or "1"; this frequency-shift keying (FSK) method can send 2,000 bits/second over a telephone line. A more sophisticated encoding method, quadrature phase-shift keying (QPSK), modifies the phase of the signal and is capable of transmitting data at four times the rate of simpler methods. When a modem is sending information, it encodes the digital data into the corresponding analog waveform; in receiving mode, it decodes the waveform and determines the bit pattern that was transmitted. This latter mode is the more difficult to implement.

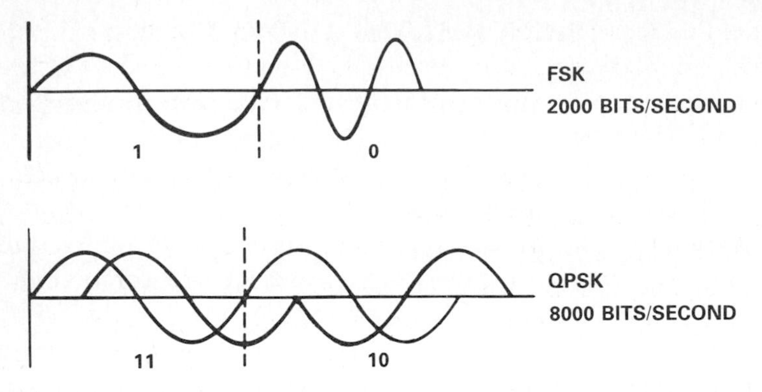

Figure 21.21. Frequency-shift keying (FSK) and quadrature phase-shift keying (QPSK).

If the transmission medium were noiseless, a modem's tasks would be limited to simple encoding and decoding—a relatively straightforward exercise. However, a phone line is a noisy transmission medium, corrupting the analog waveform. The more sophisticated the encoding scheme, the more disastrous the effects of noise and channel distortion. Therefore, a modem must effectively compensate for, or equalize, this channel distortion. To this end, high-speed modems (4,800 bits/second and above) turn to DSP for high-performance data recovery, using digital FIR filters.

Once the system channel distortion is filtered out, the modem must decide what bit pattern was originally sent. The compensation, or filtering, will rarely be perfect; that is, the amplitude, phase, and/or frequency of the filtered waveform will not generally correspond precisely to what was originally encoded. A least mean-squares (LMS) criterion (requiring fast multiplication capability) can be used to best estimate what the transmitted bit pattern is. A modem architecture for deriving the digital bit pattern from the received analog signal (and predicting the correct bit pattern in the presence of noise) can be seen in Figure 21.22. A modem using this relatively simple estimation method has a non-zero probability of bit error; more sophisticated DSP in a modem decreases the likelihood of such errors.

An additional complexity of telephone-line transmission is that its distortion properties change over time. Therefore, a modem's digital filter must be able to adapt to changes in the environment. This need to respond to a changing

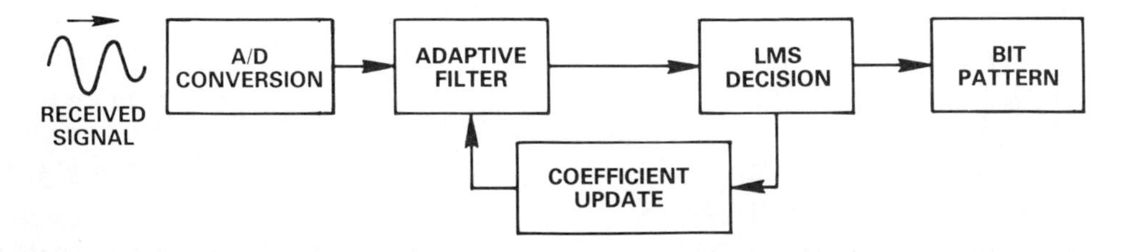

Figure 21.22. Least-mean squares modem processing architecture.

environment underscores another advantage of DSP—a digital filter's characteristics can be modified simply by changing its coefficients. Coefficient updating in a modem is determined by the observed drift of a property of the distortion in the system.

Aided by DSP, then, a modem can make it possible for high-speed data transmission to be implemented effectively. As Figure 21.22 illustrates, the DSP is the heart of a high-speed modem. The processing required generally can be handled by one digital multiplier, surrounded by the appropriate support devices (a program sequencer, an address generator). Alternatively, depending on the requirements of the modem, a single-chip processor may adequately handle all DSP requirements. The a/d conversion is an essential adjunct to the modem's signal processing. This structure allows the advantages of DSP to create a communications function that wouldn't have been possible with traditional analog signal-processing techniques.

21.3.2 STUDIO RECORDING

One of the most interesting applications of DSP is emerging in the audio processing performed in recording studios. This processing starts after the initial recording of voices and instruments in the studio; after a large number of steps, it ends with the recorded version that reaches the home stereo. Increasingly, DSP is being used to handle all intermediate steps.

The flow of activities in studio recording is complex and varied. Generally, multiple channels are used, with each track dedicated to one or more sources (instruments/voices). All channels need not be recorded at the same time. Each channel is subjected to extensive processing, including gain control, filtering, non-linear compression or expansion, reverberation adjustments, spectral equalization, and special-effects enhancements. The contributing channels are then mixed together to obtain a final arrangement with the desired overall effect.

Traditionally, channel processing and mixing were implemented entirely in the analog realm—with numerous disadvantages. Each channel's information—stored as an analog signal on magnetic tape—degrades as the cutting, splicing, and re-recording process progresses, undermining the benefits of the processing. The limited performance range available with analog processing sets a ceiling on the signal enhancement that can be attained. Also, analog circuitry can only handle one channel at a time; multi-channel mixers are expensive and difficult to control. Finally, if analog processing hardware is used, overall mixing flexibility can be achieved only through hardware modifications. In practice, this means that the mixing process loses its ability to creatively explore special effects.

Increasingly, audio processing is relying on digital techniques to improve audio quality. The first step in this transition was digital recording, which became prevalent about five years ago. Audio signals are first converted to digital form before being stored on magnetic tape. Digital recording eliminates several sources of degradation that hamper analog recordings, including the effects of non-linearities and additive noise in the magnetic materials used for recording, and wow and flutter in the tape playback mechanism.

In studio mixing applications, however, digital recording does not eliminate all complications. In the mixing and enhancement process, information is passed from one tape to another—requiring D/A and A/D conversion processes, a source of noise. These conversions are no longer necessary if all processing and mixing are handled with DSP techniques.

In DSP-based studio recording systems (see Figure 21.23), signals are converted to digital as early as possible. In fact, some implementations place a remotely controlled amplifier/converter at the recording microphone. After conversion, the audio processing is handled digitally, with high performance and flexibility. Gain factors are handled with digital multiplication. Filtering and equalization can be handled with an IIR filter that replicates the performance of standard analog filters. Alternatively, digital FIR filters can provide high-performance linear-phase filters or complex comb filters. Dynamic-range control is easily included in the system by using a multiplier for non-linear compression/expansion computations.

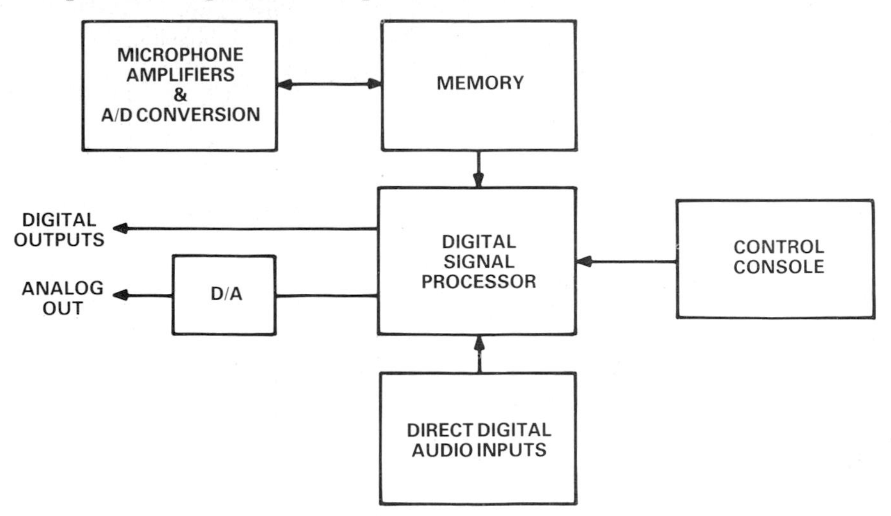

Figure 21.23. Block diagram for processing in studio recording.

The traditional mixing process is also easily implemented in a DSP-based system. Digital channels to be mixed are simply added together. Relative time delay lags can be easily introduced into the channel flows, allowing phase coherence to be explored without adding expensive delay lines to the system. An additional advantage is that the channel interconnections—which have to be hardwired in an analog processor—can be easily reconfigured in a DSP system.

In addition to improving on traditional operations, a DSP studio recording system opens up numerous new options. Unusual special effects are readily included in the system. Reverberation effects can be modeled, simulated, and integrated into the final recording. An FFT routine's spectral analysis of the signal forms the basis for frequency-domain filters that provide optimal equalization. Overall system flexibility allows the entire mixing system to be dynamically configured—processing steps can be re-ordered, mix groups and subgroups re-specified, and effects such as fading, equalization, and compression/expansion included at any juncture.

In practice, to perform the many required operations in timely fashion without introducing noise or distortion, studio recording systems face demanding processing tasks. Consequently, such designs rely on microcoded systems with numerous array multipliers. Such multipliers carry out the scaling, filtering, and FFTs required in the DSP block of Figure 21.23's block diagram.

Studio recording, then, follows the pattern of other applications using DSP. DSP techniques offers increased precision for processing steps traditionally performed with analog circuits. Of equal importance, DSP's flexibility paves the way for many new and creative processing steps. As in other areas, the DSP is shifting the role of converters; accurate ADCs and DACs are used in the system, but as close to the real-world interface as possible. The signal processing is conducted in the digital realm.

21.3.3 ULTRASOUND IMAGING

An important non-destructive imaging technology uses acoustic, or ultrasound, waves to investigate the interior of the human body. Like other medical imaging technologies, ultrasound draws heavily on the advantages of DSP for constructing and displaying images of internal organs.

In an ultrasound imaging system (see Figure 21.24), acoustic waves of a certain frequency are sent into the body. A phased array detector digitizes the reflected waves. After intensive number-crunching of the relative amplitudes and phase delays of returned signals, the shape of an internal organ can be inferred.

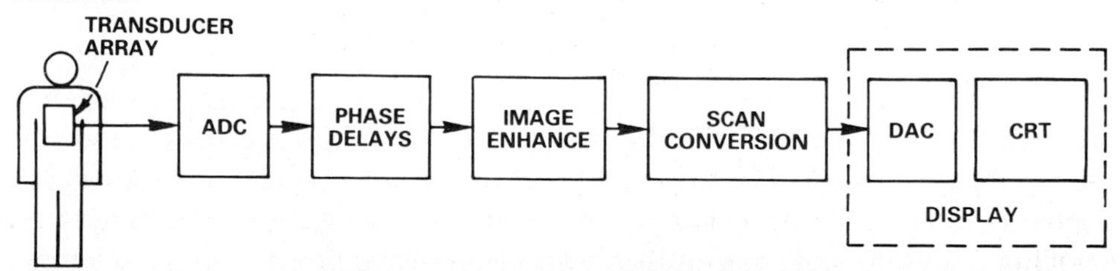

Figure 21.24. Ultrasound medical imaging.

High-speed arithmetic hardware is also used in an ultrasound imaging system to prepare the inferred image for display. The imaging data is originally expressed in polar coordinates—namely, distance from the imaging detector and angle relative to the detector. In order to display the image on a screen, a transformation from polar to rectangular coordinates must be made. This transformation—called scan conversion—is handled by fast DSP multipliers.

Since the human body is a poor transmission medium for acoustic waves, the image is degraded substantially by noise. To improve the image's clarity, an ultrasound imager performs three-dimensional FIR filtering. A one-dimensional temporal filter improves the quality of a displayed image; each pixel's displayed value is determined by a weighted average over the previous few values in time. Further image enhancement is obtained by using a two-dimensional neighborhood filter; each pixel's display intensity is based on a weighted average over its nearest neighbors in the frame. One result of a neighborhood filter is to soften edges, reducing the "staircase" effect when a diagonal line is presented on a raster-type display.

As in other applications using DSP, the presence of high-performance number-crunching hardware allows the system to perform certain tasks that would otherwise be unachievable. In ultrasound imagers, for example, it is possible to draw certain inferences about the structure of arteries. Acoustic waves reflected by blood cells will experience a Doppler shift proportional to the velocity of these cells. If an FFT is performed on the returned wave, this Doppler shift (and therefore velocity) can be computed. Then, by analyzing the velocity of blood cells over the cross-section of an artery, information is gained about the artery's resistance characteristic—specifically, whether clotting is present.

21.3.4 VIBRATION ANALYZERS

Vibration analyzers make particularly extensive use of DSP. An example of the use of a vibration analyzer is in monitoring an expensive and hard to replace turbine in an electric power-generating station. As the turbine rotates, a bearing emits a characteristic spectrum of high- and low-pitched sounds. This frequency distribution contains information about the structure and health of the turbine. The frequency spectrum will shift markedly as the bearing deteriorates; if detected in time, this information allows cost-effective preventive maintenance to be performed and catastrophic failures to be averted.

The block diagram of Figure 21.25 represents the basic functions of a DSP-based vibration analyzer. First, the system under investigation must be stimulated with a waveform. To obtain a precise waveform, DSP techniques are implemented by fast digital logic. For instance, a sine wave can be generated by using a lookup table and multiplication-intensive interpolation schemes. Then, the sine wave can be shifted to any other location in the frequency spectrum by modulating it with the output of a digital oscillator.

Stimulated with a waveform, the system will respond in a manner analyzed by the vibration analyzer. The waveform, or signal, reflecting the system's

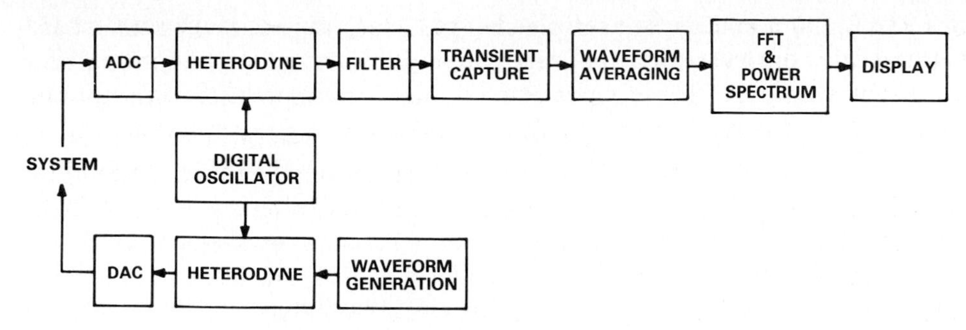

Figure 21.25. Vibration-analyzer architecture.

response is passed through a fast, high-resolution analog-to-digital converter and stored in a buffer memory. The vibration analyzer can then perform a range of processing tasks on this digitized signal in order to draw inferences about the structure of the system under test.

A key task is to filter the incoming signal. Fast DSP hardware allows all noise at undesired frequencies to be suppressed by a high-performance digital filter. Then, an FFT is performed, providing information about the spectral content of the received signal. This spectral information allows the transfer function to be estimated—information related to be basic structure of the system being monitored.

In some instances, it may be desirable to "zoom in" on a small portion of the frequency range; one technique that facilitates such zooming is to modulate the signal down to baseband, using the same digital oscillator used in stimulating the system. Then, an FFT, performed over a narrow bandwidth, provides very precise spectral resolution over a narrow frequency range.

Waveforms can be averaged either prior to or after performing the FFT; the choice may help toward improving the system's signal-to-noise ratios. With fast digital hardware in the system, sophisticated averaging schemes can be used, further enhancing system performance. Also, in displaying spectral information, certain calculations that are required can be performed, for example, power-spectrum computations. Polynomial-expansion approximations, which require fast multiply capability, can also be used by systems equipped with DSP hardware.

Once the signal has been digitized, it is a straightforward matter for the system to carry out other important computations. For example, auto- and cross-correlations can be readily calculated. Such information is useful, among others, in determining how closely the observed response matches one stored in memory. The memory-based response might correspond to a healthy system; if the correlation were to fall below a certain threshold, a failure might be likely and should be flagged.

Perhaps the biggest advantage accruing to a DSP-based vibration analyzer is its flexibility. Not only can a particular signal-processing scheme be carried out with great accuracy, but simple changes in software allow the same hardware to carry out a very different set of tasks. In this sense, vibration analyzers are representative of many types of equipment that now use DSP, not only to supersede the performance of more-traditional implementations, but to cost-effectively add new, otherwise unrealizable, capabilities to the system.

21.4 GLOSSARY OF DSP TERMS

Accumulator—an arithmetic element that adds together, or accumulates, a sequence of inputs. A DSP multiplier with an accumulator on-chip is called a multiplier/accumulator (MAC).

Algorithm—A DSP algorithm, such as the fast Fourier transform, or a finite impulse-response filter, is a structured set of instructions, and/or operations, tailored to accomplish a signal-processing task. Each algorithm has a well-defined structure; however, variations in algorithm parameters, such as the number of input points or taps, allow the same basic algorithm to perform different functions.

ALU—An arithmetic and logic unit, which performs additions, subtractions, or logical operations (e.g., AND, OR, XOR) on operand pairs.

Attenuation—The damping-out, or suppression, of signal content. Filters will attenuate the frequency content of a signal that lies in the filter's stopband.

Barrel Shifter—A device that accepts a digital number as its input and—as a function of the controls—shifts the number up or down, or rotates the word as though it were placed on a barrel. A barrel shifter is used in a system for many tasks, including scaling and normalization.

Biquad—A particularly simple recursive, or infinite impulse-response (IIR), digital filter form, often used as a building block for constructing more complicated recursive filters. A biquadratic, or biquad, section uses the three most recent input points and the two most recent output, or feedback, values to compute each output point.

Block Floating Point—A compromise between fixed-point and floating-point arithmetic. Data grouped in "blocks" is assumed to be normalized with a common exponent (but, not being attached to the data words, the exponent need not be explicitly processed with the data). In essence, the process is carried out in fixed point, with its inherent speed advantage.

Convolution—In discrete computations, a mathematical operation, defined as the summation, or integral, of a product of two functions over a range of differences in the independent variable. In the time domain, one function is the impulse response, as a set of coefficients, $h(i)$, over N time intervals; the other is the input, $f(n-i)$, as a function of the differences between the time at the instant at which the function is being evaluated, n, and the input at earlier instants, determined by the variable delay, i, from 0 to N. See equation 21.4b. In DSP, the convolution of an input signal, x, with the coefficients, h, results in the filtering of the input signal.

Correlation—a mathematical operation that indicates the degree to which two signals overlap. A high positive correlation reflects two signals that closely track each other. A negative correlation indicates that the two signals are closely related, but out of phase by roughly 180 degrees. If the correlation is close to zero, the two signals are unrelated.

Digital Signal Processing—DSP is a technology for high-performance signal processing that combines algorithms and fast number-crunching digital hardware.

Discrete Fourier Transform—the discrete Fourier transform (DFT) is a DSP algorithm used to determine the Fourier coefficient corresponding to a particular frequency.

FFT—An n-point fast Fourier transform (FFT) is computationally equivalent to performing n DFTs but, by taking advantage of computational symmetries and redundancies, can reduce the computational burden by several orders of magnitude.

FIR Filter—A finite impulse-response (FIR) filter is a commonly used type of digital filter. Digitized samples of the signal serve as inputs; each filtered output is computed from a weighted average of a finite number of previous inputs.

Fixed-Point Arithmetic—Each number is represented in a fixed arithmetic field of n bits, allowing integers in the range, 0 to $2^n - 1$, to be represented.

Floating-Point Arithmetic—Each number consists of a mantissa and an exponent, allowing wide dynamic range to be accommodated in the numbering system.

IIR Filter—An infinite impulse-response (IIR) filter is a commonly used type of digital filter. This recursive structure accepts as inputs digitized samples of the signal; each output point is computed on the basis of a weighted average of past output—or feedback—terms as well as past input values. An IIR filter is more efficient than its FIR counterpart, but poses more challenging design issues.

MAC—Multiplier/accumulator; see Accumulator.

Microcode—a set of instruction control signals stored in a program memory that govern the cycle-by-cycle operation of the various devices in a building-block architecture.

Passband—the frequency range over which a filter passes, to within some tolerance, the incoming signal content.

Pipeline—An architectural structure that allows two or more operations to be carried out simultaneously, like the stages of an assembly line. While each basic operation requires several cycles to complete, a later stage of one operation is simultaneously with an earlier stage of another operation. This structure allows the effective throughput rate for each operation to be substantially increased.

Rolloff—a measure of filter performance defined as the rate-of-change of the filter's amplitude response with respect to frequency over a transition band.

Stopband—the frequency range over which a filter attenuates, to within some tolerance, the incoming signal content.

PART V

GUIDE FOR THE TROUBLED

Chapter Twenty-Two

Guide For The Troubled

One of the hoped-for byproducts of this book is a more interesting life for data-acquisition manufacturers' Applications Engineers, because the technical inquiries they receive from our readers should become more challenging. This should be in addition to a saving of time and telephone expense for both users and manufacturers.

It is accomplished by making broadly available as much as possible of the information and advice that most commonly passes during seminars and conversations between manufacturers and users; this we have sought to do in the preceding pages.

To be useful, such information must also be accessible. There is, of course, an Index. In addition, this Chapter is intended to provide access by relating material in the book to the typical inquiries that are received, and by listing specific, recurring matters. Telephone conversations typically involve one or more of the following:

Requests for information
Requests for advice
Requests for assistance when things don't seem to work right
Urgent pleas for rescue

In order to minimize calls in all of these categories, but especially those in the latter two, listed here are a few of the most-frequently occurring topics of conversation, with comments and sources of information likely to resolve the problem. The general headings are: Frequently Asked Questions, Frequently Encountered Problems, Frequently Given Advice, and "If All Else Fails . . ."

22.1 FREQUENTLY ASKED QUESTIONS

Q. What is ESD? Should I worry about it? Why? What can I do about it?

A. *ELECTROSTATIC DISCHARGE. Yes. It can kill ICs. Handle ICs, especially CMOS, carefully. Read the 36-page pamphlet, "ESD Prevention Manual," generally available at no charge from Analog Devices, Inc.*

Q. How serious are "Absolute Maximum" ratings?

A. *They mean what they say. Ignore them, forget them, or accidentally violate them at your peril.*

Q. Is it really necessary to read the inside pages of the data sheets?*

A. *Only if you want to save time and money, get your design right the first time, and have that delicious feeling of being competent and in command of your project. A great deal of effort goes into data sheets from conscientious manufacturers, in the attempt to ensure best use of the product, anticipate the needs of the user, and answer questions that might be asked. Competently produced data sheets have product descriptions, specifications, descriptions of operation, practical suggestions for the user, and application ideas—and accuracy is a passion with the people who write them.*

Q. Is anything else that's "meaty" available to prepare me to use conversion products?

A. *There are Application Notes, Application Guides, books like this one, and serial publications, such as ANALOG DIALOGUE. In addition, manufacturers have staffs of trained sales and applications engineers, who provide seminars for users, from time to time, write articles on useful topics for the trade press, and give talks at conferences. They are naturally enthusiastic about their products, but that is generally the total extent of "hype."*

Q. What do the codes mean? What is complementary BCD? How does offset-binary relate to twos complement?

A. *See Chapter 7.*

Q. How do converters work?

A. *See Chapter 7.*

Q. How do I choose the right converter?

A. *See Chapter 11—also Chapters 13, 14, 15, 16, 17.*

Q. What are the differences between voltage- and current-switching DAC's?

A. *See Chapter 7.*

Q. What's a "glitch?" How is it caused? How can I eliminate it?

A. *See Chapters 3, 7, 11, 13, 17, 18.*

Q. How does input noise affect A/D conversion? How can I combat it?

A. *See Chapters 2, 10, 11, 12, 13, 18.*

*People really ask questions like this!

Q. Where can I find out about sampled-data systems?

A. See Chapters 2, 6, 7, 12, 13, 18, 21, and Bibliography.

Q. How can a converter have a throughput rate that exceeds 1/(conversion time)?

A. It can if it's "pipelined." For example, if it has three 50-nanosecond stages, several samples can be working internally at the same time, with a new sample accepted—and a new output delivered—every 50 nanoseconds, at a 20-MHz throughput rate. A complete conversion of one sample point nevertheless requires 150 nanoseconds.

Q. What limits the analog bandwidth in a/d conversion?

A. The sampling theorem limits the analog bandwidth to one-half the sampling rate, using equally timed samples. However, the actual bandwidth that can be handled with a given degree of resolution and accuracy may be considerably less, depending on the aperture uncertainty (See Section 2.4). A track-hold can bring about considerable improvement, even with "flash" converters (See Section 13.2.2).[1]

Q. How will my system know when a conversion is complete if my flash converter doesn't have a Data-Ready signal?

A. Set up a time delay—equal to the maximum conversion time expected—between the encode command and the time the external latch is enabled.

Q. I understand that ECL is the predominant form of logic used in flash-type video conversion. How can I convert the ECL to TTL for compatiblity with my system.

A. Probably the simplest way is to use standard ECL-to-TTL converter chips, such as the Motorola 10124 and 10125.

Q. What does a track-hold or sample-hold really do for me?

A. At a definite instant of time, it converts a rapidly changing signal into a single value that the converter perceives as dc (which is much easier to process accurately).

Q. What are the contributions of various components to errors?

A. See Chapters 11, 12, individual product data.

Q. What are the timing constraints on converters?

A. See individual data sheets.

Q. What factors affect timing of systems with sample-holds and multiplexers?

A. See Chapters 2, 3, 11, 12, 13, 18, 19.

Q. What are the requirements on power supplies?

A. See Chapters 11 and 12, individual data sheets, power-supply catalogs.

[1]See also "Flash Converters Work Better with Track-Holds," by Jerry Neal and Jim Surber, *Analog Dialogue* 18-2 (1984): 10-14).

Q. How do I use DACs and ADCs with microprocessors?

A. *See Chapters 2, 3, 4, 6, 8, 21, individual product data sheets.*

Q. How do I connect a 10-volt device for a 10.24V full-scale range?

A. *Scale-factor adjustment will usually have insufficient range, especially in high-resolution devices. Add series feedback resistance with DAC's, with due regard for resistance tempco (the loop is usually closed externally), or use attenuation ahead of the ADC input buffer (sometimes series input resistance can be used with current-summing comparators). Information on 10.24V full-scale range will be found in Chapter 7: see Table 7.3.*

Q. How is the bipolar offset circuitry connected and adjusted? How does it affect the specifications?

A. *See Chapters 7, 11, 12, 20, individual data sheets.*

Q. What are the suggested grounding techniques?

A. *See Chapters 2, 3, 10, 12, 13, 17, 20, individual data sheets.*

Q. What are the issues in low-level multiplexing vs. instrumentation-amplifier-per-channel? How do instrumentation and isolation amplifiers differ?

A. *See Chapters 2, 11, 12, manufacturers' Databooks.*

Q. What happens if my 10-V ADC input is overranged at full scale? at zero?

A. *With most ADCs, anything less than 0 reads 0, and anything greater than FS – LSB reads all-ones. Some types provide carries to flag F.S. overrange. If indication is required, accurate comparators can be used to provide flags.*

Q. What is "Differential Nonlinearity?"

A. *See Chapters 7, 11.*

Q. Can I use the analog power supply as a source of constant voltage?

A. *Yes—if it is sufficiently quiet and well-regulated to provide the desired degree of stability and accuracy. However, precision regulators and substitutes for Zener diodes are available in such numbers and variety—as well as at low cost—to suggest that the designer take advantage of their convenience and supply-independence. See Chapter 20.*

22.2 FREQUENTLY ENCOUNTERED PROBLEMS

22.2.1 GROSS MALFUNCTIONS

Power supply not connected to circuitry; power supply not turned on; *a-c line cord not plugged in.**

Wrong digital code ("Positive true" vs. complementary).

Wrong analog polarity relationship.

Using TTL levels for non-TTL-compatible CMOS devices.

*Believe it or not, we do get calls where this is the problem!

Grounds not connected (or interconnected).

Power supply not really connected, wrongly connected, or zapped.

Missing or improper connections (Study the connection diagram).

Wipeouts due to applying power to devices in the wrong order (In general, power downstream units first; avoid, buffer, or protect multiplexers that short in the power-off condition).

ICs destroyed or degraded, and thus rendered useless, after zapping by electrostatic discharges (ESD)—a result of careless handling of susceptible ICs, especially CMOS.

ICs destroyed by excessive current in latchup modes—a result of inadequate protection of susceptible devices (by such techniques as voltage and current limiting, and use of protective diodes—Schottky and otherwise) against excessive or atypical voltages applied to device terminals, or separation of grounds by more than one diode-drop. CMOS logic must not be left "hanging"; all unused inputs should be tied High or Low.

Control-logic improper (polarity, duration, timing, levels). Check logic and timing diagrams on data sheets. Data lines connected in reverse order.

Bad software.

Uncontrolled overflow in counter configurations.

Wrong diode polarity.

Bent pin that didn't go into the socket (or perhaps even broke off).

Damaged high-speed (high-power) circuit due to shorted output.

Damage due to excessive temperature rise resulting from insufficient airflow.

22.2.2 POOR FUNCTIONING

Common-mode problems in "single-ended" system (use proper grounding or difference amplifier).

Grounding problems: no ground connection, fortuitous ground connection, wrong ground connection (common analog and logic return), shields returned to wrong ground or grounded at both ends.

Pickup due to proximity of digital ground plane to analog circuits, or proximity of analog and digital wiring, in general, or poor lead dress: Keep stray capacitance low; keep analog, digital, and power leads apart; if they must cross, analog and digital leads should cross at right angles.

Intermittent behavior due to poor soldering.

Excessive load capacitance on outputs of voltage DAC's or other analog devices can in some cases cause slow response, poor settling, ringing, or oscillation ("noise").

Improper connection of built-in references (unused bipolar offset references may require grounding in unipolar applications; external use of internal Zener voltage reference generally requires buffering).

Op amp voltage offset adjustment used for zeroing anything but op amp *voltage* offsets, e.g., system offsets, can result in increased thermal drift.

Logic overloading (logic outputs may also be used for internal purposes; check actual specified loading on data sheet).

Too much attenuation because "current-output" DAC's output impedance neglected.

Nonlinearity because current-output DAC's specified maximum output voltage range exceeded.

Poor behavior over temperature because tempco of external resistance does not match tempco of on-chip resistors, when they should track.

Noisy A/D conversion, increased differential nonlinearity, and missing codes caused by widening of quantization band due to noise on input signal, or picked up in wiring.

Unanticipated "glitches" due to lack of filtering, inappropriate converter choice, marginal logic timing, limited logic slewing rates due to excessive capacitive load, stray capacitive coupling to analog circuitry.

Last one or two bits of an ADC keep flopping around even though input is "constant." (Is it *really* constant? Look at it on a scope. Check for noise pickup, either in the environment or via the power supply.)

Digital and analog "glitches" (or worse), due to insufficient power supply (and wiring) capacity to handle transient switching currents in CMOS circuitry; CMOS requires low current and dissipates little power in steady state, but may require large currents when switching between states—especially apparent when systems are clocked. Easily fixed by using power supplies and wiring with adequate capacity and regulation to handle worst-case switching situations.

Gain and offset adjustments performed in wrong order in DACs and ADCs (See Chapter 12).

Excessive thermal drifts due to: improper converter adjustment procedure; bias current flowing through resistances (MUX R_{on}, for example); use of op amp voltage offset adjustment to counteract bias-current or system offsets.

Loss of monotonicity over small or large temperature ranges: possible if a converter is specified at ± 1 LSB differential nonlinearity at room temperature—or "monotonic" at room temperature. A conservative specification of ½ LSB or "monotonic over temperature" allows variation of an additional ½ LSB with temperature.

RFI or fast pulses causing rectification that produces offsets in low-level circuits.

"Long-tailed" responses due to thermal transients (some op amp or comparator circuits), or inappropriate capacitor choice (precision capacitors should always have low dielectric absorption—polystyrene, teflon, polycarbonate are among recommended materials). Precision components should be kept at a safe distance from circuits or components that have high, or rapid changes in, dissipation.

Offsets or noise caused by leakage currents (dc) or noise currents (ac) in high-impedance circuitry due to proximity of sensitive low-level leads to power supply leads or high-level digital signals; use shielding and guarding.

In long-running projects in large organizations, problems that arise due to cost-cutting changes made to a good design (e.g., glitches, noise, and missing codes due to replacing an expensive regulated linear supply by a low-cost switcher) at a later stage by a different engineer without taking into account the thinking behind the original design. If you inherit a design that works, don't make changes unless you understand all the consequences of what you're doing.

Oscillation of voltage-output DAC, used to drive loads through long coaxial cable runs (would you drive them with a 741 op amp?) Use inside-the-loop buffer amplifier, with force-sense lead configuration, or place DAC at remote location, using serial communication.

Excessive drift in low-level circuitry due to differential "thermocouple" effects in input leads (e.g., copper-to-Kovar at IC inputs). Differential-input leads should always be close together and their junctions should be as-nearly as-possible isothermal.

When all other possibilities have been eliminated, one should not discount the possibility that the device is malfunctioning or out of specification, either innately, as a result of some recent trauma, or as a result of some "early failure" mechanism. Many manufacturers subject certain of their products to "burn-in" to eliminate innate and "early failure" problems.

By no means are all problems chargeable to the user. Manufacturers of devices and components (including Analog Devices) have been known to have made available—inadvertently, and despite considerable effort—

Data sheets with errors or insufficient data
Devices that failed, for no apparent reason, when first plugged in.

Though rare, these possibilities should not be discounted. The user of conversion devices—especially in quantity—should be prepared to perform at least simple tests on devices to verify their performance; Chapter 10 may be found useful in this respect. A user who finds information on a data sheet that raises questions will find most manufacturers quite willing to discuss them and clarify the point in question, especially as it pertains to the application.

22.3 FREQUENTLY GIVEN ADVICE

22.3.1 PREVENTIVE

First—Read the data sheet—thoroughly—especially the inside pages!

Nothing beats good initial analysis of the basic problem and conservative initial design, with double-checking to make sure that the best-available data has been used, the tolerances on resolution, accuracy, and timing are adequate, and the connection scheme is proper, and follows the manufacturer's suggestions—where appropriate. Breadboarding should be used if any aspect of the circuit or subsystem is unfamiliar or is pushing the state of your art.* The final design is not likely to succeed if the breadboard's performance is just barely in spec—and then only with tweaking and handholding. The design should include features that facilitate testing and trouble-shooting.

Be sure that common-mode, normal-mode, and induced noise problems have been considered and dealt with adequately. (Differential amplifiers, filtering, lead locations and directions.)

Be sure that grounding is proper: no ground "loops" (i.e., ground current is allowed only one path); digital and analog grounds separated; high-power and low-level signal grounds separated; One main "Mecca" point where all grounds meet, if feasible; heavy ground conductors, to avoid voltage drops in signal return leads.

Take care handling and using IC parts. Read up on and beware of electrostatic discharge (ESD). Use sockets whenever feasible, but avoid bending or breaking pins when plugging in or removing ICs. Make sure ratings are not exceeded. Use protective circuitry if necessary.

Be sure that interconnections of devices do not produce surprises as a result of (e.g.) currents and impedance levels, transient overloads during MUX switching, etc.

Despite everything we've said about the noise-rejection characteristics of digital, don't forget that digital circuits and signals are analog in nature (ohms, volts, amperes, farads, henrys). The very fast edges associated with high-frequency clocks (e.g., 10 MHz) can couple through power supplies, long leads, etc., to cause glitches and "mysterious" problems in digital circuits.

If you're using switching-type power supplies anywhere in the equipment, make sure that they are not affecting performance elsewhere, (a) via output leads, (b) reflected back through ac line, or (c) induced, coupled, or broadcast within the equipment. Don't use them for analog supplies.

*Overheard—one side of an all-too-frequent telephone conversation:
"You say you're having trouble with the data-acquisition circuit on your pc board . . .
"Well, how did your breadboard work? . . .
"Oh, you didn't build and test a breadboard? . . ."
"CLICK!"

To minimize noise from ac power supplies, power transformers should have metallic cases to eliminate radiated noise, and there should be electrostatic shielding between primary and secondary windings to minimize coupling of power-line spikes. Such transformers cost more and are often the misguided target of cost-cutting measures; if you expect noise to affect the performance of your system adversely, don't be "penny wise and pound foolish."

If the power transformer is not of adequate size to handle the worst-case load, its core is likely to saturate at the crests of sine waves; the incremental permeability vanishes, and there is no more inductance. Without this permeability in the core, the magnetic field lines can spread about as though there were no iron to channel them. In addition, the discontinuity produced by saturation can cause high-frequency transients to be radiated.

In high-speed circuitry, beware of transmission-line effects (e.g., attenuation and reflections) and antenna-like behavior of wire-wrap terminals.

After assembly, the system should be thoroughly inspected and "buzzed-out", to be sure that all connections have been made, the right elements have been plugged into the right spots, and there are no bent or broken pins.

Check the system out in small pieces and functional groupings before putting it all together. "Going for broke" often results in just that.

Remember that you can't tie two 8-bit DACs together for accurate open-loop resolution of 16 bits *unless one of the DACs has 16-bit accuracy.*

22.3.2 MEASURING DEVICES

For monitoring performance and troubleshooting, the devices that perform dc measurements should have at least twice the resolution and accuracy of the devices they are checking; the devices that perform high-speed measurements should have faster response than the devices they are checking. An oscilloscope should always be used to avoid "flying blind." A simple multimeter may be a trap (it can't see dynamic signals or oscillations; its dc resolution may be inadequate for useful measurements on the kind of high-resolution devices usually found in data systems; and its load impedance may affect the accuracy— if not the actual character— of measurements). Digital voltmeters often present a varying (and active) load to circuits being measured.

23.3.3 MEASURING AND TROUBLE-SHOOTING

First, check supply and ground voltages at terminals of pluggable devices, with the devices removed.

A useful procedure is to then perform dc, manual, and low-speed checks before performing measurements at speed. This ensures that the system is at least working properly under *some* conditions.

Try to isolate the problem.

If more than one unit of a given type is in use, an apparent failure at its location can be checked by substituting another unit. If similar units of the same kind exhibit the same problem, it is likely a design or system problem. (WARNING: If the problem is serious, involving a fault condition, the original unit and its substitute may no longer be in fit condition for further use.)

Check grounding with a simple continuity test. Use an orderly procedure. Have you localized the problem? Is it static or dynamic? Gross or subtle? Catastrophic or slightly "off?" Reproducible or intermittent? Affected by mechanical manipulation (kicking the cabinet)?

22.4 IF ALL ELSE FAILS . . . !

Reputable manufacturers want to help you solve the problem, whether it involves simple advice or the return of a unit. Although different manufacturers back their products up in differing ways, because of the sales channels they use, these guiding principles should be helpful:

If the problem seems to be related to a conversion component (either suspected or proven),

1. Prepare a summary of the problem, and outline it to the manufacturer's applications engineering staff, either at the factory or at a local sales office (or representative or distributor), over the telephone (see (4)). They may suggest some useful diagnostic procedures or put you in touch with Customer Service about having units returned for evaluation.

2. Follow the suggestions and/or instructions they give you. Ask for complete information on returns and warranty service and follow the suggested procedures. Do not return any material without receiving authorization; this procedure will enable the manufacturer or the service organization to identify the material when it arrives and process it expeditiously.

3. Be sure to include with any returned units
   A. The name(s) and telephone number(s) of the person(s) with whom the technical (and business) aspects of the problem can be discussed.
   B. Complete information on the (suspected) malfunction, and the application in which it occurred.
   C. If the unit is believed to be in warranty, the purchase date and entity from whom purchased.

4. If you are using Analog Devices products and are in a critical "bind," phone Components Group Applications, (617)-935-5565, or the Corporate Customer Service staff, (617)-329-4700, for information and action. Their U.S. addresses are:

Components Group Applications
Analog Devices Semiconductor
804 Woburn Street
Wilmington, MA 01887

Customer Service
Analog Devices, Inc.
2 Technology Way
Norwood, MA 02062

Blank space has been provided at the end of this Chapter for readers to note any useful points not covered in the text of this chapter, new discoveries, errors found in the book's text or illustrations, etc., as well as other useful information about manufacturers—Analog Devices and others: local sales-office telephone numbers, names of helpful people, Telex and cable addresses, etc. Even if, as you read these words, you are not in urgent need of help, it may nevertheless prove useful to take a few minutes *right now* to obtain and write down some of that information.

NOTES*

*If, after using this page, you consider the technical content of what you have written to be significant, the authors and editor will welcome copies of it for the benefit of future editions and readers. Send them to Editor, *Analog-Digital Conversion Handbook*, P.O. Box 280, Norwood MA 02062 USA.

*If, after using this page, you consider the technical content of what you have written to be significant, the authors and editor will welcome copies of it for the benefit of future editions and readers. Send them to Editor, *Analog-Digital Conversion Handbook*, P.O. Box 280, Norwood MA 02062 USA.

*If, after using this page, you consider the technical content of what you have written to be significant, the authors and editor will welcome copies of it for the benefit of future editions and readers. Send them to Editor, *Analog-Digital Conversion Handbook*, P.O. Box 280, Norwood MA 02062 USA.

Bibliography

SOURCES OF ADDITIONAL INFORMATION

The published references listed here have appeared in the form of books, manuals, brochures, or articles in archival publications and the trade press. Individual items have been selected because of their general or specific interest—or because of an excellent collection of further references.

Today, articles that mention converters and conversion, in relation to design, testing, system applications, new-product news, etc., are appearing in numbers that grow by leaps and bounds. It is impossible (and might in any case be undesirable) to keep track of them all. In these few pages, no pretense is made as to thoroughness; our goal is to provide a few basic sources that will supplement in greater depth the material presented here.

It is still true that much of the most-advanced, relevant, timely, and useful published material available in this rapidly growing and changing field is published by manufacturers of converters—and of systems that use them—in the course of the ordinary business of trying to sell their products. As noted earlier, Databooks, Application Guides, Application Notes, and Technical Data sheets are available from leading manufacturers in this field—generally free or at nominal cost.* Such publications are loaded with valuable information that is generally current and useful, particularly in terms of specification definitions and practical application techniques.

Magazines like *Electronic Design*, *EDN*, *Electronic Products*, *Electronic Engineering Times*, and *Computer Design* are good sources of product news and its interpretation, as well as application ideas. In general, their reporters and editors have good technical

*Publications of Analog Devices, Inc., and reprints that are available as of the initial date of publication of this book are identified by an asterisk (*); they can be obtained from Analog Devices, Inc., P.O. Box 796, Norwood, Massachusetts 02062. If a given publication is not free, the 1985 unit price (postpaid) in U.S. dollars is included in the listing. Publications not identified by (*) are *not* available from Analog Devices.

backgrounds, articles are competently written, data and specifications have accuracy as a goal, and readers are provided with various means (such as "bingo" cards, manufacturer's addresses and telephone numbers, and "hot lines") to get further information on matters that interest them. However, a cautionary word is in order: products have been known to be introduced in the press (by manufacturers in many fields) long before becoming available for purchase in reasonable quantity—and the preliminary specifications, and even pin connections, can differ from those ultimately established; such preemptive publication may aid planning but is of little help to the designer who needs accurate information for current use.

Every designer should ensure that suppliers keep him or her up to date on new products, applications ideas, and techniques. Manufacturers maintain mailing lists for that purpose. A method employed by Analog Devices since 1967 has been the publication of the technical magazine, *Analog Dialogue*—"A forum for the exchange of circuits, systems, and software for real-world signal processing;" it is mailed free to qualified interested persons.

A medium that appears to be potentially one of the most-promising services to designers employing integrated circuits—including converters and other analog ICs—is the interactive Videolog[SM] database, which is accessible to subscribers' personal computers via the telephone lines. Videolog maintains an indexed catalog of technical product information from major manufacturers; it is—in principal—always current, because its color-graphic information screens are continuously updatable by the manufacturers as new products appear and older products are de-emphasized or obsoleted. Prices are also made available. It is cross-indexed in several ways, allowing searches by device type, by manufacturer, by part number, by parameters, etc.

Despite its great potential usefulness, Videolog requires a personal computer capable of communicating over telephone lines, and the payment of fees for the service and the use of the telephone lines. Although non-interactive and not as up-to-date, such industry catalogs as *ICMaster*, published by Hearst Business Communications, Inc., are easy to obtain and access—and are quite useful.

GENERAL INFORMATION

Arbel, Arie F., *Analog Signal Processing and Instrumentation*. Cambridge, UK: Cambridge University Press, 1980.

Bice, Peter K., "Analog Prescaler Has Digitally Selected Gain," (Idea for Design), *Electronic Design*, August 16, 1980, p. 192.

Brockman, Don, and Terry Brown, "Improved Vector-Scan Displays Are Coming," *Analog Dialogue*, 16, no. 2 (1982), 14-15.

Gardner, F.M., *Phase Lock Techniques* (2nd ed.). New York: John Wiley & Sons, 1979.

Graeme, Jerald G., *Applications of Operational Amplifiers*. New York: McGraw-Hill Book Company, 1973.

———, *Designing with Operational Amplifiers*. New York: McGraw-Hill Book Company, 1977.

[SM]Videolog is a service mark of Videolog Communications, Norwalk, CT and Santa Clara, CA.

———, with Gene Tobey, and Lawrence P. Huelsman, *Operational Amplifiers—Design and Applications*. New York: McGraw-Hill Book Company, 1971.

Grebene, Alan B., *Bipolar and MOS Analog Integrated Circuit Design*. New York: Wiley-Interscience, 1984.

Higgins, Richard J., *Electronics with Digital and Analog Integrated Circuits*. Englewood Cliffs, NJ: Prentice-Hall, Inc., 1983.

Hynes, Michael J., and D. Philip Burton, "A CMOS Digitally Controlled Audio Attenuator for Hi-Fi Systems," *IEEE Journal of Solid-State Circuits*, Vol. SC-16 (February 1981), 15-20.

*Jung, Walter, Jeff Riskin, and Lew Counts, "Circuit Ideas for IC Converters," *Analog Dialogue*, 12, no. 2 (1978), 13-15.

Kester, W. A., "Design of Raster Scan Graphics Systems," *Digital Design* (August 1982).

*———, "Fast DACs Aid Raster-Scan Displays," *Electronic Design*, April 30, 1981, 123-126.

Roberge, James K., *Operational Amplifiers, Theory & Practice*. New York: John Wiley & Sons, Inc., 1975.

Soclof, Sidney, *Applications of Analog Integrated Circuits*. Englewood Cliffs, NJ: Prentice-Hall, Inc., 1985.

Sylvan, John, "Industrial Monitoring with Personal Computers," *Machine Design*, October 6, 1983.

*Toomey, Paul, "Improve Function Generators with Matched D/A Converters," *EDN*, May 12, 1982.

Watson, Dan, "Synthesize Low-Frequency Sine Waves with High-Frequency Crystals" (Idea for Design), *Electronic Design*, April 30, 1981, p. 163.

*Wynne, John, "Level-Independent Automatic Gain Control," *Analog Dialogue*, 17, no. 1 (1983), 16-17.

DATA ACQUISITION AND DISTRIBUTION

Gordon, Bernard M., *The Analogic Data-Conversion Systems Digest*. Wakefield MA: Analogic Corporation, 1977.

*Graves, Edward, "Very High Speed Data Acquisition," *Analog Dialogue*, 13, no. 2 (1979), 3-7 (includes brief bibliography).

*James, Don R., "Data Acquisition Information System," *Analog Dialogue*, 17, no. 3 (1983), 16-19.

Levreault, John E., Jr., "High-Speed Data Acquisition System Design," *Electronic Products*, March 4, 1983, 89-92.

*Melanson, Frank, and John Mills, "Sensor-Based Intelligent Data Acquisition for the STD Bus," *Analog Dialogue*, 16, no. 3 (1982), 10-11.

Ramirez, Robert W., "Digitizer Specifications and Their Applications to Waveforms," *Electronics Test* (September 1981), 74-80.

Wold, Ivar, and Fred Pouliot, et al, "MACSYM—A Complete Measurement And Control System," *Analog Dialogue*, 13, no. 1 (1979), 3-14.

Zuch, Eugene L., ed., *Data Acquisition and Conversion Handbook*. Mansfield MA: Datel-Intersil, Inc., 1979

A/D AND D/A CONVERTERS

Analogic Corporation, "Floating-Point Converters," A/DVISOR, 1, no. 1, Spring, 1975.

Bloom, Steve, and Scott Wayne, "High-Resolution A/D Applications and Methods Proliferate," *Electronic Products*, December 12, 1984, 65-72.

Brokaw, A. Paul, *Circuit Techniques for Monolithic DACs*. Norwood MA: Analog Devices, Inc., 1979. Out of print—see Chapter 9 of this volume.

———, "Input Resistor Stabilizes MDAC's Gain," (Design Idea), *EDN*, January 7, 1981, 210-212.

Bruck, Donald B., *Data Conversion Handbook*. Burlington MA: Hybrid Systems Corporation, 1974.

Burrier, Richard, *et al*, "Floating-Point Converter Uses Hardware to Get a 20-Bit Dynamic Range," *Electronic Design*, September 6, 1984, 175-186.

*Burton, D. Philip, *CMOS DAC Application Guide*. Norwood MA: Analog Devices, Inc., 1984.

DeVito, L., "Linear V-F Converter Chip Invades Module Territory," *Electronic Design*, February 23, 1984.

Dooley, Daniel, *Data Conversion Integrated Circuits*. New York: IEEE Press—Wiley, 1980.

Gordon, Bernard M., "Linear Electronic Analog/Digital Conversion Architectures, Their Origins, Parameters, Limitations, and Applications," *IEEE Transactions on Circuits and Systems*, CAS-25, no. 7 (July 1978), 391-418.

Hnatek, Eugene R., *A User's Handbook of D/A and A/D Converters*. New York: John Wiley & Sons, 1976.

Jaeger, Richard C., "Tutorial: Analog Data Acquisition Technology. Part I – Digital-to-Analog Conversion," *IEEE MICRO*, 2, no. 2 (May 1982), 20-37.

*Jung, Walter G., "Audio Application Ideas for CMOS DACs," *Analog Dialogue*, 10, no. 1 (1976), 16-17.

———, *IC Converter Cookbook*. Indianapolis: Howard W. Sams & Co., Inc., 1978.

*———, "Operation and Applications of the AD654 IC V-to-F Converter," (Application Note). Norwood, MA: Analog Devices, Inc., 1985.

Kurth, C.F., ed., *IEEE Transactions on Circuits and Systems* Special Issue on Analog/Digital Conversion, CAS-25, no. 7, July 1978.

Loriferne, Bernard, *Analog-Digital and Digital-Analog Conversion*. London: Heyden & Sons Ltd., 1982.

Michaels, Stuart R., "Getting the Best from A-D Converters," *Electronic Design*, February 18, 1982, 191-199.

*Newell, John, "Application Ideas for Multiplying DACs," *Analog Dialogue*, 12, no. 1 (1978), 16-17.

Pratt, William J., "High Linearity and Video Speed Come Together in A-D Converters," *Electronics*, October 9, 1980, 167-170.

Schmid, Hermann, *Electronic Analog/Digital Conversions*. New York: Van Nostrand Reinhold, 1970.

Stephenson, Malcolm, "CMOS DACs in the Voltage-Switching Mode," *Analog Dialogue*, 14, no. 1 (1980), 16-17.

*Wayne, Scott, "Getting the Most from High-Resolution D/A Converters," *Electronic Products*, December 12, 1983.

*Wynne, John, "Deglitching a 16-Bit Monolithic D/A Converter," *Analog Dialogue*, 16, no. 2 (1982), 16.

CONVERTERS, MICROPROCESSORS AND COMPUTERS

*Analog Devices, Inc., "Isolated Conversion from Digital to 4-to-20 mA," *Analog Dialogue*, 14, no. 2 (1980), 6.

Bibbero, Robert J., *Microprocessors in Instruments and Control*. New York: Wiley-Interscience, 1977.

Boyet, Howard, *The 8085/SDK-85* [Hands-On Volume 2] 54 Control Experiments. New York City: MTI Publications (Microprocessor Training, Inc.), 1981.

Brignell, John E., and Godfrey Rhodes, *Laboratory On-Line Computing*. Leighton Buzzard, Beds., UK: Kingswood House, Heath & Reach, 1975.

Burton, D. Philip and A. L. Dexter, *Microprocessor Systems Handbook*. Norwood, MA: Analog Devices, Inc., 1977.

*Grant, N. Douglas, "Putting the AD558 DACPORT™ on the Bus," *Analog Dialogue*, 14, no. 2 (1980), 16-17.

Lenz, James E., Jr., and Edward F. Kelly, "A Potential Difficulty in A/D Conversion Using Microcomputer Systems," *IEEE Transactions on Biomedical Engineering*, BME-27, no. 11 (November 1980), 668-669.

Lipoyski, Jack, *Microcomputer Interfacing*. Lexington MA: Lexington Books, D.C. Heath and Company, 1981.

Larsen, David G., Peter R. Rony, and Johnathan A. Titus, various books on microcomputer interfacing in the *Bugbook* and other series. Derby CT: E&L Instruments, Inc., and Indianapolis: Howard W. Sams & Co., Inc.

Lenk, John D., *Handbook of Microcomputer-Based Instrumentation and Controls*. Englewood Cliffs, NJ: Prentice-Hall, Inc., 1984.

*Mills, John, "Use Your Personal Computer for Measurement and Control," *Analog Dialogue*, 16, no. 2 (1982), p. 11.

Sargent, Murray III, and Richard L. Shoemaker, *Interfacing Microcomputers to the Real World*. Reading MA: Addison-Wesley Publishing Company.

*Schweber, Bill, "Multitasking—The Key to Effective Measurement and Control," *Analog Dialogue*, 16, no. 2 (1982), 12-13.

*Wold, Ivar, "CMOS Converters as I/O Devices," *Analog Dialogue*, 11, no. 1 (1977), 13-16.

POSITION MEASUREMENT WITH RESOLVERS AND RELATED DEVICES

*Bloom, Steve, "Advantages of 3-States in Synchro Conversion," *Analog Dialogue*, 15, no. 2 (1981), 14.

*Boyes, Geoffrey S., ed., *Synchro and Resolver Conversion*. Norwood, MA: Analog Devices, Inc., 1980 ($11.50).

Dynamics Research Corp., *Techniques for Digitizing Rotary and Linear Motion*. Wilmington MA: Dynamics Research Corp., 1976.

Herceg, Edward E., *Handbook of Measurement and Control*, "An Authoritative Treatise on the Theory and Application of the LVDT." Pennsauken, NJ: Schaevitz Engineering, 1976.

Mayer, Arthur, "Design a Multi-Purpose Network to Rotate Complex Numbers," *Electronic Engineering Times*, July 9, 1979.

———, "Low-Cost Coordinate Converter Rotates Vectors Easily," *Electronics*, Sept. 22, 1982.

Synchro Conversion Handbook. Bohemia, N.Y.: ILC Data Device Corporation, 1974.

RANDOM NOISE and SIGNAL PROCESSING

Bennett, W. R., *Electrical Noise*. New York: McGraw-Hill, 1960.

Claasen, Theo A. C. M., et al, "Signal Processing Method for Improving the Dynamic Range of A/D and D/A Converters," *IEEE Transactions on Acoustics, Speech, and Signal Processing*, ASSP-28, no. 5 (October 1980), 529-538.

*Coles, W. A., and H. J. A. Chivers, "DAC Controls Precision UHF Noise Level," *Analog Dialogue*, 14, no. 2 (1980), 15.

*DeVito, Lawrence, "Brief Bibliography on Random Noise," *Analog Dialogue*, 18, no. 1 (1984), 30.

Gordon, Bernard M., "Noise-Effects on Analog to Digital Conversion Accuracy," *Computer Design*, March-April, 1974.

Gupta, Madhu S., ed., *Electrical Noise: Fundamentals & Sources*. New York: IEEE Press, 1977. (A collection of reprints, including some noise classics.)

Johnson, Matthew, "Implement Stable IIR Filters Using Minimal Hardware," *EDN*, April 14, 1983, 153-166.

*———, "Fast, Simple Approximation of Functions," *Analog Dialogue*, 18, no. 1 (1984), 14-17.

Jung, Walter G., "Application Considerations for IC Data Converters Useful in Audio Signal Processing," *Journal of the Audio Engineering Society*, 25, no. 12 (December 1977), 1033-1038.

Oxaal, John, "DSP Hardware Improves Multiband Filters," *EDN*, March 31, 1983, 193-197.

*———, "Low-Power Digital Signal-Processing ICs," *Analog Dialogue*, 17, no. 1 (1983), 3-7.

———, "Temporal Averaging Techniques Reduce Image Noise," *EDN*, March 17, 1983, 211-215.

Rabiner, Lawrence R., and Bernard Gold, *Theory and Application of Digital Signal Processing*. Englewood Cliffs, N.J.: Prentice-Hall, Inc., 1975.

Ramirez, Robert W., *The FFT Fundamentals and Concepts*. Englewood Cliffs NJ: Prentice-Hall, Inc., 1985.

*Ryan, Al, and Tim Scranton, "D-C Amplifier Noise Revisited," *Analog Dialogue*, 18, no. 1 (1984), 3-10.

Smith, L.R., and D. H. Sheingold, "Noise and Operational Amplifier Circuits," *Analog Dialogue*, 3, no. 1 (1969), 1-16.

*Windsor, William, and Paul Toldalagi, "Digital FIR Filters without Tears," *Analog Dialogue*, 17, no. 2 (1983), 12-19. Contains a Bibliography.

Windsor, William A., II, "IEEE Floating-Point Chips Implement DSP Architectures," *Computer Design* (January 185), 165-170.

———, and Paul Toldalagi, "Simplify FIR-Filter Design with a Cookbook Approach, *EDN*, March 3, 1983, 119-128.

SIGNAL CONDITIONING

Brokaw, A. Paul, "A Monolithic Conditioner for Thermocouple Signals," *IEEE Journal of Solid-State Circuits*, SC-18, no. 6 (December 1983), 707-716.

Counts, Lew, Charles Kitchin, and Steve Sherman, "One-Chip 'Slide Rule' Works with Logs, Antilogs for Real-Time Processing," *Electronic Design*, May 2, 1985, 121-128.

Jung, Walter G., *IC Op-Amp Cookbook* (2nd ed.). Indianapolis: Howard W. Sams & Co., Inc., 1980.

*Reynolds, Dave, "Multi-Channel Signal-Conditioning Input/Output Subsystem," *Analog Dialogue*, 16, no. 3 (1982), 7-9.

*Sheingold, Daniel H., editor, *Nonlinear Circuits Handbook*. Norwood, MA: Analog Devices, Inc., 1974 ($5.95).

———, editor, *Transducer Interfacing Handbook*. Norwood, MA: Analog Devices, Inc., 1980 ($14.50).

TESTING

*Burton, D. Philip, "Checking A/D Converter Linearity," *Analog Dialogue*, 13, no. 2 (1979): 10.

Craven, Robert B., and E. Rachel Morris, "An 18-Bit Precision DC Measurement System," *Digest of Papers*, 1981 IEEE International Test Conference.

DeVito, L., "V-F Converters Demand Accurate Linearity Testing," *Electronic Design*, March 4, 1982.

*Gagne, Alfred L., "On-Line Noise Figure Test Set," *Analog Dialogue*, 12, no. 2 (1978), 16-17.

Kester, Walter, "Characterizing and Testing A/D and D/A Converters for Color Video Applications," *IEEE Transactions on Circuits and Systems*, CAS-25, no. 7 (July 1978), 539-550.

*———, "Test Video A/D Converters Under Dynamic Conditions," *EDN*, August 18, 1982.

Mohr, Steve, and Thomas Guy, "Time Window Nails Down D-A Converter's Settling Time," *Electronic Design*, July 22, 1982, 153-160.

Naylor, Jim R., "Testing D/A and A/D Converters," *Measurements and Control*, June 1981, pp. 123-130.

Perkins, D., "Fixture, Tester Whip Through Instrumentation Amplifier Tests," *Electronic Design*, February 4, 1982.

*Pouliot, Fred, Al Ryan, et al, "LTS-2000, Computerized Test System for IC Devices," *Analog Dialogue*, 14, no. 3 (1980), 3-16.

Prazak, Paul, and Tony Wang, "Superposition: The Hidden DAC Linearity Error," *Electronics Test*, July 1982, 70-79.

Ryan, Allan, "Get a Fast, Accurate Handle on FET-Amplifier Bias Current," *Electronic Design*, November 12, 1981.

*———, and John Chang, "Versatile System Console for Accurate Measuring," *Analog Dialogue*, 14, no. 3 (1980), 4-5.

Schoenwetter, H.R., "High-Accuracy Settling-Time Measurements," *IEEE Transactions on Instrumentation & Measurement*, IM-32, no. 1 (March 1983).

Souders, T. Michael, Donald R. Flach, and Thick C. Wong, "An Automatic Test Set for the Dynamic Characterization of A/D Converters," *IEEE Transactions on Instrumentation and Measurement*, IM-32, no. 1 (March 1983), 180-186.

Wilhelm, Tim, "Test A-D Converters Quickly and Efficiently," *Electronic Design*, October 15, 1981.

*———, "Testing A/D Converters Automatically," *Analog Dialogue*, 16, no.1 (1982), 12-13.

SAMPLE-HOLDS & OTHER DATA-ACQUISITION PERIPHERALS

Bolger, Steve, "Understand the Tradeoffs in IC-Analog-Switch Selection," *EDN*, September 5, 1980, 159-164.

Jung, Walter, "S-H Circuits Control Data-Acquisition Traffic," *Electronic Products*, December 12, 1984, 84-90.

Koen, Mike, "To Sidestep Track/Hold Pitfalls, Recognize Subtle Design Errors," *EDN*, September 5, 1980, 179-184.

Moscovici, Alfi, "Meet the Deglitcher," *Electronic Products*, March 4, 1983, 71-74.

*Neal, Jerry, and Jim Surber, "Flash Converters Work Better with Track-Holds," *Analog Dialogue*, 18, no. 2 (1984), 10-15.

*Whitmore, Jerry, "Behind the Switch Symbol—Use CMOS Switches More Effectively When You Consider Them as Circuits," *Analog Dialogue*, 15, no. 2 (1981), 16-17.

INTERFERENCE NOISE, ESD, and GOOD PRACTICE

Bhar, Tarak N., and Edward J. McMahon, *Electrostatic Discharge Control*. Rochelle Park NJ: Hayden Book Company, Inc., 1983.

*Brockman, Don, and Arnold Williams, "Ground Rules for High-Speed Circuits," *Analog Dialogue*, 17, no. 3 (1983), 22-23.

*Brokaw, A. Paul, "Analog Signal-Handling for High Speed and Accuracy," *Analog Dialogue*, 11, no. 2, (1977) 10-16.

*———, "An I.C. Amplifier Users' Guide to Decoupling, Grounding, and Making

Things Go Right for a Change," *Analog Devices Data-Acquisition Databook 1984*, Volume I, pages 20-13 to 20-20.

Emmens, Trevor, and Mark Lonsborough, "Use Flash ADCs Carefully to Handle High-Frequency Signals," *EDN*, March 17, 1982, 137-143.

⋆*ESD Prevention Manual*. Norwood MA: Analog Devices, Inc., 1984.

⋆Grant, Doug, and Scott Wurcer, "Avoiding Passive-Component Pitfalls," *Analog Dialogue*, 17, no. 2 (1983), 6-11.

Greason, William D., and G. S. Peter Castle, "The Effects of Electrostatic Discharge on Microelectronic Devices—a Review," *IEEE Transactions on Industry Applications*, IA-20, no. 2 (March/April 1984), 247-252.

⋆Johnson, Richard, P., "The Strange Case of the Large Offset," *Analog Dialogue*, 12, no. 1 (1978), 18.

KeyTek Instrument Corporation, *Electrostatic Discharge (ESD) Protection Test Handbook*. Burlington MA: KeyTek Instrument Corporation, 1983.

Krigman, Alan, "Sound and Fury: The Persistent Problem of Electrical Noise," *InTech* (January 1985), 9-20.

Morrison, Ralph, *Grounding and Shielding Techniques in Instrumentation* Second Edition. New York: John Wiley & Sons, 1977.

Motchenbacher, C. D., and F. C. Fitchen, *Low-Noise Electronic Design*. New York: John Wiley & Sons, Inc., 1973.

Ott, Henry W., *Noise Reduction Techniques in Electronic Systems*. New York: John Wiley & Sons, 1976.

Pease, R., "Understand Capacitor Soakage to Optimize Analog Systems," *EDN*, Oct. 13, 1982.

Prazak, Paul, and Tony Miller, "Maintaining Accuracy in High-Resolution Converters, *Electronic Products*, December 12, 1984, 74-78.

⋆Rich, Alan, "Shielding and Guarding," *Analog Dialogue*, 17, no. 1 (1983), 8-13. Also published as an Application Note in *Analog Devices Data-Acquisition Databook 1984*, Volume I, pages 20-85 to 20-90.

⋆———, "Understanding Interference-Type Noise," *Analog Dialogue*, 16, no. 3 (1982), 16-19. Also published as an Application Note in *Analog Devices Data-Acquisition Databook 1984*, Volume I, pages 20-81 to 20-84.

Travers, Don, "Preventing Electrostatic Charge Damage in Semiconductor Devices," *Microelectronic Manufacturing and Testing* (November 1984), 13-16.

Whitmore, Jerry, "Keys to Longer Life for CMOS," *Analog Dialogue*, 8, no. 2 (1974), 20.

Index

A

B

C

D

E

F

G

H

I

J

K

L

M

N

O

P

Q

R

S

T

U

V

W

X

Z